THE BIOLOGY AND EVOLUTION OF TREMATODES

AN ESSAY ON THE BIOLOGY, MORPHOLOGY, LIFE CYCLES, TRANSMISSIONS, AND EVOLUTION OF DIGENETIC TREMATODES

THE BIOLOGY AND EVOLUTION OF TREMATODES

An Essay on the Biology, Morphology, Life Cycles, Transmissions, and Evolution of Digenetic Trematodes

by

KIRILL V. GALAKTIONOV

Zoological Institute of the Russian Academy of Sciences,
St. Petersburg, Russia

and

ANDREJ A. DOBROVOLSKIJ

St. Petersburg State University,
St. Petersburg, Russia

Translated from Russian by
NATALIE V. LENTSMAN, ALEXANDRA D. KHARAZOVA,
VASILII E. STEFANOV and TATIANA A. GANF

Edited by
BERNARD FRIED and THADDEUS K. GRACZYK

KLUWER ACADEMIC PUBLISHERS
DORDRECHT / BOSTON / LONDON

A C.I.P. Catalogue record for this book is available from the Library of Congress.

ISBN 1-4020-1634-4

Published by Kluwer Academic Publishers,
P.O. Box 17, 3300 AA Dordrecht, The Netherlands.

Sold and distributed in North, Central and South America
by Kluwer Academic Publishers,
101 Philip Drive, Norwell, MA 02061, U.S.A.

In all other countries, sold and distributed
by Kluwer Academic Publishers,
P.O. Box 322, 3300 AH Dordrecht, The Netherlands.

Printed on acid-free paper

Table of Contents

Preface

The book by K.V. Galaktionov and A.A. Dobrovolskij maintains the tradition of monographs devoted to detailed coverage of digenetic trematodes in the tradition of B. Dawes (1946) and T.A. Ginetsinskaya (1968). In this respect, the book is traditional in both its form and content. In the beginning (Chapter 1), the authors provide a consistent analysis of the morphological features of all life cycle stages. Importantly, they present a detailed characterization of sporocysts and rediae whose morphological-functional organization has never been comprehensively described in modern literature. The authors not only list morphological characteristics, but also analyze the functional significance of different morphological structures and hypothesize about their evolution. Special attention is given to specific features of morphogenesis in all stages of the trematode life cycle. On this basis, the authors provide several original suggestions about the possible origins of morphological evolution of the parthenogenetic (asexual) and the hermaphroditic generations. This is followed by a detailed consideration of the various morphological-biological adaptations that ensure the successful completion of the complex life cycles of these parasites (Chapter 2). Life cycles inherent in different trematodes are subject to a special analysis (Chapter 3). The authors distinguish several basic types of life cycles and suggest an original interpretation of their evolutionary origin.

Chapter 4 features the analysis of structure and the dynamics of trematode populations and is unusual for a monograph of this type. This chapter begins with a discussion of general issues, i.e., the nature of populations of parasitic organisms and the patterns of interaction between hosts and parasites at the population level. This is followed by a detailed analysis of groups formed by free-living and parasitic stages of trematode life cycles in the environment or in the host.

The two final chapters in the book are very important. In these chapters the factual material presented in the preceding four chapters is discussed from an evolutionary viewpoint. Chapter 5 presents an analysis of apomorphic and plesiomorphic features of the main morphological characteristics inherent in the different phases of trematode development. The authors maintain an emphasis on the mother and daughter sporocysts and rediae and the hermaphroditic generation are characterized more briefly. The origin of the oral and ventral suckers and the organization of the excretory system of digeneans are of special interest to the authors. On the basis of R. M. Cable's studies and the results of their own investigations, the authors propose an original variant of the analysis of the body plan organization of the hermaphroditic generation. The authors offer an original interpretation of the events associated with trematode morphological evolution. In their opinion, evolutionary events differ as a function of life cycle stages and taxonomic groups of these parasites. The chapter concludes with a section concerned with adaptive radiation of trematodes in different ecosystems.

The concluding chapter (6) is concerned with some general questions. It consists of three distinct but closely related sections. The first section presents the authors' views on the origin and evolution of trematode life cycles. The authors of this monograph revive the idea about molluscs as primary hosts of the ancient ancestors of trematodes. They deny that two evolutionary events, transition to parasitism in vertebrates and a consequent separation of the phylogenetic branch Neodermata, were synchronous – an idea commonly accepted to-date. Special attention is focused on evolution of parthenogenetic generations in the molluscan host, and on the larvae of the hermaphroditic generation – cercariae and metacercariae. In the second section, the main evolutionary trends of trematode life cycles are discussed. Finally, the third section considers possible approaches to the elaboration of a natural classification of trematodes.

We hope that the publication of this monograph by K.V. Galaktionov and A.A. Dobrovolskij enlivens the debates about the origin and evolution of parasitic Platyhelminthes in general and Trematoda in particular. In our opinion, this monograph should be of interest to a broad circle of zoologists and parasitologists involved in studies of this unique group of parasites, the trematodes.

Dr. Bernard Fried
Kreider Professor Emeritus of Biology at Lafayette College in Easton, Pennsylvania

Dr. Thaddeus K. Graczyk
Associate Research Professor in the Department of Molecular Microbiology and Immunology and Department of Environmental Health Sciences, Bloomberg School of Public Health, Johns Hopkins University, Maryland

Introduction

Interest in trematodes has persisted for more than a hundred and fifty years. Its foundations were laid by the classical work of Steenstrup (1842), a pioneer study containing data on the life cycle of *Fasciola hepatica*. Steenstrup should also be credited with the first biological interpretation of the phenomenon described. According to him, the life cycle of the liver fluke involved metagenesis, i.e., alternation of sexual and asexual generations.

The publication of Steenstrup's (1842) research stimulated intense and versatile studies of trematodes. Researchers were concerned with every aspect: trematode fauna, life cycles, medical and veterinary significance. At the same time, faunistic studies involving vertebrate animals and molluscs were not related for a long time.

In late XIXth – early XXth Steenstrup's views were profoundly reconsidered. A discussion about the nature of flukes' life cycle arose, which has been continued until the present day. The hypothesis of heterogony was suggested as an alternative to the metagenesis hypothesis (see p. 482). Curiosity about the nature of the life cycle of flukes spotlighted on sporocysts and rediae parasitizing molluscs. In the 30s and 40s of the XXth century a series of brilliant studies by W. Cort and his followers were published, devoted to the reproductive features of these organisms.

In late 50s – early 60s trematodes figured as popular objects of electron-microscopical investigations. The number of studies applying molecular biological methods has been growing fast since the 80s, mostly devoted to general and special problems of the taxon's phylogeny.

At different times attempts were made to compile and generalize abundant and often incomparable data accumulated in the literature. We refer not so much to the systematic reviews by K.I. Skrjabin ("*Trematodes of Animals and Man*", Vol. 1-25, K.I. Skrjabin, ed., Moscow-Leningrad: Izdatelstvo Akedemii Nauk S.S.S.R. 1948-1975) and S. Yamaguti (1958), but to general monographs attempting an extensive coverage of morphology, biology and evolution of trematodes. Pioneering works in this field were those of B. Dawes (1946) and T.A. Ginetsinskaya (1968). These monographs have closed the gap in zoological and parasitological literature of that time. Reviews by J.D. Smyth (1966), D.A. Erasmus (1972) and J.D. Smyth and D.W. Halton (1983) should also be mentioned, though they were concerned mainly with biological and physiological aspects of trematodes.

Unfortunately, no monographs of the same scale have been accomplished after that. Several collective monographs have recently appeared, including "*Advances in Trematode Biology*" (B. Fried and T. Graczyk, eds. Boca Raton, New York: CRC Press 1997) and "*Interrelationships of the Platyhelminthes*" (D.T. Littlewood and R.A. Bray, eds. London and New York: Taylor & Francis 2001). They contain thorough reviews of different aspects of trematode studies and reflect the state-of-the-art knowledge of these flatworms. However, the merits of these monographs mentioned above also demonstrate that our understanding of trematodes is fragmentary and incomplete. On the whole, a paradoxical situation has occurred. An immense bulk of literature is devoted to trematode systematics, including the studies applying the newest methods. At the same time, anatomy and morphology of trematodes as zoological objects are studied insufficiently, to say the least. Trematodes have never been subjected to a rigorous comparative-morphological analysis that was applied to most turbellarian groups. Topography of the organs of the digestive system described from whole mounts can not form the basis for an analysis like this. Therefore, at present there is no solid background for detailed comparison of organization of trematodes and other parasitic flatworms and turbellaria. An attempt to establish homology between the ventral sucker of the hermaphroditic generation of trematodes and the cercomer of Cestoda and other Cercomeromorpha (see p. 437) is a good example of what happens if this fact is ignored. There is also a discrepancy between the repetitive attempts to explain the nature of trematode life cycle and the knowledge about organization and reproduction of sporocysts and rediae.

It is also difficult to discuss phylogenetic issues concerning the origin of the taxon and the analysis of evolutionary tendencies within it. Such discussions are hindered by gaps in our knowledge about representatives of different trematode groups. We know many families only from the materials of faunistic investigations conducted with dissection of vertebrate animals. There are also numerous families whose life cycles are considered as revealed. However, "revealed" usually applies to 1 or 2 representatives, and, moreover, "life cycle descriptions" often mean only that cercariae found in nature were experimentally identified with the known maritae species. Such studies mention daughter sporocysts and rediae just in passing, if at all, and leave mother sporocysts and their larvae, miracidia, completely out of the scope of attention. Finally, there are several families, mostly of evolutionary advanced and highly specialized trematodes (Schistosomatidae, Strigeidae, Diplostomidae, Plagiorchiidae, Microphallidae), whose life cycles are studied quite thoroughly, and, importantly, from many species. A similar state of knowledge presently applies to 2 families of more archaic trematodes, Echinostomatidae and Paramphistomidae. At last, there is a small group of "model" species from very different families (*Fasciola hepatica*, *Philophthalmus rhionica* and *P. megalurus*, 2-3 *Schistosoma* species). These species are studied in most detail, though not necessarily fully. Suffice it to say that a proper morphological description of *Fasciola hepatica* cercariae is still lacking. Thus, heterogeneous information obtained from a limited number of species has to be extrapolated for the class Trematoda as a whole.

The factual basis being scarce as it is, another negative tendency in recent trematode studies limits it even further. Investigators seem reluctant to draw upon the data obtained by older generations of researchers. The works of early and middle XXth century are indeed partly outdated, and yet the value of some of the data has not decreased. Even at present it is unreasonable to ignore the bulk of data accumulated by W. Cort and his school in the 1930s-40s when discussing the reproduction of sporocysts and rediae. Comparative analysis of organization and body plan of different trematode generations can not but be based upon the results of R.M. Cable's studies, mostly conducted in the 50s and 60s of the XXth century. Examples of this kind abound.

Before embarking upon the theme of the book, we would like to accent the principles we were guided by in the choice of material and in our treatment of the data, as well as to comment upon the terminology used.

Studying parasitic organisms, researchers often keep in the limelight their specific features originating from the parasitic mode of life. They seem to forget that parasites are zoological objects, to be described and analyzed from the vantage point of general zoology. There should be no essential difference between the treatment of tubellarians, on the one hand, and trematodes, on the other hand. Unfortunately, this principle has so far been never put into practice. Aspired to analyze certain structures (organization of epithelium or protonephridial fine structure), the researchers do not attempt a comparison of the body plan organization of mature hermaphroditic worms parasitic in vertebrates and that of rediae and sporocysts, or that of turbellaria or representatives of Cercomeromorpha.

Correspondingly, one of our objectives was to divert the analysis of trematode organization and biology back into the course of classical zoology. We consider the use of all data available on all life cycle phases as an important prerequisite for this goal. In our opinion, this approach is of utmost importance, since any phylogenetic scheme may prove wrong if it is based on the analysis of a single phase or even a single generation. Our principles rest on the notions of the famous evolutionist I.I. Schmalhausen (1969) who thought that evolutionary transformation of different phases of the organism's development may be different if they inhabit environments with different conditions (see pp. 215, 408, 462).

A consistent implementation of the above principles presupposed our treatment of the material. We focus our attention on those phases of trematode life cycle that are commonly mentioned in passing: miracidia, mother sporocysts, rediae and daughter sporocysts, as well as to cercariae and metacercariae. Less attention is paid to mature individuals of the hermaphroditic generation parasitizing in vertebrates, i.e., the parasitic flatworms generally thought of as trematodes. Our reasons behind such an approach are as follows.

Firstly, the bulk of literature on trematodes is devoted to the hermaphroditic generation. Analysis of the organization of these organisms (usually based on data obtained from whole mounts) is a keystone of most systematic and phylogenetic constructions. They include works of great interest and originality, e.g., a brilliant comparative analysis performed by J.C. Pearson (1992) (hindered only by the amount of information available on the topic). Many of Pearson's (1992) concepts seem convincing and we do not repeat them here. The instances where we dissent from Pearson's (1992) opinion are thoroughly discussed.

Secondly, modern morphological diversity of mature individuals of the hermaphroditic generation in our opinion reflects a relatively late stage of the trematode adaptive evolution. The main phylogenetic branches within this taxon have formed earlier. The evolutionary stage mentioned featured the colonization of new ecosystems (see p. 489) and, correspondingly, new vertebrate hosts, including the warm-blooded animals that had already existed. The colonization of new habitats in the host's organism also played an important role (see p. 311). These processes were accompanied by numerous parallelisms. Unfortunately, the time is not ripe yet for scrupulous analysis. It will only become possible when our knowledge of morphological-physiological organization of the hermaphroditic generation has passed over to a higher level.

Therefore, in writing the book, we did not set ourselves the task to conduct a detailed analysis of the hermaphroditic generation. Our main objective was to shatter some clichés that have widely spread of late. The revival of the idea about agamous nature of sporocysts and rediae is one of them. The idea that three-host life cycles were primary among modern trematodes is also doubtful. Some recent interpretations of embryonic development of trematodes, in our opinion, are very controversial, and so are recent phylogenetic schemes. Certain stereotypes dominate the discussions concerning interrelations between trematodes and other groups of Platyhelminthes and possible evolutionary ways within the taxon. These two issues are closely interlocked with another one, the determination of time and place of trematode life cycle formation.

We are far from presenting our interpretations of these issues as the universal truth. Our task is much less ambitious: to voice viewpoints different from those that enjoy popularity in modern literature and travel from one article into another, often without the necessary support.

Last but not least, the terminology used in the book calls for discussion. Both authors belong to the St. Petersburg (Russia) zoological and parasitological school. It dates back to V.A. Dogiel, a prominent zoologist and parasitologist, the founder of ecological parasitology. His school was traditionally involved in investigations of trematodes, involving morphological, biological and life cycle studies. For a long time this research was headed by T.A. Ginetsinskaya, the scientific mentor of both authors.

In the course of investigations of long standing a terminology was generated. Incidentally, its important part was borrowed from western sources. In principle, it is similar to the terminology used in modern English-language literature. However, there are discrepancies that should be pointed to from the start. We consider the life cycle of trematodes as an alternation of several generations, each generation possessing an ontogenesis of its own. The development of two generations is always accompanied by metamorphosis. One of them is the generation of the mother sporocyst, represented by two phases: the miracidium, performing the function of dispersive larva, and the parasitic phase of development, that settles in the first intermediate host (usually in molluscs). The latter phase is generally referred to in the literature as the "mother sporocyst" proper.

The second generation whose development is accompanied by a more or less complex metamorphosis is represented by individuals of the hermaphroditic generation. As a rule, 3 distinct phases are present in their ontogenesis: free-living dispersive larva, the cercaria; the phase (protected by a cyst or a capsule) that passively survives in the environment or in the second intermediate host; and, finally, the phase settling and reaching reproductive maturity in the final host (a vertebrate animal) that for a certain time actively produces eggs.

It was the latter two phases that gave rise to terminological controversy. In modern literature the larval phase is referred to as "the metacercaria" regardless of the settlement place. In accordance with an older tradition, the term "metacercaria" was reserved for larvae settling in the second intermediate host, whereas larvae forming cysts on a variety of substrates in the environment were called "adolescariae". It seems expedient to preserve these two terms. "Metacercaria" and "adolescaria" are technically the same thing, as they apply to the same stage of ontogenesis of the hermaphroditic generation. However, functional and physiological differences between them are immense. "Adolescaria", lacking external sources of energy, is a true dormant stage, at which no morphogenetical reconstructions are possible. On the contrary, metacercariae are in most instances true parasites, establishing metabolic interactions with the host. Therefore considerable morphogenetic reconstructions may occur at this phase (see section 2.2. in chapter 1). The role of "adolescariae" and "metacercariae" in the trematode life cycle is different. Thus, the use of two different terms for these two types of larvae should be recommended, as it makes unnecessary additional comment in each particular case.

In modern literature written in English the last phase in ontogenesis of the hermaphroditic generation is generally referred to as the "adult". This word implies a rather narrow meaning, "mature". However, hermaphroditic individuals settling in the final host often have to undergo a long and complex process of organogenesis, involving the reproductive system, i.e., "maturation". Therefore the neutral term "marita" that was in use in the first half of the XXth century, is more correct. The term "maritogony" follows naturally from "marita". Maritogony refers to the complex of morphogenetical transformations undergone by individuals of the hermaphroditic generation from the moment of their settlement in the final host and to the time when they reach reproductive maturity and acquire the complex of species-specific characters inherent in the maritae of a given trematode species.

The history of the term "parthenita" is also intricate. It was used during two-thirds of the XXth century for overall designation of mother sporocysts, rediae or daughter sporocysts, parasitizing in molluscs. This term had its roots in heterogony hypothesis, according to which rediae and sporocysts were parthenogenetic females that had secondarily lost hermaphroditic reproductive system. When this idea was rejected and the asexual nature of sporocysts and rediae was postulated, the term "parthenita" was also discarded. We have often argued that there are no grounds to consider sporocysts and rediae as asexual (Dobrovolskij et al., 1983; Galaktionov and Dobrovolskij, 1998; Dobrovolskij and Ataev, 2003). We give a full explanation of our opinion in the present book as well. According to our theoretical precepts the term "parthenita" should be retained and we use it widely in our presentation of material.

The use of the names "Trematoda" and "Digenea" deserves special mention. The class Trematoda used to comprise two subclasses: Digenea and Monogenea. This variant of the system was popular for a long time. In particular, it was accepted in the classical research by Dawes (1946). However, another point of view on phylogenetic interactions between the major groups of parasitic flatworms also existed. B.E. Bychowsky (1937, 1957) extracted Monogenea from the class Trematoda and affiliated them with Cestoda and other Cercomeromorpha. These ideas, with some modifications, are now accepted by most zoologists. Only some special issues are debated, concerning phylogenetic relations between lower taxa. Taxonomic status of Aspidogastrida is one of those controversial issues. B.E. Bychowsky and, later, T.A. Ginetsinskaya (1968) considered this group as undoubtedly related to true trematodes. Yet in their opinion it was distinct from them and deserved the status of a class.

Affinity between Trematoda (=Digenea) and Aspidogastrida (that was actually never questioned) has recently received additional support, including molecular data. Modern phylogenetic schemes consider Aspidogastrida as a sister group to Trematoda. As a result, the two groups are included into the same higher taxon, the class Trematoda. This solution of the taxonomical problem seems unacceptable.

While we accept both older and newer arguments in favour of a close affinity between the two groups, we hold the view that their evolutionary ways parted a long time ago. Each of the two groups underwent a unique sequence of evolutionary transformations and acquired a characteristic complex of traits. Importantly, the life cycle structure is essentially different in representatives of these two groups. They also differ considerably in comparative-morphological respect. Similarity between some aspidogastrids and trematode maritae is purely external, and by no means reflects the degree of their relatedness. In our opinion, these are two independent classes with common ancestors. A more detailed argumentation of this viewpoint deserves special investigation and is outside the scope of the present book. Here we only note that in the book the names Trematoda and Digenea are used as synonyms.

When writing the book, we encountered certain difficulties with the application of taxonomical names, mostly of family and superfamily level, arising from the absence of an accepted system of Trematoda. Treatises of K.I. Skrjabin (1947-1975) and S. Yamaguti (1958, 1971) contain an enormous amount of material, but they are outdated in many respects. However, there have been no recent publications of the same scale. An attempt to list the known trematode genera and to create keys for their identification (*"Keys to the Trematoda. Volume I"* D.I. Gibson, A. Jones and R.A. Bray, eds. London: CABI Publishing and The Natural History Museum 2002), firstly, has not been brought to the end yet, and secondly, is confined to the taxonomical boundaries from the genus to the superfamily. In order not to increase confusion in the names of well-known and widespread families, we used the names recommended in the above work where possible. The names of superfamilies were used much less often. In some cases, when the boundaries of taxa above the family level were ill-defined, we applies neutral expressions, e.g., "a group of related families". Boundaries of taxa of still higher level – orders and suborders – are so differently understood by different researchers that they are almost impossible to use.

The elaboration of a natural classification of Trematoda is a task that belongs to the future. We envisage that it would rely upon a methodological basis entirely different from the one currently utilized. Elaboration and explanation of new methodological approaches is the main objective of the book that is now presented to the readers.

Acknowledgements

This book is derived from another one, "*The Origin and Evolution of Trematode Life Cycles*", published in Russian by Nauka Publishing House in 1998. During its preparation, the English edition was radically reworked and supplemented. A new book emerged as a result, which called for a change in title. This book would have never seen the light if it were not for Dr. Bernard Fried, Kreider Professor Emeritus of Biology at Lafayette College in Easton, Pennsylvania, who suggested we publish it in English; he has given invaluable help in its editing. We are very grateful to Dr. B. Fried for his support. The publication would have also been impossible without the helpful cooperation of Peter Butler, the Publishing Unit Manager, Biomedical Unit, Kluwer Academic Publ. We are grateful to him for understanding the difficulties that arose during manuscript preparation, and for his steady encouragement of the project at all stages of its implementation.

Some of the material in the book was discussed with the late Professors Yu.I. Poljansky and S.S. Schulman, and also with Dr G.L. Ataev, Russian State Pedagogical University, Prof. M.M. Belopolskaya, Prof. V.Ja. Berger, Zoological Institute of the Russian Academy of Sciences, Dr J.O. Bustnes, Norwegian Institute for Nature Research, Dr A.I. Granovich, St Petersburg State University, Dr S.W.B. Irwin, University of Ulster, Dr E.G. Krasnodembskij, St Petersburg State University, Dr A.D. Naumov, Zoological Institute of the Russian Academy of Sciences, Dr I.M. Podvyaznaya, Zoological Institute of the Russian Academy of Sciences, Dr O.N. Pugachev, Zoological Institute of the Russian Academy of Sciences, Dr K. Skirnisson, University of Iceland, Dr I.A. Tihomirov, St Petersburg State University. We sincerely appreciate all the fruitful discussions with these people and their criticisms. We owe a particular debt of gratitude to our teacher, Prof. T.A. Ginetsinskaya who is the author of many of the ideas developed in the book. We thank all the students, post-graduate students, and the staff of the Department of Invertebrate Zoology, St Petersburg State University, the White Sea Biological Station and the Laboratory of Parasitic Worms of the Zoological Institute of the Russian Academy of Sciences and the Parasitological Laboratory of Murmansk Marine Biological

Institute for their participation in research programs that were conducted by the authors for various years. Our special thanks are due to Natalia V. Lentsman, Prof. Alexandra D. Kharasova, Dr. Vasilii E. Stefanov and Dr. Tatiana A. Ganf, who translated the Russian text of the book into English. We gratefully acknowledge the assistance of Dr. T.K. Graczyk, Associate Research Professor, Johns Hopkins University, Maryland, who helped Dr. B. Fried in correcting the English of the manuscript. We would like to express our thanks to Janet Hoffman, Manja Eliane Fredriksz, Jean van Ree and Tanja van Gaans, Assistances Publishing Editor, Biomedical Unit of Kluwer Academic Publishing for advice on technical problems during preparation of the manuscript.

We thank Professeur honoraire S. Deblock, Dr S.W.B. Irwin, University of Ulster, Dr M. Køie, Marine Biological Laboratory, University of Copenhagen, Dr E.G. Krasnodembskij, St Petersburg State University, Dr I.I. Malkova, Prof. Dr K. Odening, Dr I.M. Podvyaznaya, Zoological Institute of Russian Academy of Sciences, Dr D.H. Saville, University of Ulster, Dr I.A. Tihomirov, St Petersburg State University, Dr Z. Žd'árská, Institute of Parasitology, Academy of Sciences of the Czech Republic for kind permission to use their illustrative material in the book.

We thank the following editors and publishers for permission to use material: *Acta Parasitologica* (W. Stefanski Institute of Parasitology), *Annals of the New York Academy of Sciences* (New York Academy of Sciences), *Annals of Tropical Medicine and Parasitology* (Liverpool School of Tropical Medicine), *Annales de Parasitologie Humaine et Comparée* (Masson, Paris), *Annales Zoologici Fennici* (Finnish Zoological and Botanical Publishing Board), *Folia Parasitology* (Institute of Parasitology, Academy of Sciences of the Czech Republic), *Journal of Invertebrate Pathology* (Academic Press), *Journal of Helminthology* (CABI Publishing), *Parasitology* (Cambridge University Press), Nauka, *Parasitology Research/Zeitschrift für Parasitenkunde* (Springer-Verlag GmbH & Co. KG), Taylor & Francis, *The Journal of Parasitology* (American Society of Parasitology), St Petersburg State University Press.

Some data included in the book was obtained during our studies supported by the RFBR, programme "University of Russia – Basic Science", INTAS, Royal Society, Norwegian Directorate for Nature Management, Nordic Council, Icelandic Republic Fund and NATO.

Chapter 1

ORGANIZATION OF PARTHENOGENETIC AND HERMAPHRODITIC GENERATIONS OF TREMATODES

Analysis of each group of living organisms should start with the morphologic description of various stages of the life cycle. Of particular importance are the main trends of morphological evolution as they serve as a basis for the development of evolutionary concepts of the investigated groups. In this chapter modern data on the morphology and biology of life cycle stages of trematodes are briefly reviewed and the principal trends of the morphologic evolution of this group of parasitic platyhelminthes are traced.

1. PARTHENOGENETIC GENERATIONS AND THEIR LARVA

Trematode parthenogenetic generations are represented by a mother sporocyst and its larva (miracidium) and rediae or daughter sporocysts. The number of rediae or daughter sporocyst generations varies widely in the representatives of different digenean taxa. All parthenitae are obligate parasites of molluscs[i] and that is why they are often called intramolluscan stages.

Traditionally parthenitae are treated as sacs, packed with embryos, the so-called germinal sacs. In works on life cycles of particular species, as a rule, there are at most only a few lines dedicated to sporocysts and rediae. There are relatively few special investigations dealing with the morphology and biology of these stages in the digenean life cycle. It is not a surprise that data available on trematode parthenogenetic generations are scarce and fragmentary. This is well-illustrated by the latest reviews on digenean phylogeny. For instance, in an interesting study by Cribb et al. (2001b), which will be analyzed in later chapters of this book, digenean phylogeny is constructed on the basis of 56 morphological characters. Only 2 of the 56 have something to do with the miracidium and six – with sporocysts and rediae. All the others deal with morphology of hermaphroditic generation, mainly the adults.

In the course of prolonged co-evolution with molluscan hosts parthenogenetic generations have developed a number of morphological and functional adaptations, varying considerably in different trematode groups. This should be taken into account when the main evolution trends of the parasites are to be identified. In the material presented below we have attempted to generalize and analyze from an evolutionary standpoint the data accumulated in the literature as well as results of our own long-term studies on parthenogenetic generations of different trematode taxa.

1.1. First Parthenogenetic Generation

The first parthenogenetic generation is represented in trematode life cycle by two phases – the free-living larva (miracidium) and the parasitic imago – the mother sporocyst. According to Schmalhausen (1969) (for details see pp. 215, 408, 462), the main trends in the evolution of these two phases may differ since they exist in different environments. This is illustrated by the material presented below.

1.1.1. Miracidium

The miracidium is a free-living larva that develops into the mother sporocyst. The miracidium is the first larval phase in the trematode life cycle. Its main function consists in infection of the first intermediate host where parthenogenetic generations develop.

This process differs for different trematodes. In certain cases (Fasciolidae, Echinostomatidae, Philophthalmidae, Schistosomatidae, etc.) relatively large (80-150 μm, up to 340 μm in *Cyclocoelum microstomum*) miracidia of complex organization (Fig. 1A-D,G,J) may form. They actively swim in the water and penetrate into the molluscan host. This is an active way of infection. In other cases (Plagiorchiida, Hemiuroidea, Brachylaimoidea, etc.) somatic organization of larvae is remarkably simplified which is also accompanied by reduction of the body size (20-60 μm, up to 4-10 μm in Microphallidae, Lecithodendrioidea and Brachylaimoidea) (Fig. 1E,F,H,I). Such miracidia are not capable of active life in the environment and hatch from the eggs only in the intestines of the mollusc, which ingests them with the food. This is a passive way of infection.

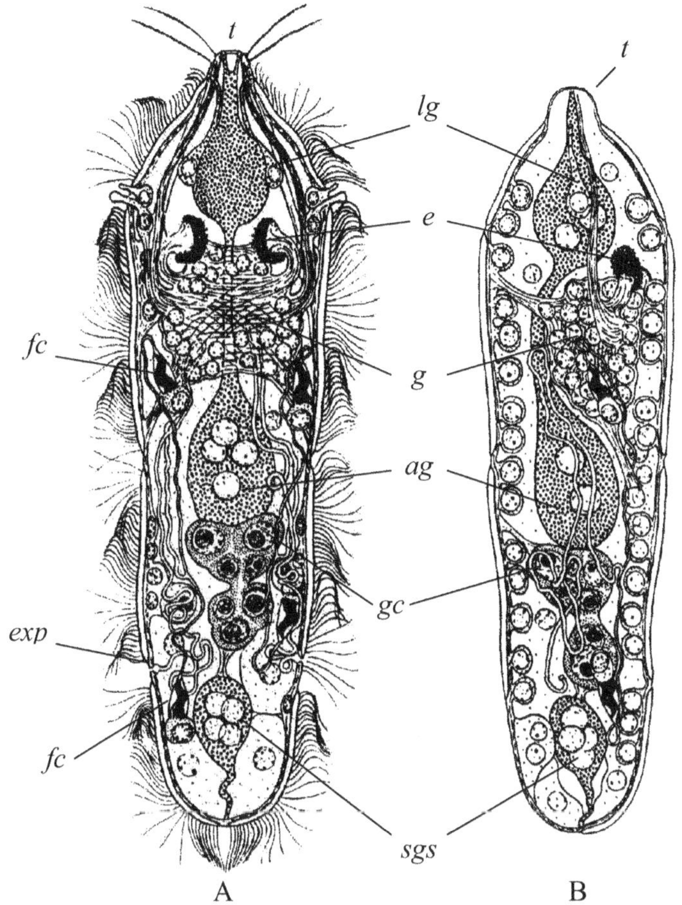

Figure 1. Morphology of miracidia. A-B, *Posthodiplostomum cuticola* (Diplostomidae), A, dorsal view, B, lateral view. From Dönges (1964), reprinted with the permission of Springer-Verlag GmbH & Co.KG; C, *Isthmiophora melis* (Echinostomatidae). From Dönges (1973), reprinted with the permission of Springer-Verlag GmbH & Co.KG; D, *Allocreadium handiai* (Allocreadiidae). From Madhavi (1980), reprinted with the permission of Springer-Verlag GmbH & Co.KG; E, *Halipegus eccentricus* (Halipeginae). From Thomas (1939) reprinted with publisher's permission; F, *Proterometra dickermani* (Azygiidae). From Anderson and Anderson (1963) reprinted with publisher's permission; G, *Sanguinicola inermis* (Sanguinicolidae). From McMichael-Phillips et al. (1992b), reprinted with the permission of Cambridge University Press; H-I, *Haplometra cylindracea* (Plagiorchiidae), H, inside of the egg, I, hatched from the egg. From Dobrovolskij (1965a); J, *Philophthalmus rhionica* (Philophthalmidae). From Tihomirov (1980) reprinted with author's permission. *Abbreviations*: *ac*, apical sensory cilium; *ag*, apical gland; *cep*, ciliated epithelial plate; *e*, eyespot; *ep*, epithelial plate; *exp*, excretory pore; *fc*, flame cell; *g*, ganglion; *gc*, germinal cell; *glc*, glandular cells; *hg*, hatching gland; *hr*, hypodermal ridge; *lg*, lateral gland; *lp*, lateral sensory papilla; *mr*, mother redia; *r*, rodlet; *s*, stylet; *sg*, secretory granules; *sgs*, syncytial glandular structure; *sp*, "spines"; *t*, terebratorium.

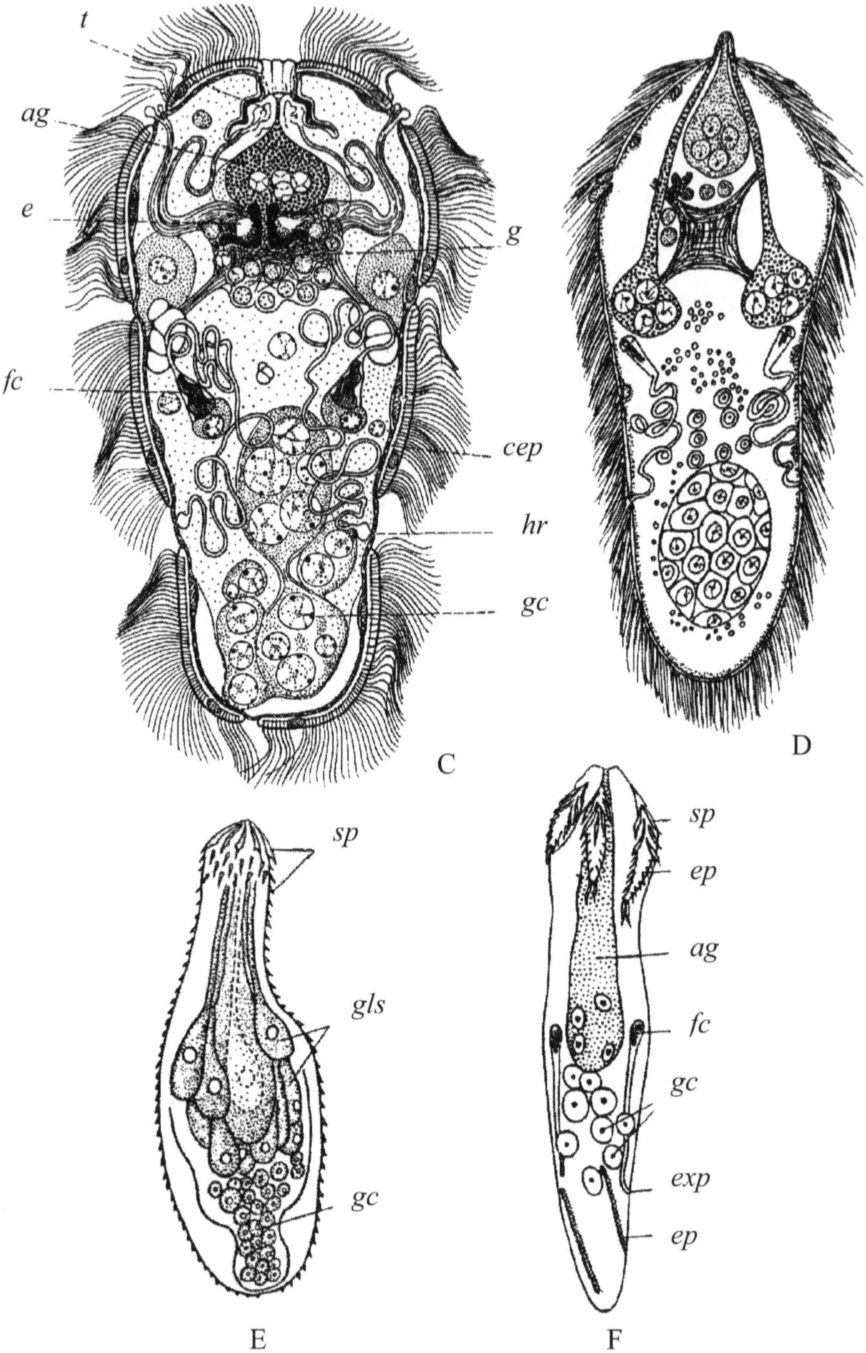

Figure 1. (Continued)

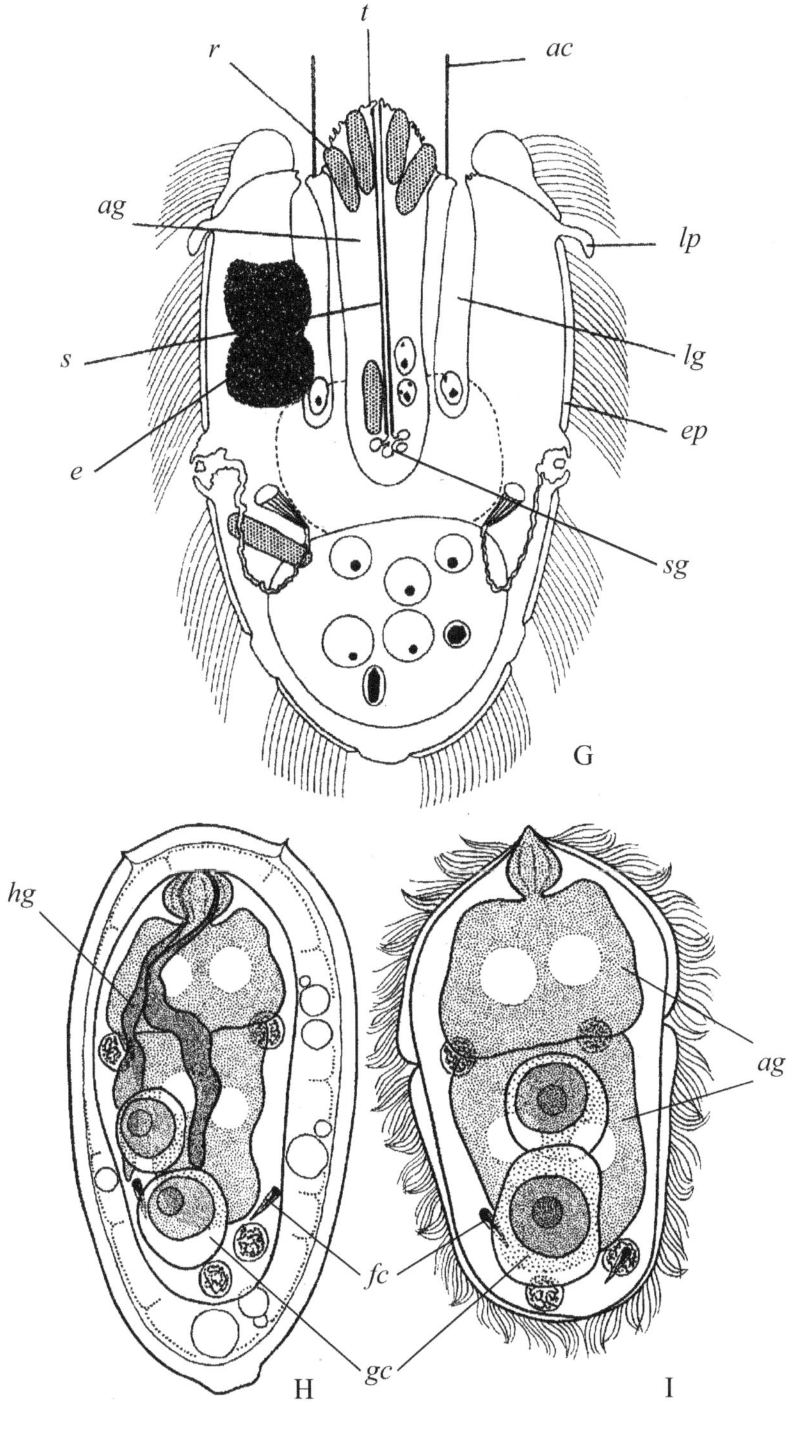

Figure 1. (Continued)

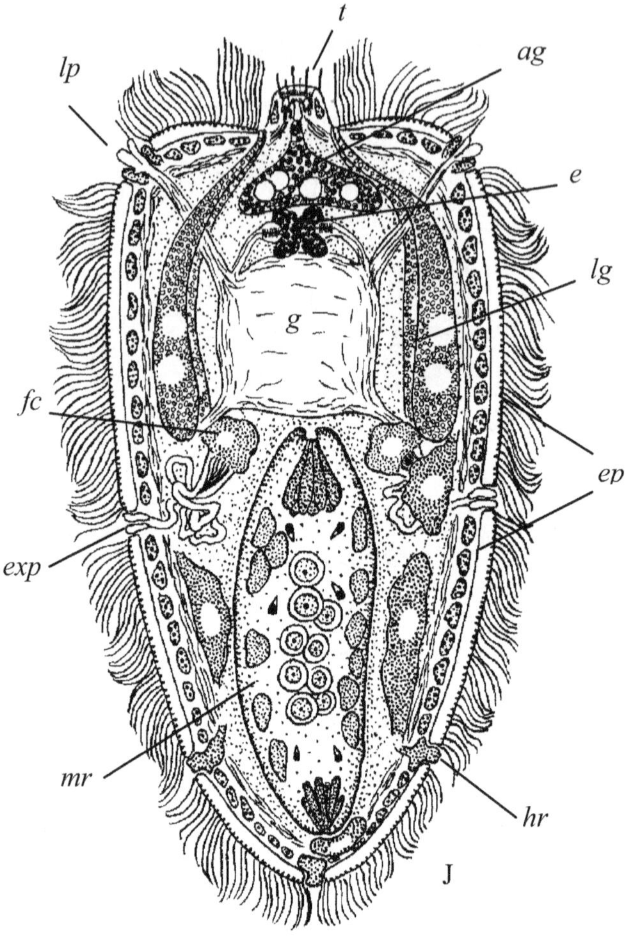

Figure 1. (Continued)

Formation of the compound egg is similar in all trematodes (Ginetsinskaya, 1968; Smyth and Halton, 1983) (Fig. 2). Primary oocytes forming in the marita (adult) ovary in the course of oogenesis pass through the oviduct into the ootype. There they are surrounded with vitelline cells, coming to the proximal part of the ootype from the common vitelline duct – vitelline reservoir. The cytoplasm of vitelline cells contains numerous protein globules utilized in the formation of the eggshell. There is also a certain supply of energy-rich materials, mainly glycogen (Irwin and Threadgold, 1972; Erasmus, 1975), whose amount varies considerably in the vitelline cells of different trematode groups depending on the type of miracidial development (Burton, 1962, 1967b; Ginetsinskaya and Besedina, 1965; Palm, 1967; Irwin and Maguire, 1979; Erasmus, 1975; Hendow and James, 1989; etc).

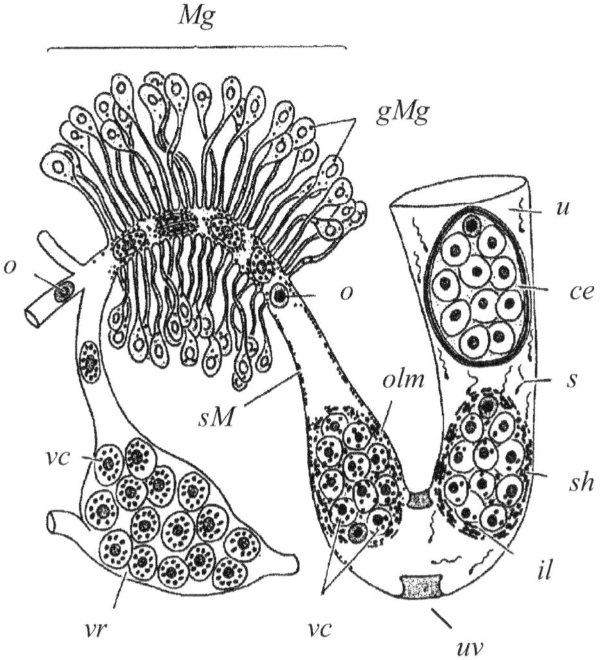

Figure 2. Formation of the trematode compound egg. *Abbreviations*: *ce*, compound egg; *gMg*, globules of Mehlis' gland secretion; *il*, forming inner lipoprotein membrane; *Mg*, Mehlis' gland; *o*, ovum; *olm*, forming outer lipoprotein membrane; *s*, sperm; *sh*, shell protein; *sM*, layer of Mehlis' gland secret on the wall of ootype; *u*, uterus; *uv*, uterine valve; *vc*, vitelline cells; *vr*, vitelline reservoir. From Clegg (1965), reprinted with publisher's permission.

"Fertilization" takes place in the oviduct or ootype resulting in the inclusion of the nucleus, mitochondrion and axonemes of the spermatozoon into the cytoplasm of the primary oocyte; the plasma membrane and cortical microtubules remaining on the surface of the oocyte (Burton, 1967b; Justine and Mattei, 1984; Orido, 1988). The channels of Mehlis' gland lead to the central and distal parts of the ootype. The Mehlis' gland may contain one or two types of cells (see reviews: Awad and Probert, 1990; Orido, 1991). In the latter case serous and mucous cells are distinguished. The function of the former is still unclear, in spite of numerous hypotheses available in the literature (Smyth and Halton, 1983; Orido, 1991; Nollen, 1997). This secretion may lower the surface tension of the shell protein globules, which in turn may facilitate fusion of these granules leading to the formation of the egg-shell (Irwin and Threadgold, 1972). The role of mucous secretion is more certain (Smyth and Halton, 1983; Orido, 1991; Nollen, 1997). This secretion covers the oocyte-vitelline cell complex with a thin layer, which presumably serves as a template to accumulate the protein secretion excreted by the vitelline cells on its internal surface (Clegg, 1965). Apparently glycoprotein secretion of the *Schistosoma* Mehlis' glands, which contain only one type of cells, has a similar function (Awad and Probert, 1990; Moczon et al., 1992; Moczon and Swiderski, 2000).

The compound egg, covered with a thin shell, is delivered from the ootype into the uterus. Here the oocyte undergoes two sequential stages of maturation division, accompanied by extrusion of two polar bodies. The forming female pronucleus is fused with the male one, which has remained to that moment in the cytoplasm of the developing oocyte. Further formation of the shell proceeds due to the tanning of its proteins, while the egg passes along the loops of the uterus (Wharton, 1983). The most usual component of the shell of the mature egg is sclerotin, though keratin can also be present (Smyth and Halton, 1983; Nollen, 1997). The only exceptions are in the Sanguinicolidae and Aporocotylidae, whose thin eggshells do not undergo sclerotization, i.e., quinone-tanning (Thulin, 1975; McMichael-Phillips et al., 1992a). Elastin was reported in the eggshell of *Orchispirium heterovitellatum* (Sanguinicolidae) (Madhavi and Rao, 1971). In many trematodes, the operculum forms at the end of the egg containing the oocyte. In *Fasciola hepatica*, a special pseudopodium of the oocyte takes part in the process. It contacts the inner side of the forming eggshell at the point of the future rupture. Consequently the shell is weakened at this point and is easily fractured on hatching (Smyth and Halton, 1983).

If all trematodes have a similar pattern of compound egg formation, the number of vitelline cells and the amount of accumulated nutrients (so called "yolk") would vary considerably. There is a correlation with the biology of miracidia. In the representatives of many families with active infecting miracidium (Fasciolidae, Paramphistomidae, Echinostomatidae, Strigeidae, etc.) division of the fertilized oocyte (and sometimes even the fusion of male and female pronuclei) takes place only after the extrusion of the egg into the environment, or only its earliest stages proceed in the marita uterus. Glycogen from the vitelline cells is the only source of nutrients for the developing larvae. That is why the number of vitelline cells in the egg may be quite large (up to 30-40). These cells normally store a considerable amount of glycogen (Palm, 1967; Ginetsinskaya, 1968; Irwin and Threadgold, 1972).

For species belonging to the families Schistosomatidae, Spirorchiidae, Sanguinicolidae and Aporocotylidae, the host blood is rich in nutrients and is an environment where larvae, contained in the eggs, develop. This is advantageous for the parasites. In the best-studied representatives of the genus *Schistosoma*, pores were discovered in the eggshell which allow for entry of glucose and a number of other nutrients required as the energy supply of the processes of embryo- and morphogenesis of miracidia (Lewert et al., 1970; Stjernholm and Warren, 1974; Bogitsh, 1983; Kawanaka et al., 1983; Swiderski, 1988). In the representatives of Sanguinicolidae and, probably, Aporocotylidae, there are no pores in the eggshell but it lacks tanning and is thin enough not to obstruct diffusion of nutrients necessary for miracidium development from the host blood (Thulin, 1975; McMichael et al., 1992a). Consequently there is no need for endogenous accumulation of considerable amounts of nutrients and the number of vitelline cells in the compound egg of the blood parasites decreases to 20-25, and even to 8-10 in *Schistosoma.*

This trend is even more pronounced in trematodes whose miracidia develop in the eggs during their passing along the loops of the uterus, delivering to the environment the eggs with fully formed larvae. This type of miracidia development is characteristic of representatives of a number of families with an active way of infection of the first intermediate hosts (Philophthalmidae, Cyclocoelidae, Zoogonidae, Gorgoderidae), as well as all trematodes with a passive mode of infection[ii]. Nutrients needed for larval formation are delivered directly from the maternal organism (Burton, 1962; Ginetsinskaya et al., 1981; Hendow and James, 1989). Therefore a sharp decrease in the number of vitelline cells in the eggs of trematode examined from this group is not a surprise. This number is equal to 3-5 in Plagiochiida, to 2 in Zoogonidae and to 1-2 in Notocotylidae, according to the data reported in Murrills et al. (1985a). The amount of glycogen accumulated in the vitelline cells also decreases and is even depleted in many representatives of Cyclocoelidae, Opisthorchiidae, Brachylaimoidea, Prosthogonimidae, Hemiuroidea, Microphallidae, etc. (Burton, 1962; 1967b; Ginetsinskaya, 1968; Palm, 1967; Coil, 1965; Hendow and James, 1989; etc).

No matter what the environment is, cleavage of the zygote contained in the egg, is always complete, non-equal and asynchronous. Embryogenesis will be discussed in the section 1.3.2. Now we shall only note that the miracidium forming in the course of embryogenesis is covered with several envelopes of a different nature. The external one, so called vitelline membrane, contacts the eggshell from the inner side. It encloses the miracidial embryo and yolk, released on the resorption of the plasma membranes of vitelline cells. The function of the vitelline membrane apparently is not the same in different trematode groups. Its ultrastructure in schistosomatids testifies to a high metabolic activity and supports the idea of its involvement in the transportation of nutrients from the host blood to the developing miracidium (Neill et al., 1988; Swiderski, 1994 a, b). The vitelline membrane is likely to have a similar function in the eggs of trematodes whose miracidia develop in the adult worm. In addition, the vitelline membrane, like any other barrier tissue, apparently plays a protective role. In schistosomes, whose eggshell has numerous pores, the membrane prevents passive diffusion of parasite antigens into the blood, thus diminishing the intensity of the host's immune response (Neill et al., 1988). In trematodes, whose eggs undergo development in the environment, the vitelline membrane is connected with special structures – "viscous cushion" or "mucoid plug", participating in hatching of the miracidium.

Data on miracidial morphogenesis are fragmentary and controversial (for reviews see: Ginetsinskaya, 1968; Dobrovolskij et al., 1983; Semenov, 1991). This can be explained by a surprising diversity of the morphology of fully formed larvae in different trematode groups. The body of actively infecting miracidia is covered with flattened ciliated epithelial cells, so called epithelial plates (Fig. 1A-D,G,J).

They form several circular rows, separated by ridges of hypodermis, which will be considered later. The number of tiers is constant within major taxonomic groups (families, groups of families). The number of epithelial plates being stable within major taxonomic groups can undergo inter- and even intraspecific variations. The latter, however, may well probe an experimental artifact (Semenov, 1991). The Fasciolidae miracidia form the greatest number of the tiers – 5; the number of epithelial plates is determined by the formula 6:6:3:4:2. In most trematode families with the actively infecting miracidium the number of rows is four: 6:8:4:2 in Philophthalmidae and Paramphistomidae, 6:9:4:3 in Schistosomatidae, Diplostomidae and Strigeidae, 6:9:4:2 in some Echinostomatidae and Lepocreadiidae, 6:6:4:2 in Opecoelidae and some Echinostomatidae. Apparently this is an optimal number for free-swimming larvae since the three-row arrangement of epithelial plates occurs rarely, e.g., 6:6:3 in Gorgoderidae.

Further reduction of the number of tiers and their ciliated cells is observed in miracidia of trematodes with a passive way of infection. Localization of the epithelial plates on the body of Opisthorchiidae, Dicrocoeliidae, Plagiorchiidae and Ochetosomatidae larvae is described by the formula 3:3 (Fig. 1H,I). In miracidia of Microphallidae and Lecithodendrioidea, which are the smallest among Plagiorchiida, the number of epithelial cells has not been counted. However, their two-row arrangement is quite distinct. The two-row arrangement of epithelial plates was also reported in small passively infecting miracidia *Prosorhynchus* (Bucephalidae) (Fig. 3A-C) and *Lecithaster* (Hemiuroidea, Lecithasteridae) (Schell, 1975). The miracidium of another representative of hemiuroid trematodes – *Genarchopsis goppo* is covered entirely with cilia, but the borders of epithelial cells can not be detected even by means of silver staining (Madhavi, 1978). The surface of Brachylaimoidea larvae is almost completely devoid of cilia. They have 2 to 6 small ciliated epithelial plates in the middle and/or posterior part of the body (Allison, 1943; Kagan, 1952; Lewis, 1969). Four ciliated cells encircling the anterior part of the miracidial body were observed in *Lecithochirium* (Hemiuridae, Lecithochiriinae) (Mathews and Mathews, 1991). A complete lack of surface cilia was reported for larvae of Didymozoidae, Halipeginae and Bunocotylinae species. On the surface of their bodies "spines" were found (Fig. 1E, 3D-F). These "spines" form 1 or 2-3 crowns on the anterior part of miracidia of some species. In *Bunocotyle progenetica* (Hemiuroidea, Bunocotylinae) they cover the whole larvae and considerably in-

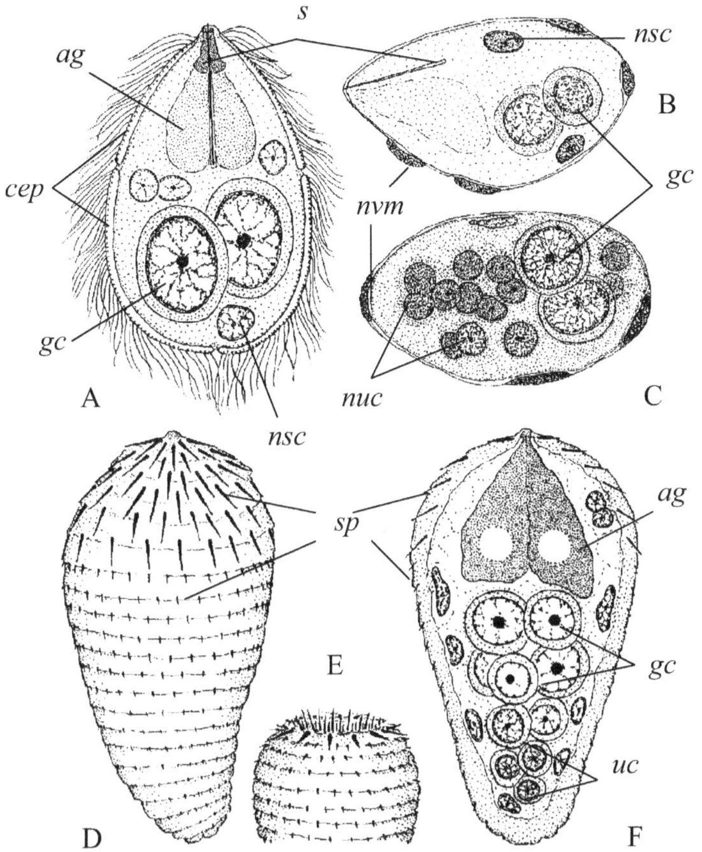

Figure 3. Structure of specialized miracidia of Bucephallidae and Bunocotylinae (Hemiuridae). A-C, *Prosorhynchus squamatus* (Bucephalidae), A, general structure, B, sagittal section of mature miracidium inside the egg (eggshell is not shown), C, longitudinal section through developing miracidium (eggshell is not shown), From Suloeva et al. (*submit*); D-F, *Bunocotyle progenetica* (Bunocotylinae), D, surface structures, E, miracidium with the anterior part of the body pulled into it, F, general structure. *Abbreviations*: *ag*, apical gland; *cep*, ciliated epithelial plate; *gc*, germinal cell; *nsc*, nucleus of somatic cell; *nuc*, nuclei of undifferentiated cells; *nvm*, nuclei of cells forming the vitelline membrane *s*, stylet; *sp*, spines; *uc*, undifferentiated cells.

crease in size from the posterior to anterior part of the body (Fig. 3D-F).

Of particular interest are Bucephalidae miracidia. According to the data available in the literature they belong to the group of actively infecting larvae. They possess ribbon-like ciliated plates, partially covering the body, and ciliated bars, protruding from the posterior end (Woodhead, 1929,1930; Kniskern, 1952; Dickerman, 1954). These observations were not confirmed in the investigations of *Prosorhynhus squamatus* miracidium (Suloeva et al., *submit*). Undoubtedly it belongs to the group of passively infecting larvae.

The *P. squamatus* miracidium is entirely covered with ciliated epithelial cells (Fig. 3A). No bars bearing cilia are visible. Neither were they shown by Woodhead (1930) in the miracidium of *Bucephalus elegans*, contained in the egg. Woodhead (1930) also points out to an extremely low number of naturally hatching miracidia, while Dickerman (1954) has not observed them at all during three month of incubation of the eggs in water. Similar results were obtained with the eggs of *Prosorhinchus squamatus* (Suloeva, 1999; Suloeva et al., *submit*). We are inclined to regard the cilia bars in miracidia in the illustrations presented by Woodhead (1929, 1930, 1931) and Dickerman (1954) as an artifact. That might be caused by observations of partially shedding epithelial plates of the posterior tier, which is quite possible because the morphologic investigations were carried out on larvae, artificially hatched from the eggs by pressing them with cover glass, quite often in hypertonic medium (solution of methylene blue) (Woodhead, 1929, 1930, 1931; Dickerman, 1954).

Ultrastructure studies of miracidia were performed for a restricted number of species with the active way of infection of first intermediate hosts (Fasciolidae – Wilson, 1969a,b,c, 1970, 1971; Southgate, 1970, etc.; Schistosomatidae – Wikel and Bogitsh, 1974; Ebrahimzaden, 1977; Pan, 1980, Eklu-Natey et al., 1985, etc.; Paramphistomidae – Dunn et al., 1987; Philophthalmidae – Isseroff, 1964; Sinha, 1978; Tihomirov, 1980, 2000; Sanguinicolidae – McMichael-Phillips et al., 1992b). Below we shall mainly refer to these data. At the same time detailed morphology of passively infecting miracidia was described only for an aberrant case of Notocotylidae (Murrills et al., 1985a) and Lecithochiriinae larvae (Matthews and Matthews, 1991). Miracidia of the entire large group of Plagiorchiida up to now has not been investigated by electron microscopy.

In the larvae of the investigated species there are short microvilli situated in the intercilia space of the surface of epithelial cells (Fig. 4). Kinetosomes are localized in the apical cytoplasm; one or two rootlets originate from each of them. There are many mitochondria with well-expressed cristae and vesicles of different size and shape, which testifies to high metabolic activity of epithelial cells. Apparently this is associated with the necessity to supply energy for locomotion. Nuclei of epithelial cells exhibit clear degenerate changes. The nuclei are not large: they often have many protrusions and vast heterochromatin zones. In schistosome miracidia epithelial cells probably lack nuclei (Pan, 1980). However, there are not well-documented data indicating the presence of nucleus-containing cell bodies (cytons) under the layers of muscles (Wikel and Bogitsh, 1974; Swiderski, 1984). However, Pan (1980), who noted the presence of cytons of epithelial plates, pointed out that they lacked nuclei. Epithelial plates of Philophthalmidae miracidia are of similar morphology. Sinha (1978) described nuclear-free projections of epithelial cells passing under the muscle layers in *Philophthalmus rhionica* miracidium. They form sac-like expansions immediately under the

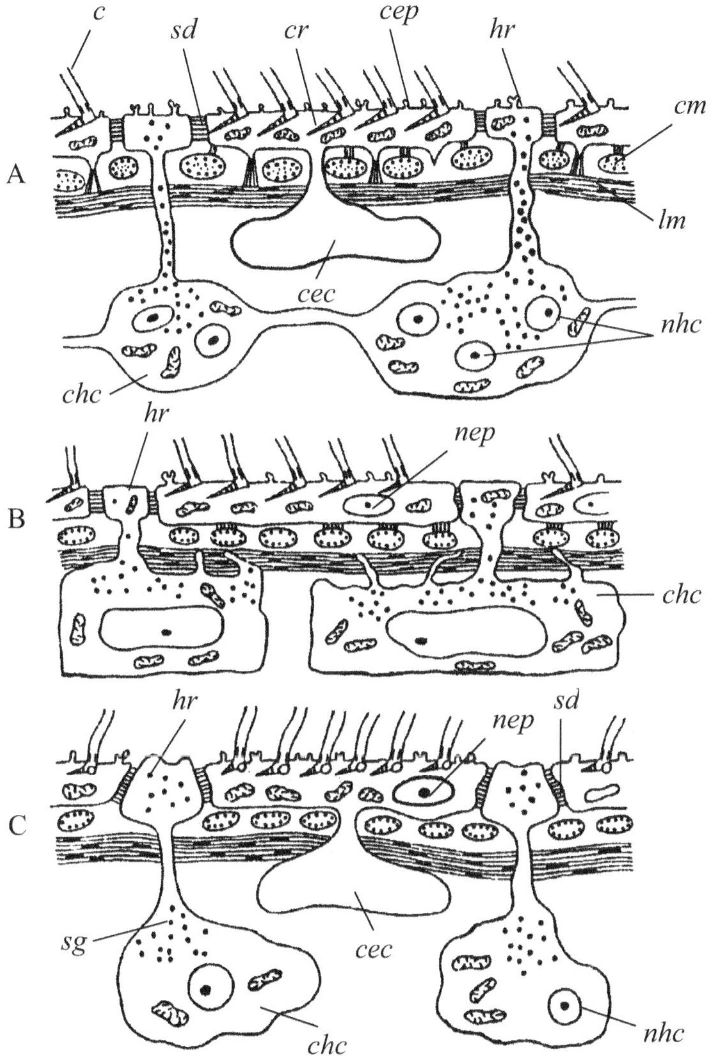

Figure 4. Diagram, showing relationship of ciliated epithelial cells with adjacent epidermal cells of miracidia of different trematode taxa. A, *Schistosoma*; B, *Fasciola*; C, *Philophthalmus*. *Abbreviations*: *c*, cilium; *cec*, nucleus-free cyton of epithelial cell; *cep*, ciliated epithelial plate (cell); *chc*, cyton of hypodermal cell; *cm*, circular muscles; *cr*, ciliary rootlet; *hr*, hypodermal ridge; *lm*, longitudinal muscles; *nep*, nucleus of epithelial cell; *nhc*, nucleus of hypodermal cell; *sd*, septate desmosome; *sg*, secretory granules. Modified from Semenov (1991), with permission.

layers of longitudinal muscles. Apparently, these structures function as a sort of anchor, which provides additional fixing of the epithelial plates on the miracidium body. Nuclei are undoubtedly absent from the epithelial plates of passively infecting miracidia of Plagiorchiida (Dobrovolskij, 1965a) and at least some Hemiuroidea (Matthews and Matthews, 1991). Under the basal membrane of the epithelial cells there is a basal lamina, tightly connected with circular and longitudinal muscles and projections of deeper localized cells (Dunn et al., 1987).

With a few exceptions contacts of the epithelial cells with the hypodermal ridges are provided by septate desmosomes (Tihomirov, 1980, 2000). The so-called hypodermal ridges are cell projections of the mother sporocyst tegument (traditionally called hypodermis), which emerge on the surface of the larva body (see p. 29). Cytons of hypodermal cells lie under the layers of circular and longitudinal muscles. In Schistosomatidae and Paramphistomidae they are interconnected by cytoplasmic projections, forming a syncytial net in the larval body (Fig. 4A, 5) (Pan, 1980; Dunn et al., 1987). The function of this net may be to provide additional fixing of the hypodermis in the miracidium body. This is particularly important during the development of the definitive tegument at the parasitic stage of the first parthenogenetic generation (see p. 31). It is quite probable that such construction of the hypodermis is common in miracidia of trematodes of different groups. Lack of relevant literature references can be explained by difficulties, encountered in electron microscopical observations of thin cytoplasmic connections between hypodermal cytons. Active nuclei, RER, dictiosomes of Golgi complex and membrane-covered secretory granules can be observed in the cytoplasm of hypodermal cells. The secretory granules play an essential role in the course of formation of the mother sporocyst tegument after penetration of the miracidium into the molluscan host (see p. 31). The surface of the hypodermal ridge has microvilli or lamellae. Secretory granules pass to their cytoplasm through the projections strengthened by microtubules.

Miracidia, lacking cilia on the body surface, possess so-called "spines" (Fig. 1E,F) whose nature still remains unclear. Semenov (1991) suggests that these might be structures resembling cirri of ciliates Hypotricha. In this case one should presume that these miracidia have epithelial cells. We believe that this is true for Azygiidae. Four and five plates covered with "spines" are observed at the anterior and posterior ends of these larvae. They can be regarded as strongly reduced epithelial plates, whereas the "spines" may represent cilia assembled in bundles (Fig. 1F). Such an interpretation is not so obvious in the case of miracidia of Didmozoidae, Halipeginae and Bunocotylinae. No structure, which could be considered as epithelial plates, was revealed there (Fig. 1E, 3D-F). However, our observations on penetration into the first intermediate host (mudsnails *Hydrobia ulvae* and *H. ventrosa*) of *B. progenetica* miracidium have shown that its spines are located on four epithelial plates, which are shed in the course of this process (unpublished data). At the same time, according to light microscopy observations, the spines of *B. progenetica* miracidia look like the tegumental spines rather than the cirri of ciliates. It is obvious that the nature of miracidia spines needs further investigations.

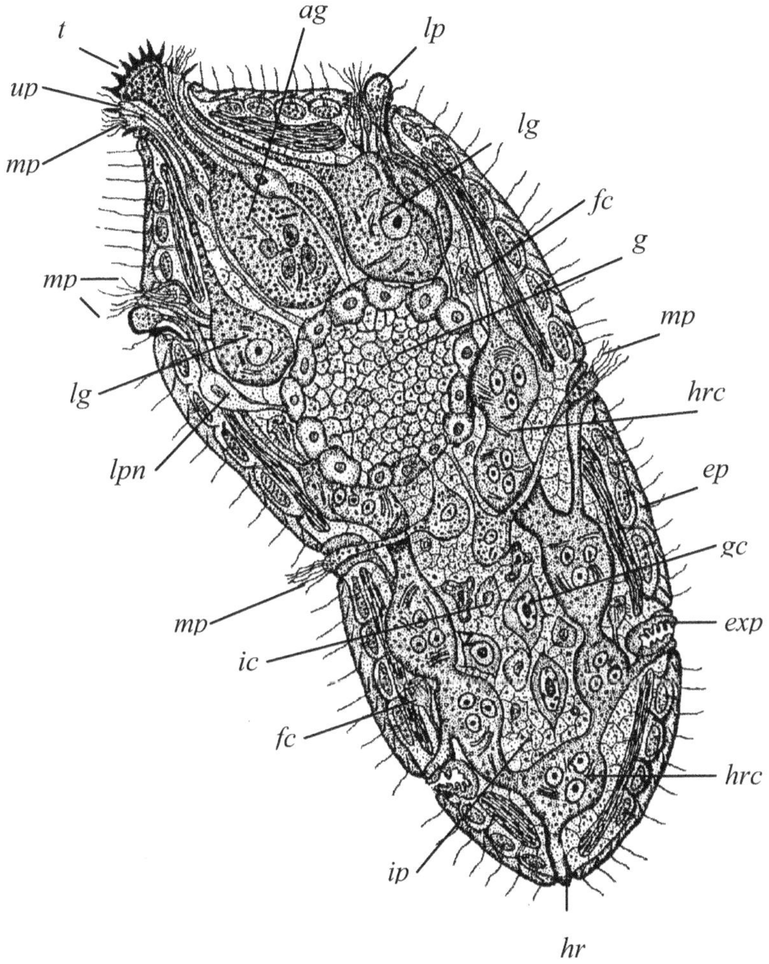

Figure 5. Miracidium of *Schistosoma mansoni*. *Abbreviations*: *ag*, apical gland; *ep*, epithelial plate; *exp*, excretory pore; *fc*, flame cell; *g*, neural mass with peripheral ganglia; *gc*, germinal cell; *hr*, hypodermal ridge; *hrc*, hypodermal ridge cyton; *ic*, interstitial cyton; *ip*, processes of interstitial cells; *lg*, lateral gland; *lp*, lateral papilla; *lpn*, perikaryon of the neuron of lateral papilla; *mp*, multiciliated sensory papillae of different types; *t*, terebratorium; *up*, uniciliated sensory papilla. From Pan (1980), with permission.

There is a terebratorium, also called apical papilla, situated at the anterior end of the miracidium. It is formed by an extension of the hypodermis and the underlying layers of circular and longitudinal muscles (Fig. 5).

There are two morphological types (telescopic and spherical) of terebratoria that can be distinguished in free-swimming larvae (Semenov, 1991). The first one found in Fasciolidae, Echinostomatidae, Cyclocoelidae and Philophthalmidae is telescopically packed in the inactive state (Fig. 6). Circular and longitudinal muscle fibers (the latter function as retractors) provide the activity of terebratorium enabling penetration into the molluscan host. The terebratorium of the second type looks like a hemispheric protrusion of the hypodermis and is encountered in the miracidia of Schistosomatidae, Paramphistomidae and Strigeidida.

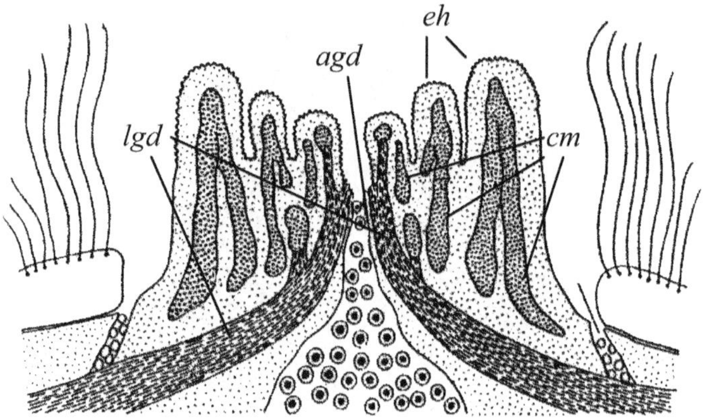

Figure 6. Telescopic terebratorium of the *Philophthalmus rhionica* miracidium. *Abbreviations*: *agd*, duct of apical gland; *cm*, circular muscles; *lgd*, duct of lateral gland; *lm*, longitudinal muscles; *eh*, extension of the hypodermis. From Tihomirov (1980), reprinted with author's permission.

The surface of the terebratoria of both types is thrown into elongate and branched lamellae, which are particularly well-developed in the terebratorium of the spherical type (Fig. 5). Interpretation of the functional role of these structures is controversial. It is suggested that the lamellae form numerous sucker-like cups which aid attachment of the larvae to the host (Kinoti, 1971; Wilson et al., 1971; Lo Verde, 1975). Numerous lamellae form a sponge-like structure on the surface of the terebratorium of *Gigantocotyle explanatum* (Paramphistomidae) miracidium and facilitate the focus of glandular secretions on a small area of the molluscan epithelium, thus avoiding waste of secretory material (Dunn et al., 1987). One can presume that by virtue of alternating contractions of terebratorium muscles during penetration of miracidium, the lamellae act as a "rasp" in conjunction with the enzymatic secretion of the glands to break down the snail epithelium (Køie and Frandsen, 1976). Apparently in passively infecting miracidia of *Bunocotyle progenetica* the same role is played by long spines at the anterior end of the body. In larvae of this species the latter can be pulled in and out like a glove finger. A small terebratorium was reported in miracidia of some trematodes with a passive way of infection, e.g., in Ochetosomatidae, Plagiorchiidae (Dobrovolskij, 1965a) and Lecithochiriinae (Matthews and Matthews, 1991). According to the electron microscopy data in the larvae of the latter it is a small protrusion of hypodermis without any lamellae at the surface (Matthews and Matthews, 1991).

A group of glands usually called penetration glands are connected with the terebratorium. Primarily it is a large apical gland located in front of the ganglion, in the anterior part of the body in miracidia of most trematodes. In larvae of Diplostomidae, Alariidae and Strigeidae the apical gland consists of a nucleus-containing part, located dorsally beyond the ganglion and a long duct (Fig. 1A,B). The latter, rounding the ganglion from the dorsal side, forms a well-expressed bulb in the preganglion zone. Usually the apical gland contains four nuclei and forms a syncytial structure. It emerges in the course of larval morphogenesis as a product of fusion of four separate cells. The duct of the apical gland reinforced by microtubules opens at the terebratorium terminal with one (Schistosomatidae, Philophthalmidae, Paramphistomidae), or several small pores (Fasciolidae, Diplostomidae). The ducts of paired lateral glands with 1-2 nuclei open at the terebratorium base. Their number varies from one (Schistosomatidae, Diplostomidae, Strigeidae) to two (Paramphistomidae, Philophthalmidae) or several (Fasciolidae, Halipeginae) pairs. In miracidia of *Fasciola hepatica* additional accessory glands are reported with openings situated at the terebratorium between the pores of the apical and lateral glands.

In miracidia studied up to now all types of gland cells have similar ultrastructure. The set of cellular organelles of these cells (well-developed endoplasmic reticulum, numerous mitochondria and Golgi complexes) is typical for secretory cells. In electron micrographs the gland secretions looks like membrane-bound granules with a heterogeneous content. In different types of cells the granules vary in shape and size. Data on the chemical nature of the granular secretions are equivocal. Most authors conclude that this is matter with sticky and/or cytolytic properties. Buzzell (1983) presumed that the secretion of the apical glands may contain an inactivated lytic factor, whereas a cofactor, activating it, is located in the lateral glands. In the apical gland of *Schistosoma* granules of α- and β- glycogen were discovered.

Apart from the penetration glands, separate gland cells were found in the body of free-swimming miracidia of a number of trematodes (Fasciolidae, Philophthalmidae and Echinostomatidae). They lack of outer pores and most probably represent specialized cells of the hypodermis, i.e., belong to the group of tegumental secretory cells. We believe that syncytial glandular structures in the posterior part of the body of Philophthalmidae, Echinostomatidae and Diplostomidae miracidia are of the same nature (Fig. 1A,B,C). In the representatives of the two latter families they are also retained in the mother sporocyst (Dönges, 1964). The function of hypodermal glands may be associated with the production of the glycocalyx at the surface of miracidia and also of the mother sporocyst.

The glandular apparatus of passively infecting miracidia undergoes some simplification. These miracidia lack hypodermal glands, while in Notocotylidae and, probably, in Didymozoidae penetration glands are also absent. The larvae of Heterophyidae, Brachycoelidae, Ochetosomatidae and Plagiorchiidae, which have been best studied, have two two-nuclear apical glands (Fig. 1H,I), while in Bucephalidae, Brachylaimoidea, Bunocotilinae and some Microphallidae only one gland is left (Fig. 3). Finally, in the fully formed miracidia of Halipeginae and some Microphallidae there are no nuclei in the single apical gland. In Cephalogonimidae, Ochetosomatidae and Plagiorchiidae miracidia, contained in the eggs, are 1-2 gland cells, called hatching glands. They open at the base of the terebratorium and are probably homologues of the lateral glands of actively infecting larvae. The glands of this type disappear after hatching of the miracidia.

Apart of the glands, stylet-like structures may be associated with the terebratorium. These structures were described in larvae of Sanguinicolidae, Brachylaimoidea, Encyclometridae and Bucephalidae (Lewis, 1969; McMichael-Phillips et al., 1992b; Suloeva, 1999; Suloeva et al., *submit*). In the miracidia of Sanguinicolidae (the only one studied by electron microscopy of the above listed species), the stylet, situated inside the apical gland, is extended along its longitudinal axis (McMichael-Phillips et al., 1992b). It is composed by three adjacent hollow tubular structures, surrounded of numerous tubes of lesser diameter. All these structures apparently consist of microtubules. Channels, connected to large secretory granules, originate from the base of the central tubules. McMichael-Phillips et al. (1992b) suggested that the stylet is capable of projecting forward, thus facilitating the penetration of miracidia through the host tissues. This process is also enhanced by the secretion, which is, apparently, delivered into the stylet lumen through the channels and excreted upon contacts with the molluscan host. Keeping in mind the location of the stylet, one may presume that it is a structure, which reinforces the duct of the apical gland. This presumption is also supported by the fact that the stylet is composed of microtubules, which usually reinforce the ducts of several types of gland cells, including the apical ones. One can not exclude the fact that the stylet of Encyclometridae plays the same role, though morphologically it resembles somewhat the stylets of virgulate cercariae with a massive bulb. It should be emphasized that the ultrastructure of the stylets of sanguinicolids, and, possibly, encyclometrids, has nothing in common with that of the cercarial stylet (see p. 121). Undoubtedly, the structures, which provide penetration of the larvae into the first intermediate host (so-called penetration complex or apparatus), include Bucephalidae stylets. Most likely, this is also true for Brachylaimoidea. The stylet of bucephalid miracidia is situated outside the apical gland and, as in the case of Brachylaimoidea larvae, is capable of moving forward and backward (Suloeva, 1999; Suloeva et al., *submit*; Lewis, 1969).

The nervous system and sense organs of the free-swimming larvae have rather complex organization. A massive cerebral ganglion (brain) is located approximately in the anterior third of the larva (Fig. 1A-D,J, 5). Nerve cords lead from this ganglion to the posterior end. Separate cords lead to the sense organs of the anterior end of the body. The central part of the brain, the neuropile, is composed of axons and dendrites. Nerve cell bodies, perikaryons, are situated at the periphery of the cerebral ganglion (Pan, 1980; Swiderski and Moczon, 1992, Swiderski et al., 1994; Halton and Gustafsson, 1996). They are fixed between processes of the adjacent cells, which encircle the brain and provide mechanical protection. In the granular cytoplasm of the neurons one can observe granules of α- and β- glycogen, lipid inclusions, smooth and RER, Golgi complexes, mitochondria, unattached ribosomes and neurosecretory vesicles of various types. The latter are particularly numerous in the peripheral nerves (Wilson, 1970; Pan, 1980), contacting muscles or indicative of specialized nerve endings.

In free-swimming miracidia of a number of trematodes, there are two simple rhabdomeric eyespots localized just above the cerebral ganglion from the dorsal side of the body. In Strigeidae, Diplostomidae, Heronimidae, Azygiidae they are situated above the lateral sides of the cerebral ganglion and widely spaced (Fig. 1A). In larvae of the other trematodes (Fasciolidae, Philophthalmidae, Cyclocoelidae, Echinostomatidae, Transversotrematidae, Allocreadiidae, etc.) (Fig. 1C,D,J), on the contrary, they are very close to each other, forming a letter "X". The pigment cell forms an eye cup, which encloses a rhabdomer – terminal part of dendrite of retinular (sensory) cell bearing numerous closely packed microvilli (Fig. 7A) (Isseroff and Cable, 1968). The nucleus is situated in the broader part of the sensory cell, immediately at the exit of the eye cup. The cytoplasm here is particularly rich in mitochondria and vesicles, which can also be found in large numbers in the base of the dendrite. The axon of the sensory cell leads back to the cerebral ganglion. With the exception of Philophthalmidae the right eyespot of all miracidia is slightly smaller than the left one, which is determined by an unequal number of their retinular cells (2 and 3 respectively) (Fig. 7A). All three retinular cells of the left eyespot associated with one eye cup, are formed by one pigment cell. The only exception known is *Heronimus mollis*. The left eyespot of its miracidium contains two pigment cups. One of them involves rhabdomers of two retinular cells, and the other one – of only one cell (Isseroff and Cable, 1968). In fact, there are two independent eyespots and the miracidium of *H. mollis* should be regarded as three-eyed.

One more sense structure found in the larvae of *Fasciola hepatica*, *Diplostomum spathaceum*, *Schistosoma mansoni* and *Echinostoma* spp. (Wilson, 1970; Brooker, 1972; Short and Cagne, 1975; Fournier, 1984; our data) is connected to the cerebral ganglion. This is a small vesicle, whose inner surface is formed by flattened endings of sensory processes of nerve cells with modified cilia (Fig. 7B). The latter are oriented inside the vesicle cavity. The structure is supposed to be a ciliary non-pigmented photoreceptor.

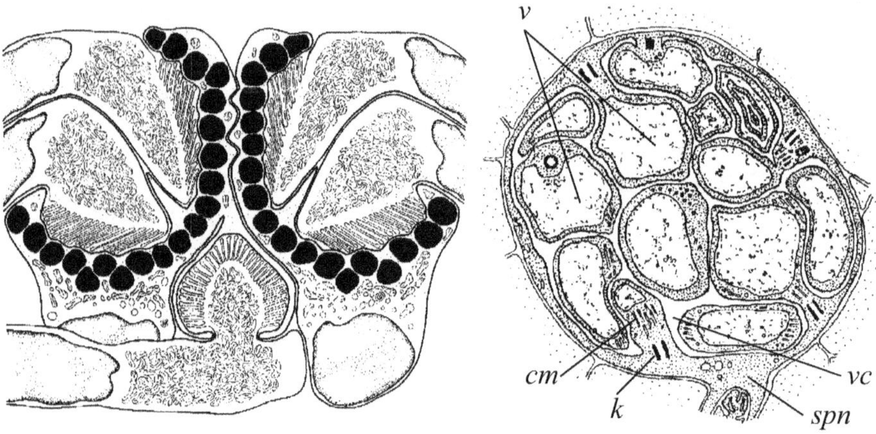

Figure 7. Diagram of miracidia photoreceptors. A, dorsal view of the eyespot of the *Fasciola hepatica* miracidium. From Isseroff and Cable (1968); B, the ciliary non-pigmented photoreceptor of the *Echinostoma* sp. miracidium. *Abbreviations*: *cm*, cilia microtubules; *k*, kinetosome; *spn*, endings of sensory processes of nerve cells; *v*, vacuole; *vc*, vesicle cavity.

Other sense structures of actively infecting miracidia are represented either by separate sensory papillae or their complex aggregations (Wilson, 1970; Pan, 1980; Tihomirov, 1980; Swiderski and Moczon, 1992; Swiderski et al., 1994). They are localized at the apex and base of the terebratorium, as well as on the hypodermal ridges between the epithelial plates of the first and second tiers (Fig. 5, 8A). The most complex organization was found in the sensory complexes of the lateral papillae (Fig. 8C), situated at the sides of the larva. In the miracidium of *Philophthalmus rhionica* the papilla is localized in the central part of the complex and represents a bulb-like nerve ending, whose cytoplasm contains orderly arranged numerous microtubules and small vesicles. From the outside papilla it is covered by a thin layer of hypodermis, with a pore on the apex. A separate papilla covered with a sponge-like layer of hypodermis is localized in front of the central papilla. The conic distal region of the nerve ending almost completely buried in the part, hardly emerging over the hypodermis surface, has three finger-like pro-

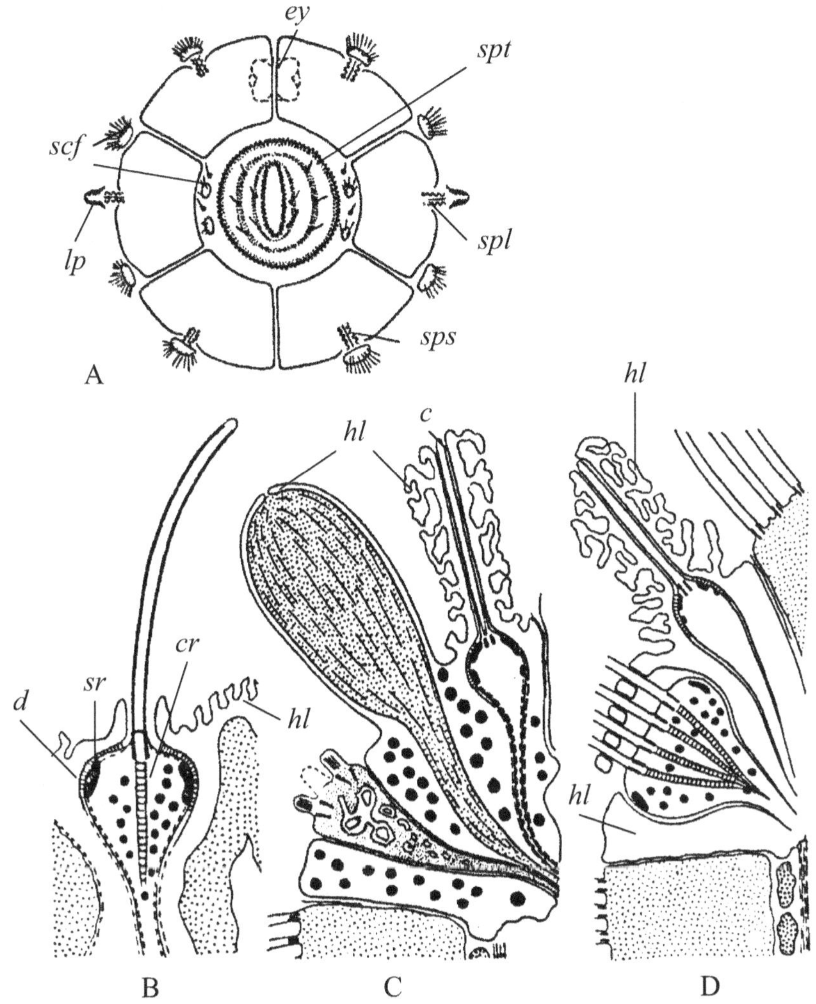

Figure 8. Structure of the sensory papillae of the *Philophthalmus rhionica* miracidium. A, location of the sensory papillae on the body of miracidium (view from the apex); B, separate sensory papilla of the terebratorium; C, sensory complex of the lateral papilla; D, sensory complex of the sensory ciliated field. *Abbreviations*: *c*, cilium; *cr*, ciliary rootlet; *d*, desmosome; *ey*, eyespot; *hl*, layer of hypodermis; *hr*, hypodermal ridge; *lp*, lateral papilla; *scf*, sensory ciliated field; *spl*, separate papilla of the lateral papilla; *sps*, separate papilla of the sensory ciliated field; *spt*, separate papilla of the terebratorium; *sr*, well-expressed cytoskeletal ring. From Dobrovolskij et al. (1983), with permission.

trusions with tightly packed microtubules.

In the miracidium of *P. rhionica*, apart from lateral papillae, sensory ciliated fields of two types are found in the hypodermal ridge between the first and the second tiers of the epithelial plates. The first one involves groups of rigid cilia at the sides of lateral papillae. The rootlets of the cilia converge into a bundle buried in the bulb-like thickening of the nerve terminal (Fig. 8D). The space between the rootlets is filled with numerous vesicles of presumably neurosecretory nature. Additionally, there are two dorsal and two ventral cilia fields of the second type that contain separate papilla similar to that of the lateral papillae complex (Fig. 8D). Its sponge-like hypodermal layer forms a protrusion covering the cilia field. It has pores for cilia.

At the terebratorium of *P. rhionica* miracidium there are cilia fields of the first type and separate single papillae (Fig. 8A,B). The latter are rigid cilia entirely deprived of hypodermal layer. The rootlet is directed from a kinetosome inside the bulbous dilation of the nerve ending, whose distal part has a strong ring of a cytoskeleton origin.

Number, location and ultrastructural pattern of sensory papillae, cilia fields and lateral papillae complexes in miracidia of other trematode species investigated (Wilson, 1970; Pan, 1980; Swiderski and Moczon, 1992; Swiderski et al., 1994) differ slightly from those described in *P. rhionica*. However, the aforementioned observations of the sensory structures of free-swimming miracidia reveal a high degree of miracidial differentiation. There are few experimental data on the function of particular structures. It is assumed that papillae and cilia fields at the terebratorium may function as mechanoreceptors, lateral papillae complex – as depth receptors, cilia fields of hypodermal ridges as photoreceptors (in schisostomes), gravitation and vibration receptors (for reviews see: Semenov, 1991). In addition, there must undoubtedly be some sensory structures functioning as chemoreceptors (Haas et al., 1995a; Haas and Haberl, 1997, see p. 231), since chemical sensitivity in miracidia has been demonstrated. Concluding the survey of the nervous system and sensory structures in miracidia we must emphasize that at present none of these receptors has been reported in the larvae of the trematode groups with the passive way of infection of the first intermediate hosts.

The miracidial excretory system is of the protonephridial type. Usually there is a pair of flame cells, but in the representatives of Schistosomatidae, Spirorchiidae, Strigeidae and Diplostomidae there can be two pairs (Fig. 1A, 5). The twisted excretory tubules emerging from the flame cells open outwards via an excretory pore in lateral sites of the hypodermal ridges, most often between the last and the next to last tiers of the epithelial plates. In the case of two pairs of flame cells, there are collecting ducts opening outwards. They are formed from the merging of the excretory tubules leading from flame cells, localized on one side of the body. In miracidia of some trematode groups (Fasciolidae, Echinostomatidae, Paramphistomidae and Diplostomidae) there is a small excretory bladder just before the excretory pore.

Electron microscopy studies of the excretory system of miracidia (Wilson, 1969b; Tihomirov, 1980, 2000) revealed its essential similarity with that described for other stages in the trematode life cycle (for reviews see: Smyth and Halton, 1983; Galaktionov and Dobrovolskij, 1987). Its main distinction consists in the formation of an excretory pore by a single cell, which is separated both from the hypodermis and excretory bladder by a septate desmosome. The excretory bladder is also formed by a single cell. Its inner surface representing the bladder lumen has lamellae and the cytoplasm is rich in active mitochondria. Short lamellae are also found on the inner surface of excretory ducts, formed as a result of rolling of flattened cells and subsequent cross-linking by septate desmosomes. The function of the excretory system consists in osmoregulation and elimination of metabolic products. It is assumed that selective absorption of substances from the incoming fluid takes place in the distal parts of excretory ducts and excretory bladder. Its preliminary filtration proceeds in the special filtration apparatus, which is formed by interdigitations or ribs of cytoplasm of the flame cell and the first tubule cell.

The excretory system of passively infecting miracidia is significantly simplified. The number of flame cells never exceeds two, whereas the excretory tubules are sufficiently expressed only in Azygiidae. In the larvae of Cephalogonimidae, Ochetosomatidae and Plagiorchiidae only flame cells are discovered. The beating of their flagella can be observed only during a short period, preceding the hatching of miracidium. No beating effect is found for inactive larvae in the egg. It is also true for the larvae, which have hatched from the egg and swim in the intestinal chyme of the molluscan host. No trace of the excretory system has been observed in the miracidia of Bucephalidae, Microphallidae and Lecithodendrioidea. According to the literature and our data there are no protonophridia in the larvae of Hemiuridae, Didymozoidae, Brachylaimoidea and Notocotylidae (Kagan, 1952; Self et al., 1963; Murrills et al., 1985a; Matthews and Matthews, 1991).

The posterior part of the miracidial body contains germinal material. The composition of the latter, though significantly variable, can be classified in two main types (Fig. 9). The first one is typical for the larvae possessing germinal cells and/or germinal balls (early embryos of daughter parthenogenetic generation) that emerge as the result of the germinal cells cleavage (Fasciolidae, Allocreadiidae, Paramphistomidae, some Echinostomatidae etc., see also pp. 33, 418). Of principal importance is the fact that the number of germinal cells (germinal balls) in the miracidial body strictly corresponds to the number of the embryos of rediae of the first generation or the daughter sporocysts (see 1.1.2.1.), which begin to develop in the molluscan host. Thus, the reproduction of the individuals of first parthenogenetic generation (the forming of germinal cells) occurs during the miracidial formation, whereas on the parasitic stage only the growth of embryos of daughter parthenitae takes place. This trend results in the appearing of so-called "viviparous" miracidiaof Philophthamidae, Cyclocoelidae, some Allocreadiidae, Clinostomidae, Echinostomatidae and Lepocreadiidae, which contain the already formed individual of daughter generation, the so called mother redia. Germinal cells and germinal balls, which produce individuals of the third parthenogenetic generation, are localized in the redia.

The data on Notocotylidae are ambiguous. The studies on *Notocotylus attenuatus* have shown that Notocotylidae lack miracidia per se. The embryo of the mother sporocyst contained in the egg is represented by two cells only – the germinal cell and the somatic cell, enclosing it (Murrills et al., 1985a) (Fig. 10). The latter is rich in glycogen and its surface, facing the inner side of the eggshell, has microvillli. A possible interpretation of this structure will be provided in 1.1.2. Still unclear is the situation with Brachylaimoidea. Our own data on the penetration of *Hasstilesia ovis* miracidia through the wall of the molluscan host intestine allow us to attribute the organization of the germinal material of these larvae to the first type. However, our data need experimental verification.

The second type of the organization of the germinal material implies the proliferation capacity preserved during a significant part of the mother sporocyst's life in the organism of the first intermediate host. The highest degree of differentiation of the germinal material is achieved in the miracidia of Strigeidida, where it is represented by a conglomerate of undifferentiated cells, capable of proliferating, germinal cells and even the earliest embryos of daughter generation (Dönges, 1964). A similar organization of the germinal material was reported for miracidia of some echinostomatids, such as *Isthmiophora melis* and *Echinostoma paraensei* (Dönges, 1973; Dobrovolskij et al., 2000). The larvae of other echinostomatids, e.g., *Petasiger* sp. and *E. caproni* have the germinal material mainly represented by undifferentiated cells and single germinal cells (Ginetsinskaya, 1968; Dobrovolskij et al., 2000).

The composition of the germinal material of *Heronimus mollis*

Figure 9. Diagram, showing transformation of germinal material in the course of digenean parthenogenetic generations development. *Abbreviations*: *M*, miracidium; *IPG*, first parthenogenetic generation (mother sporocyst); *IIPG*, second parthenogenetic generation (rediae or daughter sporocysts); *HG*, hermaphroditic generation; *IA-IB*, Fasciolidae, Paramphistomidae, Transversotrematidae, Hemiuridae and some Echinostomatidae; *IC*, Philophthalmidae, Cyclocoelidae, some Aporocotylidae and Lepocreadiidae; *ID*, Notocotylidae; *IIA*, some Echinostomatidae; *IIB*, Schistosomatidae, Sanguinicolidae and Strigeidida; *IIC*, top, Plagiorchiidae, Dicrocoeliidae and Ochetosomatidae, bottom, Microphallidae and Lecithodendrioidea; *III*, Brachylaimoidea. Germinal material (germinal and undifferentiated cells) of the first partenogenetic generation – *dotted line*; the individuals of the second (daughter) parthenogenetic generation developing from this germinal material – *dashed line*.

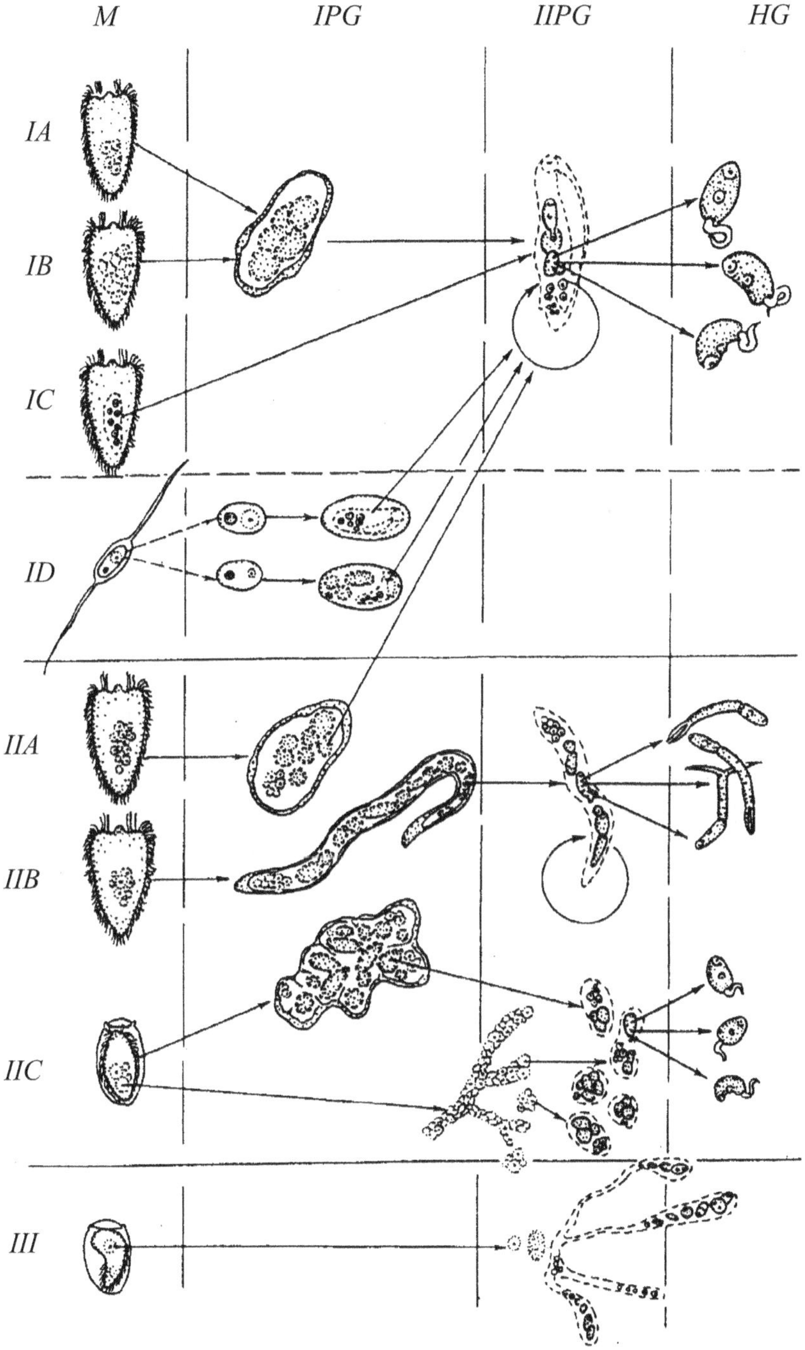
M
IPG
IIPG
HG
IA
IB
IC
ID
IIA
IIB
IIC
III

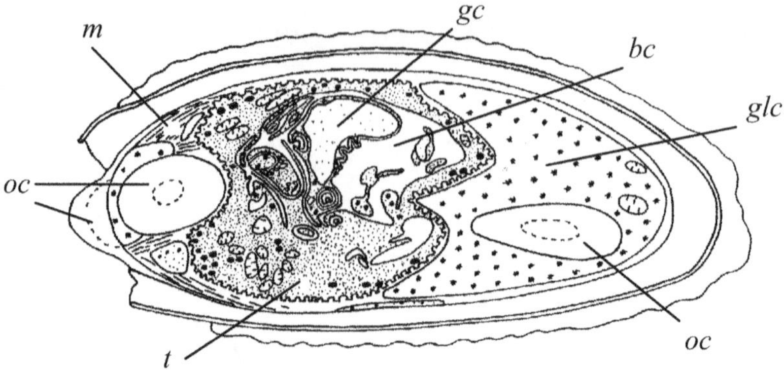

Figure 10. Section of the egg of *Notocotylus attenuatus*. *Abbreviations*: *bc*, sporocyst's brood chamber; *gc*, germinal cell; *glc*, glycogen-containing somatic cell; *m*, membranous material; *oc*, opercular cord; *t*, sporocyst's tegument. From Murrills et al. (1985a) reprinted with the permission of Cambridge University Press.

(Heronimidae) miracidia is similar to that of larvae of Strigeidida and those echinostomatids, which include *E. paraensei.* However, it forms a classical germinal mass (see p. 68), where, according to Crandall (1960) (Plate I, Fig. 1, p. 305), three zones are distinctly expressed: that of proliferation of undifferentiated cells; maturation and cleavage of germinal cells (see fig. 17A). Besides that, the germinal mass of *H. mollis* is not surrounded by tightly adjacent cells of tissue of the larvae but is rather situated in the well-expressed body cavity. In this respect the organization of *H. mollis* miracidia resembles the parasitic phase of first parthenogenetic generation (mother sporocyst), developing in the mollusc. The germinal mass is attached to the surface of the body cavity and remains in that state up to the completion of metamorphosis in the molluscan hosts (see fig. 17).

Peculiarities of functioning of Schistosomatidae mother sporocyst (see. p. 35) suggest that the germinal material of these trematodes should belong to the second type. Although it was reported to contain only germinal cells, there is also evidence that it may be similar to the second variant, described above for echinostomatids as seen from the illustrations provided by Faust and Meleney (1924) and Wu (1953). Apparently, some of the so-called interstitial cells, observed in *Schistosoma* miracidia (Kinoti, 1971; Pan, 1980), in fact, are undifferentiated cells, included in the germinal material alongside of the generative ones.

The germinal material of Plagiorchiida miracidia probably belongs to the second type (see p. 24). Its cell composition is minimal. At the most, there are 1-2 germinal cells and 1-4 undifferentiated cells (Dobrovolskij et al., 1983). Miracidia can be ordered in a series, according to the stage of ontogenesis, at which they acquire ability to infect the molluscan host. The ontogenetic stages differing in the development of germinal material (Fig. 11). In *Telorchis assula* miracidia two germinal cells are always observed. In *Haplometra cylindracea* 40% of the larvae have two germinal cells and 60% – only one. In *Macrodera longicollis* only 8% of the investigated larvae possess two germinal cells, 36% – one, the germinal material of 56% of the larvae is represented only by undifferentiated cells. This is also true for *Paralepoderma cloacicola* larvae practically lacking individuals with germinal cells.

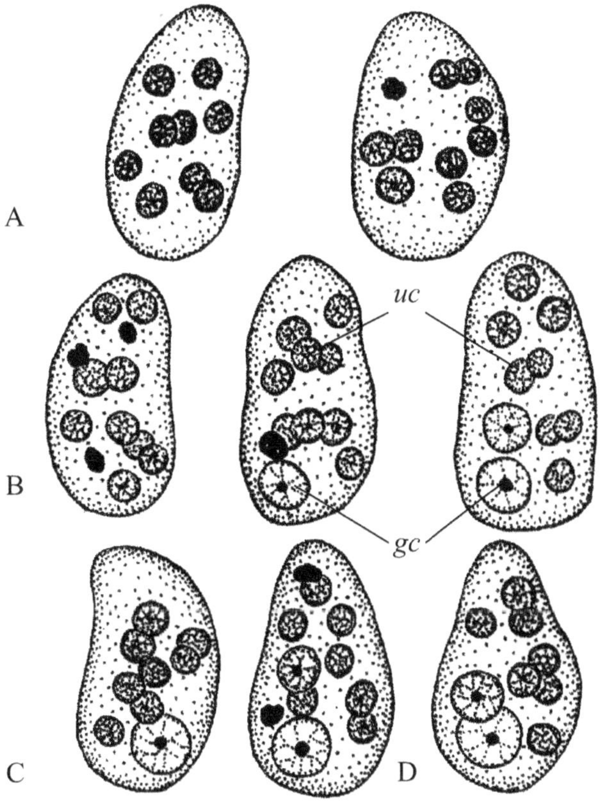

Figure 11. Cell composition of different plagiorchiid miracidia. A, *Paralepoderma cloacicola*; B, *Macrodera longicollis*; C, *Haplometra cylindracea*; D, *Telorchis assula*. *Abbreviations*: *gc*, germinal cells; *uc*, undifferentiated cells.

Concluding the morphological description of miracidia, we should emphasize that their organization is visibly simplified, when the active way of infection is replaced by the passive one. This can be demonstrated practically for all systems of organs. Primarily, the following changes can be traced: decrease in the number of tiers and total amount of epithelial plates (up to their complete disappearance), reduction of eyespots and decrease in quantity and diversity of other receptors, i.e., the structures directly related to the locomotion, spatial orientation and active search of the host. This is accompanied by diminishing of the body size, which becomes possible due to reduction of the above listed morphological structures. In turn, this causes simplification of the excretory system: the length of excretory tubules decreases, the excretory bladder and excretory pore are reduced and finally even flame cells disappear. The glandular apparatus also undergoes certain simplification, mainly owing to the loss of lateral and accessory glands. However, it should be noted that the apical gland, being of utmost importance for the process of penetration into the molluscan host, is never lost. Moreover, supplementary structures supporting this process may emerge, e.g., stylets in Bucephalidae and Brachylaimoidea, “spine” crowns at the anterior end of the body of some Hemiuroidea. Thus, passively infecting miracidia, in spite of seemingly simple organization, are completely equipped with provisional structures (primarily, penetration apparatus), which enable them to solve the crucial task of the corresponding phase of the life cycle – the penetration into the organism of the first intermediate host.

1.1.2. Mother Sporocyst

The individual of the first parthenogenetic generation parasitic in the molluscan host (however see footnote i) is called the mother sporocyst. This phase remains the least studied in the life cycle of trematodes. After a series of papers published in the 1940s – 1960s (the most distinguished of which are the works of Cort and colleagues: Cort, 1944; Cort et al., 1949, 1950, 1952, 1954; Cort and Oliver, 1943; Cort and Ameel, 1944; Ameel et al., 1949), new descriptions of mother sporocysts have appeared in literature only occasionally. As a rule, the morphological level of most of those publications is relatively sparce. Hence, mother sporocysts have not been described in representatives of a number of families including the Hirudinellidae, Ptychogonimidae, Sclerodistomidae, Syncoeliidae and Didymozoidae. No detailed description of this stage of the life cycle is available in the literature even for such a favorite model of experimental studies such as *Fasciola hepatica.* Therefore, in this chapter we refer mainly to highly reliable data.

It should be noted that the stage of mother sporocyst is missing in the life cycle of the so-called viviparous trematodes (Philophthalmidae, Cyclocoelidae, some Allocreadiidae, Clinostomidae, Echinostomatidae and Lepocreadiidae) (Pearson, 1992). In the miracidia, a young individual of a daughter parthenogenetic generation – the mother redia – is formed, which is inoculated into the molluscan host (for details see p. 53).

The appearance of the mother sporocyst is quite variable in trematodes. These may be worm-like organisms possessing a well-developed tegumentary-muscular sac and capable of migrating in the host organism (i.e., Strigeidae, Diplostomidae, Cyathcotylidae, Opecoelidae, some Echinostomatidae, Plagiorchiidae and Ochetosomatidae) (see fig. 14). They can be 10-40 mm long (*Mesostephanus appendiculatus* – Cyathocotylidae), but, as a rule, do not exceed 0.3-1.0 mm (Fasciolidae, Notocotylidae, Paramphistomidae, Echinostomatidae, Psilostomidae) (see fig. 15) (Dobrovolskij et al., 1983). Large, often branching, mother sporocysts, bag-like or of irregular shape, are seen in many plagiorchiids. In Microphallidae, Lecithodendrioidea and some other plagiorchiids practically complete reduction of the soma at the early stages of mother sporocyst formation takes place. As a result, germinal masses (parthenitae's reproductive organs) pass to independent parasitic existence (see below).

1.1.2.1. Mother Sporocysts of Trematodes with Actively Infecting Miracidia

Morphology and reproduction of mother sporocyst, their localization in the molluscan host depend significantly on "the starting position", i.e., on the morphological-biological organization of miracidia. Normally, actively infecting miracidia penetrate through the pallium, foot and tentacles. During this process most miracidia from the following families (Fasciolidae, Echinostomatidae, Paragonimidae, Cyathocotylidae, Strigeidae, Diplostomidae, some Allocreadiidae) lose epithelial plates. The events, which occur after that, were studied in detail in the larvae of *Fasciola hepatica* (Southgate, 1970; Wilson et al., 1971) (Fig. 12).

The miracidium attaches to the surface of the molluscan body by means of the terebratorium. Almost immediately (in 60 sec) the epithelial plates of the first tier are cast off, then (in 10-15 minutes) those of the second tier are lost. This is continued until the epithelial plates of the last tier are cast off. Such a sequence in shedding of the epithelial plates is a general rule for all miracidia of the above mentioned families. The mechanism of separation of the plates is not fully understood. The key element of this mechanism consists of disruption of cell junctions, connecting the epithelial plates to the surrounding hypodermal ridges and underlying structures (muscle cells, processes of the hypodermal cytons). One can not exclude the involvement of enzymes, presumably contained in lysosome-like structures of epithelial plates and activated in the course of penetration (Wilson, 1971; Wilson et al., 1971). Southgate (1970) report the emergence of large cavities between the basal lamina and basal membrane of the epithelial plates in the penetrating miracidium of *F. hepatica* (Fig. 12B), which may contribute to the shedding of the plates.

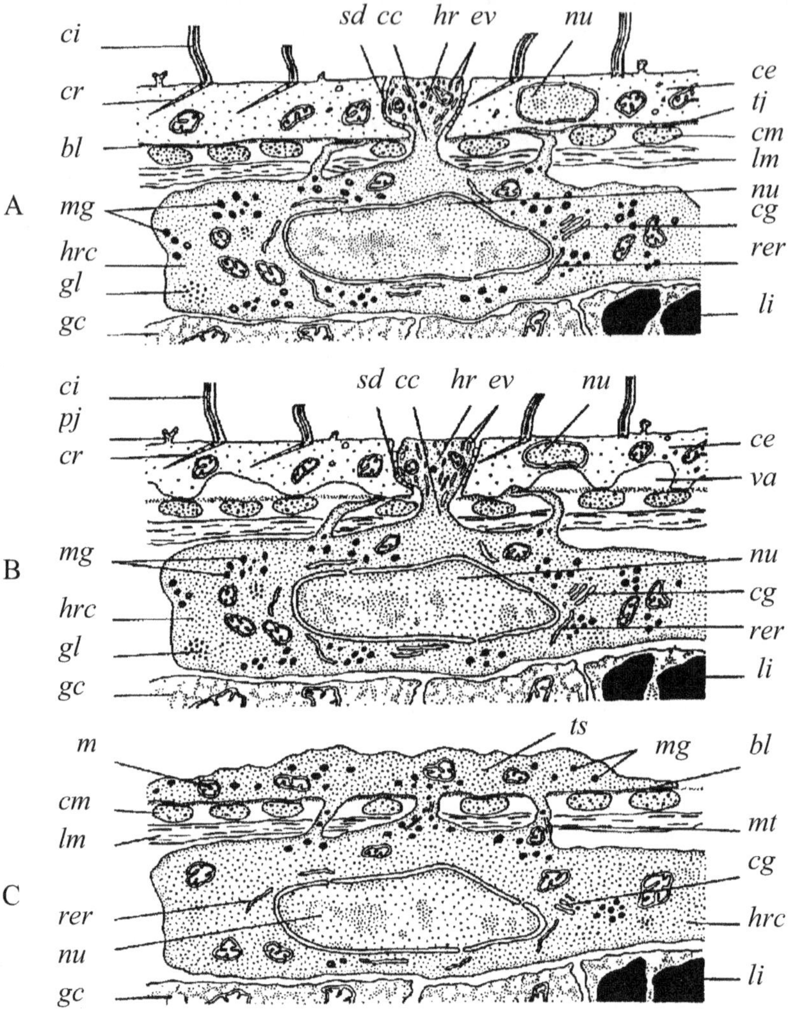

Figure 12. Diagram, showing formation of the tegument during metamorphosis of the *Fasciola hepatica* miracidium. A, intact epidermis of the miracidium; B, changes in epidermis of the miracidium during penetration through the epidermis of the molluscan host; C, the development of the tegument of the mother sporocyst inside the molluscan host. *Abbreviations*: *bl*, basal lamina; *cc*, cytoplasmic connection between hypodermal (tegumental) ridge and hypodermal (tegumental) ridge cyton; *ce*, ciliated epithelial plate; *cg*, complex Golgi; *ci*, cilium; *cm*, circular muscles; *cr*, ciliary rootlet; *ev*, elongated vesicles; *gc*, germinal cell; *gl*, glycogen granules; *hr*, hypodermal (tegumental) ridge; *hrc*, hypodermal (tegumental) ridge cyton; *li*, lipid; *lm*, longitudinal muscles; *m*, mitochondrion; *mg*, membrane bound granules; *mt*, microtubules; *nu*, nucleus; *pj*, projection; *rer*, rough endoplasmic reticulum; *sd*, septate desmosomes; *tj*, tight junction; *ts*, tegument spreading over exposed muscles; *va*, vacuoles. From Southgate (1970) reprinted with the permission of Cambridge University Press.

Along with the separation of epithelial plates, the formation of the mother sporocyst tegument proceeds due to the growth of hypodermal ridges and their spreading over the basal lamina. This process as well as the shedding of epithelial plates occurs from the anterior to the posterior end. The hypodermis starts spreading over the basal lamina immediately after the separation of the adjacent epithelial plates. Thus, either no direct contact of the basal plate with the environment takes place or this contact is very brief (Wilson et al., 1971). According to Southgate (1970), the rapid spreading of the hypodermis (=tegument) over the body surface of the penetrating larva is enhanced by membrane material, which has accumulated as elongated vesicles in the larval cytoplasm. Besides this, membranes of secretory granules are also incorporated into the outer plasma membrane of the new tegument in the course of the secretion extrusion. The latter is, apparently, used to form the glycocalyx of the developing sporocyst (Southgate, 1970; Dunn et al., 1987). Gradually the whole surface of its body becomes covered with a thin layer of tegument, which later gets thicker and acquires the appearance of the definitive state.

The above described processes look somewhat different in trematodes, whose miracidia preserve the epithelial plates during the penetration into the mollusc (Schistosomatidae, Spirorchiidae, Paramphistomidae, Gorgoderidae, Heronimidae). In the best studied miracidium of *Schistosoma mansoni* (Wajdi, 1966; Voge and Seidel, 1972; Basch and DiConza, 1974; Wikel and Bogitsh, 1974; Køie and Frandzen, 1976; Meuleman et al., 1978) the following degenerative changes in the epithelial plates were observed just after penetration: mitochondria were transformed into multilaminated bodies, outer plasma membrane was partly destroyed; extrusion of cytoplasm fragments, sometimes with cilia, took place in these sites, in some zones epithelial plates began to separate from the underlying basal lamina (Fig. 13). At the same time the hypodermal ridges grew under septate desmosomes, which connect them with the epithelial plates, and spread over the surface of the basal lamina, thus forming the tegument of the mother sporocyst. Already in an hour after the penetration of the miracidium of *S. mansoni*, all epithelial plates shed due to the combined effect of cilia beating, contracting of underlying muscles and progressive growth of the hypodermis (Meuleman et al., 1978).

As in the earlier case of *Fasciola hepatica* miracidia, a significant role in the process of separation of the epithelial plates is, apparently, played by the cavities between the basal membrane of the epithelial cells and basal lamina, which were discovered in *S. mansoni*, investigated by electron microscopy (Basch and DiConza, 1974; Wikel and Bogitsh, 1974) and *Gigantocotyle explanatum* (Dunn et al., 1987). No strict order in the shedding of the epithelial plates was observed; however, the plates of the first tier were the last to separate. This occurs after retraction of the terebratorium. After

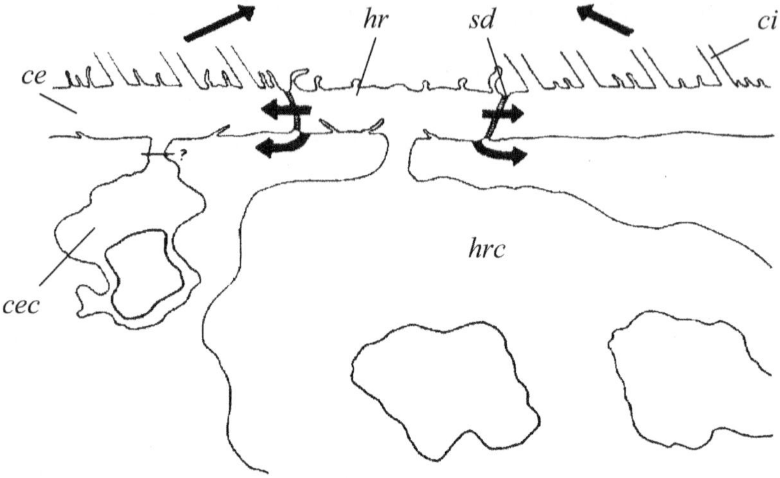

Figure 13. Diagram, showing transformation of epidermis during the metamorphosis of the *Schistosoma mansoni* miracidium. *Abbreviations*: *ce*, ciliated epithelial plate; *cec*, cyton of epithelial cell; *ci*, cilium; *hr*, hypodermal (tegumental) ridge; *hrc*, hypodermal (tegumental) ridge cyton; *sd*, septate desmosomes. From Meuleman et al. (1978), reprinted with the permission of Springer-Verlag GmbH & Co.KG.

that the tegument covers the anterior part of the body of the mother sporocyst.

Whatever the character of miracidium penetration is, its transformation into a young mother sporocyst proceeds as a regressive metamorphosis. Apart from the loss of the epithelial plates, the gland secretion is utilized, which is accompanied by subsequent degeneration of the glands; the terebratorium, cerebral ganglion, eyes (if they existed) and, at any rate, some of sensory papillae are reduced. At the same time, as noted above, the tegument thickens; microvilli are formed on its surface, the number of mitochondria in cytoplasm increases (see fig. 12). All this testifies to participation of the tegument in transportation of nutrients from the host organism (Southgate, 1970; Hockley, 1973). Unfortunately, data on the ultrastructure of mother sporocysts available in the literature are limited to the earliest stages of their development. Similar to the daughter sporocyst (see section 1.2.) one can presume that in mother sporocysts, feeding through the tegument proceeds also in the remaining period of their life activity, since mother sporocysts of all trematodes, without exception, have no rudimentary organs of digestive system. In some species, during the growth of mother sporocysts the excretory system may become more complex due to their growth and/or increasing the number of flame cells. The parenchyma volume also increases, to different degree in representatives of various groups[iii]. Later the body cavity (the schizocoel) is formed (for details see below) due to the reduction or transformation of parenchymal cells. The germinal material contained in

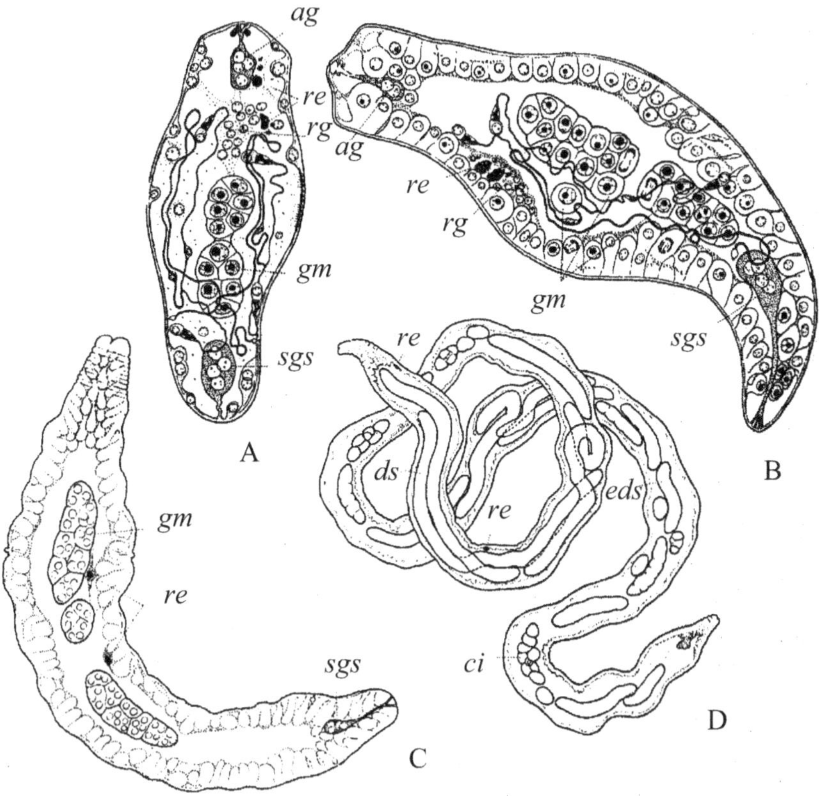

Figure 14. Development of the *Posthodiplostomum cuticola* mother sporocyst. A, 4-day-old mother sporocyst; B, 7-day-old mother sporocyst; C, 11-day-old mother sporocyst; D, 27-day-old mother sporocyst. *Abbreviations*: *ag*, apical gland; *ds*, daughter sporocyst; *eds*, embryo of daughter sporocysts; *gm*, germinal mass; *re*, resorbing eyespots; *rg*, resorbing ganglion; *sgs*, syncytial glandular structure. From Dönges (1964), reprinted with the permission of Springer-Verlag GmbH & Co.KG.

the miracidium also undergoes certain changes. It is the character of these changes that determines the features of the definitive organization of the mother sporocyst and their reproduction. In the Fasciolidae, Paramphistomidae, Transversotrematidae, Gastrothilacidae, Paragonimidae, Allocreadiidae, Clinostomidae and some Echinostomatidae, whose miracidia contain germinal cells or germinal balls formed as the result of cleavage of those cells (first type of germinal material organization, see p. 23), the fecundity of the mother sporocyst is limited by the number of these germinal elements. This is explained by the fact that mature germinal cells can not proliferate and are capable only of cleavage, forming the embryos of the subsequent generation (details on the nature of germinal cells are provided in section 1.1. in chapter 6). Mother sporocysts are small (up to 1 mm long), rounded organisms sac-like or worm-like shape. In their body cavity the embryos of mother re-

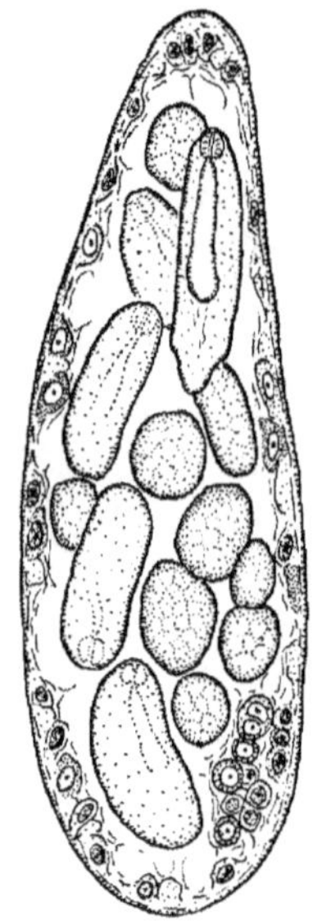

Figure 15. Mother sporocyst of *Echinostoma caproni*.

diae at different stages of development are seen (Mattes, 1949; Cort et al., 1950; Dinnik and Dinnik, 1956, 1960; Dinnik, 1961; Peters and La Bonte, 1965; Cannon, 1971; Madhavi, 1976, 1980; Cribb, 1988) (Fig. 15). The development of embryos gradually becomes uniform and is completed almost simultaneously. Young rediae penetrate the walls of the mother sporocyst, causing its death and then escape into the haemocoel. Maturation of the mother rediae embryos normally takes 2-3 weeks, but in *Bunodera sacculata* it takes more than 3 months (Cannon, 1971).

The number of individuals belonging to the next parthenogenetic generation, produced by mother sporocysts of the considered type, rarely exceeds 10, and only in *Paramphistomum microbothrium* reaches 18-20 (Dinnik, 1961). On the contrary, in mother sporocysts of *Fasciola hepatica* (Czapski, 1978), *Prototransversotrema steeri* (Transversotrematidae) (Cribb, 1988), *Clinostomum marginatum* (Clinostomidae) (Cort et al., 1950), *Paragonimus westermani ichunensis* (Shimazu and Oshima, 1983) a single redia is formed, although there are several germinal cells in the miracidium. Obviously, the best developed embryo suppresses all the others. However, it can not be excluded that this is just a random event, since the number of observations in the cited publications is very little. There are reports that at least a significant portion of embryos in the mother sporocyst of *F. hepatica* successfully accomplish their development and generate mother rediae (Skrjabin, 1948; Mattes, 1949).

The events occurring in the mother sporocyst of species with germinal material of miracidia belonging to the second type are somewhat different. In the mother sporocyst of *Echinostoma caproni*, a number of undifferentiated cells capable of proliferation are present along with the germinal cells (Ataev et al., 1997). The germinal cells considered as primary cells start cleaving practically immediately after penetration of the miracidium, often before the end of the migratory period. Proliferation of undifferentiated cells takes place during the first days postinfection in the molluscan host and results in the production of both numerous somatic cells and secondary germinal cells. The latter undergo cleavage resulting in the formation of mother rediae. It should be noted, however, that the number of rediae produced by the mother sporocyst does not exceed 20. In *E. paraensei*, primary germinal cells start cleaving in the miracidium (Dobrovolskij et al., 2000) (see p. 419).

In the young mother sporocyst, the germinal mass, containing actively proliferating undifferentiated cells (see below), is found. Nevertheless, not all newly formed germinal cells and the youngest embryos can complete their development. The mother sporocyst dies before its germinal material is exhausted. Probably mother sporocysts of some Gorgoderidae function similarly. Based on the illustrations (Fig. 16) provided by Goodchild (1948), one could presume that there are undifferentiated cells capable of proliferation in the body walls of mother sporocysts. Later these cells can specialize as germinal ones.

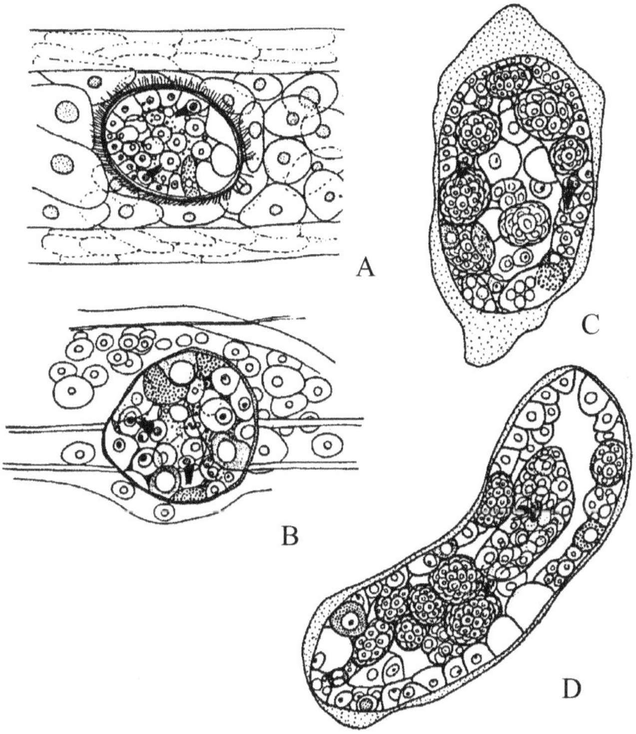

Figure 16. Development of the *Gorgodera amplicava* mother sporocyst. A, miracidium penetrating through the molluscan gill; B, 10-day-old mother sporocyst; C-D, 33-day-old mother sporocysts. From Goodchild (1948), reprinted with publisher's permission.

Representatives of Schistosomatidae, Spirorchiidae, Sanguinicolidae, Cyathocotylidae, Diplostomidae, Strigeidae and Opecoelidae, which form a special reproductive organ (the germinal mass), possess a well-developed capacity to multiply germinal cells at the stage of the mother sporocyst. Organization and possible origin of this structure in sporocysts of different generations and different trematode species are considered in section 1.2.2. Here we only note that the germinal mass consists of undifferentiated cells capable of proliferation, maturing and mature germinal cells and embryos at the earliest stages of development. All this germinal material is covered from all sides with thin laminated structures, which insure the integrity of the germinal mass.

The genesis of the germinal mass of the mother sporocyst has not been studied in all representatives of the above considered groups of trematodes. According to Cribb (1985), no germinal mass was revealed in 9-14-day old mother sporocysts of *Opecoelus variabilis* (Opecoelidae). The first individuals of the daughter generation develop in the body cavity from germinal cells, which were present in the miracidium. Later (21-days old sporocysts) undifferentiated proliferating cells, which then form the germinal mass, are grouped at the anterior end of the body. During this time it remains attached to the back wall of the body cavity by means of laminated structures. Once formed, young daughter sporocysts leave the mother sporocyst through a birth pore – a narrow channel, connecting the body cavity with the surface of the mother sporocyst. The presence of embryos of the daughter generation in the young mother sporocysts of *O. variabilis* attest to the occurrence of the germinal mass at the earliest stages of parthenita development, and possibly, even at the miracidium stage.

The transformation of the germinal material in young sporocysts of strigeidids follows a different pattern (Van der Woude et al., 1953; Cort et al., 1954; Pearson, 1956, 1961; Dönges, 1964). The cells, contained in the germinal material at the miracidium stage and considered germinal (see p. 26), actively divide and form a cellular cord in the central part of the parthenita (see fig. 14). This cord then splits into several independent germinal masses (from 3 – in *Posthodiplostomum cuticola* to 20-24 – in *Diplostomum flexicaudatum*), which are not associated with the wall of the body cavity. These are the germinal masses that produce the main part of the embryos of the daughter parthenogenetic generation. According to the figures, presented by Pearson (1961, fig.15, p. 150) a part of germinal cells, maturating already at the miracidium stage, is not included in the forming germinal masses, but passes directly to cleavage. Thus, formed embryos of the daughter sporocyst are found in the body cavity of the 5-day old mother sporocyst along with the developing germinal masses. Therefore, a certain similarity with the process of germinal material transformation, described earlier in Opecoelidae, can be traced.

Fecundity of mother sporocysts, possessing germinal masses, increases significantly, especially if there is more than one germinal mass. In strigeidids the fecundity results in the production of hundreds and thousands of daughter individuals. Consequently their worm-shaped body increases in size, becoming over 10 mm long (see fig. 14). Sporocysts have a well-developed tegumentary-muscular sac, and are mobile. There are glands, which can also be found in the body wall of the sporocysts.

Germinal masses and embryos of daughter sporocysts at different stages of development float freely in the body cavity. The excretory system becomes more complex compared to that of the miracidium. The ducts of the excretory system terminate in an excretory pore at the posterior part of the parthenita body. In most investigated species, young daughter individuals are released through the birth pore, although the latter was not reported by Dönges (1964) who thoroughly investigated mother sporocysts of *Posthodiplostomum cuticola*. Dönges, as well as Van der Woude et al. (1953), point out that germinal masses exist during a brief period of time. In the mother sporocyst of *P. cuticola*, the germinal masses disappear by the 15th day of development of the mother sporocyst in the mollusc. Later, only the embryos of daughter generation of different ages are found in the inner cavity of the sporocyst. Also, we observed many times large mother sporocysts of *Mesostephanus appendiculatus* (Cyathocotylidae), containing numerous germinal masses, germinal balls and daughter sporocysts of different ages in molluscs, which had a considerable amount of daughter sporocysts, already releasing cercariae (our data).

Mother sporocysts of trematodes with actively infecting miracidia are normally mobile and can undergo considerable migrations in the molluscan host (for details see p. 250). Most of them parasitize the body cavity and are localized in the pallium, gills, siphons and lungs of Pulmonata, i.e., in the vicinity of the site of penetration of miracidia. At the same time young parthenitae of *Echinostoma* (Echinostomatidae) migrate along the haemocoel lacunae to the heart – the site of their final settlement (Lie, 1964; Hsu et al., 1968; Ataev et al., 1997). Mother sporocysts of many strigeidids also migrate actively along the haemocoel lacunae being retained between the hepatopancreatic acini (Pearson, 1961; our data).

Concluding a brief survey of morphological and biological features of trematode mother sporocysts with actively infecting miracidia it is necessary to treat an aberrant life cycle of Heronimidae, which is not interpreted clearly in the literature (Ulmer and Sommer, 1957; Crandall, 1960; Ginetsinskaya, 1968; Brooks et al., 1985a; Gibson, 1987; Pearson, 1992). Young sporocysts of *Heronimus mollis* have eyes, a pair of protonephridia, a ganglion and remnants of the apical gland (Fig. 17). There is also a body cavity, with germinal mass attached to its wall, which is already expressed in the miracidium stage. The sporocyst rapidly develops, its body becoming an irregular multi-branched shape. The germinal mass actively functions, producing numerous embryos, filling the enlarging body cavity. These embryos do not develop into daughter parthenitae, but into cercariae, i.e., larvae of the hermaphroditic generation. There is no daughter parthenogenetic generation in *H. mollis*. Crandall (1960) and Gibson (1987) consider the described fine lining (sac – according to these authors) of the body cavity of *H. mollis* sporocyst as a remnant of the daughter parthenita. In such a case, germinal

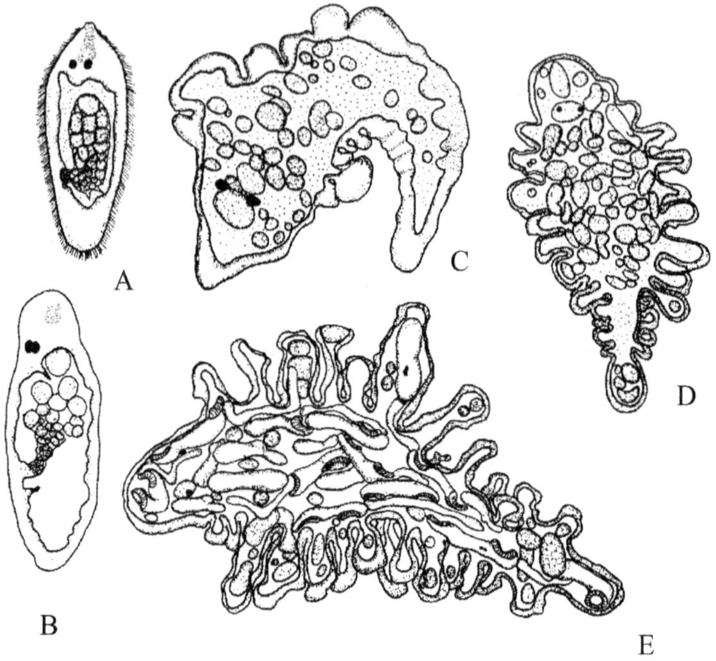

Figure 17. Development of the *Heronimus mollis* mother sporocyst. A, miracidium; B, 2-day-old mother sporocyst; C, 11-day-old mother sporocyst; D, 13-day-old mother sporocyst; E, 20-day-old mother sporocyst. From Crandall (1960), reprinted with publisher's permission.

material, which gives rise to cercariae, originates from the daughter parthenita, rather than from the mother. We do not see any good reason for such an interpretation. There is sufficient ground to consider the aforementioned sac as an endocyst, i.e., the lining of the body cavity of the mother sporocyst, which is formed due to flattening and transforming parenchymal cells (see below). The nature of the *H. mollis* sporocyst and its life cycle will be considered in detail in subsequent chapters (see pp. 436, 507).

1.1.2.2. Mother Sporocysts of Trematodes with Passively Infecting Miracidia

Morphological and biological organization of mother sporocysts of trematodes with passively infecting miracidia is even more diverse, than in the representatives of digeneans considered above. Only the first stages of mother sporocyst formation, which follow the penetration of the miracidium in the intestinal wall of the molluscan host, are common in these two groups. During this process, larvae, which have epithelial plates (all plagiorchiids and some hemiuroid trematodes – Lecithochiriinae, Halipeginae), shed them. One can presume that, in this case, alongside with this process, similar to metamorphosis of actively infecting miracidia, the parthenita tegument is formed due to extensive growth of hypodermal ridges. The described processes have not been studied ultrastructurally.

The only exception is *Lecithochirium furcolabiatum* (Matthews, Matthews, 1991). Three hours after the penetration of miracidia, mother sporocysts are completely covered with tegument (= hypodermis), which is referred to the authors as "epidermis". Small electron-lucid zones, interpreted as vesicles, can be observed in the epidermis cytoplasm in the vicinity of the basal membrane. Matthews and Matthews (1991) ascribed to them functions, similar to those of the cavities, which are revealed, when epithelial plates of actively infecting miracidia are cast off (see p. 29). Matthews and Matthews (1991) believe that the epidermis is later shed and then replaced by the definitive tegument of the sporocyst. It is difficult to accept this viewpoint. The observed electron-lucid vesicles may either be explained as a fixation artifact, or suggest the active process of trans-membrane transport. However, we believe that the aforementioned vesicles can not be regarded as a reliable proof of subsequent shedding of epidermis. According to the electron micrographs, provided by Matthews and Matthews (1991), it is most likely this tissue must be an extensively grown hypodermis, i.e., the tegument of the mother sporocyst.

Subsequent stages of mother sporocyst development vary significantly. In Halipeginae they resemble to some extent those observed in strigeidids. Multiplication of undifferentiated cells, which already existed in miracidia, occurs. Some of these cells keep on proliferating, whereas others develop into germinal cells, starting to cleave (Ameel et al, 1949) (Fig. 18). All this germinal material is grouped into several germinal masses, attached to the wall of the forming schizocoel. The sporocyst increases in size not exceeding 1 mm in length. Its body assumes a characteristic shape of a thin-walled sac, with the body cavity filled with freely floating rediae embryos of different age (Fig. 18D). The number of germinal masses gradually reduces to a single one. It is located at the posterior end of the body cavity and attached to the body wall of the sporocyst by means of a thin stem. Mature rediae leave the mother sporocyst through a birth pore, localized in the anterior end. The number of rediae, simultaneously observed in the body cavity of a mature sporocyst (one month after experimental infection of the molluscs), reaches 109. However, the real fecundity of the mother generation may exceed slightly this number, as mother sporocysts were also observed along with rediae at later stages of experimental infection of molluscs. Normally functioning germinal mass and a small number of young redia embryos were seen in the body cavity of such sporocysts.

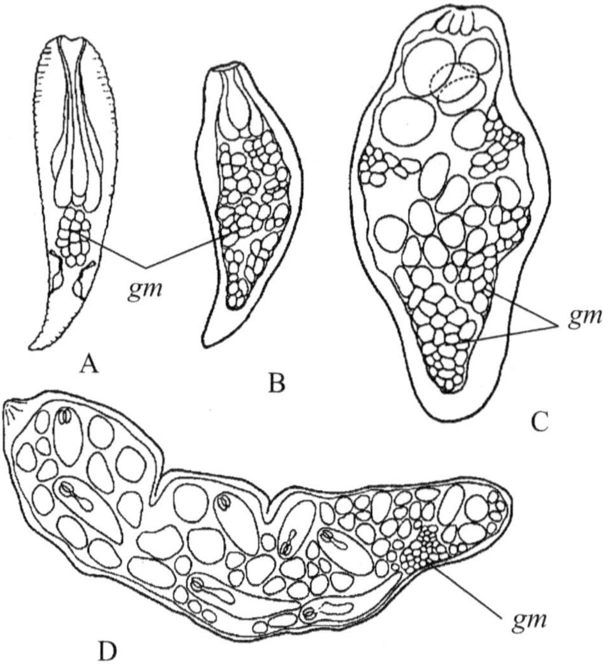

Figure 18. Development of the *Halipegus eccentricus* mother sporocyst. A, 5-day-old mother sporocyst; B-C, 11-day-old mother sporocyst; D, 11-day-old mother sporocyst. *Abbreviation*: *gm*, germinal mass. From Ameel et al. (1949), reprinted with publisher's permission.

In other hemiuroid trematodes there is no proliferation of germinal material in the mother sporocyst stage. In small (up to 1 mm long) worm-like sporocysts *Genarchopsis goppo*, localized in the molluscan haemocoel, close to the anterior part of the intestine (i.e., in the vicinity of the area of miracidia penetration), there can be 12-14 developing rediae embryos. This number does not exceed the amount of germinal cells, which can be found in miracidia of this species (Madhavi, 1978). Mother sporocysts of the aforementioned species *Lecithochirium furcolabiatum* contain only 1-3 embryos of the daughter parthenogenetic generation (Matthews and Matthews, 1991).

More fecund and morphologically diverse are mother sporocysts of plagiorchiids, which have been best investigated, though only at the light microscopy level (Dobrovolskij, 1965b, 1967, 1969, 1971, 1975; Dobrovolskij and Reichel, 1973; Dobrovolskij and Muhamedov, 1979) (Fig. 19).

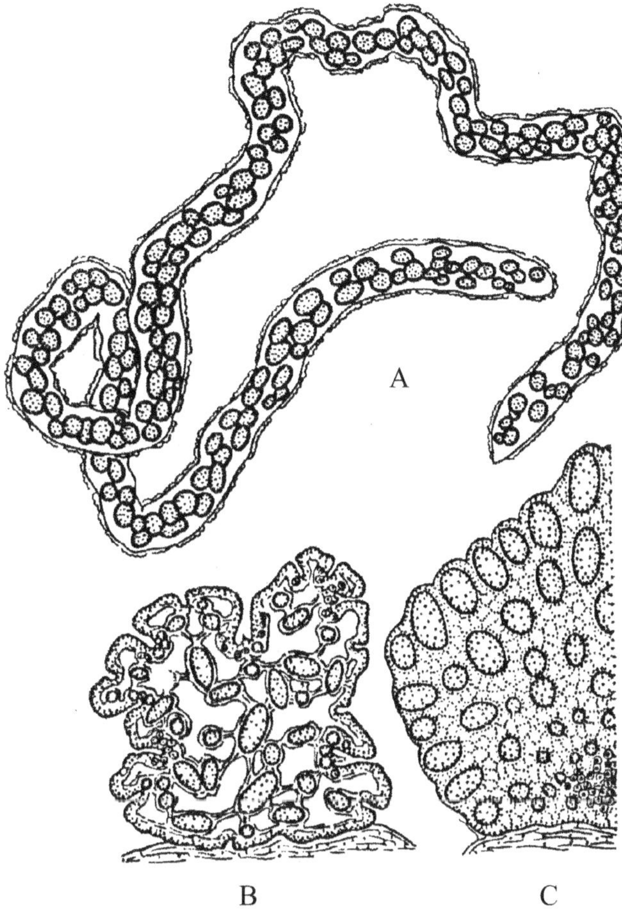

Figure 19. Mother sporocysts of the species of suborder Plagiorchiata. A, *Paralepoderma cloacicola*. From Dobrovolskij (1975), reprinted with publisher's permission; B, *Haplometra cylindracea*; C, *Telorchis assula*. From Dobrovolskij et al. (1983), with permission.

Some of them (e.g., *Paralepoderma cloacicola* of the family Plagiorchiidae and *Macrodera longicollis* of the family Ochetosomatidae) undergo relatively migrations to the place of their final settlement in the molluscan host (see fig. 95). Two distinct periods – migration and generative – can be identified in their morphogenesis (Fig. 20). The first consists in the predominant development of soma. Young mother sporocysts, which have the shape of a small oval body and contain no more than 10 cells, leave the intestinal wall, where miracidium metamorphosis took place, and escape into the haemocoel of the mollusc. Undifferentiated cells actively divide and most of them

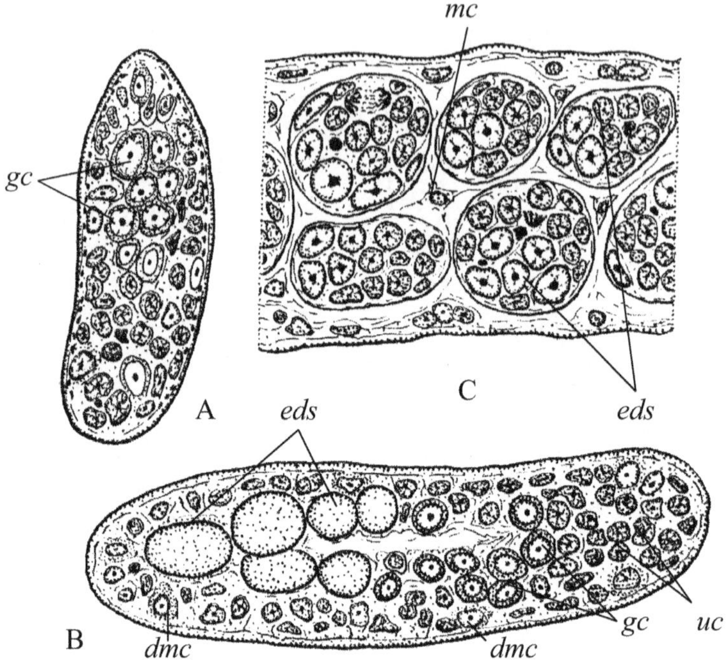

Figure 20. Development of the *Macrodera longicollis* mother sporocyst. A, 10-day-old mother sporocyst, migrating period; B, 18-day-old mother sporocyst, generative period; C, 27-day-old mother sporocyst, generative period. *Abbreviation*: *dmc*, differentiating multilaminated cell; *eds*, embryo of daughter sporocysts; *gc*, germinal cell; *mc*, multilaminated somatic cell; *uc*, undifferentiated cell.

specialize as somatic cells. They form tegumentary-muscular sac, sensory papillae and contribute to the development of the excretory system and parenchyma. Only a limited number of cells undergo differentiation into germinal ones at this stage of parthenita formation. Sporocysts increase in size and their body assumes a worm-like shape (Fig. 19A). They actively migrate in the haemocoel lacunae until they reach the place of their final settlement. In most cases, it is the lacunar system of the molluscan host pallium.

After the completion of the migration period the second stage of sporocyst development begins. At this stage most undifferentiated cells transform into germinal cells. The number of the latter rapidly increases. These cells start cleavage as their maturation proceeds. At the same time a body cavity is being formed in the sporocyst. At first its primordia are observed as narrow gaps between cells of the central cellular mass. As the sporocyst grows, the volume of the lacunae increases due to partial degeneration of parenchymal (somatic) cells, but, mostly, due to their transformation. The cells become flatter and form numerous flattened extensions, dividing the sporocyst body cavity into separate chambers. It is there that the development of embryos of daughter generation takes place (individually or in groups). Parenchymal cells (and partly cytons of the tegument), adjoining the layer of longitudinal muscles of the tegumentary-muscular sac, undergo similar specialization. Becoming flat, they form an inner lining of embryonic cavity, the so-called endocyst. The number of layers and the thickness of the latter vary significantly in parthenitae of different trematode groups.

The proliferative potential of undifferentiated cell of the germinal line in the mother sporocyst of *Macrodera longicollis* is limited. Therefore, the germinal cells, formed due to their specialization, start cleavage within a small time frame, so that the development of embryos of the daughter generation proceeds more or less simultaneously. In the course of their growth they destroy the chamber walls and freely float in the common brood cavity of the mother sporocyst. Release of individuals of daughter generation, capable of independent parasitic existence, takes place after the disruption of the body wall of the mother individual, which then dies.

The proliferation period of undifferentiated cells of *Paralepoderma cloacicola* mother sporocysts is prolonged. Groups of these cells concentrate along the wall of the body cavity in the chambers, which is being formed by transforming somatic cells. As the chambers are being disrupted by daughter sporocyst embryos, increasing in size, aggregations of undifferentiated proliferating cells appear in the body extensions, emerging in the mother sporocyst, which is extensively grown by that time. These proliferation sites constantly give rise to new cells of the germinal line. That is why early embryos and germinal cells are present in mature mother sporocysts of *P. cloacicola* along with the already fully formed young daughter parthenitae. When the mother sporocyst is completely destroyed, as a result of releasing of daughter sporocysts through its disrupted body walls, all these immature germinal elements enter the molluscan haemocoel. There is no information on the ability of immature germinal elements to complete their development in the haemocoel. However, such an issue seems probable, taking into account the information noted below on other plagiorchiids.

Unlike in the above considered cases, the brood chambers in the mother sporocyst of *Haplometra cylindracea* (Plagiorchiidae) are not destroyed in mature individuals, which become sac-shaped (Fig. 19B). The release of daughter sporocysts, developing in separate chambers does not lead to the death of the mother organism – only the emptied chamber degenerates. In other chambers normal formation of embryos proceeds. They are replenished by a sort of germinal masses (see p. 81), concentrated in chambers, adjoining the endocyst.

Development of the mother sporocyst of *Telorchis assula* (Telorchiidae), whose migration period is short, follows a somewhat different pattern. Young worm-like sporocysts are localized in haemocoel lacunae between hepatopancreas acini or directly on the outer surface of the intestine of the infected mollusc, and are washed by hemolymph. Undifferentiated cells divide intensively, which results in an increase of the parenchyma volume. The mature mother sporocyst is a multi-branched parenchymal structure of irregular shape, deprived of schizocoel (Fig. 19C). Only tegument is left of tegumentary-muscular sac, muscle fibers are not detected by light microscopy. Aggregation of reproducing undifferentiated cells is localized at one of the sporocyst poles (there is no structured germinal mass). The developing germinal cells detach from this aggregation, and are surrounded from all sides by parenchymal cells. It is just there, in the parenchyma, that cleavage of germinal cells starts and development of embryos, enclosed in separate chambers, proceeds. Mature embryos are released due to disruption of walls of these chambers, localized on the sporocyst surface, so that this process does not cause the death of the whole organism. Dying somatic cells at the sporocyst surface are constantly replaced by the new ones, whose pool is permanently renewed.

In many representatives of plagiorchiids (*Plagiorchis muris*, *P. proximus*, *Macroderoides typicus*, *Alloglossidium corti*, *Glypthelmins quieta*, *Haplometra intestinalis* – Plagiorchiidae; *Lechriorchis primus*, *Dasymetra conferta*, *Brachycoelium mesorchium*, *Pneumatophilus leidyi*; *Neorenifer georgianus* – Ochetosomatidae; apparently, all Microphallidae and Lecithodendrioidea) (Cort and Oliver, 1943; Cort and Ameel, 1944; Cort et al., 1952, 1954; Schell, 1961,1962a,b; Byrd and Maples, 1965, 1969; Jordan and Byrd, 1967; Dobrovolskij et al., 1983; Galaktionov and Dobrovolskij, 1985; Galaktionov, 1993) the active migration period is lost. Consequently there is no mobile worm-like stage, capable of migration in the molluscan body. Young individuals in the form of small aggregation of cells are localized directly at the site of penetration of the miracidium; either between intestinal epithelial cells or between the epithelial layer and the underlining basement membrane. Sporocyst growth also takes place there as the result of division of undifferentiated cells and their subsequent specialization into somatic and germinal cells.

The increase in the number of parenchymal cells in mother sporocysts of *Lechriorchis primus* (Cort et al., 1952) results in hypertrophy of their body. Its numerous branches grow into the mollusc an haemocoel. In each branch, there is a center of proliferation of undifferentiated cells, which gives rise to cells of the germinal line. Therefore, individual branches acquire a physiological independence. In other words, one can conceive that in this case a transition from unitary organization to the modular one takes place. Maturation and cleavage of germinal cells, as well as growth of embryos of the daughter sporocysts, proceed in chambers between parenchymal cells. As in the case of *Telorchis assula*, no common body cavity is formed. Chamber walls, surrounding mobile individuals of the daughter generation, already capable of independent existence, are becoming thinned and easily disrupted. Young daughter sporocysts are released outwards, without affecting the integrity of neighboring chambers, where earlier embryos and/or germinal cells are localized.

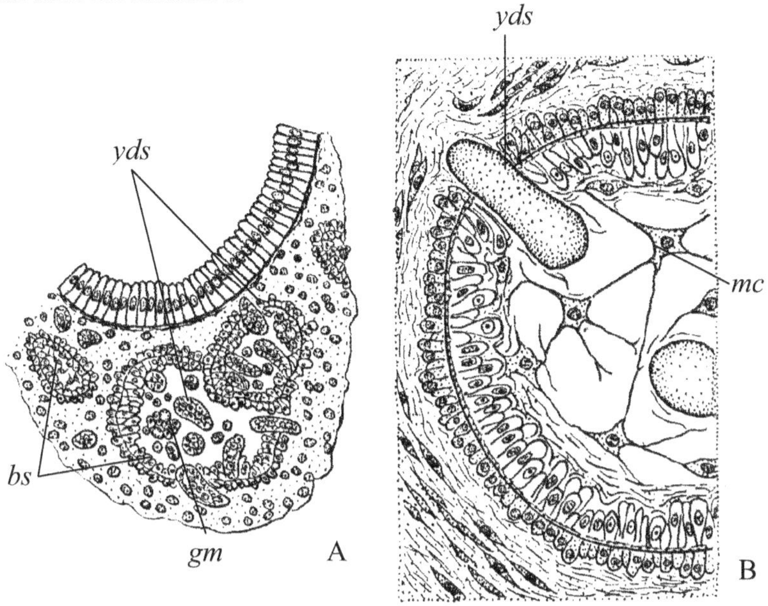

Figure 21. Mother sporocyst of *Xiphidiocercaria* sp. VII Odening. A, branched sporocyst in the tissue surrounding the molluscan gut. From Dobrovolskij and Muhamedov (1979), reprinted with publisher's permission; B, region of the mature sporocyst body. From Dobrovolskij et al. (1983), with permission. *Abbreviation*: *bs*, branches of the mother sporocyst; *gm*, germinal mass; *mc*, multilaminated somatic cell; *yds*, young daughter sporocysts.

Initial stages in formation of *Xiphidiocercaria* sp. VII Odening mother sporocysts are characterized by the fact that the majority of emerging cells specializes as germinal cells. An increase in the amount of somatic cells takes place later. Massive wide extensions (Fig. 21) are formed on the sporocyst surface, disrupting the basement membrane of the molluscan gut wall as well as connective tissue lining of the gut and protrude into the haemocoel. Eventually, the sporocyst becomes branched. As in the above described parthenitae of *Macrodera longicollis*, *Paralepoderma cloacicola* and *Haplometra cylindracea*, its body cavity is formed as the result of degeneration of parenchymal cells, which divide the schizocoel into separate chambers (Fig. 21B). The body wall is thick, although, no traces of muscle fibers have been detected by light microscopy at different stages of sporocyst development. Germinal cells, which emerged from the initial specialization of undifferentiated cells start cleavage practically simultaneously; hence, daughter sporocyst embryos develop also synchronously. Young individuals of the daughter generation emerge through local disruptions in the wall of the mother sporocyst, which immediately repair after that. However, the reproduction period of mother sporocysts is not completed with the emergence of the first group of daughter sporocysts. New germinal cells and embryos, developing from them, are generated by one or several germinal masses. This second stage of development of the mother sporocyst is characterized by sequential maturation and emergence of daughter sporocysts.

The aforementioned cases do not cover all the diversity of possible transformations of plagiorchiid first parthenogenetic generation at parasitic phase. Completing its metamorphosis, the miracidium of *Haematoloechus breviplexus* (Haematoloechinae, Plagiorchiidae), investigated by Schell (1965), enters the outer surface of the intestine and starts growing. After some time this process stops and is then replaced by degeneration of all somatic structures of the sporocyst.

Undifferentiated cells and germinal cells, undergoing different stages of maturation, get into the haemocoel. The former are still capable of proliferation for some time and replenish the pool of germinal cells. The latter start cleavage shortly and generate daughter sporocyst embryos. This germinal material is not structured. Cells and young embryos are freely dispersed in the haemocoel, many getting into unsuitable regions of host tissue, and inevitably die.

Lecithodendrioidea and Microphallidae lack a mother sporocyst. In microphallid species (*Microphallus piriformis*, *M. pygmaeus* and *Maritrema subdolum*) investigated by us (Galaktionov and Dobrovolskij, 1985; Galaktionov, 1993) (Fig. 22) in the course of metamorphosis in the gut wall of molluscan host, the miracidium loses not only provisional structures, but also all somatic structures. Only germinal and undifferentiated cells persist as a cellular conglomerate localized between epithelial cells of the intestine close to the basal parts of those cells. As the result of division of undifferentiated cells this conglomeration grows, elongates and acquires a stolon-like shape, which reveals the features of a germinal mass. Undifferentiated cells, still increasing in number give rise to new germinal cells and supporting cells, whose thin extensions permeate the stolon-like germinal mass and envelope it from the outside. This is how the integrity of the stolon-like germinal mass is insured, though its partial fragmentation can not be excluded. This is supported by the presence in the molluscan haemocoel of separate undifferentiated and germinal cells and small cellular aggregations, representing separated parts of the stolon-like germinal mass. The stolon-like germinal mass grows through the basement membrane and is localized in the connective tissue, surrounding the intestine and in the haemocoel. Apart from the stolon-like germinal mass, in the tissues, surrounding the intestine, small aggregations of germinal and proliferating cells (independent germinal masses) are observed between the hepatopancreas acini and in the gonad.

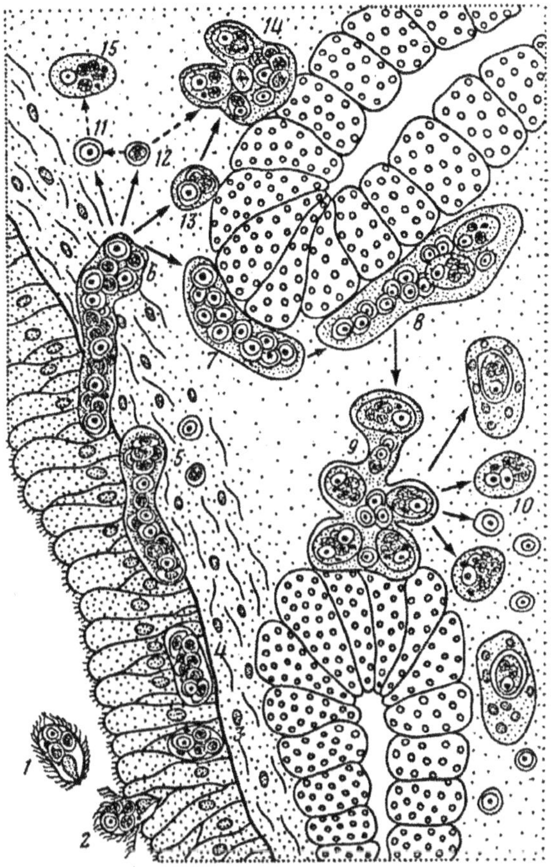

Figure 22. Diagram, showing development of the first parthenogenetic generation of *Microphallus piriformes*. *Abbreviation*: 1, miracidium hatching from the egg in the molluscan intestine; 2, miracidium penetrating the intestinal wall; 3, metamorphosis of miracidium; 4, formation of the stolon-like germinal mass; 5, the stolon-like germinal mass growing through the intestinal basement membrane and surrounding connective tissue; 6, partial fragmentation of the stolon-like germinal mass; 7-10, stolon-like germinal mass in the molluscan haemocoel: 7, phase of the germinal cells differentiation; 8, germinal cells begin to cleave; 9, phase of the formation of the daughter sporocyst embryos; 10, phase of the fragmentation of the old stolon-like germinal mass; 11, germinal cell in the molluscan haemocoel; 12, undifferentiated cell in the molluscan haemocoel; 13, separated fragment of the stolon-like germinal mass in the molluscan haemocoel; 14, independent germinal mass in the molluscan haemocoel; 15, daughter sporocyst embryo. From Galaktionov and Dobrovolskij (1985).

They are formed on the basis of separated parts of the stolon-like germinal mass and possibly individual undifferentiated cells.

In the course of its development, the stolon-like germinal mass has a stage, when all its germinal material is represented by mature germinal cells. These germinal cells start cleavage practically simultaneously, which results in synchronous formation in the stolon-like germinal mass of daughter sporocyst embryos. During their maturation these embryos destroy supporting structures of the stolon-like germinal mass and pass to the haemocoel. As no renewal of germinal material takes place, the stolon-like germinal mass ceases to exist. Apart from the stolon-like germinal mass, daughter sporocyst embryos are also produced by small independent germinal masses. The latter are also detected some time after the stolon-like germinal mass degeneration, but soon degenerate also. It is most likely that daughter sporocyst embryos may form as the result of the cleavage of germinal cells, separated from the stolon-like germinal mass.

Apparently, the development of the mother sporocyst in some Ochetosomatidae, described by Byrd and Maples (1965, 1967) and Jordan and Byrd (1967), proceeds in a similar way. According to these studies, after the miracidium metamorphosis, an aggregation of cells of the same in structure (undifferentiated cells as follows from the figures, provided in the cited works) – the so-called "brood mass", is observed in the intestine wall of the mollusc. It penetrates into the haemocoel lacunae. The authors compare the characteristics of its movement through the host tissues to the multinuclear trophozoite of protists. There, the brood mass grows intensively and acquires a multibranched shape, inside which formation of daughter sporocysts takes place. In *Neorenifer georgianus*, each cell, derived from the multiplication of the brood-mass cells, gives rise to the embryo of the daughter sporocyst. Although the figures and descriptions in the aforementioned works, can not be interpreted unambiguously, we are inclined to consider the structure, forming as the result of the brood mass growth, as a type of stolon-like germinal mass, similar to that described earlier for microphallids.

There are also data on mother parthenitae of some other trematode groups with passively infecting miracidia. However, this information can not be evaluated due to fragmentary descriptions. Thus, Dicrocoeliidae mother sporocysts are sac-like or grape-like formations, localized between the hepatopancreas acini of the mollusc (Panin, 1984). One can presume that the characteristics of their formation and functioning has much in common with that described for mother sporocysts of plagiorchiids, which have a well-developed soma.

At present, the only source of information on Renicolidae mother sporocysts is the work of Wright (1956). Twenty-four hours after infection of molluscs *Turitella communis* by the eggs of *Renicola* sp., Wright discovered small membrane-enveloped aggregations of cells on the wall of the molluscan intestine. During 72 h, the number of cells in the aggregations increases. Then these aggregations (mother sporocysts according to Wright) disappear. Wright reasonably presumed that the mother sporocysts migrate along haemocoel lacunae towards the gonad, where daughter sporocysts of renicolids are located. Subsequent stages of *Renicola* sp. mother sporocysts have not been traced. However, the presence of a large number of young daughter sporocysts at different stages of formation (up to germinal balls) without any traces of mother sporocyst (Wright, 1956; Pl. I, E) in the gonad of the mollusc in 3.5 months after infection, resembles a similar situation in microphallids at the stage following the decomposition of the stolon-like germinal mass (see above). It is not excluded that the general course of transformation of the parasitic phase of the Renicolidae mother parthenita is similar to that described earlier for Microphallidae.

Notocotylidae should also be mentioned. It was indicated in section 1.1.1. that two cells – germinal and somatic (Murrills et al., 1985a) are included in a very unusual miracidium of *Notocotylus attenuatus,* investigated by means of electron microscopy. The egg also contains the so-called opercular cord. It is assumed that this tubular structure is used as an injection apparatus for injection of the mother sporocyst's embryo into the molluscan tissues (Murrills et al., 1985b) (see fig. 10). On the 7th day after inoculation of this embryo into the mollusc's haemocoel, young mother sporocysts, containing one germinal ball (Fig. 1c, p. 163, Murrills et al., 1988), apparently resulting from cleavage of the single germinal cell, were discovered. The number of rediae, forming in mother sporocysts, varies from 1 to 8 in a few Notocotylidae species, which have been investigated in this respect (Herber, 1942; Filimonova, 1985). It is not known if this number corresponds to that of germinal cells contained in their miracidium . If it does not, then the organization of larval germinal material belongs to the second type (see p. 24). In this case one should admit that one or more undifferentiated cells are included in the miracidium instead of (or in addition to) the germinal cell. In the course of proliferation of the undifferentiated cells, proceeding after inoculation of the mother sporocyst embryo into the first intermediate host, formation of germinal, as well as somatic cells, takes place. It is quite possible that the so-called somatic cell of the miracidium is in fact undifferentiated. Some of its derivatives are specialized as germinal cells, which give rise to the second and subsequent embryos of mother rediae. In any case, by their morphology, Notocotylidae mother sporocysts belong to the same group as those of Fasciolidae, described in the beginning of this section.

Finally, one should not ignore peculiar sporocysts of Bucephalidae appearing as white or orange dichotomously branched tubules, which invade the haemocoel of the pallium, gonad and digestive gland of infected Bivalvia (Woodhead, 1931; Ciordia, 1956; Chubrik, 1966; Matthews, 1973a,b, 1974; Wardle, 1988). Cercariae are formed in these unusual parthenitae. The cercariae come out through local ruptures of the mother sporocyst body wall. There are numerous and very different viewpoints on the nature of Bucephalidae parthenitae. Sporocyst branches (Ciordia, 1956) or whole sporocysts (James and Bowers, 1967) are considered as daughter generations. In the latter case the mother sporocyst is assumed to be small in size and short-lived. That is why this mother sporocyst escapes investigators' attention (James and Bowers, 1967). The question remains unresolved until the initial stages of the development of the parthenogenetic part of the life cycle in the molluscan host become clear. It is probable that Bucephalidae parthenitae are represented by one generation only and should be regarded as the mother sporocyst.

Apart from the unusual body shape, Bucephalidae sporocysts possess other unique features, which distinguish them from parthenitae of the other trematodes. First of all the tegument morphology should be mentioned. James and Bowers (1967) were the first to describe nuclei in the tegument of *Bucephalus haimeanus* sporocysts. These data were confirmed in the electron microscopy studies of *Bucephaloides (Prosorhynchoides) gracilescens* sporocysts (Podvyaznaya, *pers com*). Nuclei either make part of the outer syncytial layer of the tegument or are localized in its outer extensions. (Fig. 23). There are no useful data on the presence of tegumental cytons in the sporocyst body wall. Therefore, the tegument appears to be inside out, i.e., the cytons are localized outside, rather than in the underlying tissues of sporocysts, which would be typical of the tegumental organization in all trematodes. However, in the other investigated Bucephalidae species *Prosorhynchus squamatus*, there are no nuclei in the outer syncytium of the tegument, while cytons are found in the sporocyst body wall. According to the light microscopy data, the tegument of *Prosorhynchus crucibulum* and *Bucephalus polymorphus* sporocysts has the same morphological structure, whereas in the outer layer of the tegument of *Labratrema minimum*, one can distinguish nuclei (Matthews, 1973a,b, 1974; Podvyaznaya, *pers com*; our data). At present, the nature of such variability in the tegument structure of Bucephalidae sporocysts remains unclear. However, it makes one think about a universal type of organization of this structure in trematodes and in Neodermata in general.

Parts, clearly distinct in their morphology and function, can be identified in the body of bucephalid sporocyst (Matthews, 1973a,b, 1974; Suloeva, 1999). In sufficiently long regions of this sporocyst the body cavity is narrow or nor expressed at all. This is also typical for young terminal parts of branches. The body wall there is thickened and separate germinal cells can be found in it. Such regions are intermixed with those, which can be defined as brood chambers. They have an enlarged lumen, containing cercaria embryos at different stages of development, and a relatively thin body wall, consisting of flattened cells. Sometimes, in addition to developing cercariae, in the young brood chambers one can discover small aggregations of germinal cells and early embryos, adjacent to the body wall. These structures can be interpreted as germinal masses (see p. 82). They are not seen in mature brood chambers where the whole cavity is filled with developing and fully

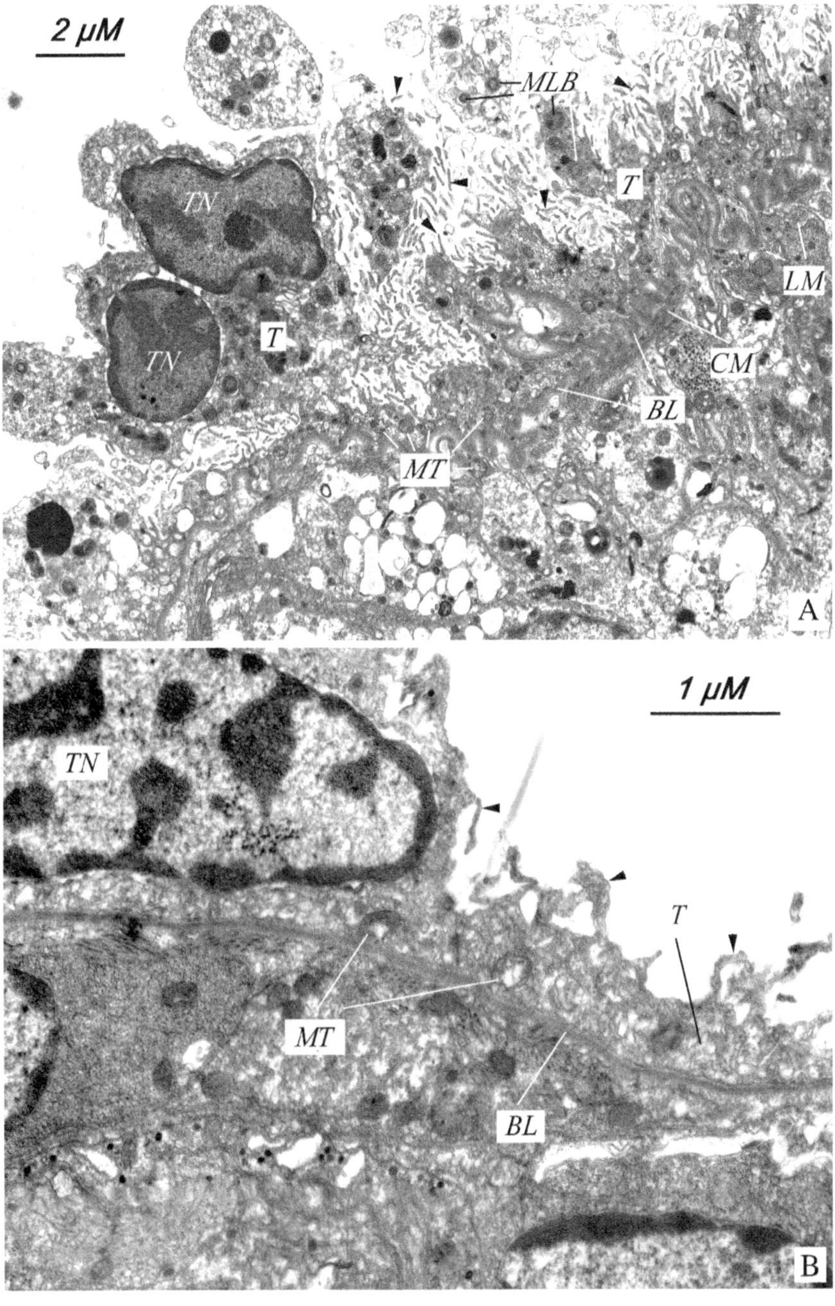

Figure 23. Ultrastructure of the body wall of the *Bucephaloides gracilescens* sporocyst. A, the region with well-expressed numerous microvilli; B, the region with the smooth tegument. *Abbreviation*: *BL*, basal lamina; *CM*, circular muscles; *LM*, longitudinal muscles; *MLB*, multilaminated body; *MT*, mitochondrion; *T*, tegument; *TN*, tegumental nucleus. Arrows indicate microvilli. Original TEM micrographs of I.M. Podvyaznaya, with permission.

formed cercariae. Finally, vast regions of the sporocyst body undergo different stages of degeneration; this process involves both narrowed parts and brood chambers. In the latter case the embryos contained in the embryonic chambers are also undergoing degeneration.

The heterogeneity in the tegument morphology suggests to functional difference between different parts of the sporocyst body. This refers mainly to the microvilli pattern. The regions, where tightly packed numerous long microvilli form a dense layer alternate with the regions, where these structures are less developed (Fig. 23B). One can presume that the absorptive function of the tegument is fully realized in the regions with numerous tightly packed microvilli in the terminal parts of the branches and, sometimes, between the neighboring brood chambers.

We suggest that transition to a modular organization takes place in Bucephalidae sporocysts. In the course of this process sufficiently independent modules (brood chambers, young parts, degenerating parts) appear in the initially integrated organism, so that degeneration of several modules, which takes place in sporocyst, does not lead to the death of the whole organism. Thus, Bucephalidae sporocysts, quite independently, achieve the same level of morphological and functional organization, as the aforementioned (see p. 43) mother sporocysts of some plagiorchiids. The same principle of modular organization is realized in some parthenitae of the daughter generation discussed below (see. p. 70).

1.2. The Second (Daughter) Parthenogenetic Generation

Individuals of the next, also parthenogenetic generation, produced by mother sporocysts, are represented by two morphological types: rediae and sporocysts, which differ from each other in morphology as well as biology.

Rediae are normally characteristic of more primitive trematodes (i.e., Fasciolidae, Echinostomatidae, Paramphistomidae, Psilostomidae, Philophathalmidae, Lepocreadiidae) but can also be found in a number of specialized groups (i.e., Halipeginae, Aporocotylidae, Heterophyidae, Opisthorchiidae). They have an elongated cylinder-shaped body with three conic locomotory extensions (Fig. 24A-B). One is localized closer to the anterior end, while the other two are situated at the posterior end of the body. Sometimes the anterior locomotory extension is replaced by a circular folding of the tegumentary–muscular sac – a collar, especially often found in Echinostomatidae rediae. The mouth opening is located terminally at the anterior end of the rediae. A small opening – a birth pore – is located subterminally. It serves to give rise to individuals of the next generation. In the digestive system, the pharynx, oesophagus and sac-like intestine can be found. There is also a protonephridial excretory system and gland apparatus. The major part of the redia body is occupied by an embryonic cavity, where consecutive stages of the development of individuals of the next generation take place.

Such a generalized description of the redia does not cover its morphological diversity. Variability, often age-dependent, is observed in the body shape as well as in the degree of development of different structures. Normally the width - length ratio of these parthenitae is 1:4 - 1:7, but rediae with a very elongated filiform body are also found. The locomotory extensions and collar can be well-expressed during the life of the redia, but can also disappear during its growth. Rediae of many Echinostomatidae, which have an embryonic cavity filled with embryos, retain only one of the posterior locomotory extensions or lose all of them. The body acquires a sac-like shape, typical for the rediae of Notocotylidae, Acanthocolpidae, Clinostomidae, Monorchiidae and Heterophyidae. However, additional locomotory extensions develop in the rediae of the specialized family Transversotrematidae. The development of digestive system is very variable. In some species, all parts of the digestive tract are well-developed so that the gut reaches the posterior end of the body. In other species the gut is short and may look like a small vesicle (Heterophyidae, Lepocreadiidae). The digestive system is most reduced in Azygiidae rediae, and in the mother rediae of Philophthalmidae it has been lost completely. It should be noted that rediae retain mobility during all their life.

Normally several rediae generations succeed each other in the molluscan host. The number of generations varies in different species. As a rule, one or two first generations, which may morphologically differ from the subsequent ones, e.g. in Philophthalmidae, give rise only to rediae, and are called mother rediae. Daughter rediae, which follow them, most often start their reproduction, giving birth to several rediae and then begin to produce cercariae. In Aporocotylidae the mother redia simultaneously produces rediae and cercariae (Køie, 1982). According to Ginetsinskaya (1954a), only one generation of redia develops in Cyclocoelidae. This redia generation gives rise to larvae of hermaphroditic individuals.

Trematode mother sporocysts, giving birth to rediae, are normally localized in the haemocoel of the mollusc pallium complex or in the mollusc's heart. From there, young individuals of the second parthenogenetic generation actively migrate along the haemocoel lacunae or, more rarely, directly through the host tissues to the site of their final settlement. Normally this is the digestive gland and/or gonad. Mother rediae of "viviparous" trematodes (see p. 23) usually take the same path. They penetrate into the tissues of the foot, pallium and tentacles directly from the miracidium, attached to the surface of these parts of the molluscan body. The mother rediae of *Philophthalmus rhionica* (Philophthalmidae) migrate to the mollusc's heart, where they settle, attached to the inner surface of the ventricle by means of tiny locomotory extensions and the posterior part of the body (Ti-

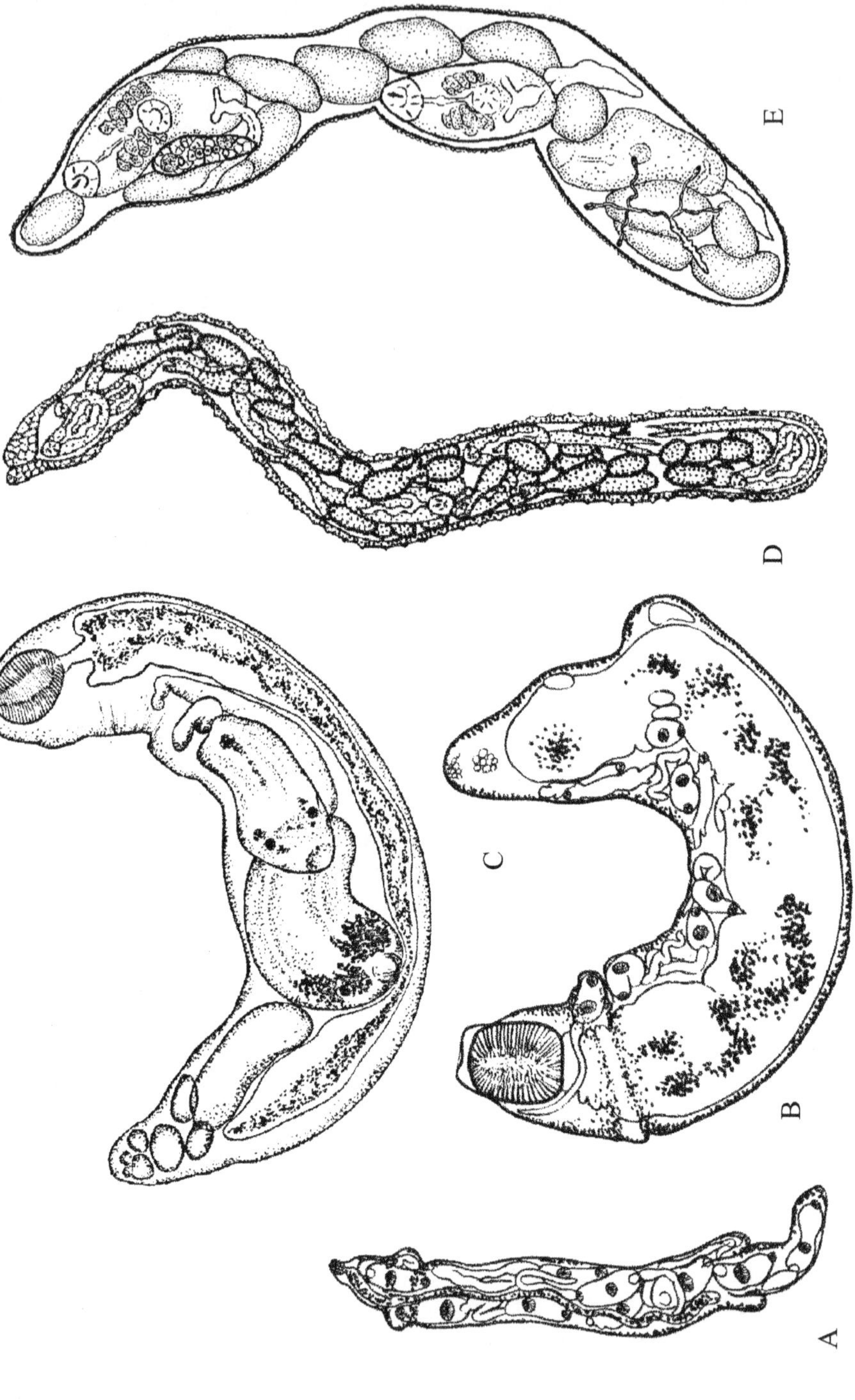
A
B
C
D
E

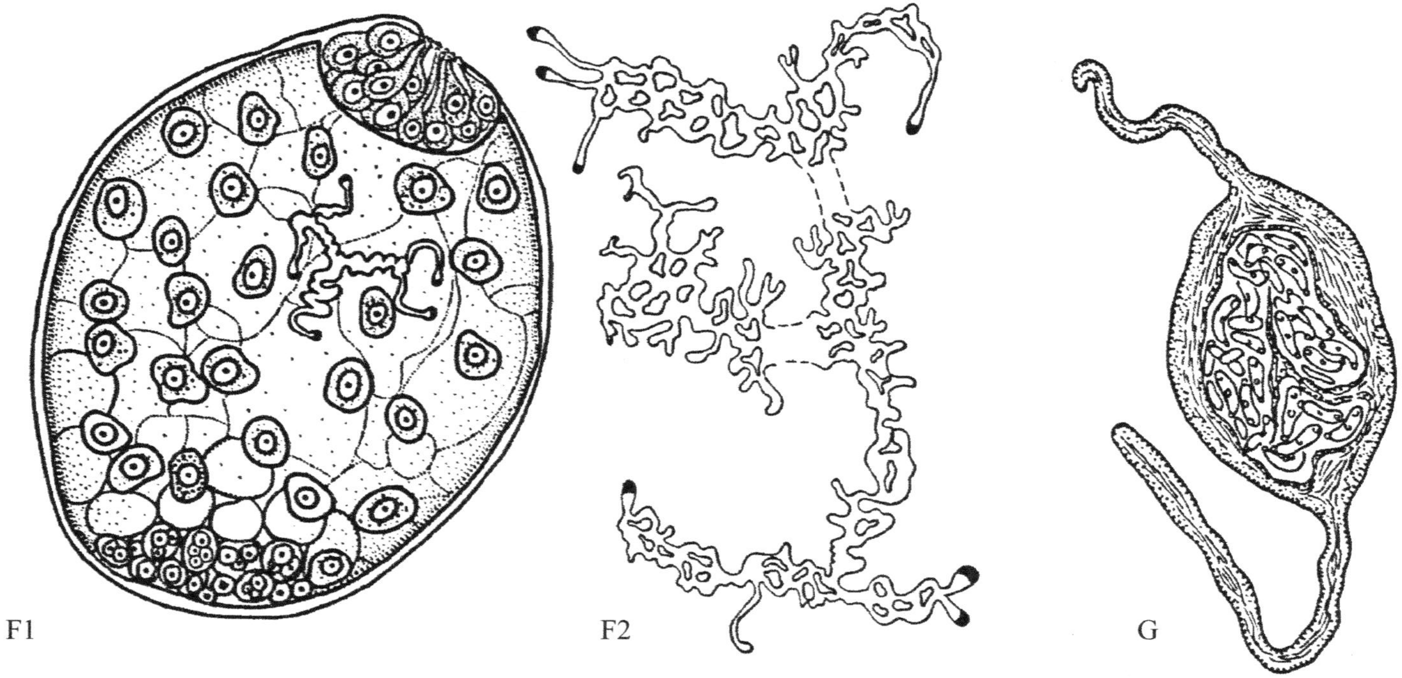

Figure 24. Morphology of rediae (A-C) and daughter sporocysts (D-G). A, mother and B, daughter rediae of Echinostomatidae gen sp. From Odening (1965a), reprinted with author's permission. C, *Cercaria fennica I* (Notocotylidae). From Odening (1963), reprinted with author's permission. D, *Mesostephanus appendiculatus* (Cyathocotylidae). From Olenev and Dobrovolskij (1975). E, *Haplometra cylindracea* (Plagiorchiidae); F, *Maritrema subdolum*: F1, young parthenite, F2, region of the excretory system of the mature daughter sporocyst. From Galaktionov (1993). G, *Eurytrema pancreaticum* (Dicrocoeliidae).

homirov, 1980). Young daughter individuals are dispersed with the flow of hemolymph and concentrate in great numbers in the pallium fold and the ctenid- ium. From there they actively migrate along the haemocoel lacunae to the visceral sac.

Daughter sporocysts are typical for the more specialized trematode groups, mainly representatives of Strigeidida and Plagiorchiida. They differ from the rediae by a much simpler organization (Fig. 24D-G). The body of strigeidid daughter sporocysts has an elongated, worm-like (Cyathocotylidae) or filiform (Strigeidae, Diplostomidae) shape. The body of plagiorchiid sporocysts is sac-like and its shape varies from sausage-like (Plagiorchiidae, Ochetosomatidae) to spherical (Lecithodendrioidea, Microphallidae). There are no locomotory extensions. The birth pore, if any, is located terminally. There is no digestive system. The glandular apparatus is simplified compared to that of the redia.

Daughter sporocysts of Gymnophallidae, Schistosomatidae, Cyathocotylidae, Strigeidae, Diplostomidae are capable of producing the next generation of sporocysts alongside of cercariae (see p. 370). In plagiorchiids presumably two sequential generations of daughter sporocysts are described only for Renicolidae (Wright, 1956). In all other species, belonging to that order, daughter sporocysts are represented by only one generation.

Mobility of daughter sporocysts is limited compared to that of rediae. The number of species, whose mature sporocysts are capable of limited migrations, is few. For this purpose some Cyathocotylidae may use annular rollers on the body surface (Fig. 24D) similar to horse fly larvae, moving in semi-fluid ground. Only rarely can mature daughter sporocysts undertake a genuine migration in the molluscan host body as is the case of *Euritrema pancreaticum* (Dicrocoeliidae) parthenitae, containing fully formed cercariae, when the former are released into the external environment (Fig. 24G). More often sporocysts are mobile only when young. This is the case of plagiorchiids, whose mother sporocysts are localized in the pallium (see p. 250). Parthenitae, produced by them, migrate along haemocoel lacunae to the digestive gland (partly using the hemolymph flow and partly due to peristaltic contractions of the body). Here they grow and soon lose or nearly lose the capacity of active locomotion. Filiform sporocyst of some strigeidids, piercing the molluscan digestive gland, may effect undulating or peristaltic movement. This generates hemolymph flow round their bodies, which delivers nutrient and oxygen. In plagiorchiids, whose mother sporocysts or free germinal masses settle on the surface of the molluscan intestine, in the digestive gland or gonad (see p. 43), daughter sporocysts are not capable of any locomotion. Young individuals may be passively carried by the hemolymph flow, but this takes place only in the molluscan digestive gland or gonads.

1.2.1. Development and Structure of Soma

Despite the aforementioned differences in morphology and biology of rediae and daughter sporocysts, the initial stages of their development are similar. In section 1.3. this process is considered in detail. Here we focus on later stages of embryo development, when the provisional tegument has already been formed and other morphological structures, typical for mature daughter parthenita, undergo differentiation. In rediae, the embryo primordial digestive system becomes detectable as membrane-bound cellular aggregations. Paired protonephridia now appear, each being represented by one flame cell and an excretory duct. The flame cell division is accompanied by capillary doubling. Non-separated parts of the ducts form collecting ducts. The latter open outside, forming independent excretory pores. In parthenitae, no bladder, typical of individuals of the hermaphroditic generation, is formed.

At initial stages of schizocoel formation, thin slot-like gaps appear practically simultaneously in different parts of the body. The volume of these gaps enlarges partly due to degeneration and partly during specialization of the cells of the central mass of the embryo, as it takes place in mother sporocysts. In the course of specialization the cells flatten and form complex branched laminated projections. They pierce the body cavity in all directions, making it look like a net-like structure and forming its multi-layered inner lining – the endocyst. In this net, embryos of the next generation undergo the initial stages of their development. In the course of their growth the walls between neighboring net meshes are destroyed and a common inner cavity of the parthenita is formed.

An extensive literature is available on the description of the fine structure of the definitive tegument of rediae and daughter sporocysts (for reviews see: Dobrovolskij et al., 1983; Smyth and Halton, 1983; Threadgold, 1984). In most investigated species the outer syncytium with microvilli or microlamellae on its surface (Fig. 25) is clearly folded. Normally there are no spines. They have been found only in the tegument of schistosome daughter sporocysts (Meuleman et al., 1980). In the cytoplasm, various structures are detected: unattached ribosomes, cisterns of the endoplasmic reticulum, mitochondria, normally concentrated in the basal part of the outer syncytial layer, often close to the basal membrane, different vesicles, including secretory granules, usually containing electron-dense material. The granules are produced by specialized tegumental cells with well-developed RER and Golgi complexes. The secretion is delivered into the outer layer of the tegument by means of cytoplasm bridges, which pass between the layers of circular and longitudinal muscles. The generalized description of the tegument of daughter parthenitae, provided above, should be extended.

Thus, branched anastomosing microvilli are most developed on the surface of the tegument of daughter sporocysts. They are described in Ope-

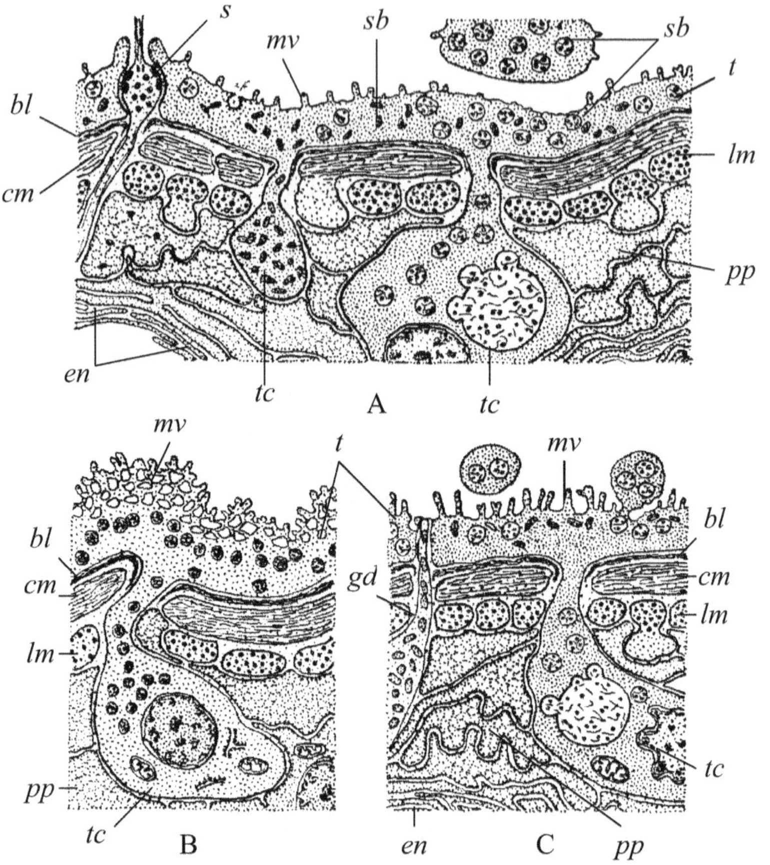

Figure 25. Diagram of the tegument structure of daughter parthenitae. A, daughter and B, mother rediae of *Philophthalmus rhionica*; C, daughter redia of *Metagonimus yokogawai*. *Abbreviations*: *bl*, basal lamina; *cm*, circular muscles; *en*, endocyst; *gd*, duct of the unicellular gland; *lm*, longitudinal muscles; *mv*, microvilli; *pp*, projections of the parenchymal cells; *s*, sensory papilla; *sb*, secretory body; *t*, outer syncytial layer of tegument; *tc*, tegumental cyton. From Dobrovolskij et al. (1983), with permission.

Coelidae (Køie, 1971b), Gymnophallidae (Al-Salman and James, 1988; Russell-Pinto et al., 1996), some Plagiorchiida (Ginetsinskaya et al., 1965; Dobrovolskij et al., 1983), Schistosomatidae (Hockley, 1973; Køie and Frandzen, 1976; Meuleman et al., 1980; Göbel and Pan, 1985), Diplostomidae (Ginetsinskaya et al., 1966), Dicrocoeliidae (Žd'árská, 1990). Microvilli are oriented perpendicular to the surface of the parthenita body and form a genuine brush border, similar to that of the intestinal epithelium of vertebrates. It was believed that only sporocysts with a birth pore have such an organization of the tegument (Popiel and James, 1978a,b; Threadgold,

1984). If there is no birth pore, microvilli are less developed or absent. The above cited authors included microphallid sporocysts in the group, lacking the birth pore. However, later this structure was found in microphallid species (see. p. 65). Our investigations of some microphallids (*Microphallus similis*, *Levinseniella brachysoma*, *Maritrema subdolum*) have shown that well-developed microvilli are typical of the tegument of young individuals (Galaktionov, 1993) (Fig. 26). Their subsequent reduction correlates with the formation of "paletot" around the sporocyst, which is formed by hemocytes of the molluscan host (see p. 256). It is just the formation of the epithelialized paletot, but not the absence or presence of the birth pore, determines the reduction of microvilli in daughter sporocysts of a number of Plagiorchiidae, including microphallids (Ginetsinskaya et al., 1965; Dobrovolskij et al., 1983). In Brachylaimoidea sporocysts, lacking the birth pore, the tegument has a well-developed layer of intermingled and anastomosing microvilli (Bakke, 1982; Žd'árská and Soboleva, 1986a; Pojmanska and Machaj, 1991).

Microvilli and/or microlamellae of the outer surface of the rediae tegument are usually less developed than in the daughter sporocysts (Rees, 1980, 1981; Køie, 1971a; Køie et al., 1977; Irwin et al., 1978; Page et al., 1980; Dunn et al., 1992a). There is a definite correlation between the tegumental structure and the digestive system. This can be detected even in different redia generations of the same species. In *Philophthalmus rhionica* daughter rediae with a well-developed intestine, ridges and microvilli are not large and are sparsely dispersed on the body surface (Fig. 25A). Mother rediae of this species, deprived of a digestive system, possess a dense net of ridge-like projections and anastomosing branched microvilli, which form a continuous fungus-like layer on the surface of the parthenita body (see Fig. 25B) (Tihomirov, 1980; Dobrovolskij et al., 1983). Rediae with a reduced intestine in Acanthocolpidae (Køie, 1971a), Heterophyidae (Irwin et al., 1978; Dobrovolskij et al., 1983) and Lepocreadiidae (Køie, 1985) have a similar structure of the tegument. In rediae of *Proterometra macrostoma* (Azygiidae), which lack the intestine, microvilli are poorly developed. However the tegument is marked by a geometric pattern of channels arranged like cylinders stacked perpendicular to the basal lamina. Each channel opens to the exterior by a pore, thereby providing an extensive increment in free surface area (Uglem and Lee, 1985). According to the estimations, provided by Uglem and Lee (1985), this accounts for the 13-14-fold increase in the surface of the redia body.

The above description of the tegument organization of daughter parthenitae undoubtedly attest to its direct involvement in the process of absorption of nutrients from the host organism. It is presumed that their transport proceeds via simple or facilitated diffusion. It has been experimentally shown that the movement of some carbohydrates into the tegument occurs

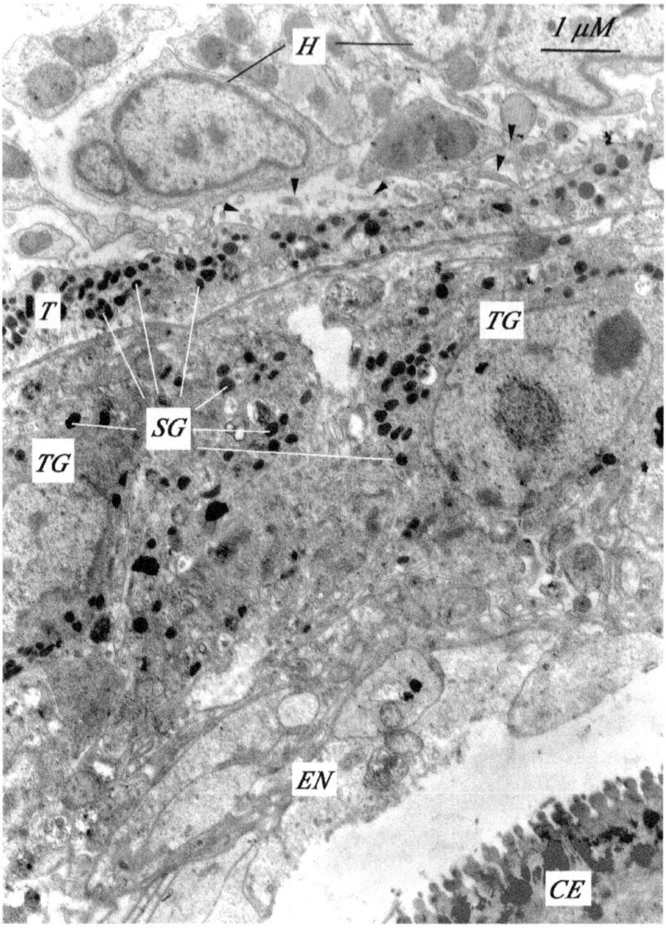

Figure 26. Ultrastructure of the body wall of the *Maritrema subdolum* daughter sporocyst. *Abbreviations*: *CE*, cercaria embryo; EN, endocyst; *H*, molluscan hemocyte; *SB*, secretory body; *T*, outer syncytial layer of tegument; *TG*, tegumental gland. Arrows indicate tegumental microvilli.

via active transport (Hoskin and Cheng, 1974; Carter and Bogitsh, 1975; McManus and James, 1975a,b; Krupa et al., 1975; Uglem and Read, 1975; Bogitsh, 1986). A model of possible consequence for the absorption of non-transported solutes was proposed by Uglem and Lee (1985) for the tegument of *Proterometra macrostoma,* described above. According to their so-called standing-gradient model the active transport of glucose by rediae is coupled osmotically with a flow of water through the channels from the exterior. Non-transported solutes would be swept (by solvent drag) into the channels where they would accumulate if they do not easily penetrate the plasma membrane. Solute accumulation could generate a concentration gradient favoring the diffusion of solute into the worm. Their model is experimentally supported by the fact that in vitro labeled mannitol (^{14}C-mannitol) penetrates into the redia only after addition of glucose to the culture medium (Uglem and Lee, 1985).

Some authors (Køie, 1971a,b; Al-Salman and James, 1989; Dunn et al., 1992a) postulate the role of pinocytosis in the process of absorption of nutrients by the parthenita tegument. They provide electron-microscopy demonstration of a process, resembling the formation of pinocytic vesicles and different stages of autophagous cycle. In the body wall of *Meiogymnophallus minutus* daughter sporocysts, Al-Salman and James (1989) describe large cells, whose basal parts form projections, lining the embryonic cavity. These cells are considered as digestive ones. According to Al-Salman and James (1989) they produce secretion, which contains hydrolytic enzymes, necessary for exogenous digestion. The secretion is delivered to the outer tegumental syncytium in the form of granules and is released at the microvilli apex. At the base of microvilli, partly digested food is absorbed via pinocytosis. Pinocytic vesicles enter the cytoplasm of tegumental cytons, where they merge with lysosomes, forming digestive vacuoles. On completion of the process of intracellular digestion the vacuoles are transformed into residual bodies. The digested food in the form of α-glycogen is accumulated in the cytoplasm of the cell and its basal projections, where it then passes to the liquid of the schizocoel, surrounding the developing embryos. Al-Salman and James (1989) distinguish several phases of activity of tegumental cells. During the synthetic phase, cisternae of endoplasmic reticulum, Golgi apparatus, mitochondria and secretory microvesicles are detected in the cytoplasm of those cells. This phase is followed by absorption of nutrients, which is then followed by the digestive phase. The latter is characterized by accumulation of digestive vacuoles and residual bodies in the cytoplasm. A cyclic transition of cells from phase to phase has been postulated.

All the aforementioned opinions of Al-Salman and James (1989) were based exclusively on transmission electron microscopy. However, according to modern cytological conceptions, endocytosis in any form, especially pinocytosis, should be proved in solid experiments. Obviously, demonstration of microinvaginations of outer plasma membrane only is not sufficient (Conradt and Peters, 1989). It should be stressed that cells in the "digestive" phase described by Al-Salman and James (1989) are similar to degenerating ones. This is supported by the fact that there are practically no intact organelles in those cells, but multilaminated bodies are present. In microphallid daughter sporocysts, we observed (Galaktionov, 1993) portions of degenerated tegumental cells and their subsequent incorporation into the endocyst. There are some data that *in vitro* sporocysts of *Cercaria linearis* and *C. stunkardi* may secrete histolytic enzymes via the tegument (Thomas and Pascoe, 1973). These enzymes are capable of converting molluscan glycogen into maltose and then glucose and decompose sucrose into glucose and fructose. Unfortunately, digestive function of the daughter sporocyst tegument is still poorly investigated and there is no reliable proof of such a function at all. This is also true for external digestion in sporocysts, whose tegument microvilli form a structure, resembling the brush border.

The protective function of the tegument is also obvious. However, its mechanisms are still unclear. It is presumed that the glycocalyx of parthenita tegument imitates that of the molluscan host tissues and, therefore, parasites are not recognized by molluscan hemocytes as non-self elements (Bogitsh, 1986). Here we deal with so-called antigen mimicry. This presumption is supported by the fact that sporocysts and rediae are not encapsulated in the specific host. The capsule is formed only around aged individuals, which have ceased functioning and whose tegument has undergone degeneration (Køie, 1987). Degeneration of the tegument in daughter sporocysts of Opecoelidae, Microphallidae and Gymnophallidae takes place long before their functional activity stops (Gibson, 1974; Popiel and James, 1978a,b; Popiel, 1978a; Al-Salman and James, 1989; Galaktionov, 1993). Organelles in the cytoplasm of the outer syncytium are transformed into multilaminated bodies, the tegument becomes thinner and may break up. In that case the parthenita body is bounded with a basement lamina, underlying the tegument. The basement lamina is formed by the layer of intermediate filaments (Abbas and Cain, 1987; Fukuda et al., 1987; Holy and Oaks, 1987). We have to presume that it plays some protective role.

The emergence and development of the digestive system proceeds only in the course of rediae morphogenesis. A mouth opening, located terminally at the anterior end of the body, leads into a cavity bound with tegumental foldings – a homologue of the prepharynx of individuals of the hermaphroditic generation. The prepharynx opens into the rounded muscular pharynx (Fig. 27, 28). The latter is composed of bundles of circular, radial and meridional muscles. Then comes a short oesophagus, which, similar to all other regions of the foregut, is lined with the extension of outer syncytial layer of the tegument. In *Parorchis acantus* rediae, this lining is present only in the anterior part of the oesophagus. Its posterior part, separated with an attachment zone, is lined with the nuclear syncytium of another origin, separated from the gut also by means of the attachment zone (Rees, 1983a) (Fig. 27). The functional significance of such a structure of the oesophagus (the presence of two separated zones) remains unclear.

The oesophagus opens into an incomplete gut, whose walls are composed of flattened epithelial cells in all rediae, investigated to date. The cells are linked into a layer by means of long septate desmosomes, located at the apical ends of the contacting cells. Additionally, in the posterior-lateral parts of the neighboring cells are other types of contacts – presumably gap junctions – as detected in the transmission electron micrographs, presented by Køie (Køie, 1971a – Fig. 7, p. 122). In the gut of *Parorchis acanthus* rediae,

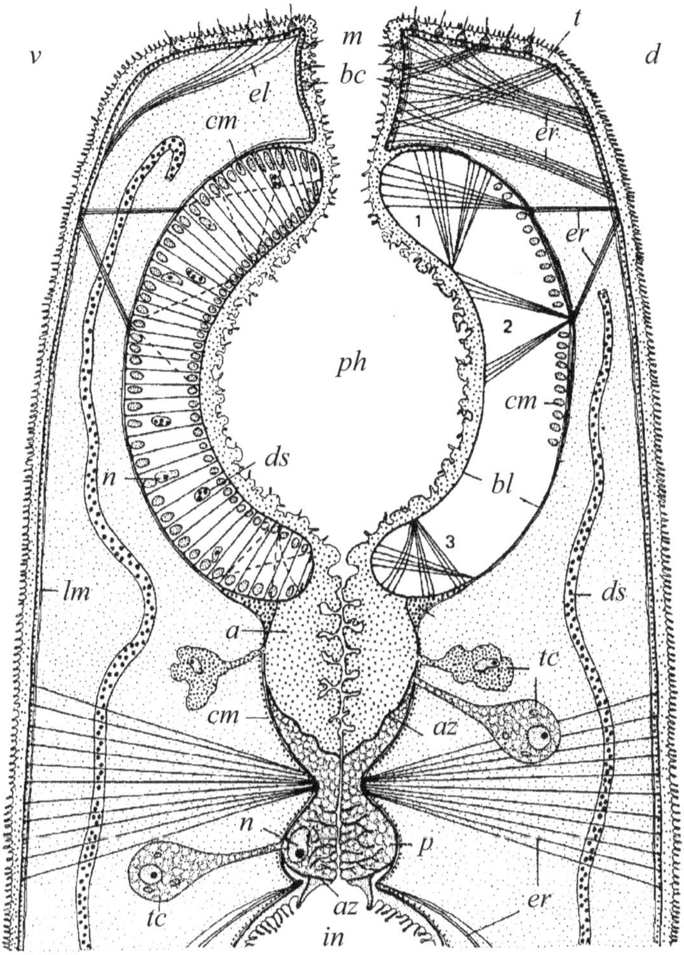

Figure 27. Diagram of the sagittal section of the fore-gut of the redia of *Parorchis acanthus*. *Abbreviations*: *a*, anterior half of oesophagus; *az*, attachment zone; *bc*, buccal cavity; *bl*, basement lamina; *cm*, circular muscles; *d*, dorsal side; *ds*, secretory duct; *el*, extrinsic longitudinal muscles; *er*, extrinsic radial muscles; *in*, intestine; *lm*, longitudinal muscles; *m*, mouth; *n*, nucleus; *p*, posterior half of oesophagus; *ph*, pharynx; *t*, outer syncytial layer of tegument; *tc*, tegumental cyton; *v*, ventral side; 1,2 and 3, anterior, middle and posterior system of additional radial muscles. Some muscles are shown dorsally, others ventrally. From Rees (1983a), reprinted with the permission of Cambridge University Press.

two types of cells – dense and lucid – are distinguished (Rees, 1983b). In the cells of the first type a well-developed RER, dictyosomes of Golgi apparatus and a big number of secretory granules with electron-dense content are ob-

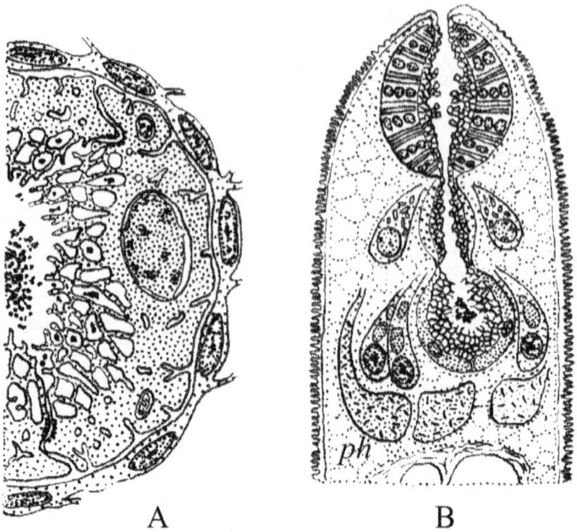

Figure 28. Digestive tract of the daughter rediae of *Metagonimus yokogawai*. A, transversal section through the gut; B, diagram showing distribution of the glands connected with the digestive system. From Dobrovolskij et al. (1983), with permission.

served. Presumably these are enzymes needed for extracellular digestion. In the lucid cells, α-glycogen is stored. In addition, in the cytoplasm of the epithelial cells of the redia intestine there are numerous vesicles, presumably of pinocytic nature (Køie, 1971a). As mentioned above, one should be cautious about such interpretations. The luminal surface of the epithelial cells has anastomosing microvilli and/or microlamellae, which are especially well-developed in rediae with shortened intestines (Dobrovolskij et al., 1983). Apparently, this enlargement of the absorptive surface to a certain extent compensates for the decrease of its linear dimensions. The basal membrane of the cells of intestinal epithelium forms numerous invaginations. In the basal cytoplasm, mitochondria are concentrated, which may suggest that active transport, connected with the transfer of nutrients to the internal tissues of the parasite, take place there (Rees, 1983b).

A channel, connecting the body cavity of rediae and some daughter sporocysts to the external environment, is traditionally called a birth pore. In the majority of species, investigated to date, this channel is lined with the extension of the outer syncytial layer of the tegument, whose surface carries microvilli/microlamellae (Rees, 1966, 1980; Tihomirov, 1980; Fournier and Theron, 1985; Fournier et al., 1986-1987). This channel is surrounded with the layers of circular and longitudinal muscles, forming a sphincter. It is assumed that the sphincter opens the lumen of the pore only momentarily, when fully formed cercariae or parthenitae of the next generation escape through it.

In the microphallid daughter sporocysts investigated by us, numerous ducts of tegumental glands pass between the bundles of sphincter muscles. The ducts are joined to the syncytium lining of the birth pore channel (Fig. 29). Large bodies of glandular cells lie behind the layer of longitudinal muscles and surround the region of the birth pore forming a hemisphere. By their structure and the type of secretion produced, these glands do not differ from tegumental cytons, connected to the outer syncytial layer of the sporocyst tegument. The birth pore and associated muscles and glands form in daughter sporocysts a sufficiently isolated morphological structure, which we call a "sucker-like" organ. It is of interest to note that it is well-developed also in daughter sporocysts of microphallids (*Microphallus pirum*, microphallids of the "*pygmaeus*" group), which do not produce cercariae. This leads to an assumption that, in addition to the birth pore function, this sucker-like organ may serve some other purposes. Its morphological similarity to the adhesive organ of strigeidid adults (obviously of convergent nature) (for reviews see: Erasmus, 1977) suggests that this may be for absorption of nutrients. This presumption is supported by a weak development of

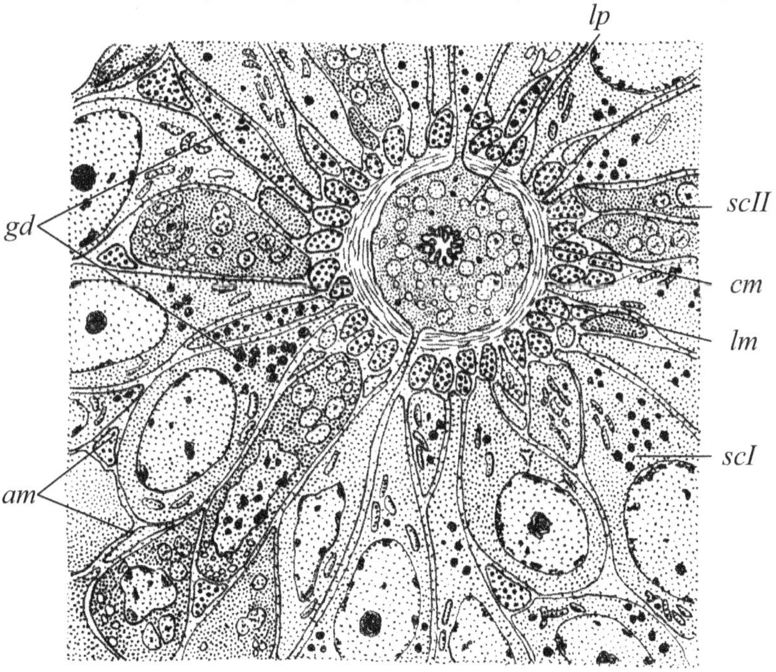

Figure 29. Diagram showing a birth pore (transversal section) of the microphallid daughter sporocysts. *Abbreviations*: *am*, additional muscles; *cm*, circular muscles; *gd*, ducts of glandular cells; *lm*, longitudinal muscles; *lp*, the birth pore lining; *sc*, secretory cells of two (I and II) types. From Galaktionov and Dobrovolskij (1998), with permission.

the tegument in daughter sporocysts of microphallids of the considered group. The outer syncytial layer is thin. Its surface is practically smooth and carries only single microvilli. It degenerates, when metacercariae, developing in the sporocysts, are at the early stages of their morphogenesis. The break up of the outer layer of the tegument of microphallid daughter sporocysts, producing cercariae, also takes place well before their functional activity stops. It is possible that in this case, at certain stages of ontogenesis, the sucker-like organ plays an important role in providing nutrients for the organism, in addition to giving birth to fully formed cercariae.

The birth pore of *Cryptocotyle lingua* rediae is organized quite differently. Excluding the most distal part (invagination of tegument) this channel is lined with epithelial cells (Irwin et al, 1978). The morphology of epithelial cells is very similar to that of intestinal cells: their luminal surface has foldings and carries microlamella, in the cytoplasm there are large active nuclei, mitochondria, ribosomes, cisterns of RER, numerous vesicles and α-glycogen rosettes. However, there are no gland cells, connected to the birth pore. Secretion of cercariae mucoid glands (see p. 124) is presumed to act as a "lubricant", needed by them to pass through the birth pore channel.

In addition to tegumental glands, connected to the outer syncytial layer of tegument and to the birth pore channel, there are other glandular structures in the rediae. The so-called frontal glands are often situated at the anterior end. Their ducts penetrate the outer syncytium and open as independent pores at the edge of the mouth opening. Specialized glandular cells (tegumental?) are localized in the locomotory projections. The secretion of these cells plays an important role in locomotion and attachment of parasites to the surface of the host body tissues. There are also several types of glands, connected to different parts of the redia digestive system that can be distinguished. There are no such glandular structures in sporocysts. However, their tegumental cells, connected to the outer syncytium, may undergo different specialization. Up to 4 types of tegumental glands are reported in daughter sporocyst of *Eurytrema pancreaticum*. The cells of one of those types are located in the so-called "warty" extensions (Panin et al., 1986). The latter are situated on the boundary between two regions of the sporocyst body – sac-like body and trunk-like appendage. It is presumed that the secretion of the glands of warty extensions provide a lytic effect on molluscan tissues when the sporocysts are released to the external environment (Panin et al., 1986).

In the course of daughter parthenitae formation the excretory system becomes more complex due to increase in number of flame cells and emergence of new loops and curves of the channels. In some species flame cell dimensions and the number of cilia, composing the "flame", increases. The excretory system of daughter sporocysts of microphallids undergoes a kind of specialization. In the course of their morphogenesis, hypertrophic growth of collecting ducts takes place: they form numerous projections, anastomosing with each other and forming a thin net, covering a significant part of the sporocyst body (Fig. 24F2).

The nervous system is better developed in rediae. Normally it is well-developed in young daughter sporocysts. Later it undergoes advanced degeneration. Central parts of the nervous system are represented by a ganglion in the anterior part of the body (in rediae it is situated in the vicinity of the oesophagus) and longitudinal nerve cords, outgrowing from it (4 in rediae and 2 in young daughter sporocysts) (Kotikova, 1971). Therefore, sensory structures are better expressed in rediae, for which several types of superficially localized sensory papillae are described (Irwin et al., 1978; Tihomirov, 1980; Rees, 1981; Czubaj and Niewiadomska, 1988). Despite certain differences, their general morphology is practically similar and corresponds to that, described for separate sensory papillae of miracidia (see p. 21). Most likely, functionally these are mechanoreceptors or chemoreceptors, which is typical for such morphological structures in invertebrates. It should be noted that in daughter sporocysts sensory papillae are observed only in young individuals and then undergo reduction.

1.2.2. Germinal Elements and Reproduction

The process of parthenogenetic reproduction of daughter parthenitae starts in the early stages of their development, preceding the formation of the schizocoel. Some of the undifferentiated cells of the central cellular mass of the embryo specialize as germinal cells (for convenience, below we call them primary germinal cells) and start cleavage. Forming embryos of the next parthenogenetic or hermaphroditic generations are initially situated in the meshes of the net, which, as mentioned above, is formed during degeneration and transformation of parenchyma cells. In the rediae of primitive trematodes (Fasciolidae, Notocotylidae, Paramphistomidae) and Paragonimidae mother rediae the reproduction as such is finished here (Cort et al., 1954) (see section 1.3.). Later, only growth and development of individuals of the next generation, produced by primary germinal cells, take place. Productivity of rediae of this type is limited by the number of primary germinal cells, which is relatively small. In daughter sporocysts of the Spirorchiidae and Gymnophallidae, during the whole period of their functional activity, new germinal cells are continuously formed in the body wall (apparently, from undifferentiated cells, localized there), which significantly increases individual productivity of parthenitae (Cort et al., 1954; James and Bowers, 1967). One can not exclude that in this case we deal with a so-called diffuse germinal mass (see p. 81-82). In the majority of sporocyst-bearing and redial trematodes, a specialized organ of reproduction is formed in daughter parthenitae – a germinal mass (see p. 75), in addition to the development of embryos from primary germinal cells.

The primordium of this structure is initially detected as an isolated group of undifferentiated cells, surrounded by parenchyma. The future of the primordium in parthenitae of different trematode groups can be different. In some Echinostomatidae the germinal mass during the whole period of its functioning remains sunk into the parenchyma at the posterior end of the redia body (Fig. 30A) (Cort et al., 1949, 1954). Normally in the rediae of Echinostomatidae, Psilostomidae, Allocreadiidae, Clinostomidae, germinal mass is displaced into the body cavity. It is tightly attached to its posterior wall, being fixed by laminated structures of the endocyst (Fig. 30B). Finally, in the rediae of hemiuroid trematodes and some species of Echinostomatidae the germinal mass is almost completely separated from the wall of the body cavity, being connected to it only with a thin stem, formed by the endocyst (Fig. 30C).

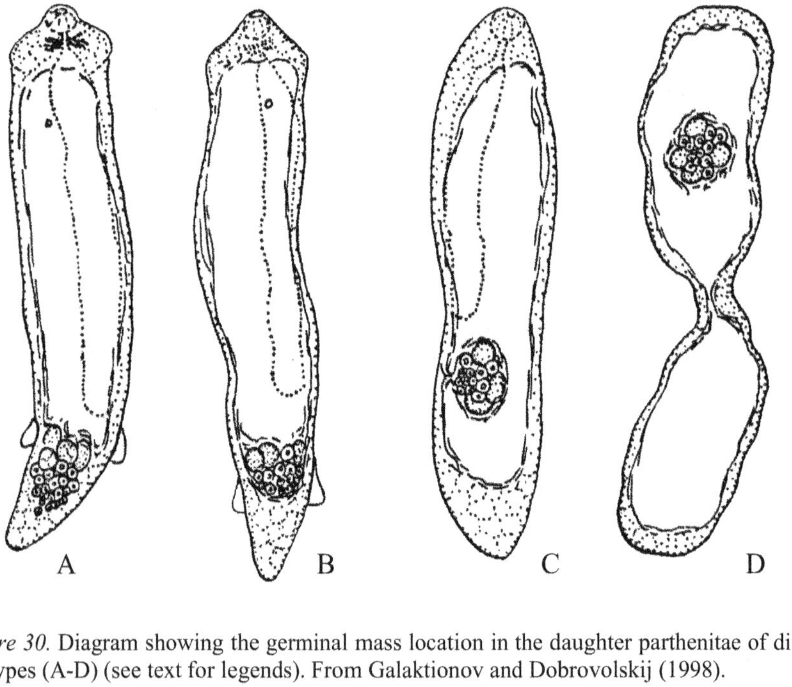

Figure 30. Diagram showing the germinal mass location in the daughter parthenitae of different types (A-D) (see text for legends). From Galaktionov and Dobrovolskij (1998).

In all the above mentioned groups a well-expressed polarity in localization of diverse germinal material, forming the germinal mass, is observed. Undifferentiated cells, capable of proliferation, are concentrated at the base, forming a proliferation zone. Then follow a zone of maturation of germinal cells and a cleavage zone. Therefore, the germinal balls are concentrated at the anterior end of the germinal mass, which is closer to the schizocoel (if the first two cases are considered). When the embryonic membrane has been formed, embryos detach from the germinal mass and are delivered to the schizocoel.

There is only one evolutionary step from the germinal mass, attached to the wall of the body cavity by the stem, to the germinal masses, freely floating in the schizocoel, which is typical for the majority of daughter sporocysts (Strigeidida, many Plagiorchiida) (Fig. 30D). The formation of germinal masses of this type has been observed by us in plagiorchhid daughter sporocysts (Dobrovolskji et al., 1983). In the posterior half of the body of the developing sporocyst, in fact, in its wall, a relatively loose group of cells, with adjoining differentiating germinal cells, is distinguished. In fact, the germinal primordium at this stage of development corresponds to the above described sunk germinal mass of archaic rediae. In the course of the formation of laminated structures of the endocyst the germinal mass becomes more and more structured and begins to evaginate into the lumen of the constantly expanding schizocoel (the stage of "attached" germinal mass).

While the net of laminated projections is retained in the schizocoel, the germinal mass remains rather loose, but it becomes more compact in the course of its isolation. The disruption of the net is accompanied by a loss of connection of the germinal mass with the wall of the sporocyst body. From this moment on, reproduction of undifferentiated cells in the germinal mass becomes autonomic. In other parts of the sporocyst body the number of undifferentiated cells rapidly decreases. A pool of cells, producing the laminated structures of the endocyst and new tegumental cells is replenished by undifferentiated cells. One can not detect undifferentiated cells in mature daughter sporocysts. The change of localization of germinal mass has probably determined the replacement of proximal-distal orientation of germinal elements in it by the concentric type of orientation: the proliferation zone is situated in the center, whereas cleaving germinal cells and germinal balls are localized at the periphery (see fig. 34B).

Attached germinal masses are encountered in some sporocyst trematodes. One of these variants is manifested in representatives of Monorchiidae (Sannia and James, 1978). In young daughter sporocysts of *Cercaria cerastodermae I*, the primordium of the germinal mass emerges in the vicinity of the birth pore. Then, laminated structures of the endocyst form a deep evagination into the schizocoel in this region. The germinal mass is therefore situated at the bottom of this evagination. This formed brood sac is connected to the endocyst of the primary body cavity by means of laminated projections. The germinal mass produces individuals of the hermaphroditic generation, which, in the brood sac, pass sequential stages of their morphogenesis up to encysted metacercariae, infective for the final host. The germinal mass of Microphallidae daughter sporocysts, like in many rediae, remains attached to the wall of the embryonic cavity during the whole period of its functioning.

This is connected with a general simplification of morphology and physiology (juvenilization) of parthenogenetic generation of these trematodes, which will be discussed in chapter 5. In microphallids of the "*pygmaeus*" group this process is even more advanced. Their germinal mass is localized in the parenchyma of the posterior end of the sporocyst body, which, as mentioned above, is typical for some species of primitive trematodes of the family Echinostomatidae.

Period and character of the proliferative activity of germinal masses of daughter parthenogenetic generation are determined by a general trend in the evolution of this phase of the life cycle of particular trematode groups (for details see section 1.3.1. and p. 408). Now it should be mentioned that these parameters may vary considerably. According to Kofiadi (1995), in daughter sporocysts of a number of Strigeidae and Hemiuridae, a stolon-like germinal mass is developed. In Strigeidae it is then completely fragmented, and thus formed small separated germinal masses and cercariae embryos pass to an independent existence in the body cavity. In giant daughter sporocysts of Hemiuridae gen. sp. Kofiadi, 1995 the stolon-like germinal mass is retained during the entire life span. In the stolon-like germinal mass new secondary germinal masses are continuously formed, replacing the ones budded earlier, which have already exhausted their proliferative potential. It is obvious that here we are concerned with a significant increase in the individual productivity of the daughter sporocyst. Another extreme case is the aforementioned microphallids of the "*pygmaeus*" group. In the newly isolated germinal mass of young daughter sporocysts of these trematodes, undifferentiated cells, capable of proliferation are absent. Therefore, a small amount of germinal cells initially included in the germinal mass, mature and start cleavage practically simultaneously. After the embryos have been released into the body cavity, the germinal mass ceases to exist.

1.2.3. Sporocysts of Brachylaimoidea

Our description of the daughter generation of parthenitae would be incomplete without considering peculiar sporocysts, typical of Brachylaimoidea. These sporocysts resemble somewhat those of Bucephalidae, described in the previous section. They are branching tubular structures scattered throughout the connective tissue of the digestive glands of their molluscan hosts (Allison, 1943; Ulmer, 1951; Villella, 1953; Mas-Coma and Montoliu, 1986, 1987, 1995) (Fig. 31). The lumen of the sporocyst branches varies considerably in width. There are germinal balls and cercariae at different stages of their formation. In the apical parts of young growing branches the youngest embryos and structures, resembling germinal masses, are detected (Fig. 31D). At the ends of the branches the birth pore is often situated. Fully formed cercariae leave the sporocyst through this pore.

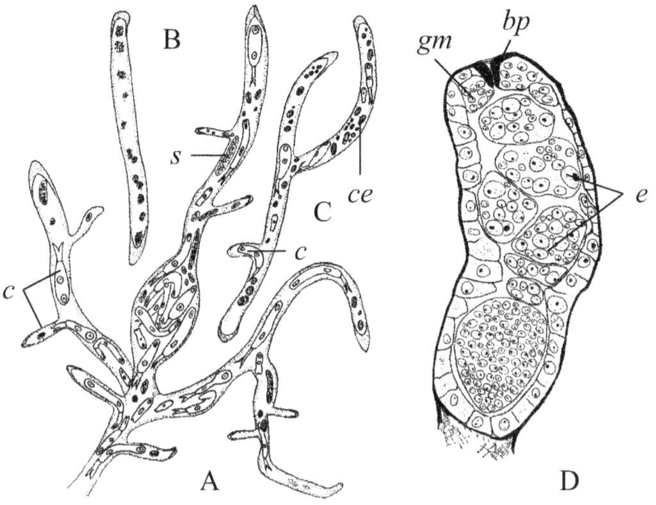

Figure 31. Sporocysts of *Leucochloridiomorpha constantiae.* A, portion of branching sporocyst; B, young sporocyst of the second generation; C, branching sporocyst of the second generation; D, tip of branching sporocyst in longitudinal section. *Abbreviations*: *bp*, birth pore; *c*, cercaria; *ce*, cercaria embryo; *e*, embryos of cercariae or sporocysts; *gm*, germinal mass; *s*, young sporocyst of the second generation. From Allison (1943).

The body of the sporocysts of *Leucochloridium* is differentiated into three parts, differing in morphology and physiology. These is a central part, colored sac-like outgrowths called brood sac, and narrowed tubes connecting the brood sacs with the central part. The central body is situated in the digestive gland of the molluscan host. It produces the embryos, which then pass through the tube to the brood sac, where they are stored and develop into the infective metacercariae. Pojmanska and Machaj (1991) demonstrated that the structure of the tegument is different in the three parts of the sporocyst body. Microvilli are best developed in the brood sac tegument. Here in the cytoplasm, numerous large mitochondria with well-developed cristae, vacuoles and vesicles of different size are observed, suggestive that active transport processes take place here. This may be associated with the delivery of nutrients from the host organism to the larvae, developing in the broodsac.

The nature of Brachylaimoidea sporocysts is still unclear. Ulmer (1951) and Villella (1953) describe two sequential generations – non-cercariogenous mother sporocysts and cercariogenous daughter sporocysts. They are quite similar in morphology and at the maturation stage they have a branched body shape, typical of brachylaimoid parthenitae. Allison (1943) demonstrated that in some cercariogenous sporocysts of *Leucochloridiomor-*

pha constantiae, sporocysts are produced in addition to cercariae (Fig. 31A,B). These are small worm-like sporocysts, which, similar to cercariae, leave the maternal organism through the birth pore and start an independent parasitic existence. Thus, an increase in the number of the initial group of sporocysts in the molluscan host takes place. Production of parthenitae, in addition to cercariae, as mentioned above, is typical of many rediae and daughter sporocysts. Therefore "mother" sporocysts of *L. constantiae* described by Allison (1943) are, most likely, daughter sporocysts. The same applies to non-cercariogenous mother sporocysts, described by Ulmer (1951) and Villella (1953). We believe that their functional role in the life cycle is comparable to that of mother rediae (see p. 53), i.e., they are the first individuals of daughter generation of parthenitae, which give rise to the whole group of parasites in the molluscan host.

This suggestion of ours is based on the data of Bargues et al. (1990) and Bargues and Mas-Coma (1991) on the early development of *Brachylaima ruminae* and on our data on the penetration of *Hasstilesia ovis* miracidia into the intestine of the molluscs *Pupilla muscorum* and *Vallonia costata*. In that instance there is no real penetration of the miracidium. It attaches to the wall of intestine and inoculates into it a cellular aggregation, concentrated at the posterior end of its body. In that aggregation 1-2 germinal cells are distinguished. Apparently other cells are undifferentiated. The cellular aggregation migrates between the cells of intestinal epithelium and several hours after its penetration can be found on the inner side of the molluscan host intestine. Unfortunately, we were not able to trace the subsequent stages of the development of the *H. ovis* parthenita. However, in the investigation of molluscs *Rumina decollata*, experimentally infected with the eggs of *B. ruminae*, at 14-28 days postinfection (dpi) cellular aggregations were detected in the host's hepatopancreas (Bargues et al., 1990; Bargues and Mas-Coma, 1991). Later, these aggregations developed into typical brachylaimoid sporocysts. The beginning of their ramification took place after 60 dpi and advanced cercarial embryos were present at 70 and 90 dpi (Bargues et al., 1990). Based on these data, one may suggest that cellular aggregation, within the brachylaimoid miracidium, is an embryo of an individual of daughter parthenogenetic generation, which gives rise to a branched sporocyst within the mollusc. This sporocyst either produces in the beginning the next parthenogenetic generation, or directly starts to produce cercariae. However, in the latter case, at least in a number of species, the cercariogenous daughter sporocyst may simultaneously produce also sporocysts of the next parthenogenetic generation. Undoubtedly the proposed scheme needs further refinement. Primarily it concerns the earliest stages of the parthenita development in the molluscan host.

1.2.4. Parthenogenetic Metacercariae of Gymnophallids

In concluding the analysis of parthenitae, one more possible variant of parthenogenetic generations encountered in trematodes, should be mentioned. We refer to peculiar parthenogenetic metacercariae, so-called germinal sacs,

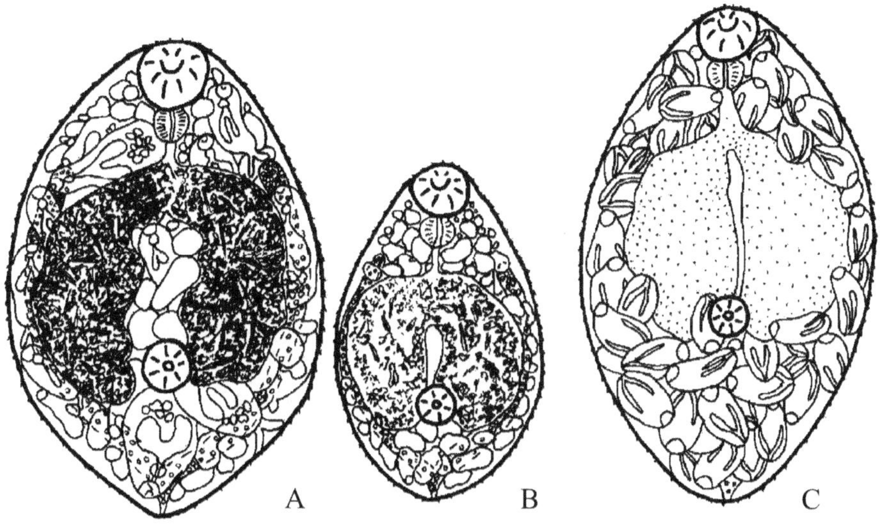

Figure 32. Parthenogenetic metacercariae of *Cercariae margaritensis* Ching, 1982. A, M1 metacercariae containing M2 metacercariae at various stages of development; B, young M2 containing embryos of M3 metacercariae; C, M2 containing fully formed M3 metacercariae (see text for details). From Irwin et al. (2003), with permission.

which up to now have been observed in some gymnophallids – *Gymnophallus australis*, *Parvatrema homoeotecnum*, *Cercaria margaritensis* (Szidat, 1962; James, 1964; Ching, 1982; Galaktionov, 1996a; Irwin et al., 2003). The life cycle of these species will be considered in detail in chapter 3. The cercaria penetrates into the second intermediate host – the mollusc, where it is transformed into a parthenogenetic metacercaria of the first generation (M1) (Fig. 32). Formation of parthenogenetic metacercariae of the second generation (M2), takes place in the M1, where, in turn, metacercariae, infective for the definitive host, develop. The processes of formation of M1 and M2 are similar. The digestive system is the most developed among definitive organs. Organs of the reproductive system are not differentiated. At early stages of the development of M1 and M2, germinal cells and young embryos of the next generation were found in the parenchyma at the posterior end of the body. Apparently, they originated from undifferentiated cells of the genital primordium, situated in that part of the larval body. Later, undifferentiated cells became dispersed over the whole body of metacercariae. They proliferated and gave rise to numerous germinal cells, which start cleavage. Parenchymal cells also undergo significant transformation. Some of them

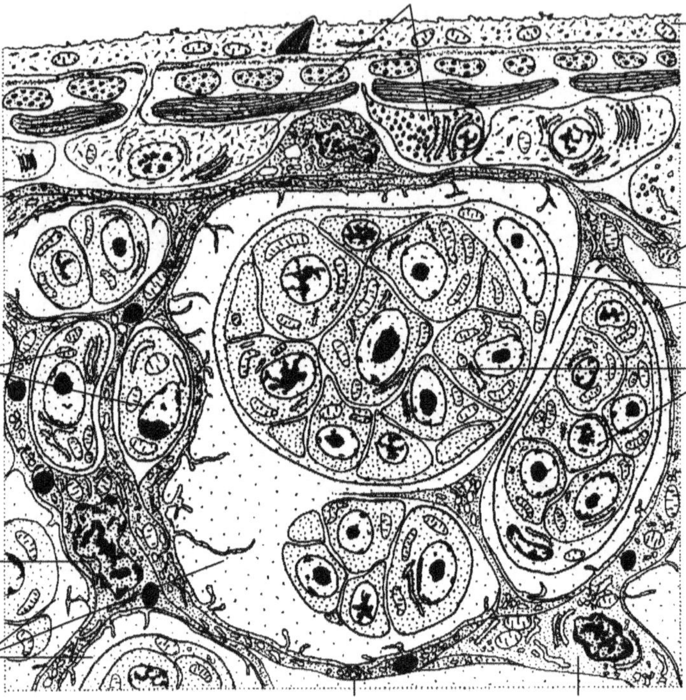

Figure 33. Diagram showing the section of M1 metacercaria of *Cercariae margaritensis* containing embryos of the M2 metacercariae (see text for details). *Abbreviations*: *bc*, brood chamber; *e*, M2 embryo; *em*, embryonic membrane; *gc*, germinal cell; *mc*, multilaminated parenchymal cell; *mls*, multilaminated structure formed by extensions of the parenchymal cells; *t*, outer syncytial layer of tegument; *tc*, tegumental cyton. From Irwin et al. (2003), with permission.

degenerate, but the main bulk of cells flattens and forms numerous laminated extensions (Fig. 33). As a result, the bodies of M1 and M2 became divided into systems of brood chambers. In each chamber a number of embryos develop. Formation of these brood chambers is similar in detail to for mation of the schizocoel and endocyst of rediae and sporocysts (see p. 57). Strictly speaking, parthenogenetic metacercariae thus formed, can be easily taken for sac-like rediae. The former differs from the latter by the presence of a two-branched intestine and the absence of a birth pore.

1.3. Organization of the Generative Apparatus and Early Stages of Development of Sporocysts and Rediae

The presence of sporocysts and rediae, described in the preceding sections, in the trematode life cycle is a unique feature, characteristic of representatives of this taxon and not found in other parasitic platyhelminths. Interpretation of the nature of these peculiar organisms, which is very important for the correct understanding of digenean evolution (see p. 482), is, to a significant degree, determined by how we treat the peculiarities of their reproduction and development. Unfortunately, these questions are not being considered at present. Moreover, much important data obtained by investigators in 20th century are ignored. Therefore we consider it important to devote a special chapter of the book to this issue.

1.3.1. Generative Apparatus of Parthenitae

The nature of germinal cells of sporocysts and rediae will be considered in detail below (see p. 486). Now we characterize the structures connected with the formation and subsequent development of germinal cells. The existence of such structures has been unknown for a long time. Plausible sources of germinal cells have hardly been discussed. Only Dubois (1929) suggested that they were formed directly from somatic cells of the sporocyst body wall.

Important work in this field included several studies by Cort and co-authors (for reviews see: Cort, 1944; Cort et al., 1954). In sporocysts and rediae of a wide range of different groups, these investigators described special structures – germinal masses, where proliferation of germinal cells takes place. In a typical case (plagiorchiids, strigeidids and many other sporocyst-bearing trematodes), germinal masses are small, most often, irregularly rounded bodies, freely floating in the body cavity of the mature sporocysts. In the works of Cort and co-authors it was noted that the germinal masses include numerous large cells and several lager well-isolated cellular aggregations. Cells with little cytoplasm and large nuclei with evenly dispersed chromatin and 1-2 nucleoli were considered as mature germinal cells, whereas cellular aggregations were considered embryos of subsequent generations. According to these authors, germinal masses were occasionally surrounded by a thin "membrane".

Supporting the views of Brooks (1930), who developed the idea of continuous germinal lineage (see pp. 84,89), Cort and his followers believed that germinal cells were isolated in the embryo in early development and that their subsequent proliferation was the result of polyembryony.

The investigation of germinal masses of mother and daughter sporocysts of several species of plagiorchiids (*Haplometra cylindracea, Astio-*

trema trituri, Paralepoderma cloacicola, Xiphidiocercaria sp. VII *Odening*, microphallids of the "*pygmaeus*" group, *Maritrema subdolum*, *Levinseniella brachysoma*) allowed us to elucidate details of their structure and the genesis of germinal cells (Reichel, 1971; Dobrovolskij and Reichel, 1973; Dobrovolskij, 1975; Gerasev and Dobrovolskij, 1977; Dobrovolskij and Muhamedov, 1979; Dobrovolskij et al., 1983; Galaktionov and Dobrovolskij, 1985; Galaktionov, 1993) (Fig. 34).

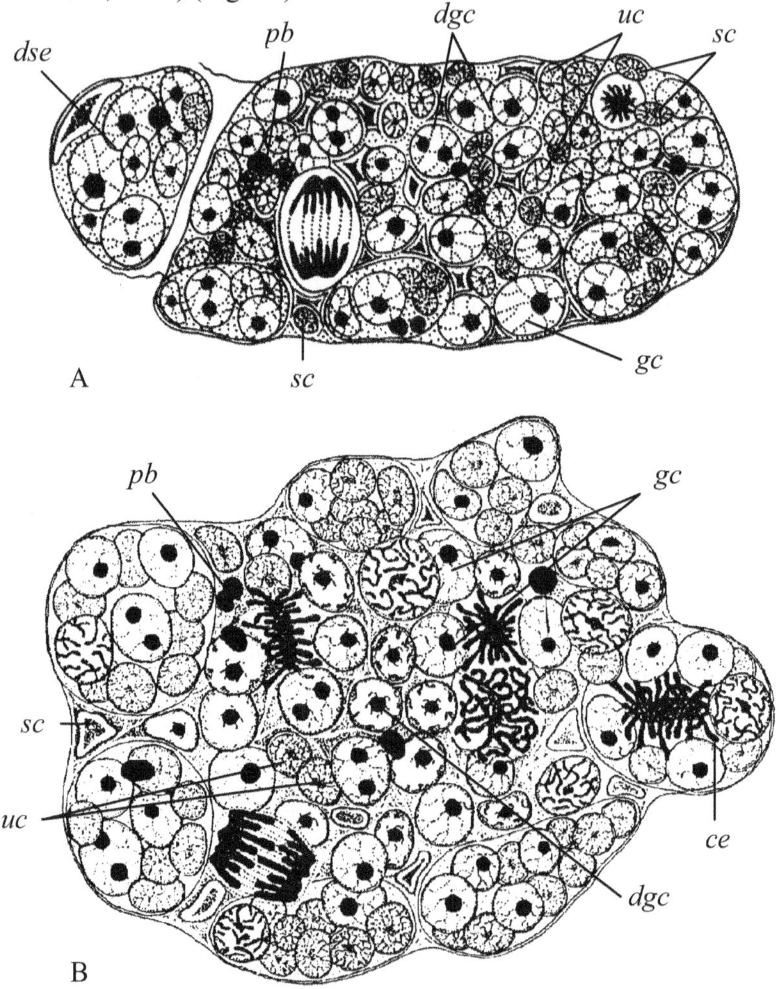

Figure 34. Diagram of germinal masses of the *Xiphidiocercaria* sp. VII *Odening* mother (A) and daughter (B) sporocysts. *Abbreviations*: *ce*, cercaria embryo; *dgc*, differentiating germinal cell; *dse*, daughter sporocyst embryo; *gc*, germinal cell; *pb*, pycnotic body; *sc*, structural cell; *uc*, undifferentiated cell. From Dobrovolskij and Muhamedov (1979) (A) and Muhamedov (1979) (B), with permission.

The central part of the germinal mass is occupied by a loose aggregation of relatively small undifferentiated cells with little cytoplasm and a nucleus with condensed chromatin. These cells are capable of proliferation, for in actively functioning germinal masses, it is in this zone that numerous mitotic figures are observed (Fig. 34). Two more types of cells are also detected there – relatively small cells with numerous branched laminated extensions and lager cells with notably dispersed chromatin in their nuclei.

We consider cells of the first type as structural ones. They form thin laminated extensions, which not only permeate the whole germinal mass but are also present on its surface, ensuring the integrity of the whole structure. Terminal parts of these extensions merge with each other and form a syncytial net (Fig. 35). The second type of cell is represented in germinal masses by maturing germinal cells. They are more polymorphous than the structural ones. Morphological differences between the cells of the second type reflect sequential stages of their differentiation.

An expressed gradient is observed in the localization of maturing germinal cells. Relatively small cells, whose nuclear chromatin begins to disperse, along with structural and undifferentiated cells, occupy the central part. The size of the germinal cells increases towards the periphery of the germinal mass. At the same time the nuclear volume increases. A large nucleolus is also observed there, heterochromatin almost disappears. At this stage of differentiation the cells become almost lucid. Their cytoplasm, lacking any inclusions can not be stained by routine histological methods. The nucleus with its evenly dispersed chromatin appears optically empty. Cells of maximal size occupy even more peripheral position. Some recondensation of chromatin can be seen in the nuclei. Heterochromatin is located mainly at the periphery of the nucleus, perinucleolar chromatin also becomes detectable. The nucleus is surrounded by a relatively thin layer of cytoplasm, which by then is basophilic. These are fully formed germinal cells, ready to cleave. The most peripheral position is occupied by several embryos, undergoing different stages of cleavage. They normally remain in the germinal mass up to the formation of an embryonic membrane on their surface (see below).

These diverse germinal elements are incorporated into the syncytial matrix, formed by structural cells. In germinal masses of microphallid daughter sporocysts, investigated by us, the matrix is well-developed in the region of aggregation of maturing germinal cells. In the cytoplasm of the matrix, mitochondria, numerous vesicles with a heterogeneous content and multi-laminated bodies are detected (Galaktionov, 1993) (Fig. 35). Apparently, a rather thick layer of matrix between germinal cells enables them to cleave freely and form germinal balls. The latter are tightly packed and only a thin layer of syncytium separates them. It can not be excluded that, along with the skeletal function, the matrix plays a certain role in nutrient delivery to the germinal elements, composing the germinal mass.

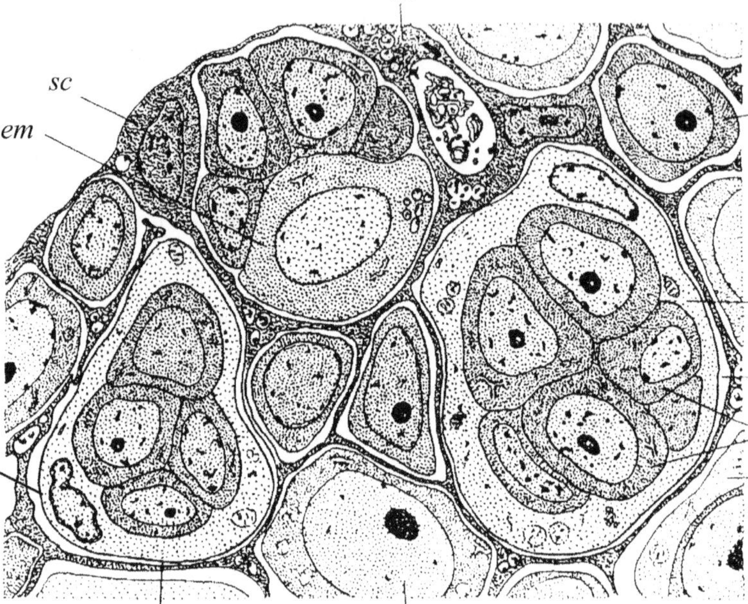

Figure 35. Diagram of germinal mass of the *Maritrema subdolum* daughter sporocyst (according to the data of transmission electron microscopy). *Abbreviations*: *dgc*, differentiating germinal cell; *ec*, embryo cells; *em*, embryonic membrane; *gb*, germinal ball; *gc*, germinal cell; *mem*, macromere forming the embryonic membrane; *sc*, structural cell. From Galaktionov and Dobrovolskij (1998).

Investigating the germinal masses of sporocysts and rediae of different trematode species, we have not observed an increase in the number of mature germinal cells, which might occur due to their direct division. Such a phenomenon could be considered polyembryony. Replenishment of mature germinal cells in the germinal mass proceeds due to the differentiation of undifferentiated cells, capable of proliferation. This process can be treated as "physiological maturation" unaccompanied by meiosis (Dobrovolskij and Ataev, 2003). Division of the mature germinal cells, resulting in formation of two blastomeres, unequal in size, represents the beginning of embryogenesis and should be considered as unequal cleavage (see p. 84).

The morphology of germinal masses is diverse and is determined by the taxonomic position of the trematode species as well as the age and physiological status of the sporocysts and or rediae. Variability in the organization of germinal masses is manifested by a number of characteristics. A well-expressed series can be traced ranging from the stationary germinal masses, attached to the body wall, up to the germinal masses, floating freely in the schizocoel (for details see pp. 69-70 and fig. 30).

The cell composition of germinal masses varies significantly. This is for the case mother as well as daughter sporocysts and is determined by the dynamics of their reproduction. In those cases, when sporocysts are characterized by a long reproductive period, which necessarily correlates with the presence of the birth pore or some other adaptations for the release of offspring into the external environment, the cell composition of germinal masses corresponds to what was described above. Undifferentiated cells are present in the germinal masses. It is from the pool of those cells that continuous replenishment of maturing germinal cells takes place.

When individuals of the next generation are not released (all embryos, formed in the schizocoel, stay in the mother organism), germinal masses contain cells practically of the same age. Proliferation of undifferentiated cells of such sporocysts terminates at early stages of their formation. Maturation of germinal cells proceeds further synchronously or with little time lag. Then, germinal cells start to cleave almost simultaneously. A similar situation takes place in daughter sporocysts of trematodes that do not release the cercariae into the external environment, as in microphallids of the "*pygmaeus*" group (see pp. 69-70) (Fig. 36) (Galaktionov, 1993).

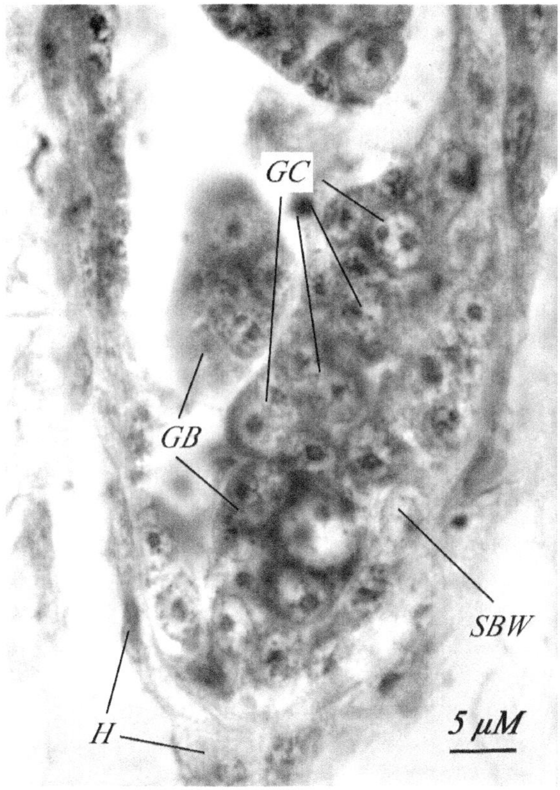

Figure 36. Section of the young daughter sporocyst of *Microphallus piriformes*. *Abbreviations*: *GC*, germinal cell; *GB*, germinal ball; *H*, molluscan hemocyte; *SBW*, sporocyst body wall.

A similar situation is observed in the miracidia of *Fasciola hepatica.* Our investigations (Dobrovolskij, unpublished data) have shown that during the last stage of miracidiae morphogenesis a group of undifferentiated cells of the embryo gives rise to maturing germinal cells. By then, when development has been completed, the store of undifferentiated cells becomes used up, and the part of germinal cells, ahead in their development compared to others, starts to cleave. Thus, in the fully formed larva, germinal material, corresponding to the germinal mass in its morphology and function, is represented only by mature germinal cells and several embryos (see p. 23). This situation in the germinal mass is retained at the first stage of the parasitic phase of the mother sporocyst of *Fasciola hepatica.* In more developed individuals, with the schizocoel still being formed, the germinal mass ceases to exist.

Apparently, a similar mode of functioning of the germinal mass is also typical of other trematodes, whose germinal material in miracidia belongs to the first type (see pp. 23, 33). In particular, Paramphistomidae miracidia have a limited set of almost fully formed germinal cells (Lengy, 1960), and mother sporocysts at the parasite phase of their development produce no more than 18-20 individuals of the next generation (Dinnik, 1961). The same type of reproductive dynamics is retained also in rediae. Already in embryos at the late stages of morphogenesis, when the digestive system has been formed, the development of germinal cells is complete. In individuals, who have left the maternal organism, there is no formation of new germinal cells. Correspondingly, there is also no germinal mass there (Van der Woude, 1951, 1954). It has already stopped functioning in the embryonic stage of the development.

In some cases an increase in the number of germinal masses, forming in the body of a single parthenita, takes place. Most often it is observed in mother sporocysts, which, on the whole, are characterized by transition from the unitary organization to the modular one (see p. 44). Fragmentation of germinal masses in mother sporocysts of a number of representatives of Strigeidida, including *Posthodiplostomum cuticola*, investigated by Dönges (1964) (see p. 36, fig. 14) can be considered as a first manifestation of this trend. Most likely, this process is connected to an elongation of sporocysts, which in many species of strigeidids becomes filiform in the course of growth and maturation. The emergence of numerous centers of reproduction of germinal cells, evenly distributed over the organism, makes possible a more effective usage of its functional abilities and may be considered a first step towards the modular organization. In fact, the parasite body is divided into a number of functional modules, autonomous to a certain degree, in which not only germinal cells are reproduced but also the development of the embryos of the next generation takes place. Similar trends are detected in the organization of daughter sporocysts of other species of Strigeidida and

Schistosomatida. The body of very long filiform daughter sporocysts (genera *Diplostomum, Trichobilharzia*) often becomes beaded-like due to numerous constrictions. Every beaded-like section functions as an independent brood chamber, where the development of the embryos takes place.

This process is more advanced in mother sporocysts of some Plagiorchiida. Many of them tend to decentralize even stronger and acquire a genuine modular structure. Growing mother sporocysts of *Lechriorchis primus* form several large well-isolated branches. Each has its own center of proliferation of germinal cells (see p. 44). These branches are functionally autonomous to an extent and formation and subsequent development of embryos proceeds there independently. In lobular mother sporocysts of *Xiphidiocercaria* sp. VII Odening, the number of relatively large and normally developed germinal masses may be considerable and they are rather evenly distributed over the entire reproducing organism (see p. 45).

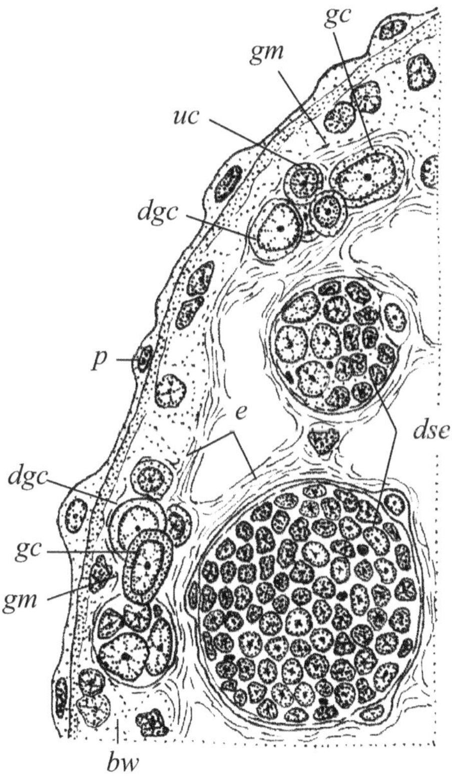

Figure 37. Diagram showing the section of the mature mother sporocyst of *Haplometra cylindracea. Abbreviations*: bw, body wall; dgc, differentiating germinal cell; *e*, endocyst; *dse*, embryo of daughter sporocysts; *gc*, germinal cell; *gm*, germinal mass; *p*, "paletot"; *uc*, undifferentiated cell.

A different situation is seen in the sac-like mother sporocyst of *Haplometra cylindracea.* On the inner surface of its body there is small, but numerous aggregations of maturing germinal cells instead of typical, well-formed germinal masses. These aggregations are not big, each contains only several (2-5) cells and 1-2 embryos in the earliest stages of cleavage (Fig. 37). Cells and embryos are attached to the wall of the mother organism by means of laminated extensions of somatic cells, forming an endocyst. These extensions also form a loose net, permeating the whole schizocoel. Large embryos are located in the meshes of this net (Reichel, 1971; Dobrovolskij and Reichel, 1973; Dobrovolskij,

1975). Such generative structures may be called "diffuse germinal masses". They originate as the result of a secondary dispersion of the germinal material during the development of the mother sporocyst in the mollusc. In the youngest individuals the material is aggregated and represents a rather typical germinal mass of "sunk" type (see p. 68). Dispersion of the germinal material starts simultaneously with the development of the schizocoel. The formation of numerous diffuse germinal masses parallels the separation of the common body cavity of the sporocyst into numerous brood chambers and ensures for the isolation of germinal material of different ages (germinal cells and embryos at different stages of their development). This enables the parasites to increase significantly the period of functional activity of germinal masses, and hence, the number of offspring (Dobrovolskij, 1975; Dobrovolskij et al., 1983).

According to our observations, the diffuse germinal mass is also typical of the multi-branched sporocysts in a number of Bucephalidae. In *Labratrema lamirandi*, e.g., individual and grouped germinal cells locate on the inner surface of the body walls of the generative zones of young, actively functioning branches of sporocyst. Apparently, the germinal material in the multi-branched sporocysts of the Brachylaimoidea (see p. 70) is organized similarly.

The above findings suggest that one should reexamine the idea that there are no germinal masses in sporocysts and rediae of representatives of different taxa and that the germinal cells lie freely in the schizocoel (Dubois, 1929; Brooks, 1930; Dobrovolskij, 1967, 1969, 1971). Cort (1944) suggested that, in all such cases, artifacts, caused by disruption of germinal masses or of young embryos, take place. The absence of large, well-formed germinal masses may be caused by two different reasons. First, there are no germinal masses in rediae and sporocysts, whose generative function has been completed in the early stages of development and the so-called "mature" organism retains only the function of a brood chamber (Fasciolidae, Paramphistomidae). Normally, in all such cases the germinal mass is easily detected in larvae (miracidia) or at other early stages of ontogenesis. Second, the absence of well-expressed germinal mass may be caused by a secondary dispersion of the germinal material, which is usually associated with the acquisition by the most specialized sporocysts of the modular type of organization.

A question arises regarding the nature of germinal masses of sporocysts and rediae. These structures functionally replace the gonad. However, it is not correct to treat them merely as an ovary, or more exactly, germarium. The germinal mass combines the functions of a gonad (reproduction and "maturation" of reproductive cells) and a brood chamber, where the initial stages of embryonic development proceed. Thus, early embryos are protected from mechanical injury, possibly caused by mature and mobile individuals. The necessity of such protection is demonstrated in investigations showing that embryos, which entered the schizocoel prior to the formation of their embryonic membrane (see below), degenerate (Dobrovolskij et al., 1983).

1.3.2. Embryogenesis and Formation of the Miracidium

The question of reproduction of sporocysts and rediae is connected with another, no less important, question – that of the nature of germinal cells and sources of their replenishment in actively reproducing trematodes. The answer to this question can be obtained only after a detailed analysis of the embryonic development of trematodes. The data on embryogenesis and morphogenesis of this group of platyhelminths are very fragmentary and controversial. The discrepancy, however, is mainly connected with the interpretation of the observed phenomena and, hence, with different terminology used.

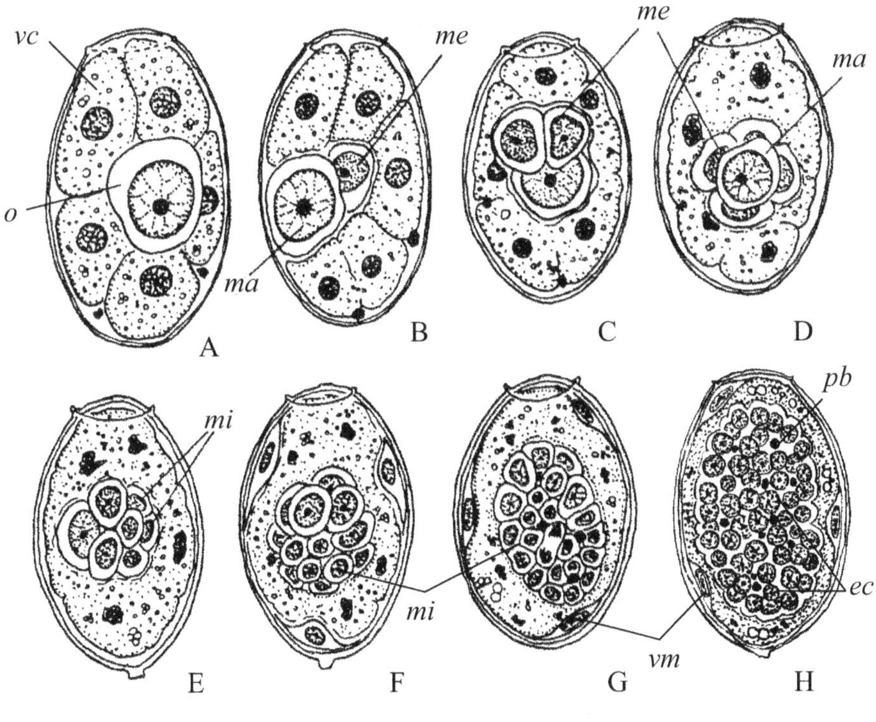

Figure 38. Successive stages (A-H) of embryonic development of the *Paralepoderma cloacicola* miracidium. *Abbreviations*: *ec*, embryo cell; *ma*, macromere; *me*, mesomere; *mi*, micromere; *o*, ovum; *pb*, pycnotic body; *vc*, vitelline cell; *vm*, vitelline membrane.

In fact, only two statements have never been discussed. First, Digenea are characterized by total, unequal and asynchronous cleavage. Second, and this is of utmost importance, (see p. 486), early embryogenesis of parthenogenetic (sporocysts and rediae) and hermaphroditic generations proceeds practically identically (Ishii, 1934; Rees, 1940; Pieper, 1953; Cheng, 1961; Dobrovolskij, 1967; Dobrovolskij et al., 1983; Galaktionov and Dobrovolskij, 1987). The first cleavage furrow cleaves the dividing cell (whether fertilized ovum in the compound egg, laid by marita, or germinal cell of parthenitae) into two blastomeres, unequal in size (Fig. 38, 39, see also fig. 47). Traditionally, these cells were called "somatic" or "ectodermal" and "propagatory" cells (Ishii, 1934; Chen, 1937; Van der Woude, 1954). The usage of such terms can be explained by the fact that an idea, assuming continuous germinal lineage in trematodes, suggested by Brooks (1930), was prevalent for years. However, as it was unambiguously demonstrated later (Haight et al., 1977a,b), neither in trematodes nor in representatives of other taxa are germinal line cells passed from generation to generation. Therefore, the terms "somatic" and "propagatory" cells (blastomeres) became meaningless. This justifies the use of terms, borrowed from general embryology: macro- and meso- and micromeres (Dobrovolskij et al., 1983; Swiderski, 1984, 1986, 1994a; Swiderski et al., 1982).

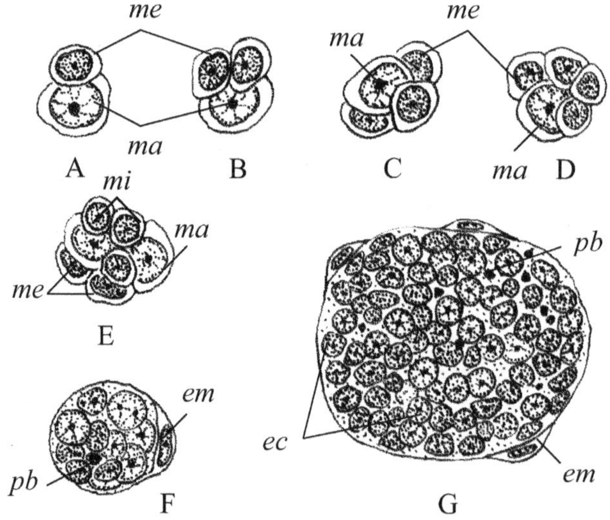

Figure 39. Successive stages (A-G) of embryonic development of the *Macrodera longicollis* daughter sporocyst. *Abbreviations*: *ec*, embryo cell; *em*, embryonic membrane; *ma*, macromere; *me*, mesomere; *mi*, micromere; *pb*, pycnotic body. From Dobrovolskij et al. (1983), with permission.

One after another, macromere yields three cells of minor dimension (mesomeres), so that the embryo becomes clearly heteropolar with a macromere, at one pole, and a more or less compact group formed by mesomeres and their derivatives, at the other (Fig. 38, 39, see also fig. 47). Normally, the first two mesomeres, yielded by the macromere, start cleavage when the third and fourth mesomeres are just forming. It is at this moment that the smallest cells of the embryo (micromeres) are being formed. During this period of time cleavage of the macromere continues. Its size gradually diminishes and it becomes hardly distinguished from other cells of medium size concentrated at one pole of the embryo. At the same time, micromeres, rapidly increasing in number, start to surround the group of larger cells (Fig. 38, see also fig. 47). The process is similar to epibolic gastrulation, but is not gastrulation (Dobrovolskij et al., 1983).

Differences between blastomeres of these three types involve not only their size, but also condensation of their nuclear chromatin. Chromatin of the nuclei of macromeres, as well as of the nucleus of the initial "propagatory" cell (whether a fertilized ovum of marita or a germinal cell of the sporocyst and redia) is strongly dispersed. Heterochromatin forms only a thin net, and in the background a large nucleolus is clearly observed. The heterochromatin of the mesomere nuclei is considerably condensed. It forms numerous relatively compact aggregations directly under the nuclear envelope, as well as in the perinucleolar zone. The nucleolus is easily detectable. In the micromere nuclei, heterochromatin is packed tightly and forms large aggregations of an irregular shape. In most cases the nucleolus is not detectable.

Cell differentiation starts during the early stages of embryogenesis. At the 15-20 cell developmental stage, the formation of an embryonic envelope (embryonic membrane) begins from several blastomeres, moving onto the surface of the embryo. Their number is very limited. In embryos of different generations of several plagiorchiid species, investigated by us, only 3 blastomeres of this type could be reliably detected (Dobrovolskij, 1965a). In schistosomes, up to 10 cells participate in the formation of the embryonic envelope around the developing miracidium (Swiderski, 1984, 1985, 1986, 1994a,c; Swiderski et al., 1982; Neill et al., 1988). Later these cells flatten and become squamous. Their future is somewhat different in embryos of miracidia (mother sporocyst generation) and in all the rest generations.

Development of the miracidium proceeds in the vitelline mass, which forms within the compound egg as the result of gradual disruption of vitelline cells. For some time their degenerating nuclei are retained around the embryo (Fig. 38). This fact accounts for the specificity of the formation of the embryonic envelope around the developing larva. It should be mentioned that these observations are interpreted somewhat differently. According to our observations (Dobrovolskij, 1965a; Dobrovolskij et al., 1983), cells of the future embryonic envelope detach from the surface of the embryo, migrate through the vitelline mass onto the inner surface of the eggshell, flatten on it and finally merge with each other, forming a continuous syncytial layer (Fig. 38F-H) This thin cytoplasmic layer envelops the vitelline mass and the embryo, which is developing in it. In many other trematodes a similar envelope is formed in the same way (Bennet, 1936; Guilford, 1961, Pearson, 1961, Neill et al., 1988).

Traditionally this envelope is considered as a typical provisional structure and is called the "vitelline membrane". The subsequent development of the miracidium proceeds directly in the vitelline mass, which is formed as the result of the disintegration of vitelline cells. The nuclei of the latter gradually become pycnotic, but normally can still be detected within the vitelline mass at all time of the miracidium morphogenesis. The vitelline membrane is retained even longer. Its remnants are easily detected in the empty eggshells, left by mature miracidia.

Quite a different interpretation was proposed by Swiderski (1985, 1986, 1994a,c), who investigated the development of *Schistosoma mansoni* miracidia. According to his view point, two embryonic envelopes (an outer and inner) are formed around the developing larva. The first is formed only by two macromeres and the "vitelline syncytium". Later, 5-6 mesomeres separate from the embryo and merge with each other to form the inner envelope. Unfortunately, transmission electron micrographs by Swiderski (1986, 1994a) are not sufficient for the step-by-step reconstruction of the formation of the embryonic envelopes. The described outer envelope corresponds to the "Reynold's layer" after Neill et al. (1988), which is treated by these authors as an acellular matter consisting of branching and anastomosing fibrils. Survey transmission electron micrographs, presented by Neill et al. (1988) and detailed row of the developmental stages studied strongly support the main conclusions, made by the investigators. These conclusions coincide with traditional views on the formation of embryonic envelopes in the course of embryonic development of trematode miracidia. The problem of the nature of these envelopes needs further investigation.

As mentioned above (see section 1.1.1.), miracidial morphogenesis is poorly understood, which makes it impossible to reconstruct the process. However, the results of some observations deserve to be mentioned. Intensive proliferation of the embryo cells ensures rapid growth of the embryo (Dobrovolskij et al, 1983). Gradually, the cells become equal in size: differences between blastomeres of different types practically disappear. Already at this stage it is impossible to trace any connection between the first mesomere ("propagatory" cell) and any group of cells, which could be treated as a "germinal primordium" (Fig. 38H, 39G).

Degeneration of a part of cellular material starts very early in development. It is accompanied by the appearance of numerous pycnotic bodies (Dobrovolskij et al, 1983; Swiderski, 1986, 1994a). The intensity of this process is illustrated by the fact that the embryo of a plagiorchiid miracidium consists of 70-80 cells, whereas the body of the mature larva contains only

10-15 cells (Dobrovolskij et al., 1983). Degenerating cells are probably used as nutrients for the embryo, as was suggested for Turbellaria by Ivanov and Mamkaev (1973). Processes of cell differentiation become noticeable almost simultaneously with the resorption of cells, which marks the beginning of morphogenesis.

Of particular interest is the formation of the miracidial epidermis (see p. 10). According to Neill et al. (1988), in the developing miracidium of *Schistosoma mansoni*, ciliated epidermal plates are formed from micronucleated cells, localized on surface (Fig. 40A). At the same time cells of the future tegument start to differentiate. These cells are situated a little deeper. First of all it concerns the nuclei-containing parts of the cells. At the surface of the embryo, there are narrow cytoplasmic projections of those cells, situated between the micronucleated cells, which give rise to hypodermal ridges in the course of subsequent differentiation.

Epidermal plates undergo a more complex transformation. At their basal surface, an extension is formed, into which the nucleus migrates. Later, this extension passes through a layer of differentiating myoblasts and ends in the central cellular mass. At the final stages of miracidial morphogenesis, nuclei are not detected in these extensions (Neill et al., 1988). Similar structures, lacking nuclei, were also described in the fully formed miracidia of *Schistosoma mansoni* (Pan, 1980).

According to our observations, the formation of epithelial plates in miracidia of representatives of other families proceeds somewhat differently (Fig. 40B-D). In this case also, several cells on the surface of the embryo begin to flatten. They cover significant parts of the embryo surface, but never contact each other directly. It is those cells, which will give rise to ciliated epidermal plates. Gaps are left between those cells, filled with smaller cells, not yet modified. Later, the nuclei-containing parts of the smaller cells start sinking into the embryo. On the surface, only nuclei-free parts are left, which later are transformed into hypodermal ridges.

Definitive formation of epidermal plates is accompanied by the development of numerous cilia on their apical surface and profound changes in the nuclear organization. However, in most known cases, the nucleus, remains in the plate (Fig. 40B-C). In many digeneans (*Fasciola hepatica*, many species of Echinostomatidae) strongly modified nuclei are retained in the epidermal plates during the entire life span of the miracidium. In other digeneans, nuclei are resorbed in the late stages of morphogenesis, as is the case of miracidia in the Plagiorchiidae and Ochetosomatidae (Fig. 40D). At any rate, there are no nuclei in the plates of fully formed larvae of all investigated representatives of these families (Dobrovolskij, 1965a).

As for sac-like projections of epidermal plates, sinking into the submuscular zone, they are usually not formed (*Fasciola hepatica,* some species

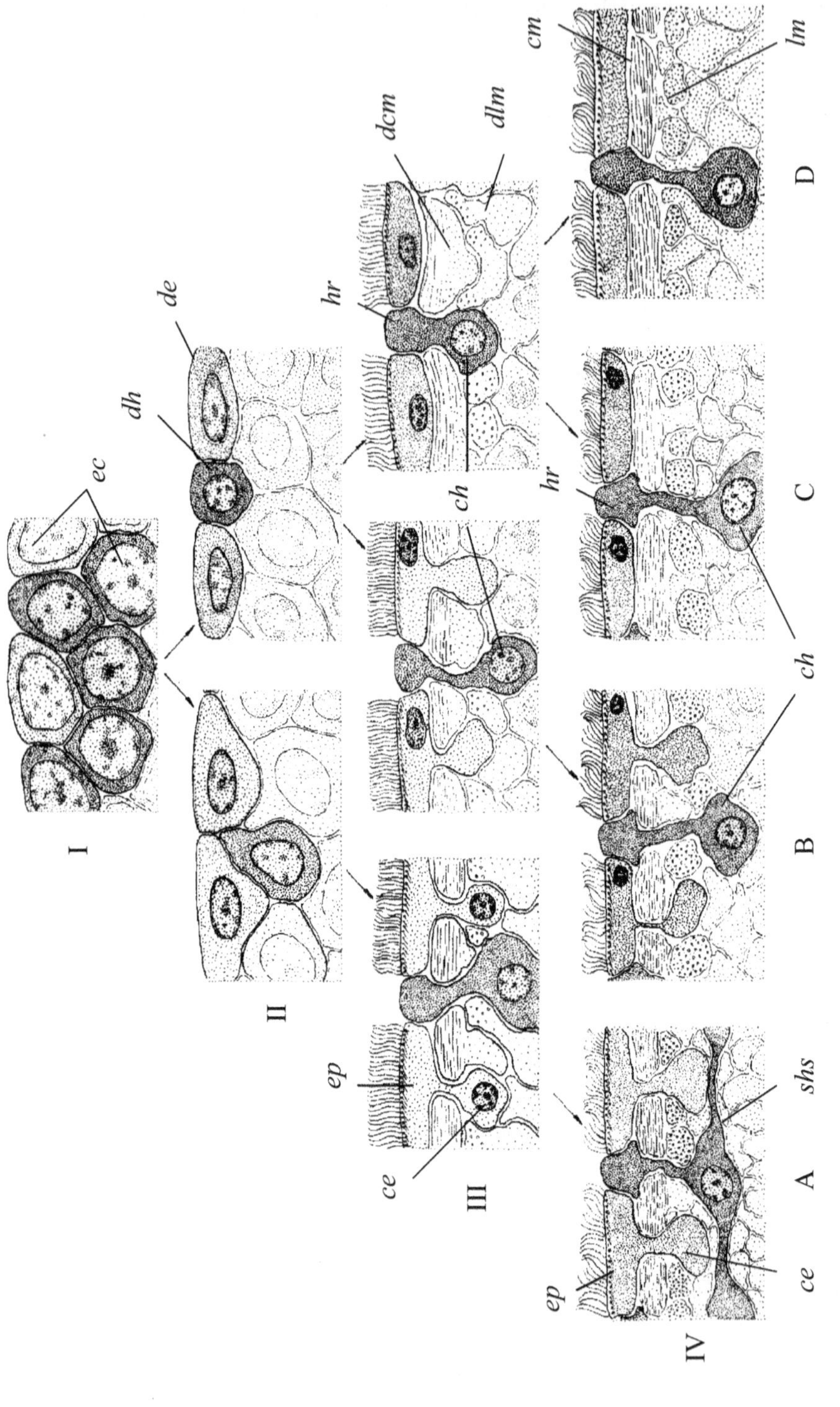
I
II
III
IV
A
B
C
D
ec
dh
de
hr
dcm
dlm
cm
lm
ch
ep
ce
shs

of Echinostomatidae). With the exception of *Philophthalmus rhionica* (Philophthalmidae). In miracidia of this species "sunk" projections have been described. Nuclei never migrate into these projections. In fully formed miracidia of this species nuclei are localized in the posterior edge of the epidermal plates.

The above described variants in the formation of miracidial epidermis differ insignificantly. This concerns the fate of the epidermal plate nuclei and details of the emergence of hypodermal ridges, which represent a primordium of the definitive tegument of the mother sporocyst (Fig. 40). All the cells participating in the formation of epidermis, are initially situated in the surface zone of the embryo. The differences between these cells appear only during subsequent differentiation.

Of no less interest is the formation of germinal elements in the developing miracidium. As mentioned above, morphological criteria do not allow one to postulate a connection between the first separated mesomere and the forming germinal cells. According to our observations, the germinal cells begin to differentiate during early stages of morphogenesis, practically simultaneously with other organ systems (see fig. 3C). They originate from undifferentiated cells, retained in the posterior part of the embryo. The "maturation" of germinal cells resembles a similar process, which takes place in germinal masses of sporocysts and rediae (see p. 75).

The dynamics of this process depends on the organization of the germinal material, characteristic of the miracidia which have completed their development (see p. 23). In those cases, when miracidia have fully formed germinal cells or even embryos of the next generation, differentiation begins during early stages of morphogenesis. Normally, in the posterior part of the embryo, where initially a rather compact group of undifferentiated cells (germinal primordium) is localized, cells, that rapidly increase in size and contain nuclei with dispersed chromatin, appear.

Later, in the miracidia with the first type of organization of the germinal material (e.g., *Fasciola hepatica*) (see p. 23) undifferentiated cells of the germinal primordium start to differentiate. When morphogenesis has been completed, the majority of cells become functionally mature. One or

Figure 40. Diagram showing a development of epidermis in the course of miracidium development in different groups of trematodes. A, Schistosomatidae; B, Philophthalmidae; C, Echinostomatidae; D, Plagiorchiidae and Ochetosomatidae; I, early embryo; II, beginning of differentiation of epidermal cells; III, differentiation of epidermal plates and hypodermal ridges; IV, fully formed miracidium. *Abbreviations*: *ce*, cyton of epithelial ciliated cell; *ch*, cyton of hypodermal cell; *cm*, circular muscles; *dcm*, differentiating circular muscles; *de*, differentiating epidermal plate (ciliated epithelial cell); *dh*, differentiating hypodermal cell; *dlm*, differentiating longitudinal muscles; *ec*, embryo cells; *ep*, epidermal plate; *hr*, hypodermal ridge; *lm*, longitudinal muscles; *shs*, syncytial structure formed by cytoplasmic projections of the of hypodermal cell cytons.

two cells start to cleave and the miracidium, which has completed its development, contains a corresponding number of embryos.

However, in trematode species, where realization of the generative function is at least partly transferred to the parasitic phase of development of the mother sporocyst (the second type of organization of the germinal material, see p. 24), only some cells of the germinal primordium undergo differentiation. When miracidial morphogenesis has been completed, a typical germinal mass is formed in the miracidium. It contains all types of cells (from undifferentiated to mature ones) and embryos at different stages of their development. A similar situation is typical for some species of Echinostomatidae (Ataev et al., 2001a,b).

When reproduction proceeds almost completely in mature mother sporocysts in the mollusc, differentiation of germinal cells in developing miracidia may not begin. In such cases, fully formed larvae (some species of Plagiorchiidae and Ochetosomatidae) contain only undifferentiated cells (see fig. 11A-B).

1.3.3. Embryogenesis and Formation of Daughter Parthenitae

Embryonic development of rediae and daughter sporocysts (as well as of individuals of the hermaphroditic generation – see section 2.1.2., p. 109) proceeds similarly, with the exception of the formation of the provisional envelope of the embryo, which in this case is called an embryonic membrane. Blastomeres, forming the surface syncytium, do not detach from the embryo, but spread on its surface (see fig. 34, 35, 39). Such a formation of the embryonic envelope, which, in this case, is called "primitive epithelium" or "investing syncytium" (see detailed discussion of the terms in Meuleman and Holzmann, 1975; Galaktionov and Dobrovolskij, 1987; Dunn et al., 1992b), is well-documented in light- and transmission electron microscopy (Tihomirov, 1980; Muhamedov, 1981; Dobrovolskij et al., 1983; Galaktionov, 1993).

There is another view point on the nature of this structure. It is based on the classical investigations of Cort (1944) and Cort et al. (1954), who were the first to describe the embryonic membrane around the embryos of sporocysts and cercariae. However, they presumed that the formation of the embryonic membrane proceeds due to the cells of the lining of the mother organism body cavity, i.e., the endocyst. According to these authors, the embryonic membrane of cercarial embryos is soon disrupted, and in sporocysts of some species is transformed into a "paletot", surrounding the body of the fully formed individuals. It was demonstrated by Schell (1961, 1962a,b, 1965), Dobrovolskij et al. (1983) and Køie (1987) that such a view point is erroneous. The "paletot" is composed of modified hemocytes of the molluscan host (see p. 256). Meuleman et al. (1980), on the basis of transmission electron microscopy of *Schistosoma mansoni* noted the connection of the embryonic membrane of daughter sporocysts with the lining of the mother sporocyst body cavity (i.e., endocyst). We believe that the electron micrographs presented in their work could be otherwise interpreted. Laminated structures of the endocyst cover separate embryos or groups of embryos, but only over their own embryonic membranes, which are formed by the cells of the embryo, localized on its surface. In microphallid daughter sporocysts investigated by us (Galaktionov, 1993), a similar character of interrelations between the embryonic membrane and the envelope, formed by laminated extensions of the structural cells of the stolon-like mother germinal mass, are clearly revealed. The nature of embryonic envelopes covering developing daughter parthenitae needs further investigation.

As mentioned above, laminated extensions of the endocyst ensure spatial isolation of developing embryos and, similar to the embryonic membrane, protect early embryos from mechanical injury. Transport of nutrients probably occurs through these endocyst extensions directly from tissues of the body wall of the mother individual to the developing embryo. Thus seems reasonable because numerous active mitochondria are found in the laminated extensions of the endocyst of parthenogenetic metacercariae of *Cercariae margaritensis* (Gymnophallidae) (Irwin et al., 2003). The transport function of the embryonic membrane seems obvious (otherwise, the embryo would be deprived of nutrients), but has not confirmed experimentally. In this connection, it should be noted that there is a direct cytochemical proof of nutrient transport through the embryonic membrane of cercarial embryos (Bogitsh, 1986).

After the embryonic membrane has been formed (the germinal ball stage), as it occurs in the case of miracidial development, an intensive proliferation of blastomeres occurs. Morphological differences between the blastomeres rapidly disappear (see fig. 39). Simultaneously an active resorption of some embryo cells proceeds, resulting in the appearance of many pycnotic bodies in the embryo. As the number of cells increases, the embryos elongate along the longitudinal axis (even in species of Microphallidae, whose mature daughter sporocysts have a spherical body). At this time, some peripherally located cells start to expand beneath the embryonic membrane. These cells flatten and merge with each other. Then their nuclei degenerate and a nuclei-free syncytium (outer layer of definitive tegument) is being formed beneath the embryonic membrane (Meuleman et al., 1980; Dobrovolskij et al., 1983). Gradually the embryonic membrane becomes thinner, degenerative changes take place there, and soon after the outer layer of the tegument has been formed, the membrane disintegrates. In the embryos at this developmental stage, large cells appear, with nuclei, containing nucleoli and dispersed chromatin. In the cytoplasm of those cells, secretory granules are usually observed. Initially these cells lie in deeper layers of the embryo body, but later all of them are found in the subtegumental layer. The cells form cytoplasmic projections, reinforced with microtubules, connected with the tegumental outer layer. Transport of the secretion, produced by tegumental glands, proceeds through these projections. The number of these cells rapidly increases, their differentiation becoming noticeable, i.e., the cytons differ from each other not only by the amount, but also by the type of the secretion they contain.

Parallel to the formation of the tegument, muscles of the tegumentary-muscular sac are formed. Myoblasts form a continuous layer, underlying the outer syncytium. In their electron-dense cytoplasm, myofibril bundles are revealed. Later, bundles of circular and longitudinal muscles, separated by projections of tegumental cells, become visible. The described type of formation of the tegumentary-muscular sac in daughter parthenitae is universal and almost identical to the similar processes, which take place in the course of morphogenesis of larvae of the hermaphroditic generation (Galaktionov and Dobrovolskij, 1987; see also p. 113).

However, the views of James and Popiel on the origin of daughter parthenitae tegument (James et al., 1966; Popiel, 1978a,b; Popiel and James, 1978a,b) were widely accepted. According to these authors, on the surface of the developing daughter sporocysts of species of Microphallidae, a nuclear syncytium is initially formed. Then it is separated into an inner nuclei-free part and outer part, containing nuclei. The separation occurs due to numerous evaginations of the outer plasma membrane. These evaginations, having passed through the nuclear layer of the syncytium, are further extended in the longitudinal direction and merge with each other. Therefore, the outer layer, containing nuclei, acquires a structure of "cellular epithelium". In parthenitae with a birth pore, this layer shortly breaks up. In parthenitae without a birth pore (it should be noted that daughter sporocysts of *Microphallus similis* and *Cercaria littorina saxatilis V* (= *Maritrema arenaria*) with a well-developed birth pore were included in this group), the layer is retained all their life. The above described type of formation of the daughter parthenitae tegument was accepted by Threadgold (1984) and supported by Al-Salman and James (1988). Such an interpretation of transmission electron microscopy findings was analyzed in detail (Dobrovolskij et al., 1983; Køie, 1987; Žd'árská, 1990). The "epithelium", covering the sporocyst tegument, is just a "paletot", forming from host hemocytes (for details see pp. 90, 256). This process has been traced in the species of microphallids, which were investigated by James and co-authors – *Microphallus pygmaeus* and *M. similis* (Galaktionov, 1993).

Interesting findings on the formation of the tegument of sporocysts, developing along with cercariae, within an earlier generation of daughter sporocysts *Meiogymnophallus minutus* (Gymnophallidae), were presented by Al-Salman and James (1988). According to these authors, the embryos of the sporocysts lack the embryonic membrane. Some cells, located in the outer layer of the germinal ball, form projections, extensively growing on its surface and merge with each other. Thus, an outer syncytial layer is formed and nuclei-containing cytons, connected with the latter by means of their projections, become tegumental cells. Therefore, it is likely that the formation of the tegument, taking place during daughter parthenitae development, may proceed at least in two ways. The process, described by Al-Salman and James (1988), resembles the formation of the hypodermal ridges of *Schistosoma mansoni* miracidia (see above) and subsequent development of the tegument of the mother sporocyst (see p. 31). At the same time, the reasoning of Al-Salman and James (1988) is not always convincing. Most likely, the young sporocyst embryo in Fig. 1 (p. 233 – Al-Salman and James, 1988) resembles a degenerating cellular aggregation (apparently, a germinal ball). This presumption is supported by "looseness" of the cellular mass and obvious degenerative changes in the cytoplasm of the cells. We repeatedly observed such structures in transmission electron microscopy investigations of parthenitae in different trematode groups. We believe that the nature of the tegument in *M. minutus* sporocysts needs further investigation.

Parallel to the formation of the tegumentary-muscular sac, inner organs and the body cavity (schizocoel) emerge and undergo development. This process is described in detail in the section 1.2. of this chapter. We note that the process proceeds similarly in rediae and daughter sporocysts. The only difference is that the digestive system is formed in rediae and is not formed in sporocysts. Moreover, the schizocoel in parthenogenetic metacercariae of gymnophallids (see p. 73) develops identically. All this testifies to the universal character of mechanisms of formation of the primary body cavity in organisms with the same level of organization as that of plathyhelminths.

2. THE HERMAPHRODITIC GENERATION

In the course of ontogenesis of individuals of the hermaphroditic generation, several stages succeed each other, each corresponding to a certain environment. Morphogenesis of the first larva, – the cercaria, proceeds in the body cavity of the daughter parthenita and may be completed during its migration in the body of the first intermediate host. Once released from the mollusc, the mature cercaria leads an active life in the external, usually the aquatic environment. The cercaria encysts on the substrate (in this case we speak of adolescaria), or actively or passively gets into the second intermediate host (invertebrates, fishes, amphibians), where, in the majority of trematode species, it undergoes certain stages of formation. Such a larva is called a metacercaria. Its final development takes place in the definitive (final) host – a vertebrate animal – and is called maritogony. After maritogony has been completed, the sexually mature individual of the hermaphroditic generation, i.e., marita (adult) – starts reproduction.

Distinct characteristics of ontogenesis of the individuals of the hermaphroditic generation makes of possible to identify a number of sequential stages that replace each other. We propose to distinguish preparatory and functional stages. In the most common case, the preparatory stages include all stages of morphogenesis of cercaria and metacercaria (up to fully formed individuals) and maritogony. The functional stages include the stages of mature cercaria, metacercaria/adolescaria and marita. At preparatory stages, a particular morphological and functional organization is created. It must provide the individual at one or another functional stage the possibilities of fulfilment of the tasks, presented by the nature of the life cycle. For example, in the case of three-host life cycle, such tasks for cercariae consist of getting into the second intermediate host, for metacercariae it is surviving in this host, for maritae – reproduction in the definitive host. The enumerated tasks can be considered as basic and primary. Their successful completion is achieved because of a complete set of adaptations, specific for each stage of ontogenesis.

In a number of cases, changes in the course of the life cycle effect significant modifications in the events of ontogenesis. Manifestation of its discrete character may increase or decrease. The first variant is encountered in strigeid trematodes of the genera *Alaria* and *Strigea*. In the ontogenesis of their hermaphroditic generation two more stages appear: preparatory (morphogenesis of mesocercaria) and functional (stage of formed mesocercaria). Both proceed in the one more (third) intermediate host, which in these trematodes gets in between the first and second intermediate hosts (see section 5 in chapter 3). An opposite situation is observed, when the second intermediate host disappears from the life cycle of a number of Microphallidae. It is accompanied by a loss of the first functional stage – the cercaria in the external environment. Morphological and functional changes, accompanying these processes, are characterized below.

2.1. Cercaria

The cercaria consists of an oval or elongated body, flattened in the dorsoventral plane, and a tail. The latter may be considerably shortened or completely lost. An oral sucker is localized subterminally at the anterior end of the cercarial body; as a rule, there is also a ventral sucker, located in the middle or posterior part of the body. In the larval body there can be seen more or less well-expressed digestive and protonephridial excretory systems, glands of different types, genital primordium and sensory organs, represented mainly by external sensory papillae and more rarely by simply organized eyespots. Tail morphology varies in larvae of different species, depending on its main function (providing locomotion and adherence or crawling on the substrate, and floating or swimming in water).

In the morphology of cercariae, two groups of characters can be clearly distinguished. The first one consists of characters, generally typical for individuals of the hermaphroditic generation – the body shape, presence of suckers, morphology of digestive, excretory and other systems, which begin to function normally only at the phase of metacercaria or marita. It is this complex of characters, in which heterochrony is observed (for details see p. 462). In most cases these systems do not function and represent just primordia, which do not affect either the viability or infectivity of the larvae.

The second group includes provisional characters, i.e., the ones, which are clearly adaptive and typical only for larvae. It is because of different provisional structures, that the main task of the first functional stage (cercaria) of the ontogenesis of most trematodes – penetration into the second intermediate host – is accomplished. Diversity of these adaptations was responsible for the variety of morphological forms, which are seen in cercariae.

2.1.1. Diversity of Cercariae

In the course of accumulation of data on morphology of cercariae, attempts to distinguish and classify their main morphological types were made. This was necessary because for a long time there were only fragmentary data available on the problem of determining which cercaria corresponds to which trematode species. In many cases, this problem still remains unsolved. Therefore, investigators of trematodes still use the system of classification of cercarial morphological types proposed by Lühe (1909), which was later defined and supplemented. The most complete version of this classification is available in the review by Dawes (1946). Below, we briefly characterize the main groups of larvae, most often encountered in the text.

Note that the morphological types of cercariae, distinguished by Lühe (1909), do not usually correspond to the categories of the modern system of trematodes, which is based mainly on the analysis of morphology of maritae. Some groups do reflect true phylogenetic relations between different trematode taxa, but there are also artificial ones, e.g., a group of microcercous cercariae (Microcerca). Assignment to this group is based on characters, which developed many times and independently in representatives of phylogenetically distant families of digeneans. They reflect similarity of

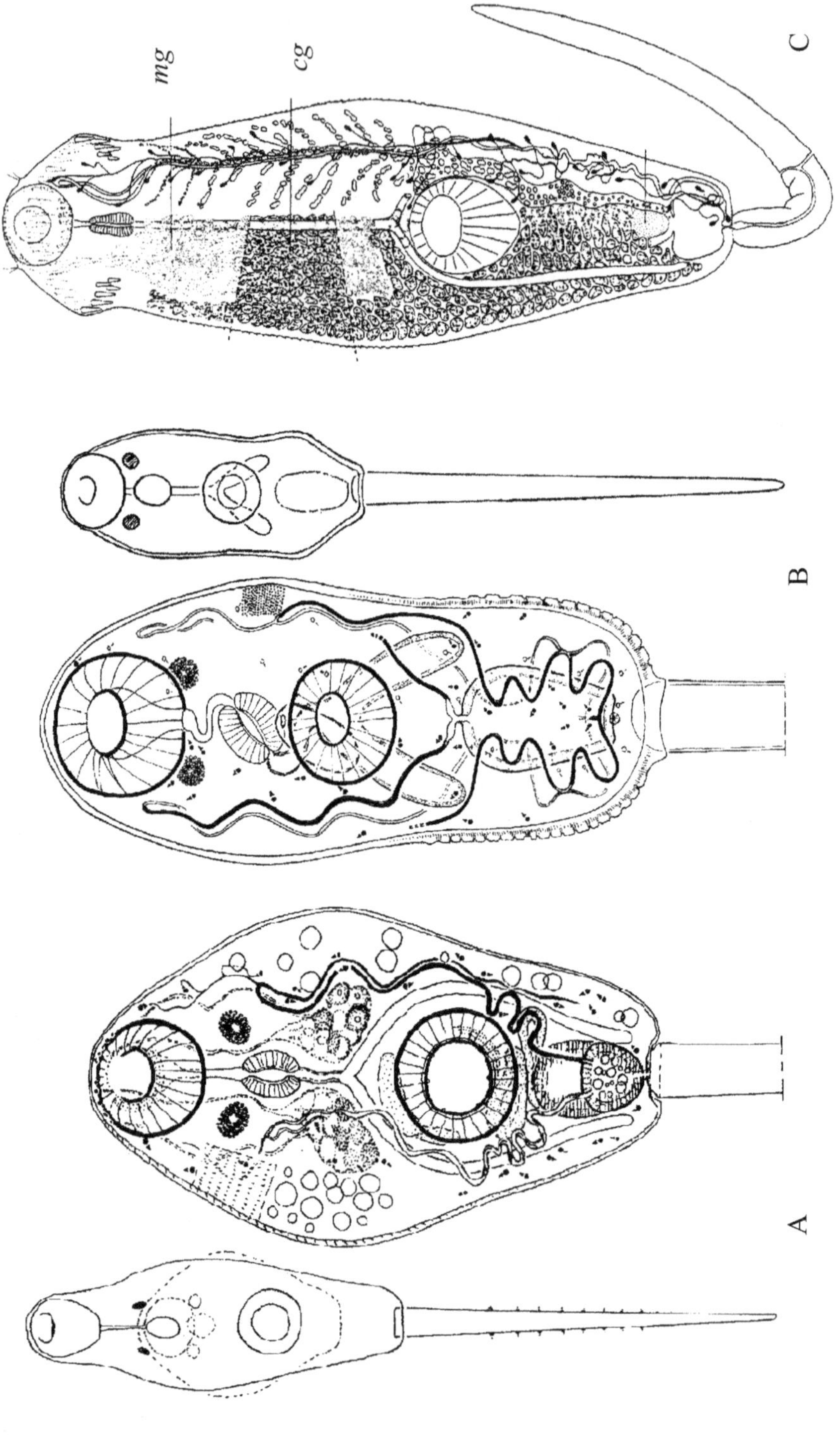
mg
cg
C
B
A

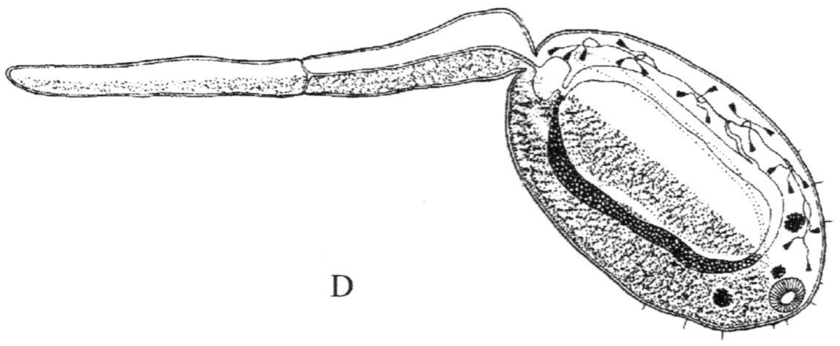

D

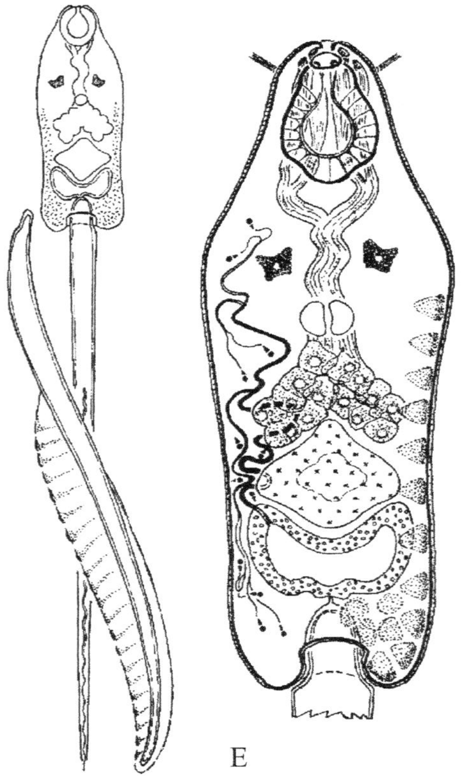

E

Figure 41. . Cercariae of Gymnocephala (A-B), Echinostomata (C), Monostomata (D) and Pleurolophocerca (E) groups. A, *Deropristis inflata* (Acanthocolpidae). From Deblock (1980), reprinted with author's permission; B, *Haploporus benedeni* (Haploporidae). From Deblock (1980), reprinted with author's permission; C, *Himasthla interrupta* (Echinostomatidae). From Loos-Frank (1967), reprinted with the permission of Springer-Verlag GmbH & Co.KG; D, *Notocotylus ephemera* (Notocotylidae). From Odening (1966), reprinted with author's permission; E, *Cryptocotyle jejuna* (?) (Heterophyidae). From Deblock (1980), reprinted with author's permission. *Abbreviations*: *cg*, cystogenous glands; *mg*, "metacercarial" glands.

ecology and behaviour of larvae, rather than phylogenetic relationships (see p. 101, fig. 138). Finally, there are groups, whose composition needs to be revised, taking into account modern data, e.g., Cercariaea.

A typical feature of the cercaria in primitive trematodes is the poor development of various provisional structures. The penetration apparatus is usually not expressed, i.e., there is no set of morphological structures used by the larva during penetration into the second intermediate host (stylet, penetration glands, specific modifications of the tail and of oral sucker). It is not surprising, that most of these cercariae encyst on the surface of the substrate, then becoming an adolescaria. Among the provisional structures present, one notes numerous cystogenous glands, a tail, and in a number of forms, e.g., eyes. If these structures are ignored, the morphology of the larvae is similar to that of young maritae. Transformation into marita does not require any substantial morphological modifications and only needs further development of tissue and organs systems of the larva.

In such cercariae several groups are distinguished. Gymnocephala includes larvae of Fasciolidae, Haploporidae, Acanthocolpidae and Psilostomidae (Fig. 41A,B); Amphistomata – Paramphistomidae, Diplodiscidae and some others. We note that Paramphistomidae cercariae already have a secondary excretory system (often called lymphatic system), which is fully developed in maritae. In the cercariae of the Echinostomata group (larvae of the family Echinostomatidae), the anterior end of the body, behind the oral sucker, forms a collar, armed with spines (Fig. 41C). Cercariae of the Monostomata group only have the oral sucker (Fig. 41D). Because they are larvae of Notocotylidae and Pronocephalidae, they do not develop the ventral sucker in the course of their morphogenesis, similar to maritae of those two families. To a certain extent, absence of the ventral sucker may be compensated for by the development of special structures. For example, in the Notocotylidae, there are adhesive pockets at the posterior end of the body, which take part in adherence of the body to the substrate (Filimonova, 1985).

Cercariae of the aforementioned groups have a rather large body (up to 1 mm long) and a tail, about the same length. They are poor swimmers, adhere to bottom water layers and encyst after a short period of active life. As a rule, cercariae of Psilostomidae and Echinostomatidae get into the second intermediate host passively, e.g., together with water flow passing into the gill cavity of fishes and crustaceans or inhalant siphons of Bivalvia. More rarely, they actively infect the host, using natural openings in its body (urethral orifices, spiracles of Pulmonata) for this purpose.

Two closely related groups of larvae of more phylogenetically-advanced trematodes are represented by pleurolophocercous and parapleurolophocercous cercariae (groups Pleurolophocerca and Parapleurolophocerca – representatives of the families Heterophyidae and Opisthorchiidae) (Fig. 41E). The only important difference between larvae of these groups is found in the position of the tail fin-folds. In the first group, the membrane consists of a median dorso-ventral fin-fold and in the second one - of a continuous lateral fin-fold. Cercariae of these two groups are significantly smaller (body length 150-200 μm) than the larvae of primitive trematodes and are characterized by less developed imaginal characters. They often lack the primordium of the ventral sucker. At the same time, the tail has a tail fin. In addition to cystogenous glands, there are numerous penetration glands, which are not yet differentiated into different groups.

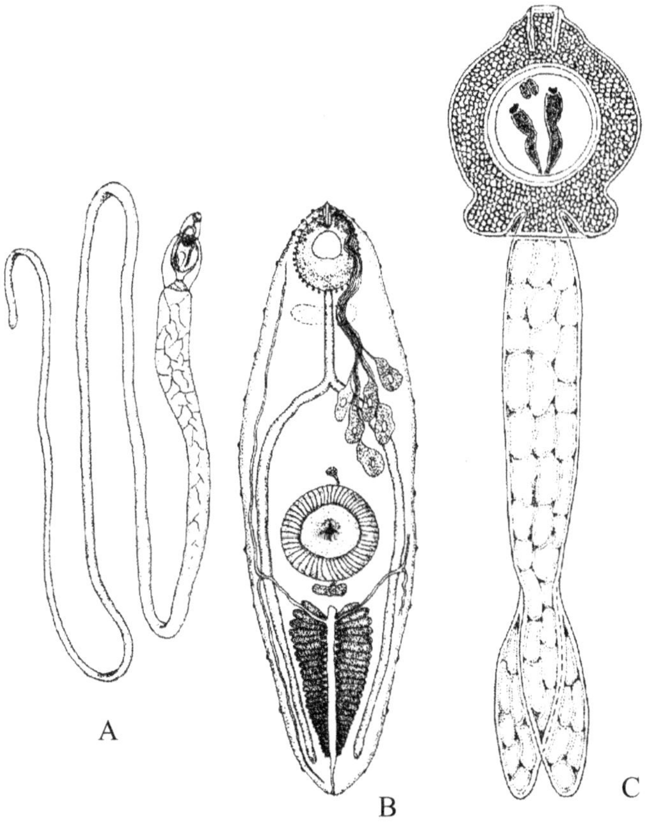

Figure 42. Cercariae with modified tails: Macrocerca (A-B), Cysticerca (C), Cystophora (D-E), Trichocerca (F-G), Microcerca (H-J). A-B, *Gorgoderina attenuata* (Gorgoderidae), A, general view, B, cercarial body. From Rankin (1939); C, *Cercaria caribbea XLVI* (Bivesiculidae). From Le Zotte (1954), reprinted with publisher's permission.

Considerable attention in the classification system of Lühe concerns the structure of the tail. Perhaps, the most substantial modifications are observed in the so-called sac tailed or cystocercous cercariae. This is an extremely diverse group, including larvae of trematode taxa, which are not closely related. However, they all have undergone a long evolution and developed excellent adaptations to get into the second intermediate host. The modified tail is a basic component of these adaptations.

In Macrocerca larvae (family Gorgoderidae) the "gigantic" tail can reach 3-4 mm long (Fig. 42A,B). Its anterior part is swelled and forms a chamber, which houses the cercaria body. Of similar structure are cercariae of Cysticerca group, typical for species of Azygiidae and Bivesiculidae (Fig. 42C). The caudal chamber is large and clearly separated from the distal part of the tail. This is especially true for Bivesiculidae. This chamber is composed of vacuolated cells, capable of swelling, when the larva gets into water. As a rule, the tail is pigmented and bifurcated at the distal end. Its movement resembles that of larvae of mosquitoes, oligochaetes and other aquatic invertebrates, i.e., organisms, on which fishes (definitive hosts of these trematodes) feed.

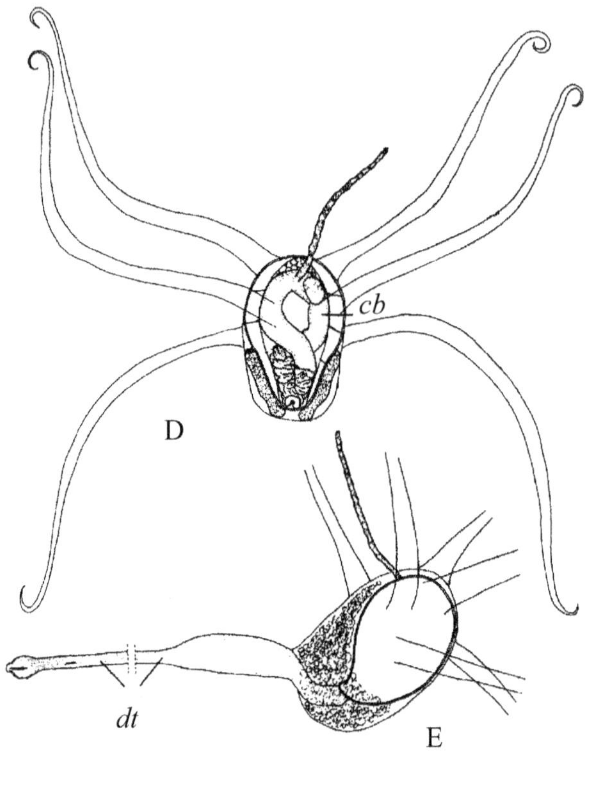

Figure 42. (Continued). D-E, *Lecithochirium rufoviridae* (Hemiuridae), D, infective cercaria, E, tail with everted delivery tube, ventral view. *Abbreviations*: *cb*, cercarial body; *dt*, delivery tube. From Køie (1990b), reprinted with author's permission.

Finally, in cystophorous cercariae (Cystophora) of superfamily Hemiuroidea, the tail is transformed into a transparent glass-like caudal cyst, which encloses the cercaria body. This cercaria is poorly developed and lacks adaptations, enabling active penetration through the epidermis of the second intermediate host – copepods (Fig. 42D,E). The caudal cyst has several appendages, often of unusual shapes, which facilitate its floating in water. It also has an apparatus of complex structure. Due to the latter, in the digestive tract of the crustacean, the larva is ejected from the caudal cyst (Skrjabin and Gushanskaya, 1960; Matthews, 1981; Køie, 1990a,b,c). The force of the ejection is so great that the larva pierces the wall of the host's intestine and gets into the haemocoel.

The groups of bristle tailed (Trichocerca) and short tailed (Microcerca) cercariae were also identified on the basis of their tail morphology. The first group includes larvae of Lepocreadiidae and some Fellodistomidae. As it is clear from the name of the group, the tail of those cercariae has regularly intermitting bundles of bristles or plates, due to which larval surface and consequently the buoyancy of the larva increases significantly (Fig. 42F, G). The group of short tailed cercariae is mixed. It includes a number of representatives of Zoogonidae, Opecoelidae, Lepocreadiidae, Philophthalmidae and even Microphallidae (see below). A common feature of organization of these larvae is modification of the tail, caused by the loss of its locomotory function, and transformation into an organ of attachment. In some species the tail stem is still retained (*Philophthalmus rhionica, Parorchis acanthus*) and only its terminal part is transformed into a sucker-like structure (Fig. 42H,I). In other species it is significantly shortened and turns into an appendix, functioning as a tail

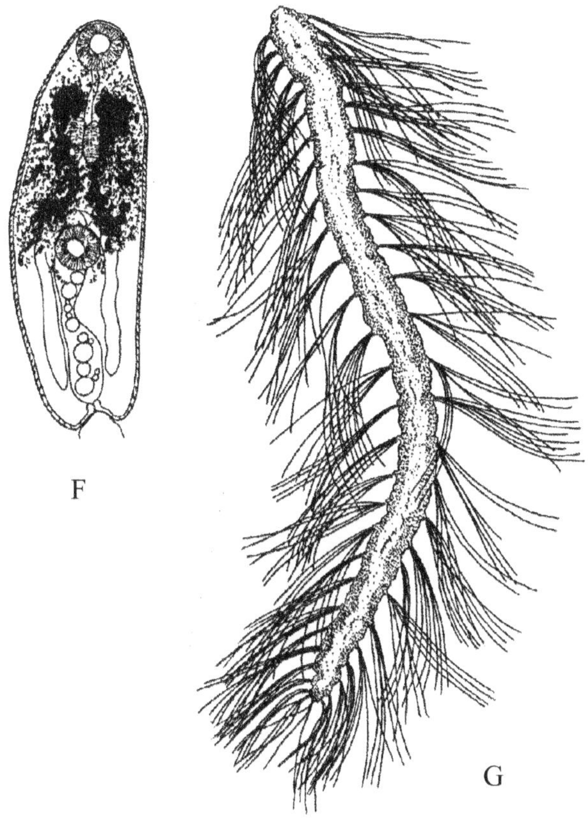

Figure 42. (Continued). F-G, *Cercaria setifera* (Lepocreadiidae), F, cercarial body, G, cercarial tail. From Bartoli (1984), reprinted with publisher's permission.

sucker (Fig. 42J) by means of which the cercaria adheres to the substrate and undertakes a complicated complex of reactions of "ambuscade" behaviour (see p. 301).

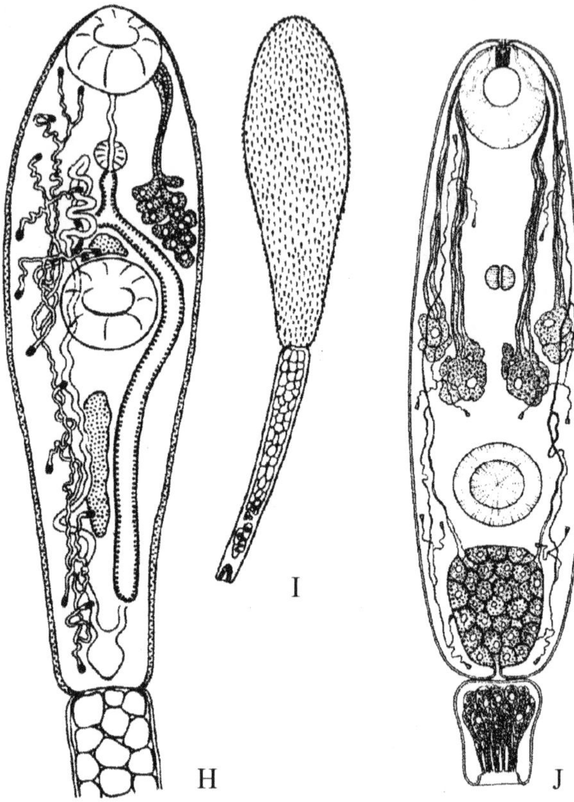

Figure 42. (Continued). H-I, *Philophthalmus rhionica* (Philophthalmidae), H, cercarial body, I, general view. From Olenev and Dobrovolskij (1975), with permission. J, *Podocotyle reflexa* (Opecoelidae). From Køie (1981), reprinted with author's permission.

Furcocercariae and Xiphidiocercariae are two major groups of cercariae. To a certain extent, these groups reflect differences in the pathways of morphological evolution of two main phylogenetical stems of higher trematodes, since all Strigeidida are included in the Furcocercariae, and Xiphidiocercariae embrace Plagiorchiida. Each group includes larvae with many primitive features (large size, poorly differentiated glandular apparatus, developed definitive systems of organs), as well as highly specialized juvenile forms with a minimal set of imaginal characters, but possessing perfect provisional structures, which provide fulfilment of the main task of the first functional stage of ontogenesis. All these questions will be discussed in detail later in the present and subsequent chapters.

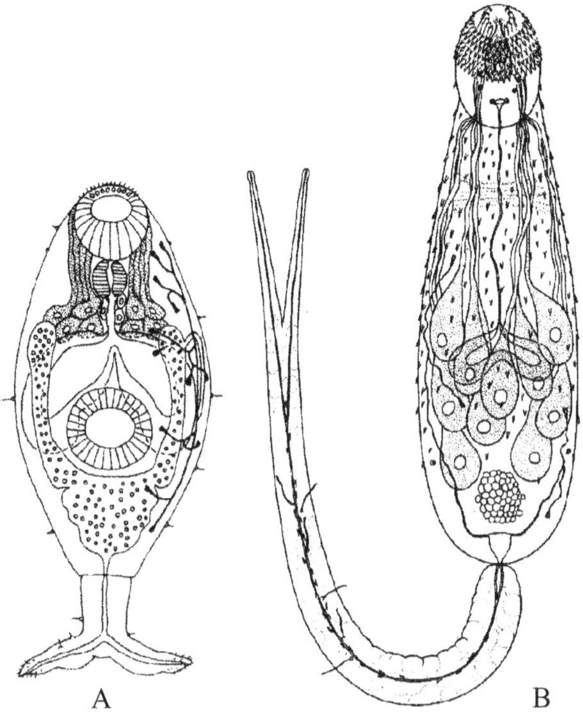

Figure 43. Furcocercariae. A, *Gymnophallus gibberosus* (Gymnophallidae) From Pekkarinen (1987b), reprinted with the permission of the Finnish Zoological and Botanical Publishing Board. B, *Aporocotyle simplex* (Aporocotylidae). From Køie (1982), reprinted with author's permission.

All larvae, whose caudal stem bifurcates at the distal end, forming two branches – a furca (Gymnophallidae, Schistosomatidae, Aporocotylidae, Sanguinicolidae, Cyathocotylidae, Strigeidae, Diplostomidae) are included in the furcocercous cercariae (Furcocercariae) (Fig. 43). The furca is used for larval locomotion and floating in water. As a rule, movement of furcocercariae proceeds with the posterior end directed forwards, due to alternating contraction of muscles of the right and left side of the caudal stem. The stroke is effected by the concave side of the stem, together with the adjacent furca branch (Fig. 44A-F). When the larva is crawling on the substrate with its anterior end directed forwards, both branches of the furca come together and passively bend to the side, opposite to the curve of the stem (Fig. 44G-L). Active swimming of furcocercariae alternates with passive floating in water, when the larvae assume a characteristic resting posture – the stem is

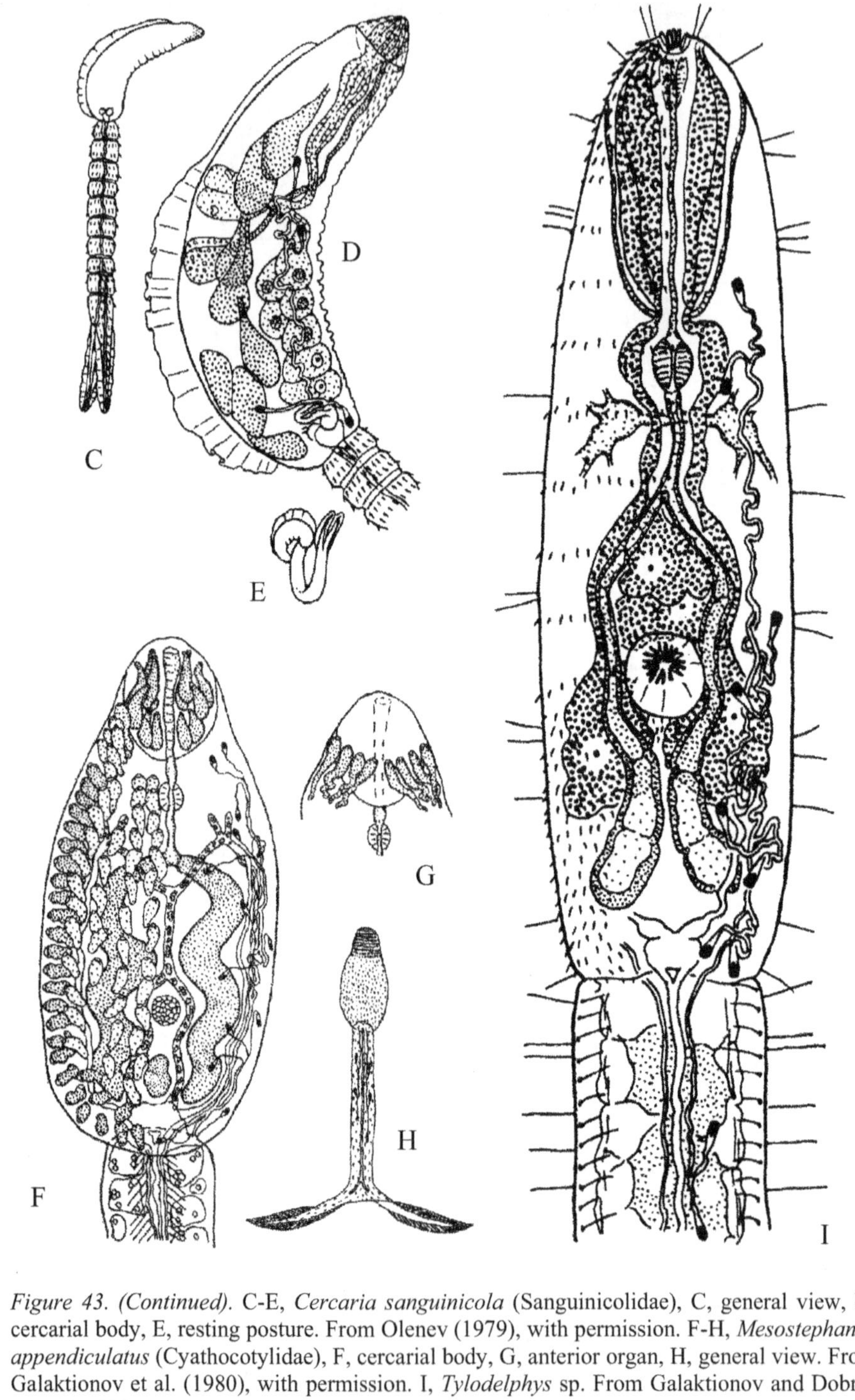

Figure 43. (Continued). C-E, *Cercaria sanguinicola* (Sanguinicolidae), C, general view, D, cercarial body, E, resting posture. From Olenev (1979), with permission. F-H, *Mesostephanus appendiculatus* (Cyathocotylidae), F, cercarial body, G, anterior organ, H, general view. From Galaktionov et al. (1980), with permission. I, *Tylodelphys* sp. From Galaktionov and Dobrovolskij (1998).

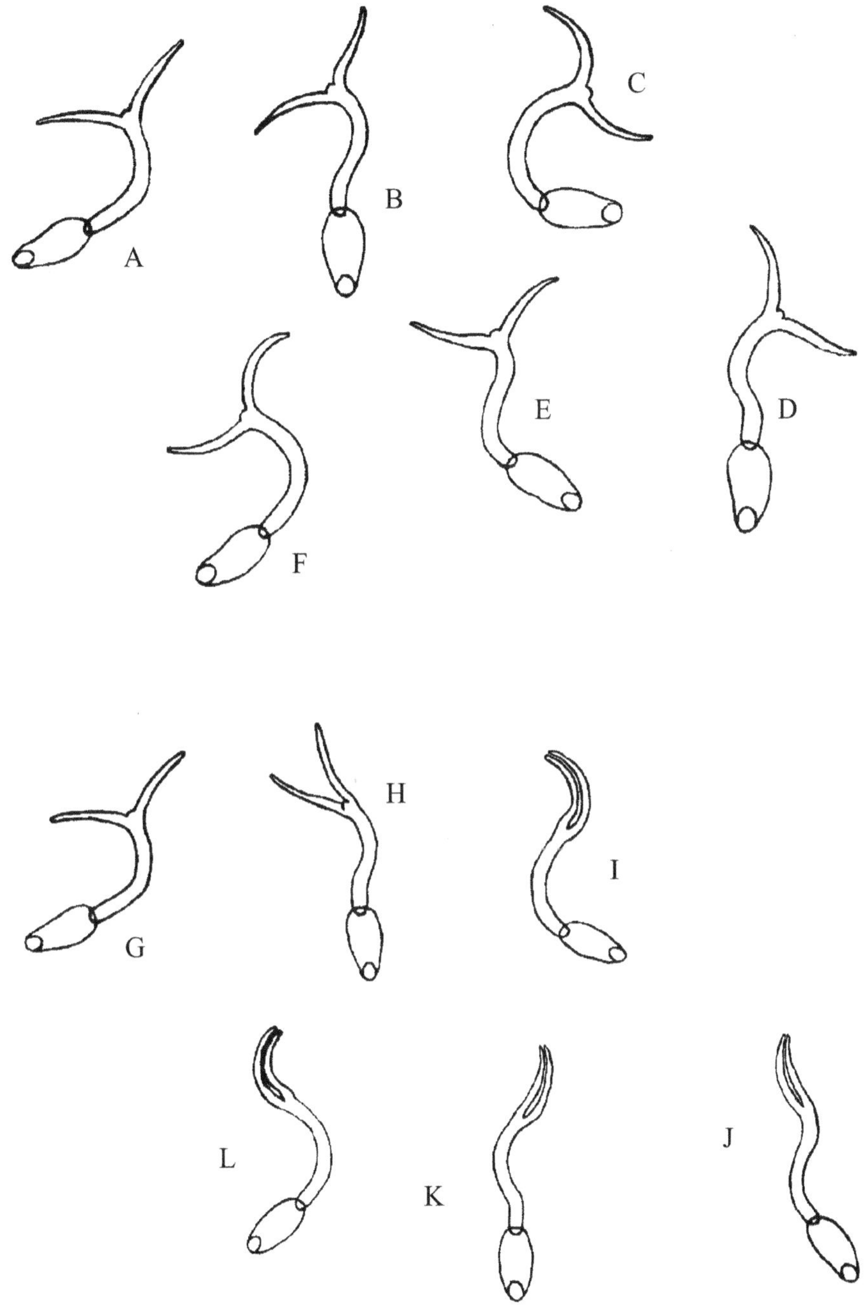

Figure 44. Postures of furcocercaria *Mesostephanus appendiculatus* (Cyathocotylidae) moving with posterior end forward (A-F, successive stages) and with anterior end forward (G-L, successive stages). From Galaktionov and Dobrovolskij (1998).

curved arc-wise; the branches of the furca are maximally diverged. This enlarges the entire surface of the body and, thus, reduces the rate of sinking (Fig. 44A,G). The resting posture of Schistosomatidae larvae significantly differs from the above described. They attach to the surface layer of water by means of the ventral sucker and so-called anterior organ. The latter is characteristic for all furcocercariae and is a modification of the oral sucker. In some cases it has rather thick muscular walls, and functionally it belongs to the glandular apparatus. Only its anterior end, capable of pulling in and out and often containing more or less developed spines, to a certain extent, compensates for the absence of a classical oral sucker.

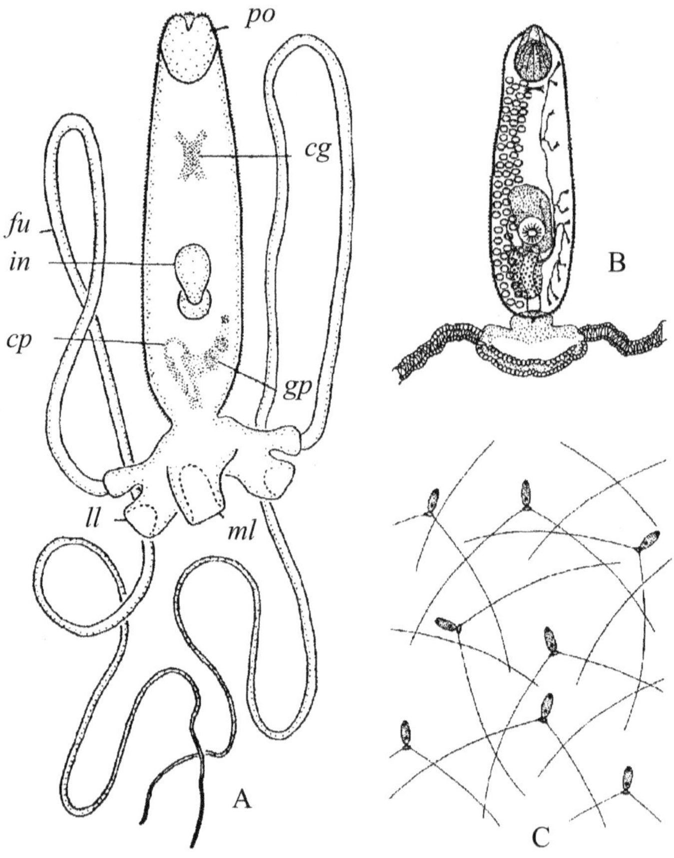

Figure 45. Cercaria of Bucephala group. A, *Prosorhynchus squamatus* (Bucephalidae). From Matthews (1973a), reprinted with the permission of Cambridge University Press. B-C, *Cercaria pleuromerae* (Bucephalidae), B, general view, C, "net" formed by cercariae in the water column. From Wardle (1988), reprinted with publisher's permission. *Abbreviations*: *cg*, cerebral ganglion; *cp*, cirrus pouch; *fu*, furca; *gp*, genital primordium; *in*, intestine; *ll*, lateral lobe; *ml*, median lobe; *po*, penetration organ.

Bucephalidae larvae are traditionally included in the group of furcocercariae, based on the formal morphological criterion, i.e., the forked tail. However, the structure and function of this formation in Bucephalidae are different, in principle, from those, described earlier for the true furcocercariae. Two long furcal rami originate from the shortened stem. They diverge at an angle of 180° (Fig. 45A,B). Due to these rami, cercariae are capable of floating in water for some time. The larvae, released from the molluscan host, may intercross with their tail rami, forming a sort of a net (Fig. 45C). This "net" has neutral buoyancy and moves freely in the water together with micro currents (Wardle, 1988). This naturally increases the chances of larval meeting their definitive hosts, i.e., fishes. Due to peculiarities of bucephalid cercariae and the fact that the Bucephalidae differ greatly from other trematodes (see p. 526), these larvae are identified as a special group of Bucephala or Gasterostome cercariae.

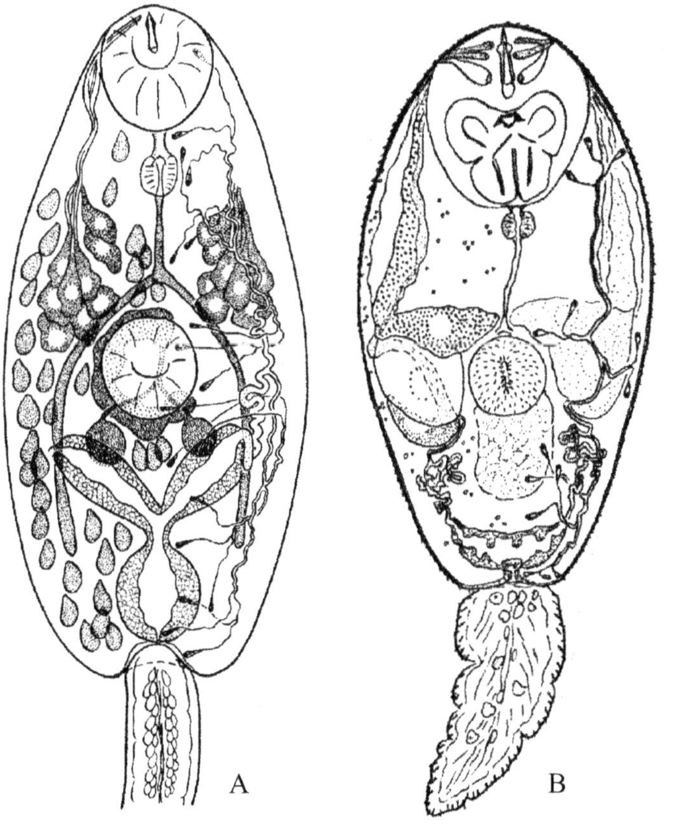

Figure 46. Xiphidiocercariae. A, *Paralepoderma cloacicola* (Plagiorchiidae). B, virgulate cercaria *Cercaria agstaphensis IV* (Lecithodendrioidea). From Galaktionov and Dobrovolskij (1998).

No less diverse are larvae, included in the group of Xiphidiocercariae (Plagiorchiidae, Pleurogenidae, Ochetosomatidae, Microphallidae, Lecithodendriidae, Allocreadiidae, Renicolidae, Dicrocoeliidae, Bunoderidae and others) (Fig. 46). The common feature of their organization is a well-developed stylet, located dorsally in front of the cercarial mouth opening. The stylet is sunk into a special pocket, and its pointed end is stuck outwards. The sides form cutting "wings". The function of this structure consists of piercing the epidermis of the second intermediate host during larval penetration. Lecithodendrioidea cercariae have a special reservoir for accumulating the gland secretion. It is formed by paired pockets, localized in the oral sucker, and is called the virgula (Fig. 46B). Hence, Xiphidiocercariae, which have this structure, are called virgulate cercariae (Cercariae Virgulae). A separate group (Ophthalmoxiphidiocercariae) includes stylet cercariae with eyes, i.e., representatives of Allocreadiidae and Bunoderidae.

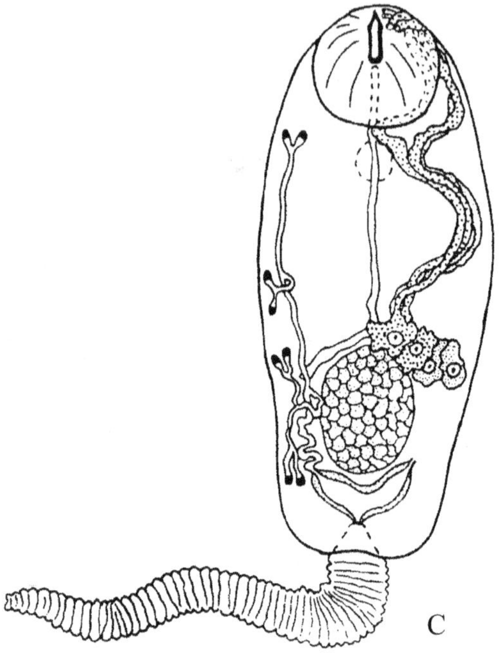

Figure 46. (Continued). C, *Maritrema subdolum* (Microphallidae). From Galaktionov and Dobrovolskij (1998).

There is an unusual mixed group of Cercariaea, grouped on the basis of the absence of the tail. For a long time larvae of Brachylaimoidea, Monorchiidae, Microphallidae, Gymnophallidae, which do not leave the daughter parthenitae, were included in Cercariaea. However, investigations of mor-phogenesis of these forms have shown that at certain stages, corresponding to the phases of development of cercariae in such trematodes, which release their larvae into the external environment, formation of the tail does take place. Its further development is more or less suppressed; the tail detaches from the body of the developing larva and degenerates. Detachment of the tail marks the transition to the stage of the metacercaria. In fact, the latter represent what was treated as "cercariaea" of the above named trematode groups. At present, only Cyclocoelidae larvae, in whose morphogenesis the tail is not formed at all, can be assigned to the Cercariaea (Ginetsinskaya, 1954a; Taft, 1986). The cercariaea of Cyclocoelidae are released from rediae and encyst in the same mollusc, forming metacercariae.

The above-description does not include the entire diversity of cercariae morphological types. A special work would be needed for a detailed discussion of this issue. Nevertheless, in the following chapters we give major characteristics of some of cercarial morphological types.

2.1.2. Development of Cercariae

Embryogenesis and initial stages of cercarial morphogenesis proceed similarly in representatives of different trematode groups (Fig. 47). Moreover,

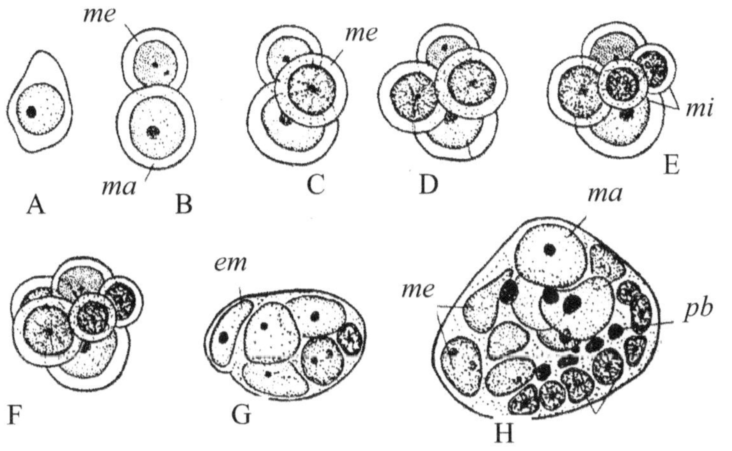

Figure 47. Successive stages (A-H) of embryonic development of *Astiotrema trituri* cercaria (Plagiorchiidae). *Abbreviations*: *em*, embryonic membrane; *ma*, macromere; *me*, mesomere; *mi*, micromere; *pb*, pycnotic body. From Gerasev and Dobrovolskij (1977), with permission.

cleavage of the germinal cell and initial stages of embryo development do not differ from these processes described above (section 1.3.) for miracidia and daughter parthenitae. As mentioned above (section 1.3.), before the embryonic membrane has appeared, embryo development proceeds in the germinal mass or under protection of endocyst envelopes of the daughter parthenitae. This stage is called the stage of "naked cell aggregates" (Cheng, 1961; Cheng and Bier, 1972). After the embryonic membrane has been formed, the embryos start to float freely in the body cavity of the parthenitae. From this moment on they are called "germinal balls" (Cheng, 1961; Cheng and Bier, 1972).

The nature of the embryonic membrane, called differently by various authors (for terminology see Galaktionov and Dobrovolskij, 1987; Dunn et al., 1992b), has provoked considerable discussion in literature (for details see Galaktionov and Dobrovolskij, 1987). The authors argue about whether the embryonic membrane is a derivative of the embryo itself or the daughter parthenitae. The second idea is supported only by authors working at the ultrastructural level (Rifkin, 1970; Meuleman and Holzman, 1975; Göbel and Pan, 1985; Halton and McCrae, 1985; Dunn et al., 1992b). However, most investigators, studying the embryogenesis of cercariae by light microscopy, describe the formation of the embryonic membrane from superficially situated macromeres (Ishii, 1934; Cheng, 1961; James and Bowers, 1967; Cheng and Bier, 1972; Gerasev and Dobrovolskij, 1977). These views are supported by a number of electron microscopy investigations (Matricon-Gondran, 1971; Hockley, 1972, 1973; Rees and Day, 1976; Tihomirov, 1980; Al-Salman and James, 1988). The results of our literature analysis (Galaktionov and Dobrovolskij, 1987) and original electron microscopy (Fig. 48) justify

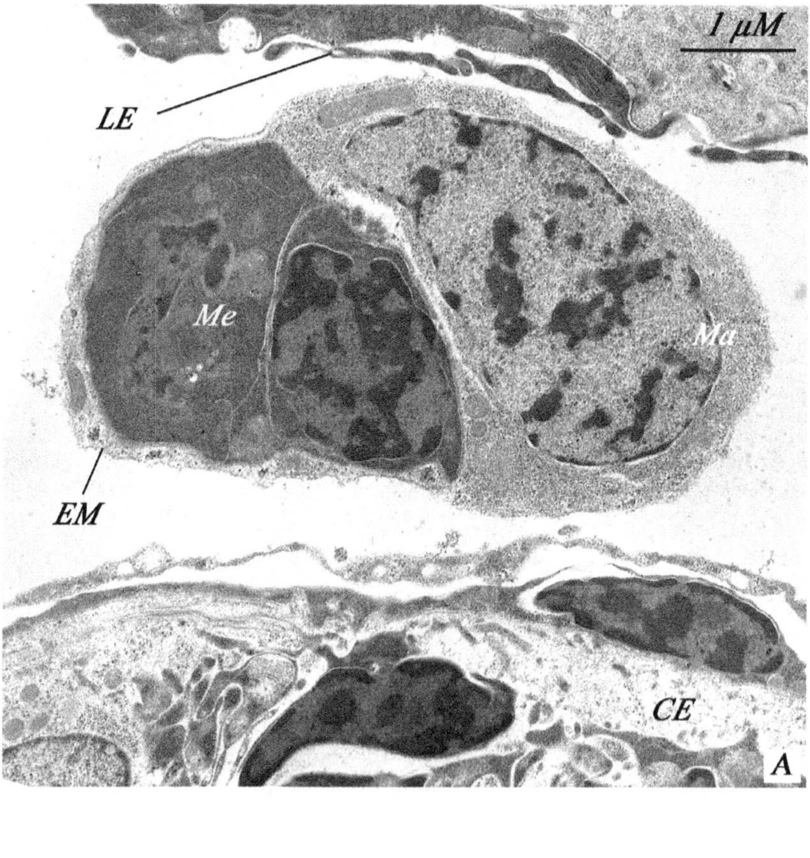

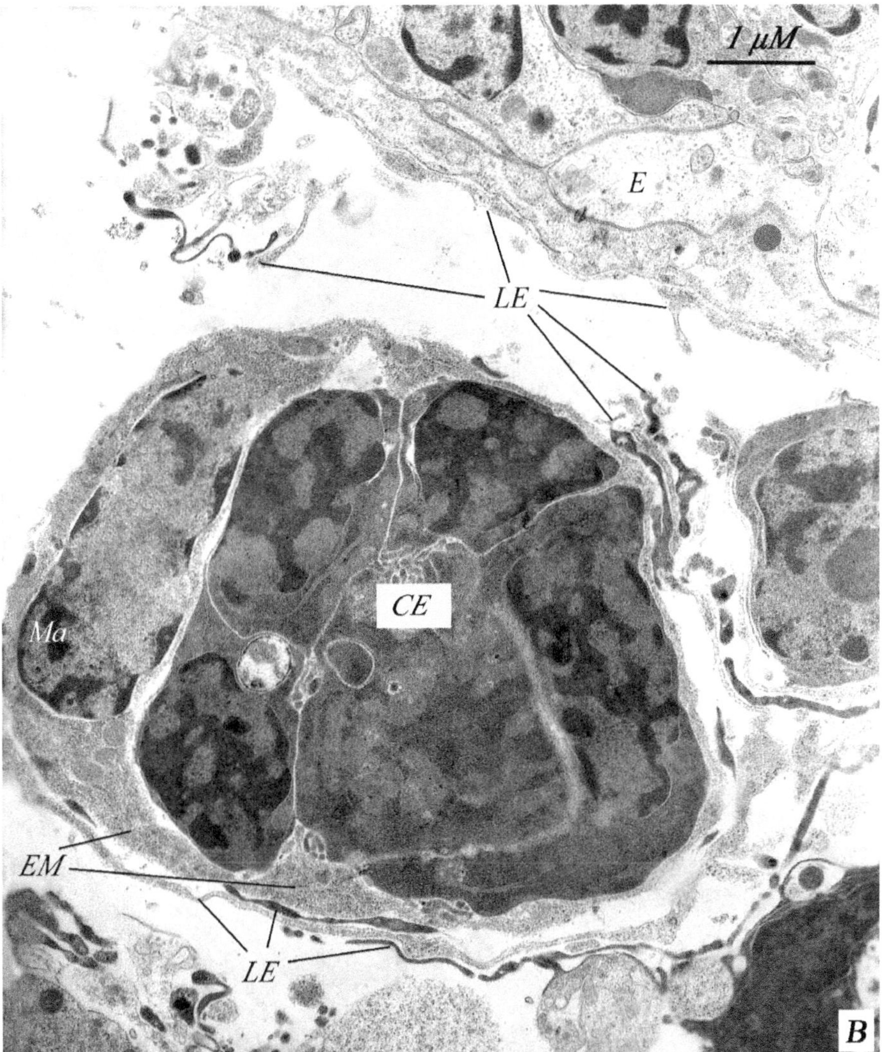

Figure 48. Successive stages (A-D) of formation of the embryonic membrane and tegument during *Microphallus pirum* cercaria development. A-B, initial stages; C, cercarial embryo surrounded with embryonic membrane; D, cercarial embryo with degenerating embryonic membrane. *Abbreviations*: *CE*, cercarial embryo; *DCM*, differentiating circular muscles; *DEM*, degenerating embryonic membrane; *DLM*, differentiating longitudinal muscles; *E*, endocysts; *EM*, embryonic membrane; *LE*, laminated structures of the endocysts; *Ma*, macromere; *Me*, mesoromere; *OT*, outer syncytial layer of the tegument.

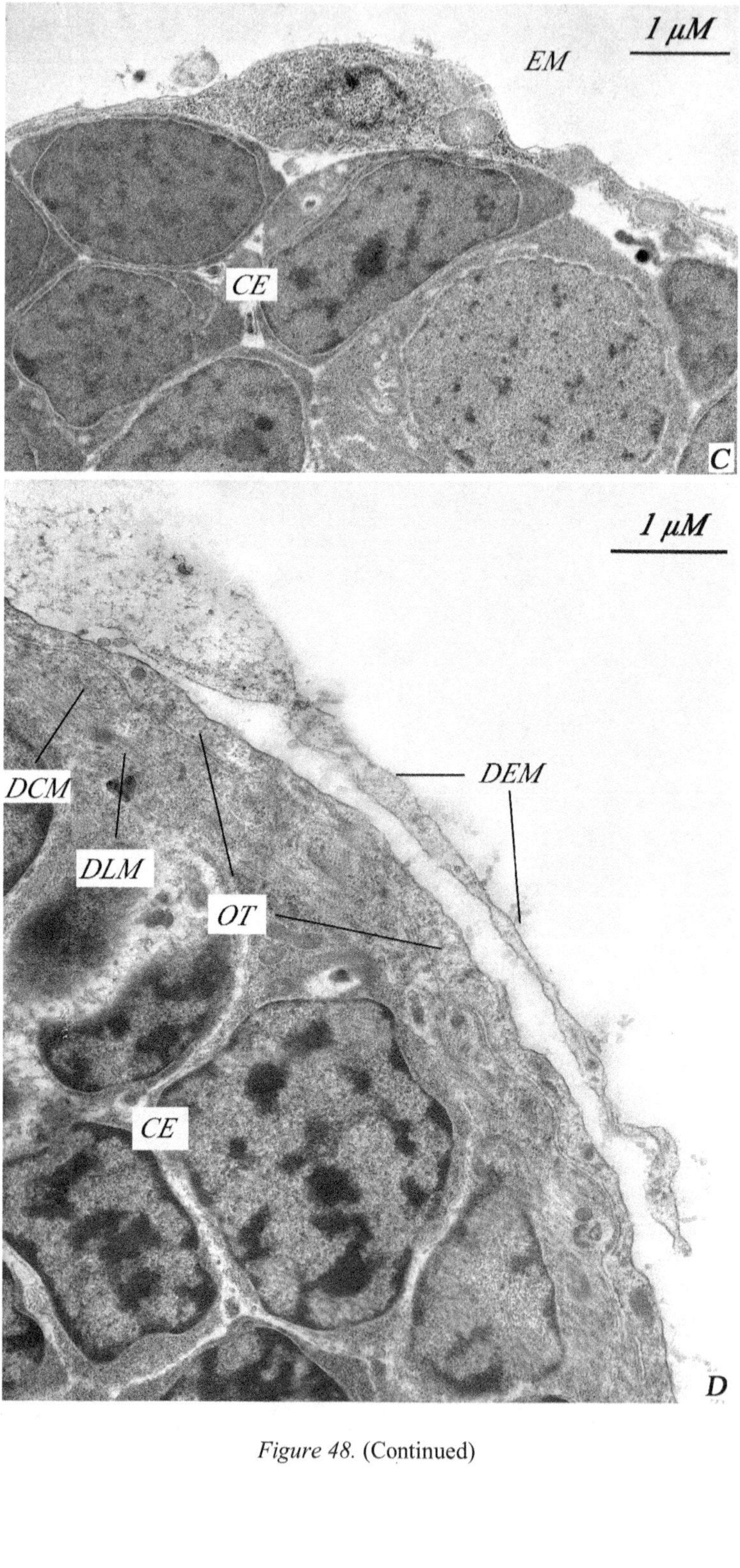

Figure 48. (Continued)

the first viewpoint. Probably the supporters of the parthenite-origin theory may have confused the embryonic membrane with the laminated structures of the endocysts (or germinal mass) of the daughter parthenita, surrounding embryos. The latter are naturally connected to the lining of the parthenita body cavity and lie (often forming several layers) over the genuine embryonic membrane.

The degree of development of the embryonic membrane and laminated extensions of the endocyst varies in representatives of different trematode groups. Thus, numerous extensions of the endocyst are observed in daughter sporocysts of schistosomatids (Göbel and Pan, 1985; Fournier et al., 1986-1987). This could have mislead the authors, working with this material, as mentioned above. The embryonic membrane, as in the case of parthenitae development, protects embryos from mechanic injury. Besides that, through this membrane, transport of nutrients to the growing embryo may take place. It functions for a relatively short period of time and soon after the outer syncytial layer of the tegument has been formed, it degenerates and is shed (Fig. 48).

Soon after the formation of the embryonic membrane, flattened cells are found on the periphery of the germinal ball. Theses are primordia of the outer layer of the tegument (Fig. 49). These cells merge with each other, forming a layer of nuclei-containing syncytium between the embryonic membrane and cellular mass of the embryo. This layer gradually thickens; the nuclei degenerate, and mitochondria, free ribosomes and elements of cytoskeleton can be found in the cytoplasm. At about the same time, the embryonic membrane is disrupted and the tegument starts to function as a boundary tissue. It is lined with a basal lamina, which is thin, with myoblasts adjoining its inner side. From theses myoblasts, circular and longitudinal muscles of the tegumentary-muscular sac are developed. Some cells from deeper layers are specialized as tegumental cells. They form extensions, which grow between muscle cells, pierce the basal lamina and come into contact with the basal membrane of the outer syncytial layer. Membranes in the zone of the contact are disrupted and the tegumental cells become connected to the outer layer of the tegument by means of cytoplasmic bridges. Their walls are reinforced with microtubules and/or microfilaments.

We have provided a generalized scheme of cercaria tegument formation, which has been accepted by most researchers (Southgate, 1971; Hockley, 1972,1973; Meuleman and Holzman, 1975; Rees and Day, 1976; Göbel and Pan, 1985; Halton and McCrae, 1985; Galaktionov and Malkova, 1990a; Dunn et al., 1992b). An exception is the study of Al-Salman and James (1988). They believe that the tegument in *Meiogymnophallus minutus* cercariae is formed by merging of extensions of superficially localized embryo cells, immediately under the embryonic membrane. The nuclei-containing cytons of these cells are found under the forming thin syncytial

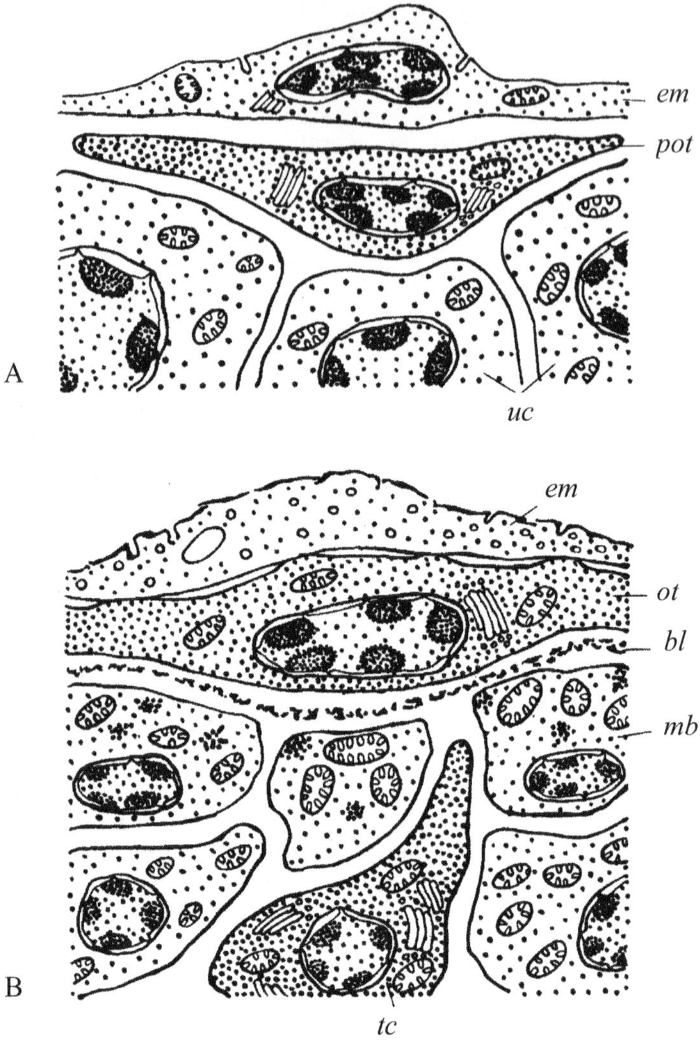

Figure 49. Diagram showing successive stages (A-D) of formation of the embryonic membrane and tegument during cercaria development. *Abbreviations*: *bl*, basal lamina; *cm*, circular muscles; *dem*, degenerating embryonic membrane; *em*, embryonic membrane; *g*, glycocalyx; *lm*, longitudinal muscles; *mb*, myoblast; *pot*, primordia of the outer layer of the tegument; *ot*, outer syncytial layer of the tegument; *s*, spine; *tc*, tegumental cell; *uc*, undifferentiated cell. From Galaktionov and Dobrovolskij (1987).

layer and become what are called tegumental cells. Unfortunately, no detailed series of electron micrographs is supplied in the cited work of Al-Salman and James (1988), which might enable us to trace accurately the process, described by these authors.

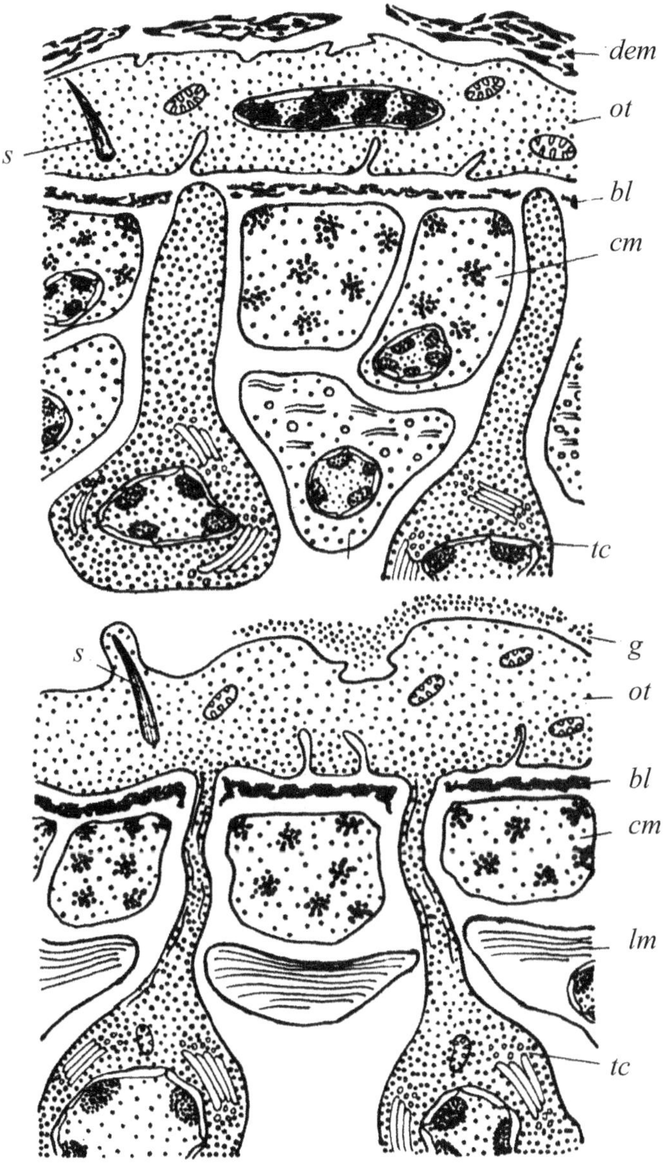

Figure 49. (Continued).

The scheme of tegument formation, proposed by Al-Salman and James (1988) agrees less with the modern views on the function of this tissue than the one, universally accepted. As shown below, at different stages of ontogenesis of the hermaphroditic generation different types of tegumental glandular cells add to the outer layer of the tegument. In fact, these two parts of the tegument are rather independent from each other. The outer syncytial layer is retained during the life of the individual and functions as a barrier tissue and reservoir for secretion. The composition of the tegumental cells is variable. There are different cells there, replacing each other, which synthesize materials, needed at different stages of ontogenesis.

Parallel to the formation of the tegument, in the cellular mass, which forms the embryo body, two protonephridia emerge. Each is represented by one flame cell and an excretory duct (Fig. 50C). Other organ systems de-

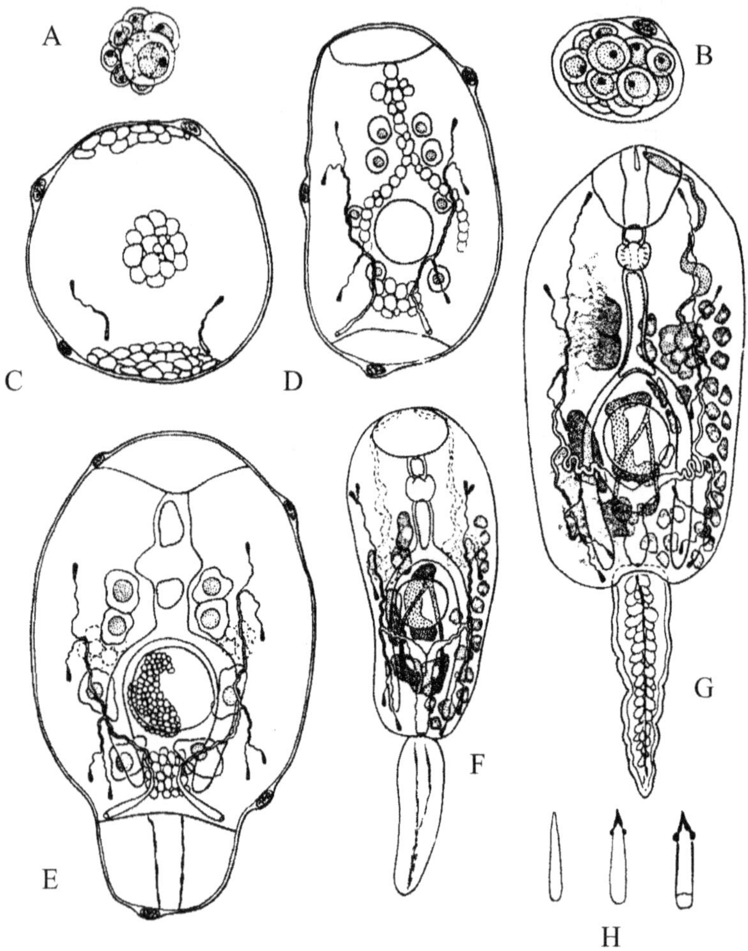

Figure 50. Development of *Opisthioglyphe ranae* cercaria (Plagiorchiidae). A, early embryo ("naked cell aggregate"); B, embryo covered by embryonic membrane; C, stage of two flame cells; D, stage of three pairs of flame cells; E, stage of four pairs of flame cells; F, stage of five pairs of flame cells; G, stage of six pairs of flame cells; H, successive stages of the stylet formation. From Dobrovolskij (1965b).

velop later, the time of their formation varying widely in representatives of different taxa. After the first flame cells have formed, the primordium of the oral sucker appears. In the species, whose cercariae have a well-developed ventral sucker (acetabulum), the primordium of the latter appears early, soon after the oral one, and then develops. In the species, whose fully formed cercariae have an underdeveloped sucker (Opisthorchiidae, Microphallidae), the sucker's primordium either emerges relatively early, and then stops development, or its formation is shifted to the last stages of morphogenesis.

The next step in the development of the excretory system consists in doubling of flame cells. At first, they are situated close to each other, and then they begin to move apart due to elongation of capillaries. Unseparated parts of the ducts become primary collecting tubules, which open at the posterior half of the embryo with independent, widely spaced, excretory pores. At the stage of 4 flame cell pairs, in embryos of many species a primordium of the digestive system is observed in the form of cellular cords (Fig. 50D). In the beginning, the oesophagus part, and later branches of the caecum, becomes visible. In the cellular mass of the embryo, big cells with vesicle-like nuclei are being differentiated. These are primordia of the first glands. Secretory granules gradually accumulate there (Fig. 50E-G). At the posterior pole of the embryo a tail primordium is being formed. There in embryos of most trematodes (Fasciolida, Echinostomatida, Psilostomidae, Notocotylidae, Opisthorchiidae, Strigeidida and others) excretory pores, as well as distal parts of main excretory ducts, are located (Fig. 51A,B). A less numerous group consists of forms (e.g., Plagiorchiida and Allocreadiidae), whose tail primordium is located entirely between excretory pores, so that no elements of excretory system can enter it (Fig. 51C).

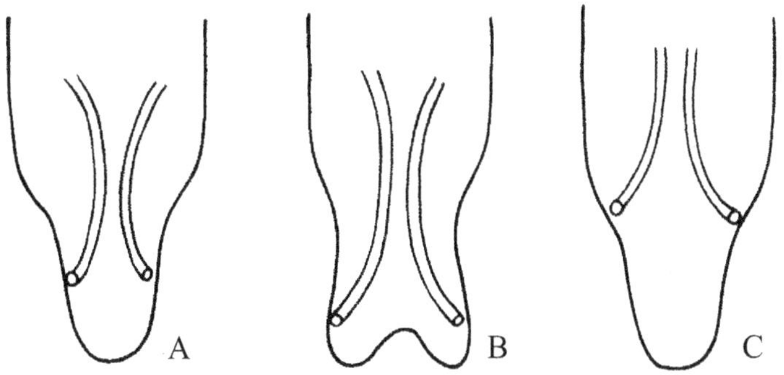

Figure 51. Diagram showing location of primary excretory pores and distal parts of main collecting ducts in embryos of cercariae of different taxa. A, Echinostomatidae; B, Furcocercariae; C, Xiphidiocercariae. From Galaktionov and Dobrovolskij (1987).

By the moment of formation of the tail primordium and appearance of the third pair of flame cells, the length of the embryo increases (Fig. 50D). Its proportions also change due to uneven growth of different parts of the body. Further, primordia of organs, which provide normal functioning of the individual at the first ontogenetic stage, are primarily developed. At the same time, the development of primordia of organs of the adult organism, which are of no importance for the cercaria life, is usually retarded. Thus, a kind of heterochrony takes place, which is expressed to a different degree in larvae of different trematode groups. Such a heterochrony is a leading trend in cercariae morphological evolution (Ginetsinskaya, 1968; Galaktionov and Dobrovolskij, 1987). We characterize this heterochrony in detail, after we have discussed subsequent stages of morphogenesis and structure of the fully formed cercariae of the representatives of different trematode taxa.

The forming tegument plays an important part in the nutrition of the embryo (Bogitsh, 1986). Through the outer syncytial layer of the embryo, nutrients are absorbed from the sporocyst's body cavity. This presumption is also supported by the abundance of large mitochondria observed in the cytoplasm of the developing syncytial layer, as well as by the occurrence of invaginations of its basal and outer plasma membranes. In embryos of some cercariae, microlamellae are found on the surface of the tegument. They disappear in the mature larvae. The active transport of glucose through the tegument of developing cercariae has shown been experimentally (Carter and Bogitsh, 1975; Smyth and Halton, 1983).

2.1.3. Tegumentary-Muscular Sac

In mature cercariae, the surface of the tegument is folded and contains the glycocalyx, which emerges at the last stages of larval morphogenesis (Caulfield et al., 1988). This is a 1-2 μm thick mesh of 15-30 nm fibrils that envelops the organism. In the chemical composition of the glycocalyx, the carbohydrate component is dominat (up to 85 % in cercariae of *Schistosoma mansoni*) (Samuelson and Caulfield, 1985; Caulfield et al., 1987). Rod-like or disc-like secretory granules containing electron-dense material are thought to produce glycocalyx. The granules are located directly under the outer plasma membrane of the tegument and are detected also in mature metacercariae and adults (Wilson and Barnes, 1974; Hanna, 1980; Samuelson and Caulfield, 1985; Caulfield et al., 1988; Weber and Sonntag, 1989) (Fig. 52). They are synthesized by special tegumental glandular cells. Apparently, the glycocalyx functions as a protective system. It may prevent filtration and selective leaching of ions through the plasma membrane (Bennett, 1963; Morris, 1971). At any rate, it was experimentally shown that cercariae of *S. mansoni*, lacking a glycocalyx, soon die in water (Stirewalt, 1963; Eveland and Morse, 1975).

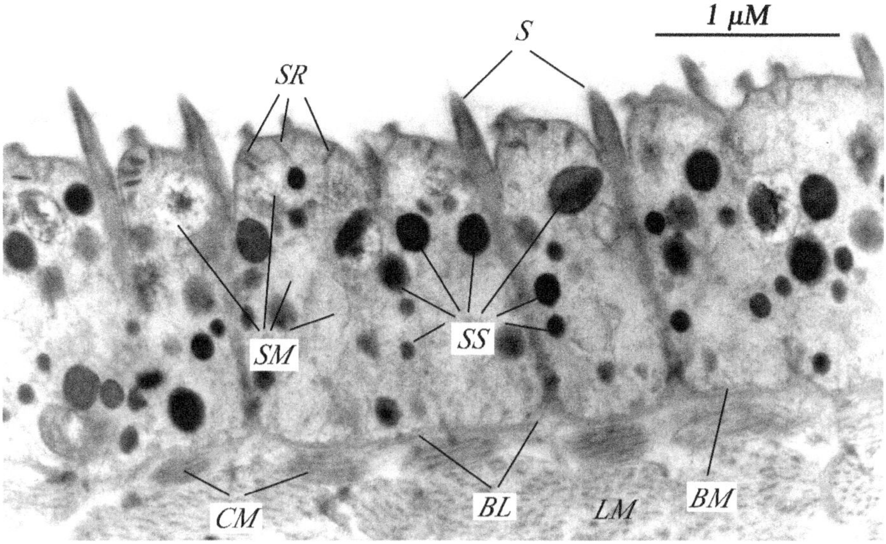

Figure 52. Tegument of cercaria *Levinseniella brachysoma*. *Abbreviations*: *BL*, basal lamina; *BM*, basal membrane; *CM*, circular muscles; *LM*, longitudinal muscles; *S*, spines, *SM*, secretory granules of mucoid glands; *SR*, rod-shaped secretory granules of tegumental glands; *SS*, secretory granules of sheath-forming tegumental glands.

In the outer tegumental layer of cercariae of many trematodes, there are spines, which are especially well developed at the anterior end of the body. They are normally of conical shape. Their pointed ends slightly protrude above the syncytial surface and are enveloped with the outer plasma membrane. In the contact site of the spine base with the tegument basal lamina, a narrow band of electron-dense material is found. Spines of microphallid cercariae *Levinseniella* and *Maritrema* have a peculiar structure. They are divided into two regions: the body of the spine, located in the apical parts of the syncytial layer, even in its special extension in *Maritrema*, and a thin long stem, which pierces the depth of the tegument and adheres to the basal membrane (Benjamin and James, 1987; Irwin et al., 1990; Galaktionov and Malkova, 1990a) (Fig. 52, 53). Formation of the spines proceeds by the thickening of actin filaments (Cohen et al., 1982; Abbas and Cain, 1987). Thus, they should be considered as a specialization of the cytoskeleton, rather than a "cuticle armament" (the term, occurred rarely in the litera-

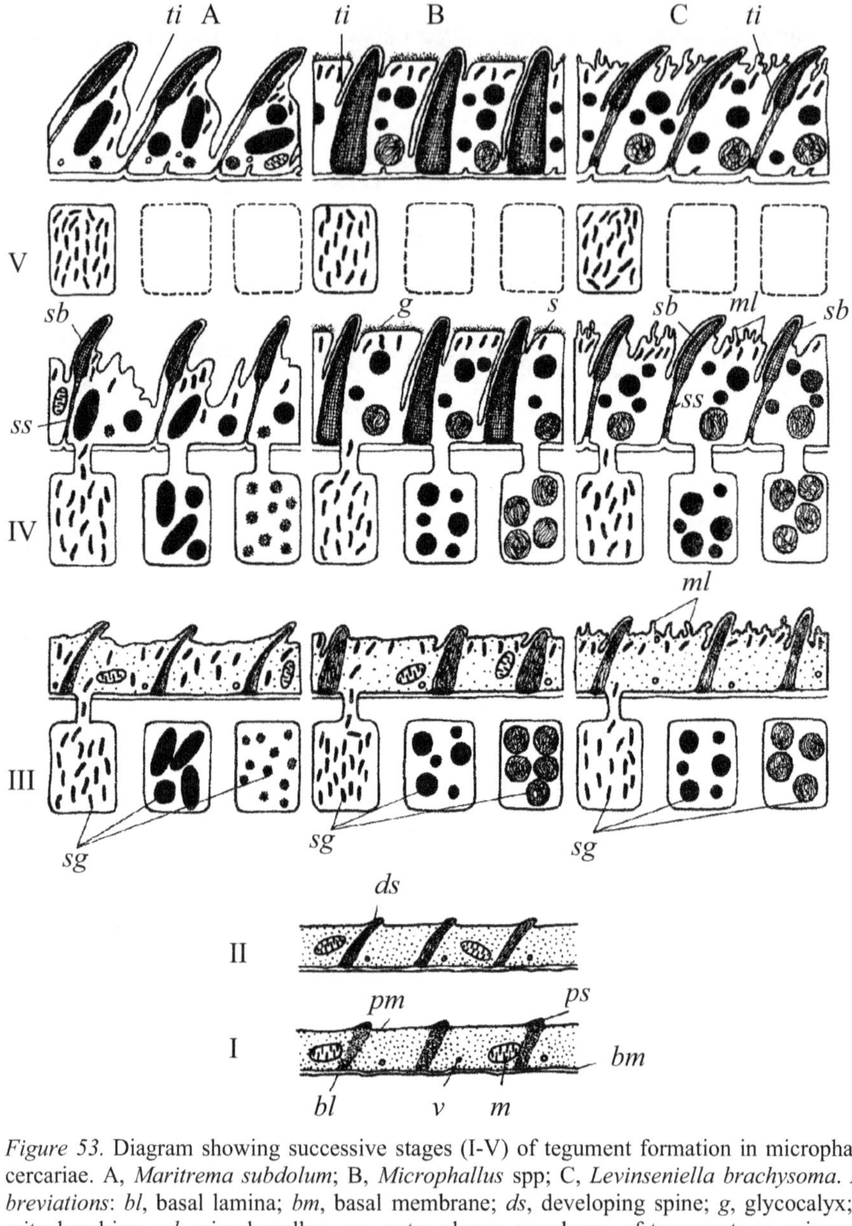

Figure 53. Diagram showing successive stages (I-V) of tegument formation in microphallid cercariae. A, *Maritrema subdolum*; B, *Microphallus* spp; C, *Levinseniella brachysoma*. *Abbreviations*: *bl*, basal lamina; *bm*, basal membrane; *ds*, developing spine; *g*, glycocalyx; *m*, mitochondrion; *ml*, microlamellae; *pm*, outer plasma membrane of tegument; *ps*, primordial spine; *s*, spine; *sb*, spine body; *sg*, secretory granules of tegumental glands of different type; *ss*, spine stem; *ti*, tegumental invagination; *v*, vesicle. From Galaktionov and Malkova (1990a).

ture on trematodes). It is well known that cuticles are extracellular structures, representing complex supramembrane formations with a chemical composition other than that of spines.

The stylet, a typical provisional structure of all Xiphidiocercariae, also represents a hypertrophic grown spine (Køie, 1971c; Gibson, 1974; Galaktionov and Malkova, 1990b) (Fig. 54). Secretion of specialized tegumental glands, which are detected in the primordium of the oral sucker at certain stages of cercarial morphogenesis, takes part in the formation of the outer light-refracting layer of the stylet (see fig. 56E). The stylet is sunk in a special pocket, formed by invagination of the tegument of the terminal part of the oral sucker. Bundles of muscle, providing its motility, with one end connected to the fibrous layer (apparently, consisting of intermediate filaments), are found in the bottom region of the pocket. At first, the muscle bundles pass along the pocket of the stylet, then, about in the middle of it, they branch into smaller ones forming a flabelliform picture. By means of the hemi-desmosomes the latter are connected with the surface muscles of the oral sucker.

Specialized circular and meridional muscle fibers are found not only in the pocket of the stylet but also in the oral and ventral suckers. In furcocercariae, special retractor muscles are developed, which enable pulling the anterior motile end of the body with spines and outer openings of glandular ducts into the pocket, formed by the walls of the anterior organ. When the muscles of the anterior organ walls are contracted, the anterior end, which was pulled in, is pulled out. Muscles of the body wall are especially developed in actively swimming cercariae, which penetrate into the second intermediate host and actively migrate in it. These are, in particular, cercariae of Strigeidida, where strongly developed subepidermal muscles are found. They consist of three layers of muscle fibers, which are relatively large compared to the size of the larva itself (Czubaj and Niewiadomska, 1997).

Tegumentary-muscular sac of the tail acquires a special structure. Unlike the rest of the body, its longitudinal muscles are represented by striated muscle fibers[iv], which are either evenly distributed on the circumference of the tail or grouped into 4 bundles (Cardel et al., 1960; Pearson, 1961; Žd'árská, 1969a; Chapman, 1973; Chapman and Wilson, 1973; Matthews, 1973a, 1974; Rees, 1977; Malkova, 1989). In Microphallidae xiphidiocercariae, which have been investigated in detail, of great importance for the functioning of the tail are circular foldings of the tegument; connection of its basal lamina in the region of constrictions with longitudinal muscle fibers by means of hemi-desmosomes and location of Z-discs of striated myofibrils in those sites (Fig. 55). Such a rigid construction enables the functioning of the tail wall as one integrated structure; at the same time, it does not impede the work of the longitudinal muscles (Malkova, 1989). During contraction and corresponding shortening of sarcomeres, the transversal foldings of the tegument thicken. On the opposite, during relaxation the foldings smooth out. Contraction of the longitudinal muscle fibers at any side of the tail results in its bending. Fast undulations of the tail, alternating in phase, provide

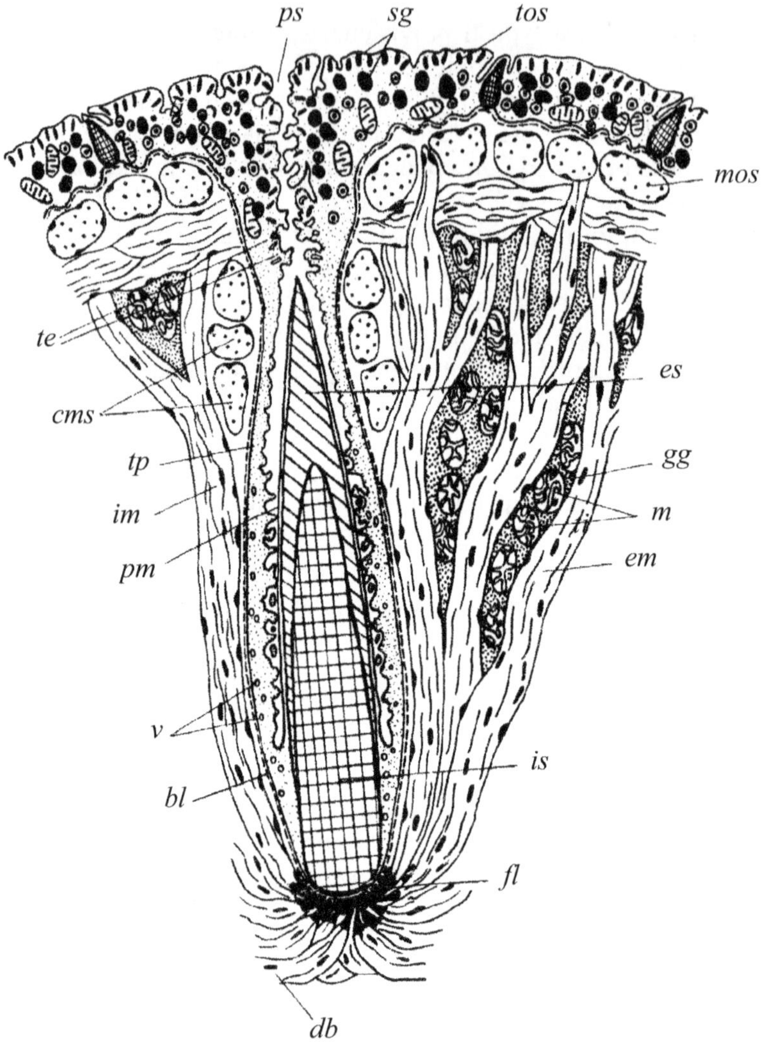

Figure 54. Diagram showing ultrastructure of the stylet of microphallid cercariae. *Abbreviations*: *bl*, basal lamina; *cms*, circular muscles of the stylet pocket; *db*, dense body; *em*, external muscle bundles; *es*, external stylet layer; *fl*, fibrous layer ; *gg*, granules of glycogen; *im*, internal muscle bundles; *is*, internal stylet layer; *m*, mitochondrion; *mos*, muscles of the oral sucker; *pm*, outer plasma membrane; *ps*, the stylet pocket; *sg*, secretory granules of the tegument; *te*, extensions of the tegument of the stylet pocket; *tos*, tegument of the oral sucker; *tp*, tegument of the stylet pocket; *v*, vesicles. From Galaktionov and Malkova (1990b).

for larval movement. Circular muscles function as skeletal structures and the liquid, filling the central part of the tail, plays the role of a hydroskeleton.

In the course of development of the tail, most nuclei of the muscle cells degenerate (Chapman and Wilson, 1973; Rees, 1971, 1977; Gerasev and Dobrovolskij, 1977; Malkova, 1989). This is easily explained, keeping

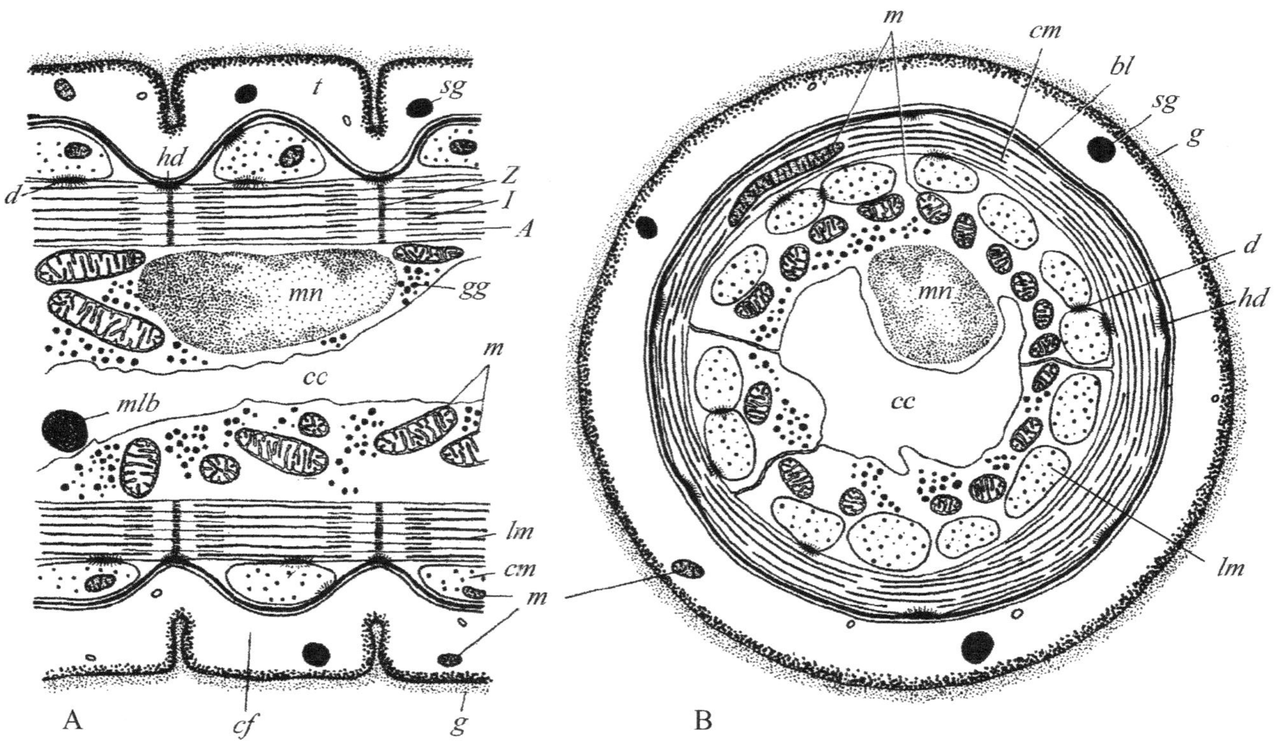

Figure 55. Diagram showing structure of microphallid cercariae tail. A, longitudinal section of the middle part; B, transversal section of the middle part. *Abbreviations*: *A*, A-zone of the sarcomere; *bl*, basal lamina; *cc*, central cavity; *cf*, circular folding of the tegument; *cm*, circular muscles; *d*, desmosome; *g*, glycocalyx; *gg*, granules of glycogen; *hd*, hemi-desmosome; *I*, I-zone of the sarcomere; *lm*, longitudinal muscles; *m*, mitochondrion; *mlb*, multilaminated body; *mn*, nucleus of the muscle cell; *sg*, secretory granules; *t*, tegument; *Z*, Z-disk. From Malkova (1989), with permission.

in mind that the tail functions as a locomotory organ only during a short period of active larval life (see below). Apparently, when cercariae are released from the sporocyst, mitochondria have already been formed in the muscle cells; glycogen and enzymes, particularly, succinate dehydrogenase, have accumulated in considerable amounts, sufficient for active work (Ginetsinskaya, 1968; Ginetsinskaya and Khari, 1975; Khari, 1975). At final stages of morphogenesis of xiphidiocercariae, along with the nuclei, cytons of longitudinal muscle cells, located in the central part of the tail, which contain these nuclei, also degenerate. In their place, the aforementioned cavity, filled with liquid, is formed. A similar situation has been described for *Parorchis acanthus* larvae (Rees, 1971). In the other trematode cercariae, investigated up to now, the central part of the tail is occupied by the bodies of muscle cells (Cardel et al., 1960; Chapman and Wilson, 1973; Rees, 1977).

2.1.4. Glands

Glands of cercariae are diverse, but, clearly form 2 large groups (Galaktionov and Dobrovolskij, 1987; Galaktionov and Malkova, 1994). The first one, most numerous, includes glands, which represent specialized tegumental cells. They are connected with the outer layer of the tegument by means of cytoplasmic projections and release their secretion into this layer. Normally, in young cells, RER and dictyosomes of Golgi complex are well-developed. The synthesized secretion gradually fills the cytoplasm, the mature gland being, in fact, a cellular reservoir with secretory material. The second group includes glands, whose ducts supplied with microfilaments and microtubules, open at the larval body surface with independent pores. Parts of the ducts, passing through the tegument, contact it by means of septate desmosomes.

Electron microscopy showed that tegumental cells are so-called mucoid cells, sheath-forming (this term was introduced by Ginetsinskaya and Bhutta, 1976, see below) and cystogenous glands, which were identified by light microscopy and histochemistry (Krasnodembskij, 1972, 1977; Ginetsinskaya and Bhutta, 1976). Mucoid glands, found in most cercariae examined, except Strigeidida, form a well-defined group (Kruidenier, 1951, 1953; Krasnodembskij, 1972; Ginetsinskaya and Bhutta, 1976; Galaktionov and Krasnodembskij, 1990; Galaktionov and Malkova, 1994) (Fig. 56E, see fig. 50G). They are formed in the embryo at the stage of 2-3 pair of flame cells, greatly increase in size in the course of accumulation of secretion, which at the final stages of morphogenesis is excreted entirely into the outer layer of the tegument. The emptied cytons degenerate and are not seen in fully formed cercariae. The secretion is metachromatically stained with alcian blue; it is of mucous nature and contains mucin. It is excreted onto the

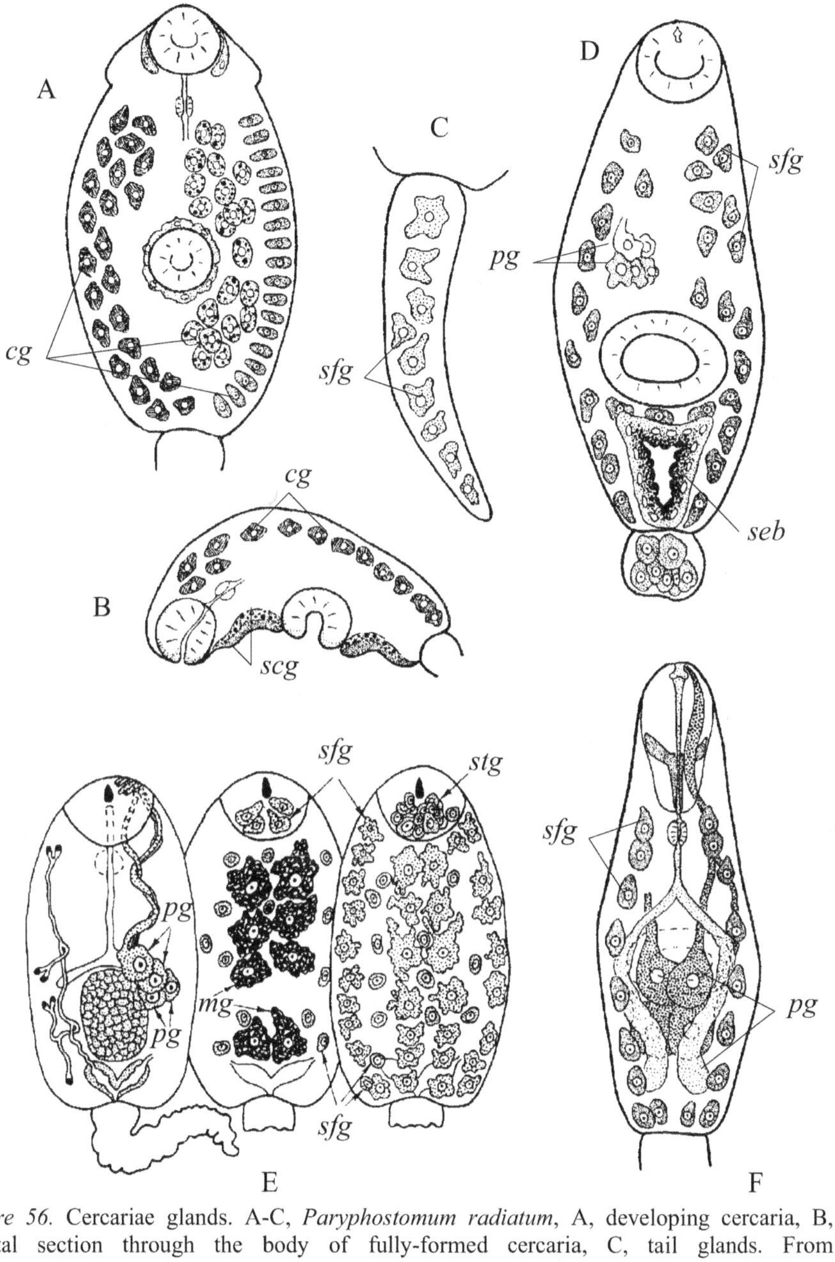

Figure 56. Cercariae glands. A-C, *Paryphostomum radiatum*, A, developing cercaria, B, sagittal section through the body of fully-formed cercaria, C, tail glands. From Krasnodembskij (1972), with permission; D, *Sphaerostomum bramae*. From Krasnodembskij (1972), with permission; E, developing microphallid cercaria. From Galaktionov and Malkova (1994), with permission; F, *Diplostomum* sp. *Abbreviations*: *cg*, cystogenous glands; *mg*, mucoid glands; *pg*, penetration glands; *scg*, secretion of cystogenous glands in the outer layer of the tegument; *seb*, secretion in the excretory bladder epithelium; *sfg*, sheath-forming tegumental glands; *stg*, tegumental glands, whose secretion is probably used to form the outer layer of the stylet.

surface of the larvae, released from the molluscan host, and, apparently, protects them from unfavourable factors of the environment. In a number of cases it enhances attachment (adherence) of the larvae to the body of the second intermediate host, thus facilitating the process of penetration (Ginetsinskaya and Bhutta, 1976)[v]. In plagiorchiids the secretion of mucoid glands participates in the formation of the outer layer of the metacercarial cyst (see p. 155).

Mucoid glands are numerous in representatives of Fasciolidae, Philophthalmidae, Notocotylidae, Opecoelidae and Heterophyidae. In more advanced forms, their number decreases (to 4 pairs in Microphallidae – Fig. 56E). Accumulation of mucoid secretion may proceed in specialized regions of the outer layer of the tegument. In microphallid cercariae it passes only into the ventral tegument, where, due to the merging of separate granules, whole "secretory fields" are formed (Galaktionov and Malkova, 1990a, 1994). In virgulate xiphidiocercariae, secretion of mucoid glands accumulates primarily in the virgula – paired thickenings of the outer layer of the tegument, deeply invaginating the oral sucker (Fig. 57A). The so-called caudal pockets of xiphidiocercariae, located at the base of the tail, have similar function and structure (Fig. 57B).

Sheath-forming glands do not compose a homogenous group. Included in this group are also tegumental glandular cells, formed in the course of cercarial morphogenesis simultaneously or a bit later than the mucoid ones, and also completing their secretory cycle before the larva is fully mature (Fig. 56E). As a rule, these glands are smaller than the mucoid ones. Their secretion is not metachromatically stained with alcian blue. Their secretion includes neutral and acid polysaccharides, their ratio, as well as that of other components varies greatly (Krasnodembskij, 1972, 1977). No less diverse is the function of the secretion of sheath-forming glands. Apparently, in some cercariae it plays the role, similar to that of secretion of mucoid glands (Ginetsinskaya and Bhutta, 1976). Tegumental cells with cystogenous secretion in Philophthalmidae, Heterophyidae, some Plagiorchiidae, Microphallidae and possibly Diplostomidae, also may be called sheath-forming glands (Thakur and Cheng, 1968; Žd'árská, 1969b, 1971a; Rees and Day, 1976; Stein and Lumsden, 1971a, b; Mitchell, 1974; Tihomirov, 1980; Benjamin and James, 1987; Galaktionov and Malkova, 1994). In addition, in forming microphallidae larvae, one more type of sheath-forming glands is observed. The cells of this type produce rod-shaped or disc-shaped electron-dense granules. As mentioned above, the content of the latter is connected with the formation of the glycocalyx. Thus, we cannot state that the group of sheath-forming glands is a natural formation.

Unlike mucoid and sheath-forming glands, cystogenous glands, which are formed at the same stages of morphogenesis or later, are retained during all free-living stage of the larva. The secretion passes into the outer

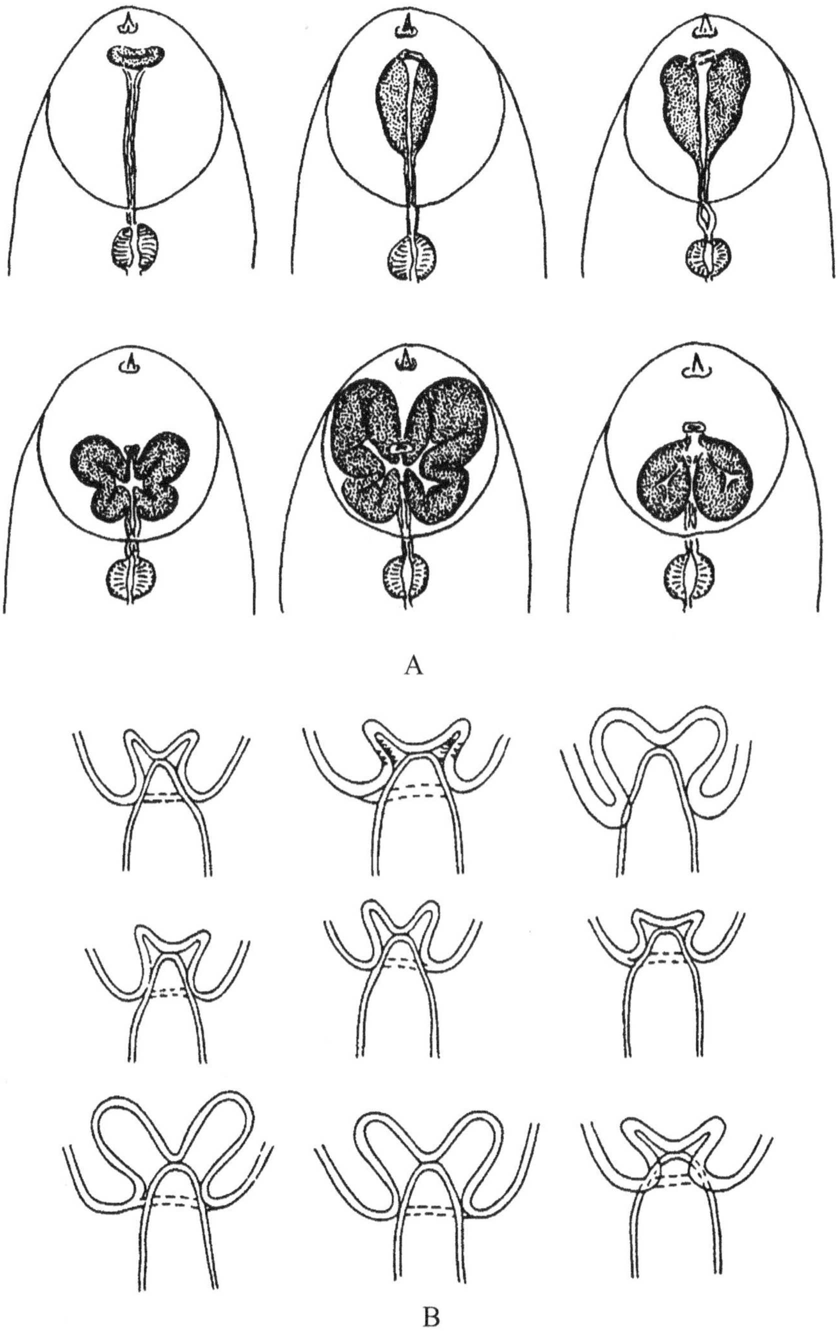

Figure 57. Spezialized structures of xiphidiocercariae tegument. A, virgulae of different types, B, caudal pockets of different types. From Bock (1980).

plate of the epidermis in the course of encystment of cercariae (Fig. 56). The most developed are cystogenous glands in larvae of primitive trematodes (Fasciolidae, Paramphistomidae, Notocotylidae), for which this phase of adolescaria is typical, or in such forms, (a number of Echinostomatidae), whose metacercariae encyst at the surface of the body and of the organs of the second intermediate hosts. In such trematodes, cystogenous glands are often represented by cells of several types, differing in structure and chemistry of the secretion. In evolutionary more advanced digenean groups, cystogenous glands can be deleted. This is associated with the transition of metacercariae to the exploitation of the host's energy potential, this transition necessarily entailing changes in the cyst structure (see for details p. 153). Typical cystogenous glands are still found in relatively primitive plagiorchiids from the families Ochetosomatidae and Plagiorchiidae, but in other representatives of the order (like in all strigeidids), they are absent. It is of interest that in Cyathocotylidae larvae, the cystogenous glands are functionally replaced by specialized unicellular glands, which open with independent pores on the periphery of the larval body, i.e., according to our classification, they are included in the second group (Galaktionov et al., 1980).

The central place in the second of the above-mentioned large groups of cercarial glands is occupied by penetration glands. They include glandular cells with long ducts, which open subterminally at the anterior end of the larva (Fig. 56). The cellular bodies are localized in two compact groups in the anterior or middle parts of the cercaria body (often in the region of the ventral sucker). Such glands are characteristic of larvae that actively penetrate into the second intermediate host, and are absent in such forms as Fasciolidae, Haploporidae, Notocotylidae, Pronocephalidae, Bucephalidae, Echinostomatidae and Hemiuridae, for which encystment in the external environment (adolescaria phase) or passive ingestion of the cercaria by the second intermediate or definite host, are typical. Penetration glands are normally formed after the tegumental ones. In developing microphallid larvae, they are formed earlier, as far as one can determine by light microscopy (Fig. 56). At the initial stages of their development, the glandular cells acquire multibranched shape. Later, one ramus elongates and is transformed into a duct. As a rule, secretion of the penetration glands is composed of proteolytic enzymes and hyaluronidase, which facilitate penetration of the larva and its further migration in the body of the second intermediate host.

The number of cercariae penetration glands decreases, along with specialization of the remaining cells, in the evolutionary series from relatively primitive forms to more advanced ones (Ginetsinskaya, 1968; Ginetsinskaya and Bhutta, 1976; Galaktionov and Dobrovolskij, 1987). In the primitive forms this number is relatively large (10-11 pairs) and the glands are uniform in morphology as well as the contents of their secretion. In more advanced forms the number of the cells is decreased (to 1-5 pairs in highly specialized plagiorchiids, schistosomatids and strigeidids). This is often accompanied by morphological and functional specialization. Thus, in *Schistosoma mansoni* cercariae, preacetabular glands contain enzymes (proteases), whereas postacetabular ones contain carbohydrates, chemically bound to proteins (Davies, 1983). In microphallid cercariae, secretion of the two anterior pairs of the penetration glands is used during penetration into the second intermediate host. In some representatives of *Microphallus* and *Levinseniella*, it is used to form the penetration cyst (see p. 275). Secretion of the two posterior pairs is used during migration of larvae in the second intermediate host. At the ultrastructural level, the secretion granules of the two anterior pairs of glands are identical in morphology and electron-density in all microphallid larvae, studied to date, whereas the secretion granules of the two posterior pairs exhibits interspecific variability (Galaktionov and Malkova, 1994).

Glandular cells, connected to the anterior organ in Cyathocotilidae furcocercariae (see fig. 43F,G) (Galaktionov et al., 1980), can be considered as penetration glands. They are represented by pear-shaped glands, located in the anterior organ, and by two groups of similar cells, situated on both sides. Obviously, these gland secretions lyse host tissues. Apparently, the same effect is caused by secretion of numerous unicellular glands, opening with independent pores at the ventral and dorsal sides of the cercarial body. Its secretion occurs, when the larva has reached its final settlement in the body of the second intermediate host (see below).

Special "metacercarial" glands are widely spread in cercariae, especially in the primitive forms. These glands are so-called because they are better seen in metacercariae, than in cercariae, though they begin functioning only in maritae (Dobrovolskij, 1965b). By morphology and localization in the cercariae, these glands resemble the penetration glands. These two types of glands are often confused (especially when larvae, penetrating into the host, are considered). All of this is also appropriate to cercariae of the Echinostomata group, whose "metacercarial" glands are most often referred to as the penetration glands (see fig. 41C). However, even in those, relatively rare cases, when Echinostomata larvae are capable of disrupting the epidermis of the second intermediate host, they achieve it not by means of the "penetration" glands, but through the release of secretions, contained in the outer layer of the tegument.

The ducts of "metacercarial" glands are normally thin. They open at the anterior edge of the oral sucker, arranged as a transversal row, but not as compact groups, as it is typical for real penetration glands. The number of glandular cells of this type varies greatly in different groups of cercariae: from several dozens in larvae of primitive trematodes (see fig. 41C) to 4-5 pairs in some Plagiorchiidae and Ochetosomatidae. The glandular cells of this type are absent in the most advanced plagiorchiids (Microphallidae,

Lecithodendrioidea), as well as in all furcocercariae of the strigeidid complex.

In addition to the glands described above, there are other more specialized types of cercarial glands. In particular, these are small tegumental cells, functioning in the oral sucker of developing microphallid cercariae. Their secretion is probably used to form the outer layer of the stylet (Galaktionov and Malkova, 1990b). Many of the glands observed belong to "glands with uncertain function" (Ginetsinskaya and Bhutta, 1976), whose elucidation is the task of future research.

2.1.5. Digestive System

The digestive system of digeneans consists of a foregut and midgut. The former includes the prepharynx, pharynx and oesophagus. The latter includes two closed branches of intestinal caeca. The only exception is the Bucephalidae, where the foregut opens midventrally and leads to an unbranched single caecum. The lining of the foregut represents a direct extension of the outer syncytial layer of the tegument and is separated from the epithelium of the intestinal caeca by means of a septate desmosome. However, there is no such a junctional complex in the Dicrocoeliidae and Brachylaimoidea (Robinson and Halton, 1983; Žd'árská et al., 1988). The cells of the forming primordial foregut become strongly vacuolated. Then, the boundaries between separate cells disappear, and cell vacuoles merge into a single cavity, which becomes the foregut cavity (Tihomirov, 1980; Galaktionov and Dobrovolskij, 1987). Studies show that the foregut is formed as an anterior invagination of the early nucleated tegument that invests the cercaria embryo (Halton and McCrae, 1985). This does not agree with studies on formation of the primordial digestive system in the form of cellular cords (Wisniewski, 1958; Cheng and James, 1960; Cheng, 1961; Dobrovolskij, 1965b, 1971; Cheng and Bier, 1972; Gerasev and Dobrovolskij, 1977; Tihomirov, 1980; Galaktionov and Malkova, 1993). Most likely, the formation of the primordial foregut and primordial outer layer of the tegument takes place simultaneously. Both initially consist of single cells; then the cells merge, resulting in formation of the syncytium.

The epithelium of the caeca may be of a syncytial or cellular nature. In the latter case, the cells are united into a single layer by means of different types of cellular contacts (Køie, 1971c, 1973; Galaktionov and Malkova, 1993). In the cytoplasm, there are large amounts of RER and many mitochondria. Golgi complexes may also be seen. This all testifies to the active state of the cytoplasm. However, there is no proof of participation of the gut in feeding of the larva, when the latter stays in the daughter parthenita (Køie, 1971c; Bogitsh and Carter, 1977). At the same time, secretory activity of the intestinal epithelium is undoubtedly present in a number of species (Køie, 1971c, 1973; Krasnodembskij, 1977; Galaktionov et al., 1980). Secretion, produced by the cells, fills the lumen of the gut and is retained in mature larvae, which have left the mollusc.

The above description is based on data, available for cercariae of species, where the digestive system is formed at that stage of ontogenesis. However, development of the primordial digestive system varies considerably in cercariae, even among representatives of closely related systematic groups. Thus, in the Plagiorchiidae there is more than one morphological series, beginning from the types, where the primordial digestive system is absent or barely expressed, to the types, where there is a well-developed digestive system with a lumen, which is fully formed along the whole caeca. There are numerous transitional types between these two extrems (Galaktionov and Dobrovolskij, 1987).

2.1.6. Excretory System

Attention of researchers has long been focused on cercariae excretory system. Its morphology and development were used for constructing taxonomic systems and evolutionary scenarios (Cort, 1919; Komiya, 1938; Hussey, 1941; Kuntz, 1950, 1951, 1952; La Rue, 1957; Odening, 1962, 1964). A detailed analysis of formation of the excretory system of cercariae of different taxonomical groups was undertaken in connection with the general trends of morphological evolution of trematodes (Galaktionov and Dobrovolskij, 1987). Below, we present a brief and generalized version of such analysis, emphasizing only those aspects, which are important for discussion of digenean evolution in chapters 5 and 6.

In the course of morphogenesis, cercariae of all trematodes pass the stage of 4 flame cells pairs, described with an excretory formula $2[(1 + 1) + (1 + 1)] = 8$ (Fig. 58). It reflects the emergence of anterior and posterior collecting tubules in the excretory system. They initiate from the point of branching of capillaries, which come directly from the flame cells, and terminate in the middle of the body, where they merge and give rise to the main collecting tubes (Fig. 58, 59, 60). In fact, formation of the system of flame cells and ducts is finished at that point in larvae of some trematodes. Further modifications are connected with the development of the excretory bladder (see below).

However, in most species the number of cells keeps on increasing and the system of ducts becomes more and more complicated (Fig. 58, 59, 60). This process proceeds differently even in closely related groups of digeneans. This difference will be considered below. In some species, 8 primary flame cells (4 from each side of the body) are doubled and give rise to

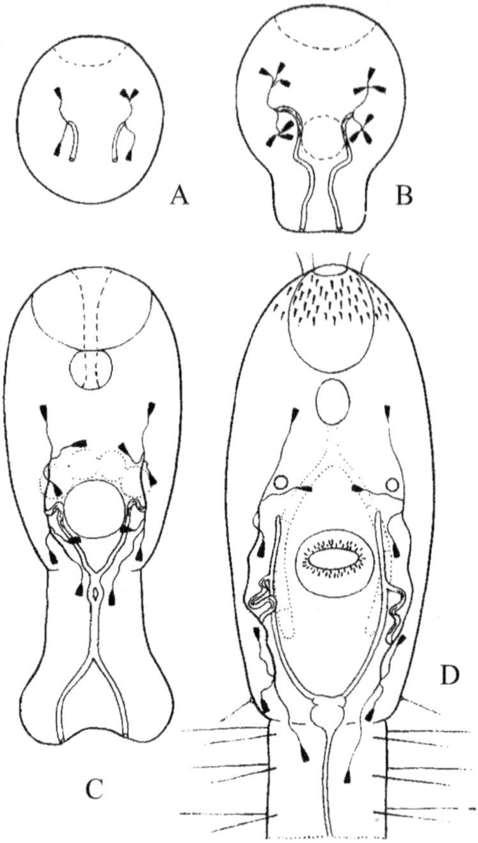

Figure 58. Successive stages (A-D) of development of excretory system of cercaria *Neodiplostomum spathoides*. From Odening (1965b), reprinted with author's permission.

8 groups of flame cells with 2 cells in each group. The excretory formula becomes as follows: $2[(2 + 2) + (2 + 2)] = 16$. Later, the number of cells may increase only within these 8 groups. No new groups are formed. In the generalized form, this can be written as $2[(N \times 1 + N \times 1) + (N \times 1 + N \times 1)] = 2 \times N \times 4$.

In other species, terminal primary flame cells divide once more. Flame cells of anterior and posterior parts of the body not being involved in this process simultaneously, two subsequent stages, are clearly distinguished in the embryo development. They can be expressed by the corresponding formulae $2[(1 + 1 + 1) + (1 + 1)] = 10$ and $2[(1 + 1 + 1) + (1 + 1 + 1)] = 12$. More rarely, the posterior terminal pair is ahead of the anterior one in its development. Then the intermediate variant assumes the following form: $2[(1 + 1) + (1 + 1 + 1)] = 10$. In most cases (with the exception of the first interme-

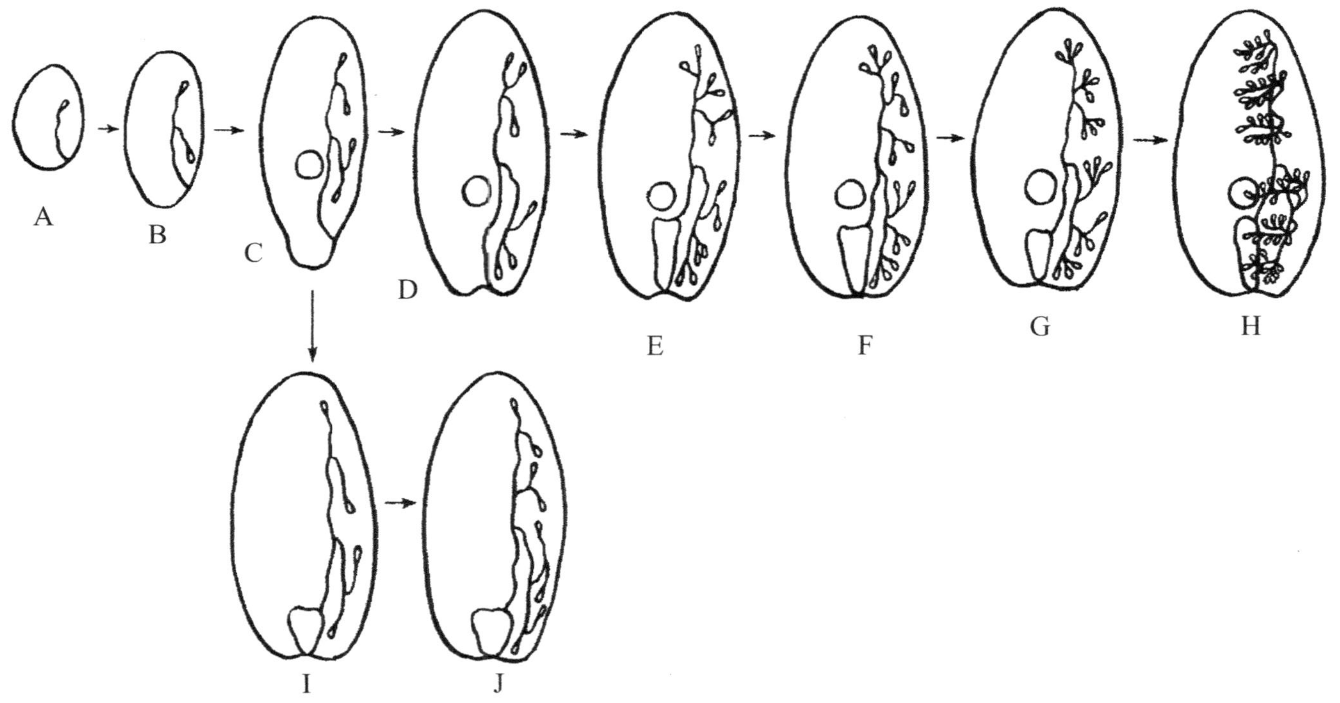

Figure 59. Excretory system of xiphidiocercariae: successive stages of development. A, 2[1] = 2; B, 2[(1) + (1)] = 4; C, 2[(1+1) + (1+1)] = 8; D, 2[(1+1+1) + (1+1+1)] = 12 – *Cercaria microcotyle*; E, 2[(2+2+2) + (2+2+2)] = 24 – Lecithodendrioidea; F, 2[(3+2+2) + (2+2+3)] = 28 – Lecithodendrioidea; G, 2[(3+3+3) + (3+3+3)] = 36 – Plagiorchiidae, Ochetosomatidae, Lecithodendrioidea; H, 2[(9+9+9) + (9+9+9)] = 108 – Lecithodendrioidea (*Cercaria rhionica XI* Olenev and Dobrovolskij, 1975); I, 2[(1+1) + (1+1)] = 8 – Microphallidae; J, 2[(2+2) + (2+2)] =16 – Microphallidae. From Galaktionov and Dobrovolskij (1987).

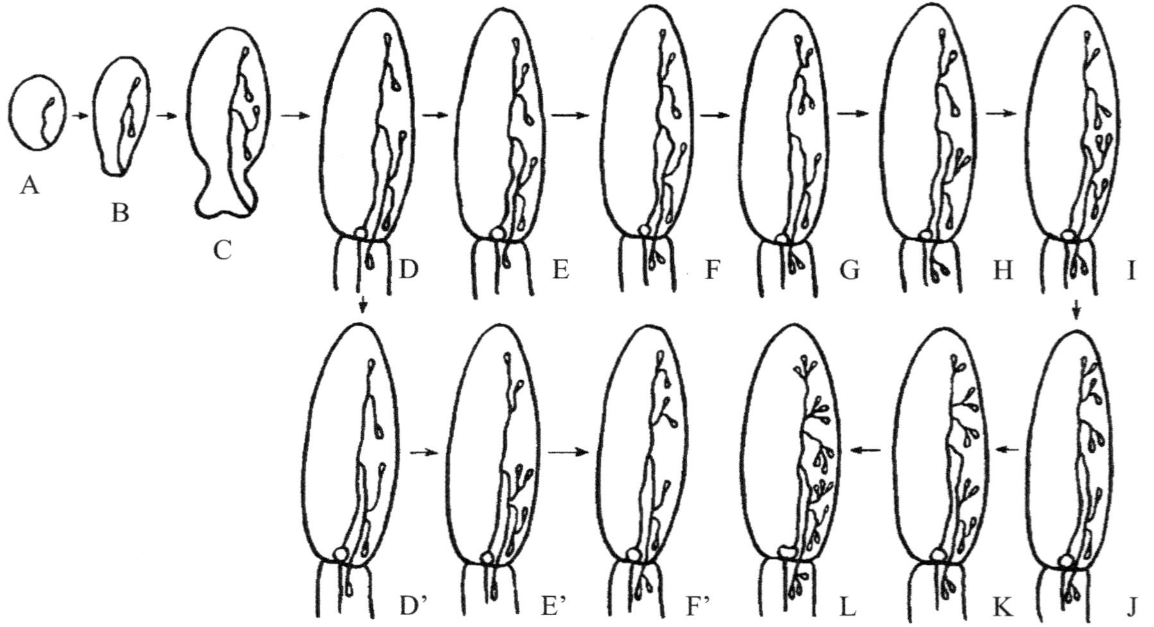

Figure 60. Excretory system of furcocercariae: successive stages of development. A, 2[1] = 2; B, 2 [(1) + (1)] = 4; C, 2[(1+1) + (1+1)] = 8; D, 2[(1+1) + (1+1) + (1)] = 10; E, 2[(1+1+1) + (1+1) + (1)] = 12 – *Fibricola texensis*, *Neodiplostomum spathoides*; F, 2[(1+1+1) + (1+1) + (2)] = 14 – *Alaria mustelae*; G, 2[(1+1+2) + (1+1) + (2)] = 16 – *Tylodelphis*; H, 2[(1+1+1) + (2+1) + (2)] = 12 – *Diplostomum spathaceum*, *D. baeri*; I, 2[(1+1+2) + (2+2) + (2)] = 20 – *Posthodiplostomum cuticola*, *P. minimum*; J, 2[(2+2+2) + (1+1) + (2)] = 20 – *Posthodiplostomum brevicaudatum*; K, 2[(2+2+2) + (2+2) + (2)] = 24 – *Holostephanus cobitis*, *H. volgensis*, *Alaria*, *Codonocephalus*; L, 2[(3+3+3) + (3+3) + (3)] = 36 – *Mesostephanus*, *Prohemistomum*, *Prosostephanus*; D’, 2[(1+1) + (1+1) + (1)] = 10 – *Strigea*, *Apatemon cobitidis*, *Parastrigea robusta*; E’, 2[(1+1) + (2+2) + (1)] = 14 – *Apatemon gracialis*, *Australapatemon*; F’, 2[(2+2) + (1+1) + (2)] = 16 – *Cotylurus*, *Apharyngostrigea*. From Galaktionov and Dobrovolskij (1987).

diate variant), the excretory system of cercariae is formed according to these primary formulae (Fig. 59, 60).

In furcocercariae of Strigeidida both versions – with 10 and 12 primary flame cells are realized. However, in both cases primary flame cells of the posterior pair grow into the primordial tail. Hence, the excretory formula is as follows: $2[(1 + 1 + 1) + (1 + 1) + (1)] = 12$ or $2[(1 + 1) + (1 + 1) + (1)] = 10$. Normally, the number of flame cells in each group increases. So, in the generalized form we have $2[(N \times 1 + N \times 1 + N \times 1) + (N \times 1 + N \times 1) + (N \times 1)] = 2 \times N \times 6$ or $2[(N \times 1 + N \times 1) + (N \times 1 + N \times 1) + (N \times 1)] = 2 \times N \times 5$. Index "N" may vary in different groups (Fig. 60).

In the majority of xiphidiocercariae of the Plagiorchiida the number of flame cells in the groups also increases: $2[(N \times 1 + N \times 1 + N \times 1) + (N \times 1 + N \times 1 + N \times 1)] = 2 \times N \times 6$. As in furcocercariae, index N may vary both in larvae of different species and in different groups of flame cells in the same larva. As a rule, groups of 2 and 3 flame cells ("diads" or "triads") are formed. Rarely, the number of cells in the group may reach 9 (Fig. 59).

Within Strigeidida and Plagiorchiida, a trend towards simplified excretory formula is clearly manifested in cercariae of the most specialized groups (Galaktionov and Dobrovolskij, 1987). In furcocercariae, the formula $2[(3 + 3 + 3) + (3 + 3) + (3)] = 36$ is reported only in some Cyathocotylidae. It can be gradually simplified to the following form $2[(1 + 1 + 1) + (1 + 1) + (1)] = 12$ in many Diplostomidae. This is even more pronounced in larvae of family Strigeidae, where the number of primary groups decreases to 5, with the maximal amount of flame cells in each group not exceeding 2. The excretory system of cercariae of the most specialized species of this family (genera *Strigea*, *Parastrigea*, some *Apatemon*) is described by the formula $[(1 + 1) + (1 + 1) + (1)] = 10$. In xiphidiocercariae, the simplest excretory formula characterizes the most advanced Lecithodendrioidea and Microphallidae (Fig. 59). Among all cercariae, the simplest version of the excretory formula – $2[(1 + 1) + (1 + 1)] = 8$ is registered for Microphallidae.

As a rule, representatives of more primitive trematode families have a more complicated excretory system. However, in some cases, the tendency towards simplification is also manifested there rather distinctly. Thus, cercariae of many Echinostomatidae have an enormous number (up to 100-200) of flame cells. Difficulties in investigation of such a complicated system resulted in a great ambiguity of excretory formulae, which sometimes count more than dozen groups of flame cells (Stunkard, 1966; Loos-Frank, 1967). Analysis of cercarial mophogenesis in such species, which we carried out on representatives of *Himasthla*, has shown that they follow the 6-members version: $2[(N \times 1 + N \times 1 + N \times 1) + (N \times 1 + N \times 1 + N \times 1)] = 2 \times N \times 6$, but the numerical expression of "N" is very large and varies in different groups. Moreover, it differs in different cercariae, released from one mollusc. From such a primitive pattern, with an unstable number of flame cells, we come to the larvae of *Echinochasmus* obeying formulae $2[(3 + 1 + 1) + (1 + 1 + 1)] = 16$ (*Cercaria rhionica VII* from *Melanopsis praemorsa*) and $2[(2 + 1 + 2) + (1 + 1 + 1)] = 16$ (*Echinochasmus* sp. from *Hydrobia ulvae*) (Galaktionov and Dobrovolskij, 1987).

The 6-membered excretory formulae are also typical for Notocotylidae, Lepocreadiidae, Allocreadiidae and Bunoderidae. In the representatives of the first family (at any rate, in the cercariae of investigated species), the number of flame cells has become stable and corresponds to the triad condition - $2[(3 + 3 + 3) + (3 + 3 + 3)] = 36$ (Odening, 1963, 1966; Filimonova, 1985). In the larvae of the other groups it varies widely. Rather common are trematodes with the 5-membered excretory formula $2[(N \times 1 + N \times 1) + (N \times 1 + N \times 1 + N \times 1)] = 2 \times N \times 5$. First of all, these are Opisthorchiidae. There the number of cells in the groups varies from $2[(5 + 5) + (5 + 5 + 5)] = 50$ (*Metorchis intermedius*) to $2[(2 + 2) + (2 + 2 + 2)] = 20$ (*Apophallus*). Other trematode groups, with rare exceptions, also follow one of the versions of typical formulae of the excretory system, presented earlier (Galaktionov and Dobrovolskij, 1987).

Distal parts of the excretory ducts also undergo certain morphological modifications in the course of cercarial development. In fully formed larvae with the excretory system of the Stenostoma type, the main collecting tubes are directed forwards from the site of the fusion of anterior and posterior collecting tubules. Then they form a loop in the region of the oral sucker, turn backwards and reach the excretory bladder, located at the posterior end of the body. Such a design of the excretory system is typical of primitive families – Echinostomatidae, Fasciolidae, Philophthalmidae, Psilostomidae (see fig. 41C, 42G). In larvae of more specialized Digenea (Notocotylidae, Heterophyidae, Plagiorchiida, Strigeidida), with the excretory system of Mesostoma type, proximal parts of the main collecting tubes do not form a loop. Instead a dense ball is formed, consisting of many small adjoining loops and bends. The tubules leave this ball and go to the excretory bladder (see fig. 41D, 43, 46). Often, especially in large cercariae, longitudinal and main collecting tubes have ciliary tufts.

Distal parts of the main collecting tubes undergo the most profound modifications at the final stages of cercarial morphogenesis. They come close to each other and finally merge. The zone of merging begins in the embryo body and terminates in the tail (in case of the forms where the main collecting tubes go into the tail – see earlier). The part of this zone, located in the embryo body, widens and gives rise to an excretory bladder (see below). In the tail, this duct preserves the tubular structure and is called a caudal duct. Its terminal regions do not merge and open at the sides of the tail (normally at the posterior boundary of its anterior third) as primary excretory pores. Such a situation is typical of mature cercariae of the most primitive families (Echinostomatidae, Notocotylidae, Heterophyidae). However, in some Echinostomatidae (*Petasiger*, *Echinochasmus*), apparently in connection with a particular specialization of the tail (Fig. 61), the caudal duct is reduced and an unpaired secondary excretory pore is formed. Such changes become obligatory for cercariae of all Opisthorchiidae. In furcocercariae, primary excretory pores are retained. They are located in the middle (Strigeidida) or at the ends (Schistosomatida and Gymnophallidae) of the furcae. In the most specialized xiphidiocercariae (Plagiorchiida) no parts of the excretory system enter the tail. The unpaired excretory pore of fully formed larvae, situated at the base of the tail, results from the merging of the primary pores.

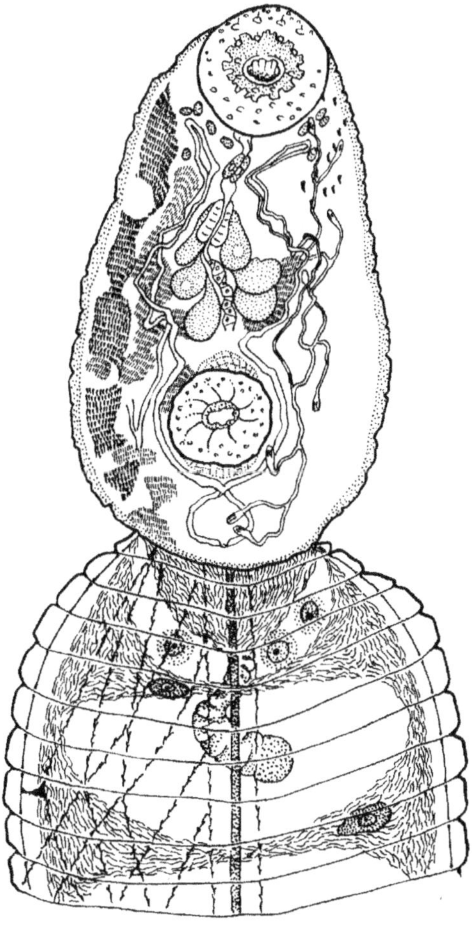

Figure 61. Cercaria *Echinochasmus* sp. From Galaktionov and Dobrovolskij (1987).

La Rue (1957) classified trematodes into two groups (Anepitheliocystidea and Epitheliocystidea), based on differences in the structure of the excretory bladders in cercariae of different taxa. For the first group (Echinostomatidae, Azygiidae, Transversotrematidae, Schistosomatida, Clinostomidae, Strigeidida) mainly small thin-walled excretory bladders with poorly developed muscle are typical, which often makes such bladders incapable of abrupt contractions. In the second group (Plagiorchiida, Allocreadiidae, Opisthorchiidae, Hemiuroidea, Heterophyidae) the excretory bladder occupies a significant part of the larval body. According to La Rue (1957), it consists of large cells and is covered with a well-developed muscle layer, providing its contraction. Electron microscopy studies led to essential corrections in these views (Ebrahimzaden and Kraft, 1971; Powell, 1972, 1973, 1975; Gibson, 1973, 1974; Rees, 1977; Popiel, 1977; Malkova and Galaktionov, 1989; Niewiadomska and Czubaj, 1996; Podvyaznaya and Galaktionov, a,b, *submit*).

It was found that the fine structure of the excretory bladder in anepitheliocystid and epitheliocystid cercariae is basically the same. In all cases (with the exception of schistosomatids, where an excretory atrium – a derivative of the body tegument (Powell, 1973) – is formed instead of the excretory bladder) it is formed by a layer of the nuclear syncytium. In anepitheliocystid cercariae it is thin and its thickness practically does not differ from that of the walls of main collecting tubes. In epitheliocystid cercariae the syncytial layer of the excretory bladder is thicker. Nuclei with the surrounding cytoplasm can evaginate into the lumen of the bladder (some Bucephalidae) (Fig 62). On the contrary, in cercariae of Ochetosomatidae and Microphallidae, nuclei with the surrounding cytoplasm are sunk in the parenchyma and retain connection with the syncytial layer of the excretory bladder by means of cytoplasmic extensions (Fig. 63). The cytons are large, adjoin each other and, when studied by light microscopy, the walls of the excretory bladder seem to be lined with cuboidal epithelium.

On the basis of the light microscopy, basic differences in the nature of the excretory bladders of anepitheliocystid and epitheliocystid cercariae were postulated. In the former group, the excretory bladder is formed by merging of distal parts of the main collecting tubes. In the latter, the walls of the bladder result from specialization of certain "mesodermal" cells, which appear in cercarial embryos in the region of merging distal parts of the main collecting tubes. Such views on the formation of excretory bladders of epitheliocystid cercariae were not supported by electron microscopy. In the larvae of Ochetosomatidae, Microphallidae and Bucephalidae, studied in this regard, formation of the excretory bladder, similar to anepitheliocystid cercariae, results from specialization of the lining of the distal parts of the main collecting tubes (Powell, 1972; Malkova and Galaktionov, 1989; Podvyaznaya and Galaktionov, a,b, *submit*). In the representatives of the two

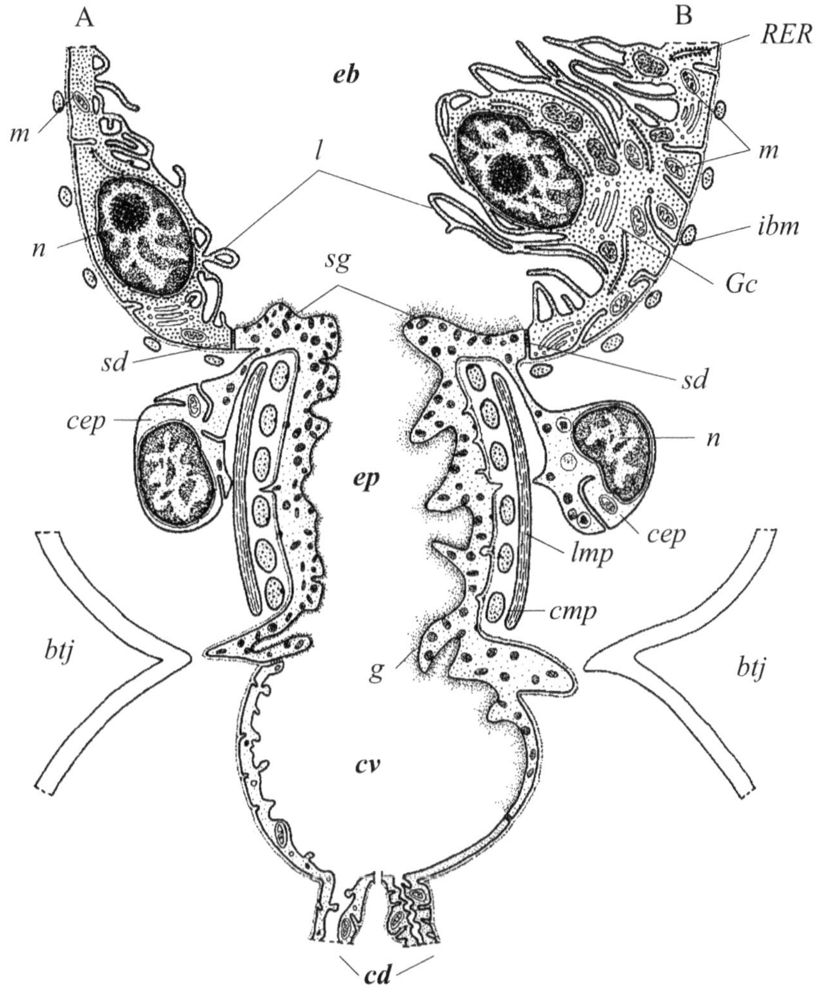

Figure 62. Diagram showing structure of excretory system of bucephalid cercariae. A, *Bucephaloides gracilescens*; B, *Prosorhynchus squamatus*. *Abbreviations*: *cd*, caudal duct; *cep*, cyton of the excretory pore tegument; *cmp*, circular muscles of the excretory pore sphincter; *cv*, caudal vesicle; *btj*, body-tail junction; *eb*, excretory bladder; *ep*, excretory pore; *g*, glycocalyx; *Gc*, Golgi complex; *ibm*, invaginations of the basal membrane; *l*, lamellae; *lmp*, longitudinal muscles of the excretory pore sphincter; *m*, mitochondrion; *n*, nucleus; *RER*, rough endoplasmic reticulum; *sd*, septate desmosome; *sg*, secretory granules. From Podvyaznaya and Galaktionov (a, *submit*).

first families nuclei-containing cytons are sunk into the parenchyma. Apparently, in earlier investigations, these perikaryons were taken for "mesodermal" cells of the primordial bladder in epitheliocystid cercariae. Falling out are data by Rees (1977), who described formation of the bladder in cercariae of *Cryptocotyle lingua* as the result of merging of "mesodermal" cells with united distal parts of the main collecting tubes. It cannot be excluded that Rees misinterpreted the process of sinking of nuclei-containing regions of the epithelium of the forming bladder. However, so far the problem of the origin of the excretory bladder in epitheliocystid cercariae is not fully elucidated, because electron microscopy data are available only for a single species.

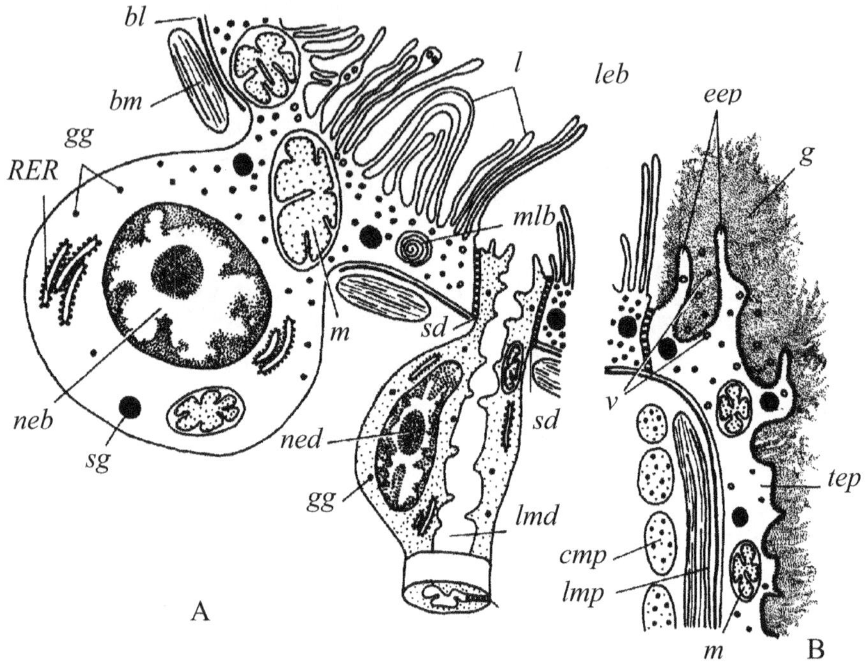

Figure 63. Diagram showing structure of excretory system of microphallid cercariae. A, excretory bladder, where it joins the main collecting tube; B, excretory pore, where it joins the excretory bladder. *Abbreviations*: *bl*, basal lamina; *bm*, bundles of muscles; *cmp*, circular muscles of the excretory pore sphincter; *eep*, finger-like extensions of the excretory pore; *g*, glycocalyx; *gg*, granules of glycogen; *l*, lamellae; *leb*, lumen of the excretory bladder; *lmd*, lumen of the main collecting duct; *lmp*, longitudinal muscles of the excretory pore sphincter; *mlb*, multilaminated body; *m*, mitochondrion; *neb*, nucleus of the excretory bladder epithelium; *ned*, nucleus of the excretory duct cell; *RER*, rough endoplasmic reticulum; *sd*, septate desmosome; *sg*, secretory granules; *tep*, tegument of the excretory pore; *v*, vesicles. From Malkova and Galaktionov (1989), with permission.

Morphology of protonephridia is rather conservative within high level taxa of parasitic flatworms. However, there are some differences between Trematoda and Monogenea on the one hand and Cestoda on the other hand (see for review Rohde and Watson, 1992; Rohde, 2001a). In cercariae, as in all other stages of the digenean life cycle, a ciliary tuft of the "flame" originates from the nucleus-containing part of the terminal (flame) cell. This tuft passes through the flame bulb and terminates in the capillary (Fig. 64). The proximal cell of the capillary wraps around itself and is joined by sep-

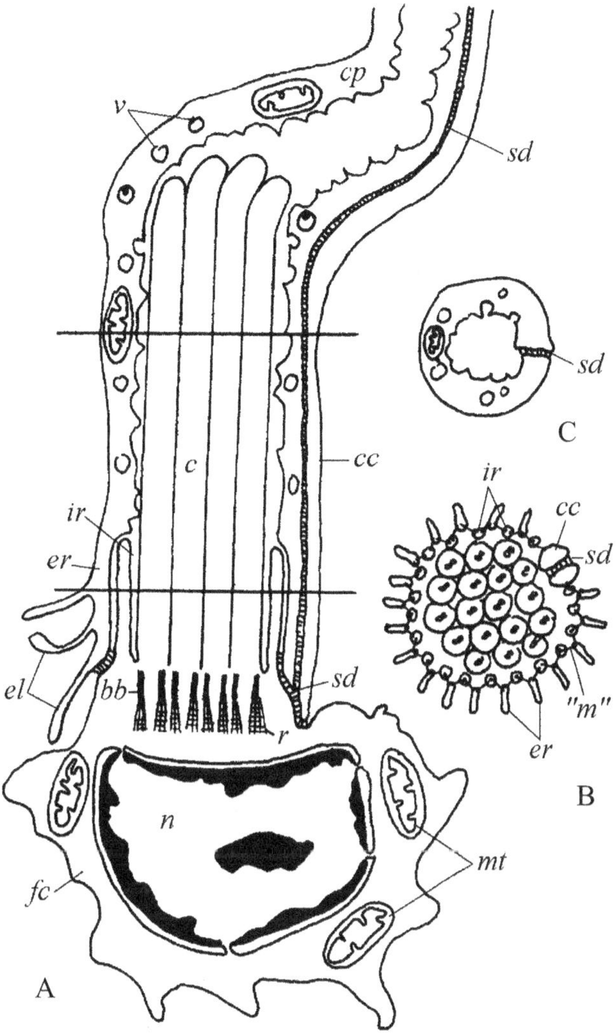

Figure 64. Diagrams showing filtering complex of cercariae of *Diplostomum pseudospathaceum*. A, longitudinal section; B, cross-section through the weir; C, cross-section through proximal capillary. *Abbreviations*: *bb*, basal bodies; *c*, cilia; *cc*, cytoplasmic cord; *cp*, capillary; *el*, external leptotriches; *er*, external ribs; *fc*, flame cell; *ir*, internal ribs; "*m*", "membrane"; *mt*, mitochondrion; *n*, nucleus; *r*, rootlet; *sd*, septate desmosome; *v*, vesicles. From Niewiadomska and Czubaj (1996), reprinted with publisher's permission.

tate junction. The capillary is joined to the flame cell also by means of the septate desmosome. Filtration apparatus or weir consists of the outer and inner layers of ribs, connected by the filtration "membrane" of apparently extracellular matrix. The outer ribs of the weir represent cytoplasmic extensions of the capillary proximal cell, and the inner ones are extensions of the terminal cell. Special cytoplasmic processes, called leptotriches, can be associated with the filtration complex. External leptotriches arise from the external ribs, and internal leptotriches arise from the internal ribs and basal cytoplasm of the terminal cell. Function of the leptotriches is not entirely clear. According to Wilson and Webster (1974), the external leptotriches keep neighboring cells away from the weir, and the internal ones keep the wall of the weir away from the ciliary tuft. Both types seem to make filtration more efficient.

As a rule, ducts of the excretory system are composed of separate cells, connected with each other by means of septate desmosomes. The cells have a prominent nucleus, situated laterally to the lumen and surrounded by a thin cytoplasmic layer. Numerous mitochondria, RER and free ribosomes are found in the cytoplasm. The luminal surface is usually increased by lamellae. The lining of the excretory bladder is connected to the main collecting tubes by means of septate desmosomes. The luminal surface of the excretory bladder may carry lamellae, especially well developed in cercariae of marine species Opecoelidae and Microphallidae (Gibson, 1974; Malkova and Galaktionov, 1989; Podvyaznaya and Galaktionov, a, *submit*). The cytoplasm of the lining is rich in large active mitochondria; also present there are cisterns of RER. In some species secretory granules are also seen in the cytoplasm, and the function of their secretion is unclear.

The fine structure of the caudal part of the excretory system, if there is any (see above), is similar to that described above for the filtration apparatus and ducts of the larval body. However, in the cercariae of Diplostomidae and Bucephalidae, the syncytial structure of the walls of the unpaired caudal duct has been described (Niewiadomska and Czubaj, 1996; Podvyaznaya and Galaktionov, a,b, *submit*). In the Bucephalidae cercariae, the proximal part of the duct widens and forms a caudal vesicle. In the region of primary excretory pores, unmerged terminal parts of the caudal ducts are connected to the tegument by means of the septate desmosome. Here the outer syncytial layer forms a small invagination, which, in fact, represents an excretory pore. In the morphogenesis of bucephalid cercariae, in the course of formation of the furcal rami, the excretory ducts in the furcae undergo reduction (Fig 65). Correspondingly, primary excretory pores also disappear. Instead, secondary excretory pores open at the surface of the caudal stem (it is more correct to call them tertiary pores, for the term "secondary excretory pore" is already reserved – see below). They also represent invaginations of the tegument, which are connected to the proximal regions of paired caudal ducts by means of septate desmosomes (Fig. 65D-G).

In most of the investigated cercariae species, the origin of the secondary (definitive) excretory pore (however, the term "pore" is not absolutely correct, for in fact this is a relatively short duct), is related to the tegument of the body (Powel, 1973, 1975; Rees, 1977; Malkova and Galaktionov, 1989).

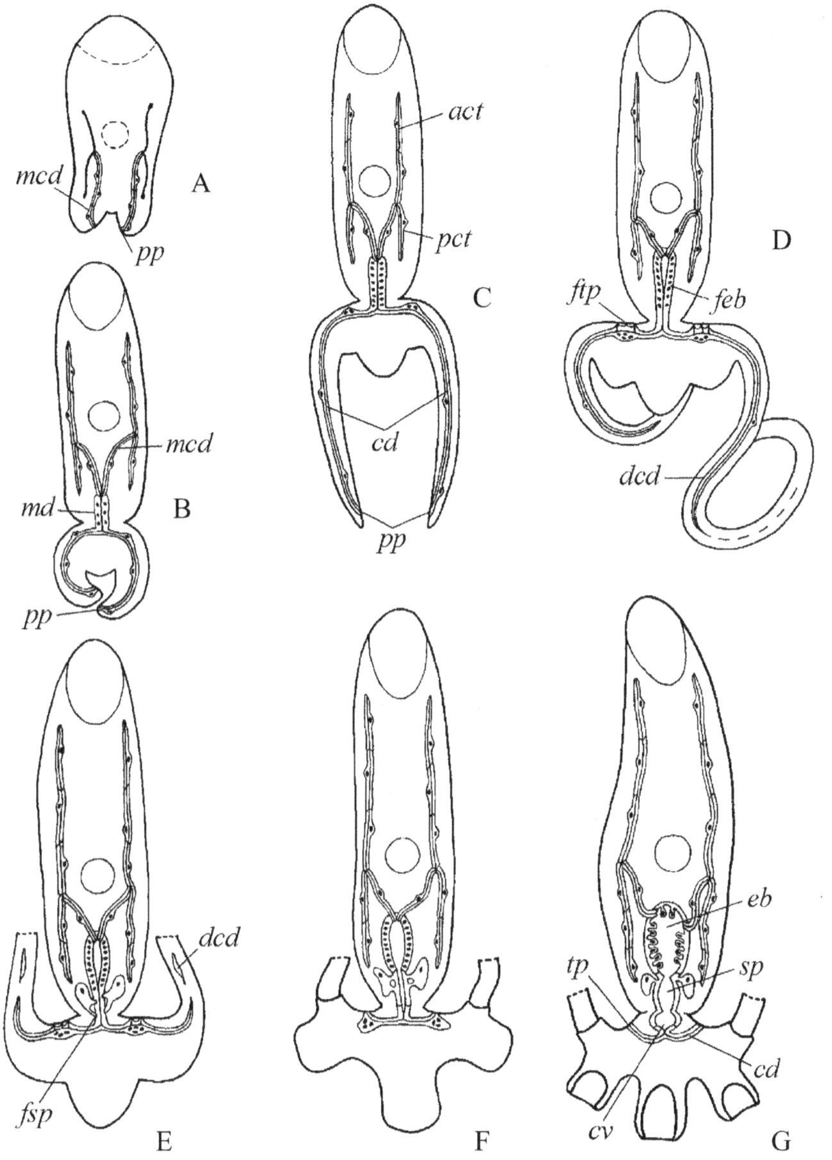

Figure 65. Successive stages (A-G) of development of excretory system of cercaria *Prosorhynchus squamatus* (Bucephalidae) (flame cells and capillaries are not shown). *Abbreviations*: *act*, anterior collecting tubule; *cd*, caudal duct; *cv*, caudal vesicle; *dcd*, degenerated part of the caudal duct; *eb*, excretory bladder; *feb*, forming excretory bladder; *fsp*, forming secondary excretory pore; *ftp*, forming tertiary excretory pore; *mcd*, main collecting duct; *md*, merged part of the main collecting ducts; *pct*, posterior collecting tubule; *pp*, primary excretory pore; *sp*, secondary excretory pore; *tp*, tertiary excretory pore. From Podvyaznaya and Galaktionov (b, *submit*).

The walls of this pore represent a direct extension of the outer syncytial layer, which here loses the spines and secretory granules of all (or nearly all) types. The second excretory pore is connected to the lining of the excretory bladder by means of septate desmosome. In microphallid cercariae (see fig. 63B), the luminal surface of the excretory pore carries large finger-like extensions, protruding into the lumen of the duct and entering the cavity of the bladder. It also carries fibrous-granular glycocalyx, which fills the entire lumen. Around the outer opening of the excretory pore, the circular muscles of the tegumentary-muscular sac form a sphincter, which provides opening and closing of the pore.

The secondary, or definitive excretory pore starts functioning, as soon as the tail has been cast off together with the caudal part of the excretory system, i.e., during transition to the stage of metacercaria. However, in cercariae, lacking the caudal part of the excretory system, as in the above described microphallids, the secondary excretory pore is formed and starts functioning already at the final stages of cercarial morphogenesis. Rees (1977) found that in cercariae of *Cryptocotyle lingua*, the secondary excretory pore, based on its morphology, is capable of functioning along with the primary excretory pores, opening at the sides of the tail. In schistosome cercariae, the tegumental atrium, replacing the excretory bladder, in fact, represents an excretory pore (Powel, 1974). It is opened at the surface of the tegument and, as in the case of *C. lingua*, functions together with the primary excretory pores.

Of a special interest is the development of the excretory pore in bucephalid cercariae (Podvyaznaya and Galaktionov, b, *submit*) (see fig. 65E-G). In the course of morphogenesis of these larvae, a group of large cells becomes noticeable in the region of the merged main collecting tubes, which directly succeed the differentiating excretory bladder. These cells form extensions, which grow through the excretory syncytium of the duct, forming septate contacts with it. Apical parts of those extensions grow extensively, merge with each other and gradually form a continuous syncytial belt, which embraces the syncytium from the outside and separates it from the basal lamina. Inside the cytoplasm of the "belt", vacuoles appear, which form cavities, when merged. At first, these cavities open into the intercellular space between syncytia and then, during the extensive growth of the cytoplasmic "belt" they increase in size and, finally, open into the lumen of the duct. The excretory syncytium is interrupted and finally degenerates, being replaced by a new lining – the tegument of the excretory atrium, which, in fact, represents, the definitive excretory pore. In bucephalid cercariae, it is not connected with the tegument of the body, but is incorporated into the excretory system as a region, connecting the excretory bladder with the unpaired caudal vesicle (see above) (see fig. 62).

Unlike the digestive system, the excretory system of cercariae is a system of organs, actively functioning at this stage of ontogenesis of the hermaphroditic individual. Its main function is osmotic regulation. Apparently, in the excretory bladder resorption of material, coming with ultrafiltrate, takes place. The active state of the cytoplasm of the syncytium of the excretory bladder testifies to transport processes taking place there. The glycocalyx of the excretory pore probably has the same function. Similar to some forms of supramembrane complexes, it may play the role of an ion trap (Zavarzin et al., 1992).

2.1.7. Nervous System and Sense Organs

The nervous system and sense organs, closely associated with it, also compose a definitive system of organs, without which the life of cercariae would be impossible. The nervous system is represented by an orthogon, which is, however, less developed than that of the marita. The ganglion is horseshoe-shaped and is located at the level of the oesophagus or pharynx. Nerve cords go from it to the anterior and posterior ends of the larva body. Normally there are 3 pairs of posterior cords – ventral, dorsal and lateral, the former being the most developed. The number of commissures varies.

Sense organs of cercariae are represented by receptors of several types, the tegumentary receptors – sensory papillae – being the most common. They are numerous, regularly distributed over the body surface and are typical, apparently, for all cercariae (at least, all cercariae, which are released into the external environment). Therefore, chaetotaxy, i.e., the distribution pattern of the sensory papillae on the larva body, is used in taxonomy as well as for constructing different phylogenetic schemes (Richard, 1971; Bayssade-Dufour, 1979). A common and widely accepted classification system of sensory papillae has been presented by Richard (1971). According to this system, certain groups of sensory papillae are connected to commissures of a conventional scheme of the orthogon. Changes, later added to Richard's system (Bayssade-Dufour, 1979) are of refining character. They do not remove weak sides of this nomenclature, which was discussed in detail earlier (Galaktionov and Dobrovolskij, 1987). These are uncertainty of connections of particular groups of sensory papillae with transverse commissures and lack of differentiation of receptors, differing in structure and function. This should be kept in mind, when one deals with this nomenclature. Nevertheless, the works of Richard, Bayssade-Dufour and their followers made it possible to trace some general principles of evolution of the sense apparatus. In the series of trematodes, from the primitive to the higher ones, the total number of sensory papillae decreases. This is accompanied by ordering of their spatial arrangement and formation of rather constant groups, localized mainly at the anterior end of the body, within the taxa of supraspecies level (Fig. 66).

As mentioned above, the term "sensory papillae" embraces formations rather different in their ultrastructure (Matricon-Gondran, 1971; Bibby and Rees, 1971a; Tihomirov, 1980; Niewiadomska and Moczon, 1982; Pariselle and Matricon-Gondran, 1985). All of them share a common morphological scheme, similar to that described above for miracidia (see p. 20). The scheme is variable. The number of sensory cilia can vary (one or more). The cilium may be replaced with a bulb-like structure, lacking the classical set of microtubules. In some cases, the cilium may lack a rootlet. The set of the cytoskeletal rings and secretory granules may also vary (Fig. 67). Finally, along with a single sensory papilla, sensory formations in the form of ciliated fields, also characteristic for miracidia, can be seen (see p. 21). It is assumed that the sensory papillae play the role of mechanoreceptors and chemoreceptors (Smyth and Halton, 1983), but there is still no experimental proof of any function, associated with any particular type of sensory papillae (Halton and Gustafsson, 1996).

In cercariae of many species (Notocotylidae, Schistosomatidae, Allocreadiidae, Lepocreadiidae, Diplostomidae), alongside with sensory papillae, there are also eyespots, situated in the first half of the body, which are closely connected to the ganglion. In some species, these are large organs of vision, with several sensory cells of rabdomeric organization and cells of pigment cups. In other species, the eyespots and pigment cup, in particular, are less developed. Apparently, in different phylogenetic branches of trematodes, the eyespots in cercariae emerged independently (Galaktionov and Dobrovolskij, 1987). It should be noted that many cercariae, lacking eyespots, such as the echinostomatids *Echinostoma*, *Echinoparyphium* and *Himsthla* spp. (Vanoverschelde and Vaes, 1980; McCarthy, 1999; Fried, 2001; our observations), also react to changes in the illumination and direction of the light beam.

2.1.8. Reproductive System

In the course of cercarial morphogenesis, the primordium of the reproductive system emerges as a group of undifferentiated cells, which are localized in the parenchyma, in the region of the ventral sucker (see fig. 50, 77). In such a form, the genital primordium is found in mature cercariae of most trematode families – primitive (Notocotylidae, Echinostomatidae, Philophthalmidae) and higher ones (Strigeidae, Microphallidae). It is a dense aggregation of cells with small nuclei, rich in heterochromatin, and a small amount of cytoplasm. Yet, no primordia of separate organs can be observed. In larvae

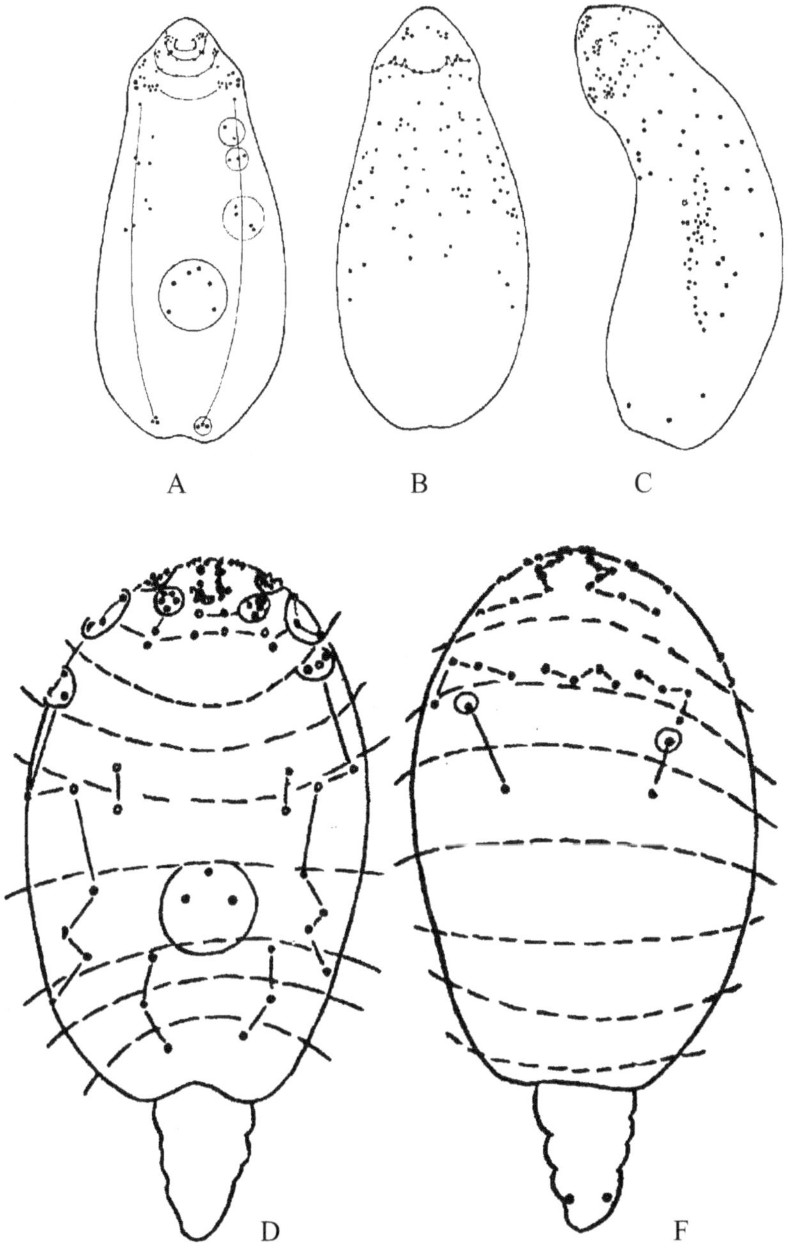

Figure 66. Distribution of sensory papillae over the cercarial body surface. A-C, *Echinostoma* sp. (Echinostomatidae) (A, ventral side; B, dorsal side; C, lateral side). From Kanev et al., 1987, reprinted with publisher's permission. D-E, *Cercaria agstaphensis IV* (Lecithodendrioidea) (D, ventral side; E, dorsal side).

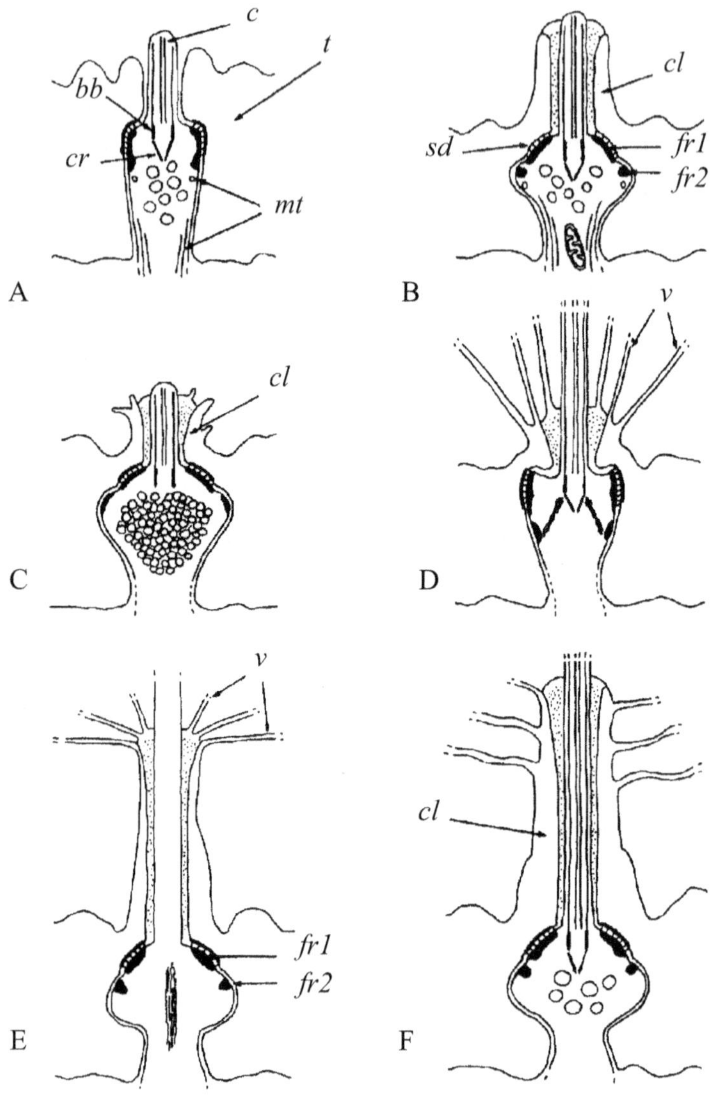

Figure 67. Ciliated sensory receptors of *Nicolla gallica* (Coitocaecidae) cercariae. A, simple receptor; B-F, sheathed receptors, the collar sheath originates from the tegument (B, smooth sheathed receptor; C, receptor with a short collar and short villi; D, receptor with a short collar and long villi; E, receptor with a long collar and long villi; F, receptor with a long branched collar). *Abbreviations*: *bb*, basal body; *c*, sensory cilium; *cl*, collar; *cr*, ciliary root; *fr1*, *fr2*, fibrous rings; *mt*, microtubules; *sd*, septate desmosome; *t*, tegument; *v*, villi. From Pariselle and Matricon-Gondran (1985), reprinted with the permission of Springer-Verlag GmbH & Co.KG.

of Strigeidae, Diplostomidae and some specialized Xiphidiocercariae, there is no distinct aggregation of primary reproductive cells.

At the same time, in other Xiphidiocercariae a clearly expressed tendency towards earlier isolation and differentiation of the reproductive primordium is revealed. This phenomenon was called "precocious development". The reproductive primordium of xiphidiocercaria has a C-like shape and lies dorsally from the ventral sucker (see fig. 50E). It is possible to distinguish there regions, which later would give rise to separate organs. The anterior, thickened part gives rise to a copulative organ and the metraterm. The middle part will become the uterus, and the posterior part, often widened at the end, becomes the future ovary. The primordia of testes and vitellaria in such larvae are not detectable.

Much rarer are the forms, where primordia of gonads, testes and ovary, are isolated from the common genital primordium (*Paralepoderma cloacicola* – see fig. 46A, *Astiotrema trituri*). Stages of ontogenesis of microphallids of the "*pygmaeus*" group, corresponding to cercariae, already have an almost fully differentiated reproductive system with all ducts formed (see fig. 77). Finally, in rare cases, e.g., in *Proterometra dickermani*, the reproductive system begins to function in the larva (Anderson and Anderson, 1963).

2.1.9. Some Trends in Morphological Evolution of Cercariae

Concluding the brief survey of formation of morphological and functional organization in cercarial ontogenesis, one can try to distinguish some common features of these processes in different trematode groups. A trend towards gradual simplification of morphological organization of larvae from primitive to the higher trematodes is distinctly pronounced. It is particularly manifested in the structure of the excretory and digestive systems, i.e., definitive systems of organs. As demonstrated above, the excretory formula of the Microphallidae and most Lecithodendrioidea corresponds to that of the embryos of plagiorchiid cercariae (see fig. 59). This is valid, if one compares Diplostomidae and Strigeidae with Cyathocotylidae (see fig. 60). A similar situation is observed, when primitive and the more specialized trematode families are compared. Among the former (Echinostomatidae, Psilostomidae, Fasciolidae, Paramphistomidae), numerous forms with many flame cells are always encountered. On the contrary, among the latter, even the initial number of flame cells, typical for their most primitive representatives, does not exceed 36 (6 pairs of "triads"). Only in some cases is the number of flame cells greater. This is an exclusion, rather than a common rule.

As a rule, simplification in the excretory system structure is accompanied by poor development of the primordia of the digestive system, which in specialized Furcocercariae and Xiphidiocercariae are represented at best by cords of small cells, slightly noticeable in fully-formed cercariae. Apparently, poor differentiation of the primordium of the reproductive system is initially characteristic for cercariae. This is suggested by the underdeveloped genital primordium in representatives of primitive families (Fasciolidae, Echinostomatidae, Notocotylidae). In more highly evolved trematodes, further development of the genital primordium can be impeded (many Diplostomidae, Strigeidae, some Microphallidae and Lecithodendrioidea). At the same time, in a number of plagiorchiids an opposite trend takes place, up to the differentiation of gonads and ducts of the reproductive system. This phenomenon, however, is not widely encountered among the species, whose cercariae are released into the external environment. However, the early formation and development of the genital primordium is, apparently, typical for ontogenesis of such trematodes, whose metacercariae are formed within the daughter parthenitae (Monorchiidae, Brachylaimoidea, Microphallidae). The aforementioned microphallids of the "*pygmaeus*" group provide a demonstrative example (see fig. 77).

The tendency towards underdevelopment of the definitive systems of organs in the course of cercarial morphogenesis is revealed not only when representatives of primitive and higher trematodes are compared, but also within rather weakly specialized families. Suffice it to compare cercariae of *Echinochasmus* with the larvae of other species of Echinostomatidae. A small size, underdeveloped digestive system, clearly simplified excretory formula; lacking or extremely poorly developed collar spines make up a set of characters, typical for cercariae of genus *Echinochasmus*. Most Echinostomatidae larvae possess alternative characters: large size, well-developed collar with large spines, well-expressed primordia of digestive system, where the lumen is often found, and, finally an extremely complicated excretory system with greater than a dozen flame cells.

Another trend is revealed in the evolution of provisional structures (e.g., stylet, glandular apparatus, tail) necessary in the long run for penetration of cercariae into the second intermediate or final host. This trend does not appear evident. In fact, in a cercarial series, from primitive to higher trematodes, cystogenous glands decrease in number and diversity and may even disappear. The penetration glands, which have emerged in the course of evolution, were initially represented by numerous small cells. In more highly evolved digeneans, their number decreases to 1-2 pairs. Nevertheless, it has nothing to do with underdevelopment of the glandular apparatus. Reduction of cystogenous glands is connected with the involvement of the second intermediate host in the life cycle and gradual development of interrelations in the host-parasite system second intermediate host-metacercaria (see p. 309). The penetration glands, as mentioned above, decrease in number, but this is accompanied by significant specialization of remaining cells. Besides, in cercariae of more highly evolved trematodes, different types of specialized tegumental glands are formed. Other provisional structures also develop and become better suited for their task. For example, this is the case with different modifications of the tail, emergence of the anterior organ, transformation of the tegumental spine into a stylet.

Thus, the obvious simplification of morphological organization of cercariae of more highly evolved trematodes is connected with underdevelopment of some definitive systems of organs (digestive, excretory and reproductive). In fact, larvae of the most specialized species correspond to embryonic stages of more primitive forms in their morphology and level of development. We call them juvenilized larvae (Galaktionov and Dobrovolskij, 1987). Juvenilization is usually accompanied by miniaturization. To be sure, it is sufficient to compare the cercarial body size in the representatives of trematode families, occupying different position on the phylogenetic tree (see the description of cercarial morphogenesis in the beginning of this section). At the same time, the process of juvenilization does not affect the provisional structures, which, on the contrary, are improved in the course of cercarial evolution and thus facilitate accomplishment of the main task of that stage of ontogenesis, i.e., entering the second intermediate or the final host.

Another trend – imaginization of larvae, i.e., approaching the organization level of the adult organism, is much rarer. Imaginization is, apparently, a secondary process. In most cases it develops on the basis of previous juvenilization. Comparing cercariae of Echinostomatidae and Psilostomidae to xiphidiocercariae, one can distinctly trace the total simplification of the latter. This also applies to the forms with the differentiated genital primordium.

2.2. Metacercaria

In the course of evolution of the trematode life cycles, the metacercaria phase was the last to emerge (Ginetsinskaya, 1968; Pearson, 1972; Gibson, 1987; for details see section 1.2. in chapter 6). Apparently, this determined the surprising diversity of metacercariae encountered in different trematode groups. On one hand, this diversity reflects the level of relations of the larvae with the external environment or with the organism of the second intermediate host. It also reflects the trends in the morphological evolution of hermaphroditic generation of a particular group of digeneans. In this connection, we must deviate from the scheme of presenting the material (generalized description of the development of organ systems in the considered phase) that has been used in previous sections of this chapter. Analysis of metacercariae will be undertaken in different groups, distinguished on the basis of common biological characteristics. To a certain extent, boundaries between these groups correspond to the main stages of evolution of hermaphroditic generation.

2.2.1. Encystment

In the life cycles of a number of trematodes (i.e., Fasciolidae, Philophthalmidae, Notocotylidae, Echinostomatidae), the adolescaria phase is retained. It is the term which is traditionally used for metacercariae, encysting in the environment. During encystment the cercaria sheds the tail and the cystogenous secretion is released outwards from its tegument. On the surface of the larva the secretion polymerizes, forming a dense envelope – the cyst. As a rule, adolescariae cysts are multilayered. Mechanically they are very solid and impermeable to water and dissolved chemical substances. Cyst layers differ in their structure and chemistry, which is explained by their formation from the cystogenous secretion, synthesized in different types of tegumental glands. In Philophthalmidae, their secretion passes into the outer layer of the tegument at final stages of cercaria morphogenesis (see section 2.1.4. – "sheath-forming" glands) (Thakur and Cheng, 1968; Žd'árská, 1971a; Tihomirov, 1980) (Fig. 68A). The secretory granules of different types do not intermingle but occupy a certain position. Normally they are localized above the site, where tegumental cells, which had synthesized the secretory granules, are located. If several types of secretory inclusions get into some part of the outer layer of the tegument, they exhibit a stratified distribution. This is associated with their successive transportation. Because of this the granules, first to enter the tegument, are concentrated in the apical cytoplasm of the syncytium. Below are the granules, which are second to appear. Thus, the number and sequence of the cyst layers corresponds to the distribution pattern of secretory granules in the outer layer of the cercaria tegument.

In adolescaria of Notocotylidae (Žd'árská, 1970; Southgate, 1971), Echinostomatidae (Žd'árská, 1968, 1971b; Laurie, 1974) and Fasciolidae (Dixon, Mercer, 1967), the cyst is formed in a different way (Fig. 68B, 69). Numerous tegumental glands of these forms develop during cercaria morphogenesis. However, the secretion of only some of these glands gets into the outer layer of the tegument. Then their cytons degenerate. This secretion contributes to the formation of the primary cyst. The remaining glands are connected to the outer syncytial layer of the tegument of the encysted larva. They then transfer the secretory granules, which they have already synthesized. The latter are released into the cyst cavity and used to form its inner layers. Thus, the primary cyst becomes an outer layer (or layers) of the fully formed cyst. The number of layers and their chemical composition in the cyst of adolescariae might differ in regions that are directed outwards and

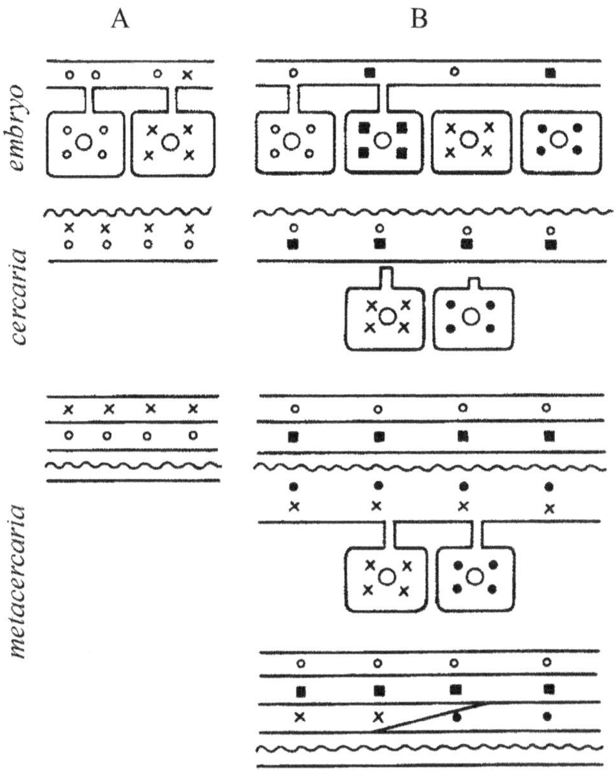

Figure 68. Diagrams showing transformations of the tegument during larva encystment. A, Philophthalmidae; B, Echinostomatidae. From Galaktionov and Dobrovolskij (1987).

facing the substrate (Fig. 69). This is also determined by the distribution pattern of secretory granules of different types in the larva tegument.

With the involvement of the second intermediate host in the trematode life cycle, the multilayered cyst is not needed any more. Moreover, in this case it prevents the growing parasite from obtaining food from the host. Therefore, in larvae of more evolutionarily advanced digenean groups, the cyst is more or less reduced, the initial stages of this process being observed already in the Echinostomatidae. The cyst becomes more thin-walled. The number of its layers decreases to 2 in the most common cases (Heterophyidae, Renicolidae and many Plagiorchiida). The shedding of the tail and encystment are shifted in time. Between these two events, more or less prolonged periods of penetration and migration (if there is any) in the body of the second intermediate host are inserted. Formation of the cyst takes places only after the larva has reached the place of its final settlement. In Heterophyidae, both cyst layers are formed by a sequential discharge of the secretion of tegumental glands from the outer layer of the tegument, where the

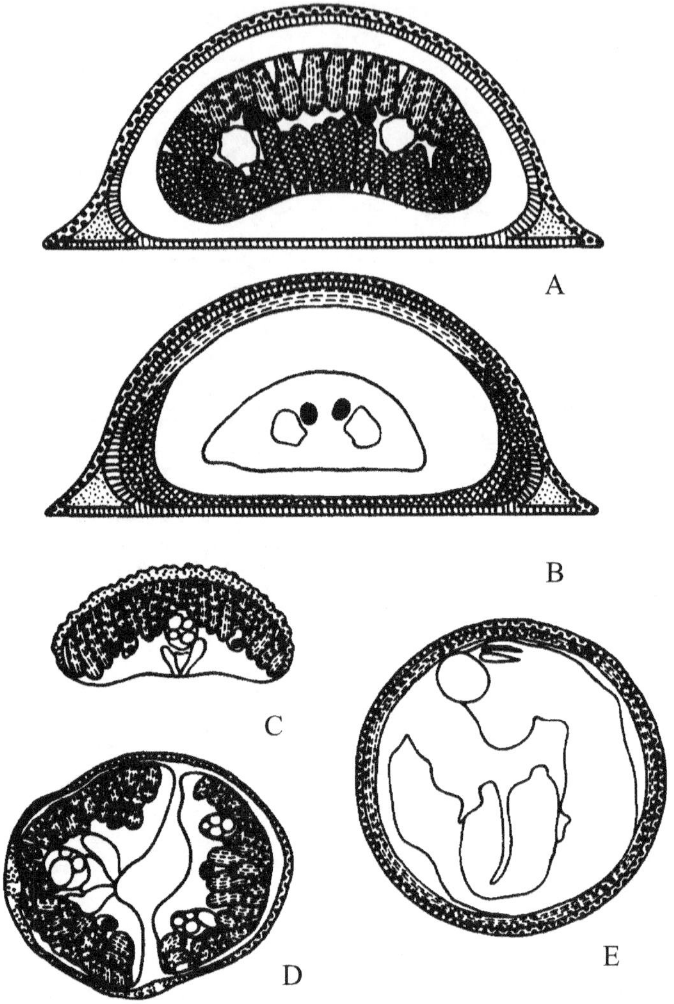

Figure 69. Successive stages of cyst formation in *Notocotylus attenuatus* (A-B) and *Echinoparyphium aconiatum* (C-E). From Žd'árská (A-B, 1968; C-E, 1970), reprinted with publisher's permission.

secretion had accumulated (Rees and Day, 1976; Day, 1976; Stein and Lumsden, 1971a,b). This resembles the above described process for Philophthalmidae with one major distinction. In Philophthalmidae, similar to adolescariae of other trematodes, every layer of the cyst is formed from the secretion, synthesized by tegumental glands of one type. In Heterophyidae, a two-layered cyst is formed by the secretion of four types of tegumental glandular cells. In other words, every layer is composed of more than one type of secretion. Thus, the principle "one type of secretion – one layer of the cyst", typical for adolescariae, does not apply. Such a way of cyst formation is characteristic of metacercariae of many trematodes.

In plagiorchiids the primary monolayered cyst is formed from the discharge of the secretion, contained in the outer syncytial layer of the cercarial tegument. This is either the secretion of mucoid glands only (see section 2.1.4. – mucoid glands), or there is also the secretory material, produced by other tegumental glands (Žd'árská, 1969a; Bock, 1988, 1989). In the latter case, unlike in cercariae, which later turn into adolescariae, the secretory granules of different types demonstrate diffuse, rather than stratified, distribution in the syncytial layer. After the outer layer of the tegument has been released from secretory inclusions, used to form the primary cyst, other tegumental glandular cells (usually of one type) become connected to it. These are cystogenous cells, according to the classification of Ginetsinskaya and Bhutta (1976) (see section 2.1.4.). Formation of the second, inner layer of the cyst, which includes proteins and carbohydrates, is associated with the secretion of these cystogenous cells. It has been proved experimentally that the functions of the outer and inner layers of the cyst are different – the former provides chemical stability of the cyst, whereas the latter is responsible for mechanical resistance (Bock, 1989).

Microphallids, one of the most specialized Plagiorchiida, secondarily acquire a multilayered thick-walled cyst. However, its similarity with adolescariae cysts is only superficial. Both sources of material for different layers and the entire process of cyst formation in microphallid metacercariae differ from those described in primitive trematodes (Strong and Cable, 1972; Alderson, 1975; Benjamin and James, 1987; Galaktionov et al., 1996, 1997). The process itself in microphallids (Fig. 70) is very prolonged. A thin monolayered cyst is found around the larvae only some days after they have penetrated into the second intermediate host, a crustacean. The cyst is formed from the secretion of one type of tegumental cells (so-called "sheath-forming" glands – see p. 126). During subsequent days, the secretion, produced by the syncytium of the excretory bladder (see below), matrix of the larval tegument, in the form of membrane-bound vesicles of different shapes and sizes (the product of tegumental cells), and copious cytoplasmic fragments produced by the tegument of the anterio-ventral region of the metacercaria enter the cyst lumen. All this material, heterogeneous in its origin, is accumulated near the inner surface of the primary cyst. Later (30-40 days after the infection of the crustacean) the material is polymerized, forming 2-3 inner layers (up to 3 sublayers can be distinguished in each layer). The latter of them, which directly adjoins the larva, exhibits a regular striation – the so-called crystalloid structure, also typical for tegumental spines. This enables one to make a solid assumption that this layer, like the spines, is formed by appropriately packed actin microfibers (Galaktionov et al., 1997).

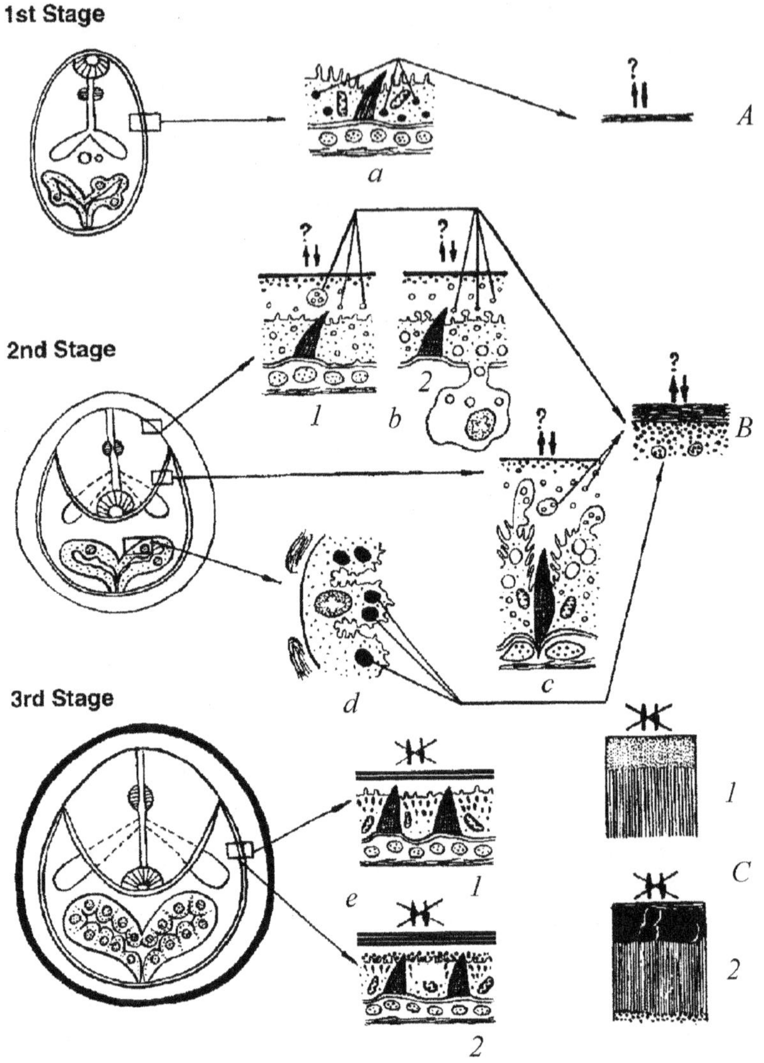

Figure 70. Successive stages (1-3) of cyst wall formation and origin of its layers during development of microphallid metacercariae. *a*, tegument of non-encysted metacercaria containingsecretorygranules; *b*, tegument of the developing metacercaria and accumulation of cystogenous matter in the cyst lumen (*1*, *Maritrema subdolum*, *Levinseniella brachysoma* , *Microphallus claviformes*; *2*, *Maritrema arenaria*); *c*, anterior ventral tegument of the developing metacercaria showing transport of cytoplasmic fragments into the cyst lumen; *d*, excretory bladder syncytium of the metacercaria in early stages of development, granules with cystogenous function are shown; *e*, tegument of fully formed microphallid metacercaria (*1*, Maritrematinae; 2, Microphallinae); *A*, primary cyst wall; B, accumulation of cystogenous matter under the primary cyst wall; *C*, cyst wall around fully formed metacercariae (*1*, Maritrematinae; *2*, Microphallinae). Arrows indicate direction of nutrient and/or excretory diffusion. From Galaktionov et al. (1997), reproduced with permission from Journal of Helminthology, 1997, p. 18 © CAB International.

Inner layers of the microphallid cyst, unlike the outer one, are mechanically resistant and impermeable to chemical substances.

The reasons, which determined the emergence of multilayered cysts in microphallids, are still unclear. It is possible that they may prolong survival of infecting metacercariae in the second intermediate host. It should be stressed that encysted metacercariae of many trematodes are surrounded by a capsule above the cyst. The capsule is formed in response of the host to penetration of non-self body (encapsulation reaction). Often, in such cases, authors erroneously refer to multilayered cysts. This is not correct, for the cyst is produced by the larva itself and the capsule is produced by the host.

In many trematodes, whose life cycle has lost the second intermediate host, the process of encystment is reduced. There is no cyst in Brachylaimoidea metacercariae, whose morphogenesis proceeds within the sporocyst, parasitizing terrestrial molluscs. However, the tegument of these larvae produces a thick glycocalyx, which must functionally replace the cyst. The glycocalyx is hygroscopic (this can be considered as an adaptation to the terrestrial mode of life), and also protects the larva from digestion by hydrolytic enzymes of the final host (Harris et al., 1974; Žd'árská et al., 1982; Bakke, 1982; Žd'árská and Soboleva, 1984). A sequential transition to the development of metacercariae within daughter sporocysts has been traced in Microphallidae (Belopolskaya, 1963; Deblock, 1977; Galaktionov, 1993). Not considering morphogenetic modifications, accompanying this process (they will be discussed below), we mention that in the larvae, encysting in daughter sporocysts, the number of cyst layers is diminished to 2-3, and in the most specialized microphallids of "*pygmaeus*" group no cyst is formed.

Inmosttrematodes with furcocercariae (Strigeidida, Schistosomatida, Clinostomidae, Gymnophallidae), the origin of protective envelopes around metacercariae, if there are any, differs from that of the larvae of the above considered digenean groups. There are no such structures in all Schistosomatida. This is, undoubtedly, associated with their transition to blood parasitism (see below). Gymnophallidae metacercariae, actively migrating in the lacunas of the haemocoel of their second molluscan host, also lack these structures. The majority of known strigeidid larvae are "coated" with a well-expressed envelope, which is nearly always called "the cyst" (Sudarikov, 1971, 1984). It can be either multilayered and thick (*Cyathocotyle*, *Strigea*, *Apatemon*), or, on the contrary, thin and transparent (*Codonocephalus*, *Cotylurus*, *Bolbophorus*). There is no cyst in a number of representatives of Diplostomidae (*Diplostomum*, *Tylodelphys, Austrodiplostomum*). Investigations on metacercariae *Posthodiplostomum cuticola* (Berezantsev and Dobrovoskij, 1968) showed that the so-called "cyst" is a typical fibrous capsule, which is formed from fibroblasts and pigment cells of the host (fish). This capsule is multilayered, its inner layer being hyalinated.

A similar structure of the cyst around metacercariae of *Clinostomum marginatum* (Clinostomidae) was described by electron microscopy (Larson et al., 1988). This formation also represents a typical capsule of connective tissue, composed of irregular layers of collagen fibers, with flattened fibrocytes between them. The inner hyaline layer has not been observed. However, in addition to *P. cuticola*, it has been detected in other strigeidids and is called "inner primary cyst wall" to distinguish it from the outer fibrous coat, i.e., the capsule itself. Formation of the primary cyst wall is thought to be associated with vesicular secretions of the tegument of developing metacercariae (Mitchell, 1974; Meade and Garsa, 1985). This may be the secretion of tegumental glands, developing in the course of metacercarial morphogenesis, as it occurs in the above-mentioned microphallids. There are no cystogenous tegumental glands in mature strigeidid cercariae (see section 2.1.4.). In metacercariae of *Mesostephanus appendiculatus* (Cyathocotylidae), investigated by us, specialized glands of the second group act as cystogenous glands (see section 2.1.4.) rather than by tegumental cells. These glands are already formed in cercariae and open at its body surface by independent pores. The secretion of these glands is discharged on the surface of the larval body soon after penetration into the fish muscles and forms a thin primary cyst. Then around that primary cyst a capsule is formed from fish fibroblasts and lymphocytes.

Thus, the envelope around strigeidid metacercariae (at least, its main part) originates from the host's tissues and should be called a "capsule". The extent of its development largely depends on the location of parasite settlement. Thick multi-layered capsules are found around parasites, localized in vertebrate tissues with a high level of immunological activity (muscles, mesentery). Larvae, settled in the material of the eye crystalline, vitreous humor, and nervous tissue (*Diplostomum*, *Tylodelphys*) are deprived of protective envelopes. The latter are also absent in metacercariae of *Pulvinifer*, parasitic in the blood vessels of oligochaetes. Relative to the cyst itself (only primary cyst wall can be regarded as a cyst), it is thin-walled and is masked by a capsule of connective tissue. Apparently, most Strigeidida either have lost the encysting capacity, or have not acquired it (see p. 494). This may be associated with transformation of the capsule into a structure, which not only provides protection of the developing parasite from immune reactions of the host, but also facilitates intake of nutrients. The capsule walls in the aforementioned larvae of *Posthodiplostomum cuticola* are densely intertwined with blood vessels, which can be considered as a foreign reaction of the host tissue to the presence of the parasite (Berezantsev and Dobrovolskij, 1968). In fact, one deals with transformation of the capsule into an organ-like structure, similar to organ-like capsules of *Trichinella* nematodes, investigated in detail (for reviews see: Berezantsev, 1974; Britov, 1982). In addition to strigeidids, in many trematode species, the capsule of connective tissue (lacking, however, any organ-like structure) is formed around encysted metacercariae as a result of the reaction of host's defense systems to the penetration of non-self object.

The formation of the capsule from fibrocytes of the host is typical not only for strigeidids and clinosotomatids, but also for bucephalid metacercariae, parasitic in fish tissues. However, in the latter case, the fibrous capsule is developed around the genuine thick-walled cyst. In Bucephalidae metacercariae, this cyst is produced by the secretion of several types of tegumental glands. Some of these glands are already developed in the course of cercarial morphogenesis (Higgins, 1980).

2.2.2. Development

Discussing the second preliminary ontogenetic stage of individuals of hermaphroditic generation in different trematode groups, it should be stressed that in the forms with adolescariae, this stage is practically not expressed. This is easily explained, keeping in mind that these larvae are deprived of exogenous sources of energy. Their tegument undergoes the most profound modifications. During encystment, granules of the cystogenous secretion disappear from the outer layer of the tegument, which, thus, becomes thinner. In the forms, whose cercariae have cystogenous glands, the latter are resorbed after the cytons have been emptied. In fact, in adolescariae of Fasciolidae, Philophthalmidae, Notocotylidae all modifications terminate at this point. However, one cannot exclude the possibility of development of some provisional structures, connected with excystment in the definitive host. Thus, in *Fasciola hepatica* an important role in this process is played by the secretion, contained in the adolescaria gut (Sukhedo and Mettrick, 1986). But it is still unclear, if its synthesis takes place at the stage of adolescaria or in the course of cercarial morphogenesis in the first intermediate host.

In any case, emergence of the second preliminary stage, as a separate part of ontogenesis of individuals of the hermaphroditic generation, is associated with the involvement of the second intermediate host in the life cycle; hence, with the emergence of the metacercarial phase. Energy resources of the host made possible the transition to this stage of some (and in a number of cases of practically all) morphological and functional modifications, proceeding in the course of adult (marita) development in the definitive host (maritogony). In addition, some provisional structures, providing normal function of the larva in the organism of the second intermediate host, began to develop.

The first steps towards the onset of the second preliminary stage in the ontogenesis of individuals of hermaphroditic generation can be traced in echinostomatids. In some of them, modifications at the metacercarial stage are similar to the aforementioned in adolescariae. However, development of the collar with spines and organs of digestive system in *Echinochasmus* already manifests some progress compared to cercariae. In species of the genera *Echinostoma* and *Echynoparyphium* considerable metacercarial growth takes place. It is accompanied by differentiation of the genital primordium, formation of ovary, testes and even vitellaria (Ginetsinskaya and Dobrovolskij, 1964). Such type of metacercarial morphogenesis, during which completion of formation of definitive organ systems sequentially proceeds, first appears in echinostomatids, and is then retained in most trematode groups, excluding strigeidids, which follow genuine metamorphosis (see below). A particular type of morphogenetic processes, characteristic for certain species, is determined by the initial level of cercarial development and final level of development of the mature metacercariae.

In most metacercariae, investigated up to now, after encystment or sometimes also prior to encystment, during migration in the host body (Diplostomidae, Microphallidae), the outer layer of the tegument gradually thickens, its surface becomes folded and may form extensions, microvilli and/or microlamellae (Bibby and Rees, 1971a; Stein and Lumsden, 1971a,b; Strong and Cable, 1972; Popiel, 1976; Halton and Johnston, 1982; Galaktionov et al., 1996; Podvyaznaya, 1999). The number of mitochondria and different vesicles in the cytoplasm increases. All these facts testify to the active transport of nutrients through the tegument, which in some cases were experimentally confirmed (Bibby and Rees, 1971a,b; Higgins, 1979; Halton and Johnston, 1982; Uglem and Larson, 1987; Larson et al., 1988). A thin-walled cyst (if any) probably remains permeable for nutrients al least during cercaria morphogenesis, since the transport is not impeded.

In microphallid larvae the maximal development of the above-mentioned cytoplasmic structures of the outer syncytium, associated with the active transport, coincides in time with the period of the most important morphogenetic modifications of internal organ systems (Galaktionov et al., 1996) (Fig. 71). Metacercariae of that age are surrounded only with a thin monolayered cyst. As mentioned above, formation of its inner layers, impermeable for chemical substances, takes place later. By this time, the outer surface of the tegument is smooth. Microlamellae either decrease in number

Figure 71. Successive stages of tegument formation during development of microphallid metacercariae. A, *Levinseniella brachysoma*; B, *Microphallus claviformis*; C, *Maritrema subdolum*; D, *M. arenaria.* For A-C: a, day-old metacercaria; b, 8-days-old metacercaria; c, 16-days-old metacercaria (1 – tegument of the mid-body; 2 – tegument of anterior ventral region); d, 30-days-old metacercaria (1,2 – same); e, 42-days-old metacercariae of *L. brachysoma* (A), 30-days-old metacercariae of *M. claviformes* (B) and *M. subdolum* (C). For D: a, metacercaria less than 18 h post infection; b-e, next stages of tegument development. From Galaktionov et al. (1996), reproduced with permission from Journal of Helminthology, 1996, p. 203 © CAB International.

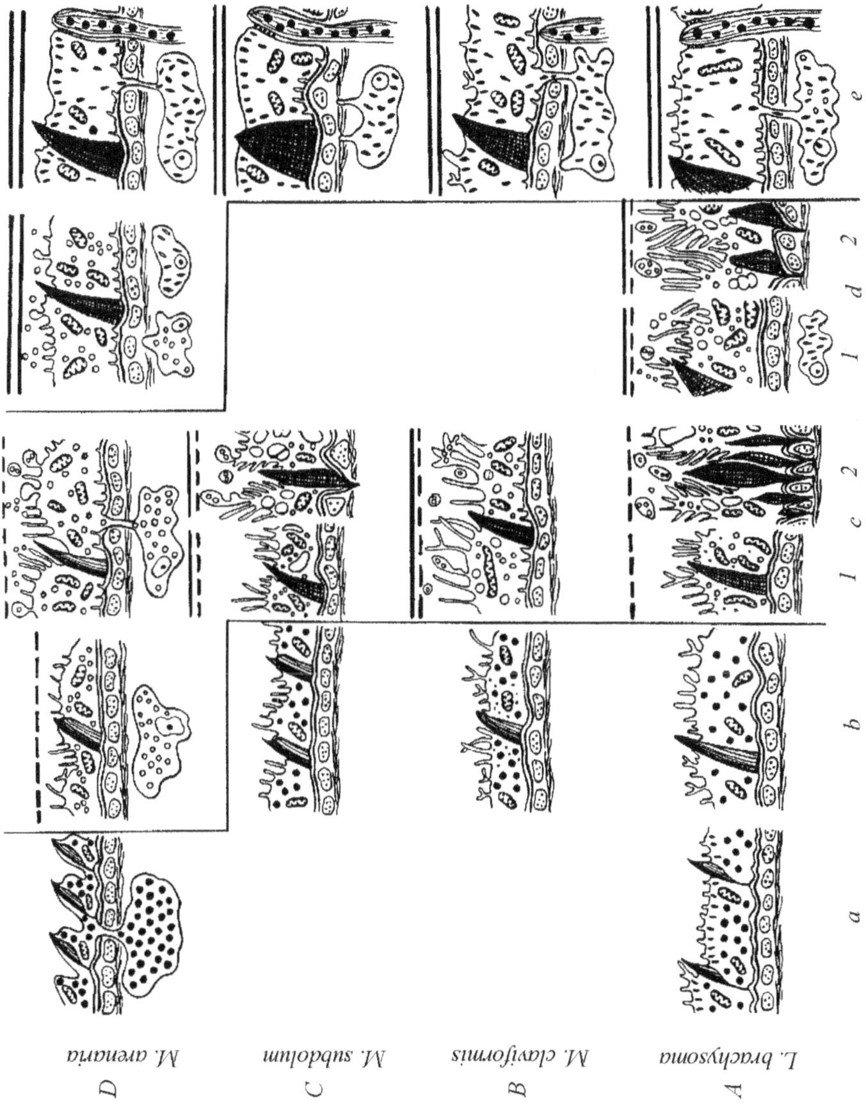
A L. brachysoma
B M. claviformis
C M. subdolum
D M. arenaria
a
b
1 c 2
1 d 2
e

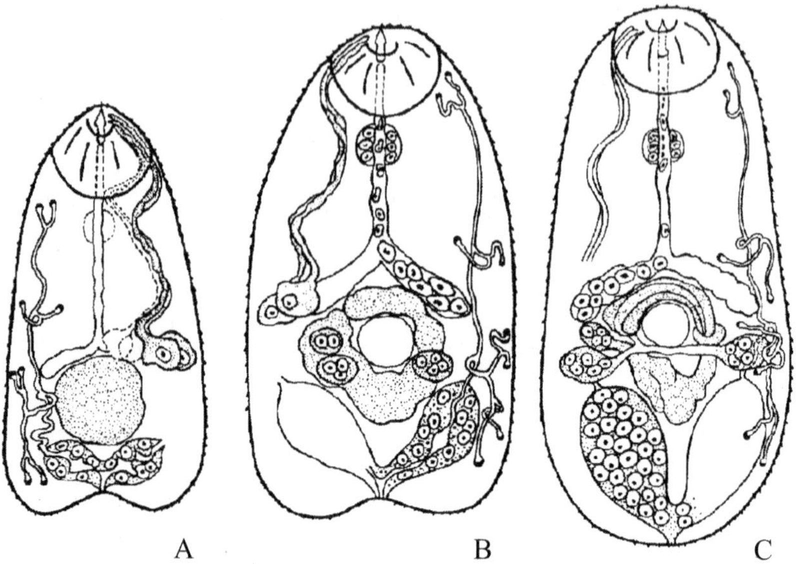

Figure 72. Successive stages of *Maritrema subdolum* metacercarial development. From Galaktionov (1993). A, day-old metacercaria; B, 9-days-old metacercaria; C, 11-days-old metacercaria.

and size, or completely disappear. The number of mitochondria in the cytoplasm diminishes and the secretion of tegumental glands, presumably associated with the formation of the adult glycocalyx, appears (see below). The tegument acquires a "marita-like" structure and movement of nutrients through it stops. A similar structure in the tegument of mature metacercariae has been reported for other trematode species.

2.2.2.1. Morphogenesis of "Smooth" Type

Morphogenesis of the "smooth" type has been investigated in detail in plagiorchiid metacercariae (Fig. 72). Those, developing from cercariae with a high level of morphological maturity (see p. 149), do not undergo any major morphological modifications (Fig. 73). They grow a little and lose provisional structures (stylet, penetration glands, cystogenous glands, virgula, caudal pockets). The digestive system changes only in those cases, when it lacks a lumen in the cercariae. Primarily, it is characteristic of juvenilized cercariae, e.g., microphallids. Their primordial digestive system consists of cords of small cells, which can hardly be distinguished from the surrounding cells. Formation of a lumen in the organs of the digestive system is connected with vacuolization of cells and/or syncytial cords (see p. 130). In microphallids, an important part in formation of the foregut lumen is also

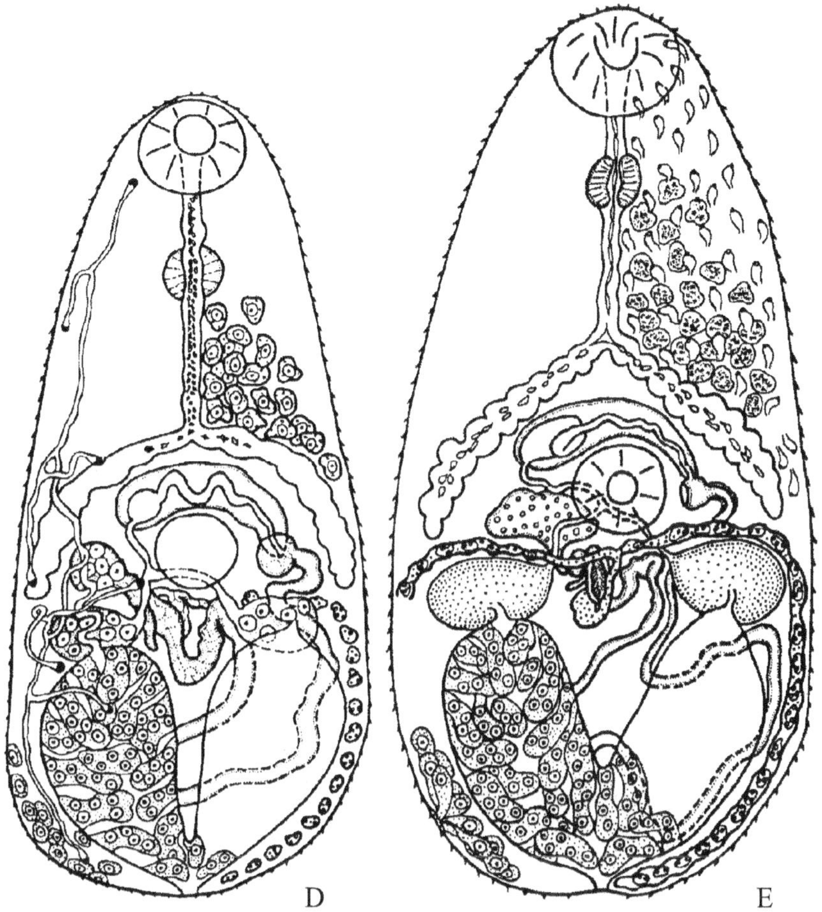

Figure 72. (Continued). D, 16-days-old metacercaria; E, fully formed metacercaria (27-days-old).

played by deep invaginations of surface plasma membrane of the tegument of the anterior part of the larval body. They grow from the front side into the central part of the syncytial cord of the primordial gut and divide it into a system of branched gaps (Galaktionov and Malkova, 1993) (Fig. 74A). A common cavity appears when the boundaries between gaps are destroyed and vacuoles, which are formed during partial degeneration of the syncytium cytoplasm, merge (Galaktionov and Malkova, 1993). Such a sequence of events in the course of the foregut morphogenesis may take place in other trematodes, but these processes have not been investigated by electron microscopy.

The formation of the lumen in the branched caeca proceeds quite differently. In young microphallid metacercariae, the primordial caeca are

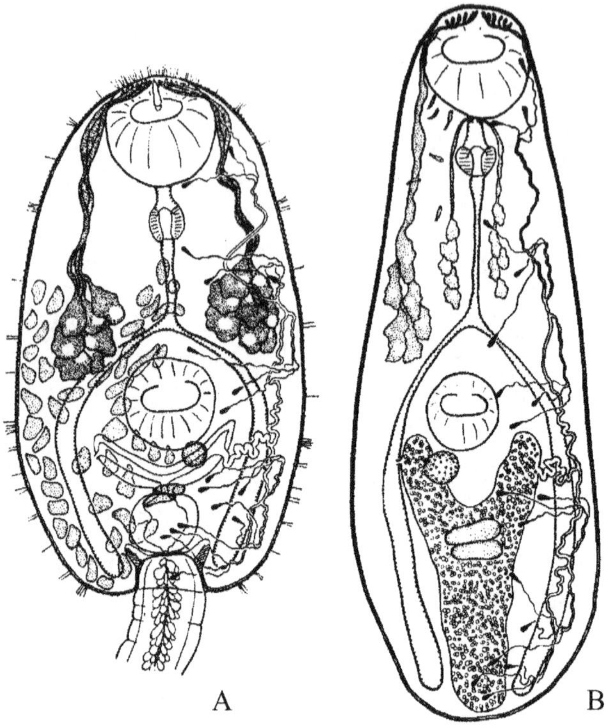

Figure 73. Larvae of *Opisthioglyphe ranae.* A, cercaria, B, metacercaria. From Galaktionov and Dobrovolskij (1987).

formed by large polygonal cells (Fig. 74B), organized in the form of a multi-row cord. Later the cell surfaces, contacting each other in the central part of the primordium, form shallow invaginations with short extensions and lamellae. Between such zones of neighboring cells, narrow cavities are revealed. During further development of the larva, they widen, merge and form a common lumen of the branched caeca. Cells, composing the caeca, are initially specialized as glandular ones. In addition to other cell organelles, they have a well-developed Golgi complex, occupying the major part of the perinuclear cytoplasm. Apparently, synthesis of the secretion, accumulating in the intestinal epithelium, is associated with this organelle. In fully formed larvae, Golgi complexes are not detected in the gastrodermal cells, and the secretion is concentrated in the apical cytoplasm. The secretory activity of the metacercarial gut is also observed in other plagiorchiids (Bock, 1989). The function of its secretion is presumably connected with the process of excystment (see section 2.3. of this chapter).

It should be noted that in the course of morphogenesis of cercariae and metacercariae of Philophthalmidae and Echinostomatidae, lumina in the

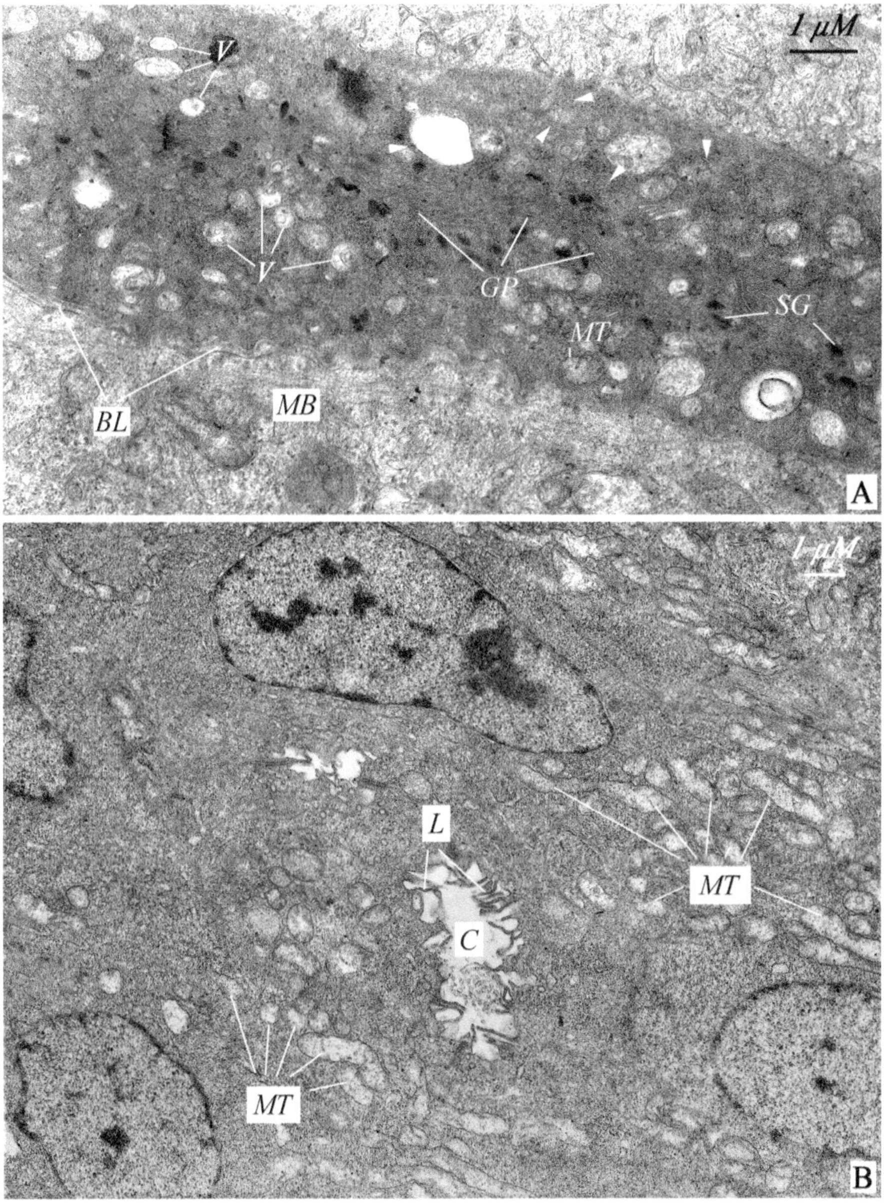

Figure 74. Ultrastructure of developing alimentary tract of microphallid metacer caria *Levinseniella brachysoma*. A, formation of the foregut lumen; B, formation of the caecum lumen. *Abbreviations*: *bl*, basal lamina; *c*, the cavity between adjacent gastrodermal cells; *gp*, gaps in syncytial layer of the foregut primordium; *l*, lamella; *mb*, muscular bundles; *mt*, mitochondrion; *n*, nucleus of gastrodermal cell; *sg*, secretory granules; *v*, vacuole. Arrows indicate microtubules.

branched caeca are formed differently (Tihomirov, 1980; our data). The primordia are represented by a one-row cord of cells, whose cytoplasm gradually becomes heavily vacuolated. The common cavity is formed as the result of merging of intracellular vacuoles and partial destruction of plasma membranes between neighboring cells. Thus, in different trematode groups, the lumen in the branched caeca is formed in different ways.

The excretory system of metacercariae of the plagiorchiid complex does not undergo significant changes, independently of the extent of morphological maturity of cercariae. Most often, the number of flame cells does not increase; their arrangement also changes only slightly (see fig. 72). The most important changes proceed in the excretory bladder. Its significant growth, as well as accumulation in the bladder of small rounded bodies, which are considered as insoluble metabolic products, can be detected by light microscopy. It is presumed that due to impossibility of excreting metabolic products from the organism after encystment, metabolism undergoes partial modifications so that insoluble final metabolites are yielded. It is possible that such events take place in a number of metacercarial species. However, electron microscopy of these larvae revealed hypertrophy of the lining of the excretory bladder, where cytoplasmic organelles, connected with biosynthesis (unattached ribosomes, cisterns of RER) are well-developed (Powell, 1977; Popiel, 1977; Smyth and Halton, 1983; Orido, 1990a; Galaktionov and Malkova, 1995). It is presumed that excreta are absorbed from the bladder cavity and transformed in its lining into a non toxic form because of the involvement of protein complexes. There is no direct proof of this view point.

Especially complicated morphological and functional modifications proceed in the excretory bladder of microphallid metacercariae (Galaktionov and Malkova, 1995) (see fig. 72, 77). At the initial stages of metacercarial morphogenesis it functions as a glandular organ and produces a large amount of cystogenous secretion. After the latter has been discharged (see above), the syncytial lining of the bladder forms numerous columnar projections, directed into the cavity of the bladder (Fig. 75). These projections grow intensively and the number of their nuclei increases. The size of nuclei decreases significantly and some of them are resorbed. Cytogenetic mechanisms of this process are unclear. Apparently, it is just this nuclear material, which is subjected to elimination that was taken for insoluble products of metabolism, accumulating in the excretory bladder. Simultaneously an intensive development of RER takes place in the projections, and numerous mitochondria appear. This suggests the synthesis of large amounts of some protein/proteins, which can be used to neutralize metabolites. Precipitation of the latter in the form of granules does not occur. Most likely, they are accumulated in non-structured form (possibly, protein complexes) in the cytoplasm of the columnar extensions. Therefore, the excretory bladder of mi-

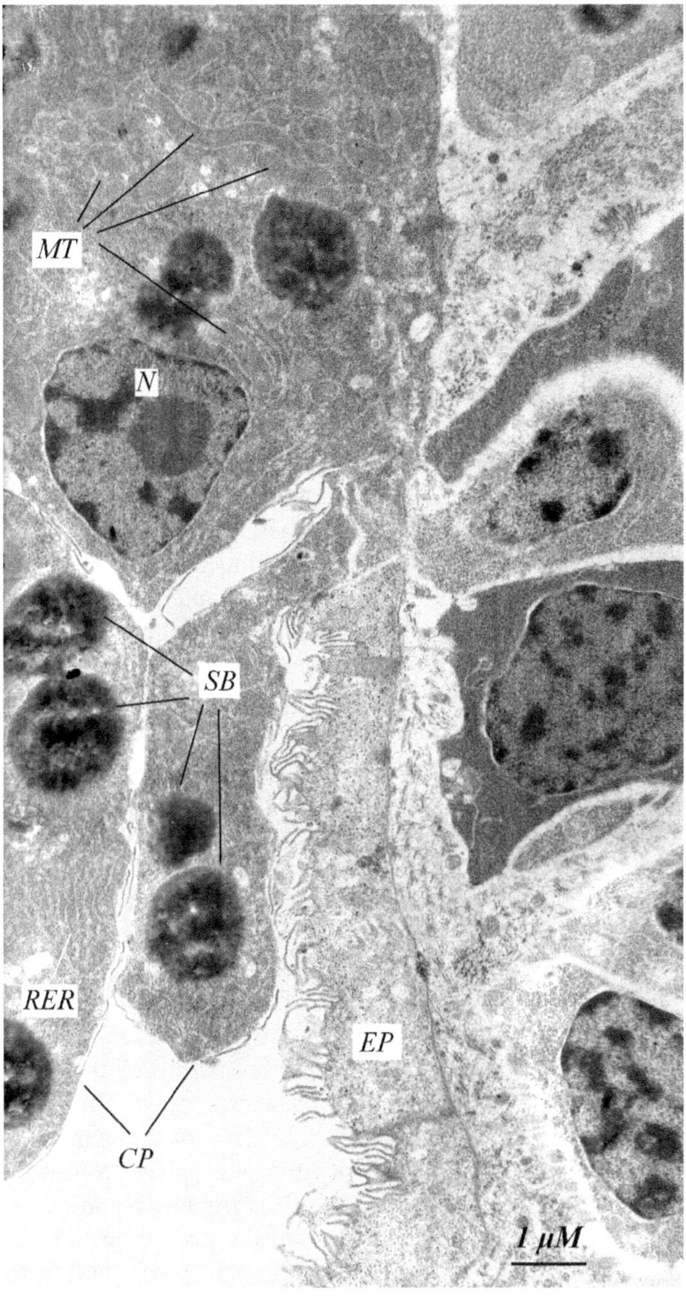

Figure 75. Columnar projections of the excretory bladder of a 16-days-old microphallid metacercariae of *Levinseniella brachysoma*. *Abbreviations*: *cp*, columnar projections; *ep*, excretory pore lining; *mt*, mitochondrion; *n*, nucleus of the excretory bladder lining; *RER*, rough endoplasmic reticulum; *sb*, spherical bodies (=resorbed component of the nuclear material).

crophallid metacercariae (and, possibly, of some other trematodes) should be considered as a sort of biological filter, where detoxication of excretory liquid takes place, so that later it can be reutilized by the organism. Apparently, this provides the possibility of osmotic regulation in encysted larvae.

Changes in glandular apparatus are associated both with the further development of such glandular complexes that has already existed in cercariae and with the appearance of new ones. The formers include the above mentioned "metacercariae" glands, which are often confused with penetration glands. It is this confusion, which accounts for the idea that penetration glands are retained in a number of plagiorchiid metacercariae. In a number of cases such a view point has a good foundation. Thus, in the course of penetration into crustaceans, microphallid cercariae use only the secretion of the 2 anterior pairs of penetration glands, whereas the 2 posterior pairs of glands are retained during the entire migration period, lasting up to 10 days. Their secretion is used gradually and, along with the stylet, apparently, it facilitates migration of the larva in the host's tissue. Glandular formations, emerging *de novo* in metacercariae, may include: often encountered groups of unicellular glands, whose ducts open into the anterior part of the oesophagus; numerous glandular cells at the anterior end of the body of microphallid metacercariae; and compact groups of glands in the posterior-lateral regions of the body (see fig. 72, 73, 77). A common feature for all these structures is that they function at the stage of marita. At the last stages of microphallid metacercariae ontogenesis tegumental glandular cells, producing small disc- or rod-like electron-dense granules, are observed (see fig. 71). They pass into the outer syncytium and are localized in its apical part. Such secretory inclusions are very typical for the tegument of trematode adults. Their secretion is presumably connected with glycocalyx formation (Smyth and Halton, 1983; Hanna, 1980; Threadgold, 1984; Sharma and Hanna, 1988).

Modifications of the genital primordium depend on the extent of its development in cercariae. In metacercariae of microphallids this process starts practically from the beginning (see fig. 72A). Its first stage is connected with transformation of small heterochromatin-rich nuclei of the genital primordium into large nuclei with evenly dispersed chromatin. Primordial testes and ovary are soon formed (see fig. 72B). In the genital primordium several zones can be distinguished: a zone, located in front of the ventral sucker, which represents a primordium of the organ's complex of the cirrus pouch. On its left there is a primordium of the genital atrium. A dense aggregation of cells originates from the primordial ovary. This is the primordium of the ducts of the female reproductive system, which, like an arc, surround the ventral sucker from behind. Some time later the primordial vas efferens and vas deferens (see fig 72C) become visible. In the primordial organ complex of the cirrus pouch the central cellular mass is bound with a thin membrane (see fig. 72C). This isolated part represents the primordium of seminal vesicle and ejaculatory duct. The cells, arranged in several layers on its surface, represent the primordial cirrus pouch. In the central part of the arc of the common primordium of the female reproductive system (future ootype) 2 extensions emerge. One of them, directed forwards, is a primordium of the vitelline reservoir, the other represents a primordial Laurer's canal. The region between the forming ovary and ootype later differentiates into the oviduct, and the region between the ootype and genital atrium – into the uterus. Approximately, at the same stage of morphogenesis, primordial vitellaria become noticeable. Subsequent growth and differentiation of the above-mentioned primordia lead to practically complete formation of the reproductive system of the metacercaria by the time, when the multilayered cyst has enveloped the larva (see fig. 72D,E).

The above described development of the reproductive system is typical for all plagiorchiids, with some species-specific features naturally being admitted. The difference lies only in association of its stages with the phases of ontogenesis of hermaphroditic generation. Very few digeneans undergo such a complete set of modifications at the metacercariae stage, as do microphallids. They include species of genus *Prosthogonimus*, whose cercariae are strongly juvenilized (see p. 149), and mature metacercariae in some cases start to produce eggs (Krasnolobova, 1962-1963). Far advanced progenesis of metacercariae, in a number of cases, leading to neoteny (see below), is also shared by other trematodes, e.g., by some species of the genus *Paralepoderma* (Dobrovolskij, 1969; Galaktionov and Dobrovolskij, 1987). However, the "starting line" turns out to be different in the cases considered. The cercariae of *Paralepoderma* possess discrete gonads and differentiated ducts of the reproductive system. On the contrary, in most plagiorchiids the development of metacercarial reproductive system is arrested at the stage preceding the formation of vitellaria. In some species this developmental level is achieved already at the stage of cercaria and the state of the primordia of gonads and the ducts of the reproductive system remains almost the same throughout the metacercarial morphogenesis (Gerasev and Dobrovolskij, 1977). If the larvae initially possess a less differentiated generative primordium, they undergo much more significant changes at the phase of metacercaria. In the end their generative primordium also reaches the level described above.

In representatives of a number of trematode families with the "smooth" type of metacercarial morphogenesis, the second intermediate host may be omitted from the life cycle (see p. 341). The entire development of the larva, starting from the stage of the germinal ball and up to the infective metacercaria, proceeds within the parthenita. In some cases, reduction of transmission pathways does not (or practically does not) affect the character of ontogenesis of the hermaphrodite individual. In daughter rediae of *Bunocotyle progenetica* (Hemiuroidea, Bunocotylinae), typical cystophorous cer-

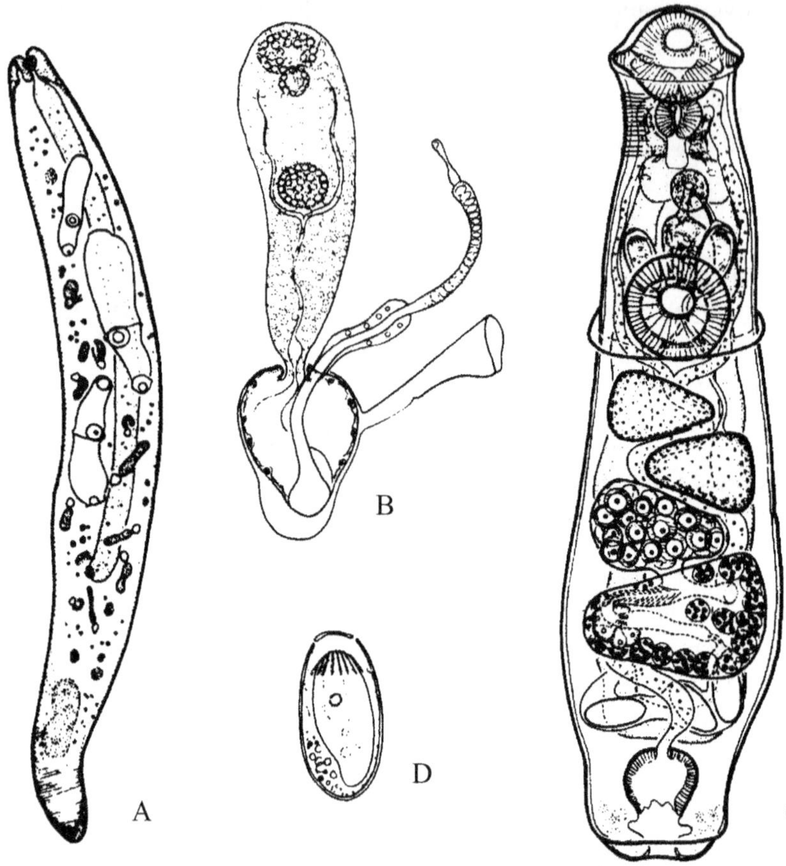

Figure 76. Life cycle stages of *Bunocotyle progenetica.* A, redia with individuals of hermaphroditic generation on different developmental stages; B, cystophorous cercaria; C, marita; D, egg with a fully formed miracidium inside. From Deblock (1974a), reprinted with author's permission.

cariae are formed, which, however, are not released into water and even do not leave the redia (Deblock, 1974a; our data) (Fig. 76). Fully formed larvae are inactive and their body is retracted into the caudal capsule. Some time later, the cercaria body is ejected from the caudal capsule, like in normal hemiuroid species. The young parasite remains in the redial schizocoel and, in fact, passes to the metacercarial stage. Its growth is accompanied by development of the whole complex of organs, typical for hemiuroid adults, and mainly, the reproductive system, which starts functioning after some time. One can distinguish the second and the third (maritogony) preliminary ontogenetic stages in morphogenesis of post-cercarial stages of this species, as well as metacercarial stage, only conventionally, comparing this process with the development of hermaphroditic generation of hemiuroids with usual life cycles.

At the initial stage of the emergence of two-host life cycles in microphallids (*Maritrema oocysta*, *M. murmanica*, *Microphallus breviatus*), cercariae retained the whole set of provisional organs: 4 pairs of penetration glands, mucoid, stylet-forming, 2 types of tegumental ("sheath-forming") glands, stylet and tail with striated muscles (Deblock, 1977; Galaktionov, 1989, 1991b) (Fig. 77A). The mature cercaria, morphologically quite similar to microphallid larvae with three-host life cycles; however, is not released into water, but encysts within the daughter sporocysts. In the course of metacercaria morphogenesis, formation of multilayered cyst and hypertrophy of the excretory bladder can be distinctly traced. These are processes, typical for microphallid larvae, developing in the second intermediate host (see above). Thus, the first and the second preparatory and functional stages of ontogenesis of hermaphroditic individual (see p. 94) are well-expressed.

More significant modifications take place in morphogenesis of the individuals of hermaphroditic generation in such species as *Microphallus pirum*, *M. scolectroma*, *M. abortivus* and *M. helicicola* (Deblock, 1974b, 1977; Belopolskaya and Soboleva, 1977; Galaktionov, 1991a,b, 1993). Developing larvae lose the stylet, penetration glands and tegumental glands, not involved in the production of the cystogenous secretions (Fig. 77B). The forming primordial tail does not develop further and soon detaches from the embryo body. We consider the stage of morphogenesis, which directly precedes this event, as equivalent of the mature cercaria of microphallids with free-swimming larva, however, calling it cryptocercaria (Galaktionov, 1991b). It is impossible to distinguish it as a separate first functional stage of ontogenesis of the individual of hermaphroditic generation. In microphallids of aforementioned species, this stage of ontogenesis is lacking. Its first preparatory stage (morphogenesis of cryptocercaria) directly turns into the second one (metacercaria morphogenesis). The absence of energy expenditure on formation of the above mentioned provisional structures facilitates the transition of differentiation of primordial digestive and reproductive systems to much earlier stages of morphogenesis – cryptocercaria embryos. At the stage of cryptocercaria, the primordial oesophagus, pharynx and branched caeca are clearly distinguished. Well-expressed are also primordial testes and ovary, primordial cirrus pouch, genital atrium and ducts of the reproductive system.

The morphogenesis scenario of the larvae of hermaphroditic generation in the microphallid species under consideration becomes smoother. It also appears to be less energetically costly since cercariae no longer have to accumulate reserve energy substances required for both the period of their free life and the morphological-physiological transformations necessary for existence and development of metacercariae in the second intermediate host.

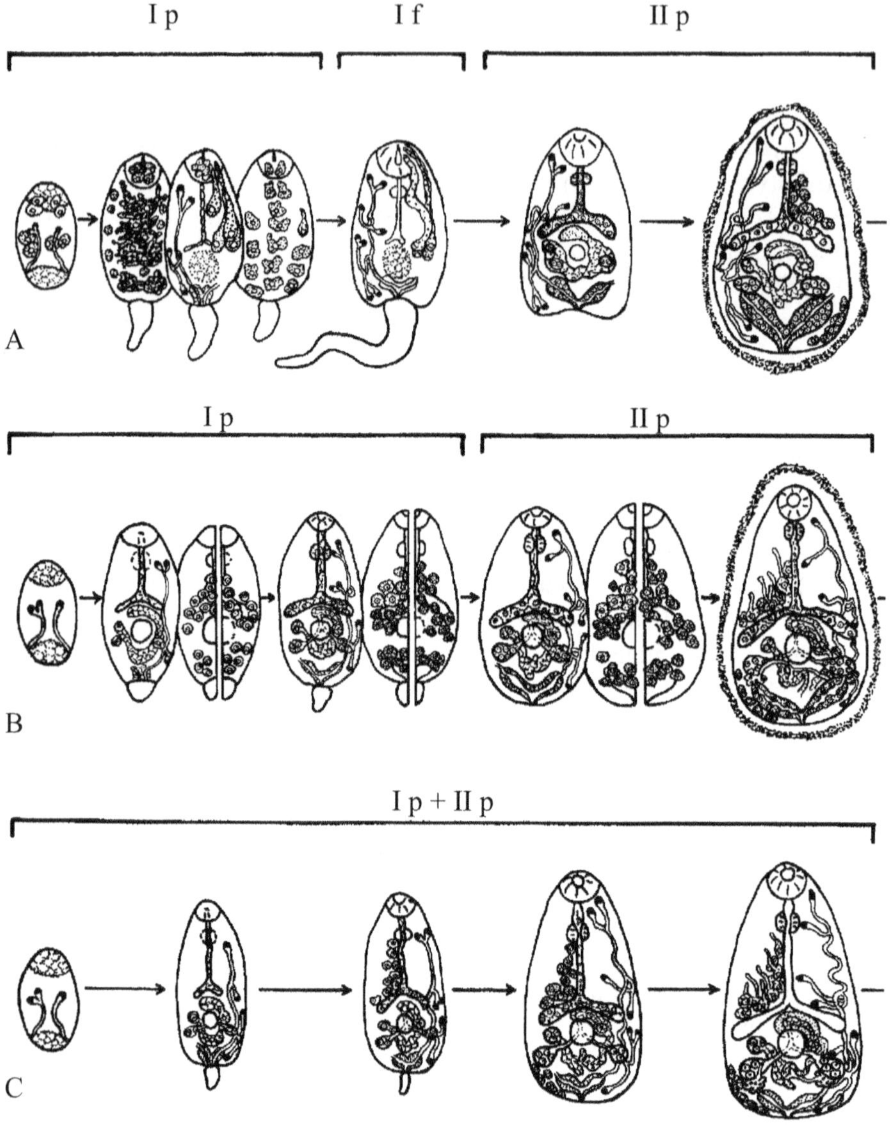

Figure 77. Ontogenesis of individuals of hermaphroditic generation of microphallids with different types of life cycles: A, three-host life cycle, B, two-host life cycle with metacercaria encysting inside daughter sporocysts (*Microphallus pirum* type), C, two-host life cycle with metacercaria developing inside daughter sporocysts without encystment (microphallids of the "*pygmaeus*" group type). I, II, III, successive stages of ontogenesis: f, functional stages, p, preparatory stages. From Galaktionov and Dobrovolskij (1998).

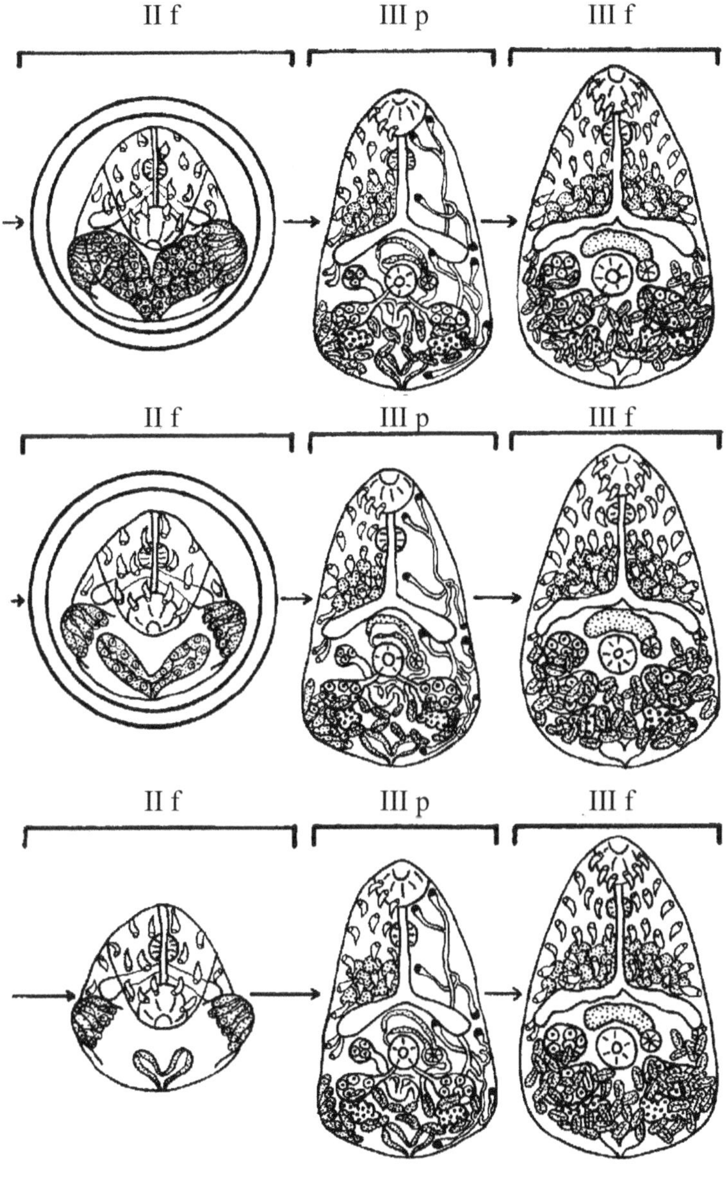

Figure 77. (Continued)

The rate of nutrients' uptake by the forming larvae seems to be more or less constant throughout all stages of morphogenesis up to the formation of a fully formed metacercaria. At least, in these larvae tegumental structures associated with nutrient transport are much less developed than in developing metacercariae of microphallids with a three-host life cycle (see above). Yet the second preparatory stage of ontogenesis in the former is rather clearly expressed, since the formation of provisional structures associated with the cyst does take place. This process involves modification of the tegument, accumulation of secretion in the caeca and successive stages of excretory bladder specialization. However, it should be noted that the lining of the latter does not form columnar extensions and no increase in the number of nuclei is observed in it, contrary to the larvae developing in the second intermediate host. It could be supposed that the shift of a considerable part of the processes of formation of definitive systems of organs to earlier stages of morphogenesis determines a relatively low metabolic activity of the encysted metacercaria. Therefore complex morphogenetic transformations of the excretory bladder (characteristic of metacercariae with a three-host life cycle – see p. 166) appear to be no longer necessary.

The final position in this series is occupied by microphallisds of the "*pygmaeus*" group. In the course of morphogenesis, larvae of their hermaphroditic generation lose all provisional structures, which were still retained by *Microphallus pirum* and similar forms (Galaktionov, 1993) (Fig. 77C). As there are no encystment and excystment, no formation of cystogenous tegumental glands or accumulation of the secretion in the lining of the excretory bladder and gastrodermis takes place. The excretory bladder remains thin-walled during all stages of development, which suggests that no metabolites are accumulated in its lining. Most likely, these metabolites are excreted into the sporocyst body cavity. The ontogenesis of hermaphroditic larvae of microphallids of "*pygmaeus*" group, from the germinal ball up to the stage of mature metacercaria, is a continuous process. Its energy supply is provided by the tegument, whose structural organization is typical for tissues with active transport function up to the terminating stages of metacercariae morphogenesis. A protective role, which usually belongs to the cyst, here is played by the endocyst of the daughter sporocyst and the fibrous capsule, surrounding it. The capsule is formed from hemocytes of the molluscan host (see p. 62).

The larvae of hermaphroditic generation of Dicrocoeliidae and Brachylaimoidea have followed a direction of morphogenesis modification, similar to that of microphallids. However, dicrocoeliids have stopped in the very beginning of this evolutionary way. Their morphological modifications are limited to shortening of the tail of cercariae, which do not leave daughter sporocysts (for details of the Dicrocoeliidae life cycle see p. 474) (Panin, 1984). At the same time, in the second group, a complete series of ontogenesis modifications of larvae of the hermaphroditic generation is observed. The series is initiated by *Leucochloridiomorpha constantiae*. Skrjabin (1947) called it "living ancestor of brachylaimoids" with the three-host life cycle, including the phase of cercaria, active in the external environment. The final position in the series is occupied by the forms of *Leucochloridium paradoxum* type. Their larvae (with reduced tail), which develop in the sporocyst, have no penetration glands. However, early formation and differentiation of digestive and reproductive organ systems is reported in the larvae (Žd'árská and Soboleva, 1986b). Among provisional adaptations of metacercariae, only the above mentioned specialization of glycocalyx (see p. 157), which functionally replaces the cyst, can be noted.

2.2.2.2. Development with Metamorphosis

Another type of development, characteristic of Strigeidida, is called metamorphosis due to its similarity (though, only superficial) to the processes of holometabolic development of insects (Ginetsinskaya, 1968; Sudarikov, 1971). The idea of metamorphic character of morphogenesis of strigeidid metacercariae (practically complete resorbtion of all internal organs, excluding nervous system and genital primordium at initial stages and subsequent formation *de novo* of the definitive organization) has been only partially confirmed by new data. However, this process is basically different from the above described "smooth" type of development and, therefore, needs a special consideration.

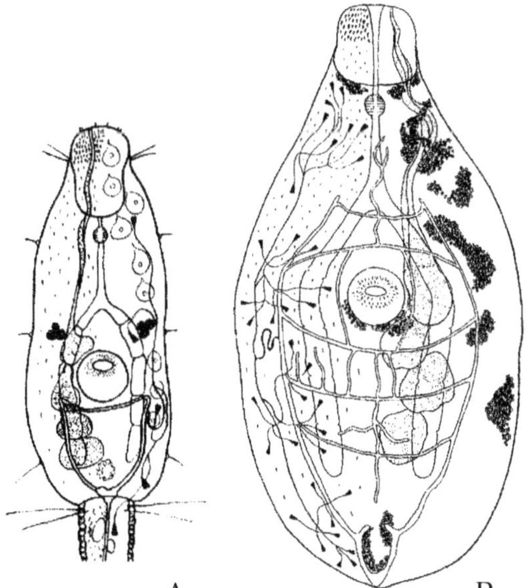

Figure 78. Cercaria (A) and mesocercaria (B) of *Strigea sphaerula*. Based on Odening (1967), reprinted with author's permission.

First of all, additional stages are inserted into ontogenesis of individuals of hermaphroditic generation in some Strigeidida (*Strigea*, *Alaria*). These additional stage are associated with the presence of mesocercaria, the larvae forming in the intercalary host which is inserted into the life cycle between the first and the second intermediate one (see p. 346). Mesocercariae lack the tail, which is shed during penetration into the host. Compared to cercariae, mesocercariae are remarkably increased in size (Sudarikov, 1971, 1984) (Fig. 78). At the same time, their tegumental spines, digestive system and genital primordium remain similar to those of cercariae. Penetration glands are also retained, though their bulk is greatly increased. Numerous unicellular glands appear in the anterior organ. In the excretory system, which is the only one among definitive organ systems that undergoes modifications, the number of the flame cells increases and formation of the secondary paranephridial excretory system starts. The mesocercaria is not infective for the definitive host and must pass in the course of its development the phase of metacercaria, which is characterized by completely different morphogenetic trends.

It is assumed that the initial stages of strigeidid metacercariae development are accompanied by degeneration of somatic structures (Ginetsinskaya, 1968; Sydarikov, 1971; Shigin, 1986). To a certain extent, this is correct. The anterior organ is resorbed, muscles of the tegumentary-muscular sac undergo drastic reorganization. Here one deals with the total degeneration of "cercarial" muscles and their replacement by the definitive ones (Podvyaznaya and Dobrovolskij, 2001) (Fig. 79). In *Diplostomum chromatophorum* (=*D. pseudospathaceum*), definitive muscles develop from myoblasts, which, in turn, originate from undifferentiated cells, situated along nervous cords of the forming metacercaria. During formation of definitive muscles in metacercariae, motility of the larvae is decreased and the body shape is not symmetric, which testifies to the functional weakness of their somatic muscles (Shigin, 1986; Podvyaznaya and Dobrovolskij, 2001). The parenchyma also undergoes significant modifications. Nuclei of its cells increase in size and have evenly dispersed chromatin. The cytoplasm is strongly vacuolated. In light microscopy, the inner structure of such metacercariae looks net-like. Electron microscopy studies have shown that the parenchyma cells become noticeably larger and most of their bulk is occupied by extensively grown RER. Its cisternae form numerous swellings, which, in the light microscopy, look like vacuoles (Podvyaznaya, 1999). Such cells can be observed during development of cercariae and metacercariae of many trematodes (Fig. 80) (our data). Judging by ultrastructure observations, these are active, protein synthesizing cells. Almost all the parenchyma is transformed into the cells of this type, which testifies to an intensive synthesizing activity.

On the 5-8th day strigeidid larvae begin to grow fast. In the tegument of larvae this period of active morphogenetical transformations is characterized by the development of the microvilli system, numerous invaginations of the basal membrane and an increase in the number of mitochondria (Podvyaznaya, 1999). This points to an enhanced absorptive function of the tegument during the time when the larva requires intensive nutrition for metamorphosis. Metacercarial growth is accompanied by division of non-differentiated cells. Myoblasts give rise to the muscles of suckers. On both sides of the oral sucker (the anterior organ is not restored) two lappets are

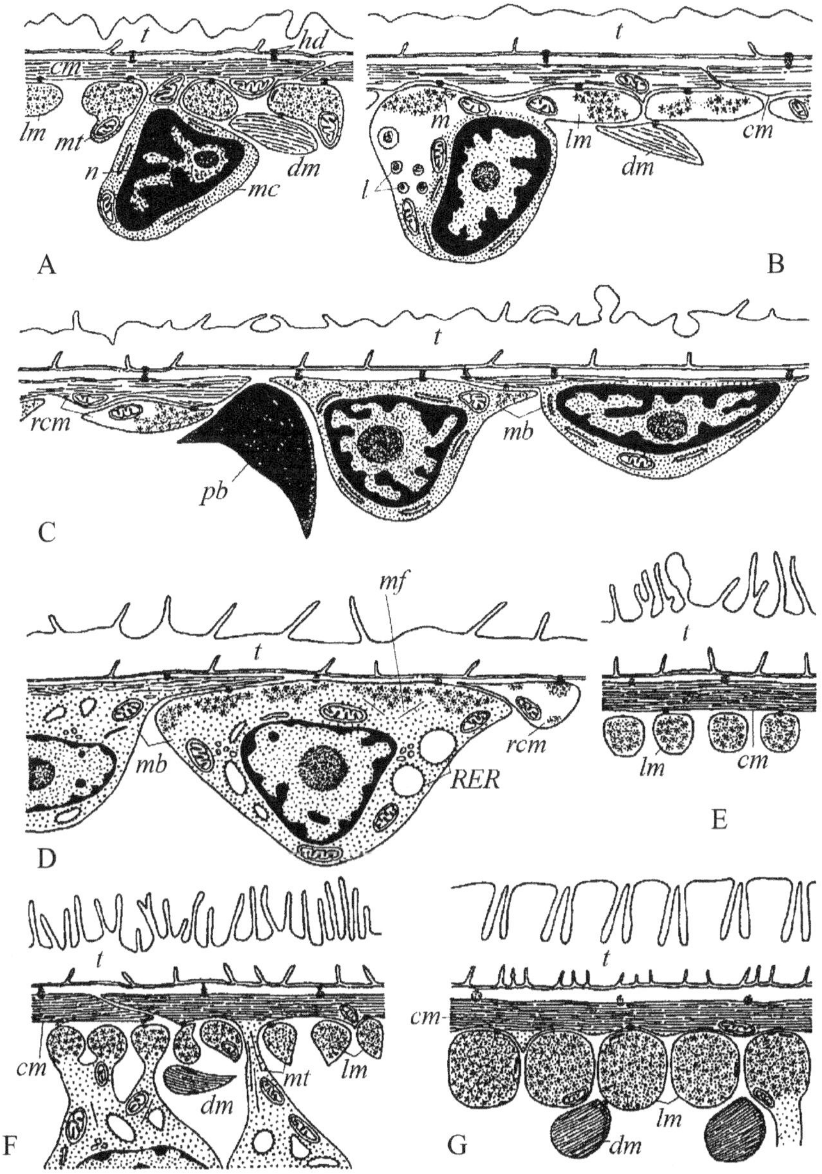

Figure 79. Diagram showing structure of body wall muscles of cercaria and developing metacercariae of *Diplostomum chromatophorum*. A, fully formed cercaria; B, 5-6-days-old metacercaria; C, 7-days-old metacercaria; D, 10-days-old metacercaria, the forebody; E, 10-days-old metacercaria, the region near acetabulum; F, 20-days-old metacercaria, the forebody; G, 40-days-old metacercaria. *Abbreviations*: *cm*, circular muscles; *dm*, diagonal muscles; *hd*, hemi-desmosome; *l*, lipid droplets; *lm*, longitudinal muscles; *m*, microtubules; *mb*, myoblast; *mc*, myocyte; *mf*, myofilaments; *mt*, mitochondrion; *n*, nucleus; *pb*, pycnotic body; *rcm*, resorbed cercarial muscles; *RER*, rough endoplasmic reticulum; *t*, tegument. From Podvyaznaya and Dobrovolskij (2001), with permission.

formed with their own muscles and a powerful glandular apparatus. Behind the ventral sucker a part of the tegument and underlying tissues specialize to give rise to the primordium of the adhesive organ and a complex glandular apparatus associated with it. The entire complex plays an important role in the feeding of the adults.

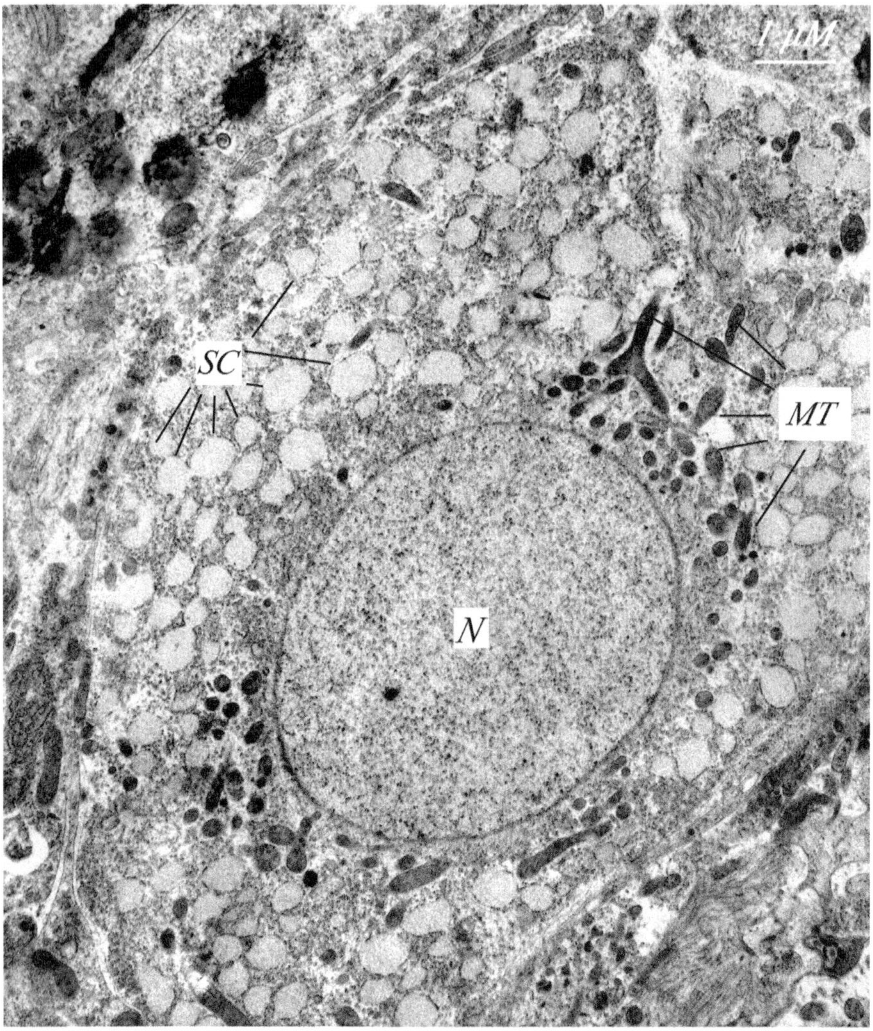

Figure 80. Parenchymal cell with swollen RER in the developing microphallid metacercaria *Microphallus pirum*. *Abbreviations*: *N*, nucleus; *MT*, mitochondrion; *SC*, swelling of RER cisternae.

The excretory system undergoes the most significant changes. Firstly, the number of flame cells increases further, sometimes quite significantly. This is also true for species with the phase of mesocercaria preceding that of metacercaria (Fig. 81). However, the type of excretory formula typical for cercaria is retained. Secondly, the paranephridial excretory system develops. In many strigeidids its primordia form as early as at the cercaria stage as transversal commissurae and closed extensions of excretory tubes. In the course of metacercarial morphogenesis they grow extensively and, joining each other, give rise to a dense net of canals. Niewiadomska and Czubaj (2000) have found that in the metacercariae of *Diplostomum pseudospathaceum (=D. chromatophorum)* canals of paranephridial system and the main excretory duct of the protonephridial system have a similar ultrastructure. The canals have a syncytial structure with few large nuclei in the walls but with no septate junctions. Typical for them is the presence on the luminal surface of numerous lamellae probably participating in resorption of nutrients and the fluid.

In the canals of the paranephridial system, excretory concretions are deposited. Their number increases with age. In a number of forms, they are localized not in the lumen of the canals, but in special side branches, terminated with closed pockets. Each pocket contains a single excretory concretion. The composition of the latter includes inorganic (calcium, magnesium, phosphorus) and organic (protein, glycoprotein and lipids) substances (Mitchell and Crang, 1976; Azevedo and Corral, 1987). Such excretory concretions are found in the excretory bladder of a number of cercariae, metacercaria and maritae of trematodes (Martin and Bills, 1964; Bennett and Threadgold, 1973; Gibson, 1973;Bennett, 1977; Powell, 1977; Orido, 1987a, 1990a). Their physiological function is not completely known. According to a number of authors (Bennett and Threadgold, 1973; Gibson, 1973; Rohde, 1989), it may be connected with their buffering osmoregulatory role.

In *D. pseudospathaceum* larvae, Niewiadomska and Czubaj (2000) observed heterocellular gap junctions between tegumental cytons and paranephridial canal walls. According to these authors, transport of nutrients between the tegument and protonephridial system proceeds due to such junctions. Thus, the paranephridial system may play an important role in the distribution of various substances in the body of developing metacercaria.

Metacercarial metamorphosis in strigeidids is accompanied by gradual changes in the shape of their initially leaf-like body. Two parts of the body characteristic of the adult are differentiated. The anterior body part includes the region bearing the oral and the ventral suckers and the adhesive organ. Ventro-lateral areas situated directly beneath the latter grow extensively. As a result the anterior body part acquires the shape of a cup with the adhesive organ on its bottom. The posterior part is much smaller than the anterior one. It comprises the distal parts of the branched caeca, the excretory bladder and the genital primordium which is usually non-differentiated. Progenesis is rarely encountered in strigeidids. It is not to be found at all in

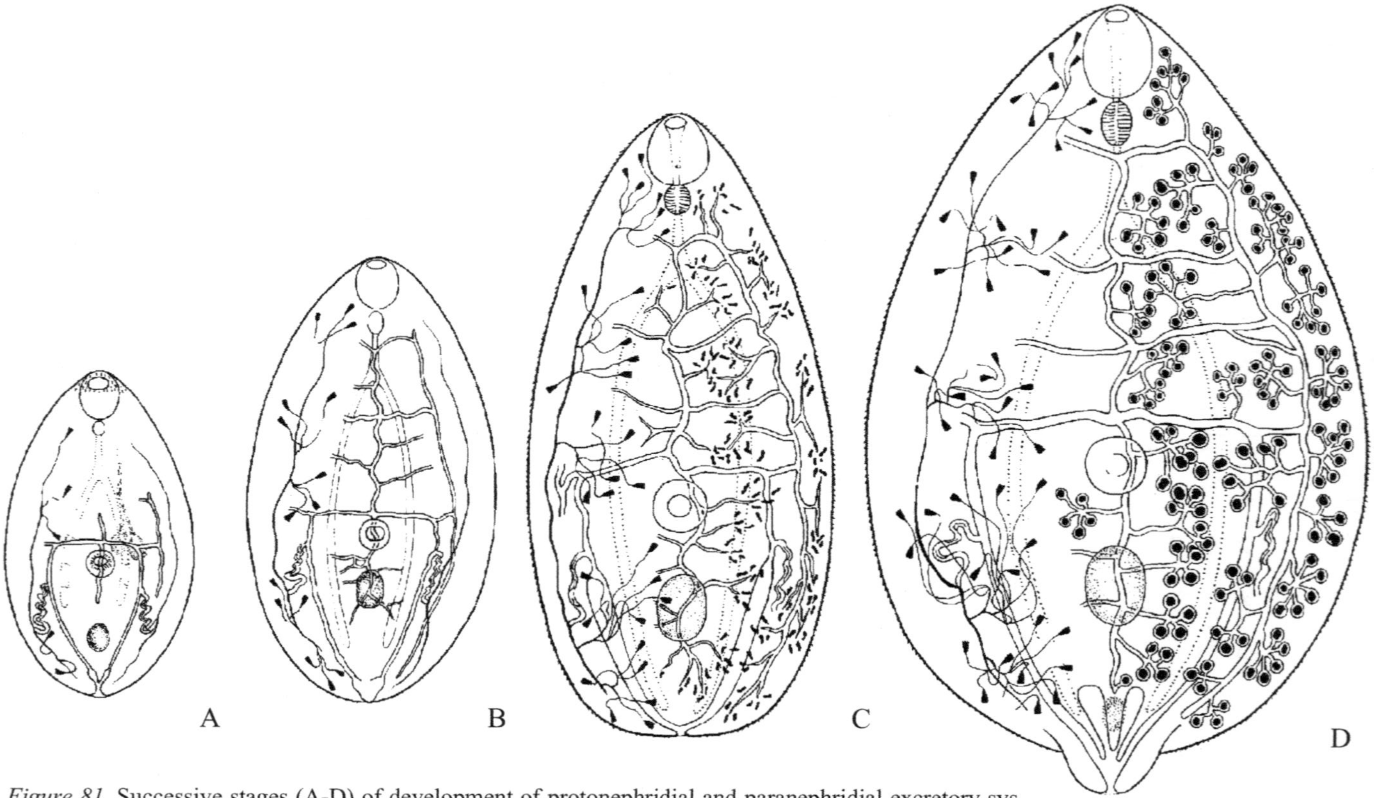

Figure 81. Successive stages (A-D) of development of protonephridial and paranephridial excretory systems in the course of metacercaria *Neodiplostomum spathoides* morphogenesis. From Odening (1965b), reproduced with author's permission.

Strigeidae. Yet in some species among Diplostomidae organs of the reproductive system (ovary, testes, genital ducts and vitellaria) are formed at the stage of metacercariae (i.e., "*Neascus hepatica*" and "*Diplostomulum minutum*"). The organs of the digestive system also do not develop almost at all during metacercarial morphogenesis in strigeidids. At its early stages the oesophagus and the branched caeca begin to look like thin cords with a barely visible lumen, this state being retained until the end of metamorphosis. The latter fact accounts for an opinion common in the literature about the resorption of the gut in metamorphosis. At the same time the gut of non-encysting diplostomid metacercariae starts its functioning a few days after the infection of the second intermediate host supplying the developing larva with nutrients alongside with the tegument (Podvyaznaya, 1999).

2.2.2.3. General Comments on Metacercarial Development

As follows from the above data, individuals of the hermaphroditic generation of different trematode groups at the second functional ontogenetic stage, strongly differ in their "morphological maturity". In adolescariae its level directly corresponds to (is exactly the same as) that achieved at the previous phase of the life cycle, i.e., cercariae. Metacercariae, with the exception of some echinostomatids, which still use the second intermediate host just as a transport facility (see p. 159), are always more advanced in their development than cercariae. Subsequent progress in the relations in the host-parasite system "second intermediate host-metacercaria" makes possible more and more significant morphogenetic modifications at the second preparatory stage of ontogenesis (see p. 93). Most distinctly this trend is pronounced in the representatives of more highly evolved trematodes of the orders Strigeidida and Plagiorchiida. Formation of their metacercariae lasts from 2-3 to 5-6 weeks (Dönges, 1969), the character ("metamorphosis" and morphogenesis of "smooth" type), as well as the final result of the larva development of these 2 taxa, being quite different.

Strigeidida metacercaria is a larva, whose morphogenesis in the definitive host should be long enough to achieve the stage of sexual maturation. In the most specialized plagiorchiids (Lecithodendrioidea, Microphallidae), metacercariae organization fully corresponds to that of maritae. Production of eggs is observed already 1-2 days after the larva has entered the definitive host.

Contrary to strigeidids, progenesis in plagiorchiid metacercariae is a rather common phenomenon. In some cases, it results in a neoteny. In the aforementioned *Paralepoderma cloacicola* with the three-host life cycle, metacercariae, starting to produce eggs in the second intermediate host, fully correspond to adults in their morphological maturity. A different situation is observed in *P. brumpti* and *P. progenetica.* In the former species, sexual maturity is reached at the stages, encysted in the second intermediate host. In the latter species, the larvae become sexually mature directly in the first intermediate host. Analysis of morphology of the sexually mature individuals of both species demonstrates that they are clearly underdeveloped compared to *P. cloacicola* adults: their suckers are less developed; the gut and the body are much smaller. As to the level of development of somatic system of organs they are more similar to the metacercaria of *P. cloacicola* and not to it marita. Thus, in this case, a real reduction of the ontogenesis due to the loss of final (adult) stages takes place, hence, one deals with neoteny. The above mentioned distinctions of the second stage of ontogenesis of the individuals of the hermaphroditic generation of Strigeidida and Plagiorchiida, reflect different pathways of morphological evolution of these digeneans, which will be discussed in chapter 6.

2.2.2.4. Aberrant Cases of Development

When describing metacercarial development we have focused out attention on the trematode groups which could serve as demonstrative examples of the main trends of evolutionary transformations of hermaphroditic generation. A detailed treatment of numerous aberrant cases is outside the scope of the present study and yet some of them should be mentioned here. In particular, we have to discuss a number of Hemiuroidea whose life cycle, similar to that of the genera *Strigea* and *Alaria,* is complicated due to insertion of mesocercaria phase.

In *Lecithochirium furcolabiatum*, the most thoroughly investigated species in this respect, mesocercariae, parasitic in marine copepods ("intercalary" host) (see p. 346), noticeably grow and differentiation of their genital primordium is observed (Matthews and Matthews, 1988b) (Fig. 82). However, the most important modifications, like in strigeidid metacercariae, take place in the excretory system. This process begins with dividing of the excretory bladder into proximal and distal parts, between which a sphincter is later formed. The lining of the proximal part does not undergo later any significant modifications. In the distal part, on the contrary, the lining greatly thickens, and numerous extensions and microlamellae are formed on its luminal surface. The lining itself represents a nuclei-containing syncytium, which is connected with the underlying nuclei-containing cytons by means of cytoplasmic extensions. This structure, the so-called "ecsoma", or "tail", may turn inside-out like a glove finger and then return to the initial state (Gibson and Bray, 1979). Histochemical tests were carried out, which showed that active transport of nutrients, especially amino acids, proceeds through the ecsoma (Matthews and Matthews, 1988b). At the mesocercarial stage, the tegument also takes part in the absorption of nutrients, but already in metacercariae this function is entirely transferred to the ecsoma. The tegument of the larva, which is formed within the fibrous capsule on mesenteries of the internal organs of fishes (second intermediate host), acquires "adult" morphology and does not possess ultrastructure organization, which is usually associated with transport processes.

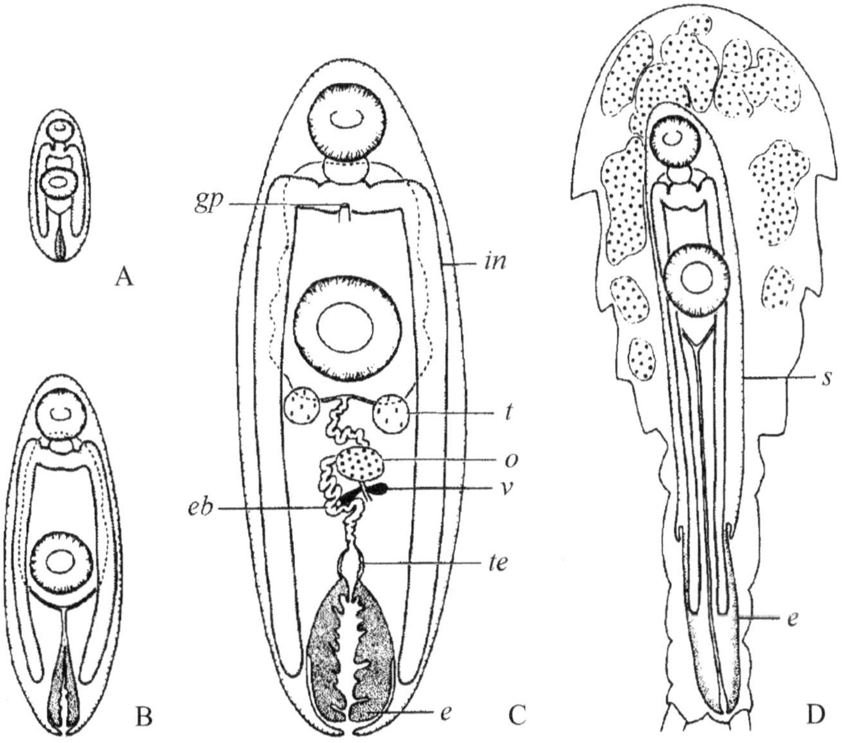

Figure 82. Successive stages of development of *Lecithochirium furcolabiatum* metacercaria in marine copepods. A, B, C, mesocercariae at 7, 14 and 21 days post-infection respectively; D, 21-days-old mesocercaria with fully extended ecsoma in the copepod haemocoel. *Abbreviations*: *e*, ecsoma; *eb*, excretory bladder; *gp*, genital pore; *in*, intestinal caecum; *o*, ovary; *s*, soma; *t*, testis; te, terminal dilation of excretory bladder; *v*, vitellaria. From Matthews and Matthews (1988b), reproduced with permission from Journal of Helminthology, 1988, p. 319 © CAB International.

We believe that formation of a large group of vertebrate blood parasites – Schistosomatida – is connected with aberrations in the course of ontogenesis of hermaphroditic individuals. Skin penetration and parasitizing in the host tissues are the main features of the Schistosomatida life cycle. These features enable us to presume that the emergence of this life cycle in the evolution is connected to the loss of the definitive host, whose role now is

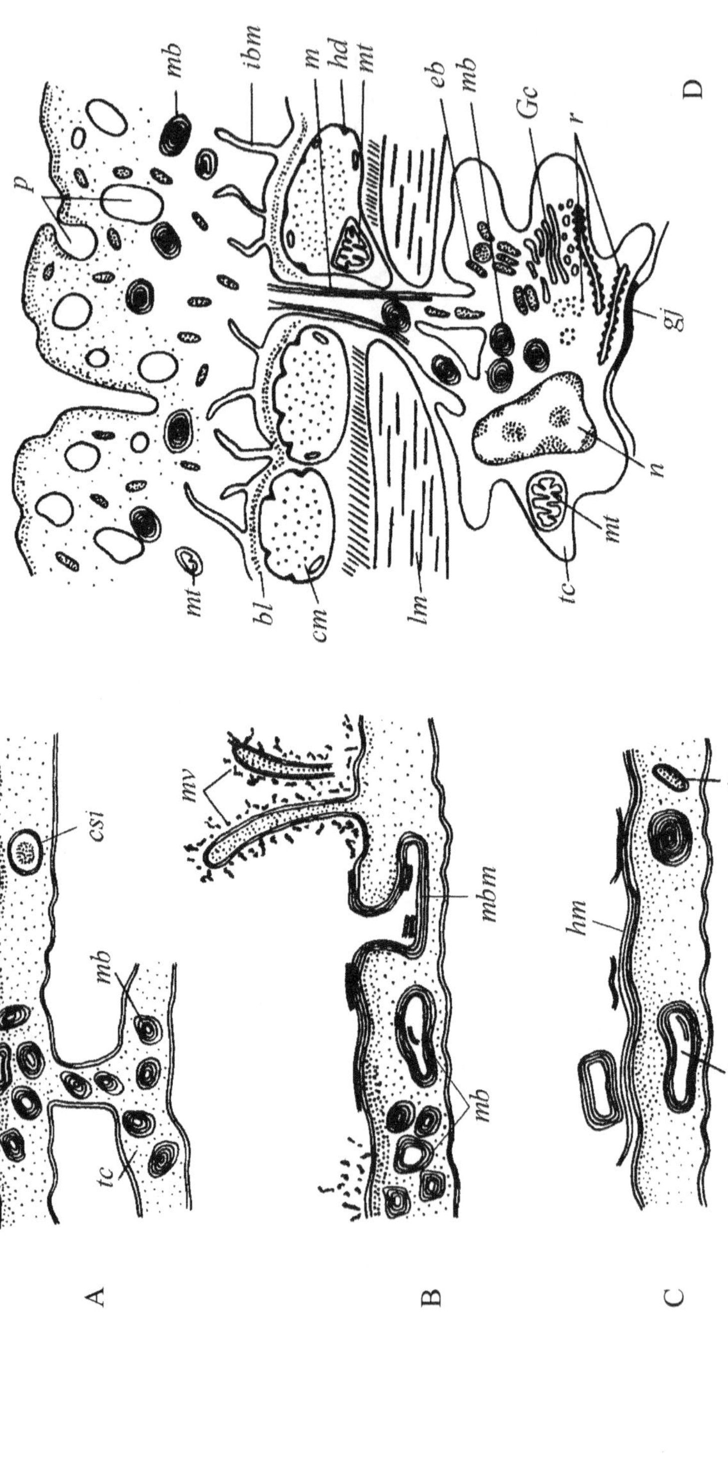
A
g
csi
mb
tc
B
mv
mbm
mb
C
hm
di
mb
D
p
mb
ibm
m
hd
mt
eb
mb
Gc
r
gj
n
mt
tc
lm
cm
bl
mt

played by the second intermediate host (for details see p. 338.). In this case, schistosome maritae should be considered as neotenic metacercariae, and development of schistosomulum (=metacercariae) – as the second preparatory ontogenetic stage, extended due to involvement of some maritogony features. The entire course of this process in schistosomes, investigated in detail in *Schistosoma* species, pathogenic for humans, and in closely related forms (for reviews see: Hockley, 1973, 1977; Stirewalt, 1974; McLaren, 1980; Smyth and Halton, 1983; Threadgold, 1984), is characterized by profound specialization to blood parasitism.

After the cercaria has entered the blood vessel through following skin penetration, the larva becomes a migrating stage, called "schistosomulum". This process, called transformation, lasts about 3 hours and is associated with drastic changes of the larval tegument (Fig. 83). In the cercaria, which has just entered the host and shed the tail, numerous microvilli are formed on the tegument surface. The zones between the microvilli soon become isolated from the external environment by a "double" or heptalaminate membrane, i.e., by two adjoining trilaminate membranes. Formation of such a structure is thought to be connected with building into the tegumental plasma membrane of membranous bodies (=multilaminate bodies), which are limited by two membranes. These membranous bodies are formed in the tegumental cytons, which sharply increase in number after the larvae have penetrated into the host.

In about an hour after penetration, all microvilli are shed and the schistosomulum body becomes fully enveloped with the heptalaminate membrane. In the course of transformation, alongside with microvilli, all "cercarial" plasma membrane and associated glycocalyx, which includes glycoproteins with antigenic properties, are also shed (Samuelson and Caulfield, 1985; Payares and Sympson, 1985; Payares et al., 1985; Caulfield et al., 1988). The above events reduce binding of antibodies and fixation of the complement of the host. Consequently, resistance of the developing schistosomulum is enhanced. This is supported by experiments on mice. Some of them were preliminarily immunized with irradiated cercariae, and others – with schistosomula. Being infected with intact cercariae, the first group demonstrated greater resistance to the infection than the second group (Sher

Figure 83. Tegument of schistosomulum and adult of *Schistosoma mansoni*. A-C, successive changes in tegument during cercaria/schistosomulum transformation (A, 30 min after penetration; B, 1 h after penetration; C, 3 h after penetration), D, tegument of adult. *Abbreviations*: *bl*, basal lamina; *cm*, circular muscles; *csi*, cercarial spherical inclusion; *di*, discoid inclusion; *eb*, elongated body; *g*, glycocalyx; *Gc*, Golgi complex; *ibm*, invagination of basal membrane; *gj*, gap junction between tegumental cyton and parenchymal cell; *hd*, hemi-desmosome; *hm*, heptalaminate membrane; *lm*, longitudinal muscles; *m*, microtubules; *mb*, multilaminate body; *mbm*, multilaminate body joined to outer membrane; *mt*, mitochondrion; *mv*, microvilli; *n*, nucleus; *p*, pits of the tegument surface; *r*, ribosomes; *tc*, tegumental cell. Based on Hockley (1973).

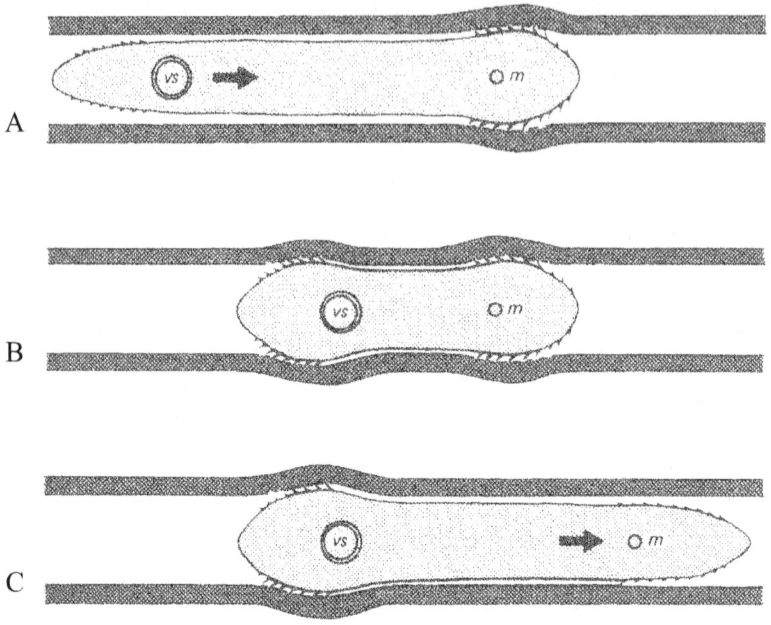

Figure 84. Diagram showing successive stages of movement of schistosomulum along the blood vessel. A, elongated schistosomulum anchored by its head spines; posterior region pulled up towards the head; B, schistosomulum at minimum length; C, schistosomulum anchored by its tail spines; anterior and mid-body extended forward. *Abbreviations*: *m*, mouth; *vs*, ventral sucker. From Crabtree and Wilson (1980), reprinted with the permission of Cambridge University Press.

and Benno, 1982). After transformation, glycoproteins are found in glycocalyx of schistosomula. However, they are not recognized as antigens in mice, previously immunized by chronic schistosomula infection or irradiated cercariae (Payares and Simpson, 1985). There is no doubt that all complicated and not fully elucidated transformations of the schistosomula surface structure enable these parasites to avoid recognition by the immune system of the host and to exist in such immunologically active environment as the blood of vertebrate animals (for reviews see: Smithers and Doenhoff, 1982; Capron, 1998; Modha et al., 1998; Wilson and Coulson, 1998).

Schistosomulum migration from the skin to the hepatic portal system is entirely intravascular (Wheater and Wilson, 1979). Tegumental spines play a rather important role in this migration along blood capillaries (Fig. 84). In the young schistosomulum, the spines disappear from the mid-body region, but are retained at the extreme anterior and posterior part of the body (Crabtree and Wilson, 1980). During elongation, the schistosomulum anchors itself in the walls of the blood vessels with the spines of the anterior end of the body and then pulls up its posterior end. Then, being fixed at the walls of the blood vessel with the spines of the posterior end of the body, the schistosomulum extends its anterior end and becomes fixed by means of the spines of the latter. The absence of spines in the middle part of the body minimizes friction between the worm surface and the capillary endothelium, thus facilitating movement along the lumen.

Soon after arrival in the hepatic portal system, the schistosomulum loses its spines. The surface of the outer layer of the tegument becomes folded and numerous pits appear (Hockley, 1973). As they become deeper, a complex system of anastomosing canals emerges, typical for the body surface of sexually mature schistosomes. Their surface undergoes a remarkable expansion, which is presumably connected with absorption of nutrients, dissolved in the plasma of the host's blood. In fact, activity of alkaline phosphatase is associated with the plasma membrane of the schistosomes tegument (beginning from the schistosomulum at the lung stage of migration or, according to other data, the liver stage of migration) (for reviews see: Smyth and Halton, 1983; Payares et al., 1985). Apparently, it is just in the migration period, that the tegument is of paramount importance in providing the parasite with nutrients, when the gut is not yet fully formed. However, in adult schistosomes, transport of glucose through the tegument was also experimentally demonstrated (Fripp, 1967; Uglem and Read, 1975; Rogers and Bueding, 1975). It is effected by means of specific glucose transporter proteins, which are integral membrane proteins (for review see: Skelly et al., 1998).

In the course of schistosomula development, the foregut is differentiated into 3 distinctly expressed zones (for reviews see: Hockley, 1973; Smyth and Halton, 1983). In the area of the oral sucker, the ultrastructure of the foregut lining is similar to that of the outer tegumental layer. It forms long anastomosing loops, which occupy the lumen of the oesophagus. In the middle part of the latter, the loops of syncytial lining do not branch and are arranged more regularly. There are secretory granules in the cytoplasm, which are produced by multinuclear oesophagus glands. Finally, in the posterior part of the oesophagus the loops of the lining are no longer regular, they become much thinner and have numerous extensions.

The function of the schistosomula foregut is ingestion and digestion of the vertebrate blood cells. The former is achieved by means of the anastomosing loops of the anterior part of the oesophagus and powerful muscles of the oral sucker. In the lumen of the middle and posterior part of the oesophagus erythrocytes at different stages of digestion are to be found, digestive enzymes being contained in secretory granules. The posterior part of the foregut is filled with non-digested remains of host's blood cells. The epithelium of the branched caeca is syncytial. Its luminal surface is plicate, ab sorption of nutrients proceeds through it. Thus, digestion processes in the schistosome digestive system occur in the foregut, which has brought about its complex morphological differentiation. Such an organization appears to have arisen in the schistosome evolution in connection with transition to blood parasitism.

In the course of schistosomula development, differentiation of the genital primordium takes place. In cercariae, it has the form of a diffuse cellular aggregation. At the liver stage of migration, schistosomes form pairs. In one individual of each pair, only male organs of the reproductive system develop, whereas in the other – only female. Up to now, formation of sexually mature individuals of the representatives of families Aporocotylidae, Sanguinicolidae and Spirorchiidae, closely related to schistosomatids, is much more poorly studied. However, the data available allow us to presume that their tegument, and possibly, organs of digestive system, undergo specialization, similar to that of schistosomes (McLaren and Hockley, 1977; Køie, 1982).

2.3. Marita (adult)

The phase of marita starts, when the metacercaria gets into the definitive host, and terminates a long pathway of ontogenetic modifications of individuals of the hermaphroditic generation. In most trematodes, this final part of ontogenesis is divided into two stages: preparatory (marita development or maritogony) and functional. At the first stage, completion of development of definitive systems of organs and/or of physiological maturation takes place. The functional stage is the period of reproductive activity of the sexually mature individual. Duration of both stages is very different in various trematode groups and largely depends on the trends of their morphological evolution.

2.3.1. Excystment

Right after the encysted larva has been swallowed by the definitive host, it undergoes a series of modifications. They are connected with its activation (one should keep in mind that the functional stage of ontogenesis of metacercariae and adolescaria are, in fact, resting stages) and excystment. This problem has recently been reviewed in detail (Fried, 1994; Irwin, 1997).

Activation of larvae within the cyst and/or capsule and their subsequent excystment proceed under the influence of external and internal factors (Fried, 1994; Irwin, 1997). The former include mechanical effects (during ingestion and mastication the food its grinding in the gizzard of birds), digestive enzymes, such as pepsin and trypsin, bile, and changes in the gas phase. The latter include the secretion, discharged from the gut of some metacercariae (*Fasciola hepatica*, *Plagiorchis* sp. I Bock, 1988, Microphallidae) or produced by singular glands, and, possibly, products, excreted from the excretory bladder. The larva becomes active and rotates in the cyst/capsule and some time later crawls out, with its anterior end directed forwards, through an opening or rupture, appearing in the envelope. The envelope is preliminarily softened either totally, or in a certain part, through which the young marita is released. It should be noted that the details of the excystment process may vary even in related trematodes.

In adolescaria of Fasciolidae and Philophthalmidae, the cyst has a ventral plug, which is softened under the influence of the secretion, discharged from the gut, and becomes passable for the actively moving larva. In Paramphistomidae, the role of ventral plug is played by a specially organized region of the cyst, which is softened under the action of exogenous factors and is disrupted by the excysting larva.

The excystment of echinostomatids proceeds in a similar way. The inner layer of their cyst lacks dense lamellae in a certain site (the so-called escape aperture) and is ruptured due to joined action of the secretion of metacercariae glands and active movement of the metacercariae. By that time, the outer layers of the cyst are partly digested by proteolytic enzymes of the host, which facilitates release of the larva. In microphallids, softening of the multilayered cyst is observed simultaneously at both poles. At one of them, the cyst is ruptured due to the pressure of the excysting individual (Saville and Irwin, 1991). The intact cyst, no structurally differentiated parts are observed (Galaktionov et al., 1997). In the course of excystment, the secretion, accumulated in the lumen of the gut, is released. Apparently, it facilitates softening of the cyst. At any rate, in *Plagiorchis* sp. I, it fulfills just this function (Bock, 1989). During excystment of the larva of this species, the inner layer (resistant to mechanic injury) of its two-layered cyst (see p. 153) is disrupted chemically in the site of release of the secretion from the gut. The outer layer, providing chemical stability, is ruptured mechanically, due to active movements of the metacercaria itself.

Disruption of strigeidid capsules proceeds under the action of digestive enzymes of the host. A rather special way of release (or rather, even expulsion) of the larva from the cyst (capsule?) was described in representatives of genus *Apatemon* (Kearn et al., 1989). After getting into the host's gut, at one of the poles of the cyst, a thin channel is formed, through which the larva is ejected due to sudden and substantial inward expansion of the inner cyst wall. Mechanisms of such excystment are not clear. It is presumed that such a way of rapid escape of metacercariae from their cysts evolved because it might be advantageous for those worms that inhabit the anterior end of the host's digestive tract.

Certain morphological and functional modifications of larvae take place already during the excystment. This may be connected with expenditure of the secretion, accumulated in the lumen of the metacercariae gut and with emptying of the glands. Leaving the cyst through the narrow opening, the larva body experiences a powerful pressure. In microphallids, when the posterior end of the larva body passes through the opening, columnar extensions, containing insoluble metabolites, separate from the base of the excretory bladder syncytium (see p. 166), and are expelled through the excretory pore into the cavity of the emptied cyst (Galaktionov and Malkova, 1995). Apparently, formation of the definitive lining of the excretory bladder proceeds due to an extensive growth of its regions, which were located between the columnar extensions and differed from them in structure. During excystment, in metacercariae of other trematode groups, liquid is removed from the cavity of the excretory bladder, together with accumulated harmful products of metabolism.

2.3.2. Development (Maritogony)

Earlier we identified 3 types of development of individuals of the hermaphroditic generation in the definitive host (Galaktionov and Dobrovolskij, 1987). The first type, apparently, the most numerous, includes species with prolonged maritogony, which includes significant morphological modifications. The second type represents digeneans with a short period of development in the definitive host, for their morphogenesis has been almost completed at the metacercaria phase. Finally, the third type, the least numerous, involves the forms, which do not develop in the definitive host. Such progenetic species with the reproductive system, which has been already formed, and sometimes even functions, use the definitive host only for disseminating eggs in the external environment.

Assignment of the individual to a certain type is determined by the difference in organization of mature metacercariae and sexually mature maritae. The larger this difference is, the longer and more complicated is the process of maritogony. On the contrary, the less the difference between the metacercaria and the marita is, the shorter is the period of time, necessary to achieve sexual maturity.

All primitive trematodes with the phase of adolescariae in their life cycle are assigned to the forms with the first type of development. The absence of significant morphogenetic modifications at this phase of the life cycle results in that maritogony takes the whole period of ontogenesis from the cercaria to the sexually mature individual. The same type of development is characteristic of maritae of many trematodes with the "smooth" type of morphogenesis of metacercariae (see section 2.2.2.1.), and also of all strigeidids. During metamorphosis of strigeidid metacercariae (see section 2.2.2.2.), crucial reconstruction of organization of the specialized, strongly juvenilized larva (cercaria) takes place, followed by re-orientation of the subsequent morphogenesis towards the imaginal development. This must finally result in formation of marita, which is also extremely specialized (see below). Apparently it is impossible to put all these complicated modifications at the ontogenetic stage in the second intermediate host, the process of development of strigeidid metacercariae (metamorphosis) already being one of the longest among trematodes (Dönges, 1969). Therefore, rather essential stages of the completion of imaginal organization proceed in the period of maritogony. In this respect strigeidid and primitive trematodes are similar, but this similarity is superficial.

Marita development in the definitive host up to sexual maturity is accompanied by allometric growth. This is manifested in uneven increase in size of locomotory and generative parts of the body, uterine and generative zones (for details see reviews: Ginetsinskaya, 1968; Galaktionov and Dobrovolskij, 1987). The latter correlates with the mode of reproduction. Forms, laying eggs with zygote, which would later develop into the larva of the next generation in the external environment (see p. 8, 331), have a short uterus (Fasciolidae, Echinostomatidae, most Strigeidida) (Fig. 85A). Growth processes dominate in their generative zone, which is connected with an intensive growth of gonads and, especially, of vitellaria. The forms, whose miracidia develop in the maternal organism (Notocotylidae, Philophthalmidae, Cyclocoelidae, Renicolidae, Bucephalidae, most Plagiorchiida), have a hypertrophically developed uterus (Fig. 85B,C). Their uterine zone is predominantly grown, whereas the generative zone remains small. The uterus, in this case, functions as a genuine uterus: it more or less provides for feeding of the embryo. Its extensive growth is connected with the necessity of simultaneous bearing of an enormous amount of eggs. Vitellaria lose their trophic function to a significant degree and are partly reduced.

The tegument of trematodes, parasitic in the gut of the definitive host, does not undergo any noticeable changes with respect to its structure in mature meracercariae. Normally, a slight thickening of its outer layer is observed, as well as extensive growth of spines, which in many maritae acquire a typical shape of serrated scales or combs (Fig. 86). Practically in all maritae, investigated up to now, small electron-dense rod-shaped or discoid granules are observed in the apical part of the outer syncytium. The content of the granules is, presumably, connected to glycocalyx formation (Køie, 1971c; Smyth and Halton, 1983; Threadgold, 1984; Hanna, 1980; Sharma and Hanna, 1988). This secretion is produced by tegumental glandular cells, appearing, in many cases, already at the metacercarial stage (see p. 160).

More complicated modifications proceed in the tegument of digeneans, whose sexually mature maritae parasitize not the gut, but in the bileducts, pancreatic gland, ureters and excretory bladder of vertebrates. Young worms migrate from the gut to the place of their final settlement in the host.

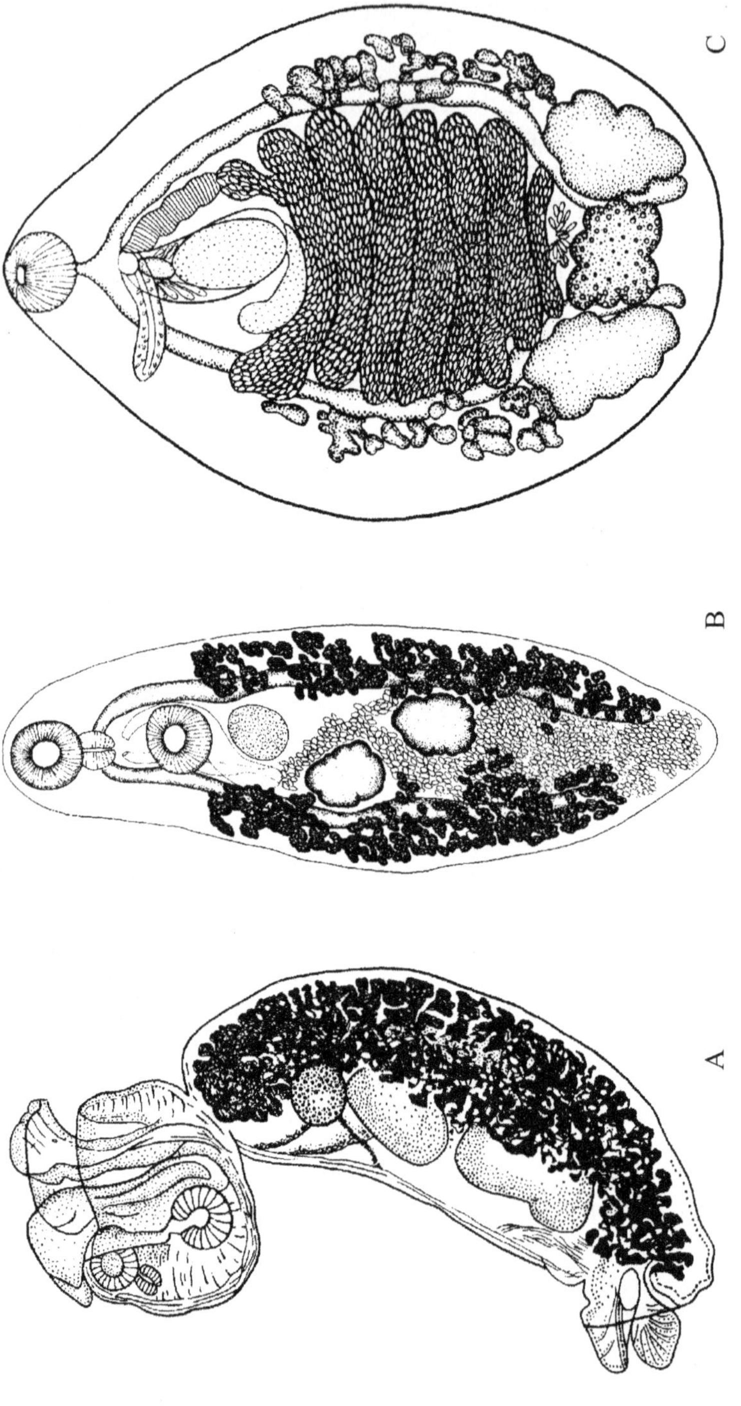

Figure 85. Trematode maritae. A, *Cotylurus cornutus* (Strigeidae) (after Dubois, 1938). B, *Plagiorchis vespertilionis* (Plagiorchiidae) (after Groschaft and Tenora, 1973), C, *Paramonostomum alveatum* (Notocotylidae) (after Filimonova, 1985)

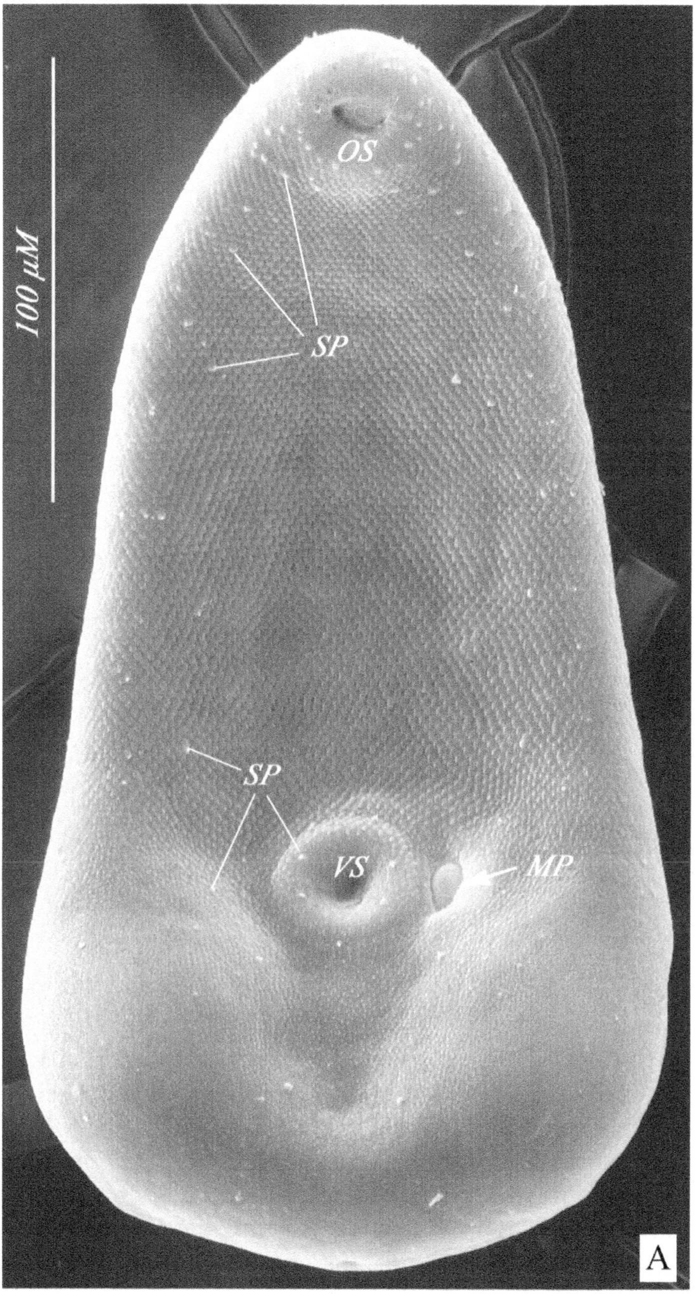

Figure 86. Surface structures of the body of young microphallid maritae. A, *Microphallus pygmaeus* with different shape of spines in the anterior and posterior parts of the body; B, *Microphallus pygmaeus*, spines of the fore-body; B, *Microphallus piriformes*, spines of the fore-body. *Abbreviations*: *mp*, male papilla; *os*, oral sucker; *sp*, sensory papilla; *vs*, ventral sucker. Original SEM micrographs of S.W.B. Irwin and D.H. Saville, with permission.

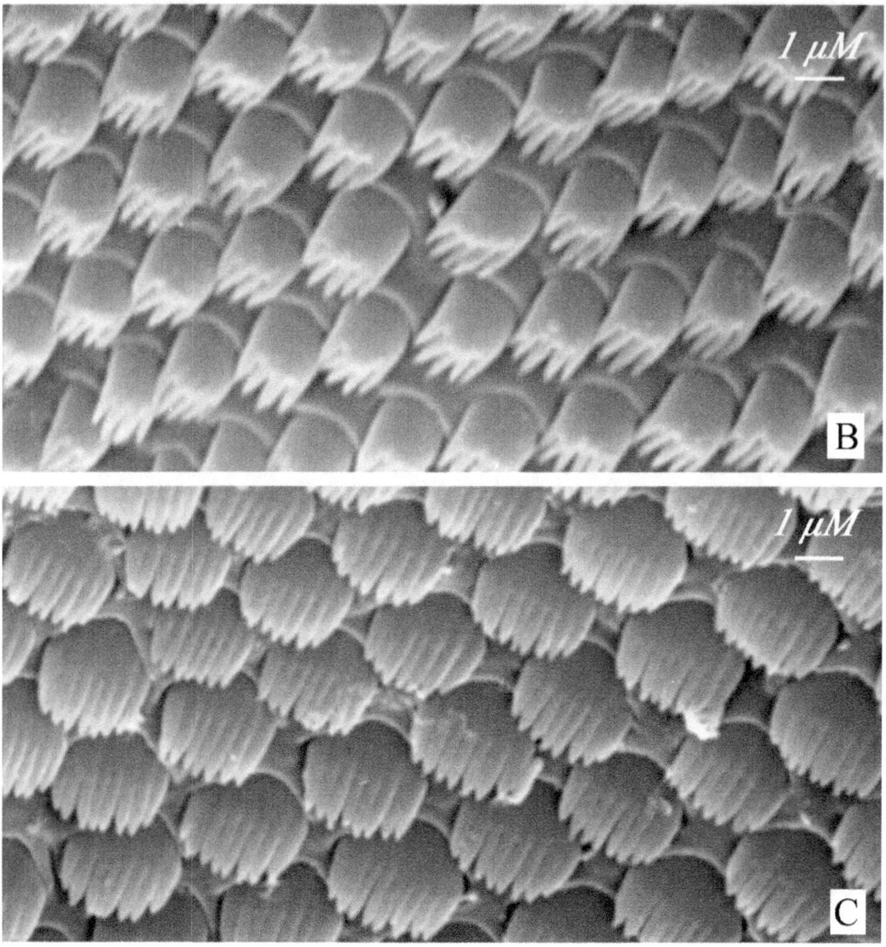

Figure 86. (Continued)

At different stages of the migration period, tegumental glands, producing different types of secretory granules, are connected with the outer layer of the tegument (Bennett and Threadgold, 1975; Hoole and Mitchell, 1983b). Their secretion is used to form the glycocalyx. The latter, in *Fasciola hepatica*, which has been investigated in this respect, inhibits proteolytic enzymes, when the juveniles are in the gut. When they migrate into the liver, the secretion contributes to the transport function of the tegument. In the bile-ducts of sexually mature maritae, inhibition of enzymes again becomes the main function of the secretion (Bennett and Threadgold, 1975). Changes in the tegument during development of the blood parasites, which we consider as progenetic metacercariae, have been described earlier (see p. 183).

The outer syncytium of the maritae tegument is underlined with a basal lamina, beneath which a layer of intermediate filaments is revealed. These filaments also enter the interspaces between fibers of the circular and longitudinal muscles and cytons of the tegument (Abbas and Cain, 1987; Fukuda et al., 1987). A similar situation takes place in cestodes. In this case, intermediate filaments are thought to play the role of integrators of the outer syncytial layer of tegument and muscles of the tegumentary-muscular sac (Holy and Oaks, 1987).

Soon after excystment, many cells of young marita are activated. We have observed the same phenomenon in metacercariae, during the first hours after penetration into the second intermediate host. However, the period of intensive cell division lasts only several days. After it has finished, cells keep dividing only in the gonads and other organs of the reproductive system, where the number of cells increases. The pause in cell division does not impede the growth of the parasite, which proceeds due to the increased cell size. Primarily, this applies to parenchyma cells, which become lobate and acquire long extensions. The latter in sexually mature maritae are so tightly packed, that only thin layers of the extracellular matrix are left between them (Threadgold and Gallagher, 1966; Matricon-Gondran, 1977; Tihomirov, 1980). The electron-lucid cytoplasm of the extensions lacks organelles, except single mitochondria. The extensions form gap-junctions with each other, as well as with the cells of the other organ systems (digestive system, tegumental cytons, ducts of the excretory system, muscles). This leads to a presumption that there exists a united transport system, which provides distribution of nutrients in the parasite body. Between extensions of parenchymal cells, extensions of myocytes are revealed. The two kinds of extensions are practically undistinguishable from each other.

It should be noted that our interpretation of parenchyma of platyhelminthes has at present significantly changed (for details see footnote iii in chapter 1). The "parenchyma" includes not only the parenchymal cells *per se*, which in some worms (e.g., in a number of cestodes) may be missing, but also cytons of the tegument, the epithelium of excretory tubules, glandular and especially muscle cells, forming numerous extensions (Conn, 1988; Conn and Rocco, 1989; Korneva, 1995, 2001). All these heterogeneous elements create parenchyma organization, typical for flatworms, functioning as the connective tissue.

In the course of marita development, new muscle cells are formed from myoblasts, resulting in strengthening of the muscle layers. The number of muscle elements also increases in the suckers, which may reach a rather large size. The opposite trends may exist in such trematodes, whose biology does not suggest their release from the host's body. Muscle apparatus of such species either remains at the "larval" level of organization, or partly degenerates. Demonstrative examples are *Brandesia turgida*, settling in the cavity of "cysts", which are formed in the wall of the frog gut, or *Astiotrema trituri*; whose maritae has a diameter, greatly exceeding that of the newt gut and are safely anchored there.

Completion of the gut development proceeds in the same way as in the forms, where this process takes place at the second preparatory stage of ontogenesis, i.e., metacercariae morphogenesis (see section 2.2.2.). Gastrodermis of maritae may be syncytial (*Gorgoderina*, schistosomes) or, most often, cellular (Erasmus, 1977; Morris, 1973; Hoole and Mitchel, 1983a; Robinson and Threadgold, 1975; Fujino and Ishii, 1978; Fujino, 1997). In the latter case, in the course of worm growth, its cells increase in number and volume. The luminal surface of the cellular epithelium or syncytium forms numerous lamellae, which may branch and anastomose with each other. The gastrodermal cells undergo cyclic transformations between secretory and absorptive forms (Robinson and Threadgold, 1975; Fujino and Ishii, 1978; 1988). In *Fasciola hepatica,* among these cells several cell types are distinguished. These are secretory cells (I type), producing digestive enzymes, cells of absorptive function (II type), and cells with unclear function (III type) (Robinson and Threadgold, 1975; Threadgold, 1978). The latter, probably, facilitate transportation of the material in the gut cavity and take a certain part in secretion of digestive enzymes and absorption of nutrients (Robinson and Threadgold, 1975).

In large maritae (Fasciolidae, Clinostomidae), the branched caeca may form diverticula and extensions, similar to those of some blood parasites (Aporocotylidae) (Køie, 1982). The above-described *Fasciola hepatica* cells of the I and II types are concentrated in the gut diverticula, where, apparently, processes of secretion and absorption proceed. The cells of the III type compose the gastrodermis of the gut (Robinson and Threadgold, 1975). It should be noted that the gastrodermis may play different roles at different stages of marita development. According to Bennett and Threadgold (1973) and Bennett (1975), in metacercariae of *F. hepatica*, which have just escaped from the cyst in the host gut, the gastrodermis is of secretory epithelium type. The secretion, it excretes, facilitates penetration of the parasite into the abdominal cavity through the intestinal wall of the host. Then, when the young maritae stays in the abdominal cavity of the host, the caecal cells undergo morphological transformation and again produce secretion, which is excreted into the caecal lumen. The apical surface of the cells forms lamellate extensions with microvilli. The basal membrane forms invaginations and contact complexes with extensions of the parenchyma cells. Apparently, these cells (type A) are capable of absorption of low-molecular compounds, which are produced in the caecal lumen during digestion of the food. In developing maritae, which have started parasitizing the hepatic parenchyma, the caecal cells (type B) gradually alter and assume the characteristics of the cells from the adult caeca.

In forming maritae, in addition to the tegumental glands, the so-called metacercarial glands complete their development and start functioning (see sections 2.1.4. and 2.2.2.). In maritae, they are normally called "head glands". In the excretory system, the number of flame cells increases. However, the structure of the excretory formula is unchanged (see section 2.1.6.). Also unchanged is the structure of the flame cells and capillaries. The walls of the collecting tubules and excretory bladder become more massive. Microlamelli of the luminal surface increase in number and size. In developing maritae, profound modifications take place in the systems, which are associated with the excretory system. These are the secondary excretory system of Strigeidida, which loses its regular geometric shape and turns into a net of irregularly arranged channels, and paranephridial plexus of Fasciolidae and some other trematodes (Pearson, 1986).

Modifications of the nervous system, observed during marita development, are connected with differentiation of the body into locomotory and generative parts. A typical orthogon with strongly developed longitudinal cords is preserved only in the first part. In the second one, the shape of the orthogon is no longer geometrically regular, and noticeably thinned cords form a neuroplex (Ginetsinskaya, 1968; Kotikova, 1971). Such differences in the organization of the nervous system in locomotory and generative parts well correspond to their different functions. At the anterior end of the body, mainly sensory papillae are located, which can be non-ciliated or uniciliated (for reviews see: Smyth and Halton, 1983; Halton et al., 1997).

Main events, connected with maritogony, take place in the reproductive system. First of all, it concerns primitive trematodes and strigeidids, where there is no significant differentiation of the genital primordium at previous stages of ontogenesis (see sections 2.1.8. and 2.2.2.). The entire process of formation and development of the primordial organs of the reproductive system, up to the beginning of their functioning, occurs during maritogony. Primordial organs of reproductive systems of many plagiorchiids, in which their differentiation at metacercariae phase stops soon after formation of primordial gonads and ducts, must also complete a complicated and long way, until their formation is fully finished (see p. 168).

At the same time, "imaginized" cercariae and metacercariae with definitive system of organs, far advanced in their development, are typical for a number of plagiorchiids (see pp. 169, 171). Maritogony of such trematodes, with the second type of development (conventionally identified by us), is shortened. Thus, in the aforementioned *Astiotrema trituri*, morphogenetical modifications during maritogony mainly consist of completion of development of the reproductive system and parenchyma (Gerasev and Dobrovolskij, 1977). Normal functioning of the remaining systems of organs (digestive system, glandular apparatus and excretory system, which in maritae remains juvenilized), formed at the previous stages of ontogenesis, enables the worms to rapidly finish formation of the reproductive system and start laying eggs. Starting of egg production does not mean the termination of maritogony. The extensive growth of the uterine loops, and, consequently, the growth of the generative part of the body continues for some time. Development of many trematode species of very different families proceeds similarly.

In some trematode species (representatives of Monorchiidae, Lecithodendrioidea, Microphallidae), processes of maritogony are more simplified, which is primarily connected with the progenetic development of the reproductive system in larval phases. In the course of metacercariae morphogenesis of such trematodes (Microphallidae, some species of *Plagiorchis*), all definitive systems, including the reproductive one, are fully formed (see p. 131). However, in the gut of definitive host, young maritae need a short period of time to complete their development, which makes up the whole third preparatory stage of ontogenesis. During that period, worms grow noticeably, differentiation of the gastrodermis is finished, reproductive systems (first male, and then female) start functioning. In microphallids, a day after infection of the host, first eggs, still abnormal, appear in the uterus. Production of normal viable eggs starts on the second or third day. In the course of accumulation of eggs in the uterus, its loops extensively grow, resulting in an enlargement of the uterine zone of the body. Therefore, the termination of the worms' growth and stabilization of their body shape proceeds, as in the previous cases, after the beginning of the reproductive period.

In species with the third type of development, both the formation of the reproductive system and the beginning of the production of eggs are shifted to the earlier stages of ontogenesis (*Paralepoderma brumpti*, *P. progenetica*). In such a case, development in the definitive host (maritogony as an independent period of life of individuals of the hermaphroditic generation) completely disappears. If sexually mature worms with the uterus, filled with eggs, enter the final host, they survive there only a short time. During this brief period, they actively lay eggs, which are disseminated by animal hosts. The role of the latter can be played by any predator, which engulfs the second intermediate host, infected with sexually mature worms.

Ultrastructure of gonads and ducts of the reproductive system was investigated in species of a number of trematode families, including Fasciolidae, Philophthalmidae, Heteropyidae, Schistosomatidae, Diplostomidae, Gorgoderidae, Hemiuridae, Bucephalidae, Plagiorchiidae, Microphallidae, Allassogonoporidae, Troglotrematidae (Burton, 1967a,b; Grant et al., 1976, 1977; Tihomirov, 1980; Otubanjo, 1980; Hendow and James, 1988, 1989; Awad and Probert, 1989, 1990; Orido, 1987b, 1990b; Podvyaznaya, 1990, 1996; Orido et al., 1994; Nollen, 1997; for review see: Galaktionov and Dobrovolskij, 1987). The male reproductive system is organized rather uniformly. Main distinctions are manifested in its distal regions, i.e., copulative apparatus and genital atrium. The number of testes varies in different species, but, as a rule, it is two. The mature testis is surrounded with a basal lamina and a thick layer of extracellular matrix with extensions of muscle cells, attached to it. They form around the testis a muscular envelope, which provides a pulsating motion that may force sperm into the vas efferens (Burton, 1960). On the periphery of the testis, multibranched somatic cells are found (Grant et al., 1976; Otubanjo, 1981; Awad and Probert, 1989; Podvyaznaya, 1996), which, in some species, form syncytium (Tihomirov, 1980; Erwin and Halton, 1983; Žd'árská et al., 1991). In the testis cavity, there are male germinal cells at different stages of specialization. Groups of such cells, especially young spermatogonia, are surrounded by cytoplasmic extensions of somatic cells. Presumably, the latter have different functions: formation of the inner framework of the testis, separation of spermatogonia of different age, providing nutrition for differentiating germinal cells and phagocytosis of the degenerating ones (Tihomirov, 1980; Galaktionov and Dobrovolskij, 1987; Awad and Probert, 1989).

Vas efferens, originating from the testes, are thin tubules, whose walls can be formed either by a cellular epithelium (*Schistosoma mansoni*, *Allassogonoporus amphoraeformis* – Otubanjo, 1980; Podvyaznaya, 1996), or by syncytium (*Philophthalmus rhionica*, *Maritrema linguilla* – Tihomirov, 1980; Hendow and James, 1988). The epithelium of the vas efferens is lined with a basal lamina, beneath which 1-2 layers of muscle fibers, sunk into the extracellular matrix, are situated. As a rule, the epithelium possesses secretory activity, whereas its luminal surface forms numerous lamellae. In *P. rhionica*, syncytium of the vas efferens forms more prominent foldings, functioning as valves, which impedes the reverse flow of testis fluid (Tihomirov, 1980). Distal parts of the vas efferens usually merge, forming the vas deferens, which empty into the seminal vesicle, localized, as a rule, in the proximal part of the cirrus pouch (Fig. 87). The latter represents a muscular sac, isolated by a basal lamina. Outwards, layers of circular and longitudinal muscles, adjacent to the extensions of parenchyma cells, are localized. Inwards, the modified parenchyma is situated. Its cells form a complex system of laminated extensions, separated by relatively wide layers of extracellular matrix.

The seminal vesicle is followed by an ejaculatory duct, which, in many species, forms several tightly packed loops in the distal part of the cirrus pouch. The structure of the walls of the seminal vesicle and ejaculatory duct may not basically differ from that of the above described in the vas efferens and vas deferens. It can be represented either by a syncytium in most investigated species (Tihomorov, 1980; Hendow and James, 1988; Žd'árská et al., 1991; Sharma et al., 1994), or a cellular epithelium (Threadgold, 1975; Hendow and James, 1988; Podvyaznaya, 1996). Only in schistosomes, is the seminal vesicle, as well as ejaculatory duct, lined with the extension of the tegument (Otubanjo, 1980; Awad and Probert, 1989). Ducts of unicellular

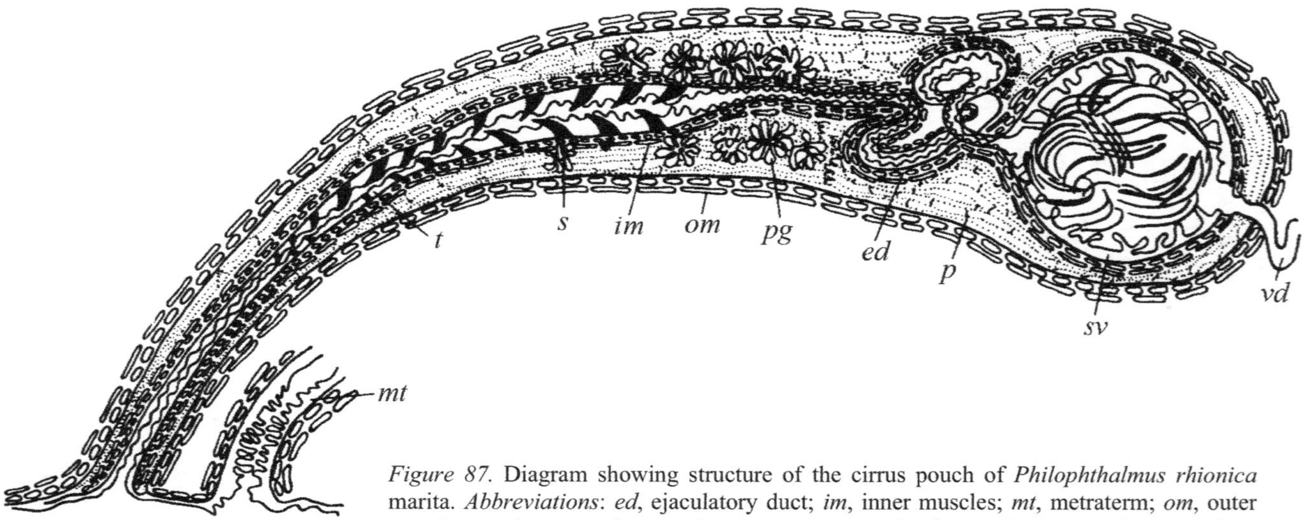

Figure 87. Diagram showing structure of the cirrus pouch of *Philophthalmus rhionica* marita. *Abbreviations*: *ed*, ejaculatory duct; *im*, inner muscles; *mt*, metraterm; *om*, outer muscles; *p*, cirrus pouch parenchyma; *pg*, prostate glands; *s*, spine; *sv*, seminal vesicle; *t*, tegument; *vd*, vas deferens. From Tihomirov (1980), with permission.

prostate glands, which may be represented by cells of one or two types, empty into the ejaculatory duct. Secretion of the glands of the first type is of glycoprotein nature; in the second type it is a phospholipid (Gupta et al., 1983; Wittrock, 1986). According to Wittrock (1986), the function of the secretion of prostate glands may consist of stimulating sperm motility and providing a fluid medium for insemination (1), supplementing the energy supply already carried within the sperm (2) and stimulating the female reproductive tract during copulation (3).

The distal part of the ejaculatory duct in all trematodes is lined with the extension of the outer syncytium of the tegument, which forms foldings and may contain spines (Tihomirov, 1980; Otubanjo, 1985). It is just this part of the ejaculatory duct, lined with the extension of the tegument, or only its distal part that represents the cirrus itself. Only this part is turned inside-out during copulation due to contraction of the outer muscle layer of the cirrus pouch. This is facilitated, first, by extra length of the twisted part of the ejaculatory duct, which is straighten up, when turned inside-out, and, second, by the parenchyma structure. Provided the extracellular matrix is well developed, the laminated structure of the parenchyma provides motility of some parts of the copulative organ, relative to the others.

In representatives of different trematode groups, the cirrus pouch is developed to a different degree. In some of them, it is reduced and, hence, the seminal vesicle and ejaculatory duct lie freely in the parenchyma. However, all reported cases of reduction require accurate investigation. It was long considered that species of genus *Microphallus* lack the cirrus pouch (Belopolskaya, 1963; Deblock, 1971). However, histological studies showed that they do have the cirrus pouch, but it is characterized by a poorly developed outer muscle layer, which is not always detectable in the stained mounts (our data).

The female reproductive system (Fig. 88, 89) begins with the ovary, whose wall structure is similar to that of the testes (Gresson, 1964; Erasmus, 1973; Tihomirov, 1980; Holy and Wittrock, 1986; Podvyaznaya, 1990). In the ovary of young maritae of *Paragonimus ohirai* and *Prosthodendrium ascidia*, supporting cells are found, that surround the germ (steam) cells and are attached to the inner surface of the basal lamina (Orido, 1987b; Podvyaznaya, 1990). Numerous extensions of the supporting cells fill in the entire space between the germ cells and form a three-dimensional net (Orido, 1987b; Podvyaznaya, 1990). In the course of development of the ovary, oogonia and oocytes form there a mass, more and more tightly packed, and displace the net of the supporting cells. Apparently, that is why they have not been described in ovaries of other species, previously investigated. In *Schistosoma mansoni* epithelial cell, lining the inside of the ovary, were observed (Erasmus, 1973). Apparently, they can be interpreted as supporting cells. The function of such cells seems to be similar to that of the supporting cells of the testes (see above).

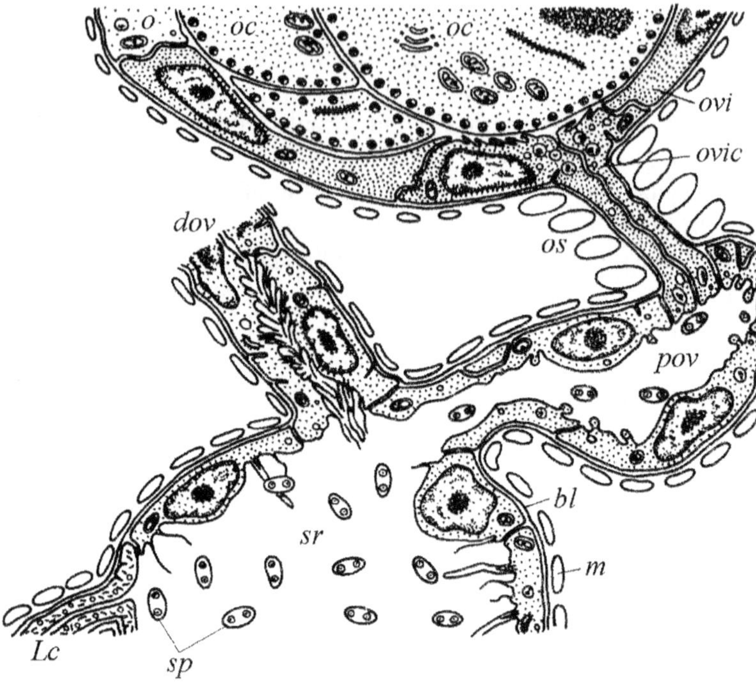

Figure 88. Diagram showing structure of the proximal part of the female reproductive system of *Prosthodendrium ascidia. Abbreviations*: *bl*, basement lamina; *dov*, distal portion of oviduct; *Lc*, Laurer's canal; *m*, muscles; *o*, ovary; *oc*, oocyte; *os*, ovicapt sphincter; *ovi*, ovicapt; *ovic*, cell forming the ovicapt canal; *pov*, proximal portion of oviduct; *sp*, spermatozoa; *sr*, seminal receptacle. From Podvyaznaya (1990), with permission.

The oviduct, originating from the ovary, begins with an infundibular region, called the ovicapt. In *Prosthodendrium ascidia,* the infundibulum of the ovicapt is closed by an annular sphincter, which regulates entry of oocytes into the oviduct (Fig. 88) (Podvyaznaya, 1990). In *Philophthalmus rhionica,* there is no muscle ring in this region. The ovicapt contains a giant cell, which, apparently, functions as a "filter", letting through only mature oocytes, capable of insemination (Tihomirov, 1980) (Fig. 90). The oviduct itself represents a long twisted tube, whose walls are lined with syncytial or cellular epithelium. The ultrastructure of the epithelium differs in different parts of the oviduct. This fact suggests that their functional role is different (Erasmus, 1973; Grant et al., 1977; Tihomirov, 1980; Orido, 1990b; Podvyaznaya, 1990; Awad and Probert, 1990). In most trematode groups, the oviduct is distinctly divided into two regions: the proximal one, lacking cilia, and the distal one, which is ciliated. In *Schistosoma mansoni* and *P. rhionica,*

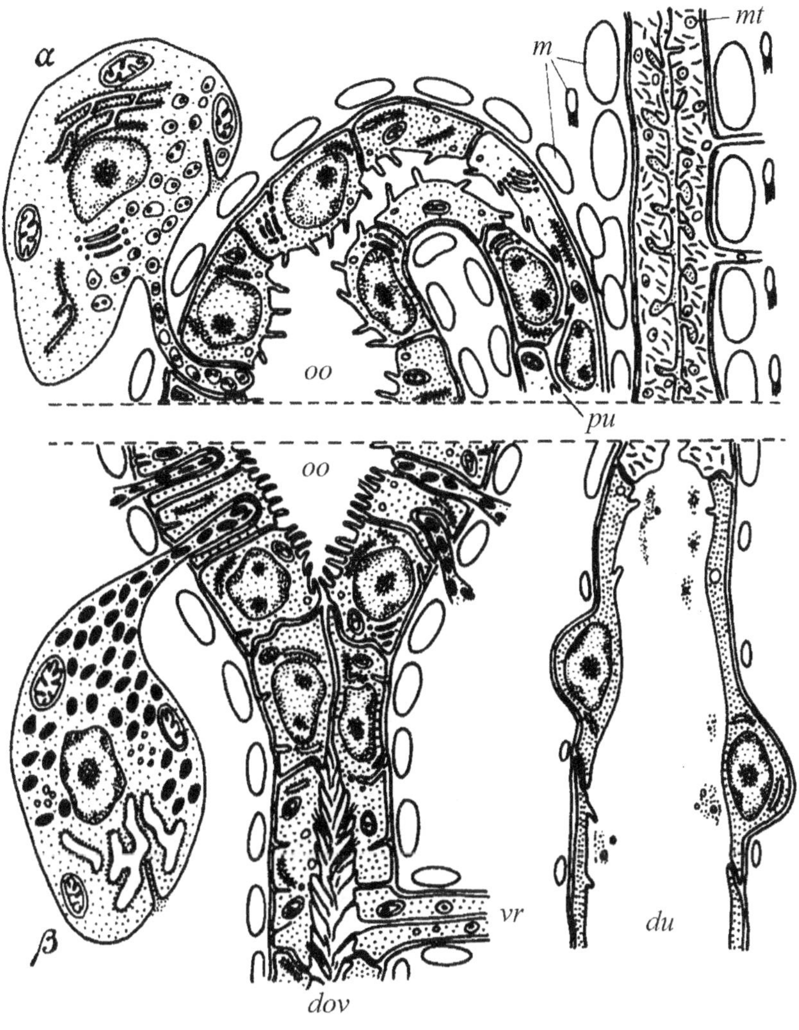

Figure 89. Diagram showing structure of ducts of the female reproductive system of *Prosthodendrium ascidia. Abbreviations*: *α, β,* α- and β-cells of Mehlis' glands; *du*, distal portion of uterus; *dov*, distal portion of oviduct; *m*, muscles; *mt*, metraterm; *oo*, ootype; *pu*, proximal portion of uterus; *vr*, duct of vitelline reservoir. From Podvyaznaya (1990), with permission.

in the proximal region of the oviduct, the luminal surface forms laminated structures (Erasmus, 1973; Tihomirov, 1980). In *P. rhionica,* these structures are numerous only in the most proximal region of that part of the oviduct. Further on, larger cytoplasmic foldings, forming infundibulum-shaped valves, appear (Fig. 90). It is presumed that the system of lamellae and valves of the proximal region of the oviduct prevents spermatozoa from entering the ovary. Possibly, the muscle sphincter of the ovicapt has the same function. In

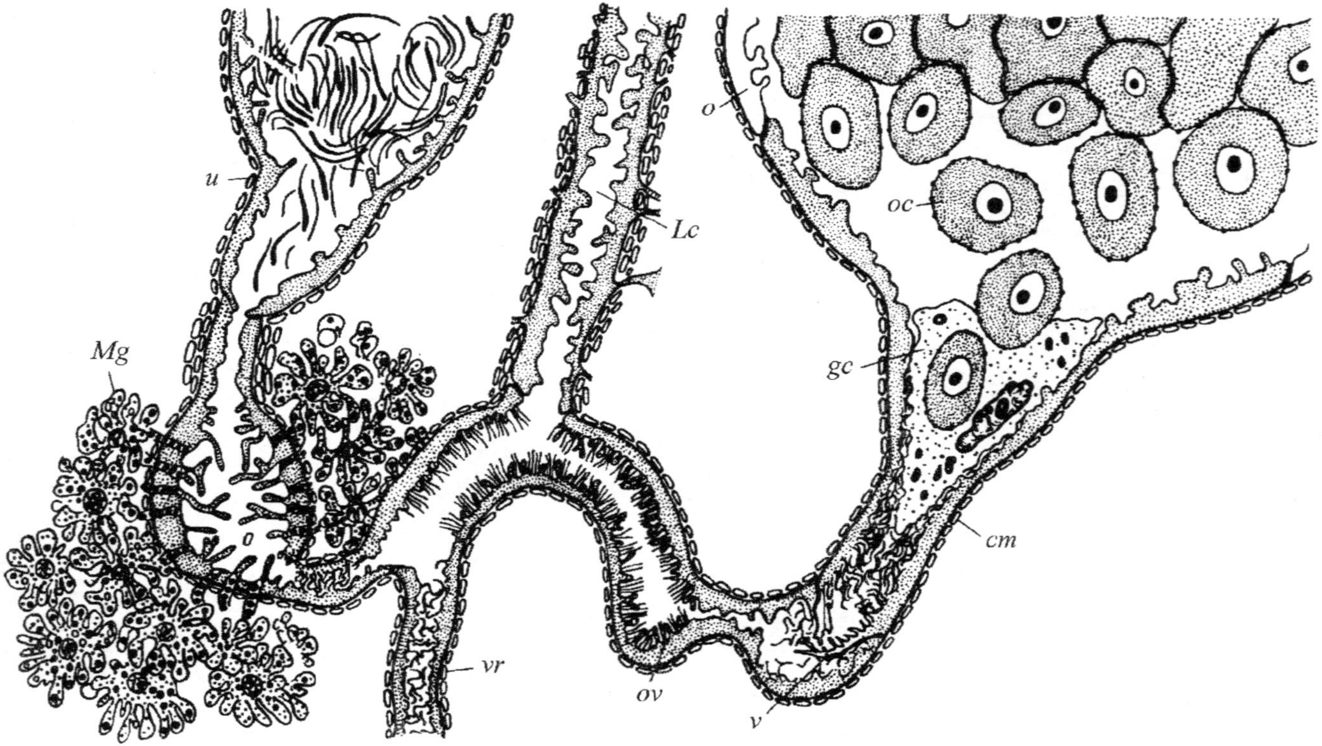

Figure 90. Diagram showing structure of the proximal part of the female reproductive system of *Philophthalmus rhionica*. *Abbreviations*: *cm*, circular muscles; *gc*, giant cell of ovicapt; *Lc*, Laurer's canal; *Mg*, Mehlis' glands; *o*, ovary; *oc*, oocyte; *ov*, oviduct; *u*, uterus; *v*, valve; *vr*, duct of vitelline reservoir. From Tihomirov (1980), with permission.

addition to laminated extensions, the epithelium of the ciliate region of the oviduct contains numerous cilia. Apparently, together with the muscles of the ovary walls, they serve to assist the movement of ova and other material along the oviduct (Awad and Probert, 1990). Erasmus (1973) considered that the oviduct wall also play a role in the lysis of degenerate spermatozoa and oocytes.

The seminal receptacle (if any) is connected to the oviduct. The former is an extensively grown part of the latter. In *Prosthodendrium ascidia*, the seminal receptacle is connected to the oviduct just at the border of its non-ciliated and ciliated regions and hardly differs from the latter in its fine structure. Laurer's canal enters the ciliated region of the oviduct or into the seminal receptacle. The lining of this canal is basically different from that of the oviduct, since it represents a direct extension of the outer syncytial layer of the tegument. If the oviduct is lined with a syncytium, the site of its contact with the walls of the Laurer's canal is distinctly marked with a septate desmosome (Tihomirov, 1980).

Still further, along the ciliate region of the oviduct, there is a site, where the vitelline reservoir enters into the oviduct. Its structure varies in different species. The luminal surface of the reservoir normally has numerous lamellae, which almost nearly completely fill its lumen. In *Schistosoma mansoni* and *Pharyngostomoides procyonis*, it also has cilia (Erasmus, 1973; Grant et al., 1977). Formation of the vitelline cells proceeds in the vitelline follicles, isolated from the surrounding parenchyma by a fibrous basement membrane. The follicles include either only vitelline cells at different stages of development (Grant et al., 1977; Holy and Wittrock, 1986), or also nurse cells (Irwin and Threadgold, 1970; Hanna, 1976; Irwin and Maguire, 1979). Immature vitelline cells are localized on the periphery of the vitelline follicle. In the course of maturation, egg-shell protein globules and "yolk" are accumulated in their cytoplasm. The cells migrate to the center of the follicle, where they break free and pass into vitelline duct (Irwin and Threadgold, 1970; Irwin and Maguire, 1979; Erasmus et al., 1982; Threadgold, 1982; Grant et al., 1977; Holy and Wittrock, 1986) (see also p. 6). Along these ducts the vitelline cells enter the vitelline reservoir.

The distal part of the oviduct is followed by the ootype (Fig. 89, 90). In many trematodes it has a shape of spherical swelling. If this is not the case, the ootype is usually considered lacking, which is not correct. It is more likely, that there are two different morphological forms of the ootype: spherical and tubular. This ootype is characterized by ducts of numerous Mehlis' glands, entering into it. The secretion of these glands plays an important role in egg-shell formation (for details see p. 7). As in the previous regions of the ducts of the reproductive system, the ootype can be lined either with syncytial, or cellular epithelium, whose luminal surface carries microvilli or lamellae (Burton, 1967a; Irwin and Threadgold, 1970; Erasmus, 1973; Grant et al., 1977; Holy and Wittrock, 1986; Podvyaznaya, 1990; Orido et al., 1994). The presence of the latter, alongside with well-developed Golgi complexes and abundant lysosomes in the cytoplasm of the ootype epithelium, allowed Orido et al. (1994) to presume that the latter not only supports the processes from Mehlis' gland cells but also plays a functional role in active synthesis and lysis in the cell cytoplasm. The digestive process in oviduct and ootype may play a part in a feedback system regulating the release of oocytes from the ovary and the supply of vitelline cells and providing some particular materials from the maturation and migration of spermatozoa in the female reproductive tract (Erasmus, 1973; Orido et al., 1994).

The distal region of the ootype narrows and opens into the tubular uterus. The length of the latter and, correspondingly, the number of its loops, depends on the biology of trematode reproduction (Ginetsinskaya, 1968). In such cases, when the development of miracidia is shifted to the period, when eggs stay in the maternal organism, different parts of the uterus may undergo different morphological and functional differentiation. Besides, in species, lacking a seminal receptacle, the function of the latter can be transferred to the proximal region of the uterus (Fig. 90).

However, in any case, the general scheme of the wall structure of the uterus is similar in all its parts (Grant et al., 1977; Rees, 1979; Tihomirov, 1980; Wittrock, 1982; Podvyaznaya, 1990). A basic difference concerns the epithelium of the uterus wall. In *Schistosoma mansoni*, along the entire duct, the epithelium is represented by the extension of the outer layer of the tegument (Erasmus, 1973). Probably such a structure of the uterus is typical for all Schistosomatidae. In all other investigated species, the uterus has its own epithelium, which is the direct extension of the ootype wall. In these trematodes, only the terminal region of the ducts of the female reproductive system, i.e., metraterm, contacting the distal region of the uterus, is formed by a special zone of the tegument. The apical surface of its syncytial layer in this site is folded. In the cytoplasm, secretory granules may occur, differing from those in the tegument of the worm body (Podvyaznaya, 1990). Circular and longitudinal muscles are particularly well-developed in the region of metraterm, which functions as a vagina.

2.3.3. Morphological and Functional Specialization

Comparing morphology of sexually mature maritae of different trematode families, several groups, differing in level and character of specialization, can be distinguished. Most primitive forms, as well as many representatives of other, more evolutionarily advanced taxa, in fact, retain the "turbellarian" organization, which is completed by some adaptations to the parasitic way of life. On the whole, the level of development of organ systems of sexually mature worms remains similar to that of neoophoran turbellarians, which testifies to relatively late transition of hermaphroditic generation to parasitism, as compared to the parthenogenetic ones (Ginetsinskaya, 1968). Initially, this process was probably accompanied mainly by modifications of surface structures. It should be stressed that we will not consider metabolic changes, which could serve a topic of special investigation. The tegument, which was, apparently, characteristic of the "turbellarian"-like ancestor, underwent specialization (Galaktionov and Dobrovoskij, 1987). Tegumental glands, producing material for the glycocalyx, which played the protective role, were formed. Formation of fixative structures – spines – was also associated with the tegument. The glandular apparatus become better developed. Certain changes, leading to intensification of the function, took place in the reproductive system. However, all these modifications did not result in any change in the level of morphological organization, achieved by ancestral forms.

It is just at this stage that we can find maritae of recent trematodes of primitive families (many Echinostomatidae, Notocotylidae, Pronocephalidae), parasitic in the vertebrate gut (see fig. 85C). The latter had been an initial place of settlement of proto-trematodes in the vertebrates (Ginetsinskaya, 1968; Pearson, 1972; Gibson, 1987). Adaptations to parasitism in the above-mentioned digeneans concern mainly surface fixative structures: appearance of collar of echinostomatids, covered with spines and of ventral papillae of notocotylids, which represent specialized glandular organs (Filimonova, 1985). There is no trace of juvenilization or, on the contrary, of any essential complication of organization of maritae of these trematodes.

Adults of digeneans, which secondarily passed over to extra-intestine parasitism (Fasciolidae, Paramphistomidae, Fellodistomidae), undergo more profound modifications. One may recollect the aforementioned paranephridial plexus, which, in Fasciolidae and closely related forms, reaches an especially high level of development. Apparently, the lymphatic system of similar origin, represented by 2 main canals and commissures, connecting them (Fig. 91), is typical for Paramphistomidae and related groups, whose representatives inhabit the stomach, rumen and bile-ducts of ruminants. The walls of the canals are formed either by a nuclei-containing syncytium, whose luminal surface carries numerous cristate extensions (Strong and Bogitsh, 1973), or by membrane-like structures (Dunn et al., 1985). It can not be excluded that these distinctions depend on the age of the parasite. The lumen of lymphatic vessels is filled with fluid, with freely moving cell elements (lymphocytes). The lymphatic system plays an important role in accumulation and transportation of amino acids, lipids and a number of other substances (Sharma and Ratnu, 1982; Dunn et al., 1985).

The ecsoma (see p. 182) of hemiuroid adults is also considered as an adaptation to inhabiting the stomach of fishes, where pH reaches 1-2, osmolarity is high and the adults may be damaged by the hard parts of freshly in-

Figure 91. Lymphatic system of *Orthocoelium scoliocoelium* (Paramphistomidae) adults. From Sharma and Ratnu (1982), reproduced with permission from Journal of Helminthology, 1982, p. 62 © CAB International.

gested prey (Gibson and Bray, 1979). Through the eversible ecsoma transport of nutrients (amino acids) proceeds, whereas the tegument of the soma is greatly thickened and has exclusively a protective function (Matthews and Matthews, 1988a). The role of the gut in Hemiuroidea feeding is still unclear.

Modifications of maritae of trematodes, inhabiting the excretory bladder of vertebrates and feeding on blood cells, present there, are mainly connected with specialization of the digestive tract. In Gorgoderidae, special oesophageal glands emerge, whose secretions have a proteolytic effect (Hoole and Mitchell, 1983a). As in schistosomatids (see p. 187), certainstagesof digestion in Gorgoderidae proceed in the oesophagus. The intestinal epithelium of Fello distomidae has separate digestive cells, which are attached to the extension of the syncytial layer of the foregut (Halton, 1982) (Fig. 92). Each cell functions as an independent "microgut". On its surface, an invagination is formed (the so-called apical cavity), which opens into the gut lumen. Into the apical cavity, blood cells enter, as well as the secretion, containing digestive enzymes and produced by the digestive cell. Thus, extracellular digestion and adsorption of nutrients take place there. As undigested remnants of food (pigment, lipid inclusions) are accumulated in the apical cavity, its content is evacuated into the gut cavity and the entire digestive cycle is repeated.

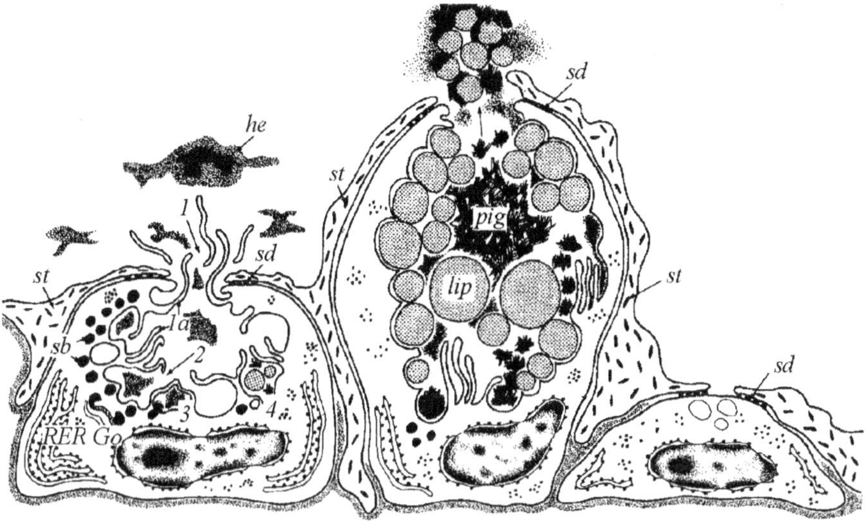

Figure 92. Diagram showing the gut structure in *Fellodistomum fellis* (Fellodistomatidae). A-C, successive stages of cyclic transformation in morphology of a single type of digestive cell (A, digestive cell; B, post-digestive cell; C, pre-digestive cell formed from B); 1, uptake of hemoglobin from semi-digested host erythrocytes in gut lumen; 2, sequestration of food by lamellae within the apical cavity; 3, extracellular digestion by enzymes packed in secretory bodies; 4, accumulation of indigestible residues, including pigment. *Abbreviations*: *Ga*, Golgi apparatus; *he*, hemoglobin; *la*, lamella; *lip*, lipid droplets; *pig*, residual pigment; *RER*, rough endoplasmic reticulum; *sb*, secretory bodies; *sd*, septate desmosome; *st*, supporting tissue (extension of the foregut syncytial layer). From Halton (1982), reprinted with the permission of Cambridge University Press.

Finally, we should discuss blood parasites of the schistosomatids (Schistosomatidae, Sanguinicolidae, Aporocotylidae, Spirorchiidae), which we consider as progenetic metacercariae (see p. 338). Existence in the immunologically active medium (blood) and mixing of II and III preparatory stages of ontogenesis of the hermaphroditic individuals are the two main factors, which resulted in the emergence of a special organization of maritae of those trematodes. The first has determined a complicated differentiation of the surface apparatus, specialization of the gut, formation of a specialized vitelline membrane of the egg (see p. 9). The second is manifested in a certain juvenilization of worms, which possess poorly developed muscles of the tegumentary-muscular sac and suckers and simple excretory system, compared to that of related maritae of strigeidids. Probably, it is just the juvenile organization of schistosomes that facilitated their transition to dioecy. The latter increases genetic variability of the parasites and, thus, helps them to successfully resist immune reactions of higher vertebrates (Basch, 1990; Combes, 1991b).

The above examples hardly encompass all the diversity of adaptive morphological modifications, developed on the basis of "turbellarian" level of organization during transition of trematodes to parasitizing the internal organs of vertebrates, other than the gut. At the same time, evolution of the intestinal forms took two quite opposite directions. One trend consists in more profound adaptations to inhabiting the gut, which caused very complicated morphological and physiological modifications, affecting practically all organ systems and even the body shape. The latter changed from the leaf-like, typical for all platyhelminthes to the cup-like. Eventually, this led to appearance of such an organization of sexually mature individuals, which can be found in strigeidids (see fig. 85A). It already qualitatively differs from that of "turbellarian"-like (especially in higher representatives of the order, such as Strigeidae and Diplostomidae), and its emergence is, primarily, connected, with formation of the adhesive organ. This resulted in division of the body into two parts and formation of a special cup by ventral-lateral sides of the body. The adhesive organ is situated on the bottom of this cup (see pp. 65, 179). Ultrastructural investigations (for review see: Erasmus, 1977) demonstrated that the adhesive organ is a special part of the ventral tegument with spines on its periphery. In the central zone, the surface is folded and may form a complex system of interconnecting canals. Tegumental cells produce secretions, which are discharged outside and take part in digestion of food, preliminarily fragmented by the spines of the peripheral zone. Absorption of these formed nutrients occurs in the region, where foldings of the central zone of the adhesive organ are situated. Thus, the above structure strongly intensifies feeding of the parasite.

In Strigeidae maritae, in addition to the adhesive organ, lappets are formed in the anterior-lateral parts of the body (see fig. 85A). In the region of their location, the tegumental surface carries finger-like extensions. It is into this space, that specialized tegumental glands also discharge their secretion, which participates in the processes of extraintestinal digestion, taking place there. Well-developed muscles of the tegumentary-muscular sac, along with the finger-like extensions, transform the lappets into a powerful organ of attachment, intended to clamp the villi of mucosa of the host's gut. The attachment role is also played by the adhesive organ, together with the cup, which is formed by the anterior part of the body. Thus, this structure appears to be multifunctional. Undoubtedly, the aforementioned secondary excretory system should be also considered as a complication of morphological and functional organization of maritae.

Such adaptations enable Strigeidae maritae to exist in the host's gut for a long time. It is very important for worms, producing a relatively little amount of large eggs per unit time (as compared to plagiorchiids), formation of such eggs being much energy-consuming process. The energy expenditure is connected with the necessity of introducing into the egg of a noticeable supply of yolk, used in the course of miracidium development after the egg has been released to the external environment.

On the contrary, the second trend in morphological modifications of maritae of intestinal trematodes is connected with their gradual juvenilization. This pathway was taken by many groups of digeneans, especially by representatives of plagiorchiid branch (see fig. 85B). In the latter case, a series of sequential simplifications in maritae morphology can be traced. This series is initiated by Opisthorchiidae and Heterophyidae with maritae, whose general organization corresponds to the "turbellarian" level, but with the function of the reproductive system significantly intensified. As in primitive trematodes, additional attachment structures may also develop, e.g., powerful spines of the anterior end of the body in some heterophyids. It should be stressed that juvenilization does not affect such organ systems, without which existence of marita is either impossible (digestive system) or meaningless (reproductive system). However, juvenilization affects different special attachment structures, muscles of the tegumentary-muscular sac and suckers, as well as the excretory system. In maritae of most Plagiorchiida, the juvenile excretory formula is retained. It fully applies to Microphallidae and Lecithodendrioidea, whose excretory system is underdeveloped in larvae (see p. 132). Juvenilization is also manifested in miniaturization of maritae, which is well illustrated by the most juvenilized representatives of the two mentioned families. Their body length rarely exceeds 500 μm. Neither long is their life span. It reaches only several days (7-10). Apparently, the juvenile somatic organization is not capable of supporting a greater longevity of the parasite. However, during such a short period of life, maritae of Microphallidae and Lecithodendrioidea produce hundreds of eggs. Because of this fact, their individual fecundity is, apparently, comparable to that of strigeidid maritae, the latter, as demonstrated above, being the most specialized among trematodes.

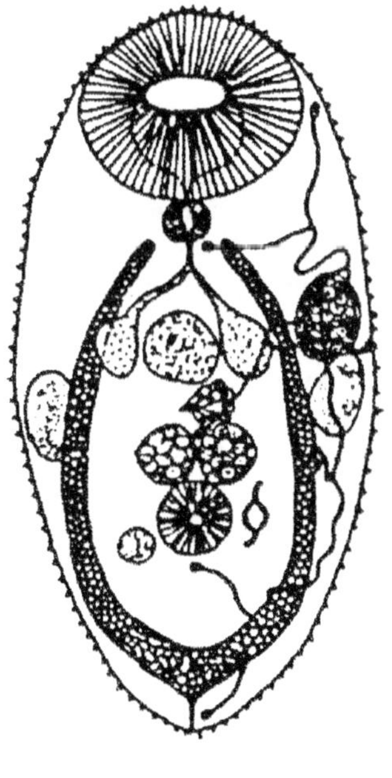

Figure 93. Marita of *Parvatrema homoeotecnum.* From James (1964), reprinted with the permission of Cambridge University Press.

Juvenililzation is not an exclusive prerogative of plagiorchiid branch of the phylogenetic tree of trematodes. It is rather widely spread among different families of digeneans, including those, closely related to the strigeidid branch. In particular, juvenilization and miniaturization are clearly expressed in some species of Gymnophallidae family, most representatives of which have relatively large maritae with the "turbellarian" level of morphological organization. Sexually mature individuals of *Parvatrema* are characterized by an excretory formula 2[(2 + 2) + (2 + 2)] = 16 or even 2[(2 + 2) + (1 + 1)] = 12, and their body length reaches 150-200 μm (James, 1964) (Fig. 93).

---

[i] The only exception having polychaetes as a first intermediate host is specialized family Aporocotylidae whose maritae (adults) parasitize in the blood of marine fishes (Køie, 1982).

[ii] It has been emphasized many times that to ensure the completion of the morphological and physiological maturation of miracidia of these trematodes, the eggs should stay for a period of certain duration in the environment. We fully support the opinion of Semenov (1991), who considers these data as an artifact of experiments. In all these experiments the initial material was obtained by pressing out the eggs from the worm's uterus or by disrupting the uterus with needles. Thus the material taken for the experiment (eggs with fully formed miracidia) was inevitably contaminated with eggs containing embryos at different stages of development (see also p. 238).

[iii] The interpretation of the term "parenchyma" with regard to platyhelminthes is ambiguous. According to some authors (Threadgold and Gallagher, 1966; Matricon-Gondran, 1977; Pedersen, 1983; Kornakova, 1989), it is a type of connective tissue, consisting of specialized cells capable of synthesis of extracellular matrix, transportation of nutrients and a number of other functions, characteristic of connective tissue. According to another point of view, tegumental cytons, muscle elements and the cells of some other organ systems should be included in parenchyma, in addition to parenchyma cells *per se* (Lumsden and Hildreth, 1983; Conn and Rocco, 1989; Conn, 1993). Analysis of literature and our own data lead us to an assumption that in trematodes one deals with parenchymal organization as a whole, rather than parenchyma as an independent tissue. We mean that in the worm body gaps between the systems of organs and tissues are filled with cell elements differing in their origin. Among them, especially at the early phases of morphogenesis of a particular stage of the life cycle, there are undifferentiated cells capable of proliferating, certain amount of specialized cells, which can be regarded as parenchymal cells, and also the cells of other cell populations such as the tegument and muscle of tegumentary-muscular sac. All this heterogeneous cellular conglomerate forms a system, which we traditionally call "parenchyma". The parenchyma composition is also discussed in section 2.3.2. (p. 195).

[iv] The only exception are data on the tail of xiphidiocercariat of *Plagiorchis laricola*, where smooth longitudinal muscle fibers were described (Žd'árská, 1969a). Žd'árská worked with light microcopy and could have missed the striation of the muscles.

[v] Mucoid secretion, excreted on the body surface of cercariae, apparently, forms a specialized glycocalyx or a more complicated supramembrane structure, which plays an important role in providing normal function of the larva in the external environment. It is not possible to state anything more definite on this subject. In fact, electron microscopy and cytochemistry of the supramembrane complex and its derivatives need special methods of fixation and subsequent treatment of the material. Unfortunately, no such investigations have been

done on cercariae, developing mucoid glands in the course of their morphogenesis (Samuelson and Caulfield, 1985; Caulfield et al., 1987).

Chapter 2

THE TREMATODE LIFE CYCLE AS A SYSTEM OF ADAPTATIONS

The trematode life cycle, which usually has a highly complex spatial-temporal structure (see chapter 4), can be completed only if the given species possesses elaborate adaptations to its environment. Viewed in this aspect, the entire complex of morphological-functional and biological traits of different life cycle phases may be treated as a united system of adaptations, which has evolved and has been genetically fixed during evolution. However, it should be kept in mind that the adaptations of free-living and dispersal stages (miracidia, cercariae, adolescariae) and those of parasitic ones (parthenitae, metacercariae, maritae) may have an essentially different character. In this respect trematodes provide a good example of the well-known idea of Schmalhausen (1969) about independent evolution of larval and imaginal phases of the development of a species in the case where these phases exist in different environments and have different functional roles in supporting the species existence.

Applied to trematodes, this means that all their life cycle phases fall into two biological and functional groups. The first group comprises exclusively free-living phases and those performing the dispersion function. These phases are, as a rule, represented by larvae. The second group comprises the parasitic adult phases (parthenitae and maritae), performing feeding and reproductive functions. It should be emphasized that there are quite definite correlations between the adaptations of dispersal and reproductive phases. Besides, it seems appropriate to especially stress one more aspect: general directions and the character of the genesis of adaptations are first of all determined by the parasitic nature of trematodes. They are subject to the general regularities of the parasites' evolution, formulated by Dogiel (1947, 1964): an increase, in many cases, in individual fecundity of reproducing individuals, a tendency to the elimination of active free-living phases and the reduction of time spent in the environment. These evolutionary changes can not be treated as adaptations characteristic only of trematodes, since they are to some extent expressed in all parasites regardless of their taxonomic status in the animal kingdom. In contrast, a concrete manifestation of these tendencies, which is sometimes drastically different in different species, is, in the first place, determined by the ecosystems where these life cycles are completed. This concrete realization is a system of true species adaptations, as they result from a precise "fitting" of the traits of a species (a life cycle) to actual conditions of existence.

These are the theoretical positions from which we will attempt in this chapter an analysis of the adaptation complex characteristic of all life cycle phases of different trematode groups.

1. ADAPTATIONS OF THE FIRST PARTHENOGENETIC GENERATION

The first parthenogenetic generation in the life cycle of Digenea, is characterized by the presence of two alternating ontogenetic phases: the free-living miracidium and the parasitic phase, the mother sporocyst.

1.1. Adaptations of Miracidia

The dispersal free-living phase in the first parthenogenetic generation development, the miracidium, possesses an elaborate complex of morphological-functional and biological features, which are traditionally interpreted as a complex of adaptations for infecting the first intermediate host. The infection of the mollusc is a complicated process, involving several successive stages: the presence in the environment, contacting the host, and, finally, the penetration into the host. Actual realization of these stages and their eventfulness vary greatly in different trematodes, which allows their division into two uneven groups (see p. 2). The first comprises the majority of trematodes, whose actively swimming larvae infect the hosts predominantly by skin penetration. The second group is formed by more specialized trematodes, whose miracidia secondarily lost the ability for active movement in the environment and infect the molluscs passively *per os*. Correspondingly, the character of the larval adaptive features turns out to be very different in these two groups. Equally different is the composition of the stages mentioned, as well as the sequence of the events.

1.1.1. Actively Infecting Miracidia

The first stage in actively swimming miracidia is the most complicated and diverse, being, in its turn, a regular sequence of phases (see reviews: Semenov, 1991; Haas and Haberl, 1997; Haas, 2000).

1.1.1.1. Hatching of Miracidia

The hatching mechanism still remains a subject of discussion. However, it should be emphasized that these discussions are mostly centered on *Fasciola hepatica*, whose miracidia have been studied in most detail in this regard. At present, three main hypotheses could be delimited. The oldest one belongs to Thomas (1883), who had thought that the larvae actively left the egg. This process in *F. hepatica* is accompanied by the thrusting of the operculum by means of muscular effort. After that the miracidium squeezes itself into the hole formed, also actively. Later these observations were not supported, and now the views of Thomas are of a historical interest only.

Much later a very thorough experimental investigation was performed by Rowan (1956, 1957), who completely denied the role of muscular activity in the larval hatching. In his opinion, environmental factors (in case of *Fasciola hepatica*, first of all, light intensity – see below) activate the miracidium, which starts to produce a special "hatching enzyme". This proteolytic enzyme attacks the "cement" material holding the operculum in place. It also causes the destruction of the yolk membrane. All this results in a change in osmotic pressure of the fluid in contact with the miracidium and with a special viscous cushion (Rowan, 1956, 1957). The emerging difference in the osmotic pressure inside the egg and in the environment results in a passive pushing of the rapidly swelling viscous cushion and the miracidium out of the egg.

The negation of the role of larva muscular activity in hatching was also the basis of the third hypothesis (Mattes, 1926). According to this viewpoint, which was later to get substantial experimental support (Wilson, 1968), larval hatching takes place by means of an abrupt increase in hydrostatic pressure inside the egg, resulting from a fast "swelling" of the cushion. The latter, according to Wilson (1968), is not simply a colloid mass but a complicated organized fibrillar mucoid-protein complex, the volume of which considerably increases during hydration. A rapid increase of the inside pressure, in its turn, causes the thrusting of the operculum and the subsequent pushing of the miracidium out from the egg. The larva often leaves the egg with the remains of the enveloping yolk membrane (see pp. 9, 83), which is finally destroyed in the environment.

This process is triggered by the external signal factors, activating the larva. It is the larva that initiates the process of cushion hydration, presumably by means of substances secreted by some glands, and alters the permeability of the yolk membrane and other barrier structures of the egg-shell. In particular, as mentioned above, special hatching glands have been described in some miracidia. These glands are not found in miracidia which have left the eggs (Lengy, 1960; Dobrovolskij, 1965a; Schell, 1961, 1962a,b, 1965). The secretion of the glands of other types, found in the larvae of some species, might also be used for this purpose (see p. 17). Many details of this process remain unstudied. Different trematode species might use various mechanisms of increasing the hydrostatic pressure inside the eggs. However, the "passive" way of hatching appears to be universal. This adaptation seems to be very important, taking into account that miracidia are absolutely endotrophic. To use of muscular activity for hatching (suggested by Thomas (1883)) involves energy expenditure, which seems to be inexpedient, miracidial energy resources being limited.

The details and dynamics of egg-hatching are essentially different in trematodes. These differences are to a large extent caused by two aspects: first, by the place and time of the larva development: in the environment, in the host organism (e.g., in its circulatory system), or directly in the marita, and the structure of the eggshell: the presence or absence of the operculum or special areas along which the eggshell is ruptured when miracidium hatches (i.e., Philophthalmidae). Summing up the available data, the following schemes of these processes can be given. If the embryonation of the egg takes place completely in the environment, the hatching of miracidia becomes possible only after they reach the stage of functional and morphological maturity. However, the very fact of reaching this maturity does not always trigger the hatching mechanism, as spontaneous "automatic" hatching of the larvae from the eggs does not occur.

In most cases, mass egg-hatching seems to be provoked by the impact of certain environmental factors or their fluctuations, e.g., rather broad fluctuations of light intensity and temperature (Smith and Halton, 1983; Xu and Dresden, 1990). Light seems to be the most likely crucial factor as was convincingly shown for the larvae of *Fasciola hepatica* and some representatives of the family Echinostomatidae. The miracidia of *F. hepatica* are released from the egg only when exposed to light. The violet and blue parts of the spectrum are the most important (Roberts, 1950). Their exclusion, as well as complete darkness, almost completely inhibits hatching. However, light appears to influence miracidia only indirectly. In shaded conditions the content of CO_2 in the environment increases. The influence of carbon dioxide upon the larvae is ambiguous. It stimulates egg-hatching at low concentrations, the concentration 2-10 x 10^{-3} M/l being optimal. The concentrations above this level start to exhibit an inhibiting impact upon hatching.

The increased range of species investigated has shown the actual picture to be less universal. Our observations demonstrate that the egg-hatching of *Echinochasmus* sp. does not depend in light intensity fluctuations. The larvae of *Echinostoma paraensei* and *E. caproni* exhibit a clear circadian hatching rhythm, with maximum hatching occurring between 11-16 h regardless of whether the eggs are exposed to light or shaded (Behrens and Nollen, 1992; Markum and Nollen, 1996; Meece and Nollen, 1996). At the same time, circadian rhythms of hatching are absent in *Echinostoma trivolvis* (Nollen, 1994).

The chemical composition of the environment also seems to exhibit a certain influence on the process of egg-hatching. According to the data of Fried and Reddy (1999), the placing of *Echinostoma caproni* eggs in the water (snail conditioned water) which previously contained *Biomphalaria glabrata* molluscs increased the hatching.

The fact that the mass hatching of miracidia is controlled by environmental factors seems to be a very important biological feature of this process. In this way a relatively synchronous release of the larvae present in the biotope is ensured, which, in its turn, conditions the temporary increase of the concentration of miracidia in the biotope. Even this fact as such may be treated as a kind of adaptation, increasing the probability of the molluscan host infection. Besides, the very nature of the above triggers – the increase in temperature and in light intensity, allows us to suggest an adaptive nature of this phenomenon, which ensures the mass appearance of the larvae in the time of the maximal activity of the first intermediate hosts, the freshwater snails of the genera *Lymnaea* and *Planorbis*.

A different situation is observed in the trematodes whose miracidia complete their development in the marita uterus (Heronimidae, Philophthalmidae, Cyclocoelidae, Encyclometridae, Zoogonidae). Data are often given in the literature on the possibility of hatching directly in the distal part of the uterus. This phenomenon has been often described in the representatives of different taxa (*Heronimus mollis, Philophthalmus rhionica, Parorchis acanthus, Encyclometra colubrimurorum, Zoogonus laevis, Lecithaster salmonis*) (Ulmer and Sommer, 1957; Crandall, 1960; Rees, 1939; Tihomirov, 1980; Schell, 1975, our observations).

Ginetsinskaya (1968) suggested to treat this larval feature as an adaptation to the infection of the first intermediate host. However, this seems to be ill-founded. All these data are likely to be artefacts, resulting from an incorrect staging of observations. Crandall (1960) showed for *Heronimus mollis* that egg-hatching in the uterus was observed only after the worms had been placed in water. This was also supported by the results of our observations on *Encyclometra colubrimurorum* from the oesophagus of grass-snakes. The distal part of the uterus of the worms freshly extracted from the hosts contained only fully embryonated eggs, the uterus walls tightly enveloping the eggs. Free-swimming larvae appear in the uterus lumen several minutes after the placing the adult in fresh water. The permeability of the tegument of the dying worms seems to be disrupted and, as a result, water starts to enter the body. This is accompanied by a swelling of the uterus; the eggs therein start to float, and soon the first swimming miracidia emerge.

Later Semenov (1991) showed that gravid maritae of *Philophthalmus rhionica* from the infraorbital cavity of birds and mammals lay their eggs in large numbers only when the tonicity of the environment decreases. This takes place only when the bird or mammals lowers its head into the water and the concentration of the lacrimal fluid drops abruptly. Thus, a considerable number of eggs simultaneously enters the infraorbital cavity, and are immediately washed into the environment. At the same time, an abrupt decrease in tonicity triggers the mechanism of egg-hatching. This process lasts a couple of minutes, and as a result numerous miracidia of about the same age enter the water.

The representatives of the family Cyclocoelidae possess a very similar mechanism of release of fully embryonated eggs (our observations). These trematodes parasitize in the pharynx and the respiratory organs of birds. In this case the active hatching of the larvae also starts only after the eggs enter fresh water, i.e., hatching is stimulated by an abrupt drop in the tonicity of the environment.

In general, the above "phenomenon" of maritae producing active miracidia simply can not exist in natural conditions: the overwhelming majority of species mentioned inhabit the digestive tract and the respiratory organs. The existence of free-swimming miracidia in these host microhabitats is not possible and the release of the larvae from such microhabitats into the environment is also not possible.

Such a mechanism of egg-hatching based on changes in the environment tonicity is also present in a number of trematodes, whose eggs are embryonated outside the adult, though not in the environment either as is the case in the Schistosomatidae. The eggs of schistosomatids get into the circulatory system of the host and embryonate there. After that they get into the lumen of the intestine or the excretory bladder through the rupture of capillary walls of these organs. The rupture appears only when numerous eggs have accumulated in the capillaries. The release of the infective agents into the environment takes place during defecation or urination. In between these acts, the eggs concentrate in the lumina of the corresponding organs. The rhythm of egg release is thus determined by the biological features of the definitive host. The host also determines the nature of the spatial distribution of eggs in the environment.

Miracidial hatching starts almost immediately after the eggs get into water as a result of an abrupt change in the tonicity in the environment[i]. Numerous local groups of larvae of the same age may occur in the biotope in this case as well (see also 3.1. in the chapter 4).

The most specialized representatives of the order Schistosomatida have acquired special adaptations to facilitate the release of eggs into water. Some schistosomatids have a characteristic egg shape, with a lateral spine, which is considered by most investigators as a morphological adaptation of this kind. It is thought that it is the presence of these spines that makes it possible for the eggs to aggregate densely in the capillaries, thus facilitating the emergence of local ruptures in the walls of the affected organs.

1.1.1.2. Morphological Adaptations of the Actively Swimming Miracidia

The organization and behaviour of actively swimming miracidia are to a great extent determined by the functions of these larvae in the trematode life cycle, and the "search" and infection of the first intermediate host. To complete these functions, the larvae should possess a strictly defined set of structures. Some of them are clear provisory structures that have been acquired secondarily in the evolution of trematodes, whereas some may be treated as recapitulations of a kind.

Functionally, the features of the organization of actively swimming miracidia are determined by the presence of four complexes of organs: locomotory, sensory, penetration and generative. Whereas the former three are always fully formed, the latter, except in a few cases, is represented by a more or less developed primordium (see p. 418).

The microscopic size of miracidia, comparable to that of most ciliates, dictates the use of locomotion by means of ciliated epithelial plates. The nature and origin of these structures will be discussed later (see pp. 408, 493). However, their presence may be considered an important morphological-functional adaptation ensuring a fast and manoeuvrable locomotion.

In our opinion, the distribution of reserve energy resources also favours the idea of the adaptive nature of the ciliated locomotory apparatus. In the majority of actively swimming miracidia (*Fasciola hepatica, Echinostoma revolutum, Echinochasmus* sp.), which shed their epithelial plates during penetration into the intermediate host, the epithelial plates contain considerable glycogen reserves. Being unable to use exogenous sources of energy, the miracidia spend their reserves on locomotion, leaving the glycogen deposited in the internal larval tissues ("parenchyma") almost untouched. The larva starts to use this glycogen only a little time before its death, when it has already lost its infectivity and is already unable of active locomotion. A purely adaptive nature of this dynamics of the reserved carbohydrate use is not to be doubted. If the larva meets a specific host, provisional epithelial plates are shed, and the parenchymal glycogen remains the last source of energy for overcoming the host tissue barriers and subsequent migration in its body.

The very act of penetration takes place by means of the activity of muscular elements of the larval tegumentary-muscular sac. The penetrated larvae also need at least a little starting quantity of glycogen. First, a young mother sporocyst starts the process of final formation of definitive covers immediately after the penetration. Until this process has been completed, there is no mention about the active feeding of the parasite. Second, carbohydrate absorption from host tissues is an energetically costly process (Phifer, 1960; McManus and James, 1975a,b) and its initiation requires at least some "initial" energy reserves in the parasite's body.

Another variant of the distribution of glycogen and the dynamics of its use is observed in the miracidia of Philophthalmidae. They bear a fully formed, yet juvenile redia. After the anchoring of the larval forebody in host epithelium, the redia makes its way inside the host itself. In this case epithelial plates are not shed (Sinha, 1978; Tihomirov, 1980) and miracidia do not start parasitizing the mollusc and do not transform into a mature mother sporocyst (see p. 418). In this case, glycogen reserves are not divided into two independent functional pools, "locomotory" one and that ensuring miracidial penetration and the initial stages of its metamorphosis. There is almost no glycogen in the epithelial plates of *Philophthalmus rhionica*, its "parenchymal" reserves being evenly spent throughout its active life (Ginetsinskaya et al., 1981).

The miracidial locomotory apparatus also includes the muscular elements of the tegumentary-muscular sac: circular and longitudinal muscles directly below the epithelial plates and hypodermal ridges. Their participation in locomotion, however, is limited. As shown by Roberts et al. (1980), the larvae change the direction of their movement, bending their forebody by means of the activity of muscle cells. Roberts et al. (1980) did not observe any changes in the metachronal wave of the cilia.

Miracidial locomotory apparatus is functionally tied in with the nervous and sensory systems. As mentioned above (see p. 19), in active larvae these systems are sometimes well developed and highly differentiated. This is not surprising since miracidia have to "solve" complex behavioural problems, associated with finding, identifying and infecting the molluscan host. The adaptive nature of these morphological-functional larval features is not to be doubted.

First of all, the relative size of the miracidial brain (ganglion) is impressive. It often occupies from ⅛ to ¼ of the body size. The larvae possess from 5 to 10 types of sensory papillae. Unfortunately, only photoreceptors (eyes) can be reliably identified functionally, the purpose of other sensory structures remaining unstudied. Knowing what it is that the larvae react to, we may only guess that the diversity of these structures included distant and contact chemoreceptors, tangoreceptors and, possible structures functionally close to statocysts. The only thing we may be certain about is that possessing this variety of receptors, the larvae obtain a lot of diverse information about the environment, which allows them to react sensitively to all external changes and disturbances.

The penetration apparatus was formed in the larvae of the first parthenogenetic generation as a joint morphological-functional adaptive complex. This apparatus is to a great extent represented by provisory structures (see p. 16). It comprises several organs: terebratorium, a set of secretory cells and, finally, the circular and longitudinal muscles of the tegumentary-muscular sac, mentioned above. The latter only to a limited extent participate in locomotion, but play the main functional role in the larval penetration into the molluscan host. Unfortunately, our knowledge about the ways of penetration of miracidia of different species is insufficient. The available data have been obtained from a limited number of species (mostly *Schistosoma* and *Fasciola*). Therefore, it is not possible to discuss the adaptive significance of the variations in the structure of the organs mentioned. It is not doubted that the terebratorium ensures the attachment of the miracidium to the molluscan host body surface. Telescopic terebratoria have been shown to function in this process as suckers (Coil, 1981). As for the details of functioning of spherical terebratoria, which seem to be characteristic of most trematodes, they remain unknown. It has been suggested that in this case the larva anchors to the body of the future host by means of adhesive secretion of special glands which open at the anterior body end (Dawes, 1960; Wajdi, 1966; Wilson et al., 1971).

Telescopic terebratoria seem to have a similar structure in all trematodes which possess them: Fasciolidae, Echinostomatidae, Philophthalmidae, Cyclocoelidae. They are less complex in fasciolids (Wilson, 1969a,c, 1971) than in other trematodes (Tihomirov, 1980), yet their functional and structural homology is not doubted. Moreover, since telescopic terebratoria are present in the representatives of the most archaic families, they are generally supposed to be the primary terebratorium type, a true plesiomorphic character. However, this is unlikely. Telescopic terebratoria most probably emerged secondarily, as the result of a profound specialization of the muscular elements of the anterior body end. Their acquisition entailed changes in the mechanism of epithelium penetration.

Telescopic terebratoria in the miracidia of archaic trematodes participate actively in the process of penetration. A short time after the attachment to the molluscan body surface, the terebratoria penetrate the epithelium and root deeply there, by means of the activity of specialized muscular elements (Køie et al., 1976; Coil, 1977). Spherical terebratoria, lacking the complex system of muscular fibers, can not perform an activity like this. In this case the main role in the disruption of the host's epidermis naturally belong to the secretory elements of the penetration complex. In general, the tendency to the elaboration of chemical ways of affecting the host epithelium seems to be common for the larval stages in the trematode development. A similar process can be clearly traced in cercariae (see p. 275). A relatively early specialization of the penetration apparatus in this direction explains retention of a relatively simple terebratorium in the representatives of specialized families. Such a mechanism of epithelium disruption is likely to be beneficial and therefore adaptive. A local application of the secreted substance lessens the traumatic impact of the penetrating parasite, thus stabilizing the host-parasite system. Besides, epithelial lysis requires essentially less energy expenditure, which is of utmost importance for any endotrophic dispersal larvae, including miracidia.

Unfortunately, our knowledge about the functioning of secretory elements is also far from complete. The researchers differ essentially in their opinion about the functions of different glandular types. Thus, some investigators consider the apical gland to be the penetration gland (Mattes, 1949; Najim, 1956; Wootton, 1957; Dawes, 1960; Wajdi, 1966; Wilson et al., 1971). It is the secretion of this gland that possesses histolytic activity, causing local disruption of the epidermis of the mollusc attacked by the miracidia. However, in a number of articles this function of the apical gland is completely denied (Goodchild, 1948; Lewert and Lee, 1954; Cannon, 1971). Its adhesive substance is supposed only to facilitate the attachment of the terebratorium to the molluscan surface.

Equally contradictory are the opinions about the possible functions of lateral glands. Some researchers think that they are true penetration glands (Goodchild, 1948), whereas others consider them as true "adhesive" glands (Wajdi, 1966; Wilson et al., 1971). Southgate (1970) has suggested that the secretion of the lateral glands functionally participates in the processes of transition from miracidial epidermis (epithelial plates) to definitive structures of the mother sporocyst (the tegument) (see p. 29).

However, despite all these contradictions, nobody doubts that all these glandular cells taken together (the apical gland, lateral glands and some others) (see p. 17) ensure a local lysis of the epithelium (Coil, 1977; Buzzel, 1983).

The disruption of the epidermis is only the first step in the process of active penetration, for the parasite has to make its way in relatively dense host tissues and, having completed a more or less long migration, to reach the place of final settlement. It can not be ruled out that the secretory elements of the penetration apparatus keep on working during this period as well. At least, it has been shown for two fasciolid species that by far not all the secretion of the apical gland is released at the moment of penetration of the epidermis, and the secretion of lateral glands causes the destruction of collagen fibers of the subepidermal connective tissue (Coil, 1981; Buzzel, 1983). Our observations of the *Echinostoma caproni* miracidia have shown that many juveniles retain all the elements of the penetration apparatus throughout migration. But the main role in the penetration process and the subsequent migration across the tissues and lacunae of the haemocoel is played by the somatic muscles of the parasite. All investigators who have observed and described the process of miracidial entry into the host body note the active work of the tegumentary-muscular sac of the larvae, which contracts and relaxes in turn (Wilson et al., 1971; Coil, 1977, 1981, Semenov, 1991, our data). It is then that glycogen reserves are used, which have been stored in the internal larval tissues (see p. 221).

The three functional complexes mentioned – locomotory, sensory and penetration, have a clear provisory nature. They are undoubtedly adaptive, having been formed in the process of trematode evolution as a complex of adaptations for successful infection of the first intermediate host, the mollusc. Nothing of this kind applies to the generative larval structures. The degree of generative primordium development does not influence either the larval behaviour or their infectivity. The generative primordium is not a provisory structure. In most cases, the functional activity of this system is shifted to another phase of the life cycle. It is this circumstance that seems to explain a tendency for the lag in development and differentiation of the generative primordium in the miracidia of more specialized trematode families (Strigeidae, Diplostomidae) as compared to archaic ones (Fasciolidae). The former have it as a primordium indeed, which gives rise to a functioning system (the germinal mass) only in mature mother sporocysts (see p. 75), whereas in the latter the germinal structures do not fall under the definition of a "primordium". Not only don't they undergo any subsequent ontogenetic transformations, but they have already completed their functioning (see p. 418).

Without embarking now upon the problem of the primary state of the generative primordium of the larvae of the first parthenogenetic trematode generation (see p. 418), we would like to note that a clear retardation of its development in the larvae of more specialized trematodes might be also treated as a type of adaptation. The shift of reproduction to a later developmental stage, especially parasitic one, makes it possible to enhance greatly the generative function of the mother sporocyst, using the energetic potential of the host (see p. 493). This, however, is an adaptation of another kind. It does not involve the properties of the larva (as noted above, the degree of generative primordium development does not play any role in the successful completion of the main miracidial function, infecting the mollusc). This adaptation alters the dynamics of the parasite's reproduction, which is expressed at the level of the life cycle as a whole functional system.

1.1.1.3. Behaviour of Miracidia in the Environment

The organization features of the free-swimming miracidia mentioned above are the morphological-functional basis of their complex and diverse behaviour in the environment. In the course of their free life, a rather rigid, genetically determined behavioural program is completed, involving several subsequent phases. The onset of each of the next stages is stimulated by the impact of strictly determined factors. If the larva is not subject to their influence during its lifetime, there is no transition from phase to phase. The interpretations of larval behavioural features by different authors vary quite significantly.

The phase starting with the miracidial release from the egg may turn out to be the only one, if the larva isn't subject to the influence of a potential host. This stage is characterized by active locomotion, so that the larvae can scan a considerable space. In this respect the abilities of the miracidium turn out to be rather significant, due to the high speed of its movement and maneuverability (the ability to quickly change the direction of movement).

Miracidia start active locomotion almost immediately after hatching. The larvae of different species move in a rather similar way. Their speed is very high. The larvae of *Fasciola hepatica* swim with a speed of 4 mm/s (Mattes, 1949). This means that they cover their body length more than 20 times a minute. Similar results have been obtained for the larvae of *Philophthalmus rhionica* (Semenov, 1991). However, it is only freshly hatched larvae that move that fast, since they have not yet used their glycogen. In the course of time, the speed of the larvae decreases significantly.

Not only the speed but also the trajectory of the larval movement changes with its age. The larvae move mainly in a rectilinear way for some time after hatching. However, later on start the larvae to make turns, with their frequency increasing gradually. This ability to change the direction of movement, increasing with age, has been noted for the miracidia of different species (Semenov, 1976; Saladin, 1979). The larvae turn even in a very "homogenous" environment balanced as to the influence of physical-chemical agents (Semenov, 1991). The choice of the movement direction is arbitrary (Plorin and Gilbertson, 1981). As a matter of fact, larval behaviour turns out to be more complicated, as they possess a species-specific set of behavioural reactions, taxes and kineses (Fraenkel and Gunn, 1961). The larvae react to the physical-chemical parameters (light, the direction of gravity, water current, chemicals) by changing the direction of their movement in a regular way. Thus, the more "heterogeneous" is the environment as to the factors mentioned, the more complex is the miracidial trajectory of movement. However, natural conditions under which miracidia exist are usually very heterogeneous, as compared to that in experimental plants.

Since the earliest investigations of miracidial behaviour, attempts have been made to interpret their species-specific orientation in the environment as complex behavioural adaptations ensuring that the larvae get into the area of the most probable contact with the molluscan host (Wright, 1959) (Fig. 94A). This looked convincing if the preference of the molluscs corresponded closely to those of miracidia. Numerous examples can be given here. As shown by Vladimirov (1962), the miracidia of *Posthodiplostomum cuticola* demonstrate a negative geotaxis which make them remain close to the water surface, and a negative phototaxis, confining them to shaded places. At first sight, this combination seems to be contradictory, as the larvae with a negative phototaxis might be expected to concentrate at the bot-

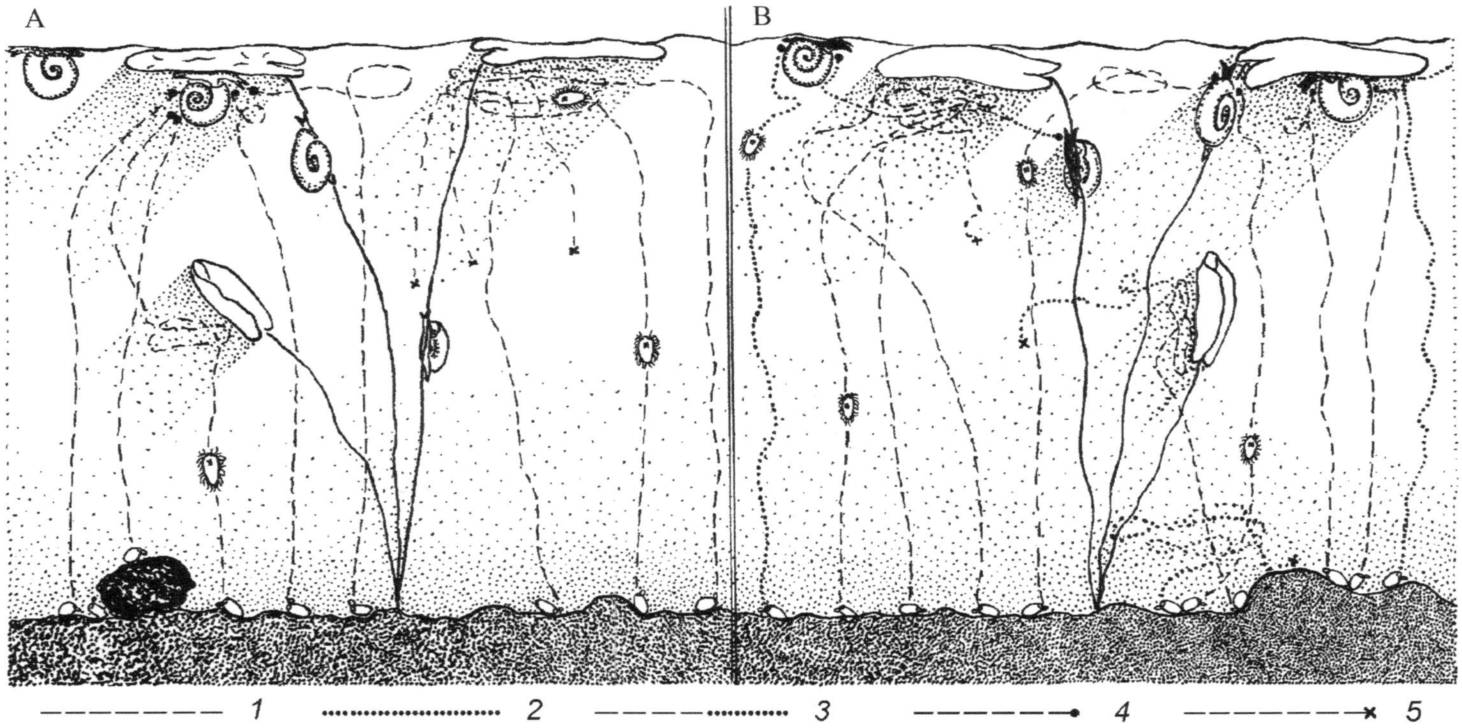

Figure 94. Scheme illustrating the miracidial movements after hatching. A, miracidia concentration in the areas of the most probable contact with the molluscan host (in accordance with Wright's (1959) hypothesis); B, dispersion of miracidia initially differing in manifestation of their taxes and exhibiting taxes changes with age (in accordance with Semenov's (1991) hypothesis). *Abbreviations*: *1*, movements of miracidia under the influence of the basical taxis type; *2*, movements of miracidia whose taxis initially deviates from the basical taxis type; *3*, age changes in the taxis sign; *4*, infection of the molluscan host; *5*, miracidial death.

tom of the water body, where the light intensity is lower than at the surface. To explain this contradiction, an idea about the possibility of dominance of one taxis over another has been put forth. The dominating negative geotaxis keeps the miracidia close to the water surface, so that they look for shady places only there. The most dense shade is given by the leaves of water lilies and other aquatic macrophytes. It is also on the lower side of these leaves that small planorbid molluscs sometimes concentrate in large numbers, the first intermediate hosts of *P. cuticola.* At first, nobody doubted that this complex of behavioural reactions is a complex adaptation increasing the probability of molluscan infection.

These ideas were developed in the monograph by Ginetsinskaya (1968). Their popularity spread, as examples similar to that mentioned above turned out to be quite numerous. The "coincidence" of miracidial responses and biological features of molluscan hosts has been also demonstrated for a number of other species such as *Fasciola hepatica* and *Galba truncatula, Schistosoma mansoni* and *Biomphalaria glabrata, Schistosoma haematobium* and *Bulinus globosus.*

In reality, events turned out to be more complicated. First of all, it was established (at first, for the phototaxis) that even the offspring of the same adult differs by this character. Thus, *Philophthalmus rhionica* miracidia of the same age (1h) hatched from the eggs laid by the same adult, predominantly demonstrated a positive phototaxis (68% of the larvae). The rest (32%) either had the negative phototaxis or did not react to the position of the light source at all (Semenov, 1991).

The general picture became even more obscure when it was shown that larval taxis changes with age: photopositive larvae become photonegative. Moreover, even a short-time impact of other stimuli, e.g., chemical ones, also resulted in the reversal of a positive reaction to light to a negative one in the miracidia of *P. rhionica.* Similar data have been obtained for the miracidia of other species (Yasuraoka, 1954; Takahashi et al., 1961; Chernin and Dunovan, 1962; Mason and Fripp, 1977; Keshavarz-Valian and Nollen, 1980). The results of all these works were twofold. Firstly, it transpired that the completion of the miracidial behavioural program is controlled by numerous factors, which were often left outside the scope of the experimental design. Hence, numerous contradictions exist in the literature, where the researchers, working with close or even the same species, got very different results. Secondly, and most importantly, the elegant and logical theory of Wright was shattered (Wright, 1959). Saladin's attempts to improve it (Saladin, 1979) can not be considered successful. In essence, Saladin's additions consist in the subdivision of the first stage of the miracidial life into two substages. The first is the dispersion, when the larvae move at a high speed and follow a rectilinear trajectory. Their infectivity at this time is supposed to be lower. As a result, the maximal possible dispersion of the larvae in space is achieved. Age changes in the taxis signs, the speed and the trajectory of movement, as well as the increase in the infectivity mark the onset of the second period, the "search" of the places where the probability to meet the molluscan host is highest. The logic of this variant of Wright's hypothesis can not be denied. Unfortunately, it does not correspond well to present data.

Semenov (1991) suggested another interpretation of the miracidial behaviour at the first stage of their life. He completely denied the adaptive nature of the larval behaviour as understood by Wright (1959) and many other researchers, as well as the existence of dispersion *sensu* Saladin (1979). Semenov (1991) thought that the actual dispersion of the miracidia in a water body resulted from the initial heterogeneity in the composition of the larvae recruited to the environment and from a complex dynamics of manifestation of their behavioural reactions (taxes and kineses), combined with the heterogeneity of the environmental conditions, which could not be measured by the present-day techniques. This very dispersion is the final aim of the completion of the behavioural program of the first stage of the miracidial existence in the environment. Semenov (1991) seems to be right, when he writes about the insignificance of the fate of an individual larva for the completion of the life cycle as a holistic system. The mechanism of compensation for the consequences of the death of large numbers of the larvae is incorporated into the life cycle structure of the digeneans, involving two stages of parasites' proliferation and ensuring an intensive recruitment of infection (see p. 328) into the ecosystem where the trematode population exists. In this respect, miracidial behaviour is adaptive, but these adaptations are revealed not at the level of an individual organism, but on the level of the whole local group (=local hemipopulation – see p. 361) of larvae, present in the water body at a given moment. In such cases we deal with group selection: it is not that every larva **must** infect a host (as a matter of fact, this would be catastrophic both for the host population and the parasites themselves), but the distribution of the group members in space, determined by the individual traits of their behaviour at every given time moment, should ensure the possible meeting of at least a part of the larvae with their potential hosts.[ii]

It should be noted that in relatively large water bodies an even distribution of larvae can not be achieved. The larvae form a kind of "cloud" with a relatively well-pronounced density gradients. The cloud itself is not constant. A redistribution of miracidia is constantly going on in the cloud, in accordance with the rapidly changing (sometimes, at the microscale) environmental conditions. The dynamic nature of such spatial structures, sensitively reacting to any change in the conditions, could be treated as a group adaptation of the larvae. The evolution of adaptive mechanisms of miracidial behaviour in response to the influence of physical environmental factors seems to have taken place by means of the selection of behavioural reactions not at the level of individual miracidia, but at the level of their local groups (Semenov, 1991).

Though we agree with Semenov (1991) in most part, it is difficult to ignore the idea that often the observed coincidence of biological features of larval behaviour and the preference of the molluscs, which are their specific first intermediate hosts, is absolutely random. In the first hours of life, when most of the miracidia still possess primary taxes, yet unmodified by the age and environmental factors, the larvae inevitably find themselves in the areas where the probability of meeting the first intermediate host is highest. If the host is there, its infection may also take place. If the host is not there, the mechanisms described above are switched on, ensuring the maximal and dynamic dispersion of infection.

In summary, the following scheme of the larval group behaviour can be suggested (Fig. 94B). After hatching, most of the miracidia move, under the influence of primary taxes, into such areas where their chances of finding the suitable molluscan host are the greatest. The primary role in this period seems to be played by photo- and geotaxis. At the same time, a lesser number of the larvae possessing initially modified taxes, occupy, correspondingly, other areas. This is a kind of a group reserve, ensuring the maximal dispersion from the very beginning. In the course of time, a considerable part of the larvae from the first group, not having met a suitable host, start changing their behaviour under the influence of age, abiotic and biotic factors. These processes are asynchronous in different individuals, which ensures their gradual redistribution in space. As a result, the second dispersion stage starts, in which all of the local hemipopulation of the same age are involved. On the one hand, in stable ecosystems the recruitment of infection into the environment takes place over a long time span. On the other hand, it is discreet due to the biological traits of the vertebrates, the definitive hosts, dispersing the infective agents. Therefore, the consequently forming miracidial groups of different ages are at different dispersion stages. There may exist in the water body both low concentrations of young, very active larvae in the areas where the probability of meeting the mollusc is the highest, and the overlapping, strongly dispersed "clouds" of older miracidia.

Free-swimming larvae enter the second stage of their life when they are in close proximity to the potential host, its presence being already perceived by the larva as an independent environmental factor. From this moment the larval behaviour is individualized, the role of "chance" is greatly decreased. The nature of the signal perceived by the miracidium has long remained a most widely discussed problem in the literature. From the beginning a logical supposition was voiced that miracidia possess chemical sensitivity, allowing them to detect and to differentiate the signals emitted by potential hosts. This idea was strongly supported by Faust (Faust, 1924; Faust and Meleney, 1924; Faust and Hoffman, 1934) and, after them, by many other researchers. Yet, opponents also appeared (Mattes, 1936, 1949; Sudds, 1960), who tried to show experimentally the absence of any kind of chemotaxis in miracidia, hence denying the role of chemical sensitivity in the search and finding of the molluscan hosts by the larvae. Following the improvement of experimental techniques used for the study of the larval behaviour, the number of researchers certain that miracidia possessed differentiated chemical sensitivity was steadily growing. Having analyzed an enormous amount of the literature in great detail, Saladin (1979) came to an unambiguous conclusion that miracidia did possess chemosensivity and used it for finding molluscan hosts. Later Semenov (1991) showed that that the conclusions of a researcher did not always correspond to the results obtained. In this respect, the analysis of the works of Sudds (1960), one of the main opponents of the idea about the presence of chemical sensitivity in miracidia, is most demonstrative. In fact, Sudds' results testified to the presence of this sensitivity, as he observed the modification of larval behaviour after the impact of the chemical factor almost in all experiments. Semenov (1991) seems to be absolutely right when he writes that Sudds' data testify only to a relatively low specificity of the substances attracting the larvae, but not to the absence of chemical sensitivity in them.

Thus, the problem of the existence of chemical sensitivity in miracidia can be considered finally solved. But this is only the first part of the problem, as it is also important to assess the role of this sensitivity in miracidial behaviour in the process of infection of their molluscan hosts. It is equally as important to reveal the nature of chemical signals perceived by the miracidial chemoreceptors.

Firstly, it may be considered proved that the miracidia of different trematode species do react sensitively to chemical substances (Reizstoffe, according to old authors, or miraxones, according to Chernin, 1970), produced by their molluscan hosts (Wright, 1966a; MacInnis, 1965; Shiff and Kriel, 1970; Chernin, 1970, 1972; Bennett et al., 1972; Roberts et al., 1979; Keshavarz-Valian et al., 1981; Kalbe et al., 1996, 1997; Haas and Haberl, 1997; Haas, 2000). This is first of all expressed in the modifications of larval behaviour.

The nature of behavioural reactions is not much argued about. Most researchers agree that these are either reactions of the chemokinesis or chemotaxis type (Fraenkel and Gunn, 1961). In the majority of cases it seems to be chemokinesis (Kalbe et al., 1996, 1997; Haas and Haberl, 1997; Hass, 2000). Under the influence of chemical signals in the environment the miracidial trajectory changes: as the concentration of the stimulating agent grows, the rate of change of direction (RCD) also increases, whereas lowering of the concentration results in a pronounced turnback swimming. In some cases, however, the larvae exhibit not a chemokinesis but a true chemotaxis, i.e., a directed movement along the gradient of the chemical signal concentration. For instance, this is characteristic of the *Schistosoma japonicum* miracidia (Haas et al., 1991). They consider this a substitution of one type of behavioural reaction by another to have been caused by the fact that

that small molluscs of the genus *Oncomelania*, the first intermediate hosts of *Schistosoma japonicum*, dwell in muddy habitats.

The possibility of the manifestation of species-specific behavioural reactions in the miracidia of different digeneans can not be completely ruled out. However, it is at present very difficult to judge the extent to which this specificity can be manifested, as the number of species investigated in this respect is so far infinitesimal.

In fact, a third variant is also possible, that is, the possibility of the change of the type of behavioural reaction under the influence of environmental conditions. The behaviour of *Philophthalmus rhionica* miracidia depend directly on the quantity and distribution of the chemical agent in space (Semenov, 1981, 1991). If the concentration of the latter is not very high and its distribution is uneven, the larval behaviour falls under the definition of a kinesis, i.e., the speed and the rate of change of direction increase. The larva in this period creates the impression of constantly searching for a stable source of chemical agent, or, at least, for the direction towards it. But if the agent's concentration is rather high and, what is most important, there is an expressed gradient in the direction towards its source, the larval behaviour fully corresponds to the manifestation of a positive chemotaxis, allowing the larvae to reach its aim as fast as possible and with minimal energy expenditure (Semenov, 1991). A greater effectiveness of chemotactic host finding as compared to chemokineses has been also noted by Haas et al. (1991).

The nature of the chemical factors modifying miracidial behaviour is also a question of great scientific and practical importance. It has never been doubted that these factors are some substances produced by the mollusc into water, excretory-secretory products (ESP). The nature of these substances, however, has long remained completely unstudied. Methodological approaches of the authors were strikingly different, which was at the root of the contradictions between the results obtained.

On the one hand, most of the researchers used in their experiments snail-conditioned water (SCW), which surely contained the ESP, on the other hand, attempts to use pure chemical substances of various nature have also been repeatedly made (see reviews: Haas and Haberl, 1997; Haas, 2000). It has also often been ignored that the concentration of miraxones in natural conditions can not be very high and is lower that that commonly used in the experiments. Even this latter circumstance alone can distort experimental results, especially concerning the degree of attractivity of chemical agents.

It has long been thought that the chemical signals are organic and inorganic products with low molecular weights, contained in the molluscan mucus. However, it was later discovered that the signal function is mediated by high molecular compounds, glycoproteins with molecular mass >30 kDa (Haas et al., 1991; Kalbe et al, 1996; Haas and Haberl, 1997). As for low-molecular products, they might play some additional role in miracidial chemo-orientation.

The demonstration that, contrary to an older view, there is a rather narrow specificity of miracidial responses to the impact of signal molecules with high molecular weight was an important achievement (Behrens and Nollen, 1992; Nollen, 1994; Kalbe et al., 1996, 1997; Haas and Haberl, 1997; Haas, 2000). Unfortunately, very few species have been investigated in this regard (*Fasciola hepatica, Trichobilharzia ocellata*, some species from the genera *Schistosoma* and *Echinostoma*). However, there are no reasons to believe that the miracidia of other digenean species behave in an essentially different way. A differential sensitivity was clearly demonstrated not only in the miracidia of different species but also of a different strain of the same species (*Schistosoma mansoni* – two Brazilian and one Egyptian strain) (Kalbe et al., 1996). Correspondingly, profound differences in the composition of the glycoproteins in the SCW from different molluscs have also been shown. In these experiments, the differences between the molluscs of different species were not essentially higher than those between the molluscs from different strains of the same species.

Haas (2000) has suggested that the glycoprotein signals are pheromones, which are needed by snails for intraspecific communication. In the course of a long co-evolution with molluscs, trematodes evolved the use of this code for their molluscan host-identification. It is this mechanism that seems to be a key to ensure the narrow specificity (often at the level of different strains of the same molluscan species) of digeneans to the first intermediate host.

In close proximity to the mollusc, the behaviour of the miracidium becomes very complex. MacInnis (1965), who studied the influence of chemical factors on the behaviour of *Schistosoma mansoni* miracidia with the use of chemical models, described 6 types of larval behavioural reactions in the zone of the influence of these factors, and 8 more types, completed after the establishment of a primary contact with the surface of the source of the chemical agents. It is a set of these reactions that is demonstrated by the larvae. In the opinion of Semenov (1991), they do not belong to a single type: some of them may be interpreted as kineses, and some as typical taxes. The larvae seem to possess a programmed behavioural repertoire, aimed at the solution of two problems at this stage: firstly, finding and establishing primary contact with the chemical agent source and, secondly, exploration of this source. MacInnis (1965) invoked a supposition about the existence of two types of chemoreception in miracidia: a distant chemoreception, responsible for the perception of low concentrations of miraxones at a certain distance from the source, and a contact one, when the larva directly explores the chemical composition of the superficial structures of the object which has attracted it. The additional ideas of Semenov (1991), who thought that larval exploratory reactions involve not only the analysis of chemical composition but also that of superficial microstructure, seems to be true as well.

The complexity of larval behaviour after the primary contact with the surface of the object (the mollusc) suggests that its receptor systems, first of all, some contact chemoreceptors, work according to the principle of accumulation of the irritant (Semenov, 1991). It is only sensitized larvae, which have reached a certain threshold level of accumulation of this information, that may pass from the second stage to the third one, the penetration into the molluscan host. There seem to be no reasons to apply this scheme to chemoreceptors only. It is not to be ruled out that the same principle is used in the work of tangoreceptors, analyzing the superficial microstructure of the molluscan body. This behavioural type appears to be not a chance one. It may be a mechanism to minimize the possibility of making a mistake: triggering of the behavioural reactions which ensure the penetration into the host makes further pattern of events irreversible. Before this has happened, alterations of larval behaviour are possible. At any rate, there are data on the addiction of the larvae to chemical agents after a prolonged exposure to them (Mason and Fripp, 1976, 1977). In such cases, the miracidia revert to their initial speed and trajectories of movement.

Thus, an elaborate complex of behavioural reactions, demonstrated by the actively swimming miracidia at their second stage in the environment and aimed, in the long run, at establishing the contact and recognizing the host, is undoubtedly adaptive. However, it should again be noted that these adaptations have an essentially different nature than those at the first stage. As mentioned above (see p. 229), at the first stage the adaptive nature of larval behaviour is mostly expressed at a group level, and the behaviour of every single larva is not individualized. In our opinion, it is impossible to find a manifestation of adaptive reactions, aimed directly at the host infection, in the behaviour of a single miracidium, especially if it possesses "reversed" taxes, as compared to those of the majority of the larvae. The whole picture alters as soon as the larva gets into the area controlled by a potential host. Already the first influence of molluscan miraxones almost completely suppresses primary taxes (a manifestation of the dominance effect, mentioned above). It is from this moment that the individual larval behaviour becomes strictly adaptive, being aimed at the infection of the molluscan host by this very larva.

The third, final, stage of free miracidial life is the penetration into the epithelium of the molluscan host. Though we do not know all the details of this process, the general picture obtained from the observations of the larvae of different species, looks quite convincing (Semenov, 1991). It seems unnecessary to reproduce it in this book in all details. However, it seems of principal importance to emphasize some specific features of this process.

Penetration is an active process of disrupting the integrity of the epithelium, involving both chemical and mechanical means of tissue de struction.

It has already been mentioned that in the trematode evolution the role of mechanical means of impact remained limited, whereas the role of tissue histolysis under the influence of the larval glandular apparatus increased. In the process of morphological evolution this was expressed in a limited distribution of motile and very contractile telescopic terebratoria, on the one hand, and the retention of spherical terebratoria in many species, on the other hand. The penetration apparatus, however, ensures only the local tissue destruction. This is followed by the next phase, the "entrance" of the parasite into the host tissues. Here, the main role in all types of the larvae is played by the somatic musculature.

The penetration process as such marks the onset of a complex regression metamorphosis, undergone by the parasite transforming into the mother sporocyst. The shedding of epithelial plates is one of the first observed features of this process. It precedes, in fact, the penetration act in some species (Fasciolidae, Echinostomatidae, Psilostomidae): a juvenile lacking these provisional structures penetrates the mollusc. In other species (Paramphistomidae, Schistosomatidae, Heronimidae) the miracidium sheds epithelial plates already in the molluscan body. Semenov (1991) paid attention to an interesting regularity. The larvae shedding the plates after the attachment to the mollusc but prior to penetration, spend from half an hour to 2-2.5 h for the penetration process, whereas those which shed the plates in the mollusc penetrate, on the average, in 10 min.

Semenov explained these differences in the duration of the penetration act by the time of shedding the plates. If the plates are shed outside the host, some time is required for the regeneration of the tegument of the mother sporocyst. Moreover, the presence of telescopic terebratorium was interpreted by Semenov in the same way: functioning as a sucker, the latter ensures a prolonged and safe fixation of the larvae at the molluscan body surface.

The shift of all these events for a later period in the host body makes the interval between the attachment and the penetration proper unnecessary. The telescopic terebratorium, as a sucker-like holdfast organ, is also not necessary in this case.

The adaptive nature of the above complex of morphological-functional features of the organization and the behaviour of the miracidia becomes especially clear when we address the problem of specificity of interactions between the parthenogenetic generations of digeneans and their molluscan hosts. Narrow specificity of the parthenitae is a well-known fact, yet the mechanisms underlying it are only now becoming the focus of close attention. When analyzing this phenomenon, it should be remembered that its first manifestations are expressed in the system "miracidium – mollusc", the existence of which was postulated by Semenov as early as in 1991. As a matter of fact, it is a subsystem of a host-parasite system of a higher level, "parthenitae – mollusc". However, it has a number of characteristic features. First of all, the specificity of relations between the partners of the system is expressed in the specific traits of miracidial behaviour, allowing the latter to solve very complicated problems of finding and infecting the molluscan host.

1.1.2. Passively Infecting Miracidia

In different phylogenetic branches of Digenea one and the same process had been occurring, independently and at different times. This process reflects general regularities of the parasites' evolution. It is the loss of active miracidia, actively infecting the molluscs. Instead, the parasites acquire the larvae which survive in the environment protected by an eggshell and which adopt the passive way of molluscan infection[iii].

1.1.2.1. Morphological Adaptations of Passively Infecting Miracidia

An abrupt change of the way of infection and hence of all major biological characteristics of miracidia has inevitably influenced the traits of their organization. Though the larvae retain all the morphological-functional complexes mentioned above (see section 1.1.1.2.), they are subject to essential modifications. As for locomotory complex, it reveals a clear tendency to simplification, first of all involving the ciliary apparatus.

There seem to be two stages in its reduction: the reduction of the number of epithelial plates (1) and a considerable decrease of their area as compared with the total area of the body surface (this is accompanied, correspondingly, by an increase of the area occupied by the hypodermal ridges, which, in fact, functionally become a definitive epidermis, the tegument) (2).

Strictly speaking, a decrease in the number of epithelial plates and the simplification of the epithelial formula up to 3:3, described for the miracidia of Heterophyida, Plagiorchioidea and Ochetosomatoidea (see p. 10) have been initially associated not with the loss of ciliated locomotion by the miracidia but with a strongly pronounced tendency to diminishing of the body size (miniaturization). Moreover, such miracidia show an increased length of the cilia as compared to the body size, which might be treated as an adaptation to active movement in relatively viscose contents of the molluscan intestine. Observations of the live miracidia of Plagiorchioidea and Ochetosomatoidea (*Haplometra cylindracea, Telorchis assula, Opisthioglyphe ranae, Haematoloechus variegatus*) have demonstrate that they combine ciliated locomotion with active peristaltic body movements resulting from the muscular activity.

A decrease of the area of the epithelial plates often seems to be accompanied by the transformation of the ciliated apparatus itself. Thus, in some miracidia (Azygiidae), narrow plates, arranged into 2 transverse rows and set widely apart, bear dense light-refractive structures resembling "spines" (see p. 14 and fig. 1F). They are supposed to be condensed cilia, forming cirri, analogous to those of some ciliates (Brooks et al., 1985a; Semenov, 1991). Of course, final conclusions will be made only on the basis of electron-microscopical investigations. Yet this viewpoint seems quite logical and may be accepted as a working hypothesis. Anyway, whatever the nature of these structures might be, the locomotory function in all such instances passes to the miracidial muscular system.

This statement seems to be even more true for the miracidia of some hemiuroid trematodes (*Halipegus, Bunocotyle, Nematobothrium*) (Rankin, 1944; Self et al., 1963; our observations). These larvae completely lack typical cilia, but their epidermis bears "spines" of obscure origin (see p. 15). Thus, the process, which has emerged as a consequence of miracidial miniaturization, seems to have soon started to generate adaptations, as it appears functionally justified to abandon ciliated locomotion for the sake of movement by means of muscular body contractions in the contents of the host's intestine.

The loss of the period of active swimming in the environment also influenced the organization of the sensory apparatus and the nervous system (see p. 22). Unfortunately, at present we know almost nothing about the fine structure of the organs of this complex. This very circumstance, however, seems to support the view that they are reduced to an extreme degree. This is biologically justified. The larvae inside the eggshells have no need of obtaining or processing information flows from the environment, as their passive life mode excludes the possibility of behavioural response.

The larval penetration complex seems to be subject to alterations in the least degree. Though its structure is clearly modified (see p. 18), the secretory elements, participating in the histolysis of the intestinal epithelium and underlying tissues of the molluscan host, are present. A certain simplification, expressed in a lesser diversity of secretory cells ensuring the penetration process, seems to be caused by a lesser complexity of tissue barriers on the way of the larvae penetrating the intestinal wall and not the body surface.

The fact that the penetration complex is retained rather fully (which is especially noticeable as compared to a considerable simplification and reduction of other provisional structures) is quite justified biologically. Both actively and passively infecting larvae finally face the same problem: disrupting the integrity of the bordering tissue structures, to penetrate into the body of the molluscan host.

The generative complex in most cases exhibits a clear tendency to retardation, which has already been mentioned above (see p. 26). Sometimes this tendency is so pronounced that the larvae lack a structured generative primordium at all: its formation and the reproductive process of this generation as such is shifted to the parasitic developmental stage – the mother sporocyst (see p. 26, 419).

The above morphological transformations of passively infecting larvae develop against the background of their extreme miniaturization. All of them are of minute size (sometimes an order of magnitude less than the size of actively swimming and actively infecting miracidia) and consist of few cells (see p. 28). Both these characteristics show that such organisms are extremely specialized and their origin is secondary.

1.1.2.2. Biological Traits of Passively Infecting Miracidia

The presence of passively infecting miracidia in the environment is characterized by a number of very specific features. First and foremost, the development of such larvae always takes place in the marita uterus up to the moment when they become infective. Literature data showing that eggs at different stages of their embryonation can be shed into the environment from the host organism and thus require some time for their final development (Styczyńska-Jurewicz, 1962; Genov and Samnaliev, 1984; Zakikhani and Rau, 1998) seem to be incorrect. We conducted a special investigation which has demonstrated that the miracidia of 6 plagiochiid, 6 microphallid, 2 heterophyid and 2 gymnophallid species fully develop in the ducts of female reproductive system of the adult. We have found only the eggs with fully formed infective larvae both in the proximal uterus parts of the trematode species studied and in the intestinal lumen, the respiratory organs and the gall bladder of the hosts, as well as in the faeces of experimental animals. Earlier erroneous ideas could have originated from misinterpretation of experimental data. Some time after they have been placed in fresh water, the maritae of many trematode species start to lay eggs very intensively, non-embryonated eggs as well. If isotonic media are used, no such phenomenon is observed. Incorrect results could also arise from the use in the experiments of non-specific definitive hosts. This must have been the case in the investigation by Zakikhani and Rau (1998). Though *Plagiorchis elegans*, the species these authors worked with, has a very broad specificity at the phase of marita (Krasnolobova, 1987), it is always birds that are its definitive hosts. Development in non-specific hosts often results in an incomplete development of the marita reproductive system, its low functional activity and the production of abortive and immature eggs.

Eggs produced into the environment might become the source of molluscan infection at once. The first two stages, active dispersion and finding and identification of a potential host, which are so characteristic of actively swimming larvae, in this case fall out. Meeting the host does not depend on the activity of the larva at all. Eggs with infective miracidia get into the host organism by chance, together with the consumed main food. Thus this process is only to a certain degree determined by the activity of the host itself. All this could have made the very fact of existence of trematodes with such miracidia rather problematic, were it not for a number of adaptations acquired by the larvae of such species in the course of evolution. These adaptations rather effectively compensate for the absence of the period of active life in the environment. The prolongation of free life of the larvae appears to be the most important of them. According to our observations, the miracidia in the eggs of some plagiorchiid species can remain infective for molluscs for 5-6 months. Similar data were also been obtained by other researchers from other trematode groups (Kagan, 1952; Rowan, 1955). Such a long period during which the larvae not simply remain alive but retain their infectivity greatly increases the probability of the egg being eaten by the host.

Quantitative aspect of the problem is equally as important. Infection success depends directly on the saturation of the environment by infective stages. This is a general regularity, applying equally well to any type of infection (eggs, cysts, active larvae). In our case, infection is represented by eggs with formed and infective larvae. The more such infective stages are there in a square or volume unit of space, the higher the infection probability is. Without increasing, in fact, energy expenses for reproduction, it is possible to produce a much greater number of very minute, few-celled and very simplified miracidia. A significantly increased life span of such miracidia makes possible their relatively high concentration in space over considerable time (up to several months). Taken together, all this can hardly be treated as an individual adaptation of the larvae. The adaptive effect in this case is expressed only at a level of the group and the life cycle as a holistic functional system. Contrary to that, an increased resistance of eggs with miracidia to the impact of unfavourable environmental factors can be certainly treated as an adaptation manifest at an individual level.

On the whole, the adaptive potential of the trematode group in question has proved to be so high that they were able to leave the aquatic environment for the life on land (Dicrocoeliidae, Brachylaimoidea), as well as to colonize ecosystems with extreme life conditions (the intertidal zone of the northern seas, the mountains) (Microphallidae, Brachylaimoidea) (see section 2 in chapter 5).

However, some barriers have proved to be insurmountable for trematodes. Having colonized arid habitats, they still did not become true land organisms. According to our observations, the eggs of *Hasstileisia ovis* (the transmission of this species takes place in conditions of alpine pastures) do not withstand even a short desiccation. They need at least some moisture to survive.

The above-listed biological features of passively infecting miracidia to a great extent determine the regularities of their distribution in the environment. Earlier it has already been emphasized that the larvae of this type lack the stage of the so-called active dispersion, the nature of which is determined by a set of primary taxes characteristic of the larvae of a given species. There can be nothing of this kind in the miracidia passively surviving in the environment. However, this does not mean that there is no dispersion. It does take place, but its mechanisms are essentially different. Out of the larval characteristics, only their longevity and the resistance of eggs to unfavourable factors influence the scale and character of dispersion. The rest is determined by biotic and abiotic factors, which are exogenous for the larvae. Being shed into the environment with the host's faeces, they form the primary focus of concentration, which is, however, very soon disrupted. The localization of the eggs is changed only passively: with the moving water masses (currents in the water basins, streams on land), as a result of the disturbance of sediment surface, caused by different factors, or the activity of different animals. In the latter case, the resistance of eggs is often of utmost importance. Being swallowed by non-specific hosts, both vertebrate and invertebrate, the eggs transit through their digestive tract without losing their infectivity and disperse in the environment again. This way of infection dissemination is the only possible one for some trematode species (see p. 341).

As a result of the impact of the mentioned and many other factors, a "spot" is formed in a certain area, whose structure turns out to be very dynamic, i.e., the infection is being constantly redistributed (Galaktionov and Dobrovolskij, 1984).

Thus, the final result of active and passive dispersion turns out to be similar: in both cases rather mobile and dynamic accumulations of larvae come to being. This similarity, however, is superficial, there being essential differences behind it. As noted above, actively swimming miracidia form "clouds", completing their behavioural programmes in three-dimensional space, whereas passive larvae are distributed on a plane (two-dimensional space) as spots. As a result, actively swimming larvae can infect the host at any point of space, the water body: at the bottom, in the water column, at the water surface. The contact with passively infecting miracidia is in fact only possible on the substrate, i.e., at the bottom of a water body. This is especially clearly seen in the so-called land species.

One more difference is associated with the microscale heterogeneity of environmental conditions, which is much more pronounced at the substrate surface. It is very important for passive larvae, unable to move by themselves. Microscale heterogeneity also exists in the water column, but there its influence upon the larval distribution is somewhat different. In the first case, heterogeneity very often results in actual or potential elimination of eggs (e.g., their accumulation in microcrevices of the sediment, inaccessible for potential hosts or the washing of the eggs into water, where they become inaccessible for land snails). As a result, the egg distribution is patchy. If new portions of infection stages cease to enter the biotope and if external conditions do not fluctuate too much, relatively stable foci of increased egg concentration can be formed within the limits of the spot. Due to the biological features of the larvae, they can survive for a long time.

In the second instance, the microscale heterogeneity also determines uneven distribution of larvae in the water body, but this unevenness is constantly being leveled due to dynamic structure of the cloud.

Free life of actively infecting miracidia starts from the moment of their hatching from the egg. For passively infecting larvae hatching, occurring after the eggs with larvae have been swallowed by a potential host, means the beginning of the transition to the parasitic phase, as it is at this moment that further events of larval life become inevitable. As a matter of fact, the larva inside an eggshell has to estimate the degree of the specificity of the host whose intestine it is in. It is the same task as the one which has to be solved by an actively infecting miracidium while investigating the surface of the mollusc it has established a primary contact with. If the conditions in the digestive tract turn out to be optimal for the larvae of a given species, hatching mechanism is triggered, whereas in case of an error the larva is destined to a quick death. If the complex of factors influencing the swallowed egg can not trigger larval activation, the egg transits the digestive tract and passes to the environment, without losing, as a rule, its intactness and infectivity.

Unfortunately we know almost nothing about fine hatching mechanisms of passively infecting miracidia. In our opinion, this problem is of immense importance, as this biological peculiarity of the miracidia represents one of the main trends in adaptive evolution of a large trematode group. *A priori* it could be supposed that the selection was aimed at the narrowing and specialization of the complex of factors triggering larval hatching, as it is only in this way that the recognition of a specific host can be achieved. At present there are no direct experimental data in favour of this hypothesis. As for indirect data, they are scarce. Unfortunately, similarly to the situation with actively infecting miracidia, the researchers have dealt with a very limited number of species (*Dicrocoelium dendriticum*, *Opisthorchis felineus*, several species of the plagiorchiids) (Ractliffe, 1968; Mitterer, 1975; German and Beer, 1984; our data). The data obtained can not even be considered contradictory – they are fragmentary and scattered. However, the general pattern of this process is described in a more or less similar way by all researchers.

The larvae in the eggs in the environment do not exhibit any signs of activity. They are absolutely immobile; their cilia, if any, do not beat, their protonephridia are inactivated. Ten-fifteen minutes after the eggs with miracidia get into the digestive system of the mollusc, the miracidia become noticeably active. The body of the larva is slowly contracting and bending, the cilia start beating, though as yet not very synchronously. Soon after that the hatching begins. First the operculum is discarded abruptly and the miracidium is ejected outside through the hole formed. Our observations of plagiorchiid miracidia show that if the energy of ejection has been insufficient for complete release of the larva, it starts trying to leave the eggshell itself by means of muscular contractions. Disrupted yolk membrane is left, as a rule, inside the shell. As noted above (see p. 18), the hatching of the larva is accompanied by considerable changes of its structure. This is associated with an almost complete disappearance of the hatching glands. No traces of these structures have been found in the miracidia actively moving in the intestine contents.

The following logical scheme can be projected onto this picture. After getting into the digestive tract of a specific host, the miracidium in an eggshell is stimulated by a complex of factors characteristic of the intestine of a given molluscan species (a group of close species). At this time not only the locomotory but also the secretory activity of the larva becomes pronounced. The purpose of the substance secreted by the hatching glands is not yet fully clear. It has been suggested that it might contain proteolytic enzymes attacking the "cement" material holding the operculum in place (Dobrovolskij, 1965a). It is not to be ruled out that it takes some part in the changing of the pressure inside the eggshell. In any case, a quick thrusting of the operculum and the pushing of the larva testify in favour of the supposition that it is not the locomotory activity of the larvae that ensures their release but a quickly increased inside pressure in the egg, i.e., the same mechanism is applied here as in the hatching of actively swimming miracidia (see p. 217).

The most interesting aspect of this problem is the nature of the factors triggering miracidial activation. It would seem logical that they could not be digestive enzymes of the host, as these higher molecular proteins are unlikely to penetrate the eggshell and to influence the larva. These factors should be rather simple and common. Chemically, they should be low molecular compounds. Ractliffe (1968) managed to demonstrate that the miracidia of *Dicrocoelium dendriticum* in the eggs reacted first of all to oxygen deficit. However, this is also supplemented by the impact of temperature: in anaerobic conditions the larvae become active in a relatively narrow temperature diapason with an optimum around 33°C. Data on the role of drying the eggs in nitrogen or in vacuum seem to be more questionable and should be tested. Our observations upon the miracidia of brachylaimids and microphallids have shown that the lack of moisture even for a very short time is absolutely fatal for the larvae. On the other hand, *D. dendriticum* belongs to a different phylogenetic branch than *Hasstilesia ovis* and the microphallids of "*pygmaeus*" group. The loss of active miracidia has taken place independently in these branches. Consequently, the adaptations of these larvae may be essentially different.

It has been demonstrated that a certain role can be played by some chemical substances (e.g., formic and capric acids), as well as by the bacterial flora of the host's intestine (Ractliffe, 1968; Mitterer, 1975). The results of observations of the eggs of *Opisthorchis felineus* add to the list of factors the necessity of a mechanical impact upon the eggs with larvae (German and Beer, 1984).

No complete picture can be deduced from these data. Yet it seems possible to make some suppositions. To begin with, the resistance of eggs with larvae to environmental factors varies. According to Panin (1984, 1986) and in agreement with our investigations, the miracidia of prosthogonimids (*Prosthogonimus ovatus*) are very sensitive to almost any kind of sudden non-specific impact (fluctuations of light intensity, temperature). They react by activation, and sometimes by rather mass hatching. We failed to observe anything like this in the numerous representatives of Plagiorchiidae, Ochetosomatidae and Microphallidae. No situations, not even extreme ones, make the larvae leave their eggs. The only thing that can be achieved, applying a slowly increasing mechanical pressure, is a slight increase in the locomotory and osmoregulatory activity of the larvae. They start moving inside the egg, exhibit a weak and often irregular metachronal wave of the cilia and an active working of flame cells. The latter circumstance makes us think that the activation of the larvae in this case is non-specific and caused by the mechanical disruption of the eggshell permeability to water. When the outside pressure reaches a critical value, the shell snaps, but we have never observed it to snap along the circular rupture of the operculum.

The same complex of experimental impacts upon the representatives of Lecithodendrioidea does not result in the slightest activation of the miracidia (our data).

Very interesting results have been obtained in our experiments when the eggs of different plagiorchiid species were cross-fed to molluscs of one and the same species, and *vice versa* (unpublished data). It turned out that the miracidia of *Opisthioglyphe ranae* hatch well in the digestive tract of some pond snails (*Lymnaea stagnalis, Radix ovata*), are weakly activated in the intestine of orb snails (*Planorbarius corneus, Planorbis planorbis*) and transit the digestive tract of viviparids (*Viviparus contectus*). It seems relevant to mention that the specific host of this trematode species is *L. stagnalis*. We have obtained similar data for *Haematoloechus similis* (specific host – *Planorbis planorbis*). It was only the ranging of the molluscs that was different: planorbid snails, pond snails (miracidia activated considerably in *Galba palustris*), viviparids.

In our opinion, these results point unambiguously at the presence of specific conditions in the molluscan digestive tract, characteristic of a definite species. The larvae of corresponding trematode species are adapted to these particular conditions. Naturally, hatching process is triggered not by some single factor but by a rather specific complex of factors.

Summing up, it can be supposed that it was not at once that the trematodes, which have started to passively infect the molluscs, acquired a reliable mechanism of larvae protection from premature hatching, both in the environment and in a non-specific host. Further co-evolution of the parasite and the host resulted in a narrowing of the reactivity of the former. The selection retained more and more specialized forms, which responded by hatching only to a very well-defined complex of factors. This can certainly be treated as an important adaptive acquisition of the trematodes with passively infecting miracidia. Interestingly, these trematodes, having lost the opportunity to choose a specific host by means of the analysis of the host's properties by contact receptors, have found another, rather reliable mechanism permitting to solve successfully the same task.

Having left the eggs, the miracidia start to penetrate the intestine wall of the host. Similar to the actively swimming larvae, penetrating the epidermis, the miracidia of the group in question use both chemical and mechanical means of influencing the host tissues. In the majority of cases, however, at the first stage of penetration they only come as far as the basal part of the intestinal epithelium. There metamorphic transformations are completed in 1-3 days, the events involved being very different in different species. However, even at this moment it is too early to speak of the completed infection of the host. Young individuals are still to undertake a migration, and, what is most important, to avoid defense reactions of the molluscan host.

1.2. Adaptations of the Mother Sporocyst

Mother sporocysts are so different in different trematode species that it is hardly possible to give their generalized biological characteristic. First of all, the parasitic phase of this generation is absent in the life cycle in a number of trematodes which are systematically rather distant (Cyclocoelidae, Philophthalmidae, Echinostomatidae, Allocreadiidae) (see reviews: Pearson, 1972; Prévot, 1971; Madhavi, 1976).

Unfortunately, the researchers, describing paedogenetic miracidia (see p. 23), do not always correctly determine the specific, and sometimes, higher taxonomic position of the trematodes studied. A well-known work by Agrawal and Pandey (1979) is a "classical" example of a misrepresentation of this kind. They undoubtedly found paedogenetic miracidia, but the attribution of the maritae which produce these miracidia to the family Telorchiidae must have been erroneous. The miracidia of true representatives of this family, as well as of all other plagiorchiid trematodes studied by now, have an essentially different morphological structure (Schell, 1962a; Dobrovolskij, 1965a). And what is even more important, their evolution followed a different tendency, not paedogenesis but a retardation in the generative primordium development (see p. 26).

All first parthenogenetic generation of species with paedogenetic miracidia is confined to these actively swimming larvae. An alternative process observed in some Plagiorchiida leads to the same thing. Early reduction and underdevelopment of the soma result in a situation when the parasitic phase is represented not by a whole organism but by its generative organ, the germinal mass (Galaktionov and Dobrovolskij, 1985), or even simply by germinal cells and embryos (Schell, 1965) (see p. 46). However, the latter are already the next parthenogenetic generation, related to the mother sporocyst as a daughter organism to a mother one.

On the other hand, mature mother sporocysts are often large integral organisms with all the main organization features of flatworms. In between these two extreme points there is a broad spectrum of forms at successive stages of morphological-physiological degeneration, up to a complete disintegration of the organism (for details see section 1.2.2. in chapter 1).

As parasites, mother sporocysts had to cope with two main functions, closely associated in most parasitic organisms. They are the feeding and the reproduction function. However, numerous facts show that the situation is not as simple as that, especially if we take into consideration the phylogenetic position of separate trematode groups.

Not all mother sporocysts reproduce at the parasitic phase (see p. 419), and even if they do, the reproduction is not always intensive. These facts would not have been surprising, were it not for one circumstance: generative function is absent or weakly expressed in the mother sporocysts of the most archaic trematodes (Fasciolidae, Transversotrematidae, many Echinostomatidae and, of course, the above-mentioned Philophthalmidae and Cyclocoelidae). Higher up the phylogenetic tree, in more specialized groups (Cyathocotylidae, Strigeidae, Schistosomatidae, Sanguinicolidae, Plagiorchiidae), a clear tendency to the intensification of the generative function is observed: the offspring of the mother sporocyst generation consists of hundreds, and sometimes, thousands of individuals.

The above considerations lead us to the only possible conclusion: the shift of the implementation of generative function to the parasitic phase, and its intensification at the expense of the energetic potential of the molluscan host took place already within the class Trematoda in the most specialized and evolutionary advanced groups. Biologically, this conclusion raises no objections: the adaptive nature of this process and its advantages for the evolutionary advanced trematodes are not to be doubted. It is much more difficult to interpret this phenomenon phylogenetically. The above data could lead us to suppose that at the time when relations between molluscs and proto-trematodes were being formed the course of evolution, the latter did not develop in the molluscs at all, but used them, at best, for the completion of the offspring development. Another interpretation is also possible: the absence of reproduction at the parasitic phase of the mother sporocyst in archaic trematodes is the result of secondary specialization on the basis of heterochrony, which is so common in trematodes. In this particular case it is manifested as progenesis, with its ultimate expression in the rise of paedogenetic miracidia. This problem will be treated in more detail in chapter 5.

Still, it is not to be doubted that it is in the specialized representatives of the class that maximal intensification of the generative function of the mother sporocyst takes place. It seems reasonable to suggest that enhanced reproduction also requires the intensification of the feeding function of the parasites. Unfortunately, we know nothing either about the fine structure of mature mother sporocysts or about the details of their feeding. It can also be supposed that the quantity of nutrients extracted from the host and the means in which this is achieved are very different in different species depending on the structural type and the reproduction dynamics.

Reproduction processes in mother sporocysts have been studied in more detail. This allows us to outline several main adaptive trends of evolutionary transformations that ensure, on the one hand, the intensification of the generative function, and on the other hand, the stabilization of relationships in the host-parasite system. Germinal mass is the main reproductive organ of mature mother sporocysts (see p. 75). There the proliferation of undifferentiated cells of the germinal lineage takes place, their maturation and the onset of cleavage. Mother sporocysts can be divided into two groups as to the dynamics of germinal masses functioning. The first group comprises the forms with a limited time of germinal mass (masses) functioning. They produce germinal cells in a relatively short time span (the number of the cells in this case is not important). The cells start cleavage only after the completion of the proliferation period. The number of embryos formed in the mother sporocyst is thus strictly determined, it can not increase after the end of proliferation and the differentiation of the germinal cells. Subsequent development of the embryos may take place asynchronously (as a rule, this is characteristic of more archaic forms producing a small number of individuals of daughter generation) or relatively synchronously (this is observed in the mother sporocysts of some plagiorchiid species, such as *Macrodera longicollis*, giving rise to hundreds of daughter sporocysts). The final result, however, is similar. After all the embryos have achieved the same developmental level, they destroy the mother organism and start to parasitize independently.

In more specialized forms the reproductive process of mother sporocysts is prolonged. A relatively rigid separation of two successive stages, the proliferation of undifferentiated cells of the germinal lineage and the development of the embryos, disappears. This poses a number of specific tasks for the reproducing parasite. First of all, it has to acquire mechanisms ensuring the discharge of mature organisms of the next generation from the mother organism. Equally as important is the necessary spatial isolation of asynchronously developing embryos of different ages. It has already been mentioned, that early embryos, yet lacking embryonic membranes, are very sensitive to mechanical impact by more developed individuals, and are easy to destroy (see p. 90). The solution of these two closely connected tasks has turned out to be achieved in two ways, which are, in fact, alternative.

One trematode group (predominantly Strigeidida and related forms) has retained mother sporocysts of a worm-like type. They are large mobile organisms with a well-developed soma, a spacious schizocoel and often with several actively functioning germinal masses, which perform not only proliferative function but also that of brood chambers (see p. 75). As a result, in the schizocoel numerous embryos of different age already can not damage each other. Besides, mature mother sporocysts of this type also have a birth pore (see p. 36). The releasing of individuals of the daughter generation is going on constantly as they mature in the schizocoel of the mother organism.

Different picture is observed in the representatives of the order Plagiorchiida. We have already mentioned quite often that in these trematodes the tasks in question are solved by means of secondary simplification, and sometimes, of disintegration of the soma of the mother sporocyst (Dobrovolskij, 1975; Dobrovolskij et al., 1983). In the mother sporocysts of this type there often appear numerous proliferation foci of undifferentiated cells of the germinal lineage (diffuse germinal mass – see p. 81), the young daughter sporocysts release at different sites of the mother sporocyst body by means of local destruction of the areas of non-specialized soma. This does not in any way tell upon the viability and productivity of the rest of the organism. This tendency finds its utmost expression in the complete reduction of the soma at the earliest stages of parasitizing, accompanied by a transition to egg and embryonic parasitism (see p. 46).

Secondary simplification and disintegration of mature mother sporocysts, which are often shaped as irregular tumor-like and branching stolon-like structures (see pp. 43-9) seem to be one of the most perfect adaptations of trematodes. The parasites obtain the opportunity to exploit more fully not only the energetic potential of the host but the space inside its body as well. Both these events facilitate the solution of the main task, the intensification of reproductive function.

It has already been mentioned that the parasites face another extremely important task. The increase of parasitic pressure upon the host, inevitably accompanying the intensification of feeding and generative functions, should not destabilize host-parasite relationship. This task may be solved in different ways. One of them, widespread in trematodes, is the parasitic castration of the molluscs (see p. 356). Mechanisms underlying this phenomenon are far from being comprehensively studied. Yet it seems certain that the parasitic phases of mother sporocysts purposefully influence the gonad from the very beginning of their presence in the mollusc. The behaviour of stolon-like germinal masses of microphallids is an example: entering into the haemocoel of the mollusc, they usually first affect the gonads (see p. 46). The released reproductive share of the host's energetic potential is in fact almost completely consumed by the parasite. The withdrawal of this share of energy does not influence the physiological state of the host, the fate of the parasite also depending upon it. Thus, the sporocysts succeed in transferring the manifestation of pathogenesis from organismic level (the level of host-parasite system) to a higher populational level. It is not the infected host itself that suffers from the parasitic castration, but the population whose reproductive potential is correspondingly decreased. Moreover, sometimes the energy released as a result of parasitic castration is sufficient not only for the full satisfaction of the parasite's requirement but also for the intensification of some functions of the host, for instance, a higher rate of the mollusc's growth (see p. 357).

The above-mentioned change in the reproductive dynamics of mother sporocysts seems to be an important adaptation aimed at the leveling of relations between the parthenitae and molluscs. Synchronous development and release from the mother organism of a limited number of embryos (in primitive trematodes: Fasciolidae, Echinostomatidae) is not especially dangerous for the existence of the host-parasite system. However, the situation changes when hundreds of juvenile specimens of the next generation are simultaneously released into the mollusc. This is what is characteristic of some Plagiorchiidae and Ochetosomatidae, the embryos of which develop in the mother sporocyst synchronously. Our experimental observations have shown a mass death of *Planorbis planorbis* snails infected by the mother sporocysts of *Macrodera longicollis* to take place from the 25th to the 35th day following infection. In synchronously staged experiments with the infection of the same host species by the parthenitae of *Telorchis assula* the death rate of infected molluscs did not exceed that in the control experiment, making up 2-5% for the same time. The difference between these two plagiorchiid species is precisely that the mother sporocysts of the former complete their development in a month; then follows their death accompanied by a mass release of the daughter generation individuals into the host's haemocoel. The parasitic pressure increases so greatly and so fast that the host simply has no time to adapt to it. The consequences of the molluscan infection by the parthenitae of *Macrodera longicollis* and other species with similar mother sporocysts in natural conditions are unlikely to be as catastrophic as that. Yet the experiments have unambiguously shown that the host-parasite system in question is not very stable.

The mother sporocysts of *Telorchis assula* start to produce the first daughter sporocysts on the 15-20 day postinfection. The parasitic pressure increases very gradually. Mother sporocysts still giving rise to the individuals of daughter generation can be found 50-60 days following the infection.

It is this circumstance that seems to allow the host to adapt to the additional pressure exerted by intensively reproducing parasites.

Other biological traits of mother sporocysts have been studied even less adequately. For instance, we know very little about the behaviour of young organisms immediately after penetration and about the localization of mature individuals in the molluscan host. It has already been noted (see section 1.1.2. in chapter 1) that after penetration and completion of metamorphosis young mother sporocysts of many trematodes migrate to the final settlement place (Fug. 95). They may localize in different parts of the host's body. Diversity of microhabitats is very large, yet three main groups may be distinguished. Mother sporocysts of some species (*Fasciola hepatica, Macrodera longicollis, Paralepoderma cloacicola*) localize in the subepidermal layer of the head, "neck" region and walls of the pallial cavity. Sporocysts of Schistosomatidae settle deeper beneath the surface in the same body areas. For instance, mature mother sporocysts of *Schistosomatium douthitti* are found in the head and the "neck" region along the oesophagus (Price, 1931). According to our observations, mother sporocysts of some *Schistosoma* species also localize in that region.

The second preferential settlement place of mother sporocysts is the ventricle and, more rarely, the aortic bulb. Mother sporocysts, e.g., of the genus *Echinostoma* can be found there (Ataev et al, 2001a,b).

Finally, mother sporocysts of some digenean species (many Strigeidida and Plagiorchiida) are localized in the visceral sac (Dobrovolskij et al., 1983). In some cases they lie freely between the lobes of the hepatopancreas and the gonad (Cyathocotylidae), whereas in other cases they lie on the surface of the gut (see p. 43).

The place of miracidial penetration does not always correlate with the localization of the mature parasite. If juveniles do not migrate, the final settlement place of parasites is not far from where the miracidium has penetrated: usually in the head and "neck" regions, the very base of the visceral sac, walls of the pallial or lung cavity. Parasites may settle in the latter owing to the fact that miracidia, apparently, often penetrate not in the open areas of the molluscan body surface but first get through natural openings into cavities or ducts and then penetrate the epithelium and get into the underlying tissues.

This behavioural feature has been noted in miracidia from different taxa. Infecting molluscs, larvae of some *Schistosoma* species often first get into the mouth cavity, the rectum or the branchial cavity (up to 57% of *S. japonicum* larvae utilize this way of infecting the host (Haas and Haberl, 1997)), though they may also infect the host via the open areas of the body surface. According to Žd'árská (1986), miracidia of *Paramphistomum microbothrium* (Paramphistomidae), on the contrary, always penetrate the walls of the lung cavity, where they get through the respiratory opening.

Thus, even if there is no migration, mature mother sporocysts may be located quite deep inside the molluscan body and not only in its open areas.

The above is also true of the cases when miracidia penetrate the molluscan mid-gut, whose loops lie in the visceral sac. If there is no migration, mother sporocysts are localized in the tissue surrounding the gut epithelium. Growing, the sporocysts remain attached to the external gut surface. This is characteristic of some Plagiorchiida (Cort et al., 1952; Dobrovolskij et al., 1983).

However, young mother sporocysts of most trematode species do undertake a more or less complex and long migration. Migrating, they use the haemocoel lacunae and sometimes also the vessels of the open circulatory system (Ataev and Dobrovolskij, 1990; Ataev et al., 1997). The migration of sporocysts in the molluscan body may be passive, achieved by means of hemolymph flow, or active, by means of the activity of muscular elements of the tegumentary-muscular sac. According to Pearson (1961), day-old mother sporocysts of *Neodiplostomum intermedium* are found in the perihepatic haemocoel. Besides, they may be observed to crawl sluggishly between hepatopancreatic acini. We recorded similar behaviour in some species of Plagiorchiida (Dobrovolskij et al., 1983).

The direction of migration is an issue of the greatest interest. Two main directions may be distinguished (Fig. 95). In the former case, juvenile mother sporocysts migrate inwards in the body of the molluscan host. Some species (*Schistosoma* species – our observations; *Schistosomatium douthitti* – Price, 1931) do not leave the head and "neck" regions, reaching only as far as the oesophagus area. Other species reach the ventricle and the aorta of the heart, using the system of lacunae and circulatory system (*Echinostoma* – Ataev et al. (2001a,b)). Long (up to 11 mm) worm-like mother sporocysts of some Cyathocotylidae lie freely in the haemocoel lacunae between hepatopancreatic acini (Olenev and Dobrovolskij, 1975, our observations). Taking into account that all the above species possess actively infecting miracidia that finally infect molluscs via the integument, we may summarize that their juvenile developmental stages, that have passed over to parasitism, are characterized by viscerotropism (migration of the parasite in the host's body towards its internal organs and tissues).

An alternative variant, when young mother sporocysts clearly demonstrate dermatotropism (migration of the parasite to the surface of the

Figure 95. Diagram showing the ways of miracidia penetration, migration and places of the final settlement of mother sporocysts of different trematode taxa in the molluscan host. *Abbreviations*: *1*, Schistosomatidae (*Schistosoma* spp., *Schistosomatium dauthitti*); *2*, Echinostomidae (*Echinostoma*); *3*, Paramphistomidae (*Paramphistomum microbothrium*); *4*, Philophthalmidae (*Philophthalmus rhionica*); *5*, Fasciolidae (*Fasciola hepatica*); *6-9*, Plagiorchiida (*6*, *Haplometra cylindracea*; *7*, *Lechriorchis primus*; *8*, *Microphallus piriformes*; *9*, *Paralepoderma cloacicola*); *10*, Cyathocotylidae; *11*, mouth opening; *12*, pharynx; *13*, oesophagus; *14*, intestine; *15*, anus, *16*, lung (pallial cavity).

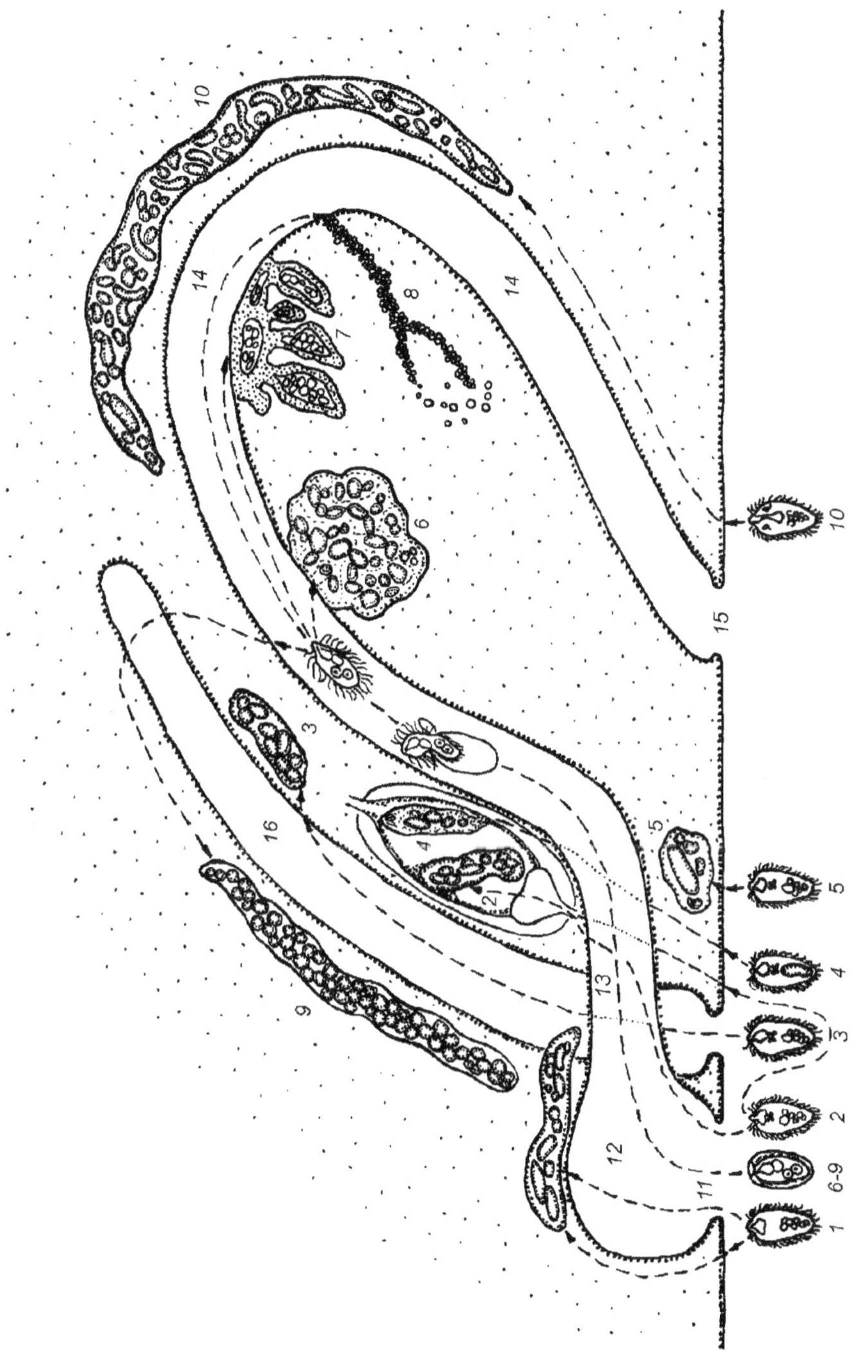
10
14
14
8
7
6
10
15
3
16
5
4
2
5
4
13
3
9
2
12
6-9
11
1

host's body, towards epidermal tissues), is also possible. Some representatives of Plagiorchiida (*Macrodera longicollis, Paralepoderma cloacicola*) are an example. Their juveniles actively migrate from the visceral sac to the subepidermal area and are localized directly below the epithelium lining the pallial cavity.

The distribution of mature mother sporocysts of different species in the body of the molluscan host is probably not random. It may reflect the sequence of evolutionary events associated with the formation of parthenogenetic generations in the digenean life cycle. These issues are analysed in more detail in the concluding chapters (see pp. 432, 492). Here we only list some of the most important aspects of this problem.

Almost all mother sporocysts known to-date are true cavity parasites, since they are localized not inside the tissues but in the haemocoel lacunae. These slit-like lumina may be so narrow that hepatopancreatic acini or gonad branches in fact lie very close to the surface of the parasites. It is probably due to this circumstance that it has often been noted in the literature that sporocysts grow into the hepatopancreas. In fact, mother sporocysts never penetrate inside the acini, they are always located between them.

However, sporocysts of the most specialized trematodes may pass over (probably, facultatively) to tissue parasitism. According to our observations, branching sporocysts of *Labratrema lamirandi* (Bucephalidae) that enter almost all parts of the host's organism, often pass inside the muscles, being situated between the muscular fibers. At the same time, we failed to observe brood chambers containing germinal cells and embryos in any of the parts passing inside the muscles.

In our opinion, it was the molluscan haemocoel that was the primary microhabitat colonized by the ancestors of modern digeneans at the early stages of their life cycle formation.

The first organ colonized by the ancestors of modern trematodes was probably the pallium and walls of the pallial cavity. A dense network of haemocoel lacunae in the areolar connective tissue of the pallial fold ensured the supply of oxygen and nutrients to the parasites. On the other hand, parasites' requirements at the initial stages of the formation of the mother sporocyst were not very large. Even in some of the modern trematodes mother sporocysts either do not reproduce at all, or their proliferative capacity is very limited (see pp. 33, 432).

The transfer of the reproductive function to the parasitic phase of development of the first parthenogenetic generation (see p. 32) required a higher rate of the trophic function of sporocysts. It was probably at that evolutionary stage that they acquired viscerotropism and the complex of morphological (retention, for a rather long time, of a soma, with actively functioning muscular elements, and the sensory apparatus) and behavioural adaptations that ensure the possibility of an active migration into the inside of the visceral sac.

The above events must have taken place independently in different digenean taxa. For instance, the transition to passive way of infecting the mollusc (by intake of an egg), when the juvenile mother sporocyst gets directly into the visceral sac, does not necessarily entail an immediate change in the localization of mature individuals. As noted above, young sporocysts of some Plagiorchiida that have just completed their metamorphosis migrate from the visceral sac to the walls of pallial cavity. Several explanations of this pronounced dermatropism may be given, the most plausible one being associated with the traits of energy exchange. Mother sporocysts of *Macrodera longicollis* and *Paralepoderma cloacicola* studied by us are probably true oxybionts. This supposition is indirectly supported by minimal amounts of stored glycogen and total lack of excretory lipids in their tissues (Dobrovolskij et al.,1983).

One more feature of mother sporocysts should be noted. As a rule, the proliferative capacity of mother sporocysts localized in the subdermal area is limited. This is true not only of *Fasciola hepatica* but also of very specialized mother sporocysts of *Macrodera longicollis* and *Paralepoderma cloacicola.* As the formation of new germinal cells in them is rigidly limited in time, their fecundity is much lower than that of mother sporocysts of other Plagiorchiida, with a prolonged reproductive period (Dobrovolskij et al., 1983) (see section 1.1.2.2. in chapter 1). Another important feature is that all mother sporocysts known to have a very high fecundity and to produce progeny for a long time, localize either directly on the gut surface (*Telorchis assula, Haplometra cylindracea, Xiphidiocercaria* sp. VII Odening, *Lechriorchis primus*), or in the hepatopancreas between its acini (Cort et al., 1952, 1954; Dobrovolskij et al., 1983).

2. ADAPTATIONS OF DAUGHTER GENERATIONS OF PARTHENITAE

The directions of the genesis of adaptations of subsequent parthenita generations are rather diverse. They are manifested both at the individual and at the group level.

2.1 Adaptations of Daughter Generations of Parthenitae Manifest at the Individual Level

Individual adaptations in parthenitae are first of all associated with the feeding mechanisms, as, on the one hand, intensive feeding of the parasite should be ensured, and on the other, the pathogenic effect should be minimized. A tendency to lower pathogenicity of parthenitae is clearly expressed. A substitution of redial forms by sporocyst-bearing ones is at the basis of this tendency. It has taken place several times independently in different phylogenetic branches of Digenea. However, the transition itself from histiophagy, the most traumatic feeding mechanism, to other, more sparing ones takes place already in the rediae. It is associated with a gradual reduction of their digestive system.

The analysis of the digestive system contents in the rediae of several heterophyids (*Metagonimus yokogawai*, *Cryptocotyle* spp., *Metorchis albidus*) has shown that no fragments of host's tissues can be observed in their extremely reduced intestine (our data). At best, single cells of the host's hemolymph can be found. The transformation of the apical surface of the tegument into a true brush boarder testifies to the functional activity of the tegument which acquires an absorptive function. With the loss of the digestive system (mother rediae of Philophthalmidae and true daughter sporocysts of all sporocyst-bearing Digenea), the feeding function passes completely to the tegument. Not only histiophagy but also hematophagy is rendered impossible.

The parasites have to obtain the nutrients directly from the hemolymph or surrounding tissues. Unfortunately, the mechanisms of this process are insufficiently studied. Apparently, the crucial role belongs to the processes of active transport of low molecular compounds (mostly carbohydrates and amino acids), and not to pinocytosis, as it had been thought (see p. 59). The nutrients are taken directly from the molluscan hemolymph. Besides, enzymatic activity of the apical surface of the tegument, noted in many works (for review see: Dobrovolskij et al., 1983), gives us reasons to think that they participate in extraorganismal (extraenteric) digestion. This supposition is confirmed by the data on glycogen hydrolysis in the areas of molluscan tissues adjacent to the sporocysts of *Glypthelmins pennsylvaniensis* in conditions of increased concentration of free glucose (Cheng and Snyder, 1962, 1963). Apparently, the parthenitae can utilize by means of distant digestion not only carbohydrates but other high molecular organic compounds as well. Thus, a transition from active histiophagy to, first, hematophagy and then to the absorption of single molecules of relatively simple chemical compounds considerably lowers the pathogenicity of parthenitae associated with mechanical damage to host tissues. However, this does not mean that there is no pathogenesis as such. The parasites do not inflict the damage that might cause the immediate death of the host. But they do incorporate themselves into the host's metabolic processes by means of enzymes of distant action and other factors secreted into the inner medium of the host by means of highly specialized tegument, and thus their presence in the mollusc has fatal consequences for the latter. However, the main goal is achieved: the mollusc can actively live for some time, required for the completion of the parasite's life cycle stage. This increases the individual stability of the host-parasite system, but it certainly does not lower the negative impact of trematode parthenitae upon the population of the hosts.

Differences in the feeding function are closely correlated with those in the structure and functions of the locomotory apparatus of parthenitae. As all flatworms, rediae and sporocysts have a typical tegumentary-muscular sac, whose muscular elements are represented by almost non-differentiated circular and longitudinal muscles. However, the degree of their development and their functional characteristics differ in different forms.

Rediae, especially young ones, have the most powerful muscular apparatus. This presupposes their locomotory activity. In full accordance with this, rediae possess a well-developed orthogon and rather numerous and diverse sensory papillae. Moreover, the locomotory activity in rediae is often enhanced by the presence of special locomotory projections situated dorsally at the anterior and posterior body ends (see p. 52). As a rule, it is such rediae that exhibit the greatest predilection to histiophagy and cannibalism. Active locomotion appears to enhance the pathogenic effect of these parasites. However, with age such rediae acquire a sac-like shape and their mobility is only manifest as contractility.

It should be specially stressed that this is not accompanied by any reduction of muscular apparatus. Its formation in the parthenitae's ontogenesis stops already in very young individuals, which continue to grow for quite a long time after that. Naturally, at some moment there appears a disbalance between the greatly increased body mass and the muscular system, whose development has stopped. As a rule, young hematophagous rediae have a sac-like body with a very short gut. Their limited motility is soon completely lost. Fully formed sporocysts of many species are incapable of locomotion, though they may generate weak undulating movements.

Having lost locomotory function, the muscular system ensures a constant flow of hemolymph around the body surface, which improves feeding and respiration conditions of the parasites. And finally, the most specialized daughter sporocysts (Sanguinicolidae, Plagiorchioidea, Ochetosomatoidea, Lecithodendrioidea, Microphalloidea and related groups) are immobile as adults, though they retain all the muscular elements of the tegumentary-muscular sac. In greater majority of cases, young individuals of the daughter generation of parthenitae (both rediae and sporocysts) are capable of locomotion. Such age dynamics of changes in motility also seems to be an adaptation: young parthenitae at the time of active settling in the molluscan body exert minimal traumatic influence upon the host tissues due to their minute size. Having reached the final site of infection and having considerably grown, they lose the ability to move, thus decreasing significantly the pathogenic effect of their presence.

Mechanisms allowing the parthenitae to penetrate the defense barriers of the host form a special group of adaptations. Schistosome sporocysts have been shown to avoid the molluscan defense system owing to so-called molecular masking (disguise) and/or molecular mimicry (see reviews: Locker, 1994; Adema and Locker, 1997). In the former case host-derived molecules are absorbed onto parthenitae surface. As a result, non-self epitopes of sporocysts are masked and do not trigger the immune response reaction. Besides, in the course of schistosome evolution the genes have evolved whose expression products are exported to the sporocyst surface as host-like determinants. This phenomenon is referred to as molecular mimicry. Both mechanisms have been so far demonstrated only for schistosomes. However, they seem to be also characteristic for other sporocysts and rediae.

In the course of evolution plagiorchiids have evolved a mechanism allowing them not only to avoid host immune response but also to modify it. Plagiorchiid sporocysts form a so-called "paletot" (see p. 90) from the hemocytes of the molluscan host (Fig. 96). The changes undergone by the host's cells in the "paletot" turn out to be so profound that they in fact change their specialization. Not only do they lose the ability for amoeboid movement, but also their ultrastructural organization, and even metabolism, is altered. In fact, functionally they become a part of the parasite's organism. The actual variety of functions of this structure is yet difficult to describe. However, one function is evident. "Paletot" cells possess a very high level of metabolic processes. They reveal a very high activity of some enzymes of the Krebs cycle, in particular, succinatedehydrogenase (SDG) (Ginetsinskaya and Khari, 1975). Besides, they accumulate much more glycogen than the surrounding cells of the host's hepatopancreas and the body wall of the parasite (Ginetsinskaya and Khari, 1975; our observations). It can be concluded that the "paletot" is very important for the feeding of the plagiorchiid daughter sporocysts.

Just how closely these cells are integrated with the parasite can be seen from the subsequent fate of these cells. Old daughter sporocysts, no long able to control their relations with the host's defense system, are often subject to hemocyte attack. The latter destroy both the "paletot" and the underlining syncytial layer of the tegument, so that the basal lamina for some time becomes the border surface of the parasite. In some host-parasite systems (e.g., *Planorbarius corneus – Xiphidiocercaria* sp. VII Odening) old dying daughter sporocysts undergo a typical encapsulation. The host's hemocytes recognize the "paletot" cells, much degraded by this time, as a part of the parasite in this case as well. Unfortunately, we know nothing as yet about the factors determining the transformation of hemocytes into "paletot" cells. However, their parasitic nature is not to be doubted.

A group of adaptations in parthenitae is associated with their physiological features, in particular, with their metabolism. Unfortunately, this aspect of life activity has been even more poorly studied in sporocysts and rediae than in other phases of the life cycle, yet some regularities of a general nature can be revealed. To begin with, the data on the nature and quantity of reserve energy substances and the activity levels of a number of enzymes of tricarboxylic acids cycle unambiguously point to a higher metabolic activity in rediae than in sporocysts, even if the latter retain their ability to move (Ginetsinskaya and Khari, 1975; Khari, 1975).

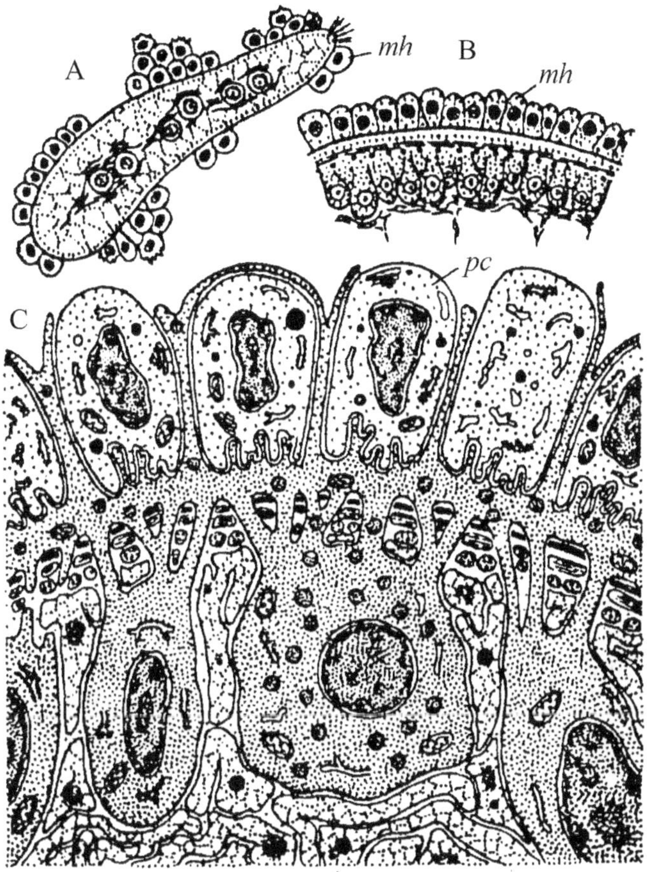

Figure 96. "Paletot" in the plagiorchiid daughter sporocysts. A, initial stage of "paletot" formation on the body surface of the young daughter sporocyst of *Paralepoderma cloacicola*; B, formed "paletot" of *Haplometra cylindracea* daughter sporocyst; C, ultrastructure of "paletot" and body wall of *Xiphidiocercaria* sp. VII Odening daughter sporocyst. *Abbreviations*: *mh*, molluscan hemocytes; *pc*, "paletot" cells. From Dobrovolskij et al. (1983), with permission.

The nature and the level of metabolism are influenced by a number of factors. Among them, the place of the parthenitae's location in the host body plays by far not the least important role. Moreover, this is true of sporocysts of one species in one and the same mollusc. The most important factor seems to be the supply of the parasite by nutrients and oxygen. The plasticity exhibited by the parthenitae in this respect has been demonstrated by the observations upon the *Leucochloridium paradoxum* sporocysts

(Ginetsinskaya and Khari, 1975). Peripheral pigmented areas, where mature metacercariae develop and accumulate (so-called brood sacs), show a high level of metabolism. On the contrary, central parts of that large sporocyst, branching in the hepatopancreas of the host, demonstrate an extremely low level of enzymatic activity. This agrees well with the data on the ultrastructure of the two above-mentioned regions of the *L. paradoxum* sporocyst (see p. 71).

2.2. Group Adaptations of Daughter Generations of Parthenitae

Biological adaptations of parthenitae, manifesting at the group level – the level of microhemipopulation (infrapopulation) (see p. 349), are first of all a number of mechanisms regulating the correspondence between the numbers and productivity of the parasite group and energy potential of the host. The importance of this kind of adaptations is not to be doubted, as any disbalance of relations between the host and the group of parasites in it inevitably results in the death of the host-parasite system as a whole. The ability to regulate these relationships efficiently is especially important in the parasites with endogenous agglomeration (see p. 320). The parthenitae of almost all trematode species are such parasites.

A host-parasite system functions according to a feedback principle. Of course, not all the response reactions of the parasites to the signals from the host can and should be treated as adaptations. A decrease in fecundity and even the death of some parasites in case of superinfection of the host can hardly be treated as an adaptive reaction, even though it can for some time stabilize the existence of the host-parasite system. However, sometimes the same factors, on the basis of the same mechanisms, result in essential reconstructions of the whole group. They include the change in the age structure of the group, re-determination of the offspring type (see p. 369) and quite regular changes in the reproductive potential. They are aimed at the minimization of the take-out of energy from the host, at giving the host a chance to stabilize its physiological status. In turn, this ensures a stable existence of the parthenitae group in the molluscan host. Similar to miracidial dispersion, when the fate of a single miracidium is not essential for infection success (see p. 229), the fate of an individual redia or a sporocyst is not important for stable functioning of the group in the molluscan host. Therefore, in this case we deal with a group adaptation.

Two main types of parthenita microhemipopulations are known at present. The first is widely represented in primitive trematodes, as well as in the orders Schistosomatida and Strigeidida. The basic feature of daughter parthenitae in these trematodes is that they are capable of re-determination of offspring type (for details see p. 369). Along with cercariae, leaving the mollusc in most cases, they can produce their like. The existence of such a group is in fact limited only by the longevity of the molluscan host, as new parthenita generations substitute previous ones. This biological trait makes it possible for the whole group of rediae (sporocysts) to react efficiently to any changes in the host. This is manifested in the fluctuations of the numbers of the group, in changes of its demographic composition and releasing of cercariae. The latter applies both to the number of released larvae and to shedding rhythm as such. All these regular fluctuations of cercariae releasing seem to play an important role in the stabilization of mollusc-parthenitae system as dynamic reconstructions of microhemipopolation itself. The point is that cercariae, being endotrophic larvae, carry out of the host a great amount of reserve carbohydrates (Shigin, 1978, 1981, 1997). All this biomass, literally saturated with energy, is created solely by means of exploitation of the host's energy potential. Therefore the rhythms of cercariae shedding, conditioned in the long run by the dynamics of parthenitae's reproductive activity, can be also treated as a manifestation of the adaptive evolution.

The second type of parthenita microhemipopulations is characterized by a rigid stability of its age structure. The number of parthenita generations is strictly limited and does not exceed two: mother sporocyst is followed by a single generation of daughter sporocysts. This is typical for the whole plagiorchiid phylogenetic branch. The numbers of the groups in this case can not smoothly fluctuate following the changes in the physiological state of the host. The number of daughter sporocysts reaches its maximum by the end of reproduction of mother sporocyst, and after its completion it can only decrease as a result of the death of some individuals. After the death of the last daughter sporocyst the microhemipopulation in the host ceases to exist, if, of course, the mollusc has lived to that time, which is by far not so common. Reproducing, all daughter sporocysts produce only the larvae of hermaphroditic generation, cercariae, which leave the molluscan host.

In the above case somewhat different adaptive mechanisms, stabilizing the host-parasite system, must be at work. The numbers of the future group are determined at the earliest stages of its formation, when the parasitic capacity of the host is far from being exhausted. The mother sporocysts of *Xiphidiocercaria* sp. VII Odening (Plagiorchiidae) are a good example here. According to the observations of Muhamedov and our own data (Muhamedov, 1981; Dobrovolskij et al., 1983), if the mollusc is infected by a single mother sporocyst of this species, 10-12 active germinal masses develop in it, producing a large number of embryos of the daughter sporocyst. The picture is entirely different in case of multiple infection: if 3-5 mother sporocysts develop in the mollusc simultaneously, the number of germinal masses in each of them is reduced up to 1-2. In case of superinfection (10-12 parasites per mollusc), mother sporocysts often contain no structured germinal masses at all. The proliferation of undifferentiated cells of the germinal lineage in them starts very early, the number of the daughter sporocysts produced being very small. It is unlikely that 10-12 young individuals, only starting their reproduction, present in the host, experience such a deficit of nutrients that it tells upon their reproductive potential. It can only be conjectured that some other mechanisms are operating, implying a direct influence of the parasites upon each other.

There are also some other data testifying to the existence of mechanisms regulating the composition and the reproductive activity of local groups of parthenitae in the mollusc. Seasonal dynamics of the composition of parthenita microhemipopulations found in a number of trematode species can only be viewed as a manifestation of rather fine adaptations to exact living conditions. It should be clearly stressed that it is possible to tell apart true seasonal dynamics from purely age changes in the composition of a local group of parthenitae only if a number of conditions are fulfilled. First of all, in host-parasite systems under investigation both the molluscs and the parthenitae in them should live for at least two years. It has been shown on a number of trematode species (Machkevskii, 1984; Rusanov and Galaktionov, 1984) that in such cases clear seasonal cycles are observed. The microhemipopulation is very active in the warm time of the year, the regeneration of its composition and mass releasing of cercariae is constantly taking place. In winter the situation is entirely different. The microhemipopulation frees itself from old individuals and cercariae production stops. In this state the parasites overwinter in the hosts. In spring the composition of the group is resumed due to parthenitae giving birth to their like, and the releasing of cercariae is resumed.

A somewhat different picture is observed in plagiorchiids, the parthenitae of which, as mentioned above, are unable of self-reproduction. They also can overwinter, if their individual lifetime permits. In winter, the most essential changes are observed in the germinal masses (Muhamedov, 1979). They enter a dormant state: the proliferation of undifferentiated cells of the germinal lineage stops completely, embryos almost entirely disappear from them, the size of the germinal masses noticeably decreases. The fate of the embryos in the sporocysts may be different.

In some species, germinal balls, which have not developed into mature cercariae, undergo resorption and might be used as an additional source of energy. In other species, embryos may be simply conserved. In spring, when, following an increase in the temperature, the infected molluscs resume the active way of life and start to feed intensively, the germinal masses are also activated and quickly regain their composition. As soon as this has taken place, the supply of the embryos to the sporocyst's body cavity is also resumed.

The stability of host-parasite system is a function of two variables. Up to now we have been speaking mainly about the numbers of the parasites, assuming the volume of the host's body and its energy potential to be more or less the same and constant[iv]. However, these characteristics of the hosts may vary very widely. Correspondingly, the parasitic capacity of the host also changes, being limited in the same way as the capacity of any other habitat. In this case the parasites also use the above mechanisms, showing a remarkable flexibility. Numerous observations show that the number of rediae or sporocysts in a mollusc depends directly upon the size of the latter (Smith, 1959; Dinnik and Dinnik, 1960; Tihomirov, 1980). At the same time, as shown on *Philophthalmus rhionica*, qualitative composition of the microhemipopulation does not depend on the volume of the host's body. The percentage of different types of rediae (young individuals which have not started the reproduction; rediae producing rediae; rediae producing cercariae; and finally, old degenerating individuals) both in numerous and in small groups in the molluscs of corresponding size remains practically the same. This is yet another consideration in favour of the idea that the productivity of parthenitae and their capacity to give rise not only to cercariae but also to their like is under constant control of the host-parasite system which has been balanced by time and selection.

3. HERMAPHRODITIC GENERATION

The development of individuals of hermaphroditic generation, similar to that of mother sporocyst generation, is a complicated metamorphosis, with several successive phases. As a rule, they are three: one free-living, the cercaria, and two parasitic, the metacercaria or adolescaria and the marita (adult). In Digenea with two-host and four-host life cycles (see pp. 334, 346) the number of ontogenetic developmental phases of hermaphroditic generation may correspondingly decrease or increase. This is often accompanied by a change in the nature of adaptations inherent to the phases.

3.1. Adaptations of Cercariae

Cercariae, the larvae of the hermaphroditic generation, possess an extremely wide complex of adaptations. The very nature of these dispersive larvae is adaptive. This is manifested in their organizational, biological, physiological and behavioural traits. The life cycles of modern trematodes are rather diverse (see p. 319). Some of them have retained a number of ancient archaic traits, others have acquired highly specialized features, as a result of complex and often many-directional evolution. Correspondingly, the tasks to be solved by means of these free-living larvae have been changing as well. One thing remains unchanged: these larvae have always performed the dispersive function, the primary and the main one.

In the most primitive trematodes this phase was intended for the completion of dispersion in its simplest form, the dissemination of infection in a certain space. Biological traits of the larvae determined the subsequent pattern of events only to a small extent. However, in the process of evolution the functional load upon cercariae has been gradually increasing. Naturally, the selection first of all supported and retained such morphological-functional and behavioural features that would, if not guarantee, but at least increase the probability of the cycle completion. These evolutionary trends have been, in part, considered earlier (see p. 149), but we will address this issue again in chapter 5 (see p. 436). Now we would only like to emphasize that the complex of properties and characters of these larvae has been changing at every stage, quite essentially but in full accordance with the tasks to be solved. As with the analysis of miracidial adaptations, we will first dwell upon the main morphological structures of cercariae, and then pass to the analysis of their biology and behaviour.

3.1.1. Morphological Adaptations of Cercariae

All cercariae may be divided into three uneven groups. The first group comprises trematodes with the most archaic, primarily two-host life cycles (see p. 335). Having left the mollusc, the larvae directly infect the definitive host. The infection, however, may take place in two ways. Firstly, free-living cercariae may be eaten by the definitive host (Azygiidae, Bivesiculidae). Secondly, cercariae may encyst in the environment prior to being eaten (Fasciolidae, Notocotylidae, Philophthalmidae, Paramphistomidae, Haploporidae). Even if these cercariae use living organisms as a substrate, their encystment is not associated with the act of penetration into the host's body through its epidermis (see pp. 152, 280). If the disruption of the host epidermis does take place, no specialized penetration apparatus is used for that. This is first of all true of the larvae from the family Echinostomatidae. Trematodes from this family demonstrate an array of possible ways of three-host life cycle formation (see p. 307).

The second group comprises the larvae actively penetrating into the second intermediate host with specialized penetration apparatus. In a surprising majority of cases these are the larvae of trematodes with three- and four-host life cycles.

Finally, the third group consists of the larvae that have lost the ability to penetrate actively into the second intermediate host. This is associated with the exclusion of the period of active existence of cercariae in the environment from the life cycle. The systematic composition of this group is heterogeneous. It comprises an array of secondarily specialized trematodes, often without any phylogenetic connections.

Our choice of biological traits of the larvae as a criterion for the delimitation of the above groups was not arbitrary. It is these traits that mainly determine the set and the character of provisional structures reflecting the directions of adaptive evolution.

Cercariae, being dispersive larvae, possess the same main functional complexes of organs as do miracidia. However, there are essential differences, conditioned by the biological traits and the nature of the larvae. Whereas all miracidia possess some kind of penetration apparatus, only the cercariae from the second group have it in the most typical form. Besides, all cercariae from the first group encysting in the external environment, a large part of those from the second group, and, finally, the least specialized forms from the third always possess a cystogenous gland complex, absent in miracidia. Attention should be also drawn to the fact that many provisional structures performing similar functions in cercariae and in miracidia have an essentially different origin and are not homologous.

The last circumstance is clearly revealed when the structure and the features of functioning and origin of the locomotory complex of the hermaphroditic generation larvae are analyzed. The tail of cercariae is one of the most demonstrable examples of provisional structures in trematodes (see p. 436). Its origin, its functional and structural details are a subject of numerous discussions in the literature. Participation in this polemics is not among the aims of the present section. Yet some arguable problems will have to be addressed.

We do not doubt an adaptive nature of this provisional organ, which has originated as a specialized organ of locomotion of the free-living dispersive life cycle stage of ancient proto-trematodes. Here we are agreement with the view of Cable (1965, 1974), who has suggested the tail to have formed as a result of specialization of the posterior body part of the free-living worm-like organisms moving in water by means of undulation. The body was divided into two parts, the vegetative-generative and the locomotory one. Subsequent functional and morphological specialization of the locomotory part must have rather quickly transformed it into one of the most perfect organs of locomotion (see p. 121) ever possessed by lower invertebrates. It was certainly to supported by selection. These views are shared by many parasitologists (see reviews: Ginetsinskaya, 1968; Pearson, 1972, 1992). It would seem that nothing here gives rise to doubt, so well this theory is supported by factual material (see p. 438). However, other viewpoints are sometimes opined in the literature. They will be addressed later in chapter 5 (see p. 436).

It has already been noted that tails of cercariae of recent trematode species are extremely diverse, not only in structure but in function as well. Constructive features of the tails and their functions are interdependent. In our opinion, this is further conclusive evidence in favour of the adaptive nature of this structure. In the process of evolution and divergence the tails of cercariae often acquire new, originally non-characteristic functions, which almost at once inform us of their structure.

The above general considerations suggest that locomotion was the original and at first the only function of the tail. This single function is often retained by the larvae with the so-called simple tails, widespread in many trematodes, both in more primitive ones with primarily two-host life cycles (Fasciolidae, Paramphistomidae, Notocotylidae), in more specialized ones showing a tendency to three-host cycle formation, and in those with obligatory three-host cycle (Cathaemasiidae, Echinostomatidae, Psilostomidae, a suppressing majority of Plagiorchiida). Though it should be always borne in mind that in terms of body plan organization the tails of plagiorchiids are very different from the tails in all other families mentioned (see p. 438), this difference does not concern their constructive and functional traits. Despite the diversity of separate details, a general structural scheme turns out to be similar.

In general, the construction of the tails (see p. 121) is somewhat similar to that of specialized nematodes with a schizocoel: the main locomotory elements – longitudinal muscular bands and a spacious schizocoel filled with liquid – are located below a relatively dense and elastic epidermis. The schizocoel houses various internal organs. Similar to nematodes, tails of cercariae can undulate in one plane. There is one, not very significant, difference: nematodes bend their body in dorso-ventral plane, whereas cercariae tails undulate in a lateral plane. This kind of structure based on the use of hydroskeleton seems to have proved the most efficient for the undulating movement in lower invertebrates.

Different as they are, organs of locomotion in cercariae and miracidia have one similar morphological-functional feature that catches the eye. (Of course, only a functional analogy is meant.) In both cases locomotory structures become the depots of reserve carbohydrates which make movement energetically possible. The division of energy reserves in miracidia into two independent pools has already been discussed (see p. 221). A similar matter is observed in cercariae. Their movement takes place first of all due to glycogen reserves in the cells of the tail, or directly in the cytons of the muscular cells, or (as in furcocercariae) in specialized cells, "caudal bodies". Even if some reserve carbohydrates are left in the body of the larva, the depletion of reserves in the tail leads to its almost total immobilization (for review see: Ginetsinskaya, 1968).

The trajectory described by the tail during movement is very different in different species, depending mainly on its length. If the tail is long (two-three times more than the undulation wave length), as it is the case in some Echinostomatidae (*Petasiger*, megalocercous species of the genus *Echinochasmus*), undulation can be very clearly observed. Such larvae move anterior end forward. Cable (1956) reported a very interesting observation. A very long tail of pleurolophocercous *Cercaria caribbea XIX* Cable, 1956 demonstrates a classical undulation during relatively slow swimming. With increasing speed, the amplitude of the undulation wave grows abruptly, the tail finally describing an "8" figure.

If the tail is shorter but its length is equal or not much less than the length of the undulation wave, then a moving larva also produces the impression of its tails describing an "8". The movement of such tails has been described in considerable detail by Chapman and Wilson (1973), who worked with *Himasthla secunda.* An actively swimming larva completely bends over to the ventral side, so that its posterior part is turned forwards (Fig. 97). The tail is thus also directed forwards. When the undulation wave is generated, the tail closes almost into a ring, starting to deviate from the longitudinal body axis. As the wave is shifted to the end of the tail, the latter straightens, but bends in such a way that its tip is directed backwards. Thus, a rather abrupt stroke on one side is made. This phase over, the wave is again generated at the tail basis. This time, however, it spreads in the other direction. Every subsequent stage of the tail movement is, as a matter of fact, a mirror reflection of the previous one, also resulting in the utmost deviation of the straightened tail backwards. The tail moves in one plane, parallel to the frontal surface of the larva's body.

It catches eye that the tail "shifts" from the right side to the left and back when a new undulation wave is being generated. At this moment the tail is bent as much as possible. In this way the resistance of water during the larval movement is minimized. Interestingly, the same principle of minimization of resistance is at work in the metachronal wave of cilia passing along a kinety in Ciliophora.

As a result of alternating one-sided strokes of the tail, the body of the larva deviates to the left and to the right in turns. In fact, during the "rectilinear" movement of the larva its body describes a sinusoid.

The swimming of cercariae is very maneuverable. They change the direction of movement (perform a turn) with the speed of 70°/sec (Chapman and Wilson, 1973). The actual speed of turning is even higher, as the experiments that have yielded these data were performed at the temperature of 5° C specially in order to decrease the speed of cercarial movement. The turning itself is made by changing the dynamics of the beating of the tail. Its trajectory becomes asymmetrical. Whereas the wave amplitude at one side remains the same as during rectilinear movement, it decreases markedly at the side the larvae is turning to. This can be achieved due to the tail starting to bend backwards much earlier, before it crosses the "line" of rectilinear movement direction. It bends less back relatively to the longitudinal axis of the larva and the stroke becomes weaker. As a result, the direction of larva's movement changes.

The beating of shorter tails of many Xiphidiocercariae (Plagiorchiidae, Ochetosomatidae, Microphallidae, Lecithodendrioidea) in principle follows the same scheme. However, when the undulation wave is being

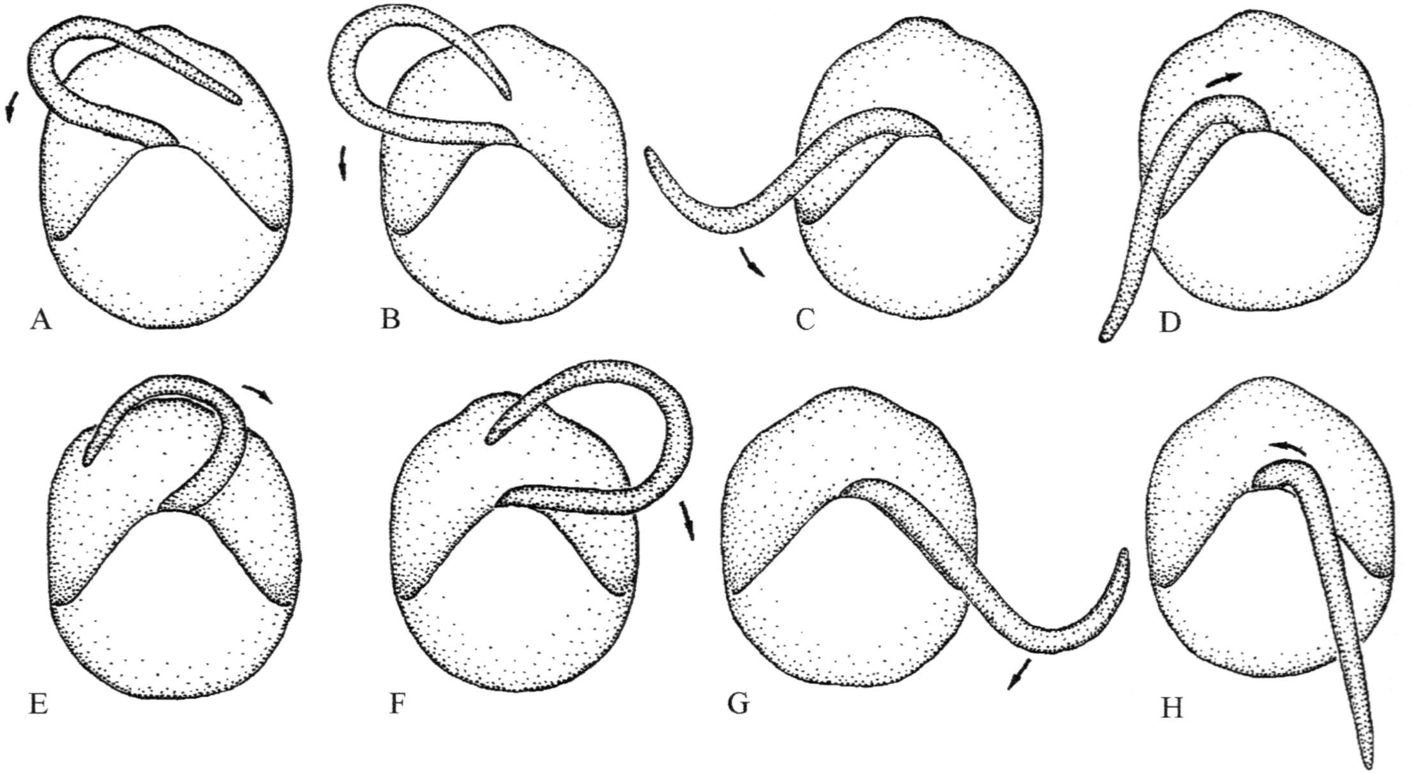

Figure 97. Diagram illustrating the movement of the cercaria of *Himasthla secunda*. A-H, successive stages of one cycle of tail movement. The large arrow indicates the direction of movement of the cercaria. The small arrows indicate the direction of movement of the tail. Based on Chapman and Wilson (1973), reprinted with the permission of Cambridge University Press.

generated, the tail is not closed into a "ring" due to its relatively short length. That is why its movement is perceived by the observer as abrupt alternating bends to the right and to the left.

Such tails usually do not perform any additional functions. When cercariae sink to the substrate and start to crawl or lie at one and the same place, elongating and contracting in turns, they drag their tails passively behind. Sometimes such tails bear additional structures for increasing the rowing surface. The above-mentioned fin-folds usually dorso-ventral and more rarely lateral. These fin-folds originate independently in different groups: some representatives of Echinostomatidae, Xiphidiocercariae "Ornata" (Plagiorchiida).

The flotation of larvae in the water column seems to be one of the first additional functions acquired by the tails. Indeed, the need for an adaptation like this was originally associated with the emergence of intermittent swimming. Large echinostome cercariae, as well as many large Xiphidiocercariae, start to sink rapidly to the bottom as soon as they stop locomotion. They have a pronounced negative buoyancy and have to spend a lot of energy, moving constantly, to remain floating in the water column. Taking into account the fact that both cercariae and miracidia are endotrophic larvae and thus have limited reserves of energy substances, their constant movement would have fast resulted into energy depletion and death. In its turn, this would have decreased the probability of infection of the second intermediate host. The emergence of periodic alternation of active movement and repose inevitably prolongs the larval life (see p. 496). However, this is meaningful only in case the larva stays at the same place where it was during active movement. If the larva sinks rapidly to the bottom, and then has to return to the area of the water body where the probability of meeting the host is high, it spends its meagre energy reserve and the adaptive significance of this behavioural type is completely lost. But if the larva is always in the zone of possible contact with the host, then increasing its life span becomes one of the leading factors influencing the successful completion of the life cycle.

This problem can be solved, first of all, by means of increasing the buoyancy of the organism, at least to such a degree that the larva would not sink too low during its repose. Different species solve this problem in different ways. It may be diminishing of the body size (miniaturization – see p. 149) (many virgulate cercariae – superfamily Lecithodendrioidea), or store in parenchyma excretory lipids which form a kind of hydrostatic apparatus (some Xiphidiocercariae – *Cercaria nigrospora* Vergun) or, finally, the development of various projections, increasing the surface area of the body and thus permitting passive flotation. This "parachuting" effect is enhanced if the larva acquires a special resting posture. Naturally, these changes first of all involve not the larval body but its provisional organ, the tail. The bifurcate tail of all Furcocercariae is such a modification, ideally suited for the solution of the task in question. It is these cercariae that have extremely di-

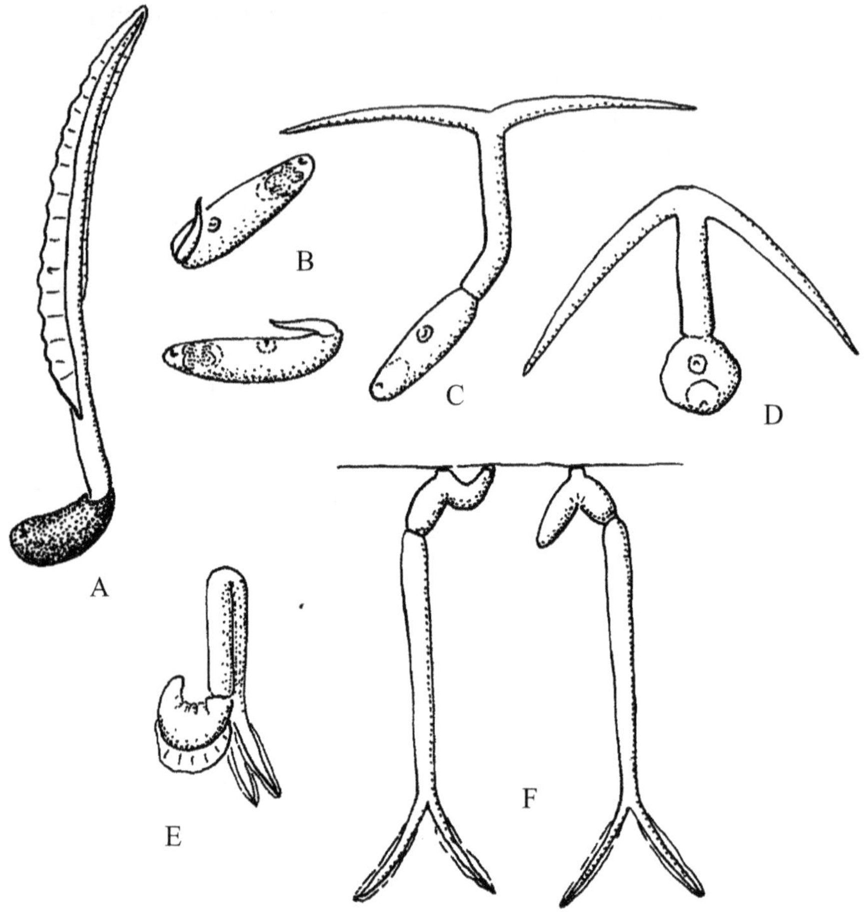

Figure 98. Resting postures of the cercariae of the different trematode taxa. A, Heterophyidae; B, Lecithodendrioidea; C, Diplostomidae; D, Strigeidae; E, Sanguinicolidae; F, Schistosomatidae.

verse resting postures. The larvae acquiring resting postures spread branches of their furca wide apart (Fig. 98). The adaptive significance of furca can be seen with special clarity if we compare their length in biologically different strigeidid cercariae (Strigeidida), belonging to the type of "longifurcate" larvae and in schistosomatid (Schistosomatidae) "brevifurcate" cercariae. The formers spend their resting periods in the water column, using their furcae for floating. The latter, ceasing to move, usually attach to submerged objects or to the water surface by means of the ventral sucker. They lack the floating phase as such and actually have no need of long furca. Furca-like appendages on the tails reach enormous lengths in the cercariae from the family Bucephalidae, which spend most of the time passively floating in the water column (see p. 107).

The above considerations do not mean that furcae are used exclusively for passive floating in the water column. They are also effectively used for swimming. The tail beating of true Furcocercariae is much more diverse than it has been described above. Furcocercariae have long been known to be capable of reversal of movement. The function of the tail during different types of active swimming has been studied in detail on *Schistosoma mansoni* (Graefe et al., 1967) and *Mesostephanus appendiculatus* (Cyatocotylidae) (see fig. 33). When the cercariae move anterior end forward, furcal ramii are closely pressed to each other. As a matter of fact, they form a true tail fin, which enhances the undulation effect of the tail stem. Chapman and Wilson (1973), analyzing the data of (Graefe et al., 1967), compare this type of movement in furcocercariae to the movement of all other aquatic animals, swimming by means of undulation of relatively long and flexible body.

When the cercaria swims posterior end forward, the tail beating is entirely different. In fact, the tail stem oscillates about two fixed points. One of them is situated at the base of the tail, the other, directly before the branching of the furcal rami. The latter are set wide apart during this kind of movement. According to the observations of Graefe et al. (1967), they are set at a constant angle relatively to the tail stem and do not make any autonomous movements. As the tail oscillates, the furca also turns alternatively to the left and to the right, thus making rowing movements.

Other structures that appear to enhance the "parachuting" effect during intermittent locomotion are the bunches of rigid bristles situated along the tail stem (trichocercous cercariae), finger-shaped projections described in haplosplachnid cercariae (at present, the life cycle of only one species from this family is known – Cable, 1954) and the above-mentioned (see p. 267) fin-folds. The latter are also found both on the furcae of some furcocercariae (Sanguinicolidae, Cyathocotylidae) and on the relatively simple tails of all Cercariae Pleurolophocerca (Heterophyoidea).

Though morphologically the tails of Heterophyoidea are rather simple, their beating during the swimming of cercariae is quite complex (Chapman and Wilson, 1973). Firstly, both the tail and the larva's body move not in one plane, as in the cercariae of the Echinostomatidae and xiphidiocercariae, but in three-dimensional space. Secondly, it is not simply an undulation wave but a true loop that moves along the tail. The loop hardly reaches the tails tip when a new one is being generated at its base. Every time the direction of the loop formation is different: it twists either clockwise or counter-clockwise. During onwards movement, the body of the larva moves along a very complex trajectory, involving clockwise and counter-clockwise turns (Chapman and Wilson, 1973).

On another topic, the emergence of alternating periods of rest and active movement also favoured the emergence of a new behavioural strategy of cercariae, aimed at the infection of the host with the lowest energy expenses possible (see p. 283).

A change in the way of infecting the host is associated with yet another direction in the evolution of tails, the emergence of the so-called macrocercous larvae. Forms with hypertrophied tails, many times longer than the ones typical for the group, appear independently in different families (Echinostomatidae, Heterophyidae, Strigeidae) (see fig. 42). They are easy to see and attract the attention of potential hosts, which swallow them readily (see pp. 282, 283). The movement of such macrocercous larvae often imitates that of other water animals, first of all, small nematodes and oligochaetes.

To enhance the attraction effect, in some representatives from, again, distant families (Echinostomatidae, Heterophyidae), the tail tips of several larvae merge to form aggregates known in the literature as "Rat-King" ("Rattenkönig") Fig. 99, see also p. 284).

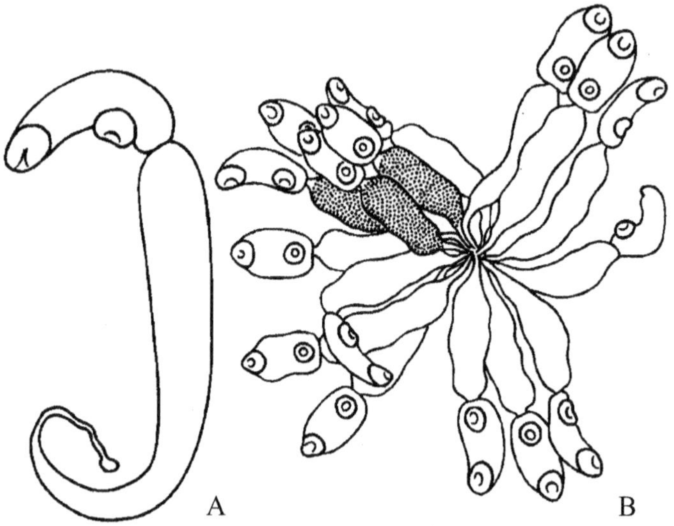

Figure 99. Zygocercous *Cercaria rhionica VII* Olenev et Dobrovolskij, 1975 (Echinostomatidae). A, general view of a single cercaria, B, cercariae aggregate of "Rat-King" type. From Olenev and Dobrovolskij (1975), with permission.

Whereas in some families macrocercous cercariae occur only in some representatives, in others they are characteristic. The emergence of giant tails must have presupposed another function of theirs, the protective one. This is typical for the so-called cystocercous and cystophorous cercariae (Azygiidae, Bivesculidae, Gorgoderidae) (see p. 99). Some researchers (Brooks et al., 1985a; Gibson, 1987) considered the submergence of the larval body into the enlarged part of a hypertrophied tail as encystment. However, in fact, no encystment takes place. The larvae lie freely in the cavity of the swelling at the basis of the tail. There is a usual connection between the tail and the body, as in all other cercariae. The elements of a functioning excretory system also come into the tail (Azygiidae) (see fig. 145).

In our opinion, the appearance of macrocercous cercariae is a secondary phenomenon, and the emergence of cystocercous and cystophorous tails even more so. This follows directly from Cable's (1974) views about the transformation of the posterior body part of proto-trematodes into a tail and is supported by the analysis of body plan organization and embryonic development of the individuals of the hermaphroditic generation (see p. 438). Large and complexly differentiated tails, whose formation is energetically rather expensive, could not have appeared originally, and moreover they are often associated with extremely specialized forms of larval behaviour in the environment (see p. 282).

An alternative direction in the evolution of cercarial tails is their secondary simplification and a more or less pronounced reduction. The tail is no longer used as a locomotory organ and, changing its function, becomes an organ of attachment to the substrate. In the beginning of this evolutionary series the length of the tail is not reduced (in *Cercaria megalurus*, the larva of *Philophthalmus nocturnus*, the tail is much longer than the body), but its internal structure is different. The tail cavity disappears, its internal volume fills with vacuolizated cells, longitudinal muscular bands are reduced, but on the other hand, a sucker-like structure with a group of secretory cells is formed terminally at the posterior end. In other representatives of the genus *Philophthalmus* (*P. rhionica*, *P. gralli*) the tail, retaining all secondarily acquired constructive features, becomes much shorter. Our observations of the larvae of *P. rhionica* have demonstrated that they can actively swim in the water column. However, the mechanism of locomotion becomes different. The larvae move bending their body abruptly. This process involves the activity of the muscles of the tegumentary-muscular sac of the body. The tail takes no part in locomotion and is passively dragging behind. According to Rees (1971), body contractions also play a crucial role in the movement of *Parorchis acanthus* cercariae. However, in the Rees' opinion, the role of the tail should not be underestimated either, as the wave initiated by the larval body also passes along the tail.

As mentioned above, the main function of the tail in philophthalmids is fixation on the substrate, including the water surface. The attached cercariae are hanging down, from time to time bending their bodies and making abrupt jerk-like movements (similar to those accompanying active swimming).

Tail reduction goes far in the representatives of the family Opecoelidae (genera *Sphaerostomum*, *Podocotyle*). These larvae completely lose the ability to swim. They can only move, poorly, upon the substrate, using the mouth and the ventral sucker. A tiny tail, fully transformed into a sucker-like organ, is used exclusively for attachment to the substrate. These cercariae spend most of their life attached to the substrate by the tail and performing smooth circular movements by the free anterior body end (see p. 301 and fig. 101).

Very similar morphological-functional changes are also observed in the cercariae from the family Gymnophallidae. The cercariae of *Cercaria margaritensis*, shed by *Turtonia minuta* bivalves, use their tiny tails with small furcae for attachment to the substrate and not for swimming (Galaktionov, 1996a).

In the above examples, forming a morphological and not a phylogenetical series, the reduction of the tail is accompanied by its specialization, its transformation into a kind of sucker. However, even that does not take place often. There is a simple rudimentation without any specialization. In some cases cercariae with rudimentary tails or without them ("cercariaea") still leave the mollusc and can even move upon the substrate (the genus *Catatropis* from the family Notocotylidae, the genus *Asymphylodora* from the family Monorchidae). In other cases the larvae completely lose the ability to lead an active life, though they still spent some time in the environment (the representatives of the families Dicrocoeliidae, Ptychogonimidae). In such forms the cercariae are either shed from the molluscs in aggregates protected with a thick layer of drying mucus (*Dicrocoelium*) or inside the sporocysts that actively leave the molluscs (the genus *Eurytrema* from the family Dicrocoeliidae, family Ptychogonimidae). And finally, a natural process of reduction of the tail as a locomotory organ is observed in a variety of trematodes whose cercariae do not leave the mollusc at all, transforming in it into metacercariae. In the trematodes with this kind of forming secondary two-host life cycle (see p. 338), the developing larvae often have the tail primordium (strictly speaking, its presence marks the cercaria stage) whose development is stopped very early. This is nothing but an atavism. The microphallids of "*pygmaeus*" group are a classical example of this kind (see p. 343). Apropos, the degree of tail reduction allows us to estimate how long ago secondary two-host life cycles have arisen in a definite trematode group. The morphological and biological features of *Heronimus mollis* cercariae which do not leave the mollusc (see p. 436) show that the reduction of transmission ways of this species in ecosystems has taken place relatively recently.

Rudimentation of cercarial tails conditioned by the loss of their original (plesiomorphic) function has taken place many times in different families from different phylogenetic branches. These events have been independent. However, every such instance was based either upon a crucial change in the individual behaviour of cercariae or upon a change in the scheme of infection transmission from the first intermediate host to the definitive one.

It has already been noted that cercariae do not always locomote by means of the tail. The tail arose as an adaptation to swimming. When the whole larval life is spent on the surface of more or less solid substrate, the larvae move by means of body musculature and the suckers: the oral one and the ventral one.

The emergence of suckers in trematodes is a subject of considerable discussions. Later on we will discuss to this problem (see p. 450). Now we would like to note that in a majority of actively swimming larvae attachment, the primary function of the suckers, is performed only after the establishment of a contact with the host. Less often (Schistosomatidae) the ventral sucker is used for attachment to the substrate or the water surface in the period of resting between two phases of active movement. In some cercariae the ventral sucker is a functional substitution of the tail of cotylocercous cercariae (see p. 301). The larvae of the genus *Asymphylodora* often attach in large numbers to the tentacle tips of freshwater snails and move together with the snail (Wesenberg-Lund, 1934). It is only rarely that cercariae can actively move upon the substrate in a manner similar to caterpillars from the family Geometridae. Brachylaimids, retaining active motile stages in their life cycle (completed on land) are especially interesting in this respect. As compared to the furcocercariae of *Leucochloridiomorpha constantiae* the larvae of *Brachylaimus aequans* retain a tiny rudiment of the tail. However, they do not use it. These cercariae develop in land snails from the genus *Macrochlamys*. Having left the host (which happens only in conditions of increased humidity), the larvae actively move in a pacing way upon the slime track left by the snails (Soboleva, 1986).

The nervous system and sense organs of cercariae are more poorly studied than those of miracidia. Moreover, whereas we have at least some knowledge of morphology of these structures, we know almost nothing about their function. The results of numerous works on cercarial chaetotaxy, widely used for systematic, populational and other purposes, give us nothing for the understanding of the regularities of sensory system function. This is regrettable for two reasons. On the one hand, ignorance of physiology makes it very difficult to study cercarial behaviour, which is one of the main factors influencing the success of infection transmission. On the other hand, some features of physiology of the nervous activity of these larvae may well be of interest for general biological problems. The point is that cercarial nervous system is in fact represented by two parts, the body orthogon and the nervous elements of the tail. The latter has both motor and sensory neurons. Pacemakers seem to belong to the former. The tails of many primitive forms (Echinostomatidae, Notocotylidae, Paramphistomidae, Heterophyidae) are known to be able to swim for some short time after the separation from the body. The tails of *Cryptocotyle lingua* have been shown to retain all their locomotory features after the separation from the body. This may happen in an especially spectacular way in macrocercous cercariae *Echinochasmus* sp., including those forming aggregates of "Rat-King" type (see fig. 99). Orientation and any manifestations of taxes (including phototaxis) are, though, completely lost. The movement of tails in the aggregate becomes chaotic and uncontrolled. Nothing of this kind can be observed in, for instance, most Xiphidiocercariae. Freshly separated tails sink at once to the bottom and are at best capable of weak contractions for a very short time. One more peculiarity catches the eye. In most actively swimming larvae from primitive groups rather large and long tails bear quite numerous sensory papillae (we are speaking about usual cercariae and not about macrocercous forms) (Richard, 1971; Olenev and Dobrovolskij, 1975). There are also numerous sensory papillae on the tails of furcocercariae. But in all xiphidiocercariae (Plagiorchiida) the tail always bears only two sensory papillae.

The above data suggest that cercarial behaviour is controlled by the orthogon, where the information received by the larva by means of photo- and numerous other receptors is processed. Functional significance of these receptor is, unfortunately, obscure. Judging by the number of sensory papillae, the tail plays an important role in receiving this information. As for the locomotory activity itself, it appears to be controlled by the motor neurons of the tail, which are, in their turn, orthogon-controlled. In xiphidiocercariae the tails are secondarily specialized (see p. 449). Their tails appear to have lost their autonomy completely, all their activity being controlled by the motor elements of the orthogon.

Penetration apparatus of cercariae is a multicomponent system comprising the organs of diverse origin and strictly defined functions. However, all these structures have to work in total accord. Their functioning is the basis of the part of genetically determined behavioural programme of cercariae which is directly associated with the penetration of the larvae into the second intermediate host.

It should be noted that the larvae may get into the host in several different ways. Not all of them are associated with penetration, i.e., with a local disruption of epidermis and the formation of a small hole through which the parasite makes its way inside the host. This is indeed performed by a special penetration apparatus. The host's epidermis can be also partly destroyed by other, less specialized means, if such an apparatus is absent. However, the larva may get into the host without any destruction at all. This can be achieved in two ways. First, the host may swallow the larva (see p. 282), second, the larva may use natural openings of the canals leading inside the body. The latter way requires no special morphological structures but complex behavioural programmes. Naturally, the cercariae utilizing the latter three ways lack specialized penetration apparatus, and so do the cercariae encysting in the environment.

The penetration apparatus in most cercariae is based on two types of structures. First of all, these are the secretory structures for chemical impact upon the epidermis and underlying tissues. They are often functionally aided by structures for mechanical destruction of epidermis.

Structures of the first type are represented by penetration glands (see p. 128). This name is very widespread in the literature, though not all the aspects of their origin and, especially, functioning can be considered fully understood. Many researchers tend to establish, however tentatively, a homology between penetration glands and frontal, or head glands of turbellarians (Ginetsinskaya, 1968; Pearson, 1992). However, one should bear in mind that not only the ducts of penetration glands but also those of the so-called metacercarial glands open at the anterior body end of many trematodes (see p. 129). The latter function mostly at the stage of marita and can be just as well considered to be the homologues of frontal glands. It should be noted that metacercarial glands are widespread in the cercariae of primitive families (Echinostomatidae, Psilostomidae), which lack true penetration glands. Taking into account the remarkable plasticity of secretory structures, two variants of the origin of penetration glands are plausible. Firstly, penetration glands could have originated as a result of morphological and functional differentiation of head glands of turbellarians. Some of the head glands cells are differentiated and start to function at the larval stage as a consequence of heterochrony (future penetration glands), whereas the others form definitively only in adult worms, as is the case with turbellarians (future metacercarial glands). Secondly, there is no biological reason why penetration glands as purely provisional structures should not appear *de novo*. In any case, their appearance should be viewed as an important adaptation of hermaphroditic generation, its origin being associated with the appearance of the second intermediate host in the trematode life cycle.

It is not to be ruled out that the acquisition of penetration glands (as well as of the second intermediate host) has taken place at least twice in the evolution of trematodes in two phylogenetic branches which separated early (see p. 497).

The above-mentioned plasticity of these structures, including functional plasticity, is clearly expressed in specialized trematodes. The differentiation of penetration glands in different trematode groups is outwardly manifested in different cells secreting substances that differ in their structure and affinity to vital and histochemical dyes (Hall, 1960; Krasnodembskij, 1972, 1977; Davies, 1983; Galaktionov and Malkova, 1994; Körner and Haas, 1998a,b; our observations). This differentiation was explained by the necessity to penetrate tissue barriers of various nature and viewed as a manifestation of adaptive evolution (Ginitsinskaya, 1968; Ginetsinskaya and Dobrovolskij, 1983). At present there are also no special reasons to doubt that. Yet in reality things are somewhat more complicated. The adaptive evolution of the cells of penetration glands is associated not only with their deeper specialization but also with a change of function. In Microphallidae and some Lecithodendrioidea the secretion produced by some penetration glands is used for the formation of a temporary penetration cyst (Fig. 100) (see p. 129) and does not directly take part in destruction of the host epidermis.

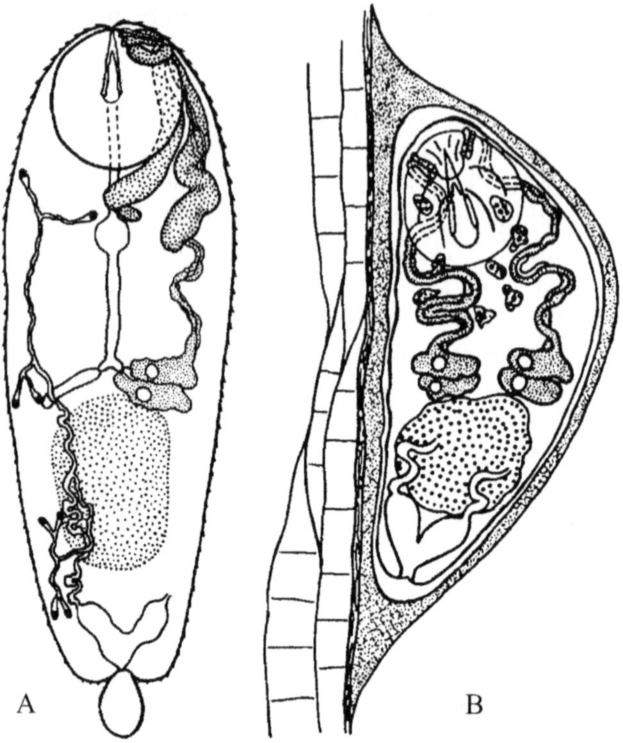

Figure 100. Levinseniella brachysoma cercaria (A) and the same cercaria inside the penetration cyst on the surface of the second intermediate host (gammarid crustacean) (B). From Galaktionov (1993).

In this connection, Pearson's (1992) opinion should be noted that penetration glands should include not only cells secreting histolytic enzymes but also cells producing different mucous protecting substances. Following this logic, the so-called mucoid glands (see p. 124) should also be attributed to penetration glands. Their mucous secretion, stored in the tegument of the larva, is partly expended in the process of cercarial penetration into the body of the second intermediate host. In our opinion, this is an unnecessary expansion of the morphologically distinct group of secretory structures, bound to result in a terminological confusion.

In contrast to head glands of other types, penetration glands are characterized by a very compact location of duct-opening pores. They must have evolved from their very emergence as an apparatus ensuring an extremely narrow point-like impact upon the epidermis and tissues of the host. This is certain to possess a biological meaning, as the reduction of traumatic impact of the penetrating larva upon a potential host is one of the factors stabilizing the evolving host-parasite system.

There are two types of structures responsible for mechanical destruction of epidermis: stylets (see p. 121) and mobile proboscides capable of turning in and out. As a rule, the latter are also armed with several rows of specialized spines (penetration spines). A close correlation can be observed here. The cercariae with stylets (xiphidiocercariae, ophtalmoxiphidiocercariae, cotylocercoid larvae from family Opecoelidae) always have a well-developed oral sucker. Its specialized muscles ensure the cutting movements of the stylet exactly at the place onto which the secretion of penetration glands are released.

In the larvae which have proboscis with penetration spines, the oral sucker is secondarily substituted by the so-called anterior organ (see p. 106) the cercariae from the orders Strigeidida and Schistosomatida demonstrating it in the most typical form. This substitution is very likely to be adaptive. The anterior organ may be viewed as a peculiar provisional structure, as it disintegrates when a cercaria is metamorphosed into a metacercaria. At its former place a true oral sucker is formed *de novo*. The functional significance of the anterior organ is twofold. The organ itself is a muscular sac, sometimes with rather thick walls (see p. 106). Most of its inner space is occupied by the enlarged parts of the ducts of penetration glands, sometimes noted in literature as duct reservoirs. To a certain extent this is true, as they accumulate large quantities of secreted material. Sometimes it contains unicellular glands of its own. In the beginning of the penetration act the contraction of muscular walls of the anterior organ results in the squeezing of the secretion onto the body surface of the potential host. The larva can regulate this process by releasing it in small portions. When the walls of the anterior organ contract, the proboscis is also turned out. It can, however, also perform some autonomous movements due to specialized muscular bundles. Its ventral surface, bearing backward-directed spines, is in turns thrust as far forward as possible and contracted backwards. As the anterior end of the larva is firmly pressed upon the host's epidermis, at this stage of movement the spines function as a milling cutter[V]. Recalling that at the very basis of the eversible part of the proboscis the pores of the ducts of penetration glands are situated in two compact groups, it becomes clear that in this case we also deal with a very advanced penetration apparatus.

The structures listed above have been acquired independently by the representatives of different trematode groups. This is true both of the stylet (it is sufficient to compare, for instance, the stylets of xiphidiocercariae and the cercariae from the family Opecoelidae) and of the anterior organs with an eversible proboscis. On the one hand, as mentioned above, anterior organs are characteristic of the larvae of Strigeidida and Schistosomatidae. On the other hand, they are encountered in the cercariae of Heterophyidae. Their structure in the latter is, however, very different from the one described above. In any case, the presence of such an apparatus unambiguously points to an extreme specialization of cercariae of the groups mentioned. To conclude, it should be noted that the lack of a penetration complex in cercariae may have a different nature. In some cases the larvae have not acquired it as yet. This is certainly true of all cercariae encysting in the open environment (Fasciolidae, Paramphistomidae, Notocotylidae). Lack of a penetration apparatus is also an important characteristic in the cercariae from the families Echinostomatidae, Psilostomidae and Cathaemasiidae. These larvae are yet in the beginning of the way to a three-host cycle and penetrate the second intermediate host either via natural openings or using non-specialized archaic means of epidermis destruction. Lack of a penetration complex is secondary in the species whose cercariae do not leave the mollusc, using it also as a second intermediate host (see p. 341). It is not to be doubted that here we deal with a case of "classical" homoplasy, as the loss of penetration complex has occurred independently in different trematode groups.

And finally, there is one more group of cercariae without the penetration apparatus. The opinion of researchers about the primary or secondary nature of this phenomenon in this group is polarized. The forms in question are two families with two-host life cycle: large cystocercous larvae are swallowed by the definitive host and reach sexual maturity in it. They are the families Bivesiculidae and Azygiidae. A two-host life cycle is also observed in Transversotrematidae, whose larvae are not swallowed by the host, the fish, but attach to it, shed the tail and crawl under the free edge of a scale, where they reach sexual maturity. In none of the instances mentioned is there any need of destroying the epidermis of the host and hence, any need of penetration apparatus. Some researchers treat its absence in these groups as a plesiomorphic state associated with their ancient age (Cable, 1974; Gibson, 1987; Pearson, 1992).

Lastly, one more instance of a total lack of classical penetration complex should be mentioned, the very specialized cystophorous cercariae of the Hemiuroidea with three- and even four-host life cycles (see p. 347). The larvae of these trematodes are very juvenalized (see p. 100). As a matter of fact, they are embryos, ejected from the caudal capsule in the digestive tract of the second intermediate host so forcefully that they penetrate into the haemocoel via the intestine wall. This adaptation is undoubtedly secondary. Nothing of this kind has been described for any other trematode group.

Cystogenous apparatus of cercariae (see p. 124) is very likely a provisional structure. It origin is more or less clear, as all structures involved in cyst formation are associated with tegumentary glands. According to the data available at present, microphallids are the only exception. In them cyst formation also involves the secretory-active walls of the bladder (see p. 155). The adaptive nature of cystogenous apparatus is also supported by a clearly expressed correlation between the degree of its development and the biology of definite trematode species. It has the most diverse and complexly differentiated elements in those cercariae that encyst in the environment. As a result, a multilayered cyst is formed, which renders the larva a reliable protection from environmental factors. The inclusion of the second intermediate host into the life cycle and a gradual establishment of increasingly close metabolic connections between the metacercariae and the host is reflected in the structure of cercarial cystogenous apparatus. It reveals a clear tendency to simplification. The quantity and diversity of tegumental glands supplying the material for cyst formation (see p. 128) are greatly decreased. And finally, some trematodes, whose metacercariae have lost the ability to encyst (see p. 338), demonstrate a secondary loss of cystogenous apparatus as well.

We have purposefully considered in more detail the cercarial structures associated with the completion of the main biological function of these larvae, infection transmission to the next host. As for the basic systems, they are represented by the more or less developed primordia of definitive organs characteristic of adult worms. However, even here two peculiarities can be mentioned. Firstly, these primordia often start to perform non-characteristic functions. Thus, intestinal branches of many typical furcocercariae (Cyathocotylidae, Strigeidae, Diplostomidae) accumulate a large amount of some secreted substance, which is almost entirely released after the penetration of the larvae into the second intermediate host. The excretory bladder in some xiphidiocercariae becomes an organ of secretion (Krasnodembskij, 1972, 1977). Adaptive significance of such transformations remains obscure. However, they do point to a deep specialization of the larvae possessing them.

The second peculiarity, applying to all trematodes, is the often-mentioned tendency to the juvenilization of cercariae (Galaktionov and Dobrovolskij, 1987; see p. 151). Juvenilization and miniaturization of cercariae, directly associated with the increased productivity of parthenitae (first of all, of the daughter sporocysts) in our opinion may be considered as an adaptation of the most evolutionary advanced trematodes to increasing the density of infective stages in the space around them.

3.1.2. Biological Adaptations of Cercariae

Cercariae, dispersive larvae with an active mode of life, possess a certain behaviour in the environment. It is very likely to be adaptive, aimed at the successful infection of the next host, intermediate or definitive. In this respect cercariae are biologically similar to miracidia (see p. 216). Yet there are essential differences between them arising from a complex of factors. To begin with, the locomotory apparatus in these larvae is different. Consequently, the ways of locomotion are different, which undoubtedly influences the behaviour. Besides, the biology of cercariae is much more diverse than that of miracidia. Not all trematodes use a second intermediate host. If they do, they may infect it in a variety of ways. And, last but not least, the range of hosts used by miracidia is relatively narrow. With one exception (Aporocotylidae – see footnote i in chapter 1), these are the representatives of two classes (Gastropoda and, much more rarely, Bivalvia) of the phylum Mollusca, i.e., mostly benthic, not very active invertebrates. With cercariae, things are entirely different. The representatives of almost all main groups of invertebrates and vertebrates, from coelenterates and flatworms to birds and mammals, can be their second intermediate hosts. This fact alone should presuppose a great behavioural diversity of the larvae infecting so broad a spectrum of hosts. However, a leveling factor is at work here: a rather broad specificity of cercariae to second intermediate hosts. Many larvae may infect taxonomically very distant hosts, from different orders, classes or even phyla of the animal kingdom. In this respect cercariae are also very different from miracidia, the latter exhibiting a very narrow specificity to certain molluscan species (see p. 233).

As a result, biological and behavioural diversity of cercariae is remarkable. As to their behaviour, several groups of cercariae can be distinguished.

3.1.2.1. Passively Infecting Cercariae

This group is very heterogeneous. The larvae belonging to it have only one feature in common: they get into the host passively. First, infection may take place *per os*. Two variants are possible here: either cercariae are swallowed by the definitive host and infect it (Bivesiculidae, Azygiidae), or, before being ingested, cercariae encyst in the environment and become adolescariae. Finally, there is one more variant: not the definitive, but the second intermediate host actively feeds off cercariae (Echinostomatidae – the genus *Petasiger*, some species of the genus *Echinochasmus*, some heterophyids – *Cercaria caribbea XVI* Cable, 1956, *Cercaria caribbea XVII* Cable, 1956) and they encyst in it in the usual way. The first two variants appear to be plesiomorphic, the last one being secondary.

Feeding on cercariae or adolescariae is a behavioural act of the host, a manifestation of its feeding activity. As a matter of fact, the host, in search of food, consumes infective larvae together with the food or instead of it. Yet, cercariae from the water column may also get into the host in another passive way, if the host's activity is expressed in other functions then feeding one. In particular, some cercariae may use the water current created by the host to ventilate the gill cavity.

3.1.2.1.1. Cercariae Encysting in the Environment

This group comprises the larvae of Fasciolidae, Paramphistomidae, Notocotylidae, Haploporidae, Megaperidae, Haplosplanchnidae. The cercariae of

Philophthalmidae also encyst in the environment, but their behaviour is closer to that of "ambuscade" type, characteristic of some more specialized cercariae. It will be discussed later (see p. 301). The larvae of this group exhibit, as a rule, a continuous swimming at a high speed. Haas (1994) paid attention to this fact, noting that large larvae with simple tails possess the best speed characteristics. It should be emphasized, however, that the speed of cercariae, as well as that of miracidia, is not constant. As glycogen reserves in the tail are being spent, the movement slows down. Young cercariae that have just left the mollusc move in the water in different directions. It some cases they seem to exhibit taxes: positive phototaxis and positive georeaction. Most of the larvae remain at the water surface, preferring well-lighted areas. However, similar to miracidia, even the larvae of the same age shed by the same mollusc show a behavioural heterogeneity (see p. 228). A small percentage of cercariae keep at the bottom of an experimental vessel, avoiding lighted areas. As the larvae age, their distribution becomes more even. Slowly swimming cercariae keep close to the bottom and appear not to react to a change in the light source position at all.

During the period of active swimming the larvae sometimes settle on the substrate for a very brief moment, moving upon it or remaining still, and then swim up and resume their movement. As they age, these passive phase in swimming become longer and more frequent. They seem to be entirely different from the pauses characteristic of intermittent swimming (see p. 286). It seems most plausible that in these brief periods cercariae search the substrate suitable for encystment. The above behavioural scheme of the larvae of the group in question is based on the results of our observations upon the cercariae from two families: Diplodiscidae (*Diplodiscus subclavatus*) and Notocotylidae (*Notocotylus* sp. from *Planorbarius corneus*). Different as fine details are (the speed and trajectory of movement, the duration of active swimming before the onset of mass encystment), the general picture has turned out to be surprisingly similar. The behaviour of fasciolid cercariae also seems to fit well into this scheme (for review see: Combes et al., 1994). Cable (1956) also reported a positive photoreaction in megaloperid and haplosplanchnid cercariae, emphasizing their remarkable energetic and highly manoeuvrable movement.

This uniformity is likely to be determined by the same tasks faced by the larvae. In fact, they simply have to disperse the infection in the environment. The larvae are not expected to exhibit any complex behavioural types. The explanation seems to be rather simple. All representatives of the families in question infect the definitive host passively, the host swallowing encysted parasites. Spatial scale of active dispersion of cercariae that are to become immobile adolescariae and their potential hosts are incomparable. Since encystment prolongs the life-span of larvae, their broad dissemination may be supposed to be more favourable for transmission that their concentration in a limited area of "host-space" (Combes et al., 1994). There is no secondary redistribution of larvae after encystment, the cysts being firmly attached to the substrate. The only possible contribution of the parasite to increasing the probability of the infection of the host is the choice of the substrate for encystment. However, in most cases there is no speaking about a strict specificity in the choice of substrate by cercariae. At best, certain groups of substrates are preferred. Pearson (1972) has noted that the cercariae of Fasciolidae, Paramphistomidae and Haploporidae encyst mostly on plants, those of Notocotylidae and Pronocephalidae often encyst on the body surface of some aquatic, mostly benthic, animals (for details see p. 335), and those of Haplosplanchnidae and some Philophthalmidae, on planktonic organisms. At the same time, according to Cable (1956), haplosplanchnid cercaria *Cercaria caribbea XXXVIII* Cable, 1956 also readily encyst on the aquatic plants their definitive hosts feed on. In experimental conditions of the absence of suitable substrate most cercariae of this type encyst on any substrate whatever.

3.1.2.1.2. Cercariae Swallowed by the Definitive Host

At present, two families are known, Azygiidae and Bivesiculidae, whose larvae are actively consumed by the definitive hosts, fishes. The cercariae of Azygiidae and Bivesiculidae belong to the morphological group of cysticercous cercariae (see fig. 42C, 145). This fact as such seems to testify to a deep specialization of these two taxa. However, there are all reasons to believe both these families to be archaic. Biology and behaviour of their cercariae have never been specially studied. There are only fragmentary data and incomplete observations (Le Zotte, 1954; Cable, 1956; our data).

The essence of morphological and biological adaptations in the cercariae of this type is aimed at making them as noticeable to the potential host as possible. As mentioned above (see p. 99), this is predominantly achieved by increasing the size of the tail.

The behaviour of macrocercous larvae is somewhat different from that of cercariae with ordinary tails, actively infecting the hosts. The character of movement and the reaction to the basic environmental factors are the same as in other furcocercariae, but, according to Le Zotte (1954) and Cable (1956), the larvae of Bivesiculidae usually swim tail forward, with the flat furca branches straddled wide apart. This is also characteristic of the *Azygia lucii* cercariae, though they may sometimes swim anterior end first. Contrary to most furcocercariae, bivesiculid cercariae demonstrate a distinct positive photoreaction. It is not very surprising, taking into account that all of them have well-developed eyes. We have failed to reveal a distinct photoreaction in the cercariae of *Azygia lucii*.

Both these cercariae demonstrate intermittent swimming. Passive phases in swimming of *Azygia lucii* cercariae are very long. Most of the time they float passively in the water column, their oarblade-like furcal branches set wide apart. However, they react to the smallest disturbances in their surroundings by abrupt leaps, including those directed almost vertically down. After a short stop, they rise back to the upper water layers, resuming passive floating. According to Cable's (1956) observations the larvae of some Bivesiculidae (*Cercaria caribbea XLV* Cable, 1956) in the passive period sink to the bottom of the experimental vessel and may lie there for a long time. They always orient themselves in such a way that their immense pigmented tails with folded furcal branches are directed to the source of light (Cable, 1956).

Such behaviour is certainly adaptive, aimed not at the search for the host but at the attraction of its attention. In this instance, an active role in infection belongs not to the parasite but to the host. The main function of dispersive larvae is not the "search" for the host but the stimulation of its searching activity through imitation of potential food objects. Ginetsinskaya (1968) called this phenomenon "mimicry", and Combes et al. (1994), prey mimetism.

Prey mimetism may be very sophisticated. Large cysticercous larvae of *Azygia lucii* morphologically and behaviourally imitate mosquito larvae with a remarkable similarity. Mosquito larvae are the favourite food of many planktonophagous fishes, including many large predators at young age.

There are data showing that infection of the final host may be achieved by the host swallowing cercariae also in some echinostomatids. According to Xiao et al. (1995, 2001) man can be infected by *Echinochasmus liliputanus* not only when eating raw fish with invasive metacercariae but also when drinking not boiled water containing cercariae. It is yet difficult to say whether it is cercariae that infect the host or the larvae that have already passed to the metacercariae phase. In experimental conditions cercariae of *Echinochasmus liliputanus* readily encysted in a variety of saline solutions including Locke's solution (Xiao et al., 2001). Infection of man can be supposed to be done not by cercariae but by metacercariae encysting under the influence of some factors in water or in the mouth cavity of a man who has drunk water containing cercariae. At any rate these data show that the relationships in the system "second intermediate host – echinostomes metacercariae" are at incipient stages of formation.

3.1.2.1.3. Cercariae Passively Infecting the Second Intermediate Host

Morphological and biological adaptations of the representatives of this group and the previous one have much in common, though there are also several important differences between them. "Swallowed" cercariae of Azygiidae and Bivesiculidae are swallowed by the host and get into the digestive tract, their final settlement place. The larvae of the group in question passively get only into the host's cavities opening into the environment (mouth and gill cavities in fishes, mantle cavity of Bivalvia, gill cavities of Crustacea, rectum of dragonfly larvae). The larvae are passive only before the contact with the host. After the primary contact has been establsihed, the cercariae actively infect the second intermediate host, as it is characteristic of trematode larvae with three-host life cycles. The definite behavioural programme of the larvae is determined by the degree of their morphological and biological specialization (see p. 497).

This group is very heterogenous both in its systematic composition (it comprises the representatives of very distant taxa: Echinostomatidae, Psilostomidae, Heterophyidae, Hemiuridae, Lecithodendrioidea) and the morphological and biological adaptations of the larvae. As to the way of infecting the host, this group falls into two rather distinct subgroups. The first subgroup consists of cercariae actively swallowed by the potential host, usually fishes and planktonic crustaceans. The second subgroup comprises the larvae which use the water currents created by the host, usually bivalves, crustaceans, the larvae of some insects.

The essence of morphological and behavioural adaptations of cercariae from the first subgroup is making them as noticeable for the potential host as possible. It is a typical prey mimetism, as in the case with the Azygiidae and Bivesiculidae larvae. It is especially patent in specialized larvae of some Echinostomatidae. Macrocercous larvae of *Petasiger* and *Echinochasmus* have enormous swollen tails. Devoid of any projections, these larvae, reaching 1 mm and more in length and undulating in one plane, look very much like small free-living nematodes and are readily swallowed by fish.

The phenomenon of zygocercous larvae, i.e., the formation of aggregates of macrocercous larvae ("Rat-King"), is one of the variants of such adaptations. As mentioned above, this phenomenon is rather widespread and is encountered in different families. Cable and McLean (1943) affiliate *Cercaria clausii* with trichocercous larvae characteristic of the families Lepocreadiidae or Gyliauchinidae. Wright (1953) thought *Cercaria buchanani* to belong to Renicolidae family. Pleurolophocercous *Cercaria caribbea XVI* Cable, 1956 certainly represents heterophyids (possibly, family Galactosomidae) (Cable, 1956). A number of forms described by different authors at various times (for reviews see: Ito, 1964; Olenev and Dobrovolskij, 1975) are from the family Echinostomatidae.

We have undertaken a special investigation of the biological and behavioural characteristics of *Cercaria rhionica VII* Olenev et Dobrovolskij, 1975 (*Echinochasmus* sp.). In laboratory conditions, the molluscs *Melanopsis praemorsa* shed from 1 to 5 aggregates per day. The number of the larvae in aggregates varied from 7 to 50. Sometimes single cercariae were also shed. Each such aggregate behaves as a functional whole. It always moves only by means of some of the larvae, forming the pole directed forwards in locomotion. Their tails undulate actively. The larvae located laterally demonstrate locomotory activity only from time to time, the group at the posterior pole being absolutely passive. This polarization of the aggregates is only functional. The aggregate does not turn in order to change the direction of movement. Another group of larvae is simply activated, forming a new anterior pole.

Freshly shed aggregates possess a clear negative georeaction combined with an also clear positive photoreaction. The former is clearly dominant, as, if the experimental aquarium is lighted from below the aggregates do not sink to the bottom. However, it is light that stimulates the locomotory activity of the larvae and determines the direction of the movement of the aggregates as a whole in the horizontal plane. As a matter of fact, the above-mentioned changes of the direction of movement are always associated with a change in the light source position and the activation of the corresponding group of larvae.

If the water surface in the experimental vessel is lighted in such a way that spots of light alternated with shaded ones, the aggregates concentrate only in the former as soon as several minutes after the beginning of the experiment. Having got into a light spot, the aggregate never leaves it. The polarization of larval activity in the aggregates is expressed along the vertical axis, the larvae in the upper "hemisphere" being active.

Directed locomotion, clearly expressed before that moment, disappears almost at once. The aggregates either slowly move along circular trajectories of a small radius or simply rotate at one place. If the light is bright, the larvae with their pigmented tails are very noticeable. Experiments have shown that if both single cercariae and aggregates are present in an aquarium, the latter are the first to be eaten by fishes.

Passive infection of the host may also occur if active forms of adaptive behaviour are absent. Cystophorous larvae of hemiuroids, with diverse projections for increasing buoyancy, are incapable of active locomotion at all. They have a zero buoyancy and float passively in the water column until some planktonic crustacean eats them by chance. It would be interesting to investigate whether there is any correlation between the vertical distribution of different species of hemiuroid cercariae and the biology of crustaceans, their intermediate hosts. Unfortunately, no such data are available. Lack of active forms of behaviour conditioned the rise of other adaptations. The life of hemiuroid larvae is very prolonged, e.g., freshwater cercariae of the genus *Halipegus* were shown to survive from three to six weeks (Krull, 1935b). Later it turned out that this was not the limit. The larvae of *H. occidualis* remain alive for several months (Shostak and Esch, 1990). All this brings to mind passively infecting miracidia, whose life span is also often prolonged up to several months (see p. 239).

The larvae from the second subgroup usually do not possess any pronounced morphological adaptations to passive infection of the host. Not that they need it. The cercariae do not attract the host's attention but make use of filtrational and breathing water currents generated by the host. In this case, behavioural adaptations seem preferable: after getting into the water current, the larva should remain passive. Turbulence should not provoke a transition from the passive phase to the phase of active movement. At any rate, it should not trigger a positive rheotaxis, which is commonly observed in cercariae. Though the behaviour of this type of larvae has not been specially studied as yet, the results of our preliminary observations favour the above suppositions. Small virgulate cercariae (Lecithodendrioidea) with a long period of passive floatation easily get into the water currents generated by insect larvae. In particular, they easily get into the rectum of dragonfly larvae, where from they penetrate into the host's haemocoel through soft, weakly cuticularized anal gills. Having got into a turbulent current, the larvae make no attempts of active locomotion in the opposite direction.

An almost identical way of establishing contact with the host is used by many very small microphallid cercariae. The currents bringing fresh water to the gills of crustaceans also carry infective larvae that use soft easily permeable gill tissues as a convenient place for penetration. Adaptations of this kind have a profound physiological significance. First of all, they are very economic from the point of view of energy expenditure. It is the host's energy, not that of the parasite, that is spent for the establishment of primary contact. Penetration through easily permeable tissues also allows the parasite to save energy, including the economy resulting from the simplification of penetration apparatus. The manifestation of such adaptations in the cercariae of the most specialized trematodes (Lecithodendrioidea, Microphallidae) seems to be associated with precisely these reasons.

3.1.2.2. Actively Infecting Cercariae

This biological group, comprising digeneans with three-host and four-host life cycles, is even more heterogeneous. Several subgroups can be delimited here, the limits between them being rather blurred. Trematodes from this group have one thing in common, the active way of infecting the second intermediate host. However, the implementation of this process can vary greatly in the representatives of different taxa. In fact, two main behavioural types of actively infecting larvae can be distinguished. The first type is characterized by a high locomotory activity: the larvae swimming relatively fast disperse in a three-dimensional space. The complex of their behavioural adaptations is directed if not at the "search" for the second intermediate host then at least at increasing the probability of meeting it. Cercariae in this subgroup usually demonstrate intermittent swimming. This type of locomotion has already been discussed (see p. 267). We denote this behavioural type as "searching".

The second behavioural type can be called "ambuscade" (Prokofiev, 1994, 1995). Cercariae exhibiting this type either do not swim at all or swim very little. The spend most of their life attached to the substrate, becoming active only when approached by a potential host.

3.1.2.2.1. Cercariae with an "Searching" Behaviour

The essence of "searching" behaviour in cercariae is hardly different from that in actively swimming miracidia (see p. 225). Under the influence of abiotic and biotic environmental factors, the cercariae distribute in space in a certain way, occupying those areas where the probability of meeting the potential host is the highest. After that, they attempt to contact with the host, often under the influence of some signal factors emitted by it.

Two groups of species are often recognized in literature, differing as to the nature of swimming. The first, not too numerous, group comprises the species whose larvae, having left the mollusc, are constantly moving. These are the so-called "continuous swimmers". The second, much more numerous group, is characterized by an intermittent type of locomotion, regularly alternating the periods of active movement and the periods of repose. They are called "intermittent swimmers".

The limits of the latter group are greatly narrowed by some researchers. Haas (1994) included into this group rather specialized cercariae with a number of adaptations for prolonged floatation in the water column (Furcocercariae, pleurolophocercous cercariae). In his comparative table the group of "intermittent swimmers" comprised 6 furcocercariae from different families and one larva of Opisthorchiidae. As for the group of "continuous swimmers" (see table 1, p. 16 in Haas, 1994,), it included the *Fasciola hepatica* cercariae (which seems justified) and the cercariae of Philophthalmidae, of all Echinostomatidae and even of Plagiorchiida. In our opinion, the inclusion of the latter three groups is unwarranted.

The cercariae of some Echinostomatidae have actually acquired the incipient forms of intermittent swimming, and many Xiphidiocercariae, even more so. Having stopped, the larva, lacking any structures for passive floatation, starts to sink very fast. These motionless passive phases are usually very short: the cercaria resumes swimming after touching the bottom or an underwater object. They do not remain absolutely immobile during the passive phase: their actively bending body contracts and elongates in turns, the tail being usually relaxed and passive. When the locomotion is resumed, the body contracts abruptly and bends to the ventral side, the tail resuming its active work. The stops are spontaneous and are not associated with the analysis of the substrate, as it is the case in the larvae encysting in the environment (see p. 281).

In echinostomatids the intermittent nature of swimming is very weakly pronounced indeed, the cercariae swimming actively most of the time. Macrocercous forms (the genera *Petasiger*, *Echinochasmus* and others) seem to be the only exception. They sink much more slowly, in all appearance owing to their large buoyant tails. This circumstance appears to cause more frequent and longer stops in their swimming.

According to our observations, intermittent swimming is much more pronounced in Xiphidiocercariae, including the cercariae of *Opisthioglyphe ranae,* mentioned by Haas (1994), than in Echinostomatidae, larval behaviour in these groups being very similar. There is, however, an important detail: if external disturbing factors are absent, Xiphidiocercariae resume locomotion under the influence of some endogenous stimuli. Even after touching the substrate, the cercaria keeps bending its body and contracting for some time and only then floats up. Our observations have shown the ratio of active and passive periods in *Xiphidiocercariae* sp. VII Odening to be 4 – 5:1, this proportion in the cercariae of *Astiotrema trituri* approaching 1:1. Small virgulate xiphidiocercariae (Lecithodendriidae, Alassogonoporidae) are in the end of this series. These larvae are extremely small and possess almost zero buoyancy. Their behaviour is much closer to that of specialized forms. The cercariae are immobile during stops, acquiring a very typical resting posture: the tail is bent ventrally, the body is elongated into a rod (see fig. 98B). They do not sink. Moreover, the weakest water current carries them across much longer distances than those covered by them during active locomotion.

The larvae of Philophthalmidae occupy a special place in the group discussed. Their attribution by Haas (1994) to "continuous swimmers" must have been a misapprehension. We have already described the organizational features of their cotylocercoid tail (see p. 101), which is not a locomotory organ at all. It serves for attachment to the substrate, including the surface water film. These cercariae spend most of the time at one place, from time to time making jerks generated by body bending. The same mechanism is used for swimming. As the larvae are still able to move actively in the water column, we can formally place them to this group. However, they should be really attributed to the cercariae with "ambuscade" behaviour (see p. 301).

There is no need to give a detailed analysis of intermittent swimming of specialized cercariae (different types of furcocercariae, pleurolophocercous larvae), as they are thoroughly analyzed in the literature (see reviews: Haas, 1994; Combes et al., 1994).

The adaptive significance of intermittent swimming is twofold. It has already been discussed that passive phase in swimming allow the larvae to spend their energy resources (glycogen reserves in the tail) more economically, especially if they remain at the same level in water. Thus, it is possible to prolong their life span and the time of infective state.

It is not to be ruled out that this was the original cause of the emergence of intermittent swimming, as we can find its first manifestations in the cercariae (some Echinostomatidae, many Xiphidiocercariae) that can make no other biological use of this locomotion type. Noteworthy, these larvae usually use sedentary benthic invertebrates (molluscs, insect larvae, flatworms) as second intermediate hosts. Events involved in the life cycle of these trematode species usually take place in limited space: either in small basins or at coastal areas (freshwater littoral and marine intertidal zone) with low depths and aquatic vegetation. The vegetation creates the effect of a multilayered substrate, allowing both the potential hosts and their future parasites to use not only the bottom layers but the water column as well (for details see p. 299).

The formation of more advanced types of intermittent swimming, accompanied by the emergence of morphological (furcae, tail fin-folds, etc.) and behavioural (special resting postures) adaptations for prolonged passive floatation, was correctly treated by Haas (1994) as a way to minimize the energy required for a constant presence of cercariae in a definite water layer. Considering that the most specialized forms spend from 85 to 98% of their active free life in passive floating (Haas, 1994), a lot of energy turns out to be saved.

Attempts to treat intermittent swimming of cercariae as an adaptation against predators seem questionable. Cercariophagy is a very widespread phenomenon (Shigin, 1978, 1981), though its true significance has never been properly evaluated. The share of cercariae consumed by different plankton-eaters can be very high (Pellegrino et.al., 1966; Christensen, 1979; Christensen and Frandsen, 1980). It has been suggested (Combes et al., 1994; Haas, 1994) that during passive floatation cercariae less attract the attention of potential predators, whereas the increase of relative speed and the frequency of locomotory cycles (tail oscillations) decreases the probability of cercariae being eaten during active locomotion. This reasoning has not been supported experimentally. Moreover, the results of our experiments poorly agree with this hypothesis. Small plankton-eaters, usually lacking a strict feeding selectivity, swallow floating furcocercariae (the genus *Alaria*) with water currents, whereas the larvae in the active phase of locomotion usually avoid this fate. As for larger organisms, e.g., the larvae of some insects, fishes (small *Alburnus alburnus* and *Leucaspius delineatus* were used in the experiments), hunting for planktonic organisms, they are first of all attracted by larger cercariae. When small furcocercariae (the abovementioned larvae of the genus *Alaria*) and virgule Xiphidiocercariae (Cercariae Virgulae group) get in the area subject to the action of currents produced by the gill operculum, they, regardless of the phase of movement, are sucked into the gill cavity. Having passed it, they are cast outside. Some larvae are damaged, but not many. Older larvae, with a markedly decreased locomotory activity, become the prey of plankton-eaters most often.

The larvae from the family Schistosomatidae form a special subgroup within the group discussed. Intermittent swimming finds in them its utmost expression. Schistosomatid cercariae spend most of their time attached to the water surface or to aquatic plants by the ventral sucker. This passive waiting occupies more than 95% of the time of larval free life. Passive floatation in them is much more rarely observed than in longifurcate Furcocercariae. Transition to active swimming is triggered either by some endogenous stimuli or by some strong disturbing external factors. In these larvae we deal with a special behavioural type, which is in fact similar to the "ambuscade" behaviour (see section 3.1.2.2.2.).

Locomotion character is only one component of that complex phenomenon, the behaviour of any motile organism. Its second component is the set of behavioural reactions to external irritants.

Unfortunately, cercarial behaviour is more poorly studied in this respect than that of miracidia. Even results obtained from close species often do not add up. It is sometimes difficult to understand whether these differences are due to species-specific behavioural features of the larvae or to a different experimental design. However, some general regularities of cercarial behaviour can be outlined rather distinctly.

Factors influencing cercarial behaviour are very diverse. Moreover, they act simultaneously, forming a very sophisticated complex, which finally determines the presence of the larva in space. Synergistic influence of environmental factors upon the behaviour of cercariae has never been studied. Experimental studies of cercarial behaviour are usually focused at the influence of one factor. The researchers try the exclude the influence of other factors or, at least, to minimize it. Chemical signals and fluctuations of light intensity and turbulence are the factors most commonly studied. The influence of gravity, a permanently acting factor, has been analyzed somewhat less often. It is with this factor that we will start our analysis of cercarial behavioural reactions.

It has been demonstrated that cercariae react to ***gravity***. It is especially clearly expressed in some "intermittent swimmers" (Furcocercariae, Cercariae Pleurolophocerca, some Xiphidiocercariae – the larvae of a number of Lecithodendrioidea). Fresh active larvae always move upwards when resuming locomotion. However, the trajectory of their movement is not rectilinear. Being in a certain water layer, they may change the direction of movement both horizontally and vertically. Vertical dispersion of larvae increases with age, most of them gradually concentrating close to the bottom.

This generalized scheme of temporal changes in the larval vertical distribution is interpreted in a variety of ways by different researchers. To begin with, the nature of the reaction responsible for the ascent of cercariae to the upper water layers is a subject of discussions. Some researchers define it as a negative geotaxis (Styczyńska-Jurewicz, 1961; Ginetsinskaya, 1968, Théron and Combes, 1983, Haas et al., 1990a,b). Others (Semenov, 1980; Prokofiev, 2002) prefer to speak of "georeaction" or "geoorientation". Without entering into details, we would like to emphasize that the very fact of existence of this reaction in the cercariae from various groups is not to be doubted.

Another problem seems to be more controversial. Prokofiev (2002), analyzing the vertical distribution of the *Renicola thaidus* cercariae, postulates the change of larval "rheoreaction" sign with age. According to his data, 1-hour-old larvae mostly remain in the upper water layers, thus demonstrating a negative geoorientation, whereas 6-hour-old ones concentrate close to the bottom, i.e., exhibit a positive georeaction. In the opinion of Prokofiev (2002), this reversal of the sign of reaction to gravity should be viewed as an important biological adaptation ensuring a considerable dispersion of larvae in the environment and preventing the superinfection of the second intermediate hosts, the mussels, at the sites where the *Nucella lapillus* molluscs shed the cercariae of *Renicola thaidus*. During the 5-6 hours when a negative georeaction is dominant, the cercariae are dispersed over large distances by inshore, tidal and wind-generated currents.

Very similar data are to be found in the work by Haas (1994), who indicated that at the first stages of their free life the cercariae of *Pseudechinoparyphium echinatum* exhibit a negative georeaction and a positive photoreaction. The author interprets these data as an adaptation ensuring the primary dispersion and allowing the cercariae to leave the habitat of the first intermediate host the larvae have developed in.

In both of the above cases the cercarial behaviour is certainly adaptive. The necessity of dispersion and its completion as a result of negative georeaction are not to be doubted. However, the interpretation of the subsequent events could be brought into challenge. Haas (1994) made no mention whatever of the subsequent fate of cercariae. Does the secondary redistribution of cercariae take place by means of the reversal of reaction sign, as it has been postulated by Prokofiev for the *Renicola thaidus* larvae? The fact of the change in larval vertical distribution as such does not explain anything. A gradual descent of the larvae may be just as well due to the expenditure of their energy reserves as to the acquisition of a positive georeaction. The former would result in the decreased speed of movement and a gradual descent. Unfortunately, Prokofiev's (2002) investigation does not answer a number of questions of principle importance. What are the dynamics of expenditure of reserved carbohydrates during the 6 hours of observation? Is the speed of cercariae and the frequency of tail oscillations constant throughout this time? What is the life span of the cercariae in definite experimental conditions? And finally, when does the experimental group of larvae pass the critical L_{50} point? Before we know the answer to these and a number of similar questions, the postulation of the reversal of geoorientation sign would be premature.

Data on cercarial reactions to ***light***, and, in particular, to changes in ***light intensity***, are much more contradictory. It has already been said (see p. 281) that many passively infecting cercariae exhibit a clear photoreaction, mostly a positive one. Thus, they concentrate in the most well-lighted areas. Sometimes photoreaction is manifested as a typical positive phototaxis, as it is the case in Zygocercous *Cercaria rhionica VII* forming aggregates of "Rat-King" type (see fig. 99).

Many other larvae of Echinostomatidae, actively infecting the second intermediate host, are also photopositive. According to our observations, freshly shed cercariae show a clear tendency to concentrate in lighted areas. Yet, they often leave these areas, and, as a result, a certain, though small, part of the cercariae is more or less evenly distributed in the whole volume of the experimental vessel. Many larvae, having left the lighted area, turn back almost immediately, thus demonstrating a typical "turnback swimming" reaction. This allows us to suppose that here we deal with a behavioural reaction of kinesis type.

The data concerning Xiphidiocercariae are contradictory. None of the larvae studied by us in this respect (*Opisthioglyphe ranae*, *Paralepoderma cloacicola*, *Haematoloechus similis*, *Xiphidiocercariae* sp. VII Odening, several species of Cercariae Virgulae – Lecithodendrioidea) demonstrated any directed reaction to the position of the source of light. However, we can not postulate a total absence of photoreactivity, as we have not registered either changes in the speed of movement or the frequency of turnings. At the same time, according to Prokofiev (2001), typical Xiphidiocercariae *Renicola thaidus* possess a clear positive phototaxis.

Apparently, locomotory reactions to the light source position, combined with geoorientation, ensure the dispersion of the cercariae which have left the mollusc and their primary concentration in the host's active space. However, cercariae (though not all) also clearly react to abrupt changes in light intensity, in particular, to shading (see reviews: Haas, 1994; Haas and Haberl, 1997). This can be treated as a response to the signals emitted directly by a potential host. The researchers failed to find a reaction like this in some Echinostomatidae (*Echinostomum revolutum*, *Hypoderaeum conoideum*, *Pseudechinoparyphium echinatum*) (Haas et al., 1995b). However, most of the intermittently swimming cercariae demonstrate a distinct reaction to sudden shading. This reaction may be different even in close species of one and the same genus, as well as very similar in the larvae from distant taxa.

As a rule, shading activates the passively floating larvae (Miller and Mahaffy, 1930; Dönges, 1964; Haas, 1969; Chapman, 1974; Feiler and Haas, 1988a; Haas et al., 1990b; Rea and Irwin, 1991; Haas et al., 1994, Haas and Haberl, 1997). The researchers have attempted to interpret the adaptive significance of this behavioural feature in a variety of ways. Shading may signal the approach of a potential host, thus triggering larval activation (reactions of kinesis type). A supposition has been voiced that shading warns the larva of the presence of a potential predator, whereas active swimming allows to avoid the predator's attack. All such reasoning is rather speculative and, as admitted by Haas (1994), often contradicts other data. A true adaptive significance of some reactions also often escapes interpretation.

This is especially clearly seen on fish infecting cercariae *Diplostomum spathaceum* (Diplostomidae) and *Opisthorchis viverrini* (Opisthorchiidae) (Haas, 1969, 1974a; Haas et al., 1990b). Shading triggers locomotory activity in passive larvae in the resting phase: they start to swim, though do not try to attach to a potential host. However, the same factor provokes an immediate stop of moving larvae. Moreover, prolonged shading in general suppressed the locomotory activity of cercariae: periods of swimming become very short. On the contrary, short shading stimulates a longer swimming period. There seem to be good grounds for treating this variant rather as an impact of an alternative signal: the end of shading is, in fact, a "flash", a sudden increase in light intensity, provoking a long period of active swimming (Haas and Haberl, 1997). The authors themselves had to admit that the adaptive significance of these reactions was obscure. If a fish stays by the larva (prolonged shading), the latter is almost completely immobilized, whereas if a fish has swiftly passed by (a short shading), the cercaria starts swimming fast and swims for a long time.

Considering that in most cases shading does not stimulate the attempts of cercariae to attach to the host and penetrate it, these reactions can hardly be viewed as adaptations to the infection of the host. It is highly probable that we simply deal here with a reaction of the organism to a disturbance. An abrupt change in light intensity signals a change in the environmental situation, which the organism reacts to by a modification of behaviour, also sudden. This is also an adaptation, yet of entirely different nature.

Whereas behavioural reactions to changes in light intensity are very similar in the two very different fish infecting cercariae described, they may also be very diverse in the larvae of different species of one and the same genus. It has been demonstrated on the species of *Schistosoma* (Saladin, 1982; Haas et al., 1987; Haas et al., 1990a; Haas et al., 1994). The cercariae of *S. japonicum* appear not to react to changes in light intensity at all. In the actively swimming larvae of *S. spindale* shading triggers a stop, whereas passively floating ones do not show any reaction whatever. The *S. mansoni* cercariae exhibit, in fact, an alternative mode of behaviour: swimming larvae do not react to shading, whereas passive ones activate, though poorly.

However, it is not to be ruled out that such arbitrary reaction could have served as a basic for the development of adaptive forms of behaviour, aimed at infecting the potential host. The interpretation of the behavioural reactions of the *S. haematobium* cercariae (Haas and Haberl, 1997) appears to be quite logical. The cercariae of this species respond to shading by active swimming, long periods of locomotion only seldom alternating with brief periods of passive flotation. As a result, the cercariae rise upwards in the water column. This is interpreted as a true adaptation increasing the chances of infection the host, man.

The duck-invading cercariae of *Trichobilharzia ocellata* react to shading in all phases: the phase of rest (when they are attached to the water surface or float in the water column) and the phase of active movement (when they swim either anterior end or tail forward) (Wright, 1974; Feiler and Haas, 1988a,b; Haas, 1992). In all the cases they respond by a long period of active swimming anterior end forward. This is accompanied by a change in photoreaction sign. Swimming tail forward, the larvae demonstrate a positive photoreactivity; starting to move with anterior end forward, they acquire a rather pronounced negative photoorientation. Moreover, shading also stimulates attempts at establishing contact with the host. This adaptation appears to be biologically justified since cercariae of this species infect ducks through the skin of their legs.

Turbulence is another factor modifying cercarial behaviour in many species of Digenea. Larval responses turn out to be very diverse in this case as well. Mention has been often made in the literature of cercariae exhibiting a positive rheotaxis (Anderson and Cable, 1950; Dönges, 1964; Ginetsinskaya, 1968). We do not doubt this fact in the least, yet we would like to note that this word combination seems to be understood differently by different authors. Most cercariae can perceive two signals of a very different nature: a current as such, i.e., a directed movement of the water masses (1) and elastic wave oscillations, readily transmitted in water yet not accompanied by its movements (2). It is only recently that the researchers started to distinguish these two factors clearly (see p. 302). Unfortunately, when analyzing literature data, it is not always possible to establish whether the generation of turbulence in the experimental aquarium was or was not accompanied by the simultaneous generation of elastic wave oscillations.

Apropos of positive rheotaxis, it does occur in a number of cercariae, penetrating mostly fishes (*Diplostomum spathaceum*, *Linstowiella szidati*, *Holostephanus volgensis*) (Anderson and Cable, 1950; Dönges, 1964; our observations). In this case we speak of directed and prolonged movement against the current. If the speed of the water current does not exceed the speed of cercariae, they either "stay" at one place or slowly move forwards. A behavioural reaction like this can hardly be treated as a direct adaptation to infecting the host. It is rather an adaptation of another kind, characteristic of many aquatic organisms: counter-movement against the current, permitting the organism to stay in its part of space (Semenov, 1991).

At the same time, many cercariae do not demonstrate any noticeable reaction to water currents whatever. Thus, we have failed to find any pronounced response in large Xiphidiocercariae (*Opisthioglyphe ranae, Haplometra cylindracea, Xiphidiocercariae* sp. VII Odening). It has already been mentioned that small Cercariae Virgulae (Lecithodendrioidea) in the resting phase, which passively get into the rectum of dragonfly larvae, do not react to the rather strong host-generated water currents at all (see p. 286). Moreover, the generated water current may sometimes trigger the transition of a cercaria from the phase of active movement to that of passive floating. A similar reaction was described by Neuhaus (1940) for *Pleurogenes medians* larvae, which are also taken up by the water currents generated by a potential host.

Haas avoids the term "rheotaxis" at all, writing only about a response to turbulence. His results make it possible to delimit one distinct regularity. The cercariae which use fishes as their hosts instantly react to the appearance of turbulent currents by a transition from floating to active swimming. In a number of cases it is also accompanied by attempts at establishing the contact with the host and attaching to it (Haas, 1969; Haas et al., 1990b). A reaction like this in the larvae of the species from the genus *Schistosoma*, infecting mammals, is either very weak or is not expressed at all (Saladin, 1982; Haas et al., 1987, 1990a, 1994). The absence of behavioural reactions is an adaptation as well: the cercariae of Schistosomatidae parasitizing in mammals are evidently adapted to a rare appearance of their terrestrial hosts. This is reflected in the lack of energy-costly reactions to a frequent impact of non-host-specific mechanical stimuli, abundant in water (Haas, 1994; Haas and Haberl, 1997).

The duck-infecting larvae of *Trichobilharzia ocellata* behave differently (Feiler and Haas, 1988a,b). When they are attached to the water surface in a characteristic resting posture, they do not show any reactions to currents. But when they are swimming or floating passively in the water column, turbulent disturbances immediately make the larva move anterior end forwards and undertake active attempts to attack the potential host. This is a typical adaptive behaviour, as ducks are much more numerous than mammals in fresh water bodies.

Cercariae, similarly to miracidia, react to ***chemical signals*** from the environment. Two types of chemical reception are traditionally delimited, distant and contact reception. In the case of the former the source of chemical signals is situated at some distance from the larva, the organism reacting to its chemical "trace" only. In the case discussed the "trace" must consist of some excretory-secretory products of the potential cercarial host. Having attached to the host, the cercaria uses contact chemical reception to analyze directly the chemical "composition" of the epidermis to be penetrated. As a matter of fact, it is the same two-stage mechanism of finding and recognizing a specific host as in miracidia (see p. 225). However, this similarity is superficial. It comes from the similar tasks to be solved both by miracidia and cercariae: the search and infection of the animal host. As for the nature of the chemical signals perceived, it is often essentially different, and the character of behavioural response, even more so. The following consideration deserves special mention here: the difference concerns the larvae of one and the same species (Haberl et al., 2000), i.e., two genotypically identical phases of the same life cycle.

Data about the presence of distant chemical reception in cercariae are very heterogeneous. This mechanism of the "search" for the host appears to be of limited spreading in cercariae. The larvae of Echinostomatidae seem to exhibit the most uniform behaviour in this respect. All the species actively infecting their second intermediate hosts which have been studied in detail demonstrate a pronounced reaction of chemical orientation (Fried and King, 1989; Hutterer et al., 1992; Haas et al., 1995a,b; Haas and Haberl, 1997). It is much more weak in different Furcocercariae. Among Schistosomatidae, a response reaction to chemical signals has as yet been reliably shown only for *Schistosoma mansoni* (Wilson, 1987; Shiff and Graczyk, 1994). Among Heterophyidae, it has been found only in *Cryptocotyle lingua* (Chapman, 1974).

According to our observations, many Xiphidiocercariae from different morphological groups (and, correspondingly, from different taxa: Plagiorchiidae, Ochetosomatidae, Lecithodendrioidea) possess no reaction of chemical orientation. Neither is it pronounced in the cercariae of *Diplostomum spathaceum* (Haas, 1969, 1974b). However, other representatives of this species-rich genus (Shigin, 1986, 1993) are almost unstudied, and so are all other Diplostomidae.

The presence of a typical chemotaxis has been reliably shown only in one species, *Hypoderaeum conoideum* (Echinostomatidae) (Haas et al., 1995b). In all other cases we can only speak of kineses. Chemokinesis of cercariae is, however, markedly different from that of miracidia. Only one variant of behavioural response has been recorded in cercariae so far: turn-back swimming in decreasing concentration of the signal substance. Noteworthy, miracidia utilize two variants: besides the above-mentioned one, they have also been shown to increase the frequency of turns (Haas et al., 1995a; Haas and Haberl, 1997).

The cercariae of *Cryptocotyle lingua* respond to a chemical signal by increased time of active swimming and a respective shortening of the periods of passive floating (Chapman, 1974).

A very important difference between cercariae and miracidia is that these dispersive larvae make use of chemical stimuli of essentially different nature. Whereas miracidia mostly react to higher molecular compounds, often species-specific of the potential host (see p. 233), chemically responsive cercariae perceive first of all lower molecular compounds. Amino acids have been shown to be the most important signal molecules for two species of Echinostomatidae (*Echinostoma revolutum* and *Pseudechinoparyphyum echinatum*) (Körner and Haas, 1998a,b). Their composition has turned out to be not so significant. It is the concentration that is important. However, different aminoacids act somewhat differently. L-Amino acids with a primary α-Amino group produce the greatest effect. The stimulation effect is even more enhanced if such excretory products as urea and ammonia are added to the amino acid mixture, acting as additional chemical signals. Amines also stimulate the locomotory activity in the larvae of *Cryptocotyle lingua* (Chapman, 1974). Small proteins have been found to act as signal molecules only in the cercariae of *Hypoderaeum conoideum*, demonstrating, as mentioned above, a true chemotaxis (Haas et al., 1995b).

This seems to be the explanation behind a rather low specificity of response reactions of most cercariae to chemical stimuli. It also agrees well with a long-known broad host-specificity of metacercariae as compared to parthenitae (for review see: Ginetsinskaya, 1968).

However, one can not postulate a complete absence of specificity. The cercariae of *Pseudechinoparyphyum echinatum* distinguish well the snail-conditioned water (SCW) of several molluscan species, markedly different in total amino acid concentration (Körner and Haas, 1998a). It is difficult to say whether this phenomenon is of any adaptive significance in the nature, as, according to the same authors, the amino acid production by one and the same mollusc varies greatly from day to day.

True specificity of host-parasite relations does exist and begins to manifest itself during the next stage of cercarial behavioural programme, the identification of the host and penetration into it. The latter is often preceded by the search for a suitable penetration site, accompanied by the migration of the larva upon the body surface of a potential host.

Distant sensitivity is followed by contact reception. Numerous observations show that the larvae analyze a broad spectrum of characters peculiar to the substrate they are in contact with. These are mechanical and chemical characteristics of the host's epidermis, its hydrophobic properties, its temperature and, very likely, some other characters that at present we can not even name.

The number of the surface properties analyzed varies greatly in different species of cercariae. These differences are caused by several reasons and are markedly adaptive. For example, the temperature plays an important role at different stages of the completion of this stage of the behavioural programme, from the establishment of primary contact to the migration, in Schistosomatidae larvae which use hot-blooded animals as hosts. Contact chemoreception is also used at the same time, the role of signals being played by different substances at different stages (Haas, 1994; Haas and Haberl, 1997). Attachment is induced by L-arginine, the establishment of firm contact after the attachment, by acylglycerol and ceramides. In the process of the larval migration upon the host body it is again L-arginine that plays an important role, whereas fatty acids become the chemical signals of the penetration as such. This general scheme is subject to essential variation in different schistosomatid species. For instance, the attachment of *Trichobilharzia ocellata* cercariae to the skin of duck's leg is stimulated by cholesterol and ceramides, yet it is its fatty acids that are the signal factor for penetration (Haas and van de Roemer, 1998). At the same time, the attachment of the cercariae of *Austrobilharzia terrigalensis*, another duck-infecting species, is provoked by free sterols, fatty acids causing but a very weak effect (Clegg, 1969). Even the excretion dynamics of the penetration glands and the chemical factors determining it may be different in different species (Haas et al., 1997).

The differences between very close species provide a very demonstrative example here. They lie not only in the complexes of signals, perceived by the larvae, but also in their behavioural programmes. The cercariae of *Schistosoma japonicum* attach to the host purely by chance (Haas et al., 1994). Migrating in search of a suitable site for penetration, they only rely on the temperature gradient. Chemical stimuli (fatty acids) start to play a certain role, along with the temperature, only during the penetration into the epidermis. The cercariae of *S. mansoni* possess an entirely different mechanism of host infection. The larvae successively use chemical signals for host identification from the very first stage, the attachment to the host. Temperature is of the least importance here (MacInnis, 1969; Austin et al., 1972, 1974; Haas, 1976, 1984; Haas et al., 1994). This difference seems to be associated with the adaptations of these two species to different habitats where the infection of the host occurs. The larvae of *S. mansoni* are adapted to the infection of the host in clean and warm water habitats, where the detection of chemical cues is not disturbed by mud components and the host body temperature is not perceived as a factor contrasting the environmental conditions. On the contrary, the adaptation complex of the *S. japonicum* cercariae could have formed in polluted cool water bodies, rich in organic compounds. There, thermoreception is more likely to ensure a successful infection that chemical reception (Haas et al., 1994).

Fish-infecting cercariae use other signal stimuli (Haas, 1974a,b, 1994; Haas and Haberl, 1997). In this case temperature does not play any role at all. The establishment of primary contact may be triggered by turbulence, and, speaking of chemical signals, by an increased content of carbon dioxide. This signal factor is, of course, non-specific. The specificity is manifest at subsequent stages. Some species of the larvae react to the presence in the epithelium of glycoproteins with sialic acid and glucoseaminoglycanes. The complex of behavioural reactions ensuring the penetration into the host may be triggered by various combinations of fatty acids, proteins and glycoproteins with sialic acid. We can not always explain why this or that combination of factors is used in a particular case. However, when we succeed in doing so, the adaptive nature of the phenomenon observed becomes evident. For example, glycoproteins with sialic acid allow the cercariae to distinguish the mucus of the fish from that excreted by the epidermis of invertebrates (Haas, 1994).

The above review demonstrates a remarkable diversity of biological adaptations of actively infecting cercariae with a "searching" behaviour (Combes et al., 1994; Haas, 1994; Haas and Haberl, 1997). However, diverse as possible combinations of behavioural reactions are, the final result is more or less the same: most of the larvae accumulate in certain zones of the water body where the host may be present. This is the above-mentioned host's active space, i.e., the section of the environment conditioned by the host.

The larvae of the species which use fishes, birds, mammals and planktonic organisms as intermediate (or definitive) hosts, concentrate predominantly in the water column or in superficial water layers, whereas the cercariae whose intermediate hosts are benthic organisms (molluscs, leeches, insect larvae, etc.) finally accumulate at the bottom of the water body. This scheme has been often reproduced in the recent literature (the already often-cited reviews by Haas (1994) and Combes et al. (1994) could be once again referred to as classical examples here). Not disputing these logical constructions, we would like to emphasize that they are made on the basis of experimental studies that are very different as to their accuracy and reliability. Besides, whereas the facts given in some articles raise no doubts, the undercurrent of their interpretation catches the eye.

Trying to prove that cercariae demonstrate the same primary dispersion as miracidia (see p. 228), Haas (1994) cites his earlier works (Haas, 1969; Haas et al., 1994). There it has been shown that the cercariae of *Diplostomum pseudospathaecum* and *Schistosoma haematobium* that have just left the mollusc demonstrate a higher locomotory activity than the larvae that have reached the host's active space. Haas (1994) considers these facts to support the idea that the cercariae leave the host's space and, hence, to prove the existence of dispersion. The facts as such are both reliable and trivial. It is common knowledge that endotrophic dispersive larvae markedly reduce both their speed and the frequency of repetition of the locomotory cycles of the tail as they exhaust their energy reserves. This is clear from the simultaneous observations of the dynamics of glycogen expenditure in the tail and the tail's activity (Ginetsinskaya, 1968). This phenomenon has no bearing on the primary dispersion whatever, and thus does not prove its existence. The above considerations apply to Haas' arguments, but not the phenomenon of primary dispersion as such. It does exist. However, the real situation is much more complicated than the above-cited reviews lead us to believe. The division of potential hosts into two groups, the inhabitants of superficial water layers and bottom dwellers, is an inadmissible oversimplification. It has already been mentioned that aquatic vegetation always creates the effect of multilayered substrate. Moreover, it produce one more effect, that of elevation. Many of the typical benthic organisms (molluscs, leeches, insect larvae) often do not stay at the bottom but rise rather high into the water column, using aquatic vegetation as a substrate. If the depth is relatively small (for macroorganisms, but not for cercariae), the spatial distribution of benthic forms along the vertical axis often turns out to be almost random. As for the pulmonate snails, the intermediate hosts of many echinostomatids, they generally prefer to remain at the upper levels, often gliding upon the water surface. These snails are often mentioned by the authors who would like to interpret any behavioural reaction of cercariae as an adaptation to the infection of the host. Yet the above feature of pulmonates' behaviour does not agree well with the data on positive geoorientation of a number of echinostomatid cercariae (Haas, 1994).

Similar examples can be also drawn from other groups of cercariae. However, in our opinion, another thing is of greater importance in this connection. The phenomenon of primary dispersion is certainly present in cercariae. As a result of the completion of a species-specific behavioural programme, based mostly on reactions of the kinesis type, the larvae distribute in the potential host's active space. It is much wider than the "space of the most probable meeting with the host" discussed in connection with the miracidial behaviour (see p. 226). This is biologically justified, as the behavioural activity of the potential hosts of cercariae is much higher than that of the molluscs, the first intermediate hosts. The secondary redistribution of the larvae in space also seems to occur on a larger scale. This is facilitated by the gaps in active locomotion: floating cercariae can be transported over long distances even by the slightest currents. The latter are constantly generated in any natural water body: convectional currents, disturbances associated with the movement of large aquatic animals. Age changes in the cercarial behaviour might also play a role in their secondary redistribution. We are now speaking not only about the temporal changes in behavioural reactions, but also about a gradual "sinking" of cercariae due to the decrease of their locomotory capacities. The infectivity of the larvae also decreases, but very gradually, so that the probability of infecting the host in another area is for some time retained.

The above considerations do not in any way contradict the data of Bartoli and Combes (1986) and a number of other researchers (see p. 389). They have convincingly demonstrated that the primary distribution of a cercarial species corresponds quite well to the biological features of their potential hosts. From this point of view, the behaviour of intermittently swimming cercariae actively infecting the host is really adaptive. But it is futile to try to interpret every movement of a cercaria as a direct adaptation to infecting the host. Dispersion is a very important adaptation. Yet, similar to the case of miracidia, it is expressed at a group level.

All this allowed Combes et al. (1994) to write that cercariae are more specific to their hosts' habitats than to the hosts themselves. It is certainly a literary exaggeration, yet it contains a large element of truth.

Generalized schemes of miracidial and cercarial behaviour in the environment are rather similar. It is associated with the similarity of strategic tasks to be "solved" by these dispersive larvae. Their tactics, however, are essentially different. Though a broad primary dispersion takes place in both cases, the mechanisms of its completion are very different. Individual behaviour of the majority of miracidia is aimed at reaching relatively small areas where the probability of meeting a not very mobile host is the greatest. Cercarial dispersion occurs at a greater scale from the very beginning, as they occupy active space of a sometimes very active host.

The subsequent behaviour of the larvae is also very different. Miracidial sense systems allow them, mostly by means of chemical reception, not only to determine the position of potential host in space but also to a certain degree to determine its specificity. It is possible because the role of chemical signals is played by the mollusc-emitted host cues, "miracidial attracting glycoproteins", which are species-specific. Low-molecular compounds, abundant in polluted organic-rich water bodies, do not hinder the identification of these molecules. Actively infecting cercariae with intermittent locomotion behave differently. The signals from the potential host usually trigger the activation of their locomotion. Chemical signals are no exception in this respect, all the more so as they are usually small molecules. Their species-specificity is low, and their action is easily masked by chemical substances from other sources. Activation of movement to a certain extent increased the chances of meeting the object which have triggered it, but still chance plays a very important role. This makes the completion of the final stages of cercarial behavioural programme all the more important, especially the identification of the host and the finding of the place to penetrate. It is at that moment that cercariae for the first time show some similarity to miracidia. The role of contact chemoreception increases in the majority of species, and, along with low-molecular substances, macromolecular substances acquire a larger importance.

There are much more adaptive peculiarities of actively infecting cercariae from the group in question than it has been listed here. Up to now we have been dealing only with individual larval behaviour, leaving outside the scope of our attention adaptations manifested at a group level. However, it is these adaptations that mostly determine the spatial-temporal structure of local hemipopulations of larvae in the biotopes. This structure, which is also markedly adaptive, will be treated in the chapter 4 (see p. 389).

3.1.2.2.2. Cercariae with "Ambuscade" Behaviour

The third biological group of cercariae comprises the larvae whose behaviour Prokofiev (1994, 1995) called "ambuscade". These cercariae are usually motile. However, they move by muscular contractions of the body, the tail being more or less reduced. Sometimes these larvae are capable of swimming (Philophthalmidae), in other cases they can only move upon the substrate (see p. 271). They spend most of the time at one place, attached to the substrate either by a sucker-like tail (Philophthalmidae, Opecoelidae) or by the ventral sucker (cercariaea of *Asymphylodora*). Though different in details in different species, the phenomenology of behaviour in these larvae is rather simple. The attached cercaria periodically bends the body, performing circular searching movements. The cercariae of Philophthalmidae make abrupt jerks in different directions, those of Opecoelidae (the genera *Sphaerostomum*, *Podocotyle*) jiggle or make smooth circular movements by the anterior body end. The organism is sensitive to signals from the environment. When the larva gets an adequate signal, its behaviour changes. West (1961), who worked with *Philophthalmus gralli*, observed that at the approach of small planktonic crustaceans, the cercariae of this species detached from the water surface and began to swim jerkily, attempting to attach to the cuticle of the host by their tail sucker. If they succeeded, they almost at once encysted on the host's surface.

We have failed to find such an association with planktonic crustaceans in a related species, *Philophthalmus rhionica*. However, the larvae react to strong turbulence in the same way. Some disagreement with the observations of West might originate from the biology of the latter species. The cercariae of *P. rhionica* can directly infect vertebrate animals (birds and mammals) if they find themselves on the mucous membranes of the nose or mouth.

Prokofiev (1994, 1995), observing marine cercariae from the group in question, revealed some of the stimuli activating larval behaviour and provoking the larva's attack upon the host. After leaving the *Hydrobia ulvae* snails they develop in, the larvae of *Levinseniella brachysoma* (Microphallidae) actively crawl upon the substrate for several minutes. This time seems to correspond to primary dispersion, though its scale is very limited. Having completed this part of their behavioural programme, the cercariae attach to the substrate by their rudimentary tail, orient vertically (waiting posture) and jiggle smoothly from side to side (Fig. 101). Every 5-10 minutes the larvae stretch abruptly, their anterior end making wide circular movements (searching reaction), and then resume the waiting posture. If a stimulus appears (oscillations of a certain frequency), larval reaction depends on the distance to the source of oscillations. If it is longer than 3-4 larval body lengths, a non-directed searching reaction is observed. If the distance is less, the cercaria shows a well-directed attack reaction. It stretches towards the source of oscillations and tries to attach to it by the oral sucker. If the cercaria succeeds, it detaches from the substrate at once and moves to the surface of the object that has activated it.

Here we also deal with purely adaptive features of cercarial behaviour. This is testified to by the nature of the stimulus. *L. brachysoma* cercar-

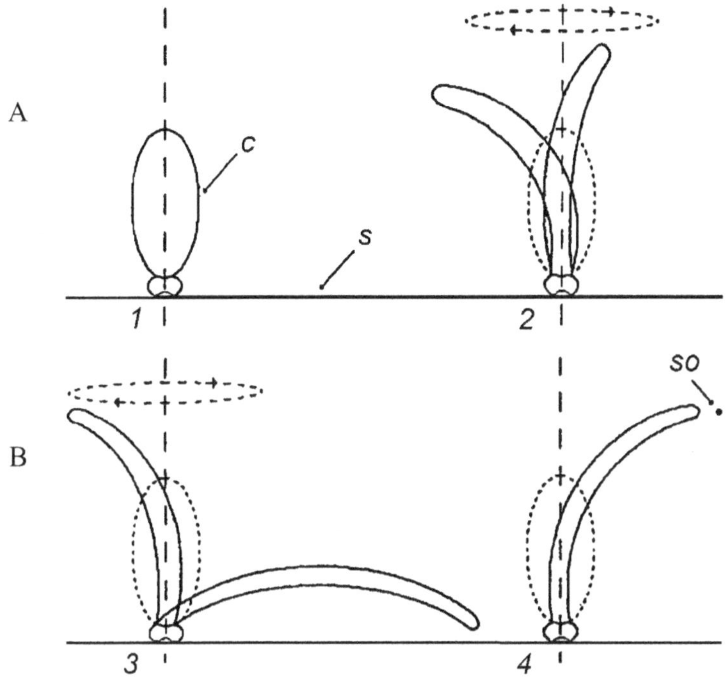

Figure 101. Diagram illustrating the ambuscade behaviour of microcercous cercaria *Levinseniella brachysoma.* A, spontaneous reactions; B, reactions in response to experimental irritation. *Abbreviations*: *1*, waiting posture; *2*, spontaneous rotation; *3*, searching reaction; *4*, attack reaction; *c*, cercaria; *s*, substrate; *so*, source of oscillations. From Prokofiev (1995), with permission.

iae react in this way only to elastic water vibrations. A generated current evokes but a brief reaction. This can be explained by the fact that the intermediate hosts of this species are intertidal gammarids. Moving, they generate wave disturbances, not directed currents.

Unfortunately, we can only conjecture the subsequent course of events. At present there are no data on the mechanisms of host identification and the stimuli triggering penetration process.

Another biological feature of cercariae with "ambuscade" behaviour also seems to be an adaptation. They often use other organisms as a substrate, in particular, molluscs (see p. 273). This seems to compensate for the absence of full-scale primary dispersion at least to some extent.

The cercariae of some zoogonid species also belong to this group. The larvae of *Diphtherostomum brusinae* are immobile (Palombi, 1930). They spend all their free life waiting passively for the host and exhibit a reaction only after a chance contact with it. The contact is always the result of the locomotory activity of the host. This seems to be the simplest, though by far not the most primitive, type of larval behaviour in trematodes.

3.2. Metacercaria

Different as the views on the origin of trematodes and the formation of their life cycles in the course of evolution are, the origin of the metacercarial phase is not very controversial. Most researchers agree that this phase was the last to emerge, as an adaptation to a more reliable transmission of infection from the mollusc to the definitive host (see p. 497). Thus, the biological essence of metacercariae is as follows: they are a transit phase between the dispersive larvae and the adult. In a typical case, the existence of the metacercaria can be divided into several successive stages.

The first stage is an intermediate one. It is started by a cercaria and completed by a young metacercaria. During this stage the larva penetrates the second intermediate host and migrates in its body to the site of their final settlement. The penetration and migration are accompanied by the metamorphic changes of the parasite. The tail is shed, the secretion of penetration glands is spent, and in some species the reconstruction of tegument starts. Transformation often involves metabolism as well.

The second stage, comprising morphogenetic changes, is the second preparatory stage of the ontogenesis of the hermaphroditic generation (see p. 94). It is at that time that the metacercarial diversity is fully expressed. Both the duration of this stage and the nature of metamorphic processes vary greatly.

Finally, the third stage, which usually starts after the larva has reached the infective state, is the period of morphogenetic rest, a relatively passive waiting in the host for a rather long time. It should be emphasized that the temporal boundary between the second and the third stages is sometimes not very clear. Many trematodes do not stop morphogenetic transformations after reaching infective state. Larval development may continue for some time (see p. 181). However, it gradually slows down and finally stops. This moment marks the beginning of physiological aging of metacercariae, which is to result in their death if the larvae do not get into the definitive host.

Thus, the host-parasite system "second intermediate host-metacercaria" performs three main functions: providing, at the expense of the host's energy potential, certain ontogenetic changes of the parasite and the reaching by the latter of infective state (1), retaining the parasite's infectivity for some time (2) and, finally, transmitting, if possible, the parasite to the definitive host (3).

The above functions of the host-parasite system are of crucial importance. The realization of each of them requires a certain set of morphological-biological adaptations of the parasites. However, the formation of these adaptations has taken place in an entirely different order, and, what is even more important, their relative significance has been changing.

The primary nature of the two-host life cycles without the parasitic phase of metacercaria is usually taken for granted. What the researchers do argue about is which types of the two-host life cycles should be considered archaic, plesiomorphic, and which should be treated as secondary. This problem is of principal importance and will be discussed in detail in chapters 3 and 6. In any case, most researchers do not doubt that the two-host life cycles with the phase of adolescaria (i.e., a metacercaria encysting in the environment) are more ancient. We will take it as a reference point in our analysis.

Trematodes are likely to have acquired the ability to encyst not from the very beginning, but after the emergence of the two-host life cycle in the form of heterogony. The infection was transmitted from the mollusc to the vertebrate host by the juveniles of the hermaphroditic generation performing the function of dispersion. They might have led an active life in the environment. This hypothesis was first suggested by Ginetsinskaya (1968). Later it has been reproduced, with changes, in the works of Pearson (1972), Gibson (1987) and others[vi].

The juvenile stages discussed appear to have acquired the ability to encyst later in the course of evolution. Dogiel (1947, 1964) proposed that the reduction and even total exclusion of active free-living stages is one of the most important evolutionary trends of parasitic organisms. This has a biological significance, as the juvenile dispersive stages are likely to have been endotrophic, with a limited span of free life. Encystment excluded the option of active search (though proto-cercariae are unlikely to have possessed the complex of morphological-functional adaptations, including advanced ways of locomotion, characteristic of true cercariae), but prolonged the time spent in the environment. Almost total immobility allowed a very economic use of limited energy resources and the cyst wall rendered a reliable protection from unfavourable conditions. As mentioned above, proto-trematodes should have possessed some preadaptations facilitating the acquisition of encystment. They must have already had tegumentary glands, which could be further specialized. The selection favoured the intensification of protective properties of the cyst, which entailed the sophistication and differentiation of cyst-forming secretory structures. In this way ancient trematodes acquired a resting stage in their life cycle, that of adolescaria. This stage is still retained in some families, which are in our opinion, archaic (see p. 499).

Passive infection of the definitive host, when the host swallows encysted larvae by chance, has both advantages and essential drawbacks. The most important drawback is that the parasite can not directly influence the success of infection. It can only be somehow increased in two ways. First, protective qualities of the cyst may be further improved, thus prolonging the longevity of the larvae. However, both these processes are limited. The amount of reserve carbohydrates, accumulated by the larvae, can not increase endlessly, if only because the larvae are too small. To create a fail-safe cyst that would retain the already starved larva is even less reasonable from energetical viewpoint. Apparently, these two parameters must have been balanced: the protection rendered by the cyst should correspond to the maximal longevity of the larva. Digeneans whose cercariae encyst in the environment demonstrate an optimal combination of these parameters. Any attempt to disbalance the system turns out to be energetically disadvantageous and is not supported by selection.

Secondly, the increase of the probability of the definitive host infection could be shifted from the phase of adolescaria to that of cercaria. Only one variant is possible here: the acquisition of such behavioural adaptations that would bring the larvae into the site with the greatest probability of meeting the host. As infection stage gets into the vertebrate host with swallowed cysts, i.e., it is associated with the trophic activity of the host, the parasite can increase the success of infection only in one way: by means of inserting itself into the trophic chain leading to the host. This appears to have been the case, as the cercariae of some trematodes with primary two-host life cycles prefer to encyst on different substrates (see p. 281). These preferences are to a certain extent adaptive. Trematode larvae parasitic in herbivorous animals readily encyst on the surface of aquatic plants (Fasciolidae, Paramphistomidae). If the host also feeds on animals, the cercariae use the chitin of arthropods or molluscan shells as a substrate for encystment (Notocotylidae, Pronocephalidae, Philophthalmidae). The choice of substrate, based on some analysis of its properties, is already a manifestation of specialization, involving not only the behaviour but also the morphological-functional features of the sense apparatus of the larvae. The importance of this acquisition can not be overestimated. First of all, it is the first step on the way to the emergence of the second intermediate host, an event of the greatest importance in trematode evolution (see p. 497). Second, this seems to be the first manifestation of adaptive behaviour, leading cercariae into the "host-space" (see p. 299).

It is from this moment onwards that two evolutionary trends appear to have become possible. The first is realized in philophthalmids. The already mentioned *Philophthalmus gralli* (see p. 302) has, as a matter of fact, a three-host life cycle, though as yet a facultative one. On the contrary, the cercariae of *Philophthalmus rhionica* appear to have secondarily lost the specificity to the encystment substrate; they usually encyst on the surface water film. The cyst of *P. rhionica* shows the features of secondary specialization, too: it is pyriform, with a characteristic "plug" closing the aperture the larva leaves the cyst through. Its protective qualities are poorer than those of strong multilayered pigmented cysts of paramphistomids and notocotylids. It is remarkable how easily the larvae excyst. Excystment can be triggered by diverse non-specific factors: changes in osmotic pressure, a slight increase in temperature and some others (Tihomirov, 1980). However, it is precisely these factors that are encountered by the larva in the oral cavity of a potential host or on the surface of the mucous membranes of the nose and eyes. Moreover, as mentioned above, the phase of adolescaria seems to be unnecessary for most philophthalmids. The cercariae can infect the definitive host just as well. Thus, the stage of metacercaria is secondarily absent in the life cycle. However, this trend has not become a leading one in the evolution of trematode life cycle.

What has become widespread is the behavioural specialization, resulting in the appearance of the second intermediate host. Such examples can be found in different families, but notocotylids seem to be the best illustrative case. After a relatively short period of active swimming, the cercariae of most notocotylid species encyst, predominantly on molluscan shells or vegetation but without any very strict association with the substrate. However, we have found the cercaria of *Notocotylus* sp. from *Melanopsis praemorsa* to show a more narrow specialization. The free life of these larvae is extremely short, usually without the period of active swimming. Immediately after leaving the mollusc they do not actively swim but crawl to the pallium's edge, shed the tail and make their way to the extrapallial cavity. They migrate at a distance of 3-5 mm from the shell edge and encyst on the inner surface of the molluscan shell. Some representatives of the genera *Catatropis* and *Notocotyloides* have advanced even further in this regard. Their cercariae have a greatly reduced tail. They do not leave the mollusc at all, but encyst directly in it. Such a change in biology presupposed not only the reduction of the tail, but also the loss of pigmented eyespots, which are characteristic of the larvae of this family (Odening, 1966; Dolfus, 1966). In such specialized notocotylids, as well as in some other trematodes (see p. 341), the mollusc combines the functions of the first and the second intermediate host.

The representatives of the family Echinostomatidae are traditionally viewed as trematodes with an obligatory three-host life cycle, necessarily with the second intermediate host. Yet in most cases they use it in a manner which is hardly different from the above examples. Many species of echinostomes get into the second intermediate host not by means of a specialized penetration apparatus (see p. 128) (it is absent in all echinostomatids), but by realizing complex behavioural programmes, characteristic of the majority of the larvae from this family. *Himasthla rhigedana* has a two-host life cycle: large cercariae of this species encyst on the surface of molluscan shells (Adams and Martin, 1963). Most echinostome larvae possess a more sophisticated behaviour, exhibiting at the same time a more narrow specificity to the "substrate", the hosts. They infect the latter through natural body openings and canals leading into the body. The cercariae usually get into the molluscs through the external kidney opening (some species from the genus *Echinostoma*) and into fishes, through the lateral line canals. In every case the larvae possess a rather narrow behavioural specialization, recalling that of many Hymenoptera (Apocrita – Parasitica). This specificity becomes the basis of the formation of a specific host-parasite relationship. A specificity like this is different from that demonstrated by more specialized metacercariae and by all other parasitic phases of the life cycle. The specificity of the latter is based on sophisticated and sometimes very individual metabolic and immune interactions of the host and the parasite. However, the most important thing has been achieved: the second intermediate host becomes indispensable, and the life cycle, an obligatory three-host one. The habitat of the larvae is now the internal environment of the host's organism. From this moment onwards another thing becomes important for the parasites: how to get into the host in a simple, reliable and energetically economic way. The *per os* possibility is no good, as the larvae are already to a certain extent adapted to the existence in a less aggressive environment (body cavity fluid, haemolymph, fluid of the excretory system ducts) than that in the digestive tract. *Per os* infection would require a total reconstruction of the adaptation system formed. Only one way is left: through epidermis. It is this method that was used by the most specialized Echinostomatidae. It seems to have been associated with a change in the functions of a number of tegumental glands. Their secretion, accumulated in the cercarial tegument, has acquired pronounced histolytic properties. Establishing a contact with the host's epidermis, the larva secretes these products from the tegument. As a result, a rather extensive focus of destruction appears, both of epithelium and underlying tissues. Having squeezed into this cellular mass by muscular efforts (we have failed to observe a cytolytic effect), the larvae encyst. These processes have been closely followed on *Cercaria rhionica VII* (*Echinochasmus* sp.), encysting on the gills of fishes (Galaktionov and Dobrovolskij, 1987).

However, it is not to be ruled out that other echinostomatids have at the same time acquired other, somewhat different mechanisms of disrupting of the host's epithelium. The cercariae of *Echinostoma trivolvis*, encysting in the planaria *Dugesia tigrina*, during penetration use the secretion of paraesophageal glands (Fried and Rosa-Brunet, 1991). Judging by the characteristic location of their ducts and external pores, these glands correspond to the already mentioned "metacercarial" glands (see p. 129). The authors have supposed that it is these cells that play the role of "penetration glands" (Fried and Rosa-Brunet, 1991). Thus, our previously voiced idea that penetration glands and metacercarial glands have emerged as a result of differentiation of a uniform group of head glands of proto-trematodes has been supported. Further evolution in this direction involved a deeper morphological and functional differentiation of secretory cells. Importantly, this process has not stopped after the emergence of specialized penetration glands. It is the latter that give rise to the cells participating in the formation of so-called penetration cyst in lecithodendrioideans and microphallids (see p. 275).

These processes have been accompanied by essential transformations both of the larval structure and of their fate in the second intermediate host. There is no longer any need for a well-developed protective cyst if the parasite dwells in the internal environment of the organism of the second intermediate host. Energy expenses for cyst formation (including those for the formation of cystogeneous apparatus) cease to pay off. This results in a partial reduction of cystogeneous apparatus and a simplification of the cyst (see p. 128). The barrier functions of the cyst are weakened and the cyst wall becomes more permeable. As a result, metabolic interactions between the host and the parasite can be established, including the use by the parasite of the energy potential of the host. Hence, the transfer of a part of ontogenesis of the adult hermaphrodite individual to the phase of metacercaria is made possible. These processes must have been accompanied by the larvae acquiring a number of special morphological and physiological adaptations.

At the same time the larvae settle in different biotopes in the host's body. The localization of the parasite often becomes a species-specific character. It can be clearly seen in the example of strigeidids parasitizing in fish. The larvae use almost all settlement places possible: epidermis and subcutaneous layer (*Posthodiplostomum cuticola*), somatic musculature (many cyatocotylids), crystalline lens and vitreous humor (the genus *Diplostomum*), the nervous system (the genus *Tylodelphis*). Such biological radiation requires numerous and diverse adaptations, first of all, behavioural (migration to the final site of infection) and physiological ones. Parasites' relationships with the host's immune system are also very diverse. Stabilization of these relationships appears to be one of the most important directions of adaptive evolution of metacercaria.

The above considerations allow us to return to the problem of the functions of the second intermediate host in the host-parasite system in question. As long as the host played the role of the "substrate" for encysting larvae, two problems could be solved. The motility of the host to some extent compensated for the immobility of the larvae. As a matter of fact, the dispersion was not only prolonged in time, but was shifted into an entirely different spatial scale determined by the behavioural features and locomotory capacities of the second intermediate host. It is equally as important that the host also met all the energy expenses for the dispersion. Besides, the definitive host is now involved into increasing the probability of successful infection as, in its active search for food, it also searches and finds potential parasites. Thus, the passively waiting parasite transfers the energy expenses to the future host. This aspect has had enormous evolutionary consequences for trematodes, increasing immeasurably the circle of potential definitive hosts and ensuring the possibility of trans-ecosystem transfer of the parasites (see for details p. 468).

With the larvae residing in the body of the host, the latter also acquired the protective function. Additional energy expenses for larvae protection inside the host appear to be amply compensated for by a longer conservation of infection in the ecosystem. However, as noted above, the juvenile stages of the hermaphroditic generation, using the internal environment of another organism as a habitat, finally establish metabolic connections with the host. The parasites could use the energy potential of the host not only for transmission tasks but also for the prolongation of their own existence in it. At first, only the longevity of juveniles seems likely to have been prolonged, as before this moment it was limited solely by the amount of their own energy resources. Later, as the mechanisms of energy withdrawal were perfected, the parasites began to get energy in such amounts that it became possible to resume morphogenesis, interrupted for a time. Thus, the host finally acquired the feeding function, ensuring a kind of energetical "priming" of the developing parasites. Primary functions of the host are certainly retained. Moreover, the parasites, inserting themselves into metabolic and physiological systems of the host, get a chance of enhancing the transmissive function of the latter. This is achieved in a variety of ways, but morphologically is always expressed in the modifications of behaviour of the animals infected. The parasites start to manipulate the behaviour of the infected hosts making them go into the host-space of the definitive host. This phenomenon is not very widespread in trematodes, but examples of this kind can be found in different groups (Diplostomidae, Dicrocoeliidae, Microphallidae) (see for details: p. 396)

3.3. Marita (Adult)

The main features of morphological-functional organization of maritae have already been discussed (see section 2.3. of chapter 1), and so was marita development (maritogony). In the present chapter we focus our attention on some aspects of the biological radiation of trematodes in the organism of the definitive host, a vertebrate.

The term "marita" is used sometimes rather conditionally. It is not always easy to establish whether a reproductively mature individual is a "true" marita or a progenetic metacercaria.[vii] In the instances when it is crucial for the analysis, we will try to provide an exact definition.

Individuals of hermaphroditic generations of recent trematode species inhabit all conceivable habitats in the vertebrate host: from the digestive system to the epidermis. This broad biological radiation points to a remarkable plasticity of trematodes, physiological one as well, and their high adaptive potential.

Most researchers agree that the digestive tract was the primary environment of hermaphroditic trematodes which started to parasitize vertebrates (Ginetsinskaya, 1968; Pearson, 1972). More precisely, it is likely to have been the small intestine, where the hydrolysis of organic macromolecules is completed, parietal digestion and resorption of nutrients take place. It seems logical to suggest that it was in the small intestine that parasites could start to feed on intestinal chyme, a kind of food concentrate, containing a large amount of partially or completely hydrolysed products.

At first sight, these hypothetical constructions are supported by the fact that a majority of maritae, from almost all main taxa of recent trematodes, settle in the duodenum and in different parts of the small intestine. Some species, e.g., *Paramphistomum cervi*, begin their existence in the vertebrate host in the duodenum and later change their localization. Most adolescariae and metacercariae also excyst in this part of the digestive tract.

Yet it is not obvious what the primary food of ancient termatodes was. Actual feeding on chyme has been demonstrated only for a few species (see for review Ginetsinskaya, 1968; Smyth and Halton, 1983). Dawes (1961, 1963a,b), who studied the feeding of *Fasciola hepatica* at different stages of maritogony, thought histiophagy to be the most ancient type of feeding in maritae. Most modern trematodes that settle in the intestine feed on epithelial cells, mucus and sometimes, blood (Smyth and Halton, 1983). Primary histiophagy, suggested by Dawes (1961, 1963a,b) may, in our opinion, testify that the hermaphroditic generation transited to parasitizing vertebrate animals relatively late, as compared to cestodes. Transition from histiophagy to feeding directly on organic macromolecules dissolved in the host liquids, accomplished by most parthenitae, is yet to be accomplished by maritae.

Further radiation of trematodes to other parts of the digestive tract took place in two directions. The maritae of *Azygia* parasitize the stomach of fishes, causing severe disruptions of the mucous membrane. *Encyclometra columbrimurorum* from grass-snakes also settles in the host's stomach, and if the infection is very intensive, also occupies the lowest part of the oesophagus.

Maritae often settle in the oral cavity. Trematodes from the genus *Halipegus* inhabit the oral cavity of frogs. Small plagiorchiids *Leptophallus nigrovenosus* localize under the lip edge of grass-snakes, the genus *Clinostomum* occupying the same habitat in herons. Maritae also expand in another direction, up to the cloaca (*Paralepoderma cloacicola* settles exclusively in the cloaca of grass-snakes and never leaves it) and even the anus area.

Relatively few species have started to parasitize tissues, without losing the connection with the digestive system. The above-mentioned *Brandesia turgida* penetrates into the host intestinal wall at the earliest stages of maritogony in such a way that the oral sucker and the closely adjoining genital atrium are at the level of the mucous membrane surface.

Further growth of the worm is accompanied by an intensive outgrowth of surrounding tissues. The body of the mature *Brandesia* marita becomes almost spherical, a dense capsule of connective tissue forming around it. The capsule can be clearly seen on the outer surface of the intestine. This process recalls the formation of galls, the outgrowth of plant tissues provoked by certain gall-forming insects and eriophyid ticks. *Chaunocephalus ferox* (Echinostomatidae) induces the formation of similar structures in birds.

The colonization of other systems of organs appears to have started rather early, taking place concurrently with the processes of settlement in the digestive tract. These other systems may be associated with the digestive system (liver, pancreas, and, to some extent, respiratory organs) or not associated with it at all (circulatory and reproductive systems, epidermis). The liver appears to be the second (after the digestive tract) favourite place of maritae settlement, especially the bile ducts and the gall bladder. The representatives of many different families can be found there (Fasciolidae, Heterophyidae, Opisthorchiidae, Dicrocoeliidae). Some species (the genus *Cyclocoelum*), which later migrate to other organs, also undergo the first stages of development in the liver (not in the ducts but directly in the parenchyma). The easiest way to reach the final site of infection is via the already existing ducts. However, this is often not the case.

The fate of juvenile maritae of *Fasciola hepatica* has been studied in detail by Dawes (1961, 1962a,b, 1963a,b,c). Freshly excysted metacercariae penetrate the host intestinal wall. They actively feed on the tissues, at the same time making their way to the abdominal cavity of the host. This is accompanied by a rather complex behaviour to overcome the tissue barriers (muscular layers of the intestinal wall) on their way. In 24 h they are already found on the surface of serous membranes enveloping different internal organs. The juveniles persist in their attempts to penetrate the organs and to get deeper into the tissues. Some worms "lose their way". They may be found in the musculature, in connective tissue, in lymphatic vessels. But only those individuals that get into the liver can complete their development.

At first the young maritae are still tissue parasites. While moving, they actively destroy the hepatic parenchyma, feeding on hepatocytes. In the course of their migration in the liver they finally reach the bile ducts and enter them. This is the final site of infection and it is there that the worms reach sexual maturity. From this moment onwards they can be formally attributed to cavity parasites. There arises a question: what is the purpose of this migration? How did this complex migration, traumatic for the host, originate? Dawes considers the primary histiophagy to have been the cause. Initially the worms had fed only on intestinal tissues, sometimes causing the perfusion of the intestinal wall. After getting into the body cavity, they remained histiophages, starting to feed on the tissues of various organs. The liver seems to have been established as a final site of infection only later. Apparently, it was facilitated by the possibility of simple and fast shedding of eggs into the external environment. This viewpoint was fully shared by Ginetsinskaya (1968), who thought that primary histiophagy might have been the cause of migrations in other species, settling in different systems of organs.

This sounds logical and may well be true. However, other causes of the origin of such migrations might have also existed. This suggestion is conveyed by a sharp pathogeneicity of the process described. The behaviour of juvenile fasciols during the first hours following excystment looks very much like a well-known "escape-reaction" of the parasite in a not-very-specific host (Dogiel, 1944, 1947, 1964). Attempting to escape aggressive and unfamiliar environment, the parasites leave the intestine and often find themselves in the most unexpected and unsuitable places. This phenomenon is referred to as "localization alteration" (Dogiel, 1944, 1947, 1964) Apparently, Fasciolidae are a secondarily specialized group. This can be seen in many features of their biology, including the use of terrestrial herbivorous animals as definitive hosts. Taking this into account, it could be supposed that their ancestors (with a similar two-host life cycle) parasitized in some other hosts. In the process of development in terrestrial mammals, the altered localization, which had appeared as the result of "escape reaction", was fixed as a characteristic of this group.

Bird-infecting cyclocoelids might be also supposed to have got into the respiratory system in a similar way. Ginetsinskaya and Saakova (1952) and Ginetsinskaya (1954a) managed to follow very closely an extremely sophisticated migration of the juveniles of *Cyclocoelum microstomum*. The adults of this species settle in the air sacs of the coot (*Fulica atra*). The juveniles leave the intestine soon after excystment (18 h following the infection all of them are found in the body cavity) and migrate into the liver. The migration takes place not by means of mechanical destruction of tissues, as it is the case with fasciols, but as a result of histolytic activity of the enzymes excreted by their numerous head (metacercarial?) glands. In the same way the young worms make the first holes in the host's liver. Young individuals of *C. microstomum* grow fast, feeding on singular intact cells. Larger individuals start to destroy the hepatic parenchyma mechanically, becoming true histiophages. Having reached a certain developmental stage, they leave the liver and get into the body cavity. The majority of individuals finally migrate into the air sacs of the host. The use of "sparing" chemical ways of destruction of the host tissues by the juveniles starting the migration seems to point unambiguously to a greater specialization of cyclocoelids as compared to fasciolids. However, the long and apparently obligate presence of cyclocoelids in the liver poses a question whether they have also had a stage of "hepatic parasitism" in their evolution. If this was the case, the settlement of the cyclocoelid trematode *Tracheophilus* in the respiratory organs of ducks could also be supposed to have taken place secondarily. However, there are also easier variants of migration of the maritae. It has been verified that *Di-*

crocoelium dendriticum, a common parasite of the bile ducts and the gall bladder of mammals, reaches its final site of settlement via the common bile duct (Panin, 1986).

The ways in which the representatives of different trematode groups have radiated into respiratory organs are sometimes obscure. Many plagiorchiids and ochetosomatids (*Haplometra*, *Haematoloechus*, *Macrodera*) inhabit the lungs of amphibians and reptiles. The species of the genus *Paragonimus* are common in mammals, including man. The representatives of a rather aberrant family Orchipedidae parasitize in the tracheae of some vertebrates. At least some of these forms (in particular, plagiorchiids) might have first settled in the oral cavity, later getting into the lungs. Regardless of the time and way of radiation into the respiratory system, all lung parasites have one thing in common: they are haematophagous. Hematophagy as a specialized mode of feeding has emerged independently in different groups of trematodes. However, in the instance under consideration it can be supposed to be the result of histiophagy specialization.

Members of the Philophthalmidae also made use of the way to the infraorbital sinuses via the nasopharynx. However, the representatives of this specialized family have developed an adaptation permitting an essential reduction of the migratory way. It is the above-mentioned ability of their metacercariae to encyst directly in the oral cavity (see p. 306). Apparently, this phenomenon should be secondary.

In their radiation in the vertebrate host, trematodes have not passed over other systems of organs. *Troglotrema acutum* (Troglotrematidae) inhabits the frontal sinuses of mammals. Representatives of different families settle in the excretory organs. *Nephrometra truncatum* (Eucotylidae) is localized in the kidneys of mammals, often destroying it almost completely. Renicolidae can be found in pairs inside special capsules in the kidney tissue and tubules of birds (e.g., waders, gulls, eiders, penguins) Species of the genus *Tanaisia* (Eucotylidae) settle in the ureters of waders and terns. Rather large maritae of *Gorgodera* (Gorgoderidae) are common in the urinary bladder of amphibians. Members of the Prosthogonimidae parasitic in birds radiated into their reproductive system. *Prosthogonimus cuneatus* and *P. pellucidus* are localized in oviducts. At the same time, *P. ovatus*, a common parasite of ducks, settled in the bursa of Fabricius. These and many other species infect the hosts *per os* and undergo a complex migration to the place of their final settlement.

However, migrations of a different kind are also known from trematodes. Young maritae of some species (e.g., *Patagifer bilobus* (Echinostomatidae), *Strigea plegadis* (Strigeidae), *Posthodiplostomum cuticola* (Diplostomidae)) from the intestine of birds demonstrate a rather peculiar behaviour (Dubinin, 1941). After excysting in the duodenum, the juveniles actively migrate to the host's rectum. After a period of feeding and growth they return to the duodenum and reach reproductive maturity there. Ginetsinskaya (1968) has treated this case as a manifestation of the "escape-reaction" (see above). It allows the parasite to adapt gradually to the environment of the most physiologically active part of the digestive tract.

A similar, but more complicated variant of completion of preceding developmental stages can be observed in some trematodes with a four-host life cycle. In a typical case, the development of *Alaria alata* involves four hosts (see p. 346). However, if mesocercariae (incorrectly referred to as metacercariae by Savinov (1953, 1954) who has described this phenomenon) get into the definitive host, a predator, they migrate through the stomach wall to the body cavity and blood vessels. Apparently using the pulmonary circulation, they get into the lungs and transform into metacercariae. After reaching physiological maturity, they ascend the respiratory tract to the nasopharynx and then return to the duodenum, their final site of settlement.

Blood parasites and species settling in epidermal and some other host tissues are somewhat different from the forms discussed. The former comprise representatives of a certain group of families (Aporocotylidae, Sanguinicolidae, Spirorchidae and Schistosomatidae). Their possible origin will be discussed later (see p. 338). Now we only note that blood parasitism, a secondary phenomenon, seems to have originated due to certain processes other than the radiation of initially intestinal forms. The so-called "adult" or schistosomatids and related groups in fact appear to be neotenic metacercariae (for details see pp. 338, 505). Blood parasitism, an extreme specialized form of tissue parasitism, is for them the primary mode of existence in the host.

The group of trematodes whose mature individuals parasitize in the epidermal tissues, mesenteries, walls of interior organs is much more heterogeneous. It comprises the representatives of two highly specialized families, Didymozoidae and Transversotrematidae. Didymozoids are often located in pairs in special capsules directly under scales, skin, gill epithelium and into the walls of interior organs of fish. Unfortunately, we know very little about their biology and life cycles. However, there are good reasons to suppose that their development followed three-host or even four-host type (Nikolaeva, 1965; Pearson, 1972; Pozdnyakov, 1996). Their definitive hosts (a wide range of pelagic and shelf fishes) are infected *per os*, which points to a complex migration from the digestive tract to the site of their final settlement. Mature individuals of didymozoids are undoubtedly true maritae.

A different picture is observed in the small and aberrant family Transversotrematidae, where mature individuals are located under the scales of freshwater and brackish-water fishes. Highly specialized furcocercariae of these trematodes demonstrate a clear tendency to progenetic development. Their reproductive system is almost fully formed, though the degree of its development may vary in the larvae of the same species (Cribb, 1988). The larvae differ from the sexually mature individuals only in their smaller size and lack of egg production. The host is infected when the larva contacts its body surface. The cercaria attaches to the host's surface and sheds its tail. Young individuals actively crawl under the free edge of a scale. In less than a week these organisms reach sexual maturity there. One detail should be noted here, seemingly unimportant but possibly of great significance. The adults have eyespots throughout their life. In our opinion, all the features of this life cycle (progenetic development, retention of eyespots, transcutaneous infection of the host) show that we are concerned not with a true marita but with a neotenic metacercaria. Metacercariae of many trematode species, using fishes as their second intermediate hosts, are localized in their epidermis, i.e., close to the body surface (*Posthodiplostomum cuticola*, many heterophyids). Having this in mind, the settlement place of adult transversotrematids may be seen as a supporting evidence that we deal with neotenic larvae and not true maritae in this case as well.

---

[i] Data available in the literature on the egg-hatching of Sanguinicolidae into the environment are very contradictory. In some articles (Schuering, 1922; Ejsmont, 1926) it has been noted that the eggs of these peculiar and highly specialized parasites of fishes hatch in the blood vessels of the gills and the miracidia are released into water via the host tissues. However, other data are also available (Lyaiman, 1949; Gladunko, 1968), supporting the universal nature of the mechanism by means of which the eggs of trematodes parasitic in blood get into water – via the rupture in the endothelium of blood vessels and bordering epithelia. The latter scenario seems more probable. Schuering's and Ejsmont's observations are likely to have been artefacts resulting from the same causes as those which lead to the observations of egg-hatching in the adult uterus.

[ii] In this case we might deal not with group selection, but with a variety of kin selection. Indeed, the reproductive traits of parasitic worms result in the emergence of an enormous number of genetically uniform larvae. In trematodes, a marita produces sibs (or clones, in the case of self-fertilization) of miracidia, whereas parthenitae, clones of cercariae (see also the footnote iii in chapter 4).

[iii] Actually, "passive" refers to the way in which the infective agent gets into the intestine of an animal that might be a potential host. If this is the case, the larva is activated and actively infects the mollusc, penetrating its intestine wall. However, the word combination "passive way of infection" has been so firmly established in the literature that we have to use it further on.

[iv] All observations of Mukhamedov (1981) upon the mother sporocysts of *Xiphidiocercaria* sp. VII Odening (see p. 45) were performed on the *Planorbarius corneus* snails of the same size and age.

[v] This mechanism of mechanical destruction of the host's epidermis recalls the work of the glossinoid type of the mouth apparatus in some representatives of Diptera Brachycera-Orthorrhapha.

[vi] In this chapter we do not discuss the whole process of the life cycle formation from its first stages. The researchers mentioned differ greatly upon this problem both with

each other and with the views of Ginetsinskaya (1968). Here we deal only with a definite, completing stage of this long evolutionary process – an acquisition of metacercarial stage.

[vii] We have already discussed this problem in the case of schistosomatids (see p. 339). It also applies to the representatives of Transversotrematidae and this instance will be discussed later on. Some terminological problems of the designation of hermaphroditic individuals should also be underlined. The term "marita", widely used in Russian parasitology, is not very common in the English literature, where a broader term "adult" is preferred. In our opinion, these two terms are not strictly synonymous. "Marita" is not just any hermaphroditic individual which has reached a reproductive state, but a sexually mature organism at a very definite stage of development. The term "adult" has a much broader meaning.

Chapter 3

THE MAIN TYPES OF TREMATODE LIFE CYCLES

One of the general features of the trematodes is the surprising diversity of their life cycles. This striking polymorphism may be caused by two reasons. On the one hand the effect of the extraordinary diversity is formed by a wide set of various morphological and biological adaptations, existing at different stages of development of digeneans, i.e., a great variety of animal hosts (both invertebrates and vertebrates) and, a broad range of biotopes and ecosystems, utilized by these parasites. Despite the great diversity of the variants, they all manifest a well-expressed biological flexibility of trematodes and, with rare exceptions, do not affect the general scheme of the life cycle, which invariably includes parthenogenetic and hermaphroditic generations[i].

On the other hand, some modifications, also connected with the properties of adaptive evolution of particular digeneans, may have certain impact on the events in their life cycle. First of all, this is manifested in the change in the number of hosts involved in the life cycle. Sometimes the set of stages composing the life cycle also varies. Naturally, these structural rearrangements should be taken into consideration in the analysis of the evolution of Digenea.

There have been attempts to classify the life cycle types of parasitic organisms. We mean a proper classification rather than constructing a particular series, which would reflect certain steps of parasite evolution. In our opinion, the best and most versatile proposal was that of Dogiel (1947, 1964). Naturally, embracing enormous and diverse material, Dogiel had to eliminate many details regarding separate groups of parasites. Moreover, at the end of the first half of the XX century, understanding of the development of parasitic organisms was limited. Nevertheless, the detailed concept, developed by Dogiel, reflects vast diversity of the life cycles of different parasites. He developed methodological approaches, which not only proved valid, but also to a great extent, together with the works by Beklemishev (1959, 1960), became the basis of the modern population biology of parasites (see p. 349).

In the analysis, ignoring details, Dogiel used a limited number of the most general characteristics: the number of hosts and characteristics of the reproduction processes of parasites and their accumulation in the host organism. In fact, for the first time in the history of the study of parasites, their life cycle was treated as a dynamic system and it was just the properties of this dynamics that made the basis of the classification proposed by Dogiel.

Main processes, determining the existence of the system and its sequential reproduction in time, are realized as strictly specified stages. The infection of the host, leading to the formation of a primary group of parasites – exogenous accumulation (*ExAc*) – makes up the first stage (*I*). The composition of the primary group (for details see p. 393) is determined primarily by the dynamics of incorporation of infective stages (eggs, cysts, active motile larvae) by the host. The number of parasites, composing the primary group, cannot exceed that of the incorporated infective stages.

Two principal types of exogenous accumulation can be easily distinguished. We shall denote them as simple and group (or block) type. In the first case the host incorporates the infective stages directly from the environment: each unit is incorporated individually (infection by eggs, resting cysts and active larvae). This can be realized by two ways – active infection via skin and passive one – *per os*. In the second case, eating the intermediate host, which has already accumulated some parasites from the environment, infects the animal. This fact significantly increases the probability of incorporation of a group of parasites of the same age into the animal host (see p. 399).

The second stage (*II*) embraces the period of time, when the parasites stay in the host. It can be realized in three different ways. In the first case only a certain phase of the development of the parasite take place in the host, i.e., the parasite undergoes some morphological and physiological changes, without reaching sexual maturity (Fig. 102A)[ii]. All these changes are accomplished within one incomplete ontogenesis.

In the second case the parasites always reach sexual maturity. The reproduction proceeds, but independently on the form of reproduction, **all** the material produced is invariably released into the environment being directed to infecting new animal hosts (Fig. 102B).

The generative function is realized in the third case as well, but the outcome of the offspring produced (any type of infective stages) is different. However only a part of the offspring is exported into other hosts, whereas the other one provides an intensive infection of the host, where the parasites originally developed. This results in the formation of a secondary group of parasites, whose number practically does not depend on the primary infection and is mainly determined by the resources of the host (Fig. 102C). The process of formation of the secondary group is called endogenous agglomeration (*EnAg*). The occurrence of the latter in the life cycle of parasites allows them to exploit the resource potential of the animal host to the maximum degree.

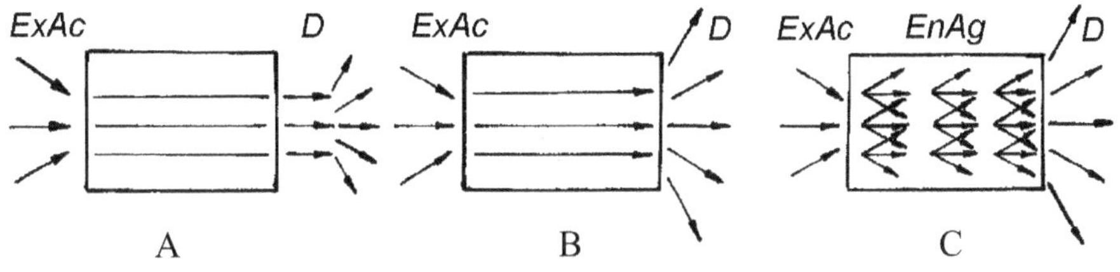

Figure 102. Three main types of parasite development in the animal hosts. A, larval parasitism – larval stages develop in the host, sexual maturation and reproduction take place in the environment; B, adult parasitism – development and reproduction of parasites take place in the host, all the offspring produced being invariably released into the environment; C, development with endogenous agglomeration – part of the offspring produced is used for the renewal of the parasite group in the host. *Abbreviations to figures 102-115*: *D*, dispersion; *DH* – definitive host; *EnAg*, endogenous agglomeration; *EnAc*, endogenous accumulation; *ExAc*, exogenous accumulation; *1IH* – first intermediate host; *2IH* – second intermediate host. Decoding of the graphical symbols used on this figure and figures *103-109* and *111-115* is given on fig. 143 (see pp. 500-1).

The third (*III*), final, stage of the parasite development is dispersion (*D*). The infective stages are dispersed in the external environment and, consequently, the possibility of infecting new animal hosts appears. Similar to the exogenous accumulation, dispersion can proceed in two ways. The most common and widely spread version is dissemination of the infective stages in the environment. However, quite often dispersion may occur without dissemination. In such a case the infected host, containing infective stages and transmitting them directly (without release of the parasites to the external environment) to the new host, plays the role of disseminating agent.

Dispersion, realized in the form of dissemination, can also proceed in several ways. The simplest one suggests that active invasive larvae pass into the external environment. In this case, dispersion is limited to primary dissemination of the infective stages by the host and a relatively short period of active redistribution of larvae in the external environment (see p. 226). These larvae can immediately start accumulating in the new host.

Similar events (with slight differences) take place when infective stages are represented by inactive larvae, fully formed and physiologically prepared to infect hosts (larvae in eggshells and cysts). The lack of an active dispersion stage, realized by the larvae, is compensated in such forms by a significantly prolonged period, that larvae stay in the environment, and possibility of secondary passive redistribution of the infective stages caused by environmental factors.

Finally, the third version differs from the first two by the occurrence of one more period (that of larvae development), which is in between the first stage (primary dissemination) and the second one (active or passive redistribution). The infective stage once it gets into the external environment is not such, strictly speaking. Some time should pass, when a certain phase of the parasite development, including morphological and physiological modifications, proceeds. As the result they acquire a capacity of infecting new animal-hosts. Obviously, in this case elimination of the infective stages, whose rate has already been quite high, will noticeably increase.

It should be emphasized that in the course of dispersion process "trans-ecosystem transfer" (see p. 467) of parasites takes place very often. Obviously, this phenomenon is especially important for heteroxenous parasites[iii]. The possibility of transition from one ecosystem into another creates excellent chances for wide spreading of parasites.

Thus, one can identify three principal typical modules, differing mainly in the events, developing during the stage *II*, when the parasite stays in the host. These events determine basic characteristic features of the parasite, which are usually typical for all representatives of the systematic group in question (see fig. 102).

Stage *I* and *III* may proceed in different ways, reflecting the consequences of the evolution of particular parasite species.

Each of these blocks can be self-sufficient for completion of the parasite life cycle. This applies to all the cases of homoxenous parasites – primary or secondary, resulting from the secondary "simplification" of the life cycle. The same blocks in different combinations may be incorporated into more complex life cycles, typical for two-, three- and heteroxenous parasites.

Obviously, all the aforementioned processes – infection (accumulation), development of the parasite in the host, reproduction and dispersion of the infective stages – make up one continuous chain, so that specific features of a preceding phase largely predetermine those of the subsequent one. In general, the parasite life cycle is a well-tuned complex system of mutually dependent and mutually complementing adaptations (Ginetsinskaya and Dobrovolskij, 1983).

Digeneans, as mentioned above, use a different number of hosts in their life cycles. Two rarely encountered variants are one- and four-host life cycles. As a rule, such life cycles occur in the representatives of very specialized groups (Hemiuroidea, Plagiorchiida, Strigeidae). Two- and three-host life cycles are more common. They include a major number of species whose biology and development have been investigated. However, comparison of these two groups shows that they differ qualitatively from each other.

Three-host life cycles are structurally rather homogenous, on the contrary, two-host life cycles form several subgroups. Distinctions between the subgroups are primarily determined by non-uniform distribution of the events of parasite development among different hosts. We believe that the latter fact testifies to independent origin of two-host life cycles in different groups of digeneans (see p. 491).

Since digeneans with three-host life cycles are numerous and widely spread, we will start our analysis with that group.

1. TRIXENOUS (THREE-HOST) LIFE CYCLES

The most general scheme of the trematode three-host life cycle is presented in figure 103.

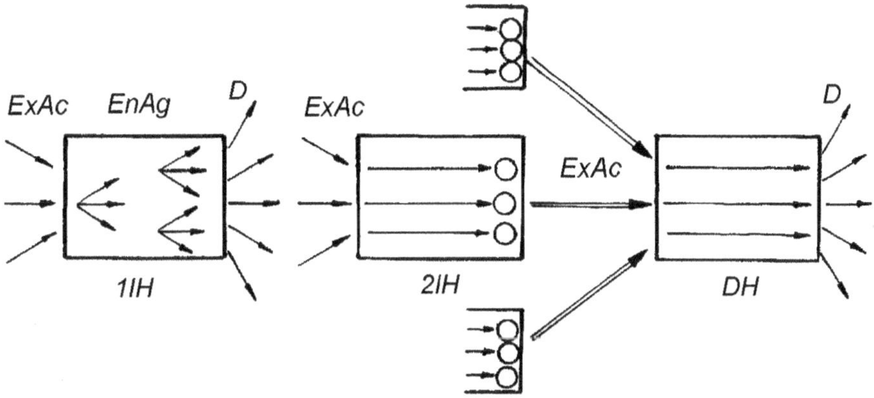

Figure 103. Scheme of the trematode three-host life cycle (for abbreviations see fig. *102*).

One should emphasize that implementation of the scheme in real life cycles of digenean species varies considerably. In our opinion, this fact reflects evolutionary flexibility of the group and ability of its representatives to adapt efficiently to existence in different ecosystems.

It should be noted that along with narrow specializations, which are characteristic of representatives of the lowest taxa, some principles, reflecting general trends in the evolution of parasitic organisms, are well expressed. It is these peculiarities that should be mainly considered in the analysis of the events in the life cycles and their classification.

A short survey of the most essential variants of stage-by-stage completion of the three-host life cycles is provided below.

Exogenous accumulation in the first intermediate host may proceed in one of two ways, which is determined by biological properties of miracidia. Despite morphological and biological diversity the latter form two distinct biological groups (see above and p. 2).

The first group includes active larvae with a well-developed and differentiated sensory system and, consequently, complex behaviour in the environment. The second group consists of morphologically simplified larvae, which survive passively in the environment being protected by eggshells. The life span of the larvae belonging to the first group is relatively short, while that of the larvae from the second group is much longer. It has been mentioned many times that the passive survival of larvae is secondary with respect to the active mode of life (Ginetsinskaya, 1968; Dobrovolskij et al., 1983; Semenov, 1991, see also p. 236). Transition from the active life to the passive one is, undoubtedly, a convincing manifestation of a trend towards shortening and eventual elimination of the mobile phases from the parasite life cycle, as was noted by Dogiel (1947, 1964).

Such a transition from the active mode of life to passive survival inevitably affected the exogenous accumulation of parasites in molluscs. This process proceeds in several stages, which only partially coincide for the two considered cases[iv].

In the actively swimming larvae, accumulation starts with the discovery and identification of the host (Fig. 104A) (see p. 225). Then the larva penetrates via the skin into the host and migrates to the place of its final location. The duration and complexity of the migration may vary significantly.

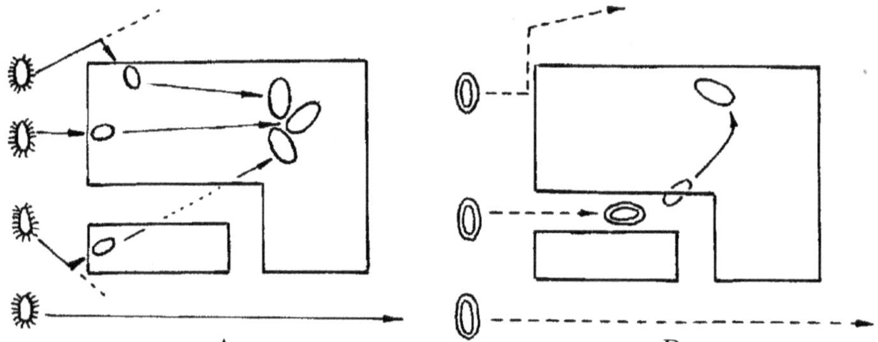

Figure 104. Scheme of the exogenous accumulation in the molluscan host. A, active accumulation – actively swimming miracidia whose behaviour is modified by mollusc; B, passive accumulation – eggs with miracidia are swallowed in the course of mollusc feeding.

In passively infecting miracidia the initial stages of the accumulation process are completely different (Fig. 104B). The entry of the infective stage (just entry, not the infection) into the host occurs randomly when the eggs with infective larvae are swallowed in the course of feeding. The first manifestations of the larvae activity become noticeable when they are still inside the eggshell. However, they are fully activated only after hatching in the intestine. Obviously, the stage of discovery of the host is completely missing, whereas identification still takes place. However, it is realized still prior to the hatching of larva from the eggshell. Then the penetration through the intestine wall (or just entry into it – see p. 238) and final migration proceed.

In the first case a successful completion of accumulation is mainly determined by the behaviour of the larvae, while the contribution of the mollusc is negligible, provided that the mollusc is active and that its body is not fully covered with the shell. In the second case the partners exchange their roles in the formation of host-parasite system. The leading role now belongs to the mollusc, for the successful accumulation is significantly determined by its feeding activity. The miracidia can affect the process of infection only via the duration of its life span (see p. 239), i.e., the period of time within which it retains its infectivity. Structural properties of the eggshells (surface microstructure, presence of filaments), enabling the eggs to gather on food substrates of molluscs, may also play a certain role. However, this problem has never been specially investigated.

Thus, two principal types of accumulation of the infective stages in molluscs – active and passive accumulation – can be distinguished.

The passive type of accumulation, has been mentioned many times, is undoubtedly secondary relative to the active one. Its acquisition by any group of digeneans should be treated as a form of advanced specialization.

There is a small group of digeneans, which combine some characteristic features of both types of accumulation considered above. This is, for example, the case of Gorgoderidae with actively swimming miracidia. However, the initial stage of accumulation is passive. First intermediate hosts of Gorgoderidae – filter feeders (bivalves from families Sphaeriidae and Pysidiidae) – pump the swimming larvae into the mantle cavity. All subsequent events are realized according to the program of active accumulation.

The development of parthenitae in the molluscan host. Endogenous agglomeration. The existence of digeneans in molluscs differs from that in other hosts by the phenomenon of endogenous agglomeration. It is realized as a result of reproduction of parthenitae (sporocysts and rediae), represented in the host by several (sometimes only two) generations replacing each other.

There is only one well documented exception - *Heronimus mollis* (Heronimidae), which lacks endogenous agglomeration. Parthenitae are represented by only one mother generation, which directly produces individuals of the hermaphroditic generation. A similar situation may also take place in the Bucephalidae (see p. 49).

Parasite groups, forming in the mollusc as a result of endogenous agglomeration, are different in their composition and life span. Here, the term "composition" does not imply the presence of rediae or sporocysts in the life cycle[v] but just the number of parthenogenetic generations replacing each other and their capacity for self-reproduction. On the base of these characteristics digeneans can be clearly divided into several groups.

Primarily, the differences are most pronounced in the first generation of parthenitae, which is somehow represented in the life cycle of all digeneans. However, its fate in various representatives of the Digenea is different.

In majority of species the miracidium, having penetrated the mollusc, develops into a mature organism – mother sporocyst – in the course of regressive metamorphosis. The latter, having reached generative maturity, starts the reproduction process, thus giving rise to the next generation of parthenitae. However, as mentioned above (see p. 28), the events may also proceed in different ways. Though these scenarios may differ in details, there is an essential similarity consisting in the fact that the mature mother sporocyst disappears from the life cycle as a separate phase. In fact, two alternative processes lead to that. First, the generative function of the mother sporocyst can be already completely accomplished at the stage of development of the miracidium. The formation of the latter is completed practically simultaneously with the formation inside it of an individual of the second generation (redia, as a rule), which becomes the first parasite phase (see p. 28).

Second, a similar result is observed in the case of the maximum simplification of miracidia, caused by the aforementioned transition to passive accumulation. Germinal material of such larvae is often represented by a minimum quantity of forming generative and undifferentiated cells, so that the generative function is practically entirely realized just at the parasite stage (see p. 26). In some Digenea (Microphallidae, Lecithodendrioidea) the development of the mother sporocyst in the mollusc sometimes is completely suppressed and only the germinal primordium of larva or even separate germinal cells (brood parasitism) start parasitizing the host. In the first case, the generative function of the first parthenogenetic generation is suppressed to a maximal degree – only one redia can form in the miracidium. The transition of germinal material (germinal masses, germinal cells) to parasitizing the mollusc does not impose such limitations and the offspring of the first generation of parthenitae in these cases is normally rather numerous.

With the exception of the aforementioned Heronimidae and Bucephalidae, endogenous agglomeration apparently takes place to a certain degree in all other digeneans. In certain species (these are probably almost all primitive redial trematodes, and some sporocyst-bearing ones such as Schistosomatidae, Strigeidae, Diplostomidae, Opecoelidae and possibly some others) parthenitae (both rediae and sporocysts) are capable of self-reproduction, i.e., along with cercariae they can produce a certain amount of individuals, similar to them, which do not leave the mollusc (Fig. 105).

Thus a parasitic group formed in the host may exist any time (see p. 369). The real life span of such a group is determined by the longevity of

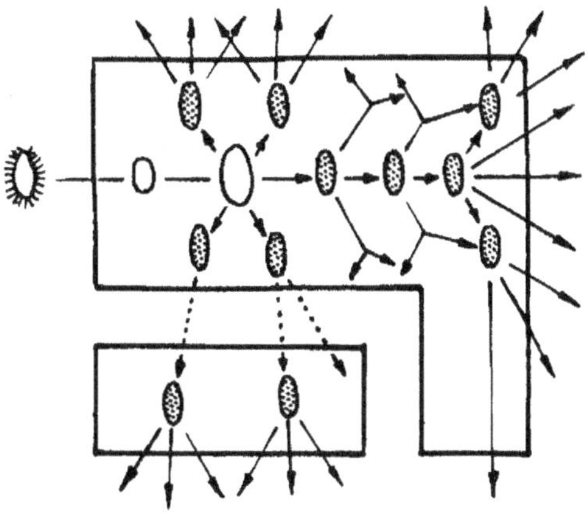

Figure 105. Scheme of an endogenous agglomeration of parthenitae in the molluscan host providing prolonged existence of the parthenitae group and simultaneous cercariae production.

the mollusc itself. The number of parasites in the host does not show unlimited growth and depends directly on host traits (for details see section 3.2. in chapter 4).

Another group consists of more specialized digeneans, with the life cycle in which there is a strictly fixed number of parthenogenic generations. This can be illustrated by species of the plagiorchiid complex (Plagiorchiidae, Ochetosomatidae, Lecithodendrioidea, Microphallidae). Their mother sporocysts give rise to only one generation of daughter sporocyst and the latter produce only cercariae (Fig. 106).

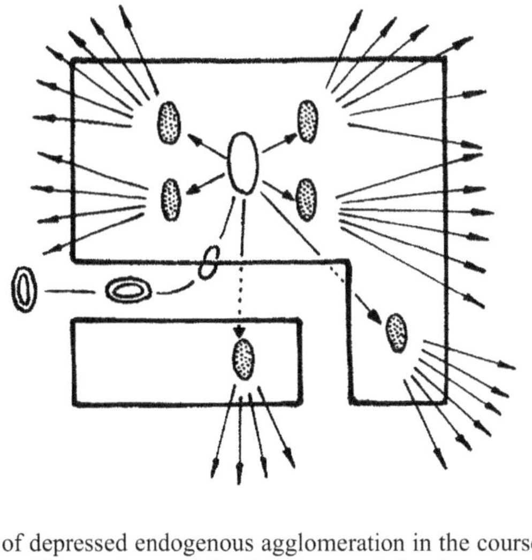

Figure 106. Scheme of depressed endogenous agglomeration in the course of which only two parthenitae generations (the mother and the daughter one) develop, the daughter parthenitae can produce only cercariae.

Therefore, in this case the process of endogenous agglomeration is strongly suppressed. The number of parasites, having reached a certain level, does not change any more, and the group of parasites in the host cannot be replenished. This trend is even more pronounced in Brachylaimoidea, whose miracidia inject several (apparently, 1-3) germinal cells in the course of exogenous accumulation. Provided that all the latter successfully develop, they give rise to a corresponding number of large branching daughter sporocysts, which grow mostly into all host tissues and organs, including the tentacles, as in the case of species of genus *Leucochloridium* (see p. 70) (Fig. 107).

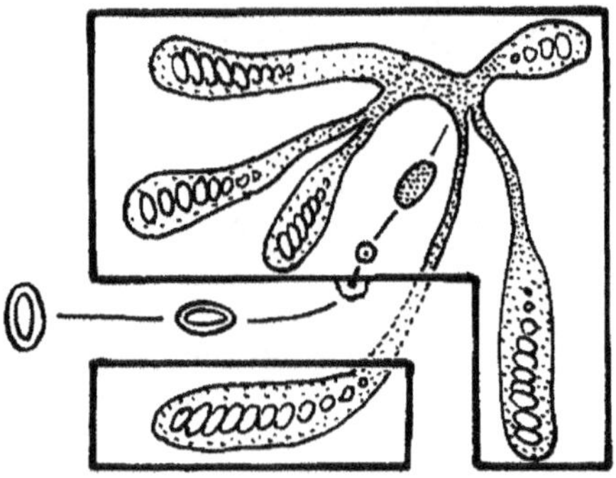

Figure 107. Scheme of practically total depression of the endogenous agglomeration in the Brachylaimoidea parthenitae – the small number of daughter sporocysts are compensated by their intensive growth and branching.

As mentioned above, the prolonged endogenous agglomeration, based on the capacity of parasites to self-reproduce during several generations, provides a high density of parasites in the host and, correspondingly, an efficient exploitation of host resources. To a significant degree a similar result can be achieved by species with a limited period of endogenous agglomeration. In plagiorchiids, the offspring of highly productive mother sporocysts is usually enough to exhaust the mollusc's resources completely. In brachylaimoids this is due to abnormally high growth of one or several branching sporocysts.

The first dispersion, starting with the emission of cercariae into the environment, is a necessary element of trixenous life cycles. This process varies widely in different species. This refers to the number of the produced cercariae and the rhythms of their shedding. Biological traits of larvae of different species (e.g., life span, behaviour) are also very diverse. All the above, together with the environmental factors, determine the spatial distribution of cercariae and, eventually, the probability of infection of the second intermediate host (for details see: section 3.1. in chapter 2 and section 3.3. in chapter 4).

The stage, that follows dispersion, is the **exogenous accumulation** of the infective stage (cercariae) **in the second intermediate host**. The ways of entry of larvae into its body are also very different, being determined by the biology of a particular digenean species (see p. 279).

The parasite stay in the second intermediate host is of particular interest, because it is just this stage of the development of the hermaphroditic generation largely determines subsequent events. As mentioned above (see p. 159), due to the involvement of a second intermediate host in the pathways of digenean circulation some ontogenetic transformations could be transferred to the metacercariae stage. The advance of different digenean species in this direction is not uniform, so that even closely related species may differ significantly in that respect. All this makes it practically impossible to distinguish any distinct groups, for the boundaries between them are poorly defined. Therefore, we identify very conventionally several levels of metacercariae specialization, which, in our opinion, reflect the main tendency of evolution of this stage of the development of the hermaphroditic generation.

The first level embraces species (Echinostomatidae, Monorchidae, Renicolidae), whose metacercariae are isolated from the host by multi-layered protecting cysts and, hence, no close metabolic contact between parasites and the host is possible. This fact largely determines the future of larvae, which have penetrated into the host. They do not undergo any significant metamorphic changes, retaining many features of cercariae.

The second level of specialization is represented by larvae with secondarily simplified cysts. They have relatively thin walls and most often consist of two layers (Heterophyidae, Plagiorchiidae, Ochetosomatidae, Lecithodendrioidea). Sometimes these larvae initially form a very thin provisional cyst, which is later replaced by a thicker definitive one. In some cases the larvae do not encyst immediately after penetration. There is an obligatory time lag between the penetration and encystment during which the metacercariae can freely move in the tissues and cavities of the host body. Apparently, complex relations between the parasite and host, including merely trophic ones, are established in the course of realization of the above-mentioned variants (see p. 159). Larvae gain access to the energy resources of the host, so that some morphogenetic modifications, that earlier took place at the stage of marita, are transferred to this phase of the life cycle.

Finally, the third level of specialization is achieved by metacercariae, completely incapable of encystment (Gymnophallidae, Strigeidae, Diplostomidae). In most the host cells form a capsule of connective tissue surrounding them. The development of such metacercariae in the second intermediate host is often characterized by the most complex transformations (metamorphosis) (see p. 175).

In most cases, achievement of the infectivity by metacercariae implies a transition from the state of morphogenetic activity to that of inactivity. A fully formed metacercaria ceases to develop at that stage, from which the process of maritogony starts in the definitive host.

The dispersion of parasites after staying in the second intermediate host acquires rather specific features. The process of dissemination of infective stages in the environment is skipped altogether. The hosts accumulate metacercariae and maintain them during their entire life. Naturally, such formed parasite groups can be rather heterogeneous in their age and physiological status (see p. 393). The dispersion is realized in this case by the intermediate hosts themselves. Its particular manifestations are determined, on one hand, by the biology of animal hosts, and on the other, by the relations in the host-parasite system (see p. 358). It is these factors that mainly determine the principles of spatial distribution of the infective stages and the probability of infecting the definitive hosts.

The elimination of dissemination determines specificity of the **exogenous accumulation** of parasites **in the definitive host**. As a rule, the latter is infected not by a single individual (infective stage) but by a group of parasites or a block. We have called this type of accumulation "group or block exogenous accumulation" (see above). Under favourable conditions a rather numerous group (local hemipopulation – see p. 399) of sexually mature parasites may form in the definitive host even if infection occurred only once.

The specificity of **parasite stay in the definitive host** is determined by the fact that the generative phase of the hermaphroditic generation develops there, i.e., maritae reach maturity and start reproduction. The initial stages of forming of maritae functionally are very similar to metacercariae development. Juvenile maritae undergo a more or less complex migration, and having reached the place of final settlement, resume their development. Sometimes morphogenetic processes may start even during the migration period. Most often, this takes place, when the migration path is very complicated and lasts relatively long, because parasites have to move from the intestine into other organs – lungs, bladder, gall bladder, liver, etc. (see p. 310).

The most important element of the development of maritae is formation of the reproductive system. Sexual maturity does not always coincide with the completion of morphogenetic changes. The latter may continue for sometime even after the production of eggs has begun. Usually it is just in this period that the maritae acquire a complex of species-specific morphological characteristics.

It should be emphasized that intensive reproduction of parasites does not result in an increase in the numbers of the group of parasites in the host. No elements of endogenous agglomeration can be observed in this case. All the eggs produced by maritae are passed to the external environment. The parasite group in the host can be replenished only at the expense of repeated infection with metacercariae (exogenous block accumulation).

The final stage of the trixenous life cycle is the **second dispersion**, which always proceeds via dissemination of eggs in the environment. Three main types of such processes can be distinguished. They differ in two principal characteristics – the time spent by the infective stage[vi] in the environment and the way of infection of molluscs (see p. 2).

There is a typical situation, which is apparently often encountered in trematodes, both primitive and highly specialized. The eggs, laid by the adult, need a certain period of time (t') to reach the infective state. Here one deals with eggs undergoing their embryonation, partly or fully in the environment (Fig. 108).

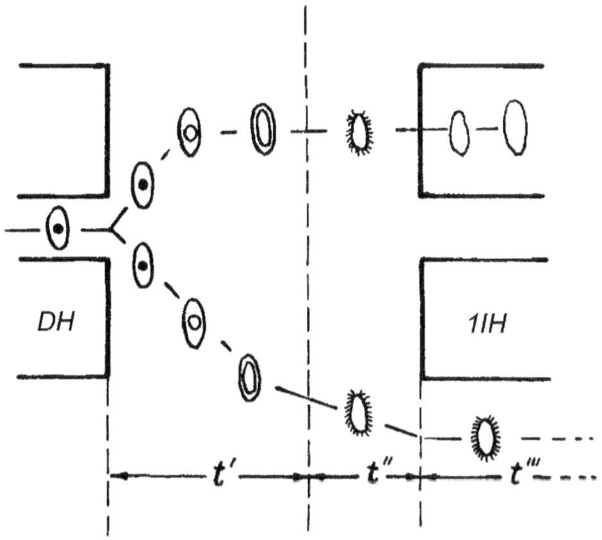

Figure 108. Scheme of dispersion of the eggs undergoing their embryonation in the environment (see text for explanations).

A fully formed larva hatches from the egg and has to infect a mollusc during a relatively short period of time t'' (this time is determined by a combined effect of endo- and exogenous factors - see p. 225). As a rule, the age-dependent loss of infectivity does not coincide with the physical death of the larva. The latter, not being infective, can still survive in the environment for some time (t''').

The second version of dispersion (Fig. 109A) is also characterized by a short t'' period (retaining infectivity) and an active way of infection of the host. The only difference is that all the development of miracidia proceeds in the mother organism, so that the t' period is completely lost.

In the case of dispersion of the third type (Fig. 109B) the entire embryonation of the eggs also proceeds in the marita uterus (t' period is skipped). But, unlike in the first two situations, in this case t'' period is sig-

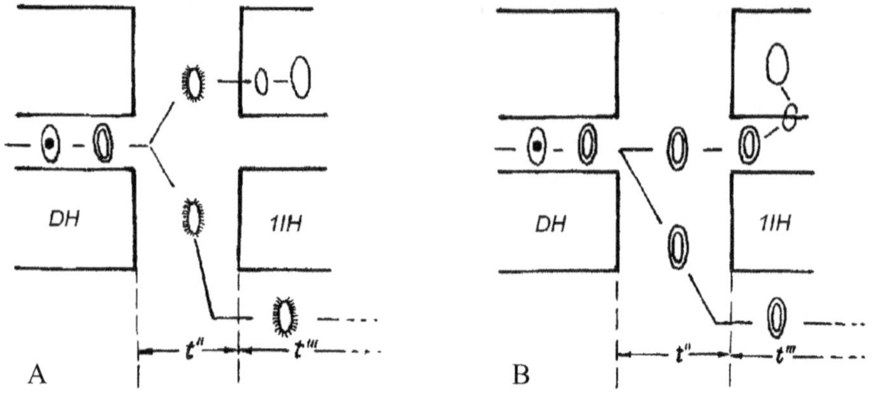

Figure 109. Scheme of dispersion of the eggs, whose entire embryonation proceeds in the marita uterus. A, actively infecting miracidia; B, passively infecting miracidia (see text for explanations).

nificantly prolonged due to transition to the passive way of infection of molluscs. Miracidia, as mentioned above, do not leave the eggs and survive there for a long time, completely retaining their infectivity.

2. TRIXENOUS (THREE-HOST) LIFE CYCLES WITH TWO ENDOGENOUS AGGLOMERATIONS

In some of Gymnophallidae, belonging to specialized and evolutionary advanced family of trematodes, the typical three-host life cycle becomes more complex due to the involvement of the second stage of endogenous agglomeration (Fig. 110).

All the events, connected with the infection of the first intermediate host, development of parthenitae there, dispersion of cercariae and their accumulation in the second intermediate host, are similar to those in most other trematodes (Fig. 111). The situation drastically changes upon the infection of the second intermediate host. In some gymnophallids metamorphosis of cercaria into metacercaria acquired unique traits, not encountered in other cases. It is strongly regressive, so that the organization of larvae is not similar to that of marita, but on the contrary, acquires features, rather characteristic of parthenitae (see p. 73). A large body cavity divided into a system of brood chambers is formed. Each brood chamber contains germinal material which apparently, originated from the undifferentiated genital primordium of the larva (see p. 74). The body of this metacercaria grows and acquires a spherical shape, resembling in its appearance a large sac-like redia. However, it

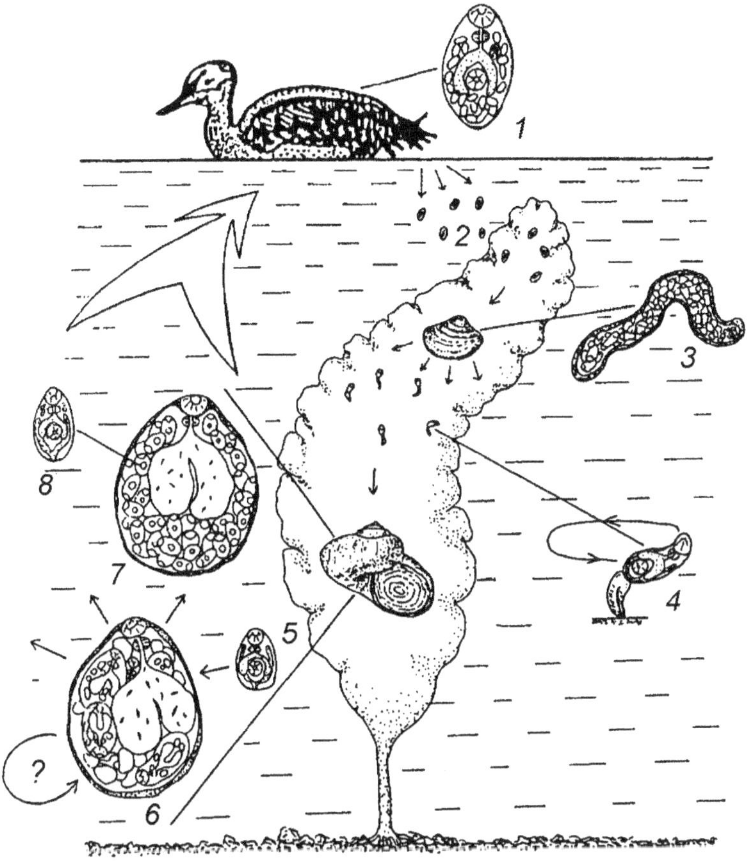

Figure 110. Life cycle of *Cercariae margaritensis* Ching, 1982. *1*, marita in the common eider (*Somateria mollissima*) intestine; *2*, egg with miracidium; *3*, daughter sporocyst in the first intermediate host – bivalve *Turtonia minuta*; *4*, cercaria; *5-8*, stages of development in the second intermediate host – gastropod *Margarites helicinus*; *5*, young metacercaria of the first parthenogenetic generation (M1); *6*, fully formed M1 containing developing metacercaria of the second parthenogenetic generation (M2); *7*, fully formed M2 containing metacercariae (*8*) invasive for the definitive host. From Galaktionov (1996a).

differs from a typical redia by the occurrence of a two-branched intestine, characteristic of the hermaphroditic generation. Similar to rediae, these metacercariae reproduce by parthenogenesis (see p. 73). The first generation produces one more parthenogenetic generation. However, there may be also several generations. Such an assumption is based on the material on the development of *Parvatrema homoeotecnum* in the molluscs *Littorina saxatilis* in the Barents Sea, which we have obtained. These data, however, require certain refinement.

The above-described process fully agrees with our concept of endogenous agglomeration. The latter, however, is limited in time, for at a

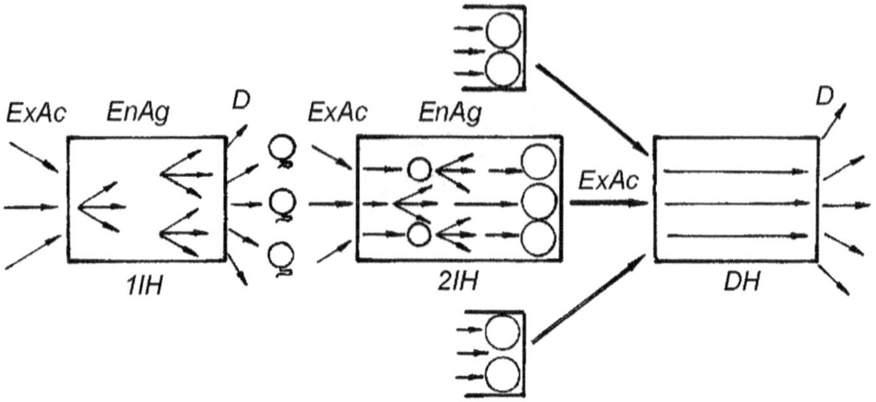

Figure 111. Scheme of the life cycle with two endogenous agglomerations (for abbreviations see fig. *102*).

certain moment these redia-like parthenogenetic metacercariae start to produce normal metacercariae, which accumulate in their body cavity in large numbers. Normal metacercariae are not capable of reproduction and can only infect the definitive host. Their number increases as long as reproduction of parthenogenetic metacercariae goes on. As soon as it is finished the quantitative composition of the group stabilizes. This process is similar to what is observed in the course of usual infection of the second intermediate host by metacercariae, as far as the final result is concerned. The only difference is that in this case the infective stage does not come from the environment but is produced by preceding parasite phases. This allows us to interpret the observed phenomenon as **endogenous accumulation**.

From the functional viewpoint, the host, so far traditionally called second intermediate host, can be treated that way only after the first genuine metacercariae appear in it. Dispersion, like it takes place usually in trixenous life cycles, proceeds without dissemination. Infection of the final host and development of maritae in it obey a general scheme, common for most trematodes.

3. DIXENOUS (TWO-HOST) LIFE CYCLES

The two-host life cycle in trematodes is a very common phenomenon, occurring even in phylogenetically distant taxa. This fact alone makes one presume that the nature of the dixenous life cycles and their origin may be completely different. In some cases, apparently, these are archaic life cycles, preserving their primary structure. On this basis, the aforementioned trixenous life cycles, encountered in most trematodes, secondarily evolve. In other cases we deal with different tertiary modifications. Their emergence is connected either with the shortening of the life cycle itself (see p. 181), due to the elimination of some phases of development, or with the reduction of the parasite transmission pathways in ecosystems (see p. 344). In this case all developmental phases are retained, though sometimes strongly modified, whereas some hosts can be lost (Galaktionov and Dobrovolskij, 1989).

3.1. Primary Dixenous Life Cycles

The life cycles of this type are characterized by lack of the second intermediate host, though in many digeneans there is a corresponding phase of the ontogenetic development of the hermaphroditic generation. Trematode taxa with primary dixenous life cycles significantly differ in the degree of specialization. Both, archaic groups (Fasciolidae, Paramphistomidae, Azygiidae) and families with various traits of advanced specialization (Philophthalmidae, Parorchiidae, Notocotylidae, see pp. 24, 28) can retain primitive two-host cycles.

Similar to trixenous life cycles, exogenous accumulation is achieved either via skin by actively swimming miracidia (Fasciolidae, Philophthalmidae, Paramphistomidae), or passively *per os* (Notocotylidae, Pronocephalidae). The occurrence of rediae is a common archaic feature of all trematodes with primary dixenous life cycles. Endogenous agglomeration as in most redial digeneans proceeds with an unlimited number of generations: rediae are capable not only of producing cercariae, but can also self-reproduce (see p. 370).

Compared to other types of life cycles in the species with primary dixenous life cycles the first dispersion is very complicated. It includes three phases: emergence of the fully-formed cercariae into the environment, period of active swimming (these two events ensure the dissemination of the infective stages in the environment) and, finally, encystment on the surface of underwater objects followed by a more or less prolonged survival of the larvae protected by the cyst. Realization of this stage of the life cycle in various digenean species is subjected to rather significant variations (Fig. 112).

Of major importance are variations in the period of cercarial active life. It can be long (up to several hours) (Fig. 112A), though it can also be significantly reduced and even eliminated (Fig. 112B,D). Also variable is the time, when larvae acquire infectivity. In some cases they must encyst and transform into adolescariae, because only after that can they infect the definitive host. In other cases the encystment is not a compulsory phase of metamorphosis, so that the definitive host can be infected not only by adolescariae, but also by free-living cercariae (Fig. 112E), as in the Philophthalmidae (see p. 302).

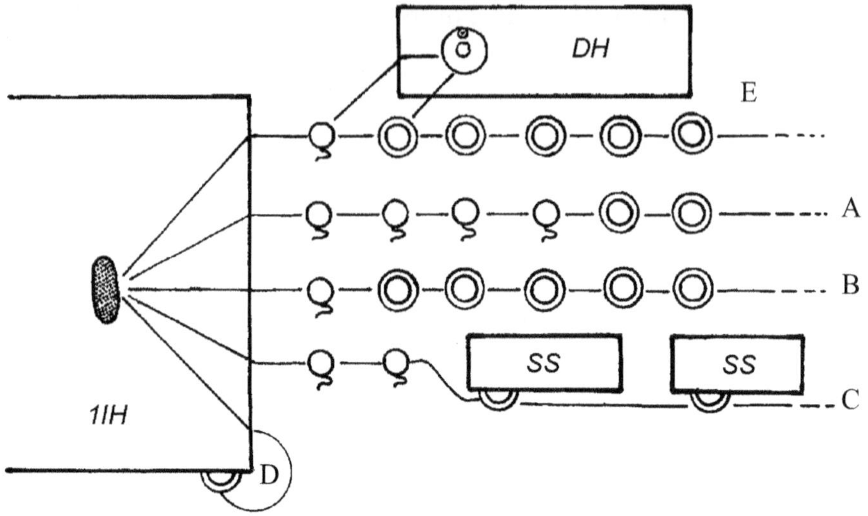

Figure 112. Possible transmission ways of the primary dixenous life cycles with cercariae encystment in the environment. A, species with prolonged period of cercariae active life (Fasciolidae, Psilostomidae, some Echinostomatidae); B, species with reduced period of cercariae active swimming; C, species with cercariae encysting on specific substrate (some Notocotylidae); D, species, whose cercariae loose the period of swimming in the environment and encyst on the body of the same host, where they have developed (some Notocotylidae); E, species, whose free-living cercariae can optionally infect the definitive host without encystment (Philophthalmidae). *Abbreviations*: *SS*, specific substrate (for other abbreviations see fig. *102*).

And, finally, the last group of distinctions is connected with the type of substrate chosen by the larvae for encystment. Larvae of a number of species do not show any substrate specificity (Fig. 112A,B,E; see p. 280). Others exhibit substrate preference such as mollusc shells or the chitin cuticle of aquatic arthropods (Fig. 112C). Lately, there is a group of larvae usually encysting on the body of the same host, where they have developed (Fig. 112D). In this case cercariae may often effect complex migration (see p. 307). Here, one deals with gradual complication and ever increasing specialization of the cercarial behaviour (undoubtedly, secondary in their nature), which testify to smooth transition to trixenous life cycles. In fact, these are initial stages of involvement of the second intermediate host into the pathways of parasite transmission. The emergence of this host facilitates the transmission of the infective agents from molluscs to the definitive host – a vertebrate animal. In those cases, when such a specialization in larval behaviour is obligatory (some Notocotylidae and many Echinostomatidae), one deals with very primitive and archaic trixenous life cycle.

Life cycles, where the dispersion of infective stages (cercariae), produced by the first intermediate host (mollusc) is replaced by exogenous accumulation of the larvae in the host, which is undoubtedly definitive

(Azygiidae, Bivesiculidae), should be considered as primary dixenous ones. The infection proceeds *per os* and the ingested larvae immediately start maritogony in the specific sites of their final settlement (Fig. 113). In this case, life cycles of highly specialized parasites are realized according to an archaic scheme. This specialization is primarily connected with peculiarities of dispersion of cercariae, produced by the mollusc and their subsequent accumulation in the second host.

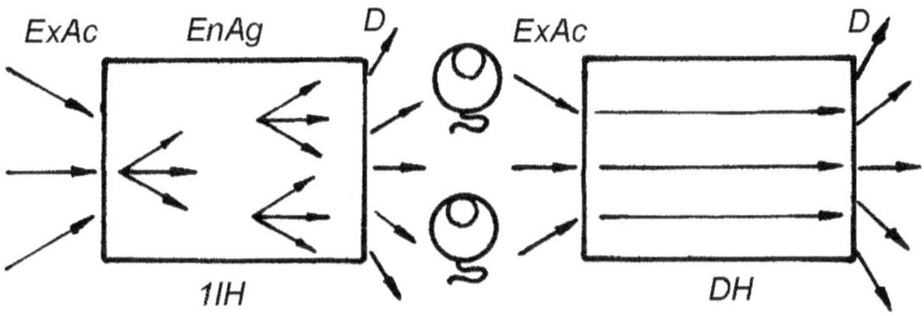

Figure 113. Scheme of the transmission ways of the primary dixenous life cycle without the phase of adolescaria (Azygiidae, Bivesiculidae) (for abbreviations see fig. *102*).

Dispersion in a life cycle of this type is a one-stage process, i.e., cercariae, leaving the mollusc, should be immediately eaten by the definitive host. Their life span is limited, as in overwhelming majority of cercariae, and they are not capable of encystment in the environment. The larvae of these digeneans are often large enough (this is achieved exclusively due to provisional structures, primarily a tail); sometimes one can observe even their mimicry (Azygiidae).

The accumulation of infective larvae by the host is also characterized by a number of specific features. In fact, it should be considered as a passive type of infection like any case of *per os* infection. However, this formal approach gives scarce information on the essence of the process. Specific behaviour of the larvae, often imitating the behaviour of typical pray of the host, attracts its attention. The meeting of the parasite with the host is largely provided by the activity of the feeding behaviour of the latter. However, this activity is to a certain extent stimulated by the infective stage itself.

Apparently, such an activation processes of infection of the definitive host must compensate for the absence in the life cycle of a prolonged phase of passive survival, which is typical of adolescariae. It is evident that morphological adaptations of cercariae are secondary. Their strict correspondence to biology and behaviour of the host results from secondary specialization.

3.2. Secondary Dixenous Life Cycles

Other variants of dixenous life cycles are clearly of secondary nature. Their emergence, as mentioned above, was caused either by changes in pathways of parasites transmission or shortening of the life cycle itself. Classical examples of two-host life cycles are observed in a large group of blood parasites – Aporocotylidae, Sanguinicolidae, Spirorchiidae and Schistosomatidae, which explored a large number of vertebrates (from fishes to humans) as hosts.

These cycles appear very simple. They include only two hosts, two phases of exogenous accumulation (primary simple, not block), one endogenous agglomeration and two dispersion phases, realized as dissemination (Fig. 114).

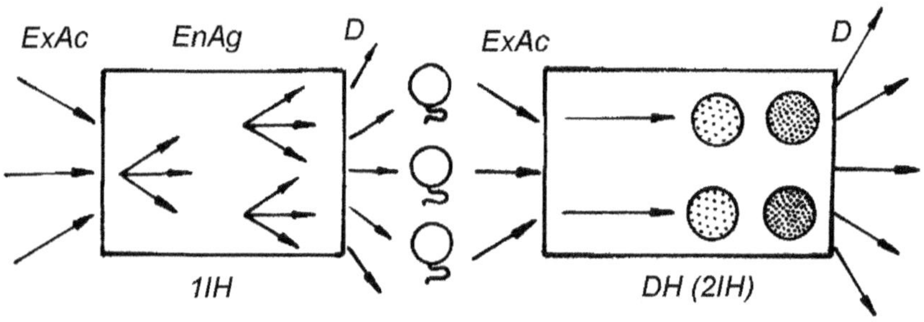

Figure 114. Scheme of the transmission ways of the secondary dixenous life cycle lacking the definitive host (Sanguinicolidae, Aporocotylidae, Spirorchiidae, Schistosomatidae) (for abbreviations see fig. *102*).

The first exogenous accumulation is an active process. Endogenous agglomeration involves several generations of parthenitae, which follow each other and form a genuine microhemipopulation (see p. 369). Distinctions are observed only in the type of daughter and subsequent generation of parthenitae. In some cases, these are rediae (Aporocotylidae), and in all the others (Sanguinicolidae, Spirorchiidae, Schistosomatidae) these are sporocysts.

Total fecundity of parthenitae, primarily, of sporocysts, is normally very high. This ensures an intensive dissemination of cercariae in the environment. Having established contact with the next host, larvae actively penetrate the host's skin. This usually takes place in the typical trixenous life cycle upon the infection of the second intermediate host. However, after the second exogenous accumulation has been completed, the subsequent events develop quite differently. Having finished migration, juveniles, which now inhabit strictly defined regions of the blood circulatory system of the vertebrate animal, complete their ontogenesis and reach maturity.

The dispersion period of all the representatives of this ecological group consists of two distinct phases. During the first phase eggs are being accumulated in capillaries of different organs (as a rule, directly connected with the external environment – intestine, bladder, etc.). It is during this time that embryonation of eggs takes place. Dissemination proceeds due to disruption of the endothelium of injured capillaries. The eggs get into the lumen of the corresponding organ and from there to the external environment. There miracidia almost immediately hatch from the eggs (period t' is missing) (see fig. 109A and p. 331).

The nature of such life cycles has not produced much discussion. Most research suggests that these life cycles have lost the second intermediate host. Vertebrate animals, parasitized by sexually mature worms, homologous to maritae of other trematodes, are the normal definitive hosts. Consequently, the metacercarial phase vanishes from the life cycle together with the second intermediate host. Having penetrated into the vertebrate animal, juveniles immediately start a maritogony. Similar to archaic (primary) dixenous life cycles, this process includes a whole set of morphogenetic transformations from the fully formed cercariae to the reproducing adult.

However, there is also exists a different view point (La Rue, 1957; Shoop, 1988; Combes, 1991b). Infection of the host via the skin, very common in trematodes, is clearly of secondary origin (see p. 308). The onset of this phenomenon in the evolution of digeneans, is, most likely, connected with the involvement of the second intermediate host in their life cycle (see p. 307). As for definitive hosts, in most cases, they accumulate the infective stages *per os*, ingesting adolescariae or second intermediate hosts infected by metacercariae. It should also be remembered that in the normal definitive host, maritae are usually localized in the intestine or organs connected with the digestive tract (gall bladder and bile ducts, lungs and respiratory system). More rarely they effect complex migrations settling in the kidneys or excretory bladder. Finally, even more rarely can they settle in tissue capsules (i.e., *Brandesia*). However, even in this case the capsule cavity is not isolated, but rather connected either to cavities of some organs, which have access to the external environment, or directly to the latter. Undoubtedly, the formation of such capsules in most cases can be considered as a secondary adaptation of parasites, which infect the host *per os* and necessarily undergo an initial phase of cavity parasitism.

Representatives of the aforementioned group of blood parasites with skin penetrating mechanism of infection (their cercariae have one of the most complex and specialized penetration complex – see p. 129) are, in fact, specialized tissue parasites. Blood is not just a fluid, but first of all, a tissue of internal medium, which is characterized by a high degree of immune reactivity. Thus, neither by the mechanism of infecting the host nor by localization in it, sexually mature worms of the trematode group in question differ from the metacercariae, and the host, they settle in, does not differ from the second intermediate host.

Nevertheless, there are clear distinctions: first, these parasites achieve sexual maturity; second, they do not encyst. However, in our opinion, these distinctions are not of key importance. Progenetic development of metacercariae up to the appearance of genuine neotenic individuals is rather common in various groups of trematodes (see pp. 149, 169, 345). As a matter of fact, that might be facilitated by transition to blood parasitism. Such a transition creates for parasites very favourable conditions within the host. Of no less importance is the fact that absence of encystment or, at least, its delay is quite frequent among digeneans. Reduction of cystogeneous apparatus and suppression of cyst formation (see p. 157) are normally connected with a transfer of morphogenetic activity to the metacercarial phase. "Metacercariogenesis" to a great extent contributes to the maritogony. The organism inhabiting the lumen of the vessel and constantly subject to the influence of bloodstream should not lose mobility. In this case, encystment is not merely useless, but even harmful. Enhancement of the trophic function of the tegument is also, undoubtedly, beneficial, which was demonstrated many times in the case of shistosomes (see p. 183). It is through the tegument that the uptake of nutrients to the developing metacercariae of many other trematodes takes place. Participation of the digestive system in these processes in most cases is excluded. In those cases, when metacercariae do not encyst, they may pass to active histophagy after formation of the digestive system has been completed (see p. 181).

However, there is a different interpretation of the absence of cyst formation in all representatives of the blood digeneans. The absence of cystogenous apparatus and incapability of metacercariae to form cysts are encountered in most trematodes with genuine furcocercariae (Strigeidae, Diplostomidae, Gymnophallidae). There are good reasons to believe that, in this case, one deals with a plesiomorphic character, inherited from ancestral forms (see p. 498), rather than with a secondary lost of the cyst.

Thus, addressing again the problem of the nature of heterosexual (Schistosomatidae) and hermaphroditic (all the remaining) individuals of the discussed trematode groups, we suggest that they should be considered as neotenic (progenetic) forms. In all these specialized blood digeneans the emergence of dixenous life cycles must have been accompanied not only by a loss of the genuine definitive host (shortening of pathways of parasite transmission in the ecosystem), but also by a reduction of the life cycle itself due to the elimination of the phase of the primary marita.

While the interpretation of the life cycle of obligatory blood digeneans can still be argued, the nature of dixenous life cycles of some Plagiorchiida is unambiguous. A very convincing proof is provided by the species *Paralepoderma brumpti*, which completes its development and reaches sexual maturity in the second intermediate host in the metacercarial stage (see

p. 169). However, in this and other similar cases, trematodes experience certain difficulties with their dispersion, since metacercariae, localized in the body cavity of the host, are encysted and, moreover, are often encapsulated. In this case, dispersion of the infective eggs in the environment can proceed in two ways.

First, if the host dies (see fig. 143,Vd). However, this way can be efficient only if the host's life span does not exceed the time needed by the parasite to reach sexual maturity. In the case of *P. brumpti*, whose metacercariae develop in amphibians, this way can not be reached in nature. In large frogs *Rana ridibunda* and *R. esculenta*, often we could observe encapsulated calcificated larvae with already non-infective eggs

The second way of dispersion requires the presence of a non-specific predator, which would eat the host, infected with sexually mature metacercariae (see fig. 143,Ve). Trematodes do not live in such a host, and eggs, released after their death, transiently pass the digestive system of the casual host-predator and pass outside.

Two more versions of dixenous life cycles are realized only due to shortening of the pathways of parasite transmission resulting from the loss of the second intermediate host. Functions of the eliminated host pass to the first intermediate host (mollusc) or definitive host.

The most typical case of the first version is rather common in trematodes and can be encountered in representatives of various groups (Microphallidae, Brachylaimoidea) (see fig. 143,Va,Vb). As a rule, these are representatives of highly specialized families, adapted also to conditions, which are extreme for trematodes (intertidal zone, different types of terrestrial ecosystems). This factor alone is sufficient to exclude any possibility of active exogenous accumulation. As mentioned above, the infection of molluscs proceeds passively, when ingesting the eggs, which often possess solid protective envelopes and contain fully formed infective larvae.

At the earliest stages of endogenous agglomeration in the mollusc, the type of specialization of these digeneans is already well-developed. The most pronounced changes occur in the mother sporocyst. Miracidia, with all its typical characters, are retained everywhere, whereas the parasitic phase of mother sporocyst development (considered as an independent organism) is, apparently, very often secondarily lost. At any rate, in representatives of the aforementioned families, only germinal elements (germinal mass, separate germinal cells), directly producing individuals of the second (last) parthenogenetic generation, but not the entire organism, gets into the mollusc (see. p. 146). But even in this case, several variants may exist. In microphallids this germinal material is capable of active proliferation, resulting in a great amount of forming daughter sporocysts (during intensive infection, practically the whole hepatopancreas can be impaired by parasites). In the most advanced Brachylaimoidea, germinal cells inoculated into the mollusc give rise to a corresponding number of sporocysts. However, when growing, the sporocysts branch intensively and infiltrate the entire visceral sac (see p. 70). The parasite group, which is being formed in the mollusc, cannot be regarded as a genuine microhemipopulation (see p. 366), for the individuals, composing it, totally lose their capacity for self-reproduction. In fact, the single generation of sporocysts (or even one sporocyst) produces larvae of the next hermaphroditic generation, rather than a generation of parthenitae. Still remaining in the sporocyst, the larvae keep on developing, so that corresponding phases could be identified. This is well-expressed in microphallids, in whose sporocysts cercariae are initially formed. This is supported by the presence in the larvae of a rudimentary, though distinct tail. The shedding of the latter marks the transition from cercaria to metacercaria, which after a period of morphogenetic transformations, passes to the dormancy.

The period of proliferation of germinal material in such sporocysts being limited in time, replenishment of the group of young larvae of the individuals of the next generation ceases. Thus, eventually, younger individuals match in their development the older ones and all larvae become functionally the same age. They all are infective and are at the same morphogenetic stage. In this state, the larvae can remain for a long time inside the sporocyst which produced them. In fact, accumulation of the infective stages of the next generation (metacercariae) takes place in the host due to reproduction of preceding generations (parthenitae). Once in the definitive host, metacercariae immediately start maritogony. Thus, parasite development in the mollusc ends in accumulation of metacercariae, rather than in dispersion of cercariae. As this process has nothing to do with the entry of larvae into the host from the external environment, we call it endogenous accumulation, as suggested by Dogiel (1947, 1964). Initially being a first intermediate host, the mollusc, functionally becomes a second intermediate host, providing transmission of the infective stage to a definitive host along the food chains.

The aforementioned trend to suppressing the endogenous agglomeration is manifested in representatives of phylogenetically very distant taxa, e.g., Microphallidae and Brachylaimoidea. The mollusc's function as a host in the completion of the trematode life cycle having significantly changed (see pp. 474-7), the ratio of the two stages of parasite development in it also changes notably. The main role goes to endogenous accumulation of the larvae, which are infective for the definitive host. Endogenous agglomeration is preserved only to an extent, which is necessary for accumulation of the maximum possible number of metacercariae.

We have considered above the most specialized obligatorily dixenous life cycles as examples. In the course of their evolution, metacercariae completely lost the capacity of encystment. Some of them even acquired a well-expressed tendency to progenetic development (e.g., microphallids of "*pygmaeus*" group). More and more prolonged periods of marita morphogenesis are shifted to the time of metacercarial formation. However, among representatives of various groups of digeneans rather frequently there are species with obligatorily dixenous life cycles, retaining a number of archaic features, which testify to the fact that these species originated from the forms with typical three-host life cycles.

Very demonstrative in this respect are the Microphallidae, where dixenous life cycles are widely spread (see fig. 77B,C). In more archaic species with dixenous life cycles the development of individuals of the hermaphroditic generation in the mollusc, still retains all phases of ontogenesis of species with trixenous life cycles. In the beginning, a typical cercaria with a complete set of provisional structures, originally characteristic of that phase of development is formed (see p. 171). They then encyst without leaving the daughter sporocyst. Finally, a typical metacercaria, corresponding in its development to the larva, developing in the second intermediate host, is formed within the multi-layered cyst (see p. 171).

In representatives of the second group of microphallids with dixenous life cycles a clear trend to simplification of the cercariae is seem. Primarily they lose provisional structures, which allow larvae to exist in the environment and thus infect the second intermediate host. However, they still retain the capacity of encystment and, therefore, the cystogenous apparatus. The aforementioned microphallids of the "*pygmaeus*" group comprise the third group of species, strongly influenced by transition to dixenous life cycles.

Once again we considered microphallids in detail in order to emphasize two aspects, which we consider to be of great importance. First, reduction in the number of animal hosts, involved in parasite transmission, always precedes any rearrangement of the life cycle itself. The pattern of parasite development is more conservative than the pathways of parasites' transmission in ecosystems. Second, even within the life cycles of the same type, the events, connected with the parasite development, may differ significantly, which, in our mind, reflects differences in the degree of specialization of species.

In some trematodes, the facultative dixenous life cycle is encountered along with the obligatory one, as in the case of some Echinostomatidae. During completion of the typical trixenous life cycle at certain conditions (especially, unfavourable), shedding of cercariae practically ceases, and newly formed larvae are encysted in the mollusc, without leaving it[vii].

A very peculiar version of facultative dixenous life cycle is encountered in the plagiorchiid trematode *Opisthioglyphe ranae*. Normally the life cycle of this parasite of the frog intestine involves three hosts: molluscs, tadpoles and adult amphibians (see fig. 103). Large frogs, *Rana ridibunda*, are infected, eating tadpoles of the same species infected by metacercariae. In frogs of other non-cannibal species the life cycle becomes dixenous (Fig. 115). Cercariae actively infect tadpoles. Encysted metacercariae are predominantly localized in the tail. In the course of metamorphosis, when tail muscles are being resorbed, the larvae excyst and migrate into the intes-

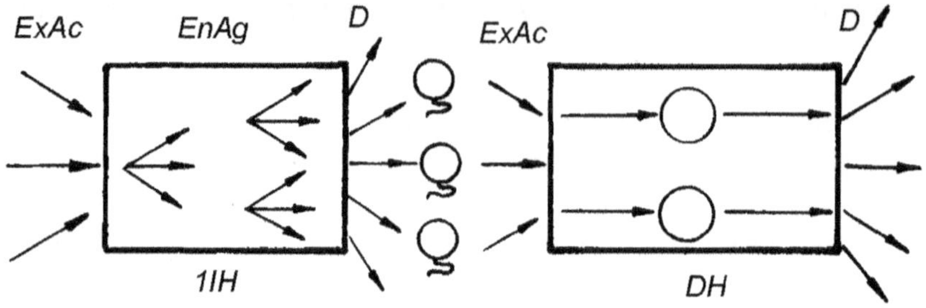

Figure 115. Scheme of the transmission ways of the secondary dixenous life cycle lacking the second intermediate host (*Opisthioglyphe ranae*) (for abbreviations see fig. *102*).

tine, where the regular process of maritogony takes place. In this case no reduction of the life cycle occurs – the full program of the development of hermaphroditic generation is realized. Only the number of hosts, explored by the parasite, decreases. The reason is that in this case the vertebrate functions as a second intermediate host and later as the definitive host.

4. HOMOXENOUS (ONE-HOST) LIFE CYCLES

In trematodes, homoxenous life cycles are relatively rare and, unlike dixenous, all of them belong to the same type (see fig. 143,VIa,VIb). They result from reduction of transmission ways of parasites in the ecosystem due to loss of the second intermediate and definitive hosts. The life cycle itself in that case may remain practically complete without undergoing any substantial structural rearrangements. However, it can also be reduced because of elimination of certain stages and mainly of the final stages of ontogenesis of individuals belonging to the hermaphroditic generation.

The first option is best illustrated by the development of *Bunocotyle progenetica* (Hemiuroideae, Bunocotylinae) (Deblock, 1974a; our data). The whole life cycle of this species is realized in mudsnails, *Hydrobia ulvae* and *H. ventrosa*. Exogenous accumulation leading to infection of molluscs is succeeded by limited endogenous agglomeration (reproduction of mother sporocysts and formation of, most often, one redia generation), which is in turn followed by endogenous accumulation of individuals of the hermaphroditic generation. The ontogenesis of the latter is preserved almost completely: developing in rediae are cystophorous cercariae with a caudal chamber and appendages, typical of hemiuroid trematodes. Still remaining in the redia, cercariae are transformed into metacercariae, which not only achieve sexual maturity, but also acquire a whole set of characters, typical of gravid hermaphroditic adults of Hemiuroidea.

It should be emphasized that in the above-considered case, one can not detect any trace of progenesis (premature development). The generative function in *B. progenetica* is accomplished by marita rather than by metacercariae or cercariae.

As has been mentioned above, the onset of both homoxenous and dixenous life cycles can be accompanied by a reduction of ontogenesis of individuals belonging to the hermaphroditic generation, which takes place due to loss of final phases – marita and even metacercariae. However, (see above and p. 181) each case needs a special analysis. Finding of marita-like individuals with eggs in the second or even first intermediate host is usually interpreted as a manifestation of progenesis. Yet, it can not be considered as a reduction of the life cycle itself.

There are publications providing descriptions of sexually mature cercariae belonging to various groups of trematodes, i.e., *Genarchella genarchella*, *Parahemiurus bennettae* (Hemiuroidea), *Sphaerostomum* sp. (Opecoelidae), *Plagiorchis* sp., *Paralepoderma progenetica* (Plagiorchiidae), *Proterometra* spp. (Azygiidae), *Asymphylodora* spp. (Monorchidae) (Szidat, 1956; Biguet et al., 1956; Shahmatova, 1961; Chernogorenko-Bidulina and Bliznyuk, 1960; Anderson and Anderson, 1963; Jamieson, 1966). Some of these trematodes exhibit facultative homoxenous life cycle, e.g., *Genarchella genarchella*. For other species, e.g., *Proterometra dickermani*, it has become obligatory.

Dispersion of eggs may proceed in various ways. Trematodes may benefit from the host's death or the host being eaten by non-specific predator and even from active emergence of infective stages into the environment. The latter situation takes place in *P. dickermani,* whose rediae, together with the enclosed cercariae and eggs, laid in their cavity, leave the mollusc and die, contributing to dissemination of the infective agents. In the aforementioned cases, the sexually mature stages are normally treated as cercariae for two reasons. First, they are always found in molluscs and most often in parthenitae. Second, all of them, with the exception of *Asymphylodora* spp. cercariaeum, have more or less expressed tails. In some cases these specific provisional organs are as well developed as the corresponding structures in actively swimming larvae of closely related forms. In other cases they have already undergone considerable reduction and secondary simplification.

However, the fact that they are located inside parthenitae and possess a tail or its rudiment, does not prove that these are real cercariae. We have already emphasized that it is necessary to thoroughly evaluate the level of general morphological development, achieved by a particular larval stage. This is well-illustrated by *Paralepoderma progenetica*.

The morphology of sexually mature individuals of this species, observed in molluscs, practically fully (not mentioning their tails, which are obviously reduced) corresponds to that of the metacercariae of the aforementioned (see p. 181) closely related species *P. cloacicola* and *P. brumpti*, rather than their cercariae. Therefore, no reduction of ontogenesis of individuals of hermaphroditic generations in *P. progenentica* takes place compared to that of *P. brumpti*. They reach the stage of generative maturity at the same developmental phase, corresponding to a normal metacercaria of the trixenous life cycle of *P. cloacicola*. Compared to maritae of the latter species, sexually mature individuals of *P. brumpti* and *P. progenentica,* which have not completed their morphogenesis, remain morphologically "underdeveloped". These two species completely lack this distinction.

Unlike in *Bunocotyle progenetica*, whose meracercaria phase is well "marked", in *Paralepoderma progenetica* it is just a stage in the continuous ontogenesis of individuals of the hermaphroditic generation, which has secondarily lost its discrete pattern. All the above, apparently, refer to *Plagiorchis* sp., described by Shahmatova (1961).

Therefore, one can judge the real nature of sexually mature individuals in homoxenous life cycles only on the basis of a detailed comparison of their organization with the corresponding developmental phases of closely related species possessing genuine trixenous life cycles.

5. TETRAXENOUS (FOUR-HOST) LIFE CYCLES

There is an alternative trend, which is realized in trematodes more rarely than the reduction of the life cycle and transmission ways in ecosystems. An additional phase appears in the life cycle and, hence, one more host (third intermediate) is involved (see fig. 143,VII). To date such examples are known only in two groups, i.e., the Strigeidae and Hemiuroidea.

Considered changes are not connected with the parthenogenetic part of the life cycle, which remains the same as in all other representatives of each mentioned taxon. The development of individuals of the hermaphroditic generation, however, becomes more complex. One more parasitic phase – mesocercaria (see p. 175) – appears in between the free living cercaria and the parasitic metacercaria. Apparently, morphologic processes, accompanying formation of mesocercariae in strigeids and hemiurids, differ significantly. This is not a surprise, if one considers the fact, that the transition to tetraxenous life cycle in these groups took place independently, whereas phylogenetically they are distant.

Having penetrated into the second intermediate host, strigeid cercariae (genera *Alaria, Strigea*), do not undergo metamorphosis, typical of most metacercariae. Retaining practically all the features of cercarial organization, including all provisional structures, except the tail, they show a noticeable increase in size. This can be interpreted as a complexity of their organization, because it is accompanied by an increase in the number of flame cells. Further development proceeds in the next – third intermediate host, where mesocercariae, having reached the place of their final settlement, turn into infective metacercariae, due to a complex metamorphosis, typical for that stage of strigeid development. The infective metacercariae achieve sexual maturity in the fourth host, which is the definitive one.

A different sequence of events takes place in the development of the hemiuroid trematode *Lecithochirium furcolabiatum* in the second and third intermediate hosts (Matthews and Matthews, 1988b). Here, the main morphogenetic changes of the parasite, connected with the development of ecsoma (see p. 182), occur just in the second host. In the third host, morphogenetic changes are minimal – metacercariae are rapidly encapsulated and pass to a resting survival stage. Functionally such a host corresponds to the so-called paratenic hosts, often used by certain cestodes and acanthocephalans. Parasites do not undergo any noticeable development there, but are often accumulated in large numbers, which increase the block accumulation of infective stages in definitive host.

However, it is impossible to associate the third host of *L. furcolabiatum* with paratenic hosts of other parasitic worms. The paratenic hosts are facultative, but in some cases when they fill ecological gaps in the parasite transmission (e.g., in the case of *L. furcolabiatum* when infection can be transferred from the second intermediate host, plankton copepods to the definitive one, carnivorous fishes only by means of third host, small plankton-eating fishes), their participation in the completion of the life cycle, in fact, becomes obligate. In *L. furcolabiatum* the third host is obligate and, therefore, should be considered as a real intermediate host.

Hemiuroidea also demonstrate a life cycle with a true paratenic host (Zelmer and Esch, 1998). In ostracods, *Halipegus occidualis* larvae develop into metacercariae, infective for definitive host (green frogs). Odonate naiad, where no further development of the metacercariae is observed, serves as a paratenic host. Both the metacercariae from odonates and the metacercariae from ostracods are infective to the frog definitive host.

Concluding a consideration of tetraxenous life cycles, one should emphasize that the emergence of one more host in digenean transmission, is not accompanied by additional proliferation of parasites, or by an increasing number of disseminations of infective stages in the environment. From this viewpoint, the addition of the mesocercaria phase and, hence, of one more (third) intermediate or paratenic host can be interpreted as a quantitative rather than qualitative modification of this life cycle. The biological meaning of this modification has been analyzed in chapter 5 (see p. 471).

---

[i] The exceptions, discussed below, are rare and provide material of interest for the analysis of the evolution of a complex biological phenomenon such as the "typical" digenean life cycle.

[ii] Hereafter we will use graphic interpretation of life cycles suggested by Dogiel (1947, 1964).

[iii] In modern parasitological literature terms "mono-, poly-, homo-, heteroxenous parasites" are not always correctly used. Parasites, whose life cycles involve only one host, should be called homoxenous parasites. If the development of parasites requires a sequence of hosts, replacing each other, one should use the term heteroxenous parasites. One or several intermediate hosts and definitive host form the so-called metaxenous series. These are metaxenous hosts. However parasites of the same species (or sometimes one and the same phase of the life cycle of particular parasite species) can use different species of animals as hosts. If the parasite species (or some stage of the life cycle) is of narrow specificity and develops in one host species, one deals with monoxenous form. If the species is widely specific and can use animal-hosts of different species (paraxenous series, paraxenous hosts), it should be treated as polyxenous form (for review see: Granovitch, 1999).

[iv] Naturally, we are analyzing a situation when the partners of the host-parasite system are mutually specific. If the specificity factor hinders the completion of parasite behaviour program providing the successful infection of the host, the process can be stopped at any stage – finding of the host, its identification, penetration, migration in the host. That is why we suggest that the successful completion of exogenous accumulation implies that the parasite has reached its specific locality in the molluscan host and morphogenetic processes, transforming migrating larva into a young mother sporocyst, have resumed.

[v] The morphological type of parthenitae is of no importance for our analysis. Moreover, we believe that the significance attributed to the occurrence of rediae or sporocysts was exaggerated (Odening, 1961). We share the viewpoint of the authors who suggested that the sporocysts may have appeared several times quite independently, whereas the two parthenite morphotypes are homologous (Ginetsinskaya, 1968; Pearson, 1972).

[vi] The "infective stage" in this case implies only mature miracidia, which have already acquired or, on the contrary, still retain the ability of infecting molluscs. This period is much shorter than the period of time they spend in the environment, which often includes also embryonic development (for details see p. 83).

[vii] The fact, that metacercariae of the same species, as shed cercariae, were found in the mollusc, does not prove the existence of facultative dixenous life cycle. Quite often in the life cycles, involving the mollusc as a second intermediate host, cercariae, having left the mollusc, may immediately penetrate it again and encyst there. Especially often it is observed in the experimental conditions in small jars. There is no reason to consider such cases as a manifestation of facultative dixenous life cycles, since in nature such a self-infection rarely takes place. According to our observations, molluscs spontaneously infected with parthenitae of *Xiphidiocercaria* sp. VII Odening in nature are infected with metacercariae of the same species rarely (1-2,5% at intensity 5-12 larvae per host). After a few days (up to a week) of maintenance in aquarium prevalence with metacercariae reaches 100%, and the intensity varies from several hundreds to several thousands. This is another evidence of how cautious one should be, when interpreting data, obtained in the experimental conditions and extrapolating them to natural conditions.

Chapter 4

SPECIFIC TRAITS OF POPULATIONS FORMED BY TREMATODES

This chapter is concerned with some special traits of formation, functioning, spatial and temporal distribution of the groups formed by different phases of the trematode life cycle. It is these population parameters that appear to us to be the most significant for understanding the ecological evolution of trematodes, i.e., for revealing the causes which enabled this or that species/group of species to penetrate into the ecosystems of a certain type and to successfully complete their life cycles there.

This approach is based on the teaching of Beklemishev (1959, 1960) that the parasites are co-members of ecosystems to the same extent as their hosts, and that the parasite-caused diseases are normal ecological phenomena. Therefore, it is impossible to treat the evolution of parasites, including trematodes, out of the ecological context. Larvae and mature individuals exist in ecosystems not independently but as members of groups formed by the given phase of the life cycle, free-living or parasitic.

1. ON THE NATURE OF TREMATODE POPULATIONS

Before presenting the material, some terminological problems should be discussed. First of all, it concerns the terms "phase" and "stage", which are often equated. We consider a phase to be a larger division than a stage. It defines a significant span of the life cycle[i], essentially different both from the previous and from the following one, and passing, as a rule, in a different biotope than the previous and the following phases. For example, in the case of benthic molluscs and polychaetes, we may speak about the phase of planktonic larva and the phase following the settlement of the young. For trematodes these are the phases of the miracidium, mother and daughter parthenitae, cercaria, metacercaria and marita. The stage characterizes a much less significant span of ontogenesis and the choice of its limits is rather arbitrary. For example, Cheng and Bier (1972) delimited as many as 7 consecutive stages in the development of the cercariae of *Schistosoma mansoni* in the daughter sporocysts, from the germinal cell to the fully developed cercaria.

It is the analysis of the phases that is meant by population analysis of trematodes, each of the phases forming a definite group of intra-population level. The classification of these groups is the second terminological problem calling for a discussion. In the Russian literature, the terms proposed by Beklemishev (1959, 1960) became popular. They were subsequently elaborated (Kontrimavichus, 1982; Galaktionov and Dobrovolskij, 1984; Granovitch, 1999). These terms are convenient to use, which is also due to the fact that they are equally suitable both for free-living and for parasitic animals. Besides, this terminology corresponds with Dogiel's views on the structure of the parasite's life cycle, discussed in chapter 3. However, Beklemishev's classification remained unknown abroad, where an independent classification of parasitic populations developed (Esch et al., 1975; Margolis et al., 1982; Bush et al., 1997). Detailed juxtaposition and analysis of these terminologies, as well as the description of the population structure of parasites, has recently been done (Granovitch, 1999). In the paper (Granovitch, 1999) a detailed glossary is given. This simplifies our task and allows us to dwell on the terminology accepted in this book only briefly.

According to Beklemishev (1959), populations and subpopulations (parts of a large population) have one feature in common: they occupy "certain parts of the Earth's surface, i.e., they are associated with certain biotopes or complexes of biotopes" (p. 216). In contrast to these populational units, micropopulations occupy microbiotopes, which "are not so much parts of the Earth's surface but rather are situated on this surface or close to it" (Ibid, p. 216). It may be dung or corpses of animals, holes and nests, and also, for parasites, separate host individuals. A characteristic feature of micropopulations is their ability to sustain their existence by means of self-reproduction, and their brevity, defined by the brevity of the microbiotope itself. Among parasitic organisms, micropopulations are formed by many arthropods (sucking lice, chewing lice and others), which live on their hosts throughout their lives.

All the animals, including parasites, which pass different phases of their life cycles in different habitats, form hemipopulations, which are the settlements of separate phases of the life cycle (Beklemishev, 1960). Hemipopulations form more or less numerous local hemipopulations, the term used for a group of a certain phase of the parasite's life cycle in an individual host, or for free-living phases in a microhabitat (cercariae in a marine pool, a small pond, etc.) (Galaktionov and Dobrovolskij, 1984). Hemipopulations, in contrast to subpopulations and micropopulations, do not self-reproduce. However, if they do possess the ability to self-reproduction, then the group of self-reproducing phases of the life cycle in a microhabitat (e.g., in a host specimen) should be called a microhemipopulation.

The term hemipopulation corresponds to component population as used by Esch et al. (1975), Margolis et al. (1982) and Bush et al. (1997), local hemipopulation for parasitic phases of the life cycle coinciding with infrapopulation, following the terminology of these authors (for synopsis see: Granovitch, 1999). At the same time, there is no term corresponding to the local hemipopulation for free-living phases of the parasites' life cycle and for the denotation of micropopulation and microhemipopulation. The latter, though, may be called self-reproducing infrapopulation. The dynamics of self-reproducing phase groups of parasites essentially differs from those lacking this ability. In reference to trematodes, this will be discussed later. Here we note that the introduction of a special term for the denotation of such a situation does seem justified. The above considerations led us to accept Beklemishev's classification of intrapopulation groups, which is used in this book.

Each phase of the life cycle of trematodes, which are phase parasites, is, as a rule, a well-defined hemipopulation, consisting of local hemipopulations. In a number of cases true microhemipopulations will be considered. Such groups are characteristic of rediae and daughter sporocysts of the representatives of many families, e.g., Fasciolidae, Notocotylidae, Philophthalmidae, Echinostomatidae, Schistosomatidae, Opecoelidae. Rediae and sporocysts of these forms, parasitic in molluscs, can produce their like, i.e., the generations of grand-daughter, grand-grand-daughter and so on rediae and sporocysts, apart from producing cercariae. A more detailed analysis of parthenita microhemipopulations will be given below.

The life cycle of trematodes, as often noted in the previous chapters, is a regular alteration of ontogeneses of the individuals belonging to parthenogenetic and hermaphrodite generations. Correspondingly, phase groups of these parasites are characterized by their own demographic and spatial structure, which undergoes obligatory temporal changes. As a matter of fact, each such group occupies a certain niche. A sum of niches occupied by separate hemipopulations is the common niche of a certain trematode species.

This reasoning requires some elaboration. The fact is that the phenomenology of the life cycle, expressed in definite transmission pathways, in some cases turns out to be more important for population analysis. As shown above (see p. 341), these circulation pathways do not always correspond to the series of phases. The species *Bunocotyle progenetica* (Hemiuroidea) (see p. 344) is one of the most striking examples. In this species the whole ontogenesis of the individuals of the hermaphrodite generation, from the germinal ball to the stage of gravid marita, takes place inside rediae. In this case it is impossible to delimit the hemipopulations of cercariae, metacercariae and maritae, though all these phases are morphologically distinct (see p. 344).

Here we can only distinguish the hemipopulation of parasites in a mollusc (the phases of mother sporocyst and rediae containing developing and mature specimens of the hermaphrodite generation) and the hemipopulation of eggs with miracidia located in the environment. It seems that the former group should be called a complex, or compound, hemipopulation.

We often come across a similar situation in case of the differentiation of groups of mother and daughter generations of parthenitae. If the former and the latter exist and function in one mollusc at the same time (many Strigeidida, Plagiorchiida, see pp. 36, 43), they have to be united and called a complex, or compound, hemipopulation. However, it should be remembered that this association is artificial, as mother and daughter parthenitae are different phases of the life cycles (see pp. 2, 52).

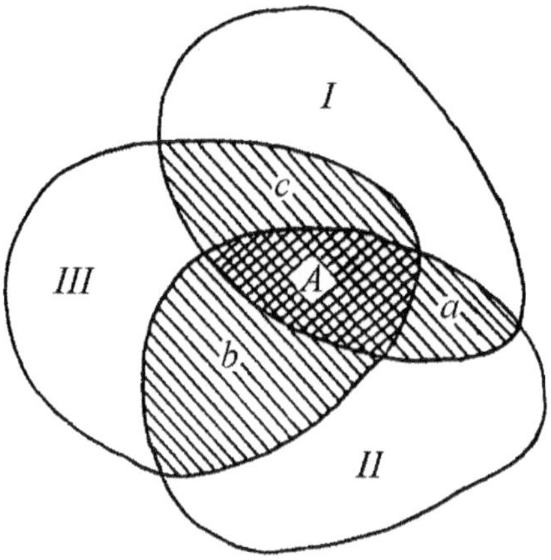

Figure 116. Diagram showing a possible variant of correlation between the distribution area of a parasitic species with a three-host life cycle and the distribution areas of its hosts. *Abbreviations*: *I*, *II*, *III*, distribution areas of the first intermediate, second intermediate and final hosts, respectively; *A*, the distribution area of the parasitic species; *a*, *b*, *c*, zones of occurrence of the parasitic species.

When discussing the spatial structure of the parasites populations, it should be remembered that it is more complicated that that of free-living organisms. This is due to the fact that its formation is influenced by the spatial distribution of the intermediate and definitive hosts (Fig. 116). For free-living animals the zone of occurrence coincides with the species distribution area, excepting the cases of an accidental occurrence. In parasites, the situation is different. The occurrence of the larvae or mature individuals of a parasitic species in a host or its larval stages in the environment in some geographic region does not necessarily indicate that its whole life cycle is completed there. It is the case only in the zone of overlapping of the distribution areas of all hosts involved into the given life cycle (one or several intermediate and the definitive host). This zone is the distribution area of the parasite (*A*). Other parts of the distribution areas of hosts (*I*, *II*, *III*) are only the zones of occurrence for the parasite, where separate phases of its life cycle can be found and/or some of its stages can be completed. The latter takes place, e.g., in case of the three-host life cycle, which is the most common in trematodes, in the regions, which two of the three hosts (*a*, *b*, *c*), necessary for the completion of the cycle, inhabit together.

Actual size of the occurrence zones depends on the motility of the host and its ability to migrate for certain distances from the parasite's distribution area. Maritae and metacercariae in vertebrates (definitive and second intermediate hosts) may occur, in extreme cases, throughout all their distribution areas. In case of parthenitae and metacercariae in invertebrates (first and second intermediate hosts), incapable of long-distance migrations, the zones of occurrence are unlikely to extend far beyond the distribution area borders of the trematode species (*A*). Correspondingly, the chances of the completion of a part of the trematode life cycle are very slim in the overlapping zones of the distribution areas of the first and the second intermediate hosts and the second intermediate and the definitive hosts, but they are very good in the zone where the definitive host meets the first intermediate host.

Besides the distribution of the hosts, a significant role in the delimitation of the distribution area borders of parasites is played by environmental factors, limiting the transmission possibility of free-living larvae. This can be illustrated by the results of the investigation of the distribution patterns of trematode intramolluscan stages in the intertidal snails *Littorina saxatilis* and *L. obtusata* along the Barents Sea coast (Galaktionov and Bustnes, 1999). Despite the presence of the second intermediate (littoral snails, crustaceans and fishes) and definitive (marine and coastal birds) hosts, a gradual disappearance of species with the free-swimming miracidium and/or cercaria (the species of the genera *Himasthla*, *Renicola*, *Cryptocotyle lingua*, *Microphallus similis*, *Maritrema arenaria*) is observed in the eastward direction (in this direction the climate acquires more and more pronounced arctic features). In the arctic intertidal zone of the islands Novaya Zemlya and Vaygach, the snails investigated were infected almost exclusively by the microphallids of "*pygmaeus*" group, which lack actively swimming miracidia and cercariae in their two-host life cycle.

Individuals belonging to different phases, forming separate hemipopulations, are distributed unevenly on the territory occupied by a trematode population. They form spatially separated groups, and it is these groups that are local hemipopulations or microhemipopulations. This is true both of free-living phases (developing eggs, miracidia, cercariae and adolescariae) and parasitic ones (parthenitae, metacercariae and maritae). The uneven distribution of the former is determined, on the one hand, by the heterogeneity of environmental conditions, and, on the other hand, by the unevenness and discreteness in the dissemination of infective stages by animal hosts. As a result of a joint action of these factors, even in small areas (of a water basin, intertidal zone or land), the zones of high concentration of trematode larvae alternate with the zones of their total absence. As for parasitic phases, there the isolation of separate groups is pre-supposed by the discrete nature of the environment itself, represented by separate individuals of animal hosts.

2. HOST-PARASITE INTERACTIONS AND THEIR MANIFESTATION ON POPULATIONAL LEVEL

For the objectives of populational analysis, it seems very convenient to classify the nature of the influence exerted by the parasite on its host. To do so, it is not necessary to know the definite mechanism of the influence. It is quite sufficient to define its external manifestations, i.e., what happens to the host infected with a given intensity by this or that parasite (Galaktionov and Dobrovolskij, 1989). Adopting this approach, all parasitic influences may be subdivided into three large groups: those resulting (directly or indirectly) in the host's death, those influencing its reproductive function and those altering its phenotype (Galaktionov, 1993). The former two groups provide information about the abundance and demography of the hosts' populations, the latter, predominantly on their spatial-temporal structure. It should be noted from the very beginning that this division is arbitrary, as the borders between the groups are blurred. Thus, the influence causing the weakening of the host's organism can by this fact alter its behaviour and make reproduction processes impossible. Disruptions of the host's behaviour induced by parasites often result in the increased accessibility of the infected animals to predators. More examples of this kind can be given. Yet the classification suggested seems to be of utility as it makes it possible to circumscribe more precisely the sphere of the ultimate causes of the phenomena encountered during the analysis of host-parasite interactions at the population level. Below we present their brief characteristics, using examples mainly from the trematodes.

1. Influences on the survival of the host. This is a vast group, within which several gradations may be outlined, in their turn. *The first* includes the influences that normally result in the death of the host. They are caused by the life activity of those parasites for which the death or the illness of the host is a necessary or favourable condition for passing to the next phase of the life cycle (e.g., myxozoan and *Ligula* infection in fishes, echinococcosis in mammals).

Into *the second* gradation we group the influences resulting in the death of the host if the balance of the host-parasite system is disrupted. For most parasites the host's death is "not advantageous" as it results in their own elimination, and thus in the shortening of reproductive period or impossibility to continue the life cycle in case of the intermediate host. Infection by such parasites normally does not cause any noticeable disorders of the host's activity. However, if the infection is very intensive, pathogenesis increases, which takes place in case of epizooties. The latter usually occur in conditions of increased abundance and/or concentration of the hosts due to natural or anthropogenic factors (for reviews see: Lauckner, 1984, 1985; Karpovich, 1987; Bustnes and Galaktionov, 1999; Bustnes et al., 2000). Cases of acute course of the disease in hosts new for a given parasite should be attributed to the same gradation. Mass death of salmon young in the rivers and fish farms in Norway due to the invasion of *Gyrodactylus salaris* may serve as one of the latest examples of this kind (Johansen and Jensen, 1986). This monogenean invaded there from the Baltic Sea basin, where it is a typical component of the parasitic fauna of salmon and does not exert any pronounced pathogenic influence on the fish.

The third gradation comprises the influences altering the host's resistance to environmental factors. This seems to be the most widely spread variant of parasitic influence. It can be clearly observed in animals from ecosystems with greatly fluctuating environmental conditions, e.g., the intertidal zone. A lesser resistance of littoral snails infected by the trematode parthenitae, as compared to non-infected individuals, to desiccation, extreme temperature and salinity fluctuations was described by many authors (Vernberg and Vernberg, 1963; Tallmark and Norrgren, 1976; Berger and Kondratenkov, 1974; Berger, 1976, 1986; Sergievskii et al., 1986; Galaktionov, 1990).

However, data of another kind have been accumulating recently (Sousa and Gleason, 1989; Galaktionov, 1990, 1993; Jensen et al., 1996; Berger et al., 2001). Experiments with marine littoral snails *Littorina saxatilis* have shown that the survival rate of non-infected individuals and those infected by the parthenitae of the microphallids of the "*pygmaeus*" group in conditions of desiccation and extreme temperatures is approximately the same (Galaktionov, 1990, 1993; Berger et al., 2001). Moreover, infected snails turned out to be even more resistant to freezing, and in some experiments, also to exposure to high temperatures. At the same time, their increased mortality rate, as compared to non-infected specimens, was observed when they were kept in fresh water. Taking into account that the resistance mechanisms in molluscs to different factors vary greatly (Berger, 1986; Berger and Kharazova, 1997; Sokolova et al., 2000a,b), it may be suggested that trematode parthenitae exert a differential influence upon them.

The nature of interrelations in a particular host-parasite system also plays an important role. The degree of pathogenic influence exerted upon the molluscan host depends on many factors. Some of them are associated with its own life-history (e.g., life span, character of reproduction), whereas others are determined by the characteristics of migration, localization, growth, development and feeding of parthenitae, the absence or presence of free-living cercariae, the ways of cercariae migration in the host organism and the rhythm of their emission. The data accumulated to date lead us to conclude that the trematodes with the phase of free-swimming cercaria in their life cycle exert a greater negative influence on the survival rate of the molluscan hosts (Meuleman, 1972; Lo, 1972; Christensen et al., 1980; Makanga, 1981; Woolhouse, 1989; Zakikhani and Rau, 1998; Jokela et al., 1999; Berger et al., 2001; Galaktionov et al., 2002). It should be kept in mind that the resistance of infected molluscs to environmental factors may fluctuate depending on their physiological state. Thus, on the Barents Sea coast mass death of periwinkles infected by microphallids of the "*pygmaeus*" group is observed once a year – in May-June, when the snails, weakened after a long winter starvation, begin an active life (Galaktionov, 1992).

II. Influences on the reproductive function of the host. It has already been mentioned that the parasite-induced weakening of the host may lower its reproductive potential or completely exclude it from reproductive processes. Parasites, whose development in the host results in its full or partial castration, are in a special category. The phenomenon of parasitic castration is very widespread. It is treated as a strategy of parasites, aimed at providing better conditions for their survival in the host and the completion of their life cycle (Kuris, 1974; Baudoin, 1975; Dobson, 1988). This effect does decrease or completely exclude the reproductive efforts of the host, thus releasing considerable energy resources which are used by the parasites.

Parasitic castration of molluscan hosts is a frequent result of the development of trematode parthenitae (see reviews: Ginetsinskaya, 1968; Wright, 1966b; Thompson, 1990, 1997; Thompson and Kavaliers, 1994; Poulin, 1998; Gorbushin, 2000). It may be provoked by mechanical (rediae feeding on generative cells, tissue necrosis resulting from the pressure of growing parthenitae) or chemical causes. In the latter case, parasites secrete certain substances which act negatively upon the reproductive function of the molluscan host, i.e., inhibiting spermatogenesis or oogenesis (Pearson and Cheng, 1985; Sullivan et al., 1985; Thompson and Kavaliers, 1994).

In recent years the data on the hormonal basis of reproductive inhibition in infected molluscs have been accumulating (Schallig et al., 1991; De Jong-Brink, 1992, 1995; Thompson and Kavaliers, 1994; Thompson, 1997). In the hemolymph of *Trichobilharzia ocellata*-infected *Lymnaea stagnalis* the presence of a peptide, schistosomin, was reported. Schistosomin is produced in the snail's central nervous system in response to parasitism and interferes with the host's neuroendocrine system. It suppresses the stimulating action of several hormones on the albumin gland, the organ synthesizing a liquid, rich in nutrients, in which the molluscan eggs and embryos develop. It is of interest that schistosomin production coincides with cercarial maturation in daughter sporocysts.

Increased resistance or survival rate of the trematode-infected molluscs, described by some researchers, is regarded as a consequence of parasitic castration (see reviews: Dobson, 1988; Galaktionov, 1993; Berger et al., 2001). Numerous authors tend to consider this phenomenon as a special means of "managing the resources" of the mollusc, which turns out to be mutually beneficial both for the parasites and the hosts: the former get intensive feeding, and the latter, a chance to reproduce (Minchella, 1985; Dobson, 1988; Théron et al., 1992a; Théron and Gerard, 1994).

This event may work in different ways. In some cases, which seem to be quite rare, the mollusc regenerates normal functioning of the gonads after the death of the parthenite group parasitizing in it (Minchella, 1985; Dobson, 1988). Sometimes an abrupt increase in fecundity during the early stages of infection can be observed, which is followed by parasitic castration (Minchella and Loverde, 1981; Minchella et al., 1985; Thornhill et al., 1986; Schallig et al., 1991; Théron et al., 1992a). A high resistance to trematode infection was noted in the periwinkles *Littorina littorea* in the first reproductive season (Robson and Williams, 1971). It is these individuals that mostly ensure the reproduction of the periwinkles, older snails being heavily infected by different trematode species (Robson and Williams, 1971; Hughes and Answer, 1982; Lauckner, 1987). In bivalves *Transennella tantilla*, partial parasitic castration of females was seen. Its degree is determined by the intensity of infection by a monorchiid trematode *Telolecthus pugetensis* (Kabat, 1986), which is exponentially connected with the age of infection. Because of this, *T. tantilla* can reproduce, in spite of almost 100% infection with *T. pugetensis*. Diverse mechanisms, compensating for reproductive losses due to parasitic castration, were found in the populations of the periwinkle *Littorina saxatilis* (Galaktionov, 1985, 1993; Granovitch, 1992; Sokolova, 1995).

The phenomenon of an increased growth of molluscs infected by the trematode parthenitae, so-called gigantism, is also closely associated with parasitic castration. It is supposed to take place by means of an energy release, this energy in a healthy mollusc being spent on reproduction. In the past, the problem of gigantism was often raised in the literature. Some authors (Fernandez and Esch, 1991) tended to view it as an experimental artifact, caused by very favourable feeding conditions of the experimental molluscs. However, in recent years an increased growth of molluscs infected by the parthenitae of some trematode species was also demonstrated in field conditions (Mouritsen and Jensen, 1994; Huxham et al., 1995; Gorbushin, 1996, 1997, 2000; Gorbushin and Levakin, 1999). It is evident that this phenomenon depends on many factors, an important role being that of the life history of the molluscan host (Sousa, 1983; Minchella, 1985; Mouritsen and Jensen, 1994; Gorbushin, 1997, 2000; Gorbushin and Levakin, 1999; Curtis et al., 2000).

An analysis of the data from the literature has led Gorbushin (1997, 2000) and Gorbushin and Levakin (1999) to conclude that gigantism is common among molluscs with medium life span (3-4 years), while it occurs only as an exception in short-lived species (0.5-1.5 years). The growth rate of infected long-lived molluscs (6-27 years) does not change, or sometimes even decreases. However, a certain combination of factors may result in its increase in long-lived molluscs as well. A possible explanation is provided by the analysis of the trade-offs between energy costs of reproduction and the struggle against parasites in molluscs with different life spans (Gorbushin, 1997, 2000; Gorbushin and Levakin, 1999). Energy costs for the struggle against parasites in long-lived molluscs are higher than in short-lived molluscs. The latter invest actively in reproduction, sacrificing the efficiency of reparation mechanisms. Therefore, it may be supposed that the parthenitae of one and the same trematode species will be more pathogenic for them than for long-lived molluscs. Thus, energy released during parasitic castration does not result in an increased growth of short-lived molluscs because of a strongly pronounced pathogenic effect of the parasites. On the contrary, in some cases a slower growth was noted (Raymond and Probert, 1993; Gérard et al., 1993). The antagonism between the same parasitic species and a long-lived mollusc would be less. However, the investment of the latter into reproduction is lower and the energy released during parasitic castration is, as a rule, insufficient for ensuring an increased growth. Gigantism, according to the model proposed, should be observed in molluscs with a medium life span, which agrees well with the facts.

III. Influences on the host's phenotype. These types of parasitic influences have been attracting close attention, especially in connection with their possible role in the evolution of the hosts (for synopsis see Dawkins, 1982; Zuk, 1992; Goater and Holmes, 1997; Hudson and Dobson, 1997; Poulin and Thomas, 1999). In trematodes, the cases of gigantism of molluscs infected by parthenitae and the hypertrophy of tentacles in *Succinea* snails infected with the sporocysts of *Leucochloridium paradoxum* can be seen, as well as numerous cases when trematode parthenitae and larvae manipulate the behaviour of the first and the second intermediate hosts (for reviews see: Lafferty, 1992, 1999; Thompson and Kavaliers, 1994; Poulin, 1994, 1995; Moore, 1995; Lafferty et al., 2000). Two gradations may be delimited within the influences of the third group.

Changes in the host phenotype, non-adaptive for the parasite (at least, seen as non-adaptive at present) fall into *the first* gradation. They appear only as a side-effect of parasitic infections (Poulin, 1995, 1998). Here, the changes of growth rate in molluscs infected by trematode parthenitae (see above) may be recalled, as well as the cases, frequently described in the literature, of the absence of seasonal migrations of infected snails in the intertidal zone (see p. 375).

To *the second gradation* such alterations of the host phenotype are attributed which are favourable for parasites and increase the probability of trophic transmission of the infection to the definitive, or, more rarely, to the second intermediate host. Combes (1991a) suggested a French word *favorisation* for phenomena like these. The influences of parasites in this case may be both specific and non-specific.

In the first case, modifications of the host's behaviour are the result of a direct parasite-inflicted damage of the organs associated with movement, feeding and breathing. Influences of this kind are often exerted by the metacercariae of Gymnophallidae and Echinostomatidae, parasitizing in burrowing bivalves of marine shores (for reviews see: Lauckner, 1983; Thomas and Poulin, 1998). In these molluscs, intensive infection by metacercariae causes pathologies of the foot and/or the siphons. As a result the molluscs loose their ability to burrow and during low tide remain on the surface of the sediment, where they become easy prey for birds. Interestingly, oystercatchers, common at the shores of European seas, do not feed on those *Macoma balthica* molluscs which are lying on the surface of the sediment and are, as a rule, infected by the parthenitae and metacercariae of a gymnophallid *Parvatrema affinis*. They feed on the molluscs buried in the sediment, where they are not as much infected by the parasite (Hulscher, 1973, 1982; Kersten, 1996). It may be that here we are concerned with a case of counter-adaptation of the host, which appeared as a response to the corresponding non-specific influence of the parasites on one of their most important food objects. We have given this example as an illustration of ambiguities in the interpretation of the data on parasite-induced alterations of behaviour. Before postulating their adaptive nature, it is necessary to conduct a thorough investigation of the nature of relations in specific host-parasite systems.

In case of specific parasitic influence, parasites exert a well-directed influence upon the nervous system of their hosts. This seems to be true in numerous instances when the behaviour of crustaceans and insects is altered under the influence of parasitizing acanthocephalan larvae (for reviews see: Moore, 1983; Moore and Gotelli, 1990; Thompson, 1990; Thompson and Kavaliers, 1994). It has been shown that in the host-parasite system "cystacanth *Polymorphus paradoxus* – amphipod *Gammarus lacustris*" the parasite's presence stimulates the secretion of serotonin, a neuromodulator, the increased concentration of which provokes the clinging behaviour, characteristic of this shrimp only during the reproductive period. Infected individuals rise to the upper water layers, where they attach to any object they can find (Helluy and Holmes, 1990). Thus, the chance of them being eaten by the definitive host (ducks) increases. Specific influence on the behaviour of the infected host seems to be wide spread in trematodes as well.

Some examples of this kind will be given in sections 3.2. and 3.4. The mechanism of the influence on the nervous system of the host has been described only for the heterophyid *Euhaplorchis californiensis*. Metacercariae of this species parasitize the gills of the killifish *Fundulus parvipinnis*, which results in the infected fish becoming more conspicuous: they surface abruptly, jerk suddenly, vibrate, bend at right angles, and rub themselves on the substrate (Lafferty and Morris, 1996). The metacercariae secrete fibroblast growth factors, which are known to be associated with neural pathology, including Parkinson's disease in humans (Lafferty et al., 2000).

3. PHASE ANALYSIS OF TREMATODE POPULATIONS

3.1. Hemipopulations of Mother Sporocyst Larvae

The structure of hemipopulations formed by mother sporocyst larvae varies greatly in different trematode groups. This is determined by their morphological-functional organization and biology (see chapters 1 and 2). The adults of the representatives of archaic families and of Strigeidida lay compound eggs containing only the zygote (see, however, the footnote ii in chapter 1). Eggs pass to the outside and undergo their entire embryonation in the environment. On its completion, a fully formed miracidium hatches from the egg and starts an active life. Thus, two independent phases can be clearly revealed, forming, respectively, two hemipopulations: embryonating eggs and free-swimming miracidia.

More specialized forms are characterized by a secondary shift of development to the period when the egg passes the uterine bends of the adult worm. Eggs pass to the environment already with a fully formed larva. Thus, the hemipopulation of embryonating eggs falls out of the phase structure of the population. As for the hemipopulation of mature miracidia, in some species it is represented, as in the previous case, by active larvae which hatch immediately after the eggs reach water (e.g., Philophthalmidae, Cyclocoelidae, Schistosomatidae). In other species (e.g., Notocotylidae, Gymnophallidae, Plagiorchiida, Hemiuroidea), miracidia don't hatch in water but only after the ingestion by an appropriate molluscan host.

The composition of hemipopulations of eggs with developing miracidia turns out to be heterogeneous. This results both from the input of new portions of eggs into the environment (if the definitive host is present in this biotope for a long time) and from the micro-scale heterogeneity of environmental conditions experienced by the eggs. Depending on the combination of the latter, the larvae formation may slow down (e.g., in shady places with cool water) or, on the contrary, accelerate (e.g., in well-warmed shallows). Elimination occurs during the hatching of miracidia, which either form or recruit their own hemipopulation (see below), or as a result of the death of embryos in the egg due to desiccation or washing down to greater depths. On the whole, if there is no recruitment, the hemipopulation of embryonating eggs exists for a limited time, which is determined by a relatively short period of miracidia development (5-20 days).

Hemipopulations of fully embryonated eggs formed by the trematodes with passively infecting larvae are more homogenous, as it is only eggs with mature larvae that are released into the environment. Their infectivity is retained for a long time, months for Plagiorchiida (see p. 239). Desiccation appears to be an important eliminating factor for the individuals forming the hemipopulation of this type. It results in an immediate death of the miracidia contained in the eggs (see p. 240). With time, the larvae loose their infectivity, so the group of eggs with fully formed miracidia contains a certain percentage of individuals unable to infect the definitive host.

Parameters of hemipopulations of free-swimming miracidia are subject to considerable variations. Yet, two large groups can be outlined here, the border between them being determined by the nature of miracidia development (see p. 9). If the larvae are fully formed in the marita uterus (e.g., Philophthalmidae, Cyclocoelidae) or in the definitive host organism (Schistosomatidae), they hatch immediately after getting into water (see p. 220). Here, the peculiarities of dissemination are such that they ensure an immediate output of numerous eggs into the environment (e.g., during the contact of eyes of mammals and birds with water in case of philophthalmids, or when schistosome hosts urinate while drinking). A local hemipopulation of miracidia emerges, all of them of the same age. It is their life span that limits the time of the existence of this group in the environment. The situation may become more complicated under certain conditions (e.g., in crowded watering-places), as the group may consist of the larvae hatched from the eggs that have reached the water at a different time and from different host individuals.

Local hemipopulations formed by the miracidia of the trematode groups in question may be either "ephemeral" or "pulsating". In the former case we deal with, e.g., a local group formed by philophthalmid miracidia which reached a certain area of the water during an accidental stop of the infected bird or mammal. Renewal of such local group at the same place depends only upon an accidental series of events. Groups of the second type are resumed periodically. Thus, animals in tropical countries drink water at permanent sites and at the same time of the day, different for different animals. It is during this time that schistosome eggs get into the water in large numbers, forming a local hemipopulation of miracidia with a heterogeneous age structure (it should remembered that for short-lived larvae even several minutes difference in hatching is significant). They live for several hours, after which the group disappears completely, to resume again during the next water drinking. Naturally, numerous transitions are possible between the two extreme variants discussed.

In another case, when a local hemipopulation of miracidia arises from a group of eggs with embryos of different ages (see above), its age structure is heterogeneous, as new individuals are more or less regularly recruited. In theory, such a population can be imagined to be permanently present in the environment. However, this is unlikely to happen in reality. Egg hatching is triggered by a certain combination of environmental factors (temperature, light intensity – see page 218), subject to temporal diversity at different scales. Thus, when a local group of eggs if sufficiently dense, mass hatching occurs within a limited time and the nature of the miracidia group formed is closer to the "pulsating" type described above.

Finally, we have to discuss what seems to be a quite common situation, when we can only conditionally speak about the presence of local hemipopulations both of miracidia and of embryonating eggs. We refer to the trematodes which increase transmission success by means of the strategy when a long-lived marita produces, in a unit of time, a relatively small number of large eggs rich in yolk (e.g., Fasciolidae, some Echinostomatidae and Strigeidida). In this case, it is often only several eggs, or even one egg that gets into some area of the water. Thus, it is possible to consider only hemipopulations of embryonating eggs and miracidia in an extensive enough part of the distribution area of the definitive host(s), but not separate local groups. It should be noted that the opposite strategy of increasing transmission effectiveness, used by the juvenalized forms, producing many small eggs with formed miracidia during the short life span of a marita (e.g., Microphallidae and Lecithodendrioidea) almost always results in the formation of numerous local groups of eggs with fully formed larvae. This is due to a large number of infective agents which get into the environment during every defecation act of the host. In some species the latter circumstance is to a great degree facilitated by the features of the structure of local hemipopulations of maritae, which will be discussed later.

The spatial structure formation of the groups of mother sporocyst larvae takes place in three stages. The first is associated with the activity and biological features of the definitive hosts. It is this factor that first of all influences the distribution of local hemipopulations on the whole territory occupied by the population of a given parasitic species. High concentrations of infection can be expected in places frequented by the hosts (e.g., feeding and watering places). Therefore, the first stage of dispersion to a great extent determines the focal ("spotty") character of infection distribution. The formed foci of infection actually corresponds to separate local hemipopulations. They may be small, or they may occupy a considerable area and have a complex and dynamic structure.

The second dispersion stage is mostly associated with the impact of abiotic factors, ensuring the dispersion and re-distribution of eggs on a limited territory around the primary "spot". Eggs with developing or mature miracidia move from one place to another under the influence of different factors (most often, water currents). Eggs accumulate in some microhabitats and are carried away from others. A secondary spottiness appears, which is especially strongly pronounced if environmental conditions are highly heterogeneous. In closed sheltered areas eggs with long-lived miracidia survive well and may accumulate in large quantities. Open exposed areas are soon cleared of them by means of elimination (e.g., by desiccation) or the influence of other factors (e.g., washing to greater depth). Thus, the ecological heterogeneity of the environment ensures, on the one hand, infection dispersion, and, on the other hand, its concentration in certain areas. It should be noted that the second dispersion stage is almost absent in the trematodes the miracidia of which hatch immediately after they get into water.

The third stage of dispersion is associated with the activity of the larva itself. Naturally, it is lacking in the trematodes which passively infect the first intermediate host. Spatial distribution of miracidia of different trematode species is determined by their systems of species-specific behaviour (see p. 226). This makes a generalized description difficult and requires the analysis of each separate case. Yet, a discussion concerning some general considerations seems appropriate. All of them refer to the trematodes which form local hemipopulations when the eggs get into water (e.g., Schistosomatidae, Philophthalmidae, Cyclocoelidae). It is the larvae of these trematodes that gave the bulk of the literature data on the miracidial biology.

Within a local group, which in the most simple case is formed by the larvae hatched from the eggs produced by one marita, there are individuals differing by the sign of taxes and kineses, these signs sometimes reversing as the larva grows (see p. 228). As a result, miracidia distribution in the environment turns out to be not monotonous but discrete. A complex and dynamic system of microaggregations (clouds) of miracidia is formed in the area of primary focus, and it is this system that appears to ensure the greatest probability of infecting the first intermediate hosts (see p. 230). Taking into consideration the facts discussed in chapter 2 (see section 1.1.1.3.), it can be supposed that a "dominant" cloud is formed during the primary period of the group existence, which includes individuals with the taxis signs dominant for the species, and much less numerous "satellite" clouds consisting of individuals with another taxes sign (Fig. 117).

It should be noted that the signs of dominant taxes may be determined by environmental factors. For instance, the miracidia of *Schistosoma haematobium* possess a negative phototaxis when the water temperature is above 15°C, which is reversed in the water with lower temperature (Shiff, 1974). In any case, the "dominant" cloud is associated with the water body areas where meeting the molluscan host is more probable. In the above example with *S. haematobium*, the reversal of the dominant phototaxis sign is closely connected with the molluscan host behaviour: the molluscs migrate to the depth during the summer warming of the water and migrate to the surface when the temperature cools to about 15°C (Schiff, 1974). As the group

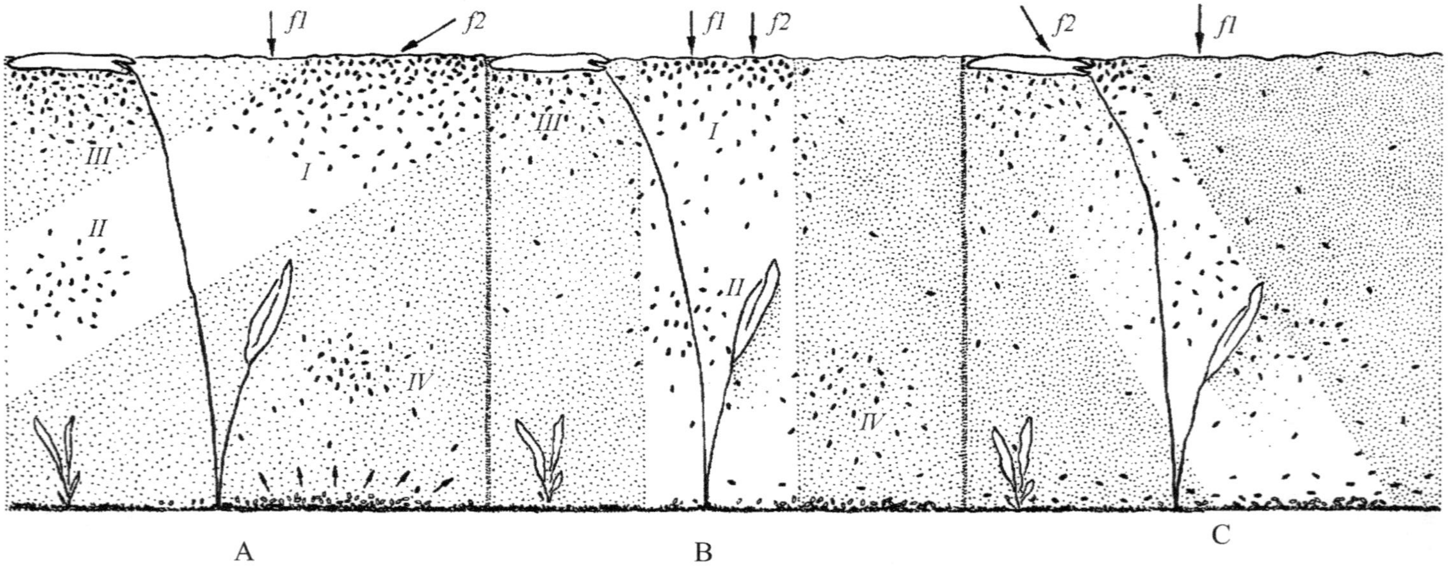

Figure 117. Diagram showing possible dynamics of redistribution of miracidia in a water body. A-B, successive stages of changes in position of miracidial "clouds"; C, disintegration of the primary "clouds" with ageing of miracidia. *Abbreviations*: *f1*, standing factor; *f2*, factor, whose direction and strength change in time; *I*, dominant "cloud", miracidia showing negative response to *f1* and positive to *f2*; *II*, satellite "cloud", miracidia showing positive response to *f1* and positive to *f2*; *III*, satellite "cloud", miracidia showing negative response to *f1* and negative to *f2*; *III*, satellite "cloud", miracidia showing positive response to *f1* and negative to *f2*. Arrows show direction of the factors.

ages and the taxis signs of its miracidia reverse, the borders of the clouds become blurred and the spatial dispersion of the local hemipopulation reaches its maximum. The adaptive nature of such spatial distribution has already been discussed (see p. 230). To conclude, it is noted that the fixation of the strategy of the "satellite" clouds formation by means of natural selection may have been determined by the fact that in the host populations there always are individuals or groups of individuals, which exhibit behavioural reactions different from those of the majority.

It is difficult to say to what extent the above scheme can be used for the description of the spatial structure of miracidia hemipopulations which form from local groups of eggs embryonating in the environment. It is quite plausible in the places where the concentration of the latter is high, taking into account the "trigger" mechanism of the onset of egg hatching. As noted above, the parameters of the local miracidia group formed in this case resemble those described for larvae which hatch when eggs contact water. However, starting a discussion like this we find ourselves on dangerous grounds, since the material which could be used to support or to disprove such theoretical constructions is infinitesimal. Until we have obtained reliable data on the miracidial behaviour of the group in question, their temporal dynamics and other aspects of miracidial biology, further development of the topic started would be speculative.

In the areas with pronounced seasonal variations of the environment, miracidia groups, as is clear from the above discussion, may exist in the environment only in the warm season, and, naturally, if active definitive hosts, disseminating the infection, are present. When it gets colder, hemipopulations of miracidia and embryonating eggs disappear from the biotope. The groups of eggs with mature miracidia, capable of withstanding unfavourable conditions, however, prolong their existence and, probably, are capable of overwintering. In the areas with warmer climate, seasonal dynamics of the abundance of definitive hosts (e.g., overwintering birds) is the main limiting factor (see section 3.5.).

3.2. Hemipopulations (Infrapopulations) of Parthenogenetic Generations

Penetration of a miracidium may be considered as the beginning of the development of parthenite group in the molluscan host. The nature of this process differs greatly in actively infecting larvae and passively infecting ones. But whereas it is almost unstudied in the latter, for the former its dependence both upon the morphological-functional organization of miracidia and upon environmental factors and the state of molluscan host has been shown (see pp. 29, 234). Further success of the penetrated larvae is determined by its mined by its ability to exist inside the host. The first thing it encounters is the molluscan plasma, which may contain substances toxic for the forming sporocyst (Bezerra et al., 1997; Sapp and Loker, 2000a).

The nature of plasma factors remains unclear. They may be a part of the internal defense system of the snail, or they are simply "housekeeping molecules" that are fortuitously toxic to exotic digeneans (Sapp and Loker, 2000a). Proper internal defense system of molluscs consists of both cellular and humoral components. Their role in the determination of the host's immune response to the penetrated digenean larvae is now being actively investigated (for reviews see: Adema and Loker, 1997; Horák and Van der Knapp, 1997). An especially important role is attributed to hemocytes, cellular elements of molluscan hemolymph. The functional significance of these cells is similar to that of monocyte-macrophage lineage of vertebrate immune cells. They can bind to and kill digenean intramolluscan stages by phagocytosing the syncytial tegument, or releasing cytotoxic compounds, or both. Hemocytes appear to be the last filter on the way of the successful development of the mother sporocyst in a given molluscan host (Sapp and Loker, 2000b).

The efficiency of the barriers mentioned can be rather high. These barriers stop not only the miracidia which penetrated by mistake into a non-specific host, or a specific host belonging to a resistant strain, but also a considerable part of the larvae in highly susceptible molluscan strains. The experiments of Ward et al. (1988) demonstrated that only 25% and 50% of the penetrating miracidia of *Schistosoma mansoni* developed into mother sporocysts in snails *Biomphalaria glabrata*, exhibiting the lower and higher rate of susceptibility correspondingly. Théron et al. (1997) showed that 35% of subadult and 75% of adult snails *B. glabrata* of the Brasilian strain were resistant to *S. mansoni* infection. It can only be guessed to what extent these figures characterize the situation in other digenean-snail associations. The processes connected with the resistance of molluscs to trematodes have been investigated almost exclusively in the host-parasite systems formed by the Pulmonata (mostly *Biomphalaria, Bulinus, Lymnaea*) and the parasitizing representatives of the genera *Schistosoma*, *Trichobilharzia*, *Echinostoma*, *Fasciola* and a few others (for review see: Adema and Loker, 1997). It should also be noted that investigations of this kind are usually carried out on a limited number of laboratory strains of molluscs. Therefore it is with great caution that the results of these investigations should be extrapolated to natural populations.

In this connection, the point of view of Shiff (1994) seems to be well-based. He postulated that among compatible host-parasite associations susceptibility in the field is general and the molluscan hosts are actively exploited by trematodes. Molluscan response to trematode infection is to a greater degree shifted to populational level. It is expressed in the increase and development of an *r*-type reproductive strategy with a rapid population growth. In the opinion of Shiff (1994), a response like this requires essentially lower evolutionary costs than the competition for internal resources with the parasites on an individual basis. Besides, this response seems to be universal, as it is in this way that molluscan populations react to the pressure of predators and unfavourable environmental factors. These views are supported by the data on increased individual fecundity of the females of periwinkles *Littorina saxatilis* in the populations heavily infected by microphallids of the "*pygmaeus*" group, as compared to lightly infected ones (Granovitch, 1992; Sokolova, 1995).

In an overwhelming majority of the cases, a single miracidium penetrates into the host, and one mother sporocyst develops in it. In other words, there is no reference to a group at this stage of infection development. At the same time, the application of molecular biological methods in parasitology made it possible to ascertain the possibility of multiple infection of molluscs in the nature. Minchella et al. (1995) showed, using a repetitive polymorphic DNA element, that 57% of the molluscs *Biomphalaria glabrata* from the Brasilian population studied infected by *Schistosoma mansoni* harboured multiple infections, with some snails carrying at least 9 parasitic genotypes. Here we already deal with groups of mother sporocysts.

It should be noted that Minchella et al. (1995) worked in places with a high prevalence of schistosomes in molluscs, varying from 11.4% to 26.0%. It may be supposed that in the water bodies studied by them the densities of miracidia "clouds" were high, which increased the possibility of multiple infections. Sire et al. (1999) found that in the populations of *B. glabrata* from Guadeloupe with low prevalence of *S. mansoni* (0.21-4.76%), most of the infected snails (88.4%) harboured a single recognizable parasite genotype and parasite intensity did not exceed 3 genotypes per snail. Besides schistosomes, the probability of multiple infection of molluscs seems to be rather high in some other trematodes. First of all, this is the case for philophthalmids and cyclocoelids, the formation of miracidia groups in which is similar to that in schistosomes (see p. 361). In some trematodes with passively infecting miracidia, for example, in microphallids, dead maritae full of eggs with invasive miracidia are shed from the definitive host together with faeces. It is clear that if a mollusc eats a worm like this, it will result in multiple infection. This was demonstrated experimentally (Galaktionov and Dobrovolskij, 1985).

The number of individuals in the group of mother sporocysts in the molluscan host may also be regulated in some cases. First, the mollusc may acquire resistance preventing its delayed reinfection as recently demonstrated for the *S. mansoni/B. glabrata* combination (Sire et al., 1998). If a mechanism like this is absent and several miracidia penetrate successfully into the molluscan host, some of the developing mother sporocysts may die within the first few hours following the infection (Pan, 1965; Richards, 1975). Studying the same system *S. mansoni/B. glabrata*, Niemann and

Lewis (1990) observed that in large snails only one mother sporocyst developed after the exposure to 7 miracidia. We observed degeneration and death of a part of stolon-like germinal masses (see p. 46) in the first 5-10 days after plurimiracidial infection of periwinkles by microphallids of the "*pygmaeus*" group (unpublished data).

All this points to the existence of a density-dependent regulatory process. It may be determined by direct interactions between mother sporocysts or/and to be snail-mediated. The presence of such a regulatory process seems to be determined genetically. Théron et al. (1997) demonstrated that when *B. glabrata* was experimentally infected by a Brazilian strain of *S. mansoni*, the average number of mother sporocysts in the mollusc grew exponentially with the increase in the miracidial dose. Thus, there is no any regulation here. At the same time, the molluscs infected by a Guadeloupean strain of the parasite soon demonstrated saturation and further increase of the miracidial dose was not followed by an increased infection intensity. Histological studies showed that "extra" mother sporocysts in this case were encapsulated and dead in 48 h or degenerated without host reactions (Théron et al., 1997).

The parameters of the forming group of the daughter parthenogenetic generation are determined both by the developmental history of mother sporocyst/group of mother sporocysts and by host-dependent factors. The most detailed investigations of interactions of these two components of the host-parasite system were done on the *S. mansoni/B. glabrata* system. In susceptible molluscan strains, the numbers of daughter sporocyst groups and the total production of cercariae depend directly on the initial number of mother sporocysts (Ward et al., 1988). In case of plurimiracidial infection, several mother sporocysts develop in the mollusc, giving rise to a more numerous group of daughter parthenitae than one mother sporocyst in case of monomiracidial infection. Correspondingly, the total production of cercariae by the first group is higher than that by the second group.

Further investigations showed that the numbers of the group of daughter sporocysts and the volume this group occupies in the digestive gland of the mollusc depend not only on the number of miracidia used for infection but also on the initial size of the molluscan host and its growth rate (Théron et al., 1992b; Gérard et al., 1993). In larger (mature) infected snails the volume occupied by the sporocyst group is always larger than in smaller (immature) infected snails. However, in immature snails exposed to a large number of miracidia, the volume occupied by the resulting sporocyst group is significantly smaller than that in case of monomiracidial infection. It seems to be associated with the fact that the growth rate of molluscs exposed to 8 miracidia slowed down significantly in comparison with those infected with 1 miracidium. Correspondingly, the volume of digestive gland which could be colonized by the sporocysts also decreased.

These data suggest the existence of certain regulatory mechanisms in the host-parasite system discussed, which confine the use of host resources to the maximal possible level which does not result in the host's death. Théron et al. (1992b) and Gérard et al. (1993) suggested that two types of regulatory mechanisms can intervene: the density-dependent type, which affects the number of sporocysts through demographic process, and the biomas-dependent type, which affects the size of the individual daughter sporocysts. The former mechanism will be discussed later (see p. 370). The presence of the latter is supported by an increase in the size of individual daughter sporocysts that are developing within larger snails in which more energy is available (Gérard et al., 1993). These larger sporocysts also produce larger cercariae in greater numbers than smaller sporocysts from smaller molluscs (Gérard et al., 1993).

Embarking on the description of demographic processes in sporocyst groups from the molluscan host, it should be noted that, as mentioned above (see p. 352), it is practically impossible to delimit the groups of mother and daughter parthenita generations if they are simultaneously present in the molluscan host. This makes us consider them as a part of a complex microhemipopulation. Formation and functioning of such groups has already been discussed (see pp. 248, 258). Here we are only concerned with aspects which were either not discussed or only mentioned in passing in chapter 2.

One of them deals with the age structure of local groups. In case of microhemipopulations with a determined number of generations – mother and one daughter generation (e.g., Plagiorchiida, Dicrocoeliidae, Gymnophallidae), its structure is determined by the traits of functioning of germinal masses of a mother sporocyst (see pp. 42, 246, fig. 9, IIC). If their proliferative activity is limited in time and the maturing of germinal cells is synchronized, as is the case in some Renicolidae, Plagiorchiidae and Microphallidae, the age structure of daughter sporocysts turns out to be homogenous. It becomes especially close to the age of the group, when the parthenitae lagging in their development either overtake more advanced ones or are suppressed by them. However, in the majority of Plagiorchiida, the reproductive activity of the mother sporocyst is sustained for a long time, which results in the formation of microhemipopulations, the age structure of which is quite diverse. In both cases, the group functions as long as the parthenitae live, if the molluscan host does not die previously.

More complicated is the structure of microhemipopulations with several generations of daughter parthenogenetic generation. At different stages of its formation, such a group may include a mother sporocyst ("founder individual"), parthenitae producing their like (replicating parthenitae), young parthenitae with cercariae embryos, mature individuals "exporting" cercariae and so on – all this in various combinations. The nature of the changes in the structure of microhemipopulations like these is determined by internal and external factors. The production by the parthenitae of their like is one of the most important internal factors, along with the nature of relations in digenean-mollusc association, mentioned above and in chapter 2. External factors may be also associated with the seasonal changes in the environment, which will be discussed later.

For instance, daughter rediae and the following rediae generations of *Philophthalmus rhionica* give rise to rediae in the beginning of their reproductive period (Fig. 118) (Dobrovolskij et al., 1983). As for the number of young individuals recruited to the microhemipopulation, in this instance it is determined by the total abundance of the group and the physiological state of the mollusc. During the period of early microhemipopulation formation, daughter rediae produce almost exclusively rediae too, subsequent generations producing mostly cercariae. Only the youngest individuals, starting reproduction, give rise to 1-2 rediae. Such reproductive dynamics is the cause of clearly expressed age differences in the parthenita microhemipopulation. It always includes both young and old individuals, belonging to different generations.

The dynamics of sporocyst microhemipopulation of *Schistosoma bovis* resembles the one described above (Touassem and Théron, 1986). At the initial stages of its formation, a certain part of daughter individuals (Sp II), produced by the mother sporocyst, give rise only to the next sporocyst generation (Sp III) (1). Another group of Sp II starts to produce cercariae, at the same time producing a certain number of Sp III (2). Finally, inside the rest of Sp II, first cercariae are formed, and then this process is arrested, the remaining cercariae embryos degenerate, and the sporocysts begin producing Sp III (3). Thus, there are three variants of parthenitae producing their like, the first evidently ensuring the fastest expansion in the host organism during the formation of the group, with its further sustenance provided for by the 2nd and the 3rd variants.

In *S. haematobium*, in the course of initial formation of microhemipopulation, the 1st variant is accomplished, followed directly by the 3rd (Kechemir and Théron, 1980). In *S. mansoni*, the local sporocyst group is reinforced only according to the third variant at all stages of its existence (Théron and Jourdane, 1979). This process is regulated in a certain way, so that the periods when the bulk of the sporocysts produce cercariae "for export" (during this time, cercariae emission from the mollusc reaches its maximum) are regularly alternated by the periods when the group composition is dominated by degenerating individuals, or by those reconstructing their reproduction according to the 3rd variant (Fig. 119) (Théron, 1981a,b). Unfortunately, the examples given are, in fact, all that we have in the way of full descriptions of age dynamics of the microhemipopulation type discussed from the moment of the molluscan infection. This does not allow us to make any far-reaching conclusions about the traits of these processes in different trematode groups.

The spatial distribution of parthenita hemipopulations is heterogene-

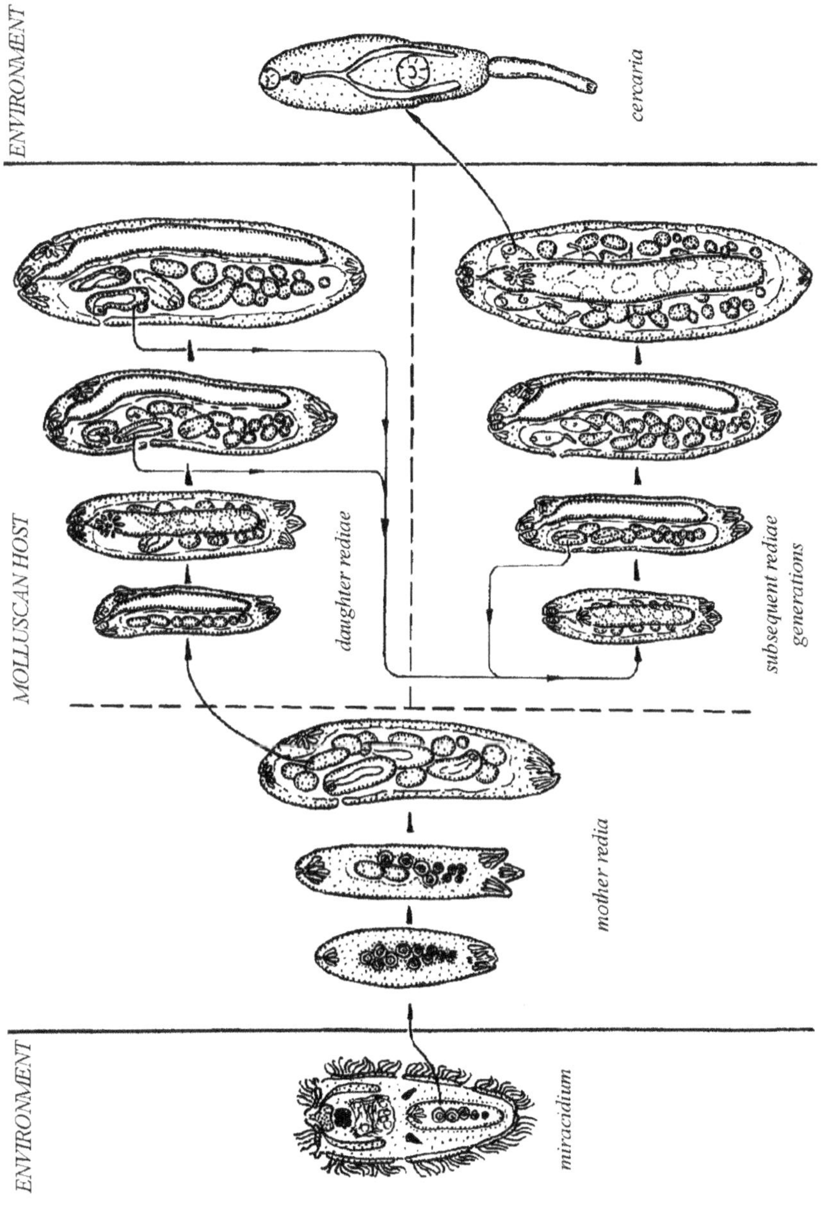

Figure 118. Diagram showing succession in development of rediae generations in *Philophthalmus rhionica*. From Dobrovolskij et al. (1983), with permission.

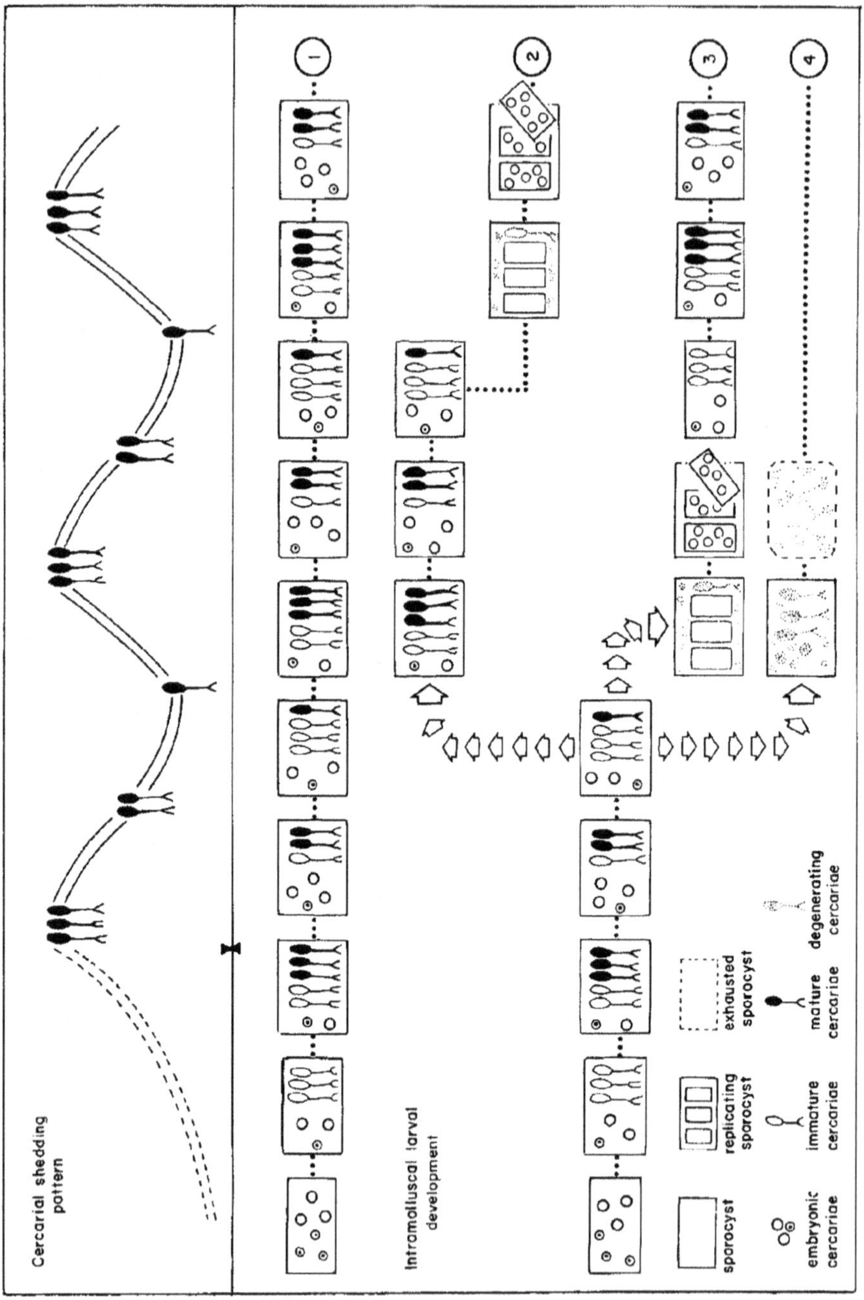

Cercarial shedding pattern
Intramolluscal larval development
1
2
3
4
sporocyst
replicating sporocyst
exhausted sporocyst
embryonic cercariae
immature cercariae
mature cercariae
degenerating cercariae

ous. This heterogeneity expresses itself in the existence of foci of the molluscan infection by their local groups (Ginetsinskaya and Shtein, 1961; Curtis and Hurd, 1983; Galaktionov and Dobrovolskij, 1986; Kuris, 1990; Sousa, 1990; Williams and Esch, 1991; Huxham et al., 1995; Galaktionov and Bustnes, 1995; Granovitch and Johannesson, 2000). This "spottiness" is caused by two kinds of factors. On the one hand, the distribution of hosts themselves is uneven. The density of molluscs may be different even within vast continuous populations, often formed by them, to say nothing of the fact that the parasite's population may cover several isolated populations of the first intermediate host. On the other hand, the distribution of infection in the environment – local miracidia hemipopulations, eggs with developing and mature larvae – is also "spotty" (see p. 362). Accordingly, these foci are retained for a long time even in the areas with an even distribution density of the first intermediate hosts, as their mobility is limited.

The spatial heterogeneity of hemipopulations of trematode parthenitae has its macro- and microscale components. It is associated with the fact that environmental factors as well as biological peculiarities of different phases of definite life cycles and the hosts involved in their completion specifically manifest themselves both on large territories and on rather small areas. As an example, it seems convenient to discuss the distribution of infection by trematode parthenitae in littoral molluscs from the northern seas, which have been studied by us (see reviews: Galaktionov and Dobrovolskij, 1986, 1989; Galaktionov, 1993; Galaktionov and Bustnes, 1995).

It was discovered that the maximal density of parthenita hemipopulations in littoral molluscs is reached in sheltered inlets, where the definitive hosts of the parasites, sea birds and the young of the fish, concentrate. Hydrodynamics is also much less pronounced there, which favours the survival of hemipopulations of eggs with developing and formed larvae and promotes, as far as it can be in severe conditions of the northern coast, a successful infection of the hosts by local miracidia groups. In the more exposed areas, parthenita hemipopulations are significantly dispersed, many species being simply absent, especially those with the phase of free-swimming miracidia in their life cycle (e.g., echinostomes of the genus *Himasthla*). The same picture is observed in the areas of seabird cliffs, where, on the one hand, the concentration of definitive hosts reaches its maximum, but, on the other hand, the exposure to wave action is strong, as the colonies of sea birds are situated on rocky open islands. As a result, the average prevalence of such widespread bird parasites as *Cryptocotyle lingua*, *Microphallus similis*

Figure 119. Model for development of sporocyst microhemipopulation of *Schistosoma mansoni* in *Biomphalaria glabrata*. *Abbreviations*: *1*, the daughter sporocyst produces several generations of cercariae; *2*, *3*, the daughter sporocyst ceases production of cercariae after the 2nd or 1st generation and produces the next sporocyst generation (Sp III); *4*, the sporocyst degenerates. From Théron (1981b), reproduced with the permission of the Liverpool School of Tropical Medicine.

and others in molluscs does not greatly exceed there the values recorded for coastal areas far from the colonies. This is the macroscale.

However, even on a small island with a bird colony the prevalence in *Littorina saxatilis* fluctuates between 0.23% in an exposed rock crevice and 6.89% in a more sheltered locality (Podlipaev, 1979). Moreover, even at a certain site of the shore the density of parthenita hemipopulations turns out to vary at different littoral levels, these differences being quite significant. For example, on the same coastal area of Malyi Aynov island (the Barents Sea), the prevalence of *M. similis* in *L. saxatilis* in the lower level was 58.6 ± 4.83%, further up the intertidal zone 9.7 ± 3.07%, then 3.9 ± 1.02% and, finally, 0% at the supralittoral edge (our data).

The vertical heterogeneity of the prevalence of digeneans in littoral molluscs was first noted by James (1968), who supposed that its nature was defined by the presence of free-living larvae in the life cycles of the parasites. The species with free-swimming miracidia and cercariae can be found in the molluscs from the lowest levels, those with free-swimming cercariae and passively infecting miracidia, in the middle levels, and those with no free-swimming larvae at all, in the upper levels. James' hypothesis seems to be true only in its first part. Indeed, the species with the phase of the free-swimming miracidium in their life cycle (e.g., the above-mentioned representatives of the genus *Himasthla*) are usually associated with the lower levels of the intertidal zone. The period of desiccation here is shorter, and the change of environmental conditions in the course of the tidal cycle is not as drastic as at higher levels. Therefore, the probability of the mollusc's infection by a delicate short-lived larva such as a miracidium is greater. At the same time, the very fact of presence or absence of the phase of free-swimming cercariae in the life cycle as such cannot influence the distribution of the parthenitae hemipopulation density relatively to the littoral level (given, of course, that the parasite does not modify the behaviour of the infected mollusc – see below). For example, at many coastal sites of the Barents Sea the maximal prevalence of *Microphallus pygmaeus*, which has neither the phase of free miracidium nor that of free cercaria, in periwinkles is seen at the lower littoral level (our data). This is due to the fact that it is there that the density of local hemipopulations of eggs with miracidia seems to be the highest, which is, in its turn, associated with the biology of the definitive host of the parasite – the common eider.

The microscale spatial structure of parthenita hemipopulations of the littoral trematodes with the phase of the passively infecting miracidium in their life cycle is determined by a joint action of numerous factors, both biotic and abiotic. The spatial heterogeneity of molluscan host populations also should not be underestimated: individuals of different sex and age, polymorphic by different genes prefer different microhabitats. Besides, the molluscs tend to aggregate in places with the most favourable microclimate, where shelter is available and food is abundant. All this results in the formation of an extremely heterogeneous microscale structure of parthenita hemipopulations, their density being different even on different substrates of the same level. The data on the microphallids of the "*pygmaeus*" group provide a demonstrative example (Fig. 120). In this instance, however, a certain role in the formation of the spatial structure of hemipopulations is also played by the parasite-induced alterations in the behaviour of infected molluscs (see pp. 359, 396).

The importance of this latter factor for the dynamics of the spatial structure of parthenita hemipopulations is as yet difficult to assess. It has already been discussed that parasitic influences of this kind are best pronounced when the infection is transferred from the definitive to the intermediate host by the trophic chain (in trematodes, from the second intermediate to the definitive host) (see, i.e., Keymer and Read, 1991, and also p. 358 of this book). Consequently, we come across the manipulation of the molluscan host behaviour mostly when it is the first and the second intermediate host at the same time, as is the case in microphallids of "*pygmaeus*" group (see p. 396).

At least in one instance we are concerned with a specific adaptive influence (Curtis, 1987, 1990). In high tide, the molluscs *Ilyanassa obsoleta* infected by *Gynaecotyla adunca* crawl to the water edge, where cercariae are emitted. There the second intermediate hosts of this parasite occur, the fiddler crab and some amphipods which are not to be met in the lower levels, where *I. obsoleta* normally lives. It should be emphasized that these migrations are timed mostly to such high tides which are followed by the nocturnal low tides. Thus, when the night-time low tide comes, the infected snails are high up on the shore. It coincides with the time of the activity peaks of the second intermediate hosts. Besides, at night cercariae have a better chance to survive desiccation and excess heat than in the daytime.

It is not to be ruled out that the influence of the parthenitae of *Plagiorchis elegans* upon *Lymnaea stagnalis* snails, described by Lowenberger and Rau (1994), is of a similar nature. Infected snails spend more time at the water surface, and it is there that cercarial release takes place. The larvae emitted here have better opportunities for passive dispersion (cercariae of *P. elegans* are poor swimmers) before reaching maximal infectivity, which happens 2-3 h after the shedding. Finally, it should be noted that non-adaptive changes of molluscan host behaviour due to pathologies of various kinds, caused by parthenitae, appear to be rather widespread (Sindermann and Farrin, 1962; Lambert and Farley, 1968; James, 1969; Tallmark and Norrgren, 1976; Curtis, 1985; Moukrim and Rondelaud, 1992).

Turning to the consideration of seasonal aspects of the variability of parthenita hemipopulations, it should be emphasized that there are few, if any, literature data suitable for such an analysis, despite the abundance of material on the so-called "seasonal dynamics on molluscan infection". The point is that in the overwhelming majority of works in this field the authors

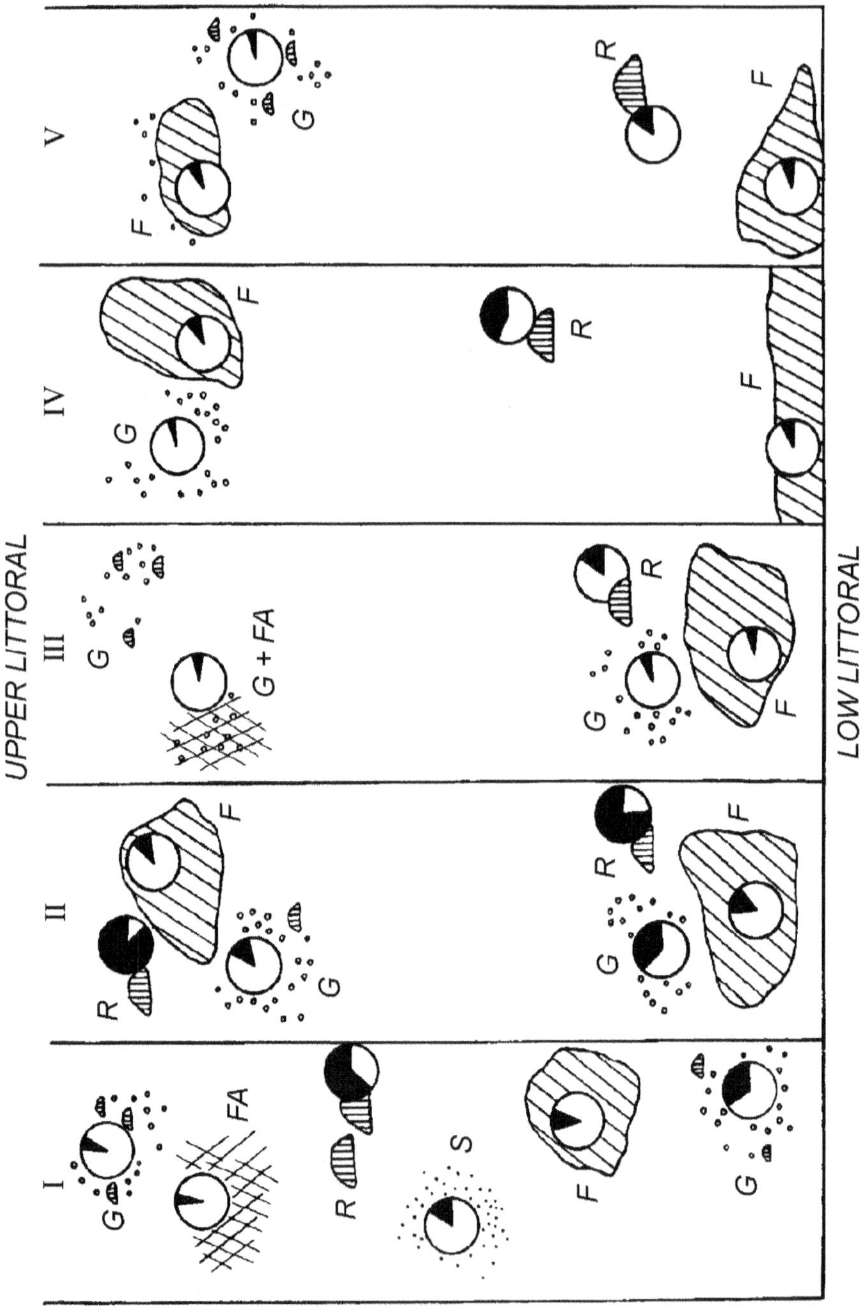
UPPER LITTORAL
LOW LITTORAL
I
II
III
IV
V
G
FA
R
S
F
G + FA

note solely the fact of the molluscan infection, leaving outside the scope of their attention both the characteristics of the dynamics of molluscan hosts populations and the age structure of the parthenita groups parasitizing in them (Lemly and Esch, 1984; Fernandez and Esch, 1991; Williams and Esch, 1991; Galaktionov, 1992). In particular, the infection by young parthenitae is often simply ignored by the researchers.

As a result, some completely erroneous notions can still be found in the literature. One of them is the "disappearance" or "abrupt fall" of the prevalence of digeneans in molluscan populations in the climatically unfavourable season (i.e., autumn-winter for the northern regions). In reality, only a partial elimination of individuals infected by mature parthenita groups may occur, or else such microhemipopulations may "realign" from producing cercariae to the production of the next generation of parthenitae, accompanied by the death of a part of old rediae/sporocysts. However, a significant part of the parthenita hemipopulations is formed by young local groups, which frequently escape the researcher's attention. Objective estimates of the seasonal dynamics of molluscan infection by trematode parthenitae are only possible after a thorough analysis of the age structure of local parasitic groups concurrently with the investigations of populational parameters of the molluscan hosts (Dobrovolskij et al., 1983; Galaktionov and Dobrovolskij, 1984; Lemly and Esch, 1984; Fernandez and Esch, 1991; Williams and Esch, 1991).

Such methodology reveals a rather complex picture of the seasonal variability of the composition of parthenita hemipopulations, as in real natural environment the variants of the formation local group, described earlier in this chapter (see p. 369), are influenced by season-dependent variations of molluscan activity and physiological state and by environmental factors which can accelerate or, *vice versa*, slow down reproduction in parthenitae (see p. 260). Besides, when hemipopulations of miracidia or eggs with formed larvae exist in the environment, new molluscs are being infected and young, newly formed local groups are being recruited into parthenita hemipopulation. Naturally, the influence of all the factors mentioned manifests itself in a variety of ways in the groups where rediae/sporocysts can give rise to their like and with a determined number of parthenogenetic generations.

Parthenita groups of *Podocotyle atomon* studied by us in the periwinkles *Littorina saxatilis* on the Barents Sea shore belong to hemipopulations of the first type (Rusanov and Galaktionov, 1984) (Fig. 121). Mass in-

Figure 120. Prevalence (black sectors) of the "*pygmaeus*" microphallids in periwinkles *Littorina saxatilis* at different levels and substrates of some littoral sites (I-V) of the White Sea. *Abbreviations*: *F*, fucoid algae; *FA*, filamentous algae; *G*, gravel; *R*, rock; *S*, sand. From Galaktionov and Mikhailova (1990), with permission.

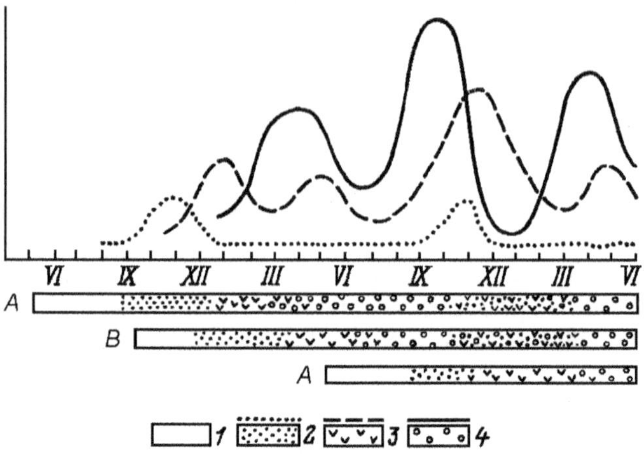

Figure 121. Seasonal dynamics of parthenitae microhemipopulations of *Podocotyle atomon* in periwinkles *Littorina saxatilis* at the Barents Sea coast. *Abbreviations*: *1*, mature mother sporocysts; *2*, microhemipopulations of young daughter sporocysts; *3*, microhemipopulations which include daughter sporocysts of small and medium size, containing single fully formed cercariae; *4*, mature microhemipopulations shedding cercariae; *A*, dynamics of microhemipopulations in molluscs infected in spring; *B*, dynamics of microhemipopulations in molluscs infected in autumn. From Rusanov and Galaktionov (1984), with permission.

fection of periwinkles takes place twice a year, in spring and in autumn, due to the nature of infection dispersion by the definitive hosts, the fishes. The snails infected in April-May harbour developing mother sporocysts (1) throughout the summer, and by September-October microhemipopulations of young daughter sporocysts (2) become apparent. As they grow, groups are formed which include sporocysts of small and medium size, containing in their brood cavity several fully formed cercariae as well as embryos (3). In winter months the development of these groups slows down, as does the development of all other groups, but it resumes early in spring and by March-April they are transformed into mature microhemipopulations shedding cercariae (4).

The second, more intensive, infection of periwinkles occurs in late summer-autumn. In winter the snails harbour either mother sporocysts (1) or groups at the 1st stage. In spring they reach the 3rd stage of development, and by late summer – autumn, exactly in a year after the infection of the snail, the 4th. Thus, in autumn mature microhemipopulations (4) are present both in the snails infected last spring and in those infected last autumn. Two infection waves interfere and the groups at the 4th stage dominate in the hemipopulation. This state, characteristic of many parthenita hemipopulations in the northern regions, is referred to in the literature as "the autumn maximum of infection". However, as explained above, it is not followed by a "disappearance" of parthenita hemipopulation. A decrease in the prevalence

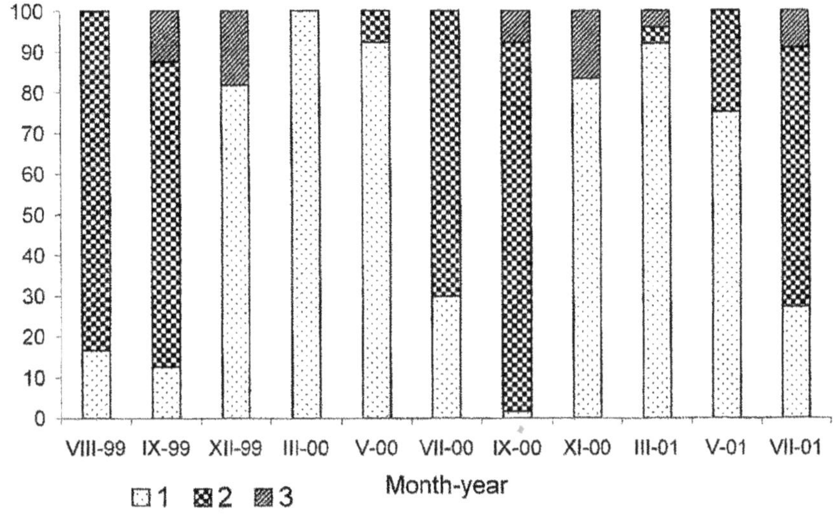

Figure 122. Ratio (%) of molluscs containing different rediae groups in the total infection of mudsnail *Hydrobia ventrosa* with *Paramonostomum anatis* at the White Sea (Suhaya Inlet, Kandalaksha Bay) in 1999-2001 (Galaktionov, Nikolaev, Kofiadi and Prokofiev, unpublished data). *Abbreviations*: 1, the groups, containing young rediae; 2, the groups containing rediae and numerous cercariae maturing in the snail; 3, the groups containing young rediae and single cercariae maturing in the snail.

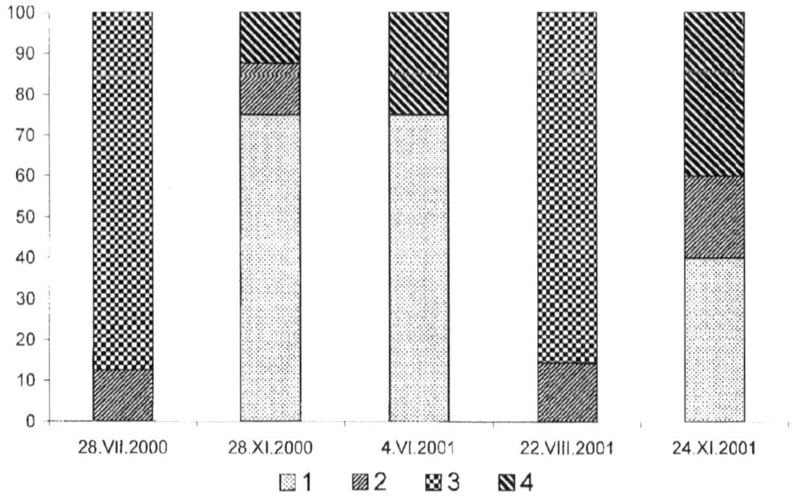

Figure 123. Ratio (%) of molluscs containing different rediae groups in the total infection of periwinkles *Littorina saxatilis* with *Himasthla elongata* at the White Sea (Kem'-Ludi archipelago, Kandalaksha Bay) in 2000-2001 (Galaktionov, unpublished data). *Abbreviations*: 1, the groups, containing young rediae with germinal balls and cercariae embryos; 2, the groups containing rediae of different age, some of them contain single fully-formed cercariae; 3, the groups containing rediae of different age, most of them contain fully-formed cercariae; 4, the renewing redia groups containing old degenerating rediae along with very young ones.

of the groups at the 4th developmental stage in molluscs, manifest in late autumn-winter, is only partially determined by the death of infected snails after the onset of unfavourable conditions. In most of the periwinkles the nature of parthenitae reproduction changes: some individuals start replicating, with many old sporocysts dying. As a result, the groups that were at the 4th stage in early autumn, revert to the 3rd stage by winter. Reconstruction processes appear to be triggered by a considerable drop of temperature as early as in September, which also entails changes in the physiological state of molluscan hosts.

The process described, together with the development of daughter sporocysts in the periwinkles infected in spring, is the cause of the considerable increase of the portion of the 3rd stage groups in parthenita hemipopulation during winter months. The highest death rate of infected (mostly by the 4th stage groups) snails is observed in early summer, when periwinkles, weakened after long winter starvation, are subject to the influence of numerous stress factors, first of all, a considerable water freshening caused by melting ice and snow.

Seasonal reconstructions of the character of parthenita groups reproduction appear to be widespread, which is testified to by the results of the studies of parthenita hemipopulations dynamics of a number of trematode species in the littoral snails of the White Sea. Parthenita groups of notocotylids of the genus *Paramonostomum* in the mudsnails *Hydrobia ventrosa* and *H. ulvae* in summer are represented by rediae of different ages (young rediae with germinal balls and cercariae embryos, rediae with embryos as well as almost fully formed cercariae, rediae with 1-2 young forming rediae alongside with developing cercariae) and cercariae maturing in the snail (Fig. 122). The latter leave the snail as they mature. In autumn (October-December) such groups disappear. All cercariae are released from the molluscan host. There is no recruitment, as some of the old rediae die and in young individuals embryo formation is arrested. Throughout the winter-spring period the groups of young rediae are to be found almost exclusively in the snails. In May-June, when water temperature increases, the functioning of germinal masses and embryonic development in rediae is resumed and by July most of the notocotylid groups in infected snails reach their mature summer state.

Similar reconstructions are experienced by rediae groups of echinostomes *Himasthla elongata* in the periwinkles *Littorina* spp. (Fig. 123). Contrary to notocotylids, rediae in winter groups remain of different ages: there are very young rediae together with considerable numbers of larger and more mature individuals, as well as old degenerating ones. Common for all these rediae of different ages is an almost total absence of formed cercariae ready to leave the host. Shedding of these larvae from the infected molluscs ceases in autumn, when water temperature drops below 1°C. At this temperature the development of cercariae embryos is blocked and maturing of new larvae does not resume until spring. Summer groups are almost always represented by rediae of different ages. This points to the fact that the production by rediae of their like is not timed for a certain season but takes place constantly when the group is functionally active. Indeed, rediae which have young rediae of the next generation as well as cercariae can be found in all seasons.

In contrast to *Himasthla elongata*, in winter groups of rediae of *Cryptocotyle lingua* and *C. concavum* (Heterophyidae), parasitizing, respectively, periwinkles and mudsnails, mature cercariae are found (Fig. 124). Their shedding is observed even at 2-3°C, but only single cercariae are released. On the whole, in winter groups of the heterophyids studied, young rediae prevail, the number of fully formed cercariae in mature rediae being less than in summer. In spring, when the water warms up to 8-10°C, the release of cercariae from the molluscs infected by these parasites quickly returns to summer values.

Similar to heterophyids is the situation with the groups of daughter sporocysts of renicolids and microphallids that belong to the second type, i.e., their daughter sporocysts are incapable of self-reproduction (Fig. 125). Low temperatures almost completely block the release of cercariae from molluscan hosts. Functional activity of germinal masses in daughter sporocysts also seems to be arrested. Anyway, in wintering daughter sporocysts the ratio of developing embryos and mature cercariae is markedly shifted towards the latter as compared to the summer state. Cercarial shedding in spring is resumed quickly enough, as soon as the water warms up to 4-6°C. Though many molluscs with old groups releasing cercariae die in early summer, by July-August this stage of maturity is reached by daughter sporocyst groups that spent the winter in the stages of mother sporocysts or young daughter sporocysts.

Seasonal composition changes of parthenita groups of microphallids of the "*pygmaeus*" group, investigated for several years at the Barents Sea (Galaktionov, 1992), have a special nature. These trematodes are characterized by metacercariae development inside daughter sporocysts. In this instance, the situation is complicated by the fact that during the period when the periwinkles (the first intermediate hosts) are active, the hemipopulation of microphallids of the "*pygmaeus*" group spatially (in a microscale aspect) falls into two heterogenous groups: the snails infected by sporocyst groups with old invasive metacercariae (OIM) or with young invasive metacercariae (YIM) mostly aggregate on the substrate surface (rocks, seaweed fronds), whereas young sporocysts with developing larvae (DM) are mainly localized in sheltered microhabitats (e.g., in the mass of seaweed fronds and on the ground under them, at the bases of stones and boulders). Such spatial heterogeneity appears to be caused by specific influence on the behaviour of infected molluscs, the adaptive nature of which will be discussed later (see p. 396).

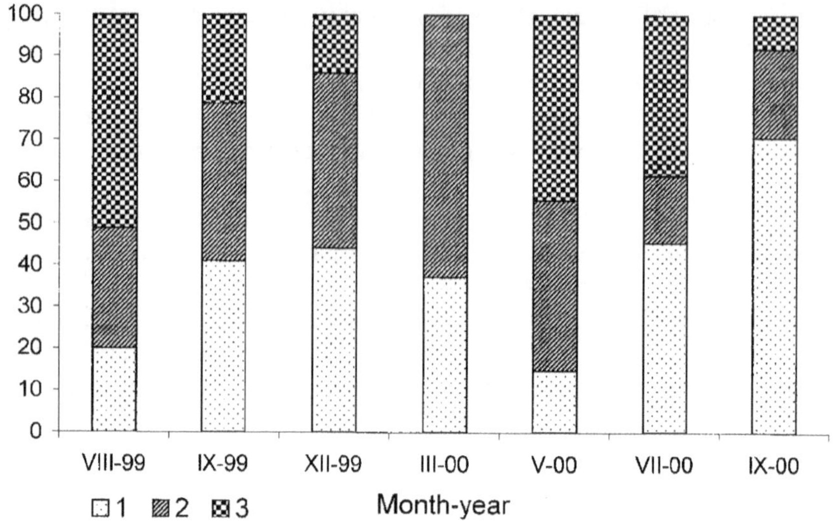

Figure 124. Ratio (%) of molluscs containing different rediae groups in the total infection of mudsnail *Hydrobia ulvae* with *Cryptocotyle concavum* at the White Sea (Suhaya Inlet, Kandalaksha Bay) in 1999-2000 (Galaktionov, Nikolaev, Kofiadi and Prokofiev, unpublished data). *Abbreviations*: 1, the groups, containing young rediae with germinal balls and cercariae embryos; 2, the groups containing rediae of different age, some of them contain single fully-formed cercariae; 3, the groups containing rediae of different age, most of them contain fully formed cercariae.

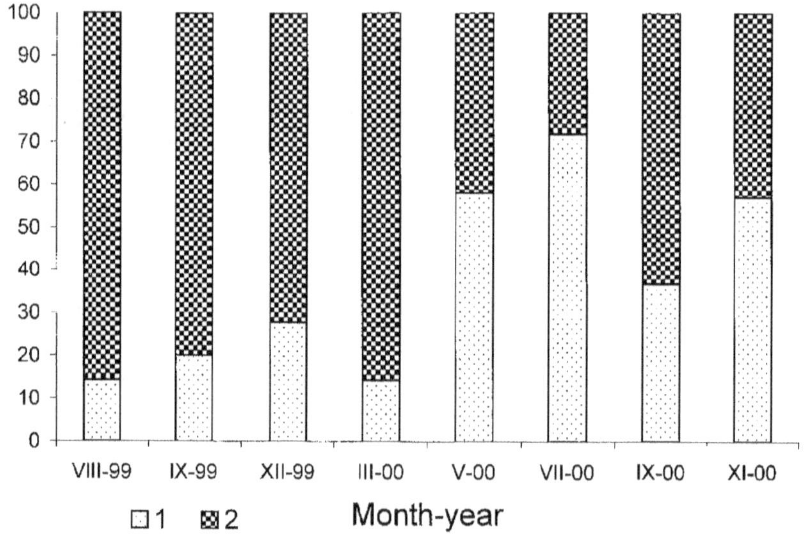

Figure 125. Ratio (%) molluscs containing different daughter sporocyst groups in the total infection of mudsnail *Hydrobia ulvae* with *Maritrema subdolum* at the White Sea (Suhaya Inlet, Kandalaksha Bay) in 1999-2000 (Galaktionov, Nikolaev, Kofiadi and Prokofiev, unpublished data). *Abbreviations*: 1, the groups, containing young sporocysts with germinal balls and cercariae embryos; 2, the groups containing sporocysts with germinal balls, cercariae embryos and fully formed cercariae.

Periwinkles are infected at the greatest intensity in sheltered microhabitats, where, due to the factors described above (see p. 362), the probability of increased density of local hemipopulations of eggs with formed miracidia is the greatest. The latter can survive in the environment for a long time (see p. 239), which makes the infection of molluscs possible throughout a considerable period of their activity. This is probably the reason why the fluctuations of the percentage of DM groups in the part of the hemipopulation of the parthenitae of "*pygmaeus*" group aggregated in sheltered microhabitats are not large, though they are subject to a certain regularity associated with seasonal changes in the behaviour of the definitive host – the common eider.

More dramatic are the events occurring in the part of the hemipopulation on open substrates. In April, the snails infected mainly by OIM appear here, only to die for the most part in early summer of unfavourable environmental factors described in the example with *Podocotyle atomon* (see above). In June, the percentage of OIM is the lowest, but there is an abrupt increase in the share of YIM, supplied by the "young" part of the hemipopulation located in shelters. By late summer – early autumn they mature and become OIM, which overwhelmingly predominate at this time in the "open substrate" group. After the onset of the autumn cold, the snails leave open substrates and aggregate under stones. At this time the two parts of the hemipopulation mix and there is noted an abrupt drop in the infection rate of the snails. However, this drop is by no means caused by the death of OIM-infected periwinkles (it is these groups that are usually recorded in a routine parasitological research). The point is that in summer sheltered microhabitats harbour about 80% of the numbers of *L. saxatilis* population which are infected by parasites with a relatively low extensity, mainly by DM and YIM groups. Because of that, when in autumn individuals from open substrates, infected mostly by OIM, enter common aggregations they, so as to say, "dissolve" in the weakly infected part of the population. Recalculated to total numbers of molluscan population, the composition of the parthenita hemipopulation of the "*pygmaeus*" group turns out to be about the same both in summer (except in the beginning) and in winter.

To conclude, it should be noted that, while writing this chapter, we had to be content mainly with the descriptions of concrete examples. The reason is because of an extreme scarcity of investigations that would analyse not only percentage of the molluscs infected mainly by cercariae-releasing sporocysts, but the microhemipopulations formed by the parthenitae of different generations. The time for generalizations has not yet come. It should be noted that the processes determining the functioning of parthenita groups of different trematode species are very complex and diverse. They are determined by both components of the host-parasite system and are subject to a certain canalization under the influence of external as well as internal factors. No doubt they are highly adaptive, which could happen only in the course of prolonged co-evolution of the trematode parthenitae – mollusc associations.

3.3. Hemipopulations of Cercariae

The cercaria, the second free-living phase of the trematode life cycle, possesses more pronounced morphological-functional and behavioural adaptations (see section 3.1. in chapter 2) than the miracidium. These adaptations are also expressed at the population level. Already the first dispersion stage, i.e., the emergence of cercariae from the infected mollusc, is adaptive, at least, for the larvae whose further development involves their penetration into a host. This process is not stable and undergoes periodic fluctuations which are genetically controlled (Théron and Combes, 1988; Théron, 1989; Pages and Théron, 1990a) and are referred to as "cercarial emergence rhythms". They are divided into circadian rhythms (the interval between the peaks of larvae shedding is 24 h), ultracircadian (two or more peaks during the day and the night) and infracircadian (the interval between the peaks is more than 24 h) (Combes and Théron, 1981).

Numerous investigations (see reviews: Ginetsinskaya, 1968; Smyth and Halton, 1983; Rea and Irwin, 1994; Combes et al., 1994) have shown the cercariae emergence to occur, usually, at the time of the day when the chances of meeting the host are maximal, i.e., it occurs at the "host-time", following the terminology of Combes et al. (1994). The representatives of the genus *Schistosoma* studied in most detail are a good example. The maximal output of *S. curassoni* and *S. bovis* cercariae are timed for early morning, when the definitive hosts of these parasites, the bulls, come to drink water (Pages and Théron, 1990b; Mouchet et al., 1992), those of *S. haematobium*, *S. intercalatum*, and *S. mansoni* cercariae, for the daytime, when people are active (Théron, 1984; Pages and Théron, 1990b) and those of *S. rodhaini* cercariae, for the evening dusk, when murine rodents, the definitive hosts in the life cycle of this species, leave their shelters (Théron, 1989). *S. margrebowiei* has an ultracircadian cercarial emergence rhythm with the peaks at dawn and dusk, the former coinciding with the visits of antelopes to watering places, the latter, with those of the waterbucks (Raymond and Probert, 1991). And, finally, while the molluscs, infected by the *S. mansoni* of the strain which uses man as the definitive host, shed the cercariae, as stated above, mostly during the daytime, the strain of murine origin has its shedding peak in the evening (Théron, 1984).

The above examples are concerned with those cercariae which penetrate the host via the skin. However, the adaptive nature of the emission rhythm can be also revealed in the group of the larvae eaten by the host (see p. 282). In particular, it has often been demonstrated that the peaks of cercariae shedding by different species of the genus *Proterometra* (Azygiidae) occur during the periods of activity of those fish species which are their definitive hosts (Horsfall, 1934; Lewis et al., 1989). So far the short-lived cercariae, most widespread among the trematodes, were discussed. In the long-lived larvae, the adaptive nature of the emission rhythms may be different. This aspect was investigated in the long-lived cercariae of *Halipegus occidualis* (Shostak and Esch, 1990). Their life span lasts for weeks and sometimes for months. Their second intermediate hosts, the benthic crustaceans, are active during the daytime. However, the maximum of cercarial emergence occurs at night. It has been suggested (Shostak and Esch, 1990) that here we are concerned with an adaptation that lowers the risk of the emitted cercariae being eaten by the commensal oligochaetes *Chaetogaster limnaei,* which search for their prey by sight. It appears that for the long-lived larvae this factor (the risk of being eaten by predators) may be crucial for the formation of the rhythm of their shedding from the molluscan host, considering that the coincidence of the emission peak with the "host-time" is less important for the larvae of this type than for the short-lived ones. Taking into account a high nutritive value of cercariae, which are, as a matter of fact, reservoirs with energy-rich glycogen (see p. 289), it can be supposed that the factor of predators may determine the emergence rhythms of some cercariae to a greater degree than it is commonly thought.

It is more difficult to understand the reasons behind the cercarial emergence rhythms, when the first and the second intermediate hosts are in the same biotope all the time and/or when the range of the second intermediate hosts is broad. For instance, a freshwater pulmonate gastropod *Gyraulus chinensis* is the first intermediate host of the echinostomatid *Euparyphium albuferensis*, this and other freshwater snails also serving as the second intermediate hosts (Toledo et al., 1999). Cercariae shedding takes place strictly in the afternoon with the peak between 1 and 2 o'clock p.m., i.e., when the water is the warmest. As the infectivity of cercariae increases with temperature, the formation of dense larval "clouds" at this time increases the probability of the successful infection of the second intermediate hosts (Toledo et al., 1999). It seems reasonable to agree with the opinion of Toledo et al. (1999) that here the adaptive nature of the cercarial emergence rhythms is expressed not only and not so much in the increased probability of meeting the host but rather in the timing of cercariae appearance for the period when the conditions are the most favourable for successful parasite-host contact.

A somewhat different scenario of adaptation to transmission is demonstrated by the cercariae of the already mentioned species *Plagiorchis elegans* (see section 3.2., p. 375). They also possess a wide specificity to the second intermediate hosts (Lowenberger and Rau, 1994). The researchers failed to reveal the timing of the emergence rhythms of these larvae for the "host-time", the hosts (benthic larvae of aquatic insects) being available almost always. It is more important for these larvae to avoid the superinfection of the host, which may result in its death and thus in the death of the penetrated larvae. The cercarial emergence occurs when the infected molluscs rise to the water surface. The cercariae of this species reach their infectivity maximum in 2-3 h after shedding. During this time the larvae disperse passively (by water microcurrents) and settle on vast areas of the bottom. This decreases the probability of the superinfection of the second intermediate hosts, thus facilitating the successful completion of the life cycle. It remains to be said that in some species, the cercariae of which encyst in the environment and form adolescariae, have no pronounced emission rhythms (Kendall and McCullough, 1951; Ginetsinskaya, 1968, 1983; Bouix-Busson et al., 1985; our data).

The synchronization of the cercarial emergence rhythms is achieved by means of control by environmental factors which change in a regular manner during the day. These factors are first of all associated with photoperiodism (see reviews: Ginetsinskaya, 1983; Smyth and Halton, 1983; Rea and Irwin, 1994; Combes et al., 1994), with light intensity and water temperature being the leading ones. However, it must be taken into account that the changes in all these factors are closely correlated and not all the investigations which postulate the domination of this or that factor are based on the experiments which allow us to consider a postulation like this to be correct (Prokofiev, 1995). As a rule, the changes in light intensity play the paramount role (see, e.g., Smyth and Halton, 1983). However, in a series of laboratory experiments conducted under the conditions of the artificial lack of photoperiodism, it has been convincingly shown that the cercarial emergence rhythm may be also determined by the experimentally set temperature fluctuations (Valle et al., 1973; Lewis et al., 1989; Prokofiev, 1995). This factor acquires special significance in the conditions of the polar day. In a demonstrative series of experiments, Prokofiev (1995) showed that it was the temperature that regulated the cercarial shedding in a number of trematode species (*Podocotyle atomon*, *Cryptocotyle lingua*, *Maritrema subdolum*, *Levinseniella brachysoma*, *Renicola* sp.) from littoral snails at the Barents and the White Seas, where daily changes in light intensity in summer are not very pronounced.

The short-term cercarial emergence rhythms described above are superimposed by temporal oscillations on a larger scale: the values of daily peaks decrease periodically, to reach their maximum again later. In the case of the microhemipopulation of the parthenitae, capable of periodic rejuvenation (see p. 370), it is these processes that determine the long-term changes in the daily cercariae production (Théron, 1981a,b; Dobrovolskij et al., 1983; Niemann and Lewis, 1990; Toledo et al., 1999). In schistosomes the number of cercariae released daily periodically decreases until there is no emission at al. It is associated with the fact that some of the daughter sporocysts degenerate and the others "switch" from the production of cercariae to the production of sporocysts (see p. 370 and fig. 119). Cercarial shedding starts to increase again as the microhemipopulation is being renewed, to reach its maximum after the maturing of all newly formed sporocysts. A cycle like this in schistosomes takes 30-40 days (Théron, 1981a,b). We have also observed long-term fluctuations in the number of the larvae shed by a snail infected with plagiorchiid trematodes which have a strictly fixed number of parthenita generations (mother and daughter sporocysts – p. 369). The reasons behind these fluctuations remain obscure. They are likely to be associated with the dynamics of interrelations in the definite molluscan host – trematode association.

The emergence patterns and the life span of cercariae determine what Combes and Théron (1981) call "the rhythm of presence", i.e., temporal changes in the density of the larvae in water. In our terminology, it is the dynamics of local hemipopulations of cercariae. As noted above, most cercariae have a short life span, not exceeding 24 h (Combes et al., 1994). In case of an infracircadian emission rhythm (or a circadian rhythm, with a short maximum), a group of such larvae is practically of the same age. At a certain time of day, it appears in the environment, to disappear rather soon. The latter circumstance results not only from the short longevity of the larvae but also from their elimination by predators and their death under the influence of unfavourable environmental factors. Such groups may be, though conditionally, called "pulsating" (drawing an analogy with miracidia – see p. 361).

The point is that cercariae shedding from the mollusc seldom falls to zero, a certain number of larvae almost always being emitted between the two neighbouring maxima. However, their number is relatively low and the term "pulsating" is quite suitable for characterizing the nucleus of the local hemipopulation. Besides, some individuals emitted during the previous peak of emission seem to be capable of surviving till the next one. In this instance, the age structure of the local group turns out to be heterogenous. However, old individuals are unlikely to be of any importance for the parasite transmission, as cercariae loose their infectivity long before the end of their life. "Pulsating" local hemipopulations occur in their purest form in the habitats with frequent changes of environmental conditions, e.g., in the intertidal zone.

There, the time of existence of cercariae groups of many species is determined by the tidal cycle. Thus, the emergence of *Maritrema subdolum* larvae from *Hydrobia ulvae* mudsnails in the White Sea takes place during the daylight time at low tide (the emergence is temperature-controlled) in small marine pools (puddles in the depressions of the sediment, rocks, etc., left after the water recedes). It is there that the infection of the second intermediate hosts, gammarid amphipods, occurs (Ginetsinskaya, 1954b; Prokofiev, 1995). At high tide these microhabitats disappear, together with the local hemipopulations of *M. subdolum* cercariae. The formation of new local groups takes place only 24 h later, at the time of the next day's low tide.

Another picture is observed in case of an ultracircadian cercarial emergence rhythm, or a circadian rhythm with a smooth, i.e., a more or less prolonged, peak of the larvae shedding. In this instance, the age structure of the local hemipopulation of cercariae is heterogeneous. Such groups, which can be termed "permanent", may sustain their existence in the environment for a long time, by means of an almost constant recruitment of new individuals. The local hemipopulations formed by relatively long-lived larvae regardless of their emergence rhythm should also be attributed to this type.

Up to now we have not been considering the problem of the numbers of local groups formed by cercariae. It is difficult to perform a comparative analysis of different trematode taxa, despite a vast amount of literature data on the "daily production" of cercariae by the infected molluscs. The difficulty is that one has, in fact, to compare the incomparable: the number of larvae emitted by the molluscs of different size and age, i.e., with different tissue reactivity and the volume of the visceral sac, infected with parthenita microhemipopulations on different stages of their formation. As a general tendency, the increased "daily production" of cercariae in the representatives of the higher trematodes with juvenalized larvae, e.g., Strigeidae, Diplostomidae, Microphallidae, Lecithodendrioidea, can be noted. Here, we often deal with the orders of magnitude up to ten or hundred thousands. In Echinostomatidae, Fasciolidae, Notocotilidae and others, the number of cercariae shed during the day varies, as a rule, from several dozens to several thousands. These figures give us only an approximate idea of the actual state of things. Even species within one taxon (often within a genus) may differ in the production of larvae, which may be associated, among other things, with their behaviour (see section 3.1.2. in chapter 2).

The "daily production" of cercariae with an "ambuscade" behaviour turns out to be relatively low (see p. 301). It is about 20-100 for *Levinseniella brachysoma* larvae, whereas the daily production of the cercariae of *Maritrema subdolum* (also a microphallid) is 3000-4000, the molluscs being of the same size and species (*Hydrobia ulvae*) (Prokofiev, 1995). We have also obtained similar data on the gymnophallids, the first intermediate host of which is a small bivalve *Turtonia minuta*, common on the upper intertidal level of the Barents Sea. The molluscs infected by the species *Cercaria dichotoma* Chubrik, 1966 produce on the average 500-2000 cercariae a day, whereas those infected by *Cercaria margaritensis* Ching, 1982 produce only 10-300 individuals, which exhibit an "ambuscade" behaviour (our observations). Interestingly, the larvae of the former species are more than 1.5 times larger than those of the latter, i.e., in this instance low cercariae production cannot be associated with their larger size. At the same time, it is this factor that appears to have played a crucial role in determining an extremely low "daily production" of gigantic macrocercous and cysticercous larvae. It is sufficient to recall that the molluscs infected by *Proterometra macrostoma* shed on the average 1 larva in 3 days, and those infected by *P. edneyi*, 1 in 2 days (Lewis et al., 1989).

The spatial structure proper of cercariae hemipopulations is formed at the second dispersion stage, the secondary redistribution of the larvae shed by the mollusc. Its character is determined by the complex of behavioural reactions of the cercariae, as well as by the joint influence of environmental factors upon them. If these factors are not catastrophic, the group of cercariae becomes dispersed in the places where the presence of the host is possible, i.e., in "host-space", following the terminology of Combes et al. (1994). Contrary to the case for miracidia (see p. 228), this is undoubtedly the result of the realization of the cercarial behavioural reactions, which has been demonstrated by many researchers. At the same time, it should be noted that only a few authors extended the traits of individual larval behaviour to a group level, using them to characterize the spatial structure of the group.

Classical studies in this field were carried out by the French scientist P. Bartoli and his colleagues (Bartoli, 1983, 1986; Baroli and Combes, 1986; Bartoli and Prevot, 1986; Bartoli and Holmes, 1997) on the cercariae of different trematode species released by infected benthic molluscs of the coastal fauna of the Mediterranean Sea (Fig. 126). They found out that the cercariae of the trematodes from different taxa (Zoogonidae – *Diphtherostomum brusinae*, Opecoelidae – *Cainocreadium labracis*, Gymnophallidae – *Gymnophallus choledochus*, Lepocreadiidae – *Holorchis pycnoporus* and *Lepocreadium pegorchis*, and others), which have bottom invertebrates (polychaetes, amphipods, bivalves) for the second intermediate hosts, disperse only horizontally after leaving the molluscan host, forming a "cloud" in the bottom water layer. The hemipopulation of the larvae of *Cardiocephalus longicollis* (Strigeidae) disperse in the water column closer to the surface, where their second intermediate hosts keep, the garfish *Belone belone*. Shed by the molluscan hosts, the cercariae of *Renicola lari* (Renicolidae) first swim upward, reach the superficial stratum of the water, and there disperse (both actively, and passively by means of water currents), forming more or less extensive "clouds". It is there, close to the surface, that the second intermediate hosts of *R. lari*, the atherinid fish, are localized.

In the two latter cases the distribution of local groups in the "host-space" does not take place at once. It is preceded by a migration period. It complicates the spatial structure of the local hemipopulation: its nucleus is localized in the habitat where the contact with the host is possible, the rest of the larvae following the migration ways from the molluscan host. The latter scenario is widespread in the trematodes, as their first intermediate hosts are benthic animals, and their second intermediate hosts are often, if not predominantly, pelagic invertebrates and fishes.

Even more complex is the spatial structure of the local cercariae groups, when the first and the second intermediate hosts inhabit not only different habitats but a different environment. Thus, the parthenogenetic gen-

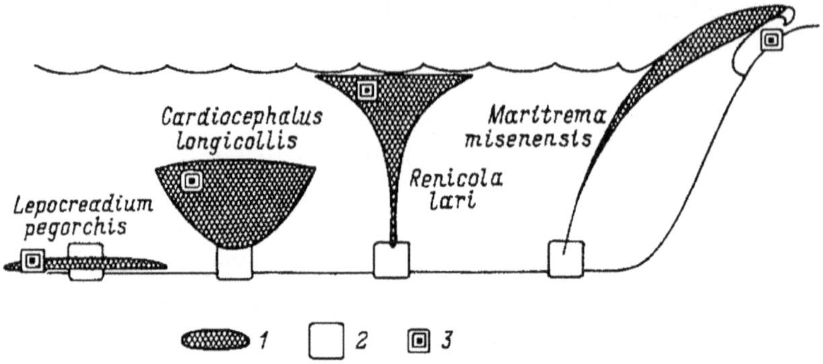

Figure 126. The dispersion of cercariae of various trematodes in water, correlating with the usual position of their target (second intermediate) hosts. *Abbreviations*: 1, distribution of cercariae in the water column; 2, benthic molluscs which release cercariae; 3, cercarial targets – second intermediate hosts. Based on Combes et al. (1994), reproduced with the permission of Cambridge University Press.

erations of the microphallid *Maritrema misenensis* develop in the sublittoral mediterranean molluscs *Cerithium rupestre*, their second intermediate hosts being the crustaceans from the genus *Orchestia*, which live among the rotting seaweed washed ashore (Bartoli, 1986). As the cercariae shed from the molluscs possess a negative geotropism, they swim actively to the surface of the water where they cling to the air-water interface by means of their oral sucker. The group of larvae formed at the water surface drifts to the surf zone under hydrodynamic influences and is transferred (partially, of course) to the "host-space" with splashes of the seawater (see fig. 126). There, in the humid environment of the masses of rotting seaweed, the larvae accumulate, which promotes the infection of the second intermediate host.

Local cercarial hemipopulations with a complex spatial structure may be formed not only when the habitats of the first and the second intermediate hosts do not coincide. For the renicolid species *Renicola thaidus* and *Cercaria parvicaudata* Stunkard & Shaw, 1931 studied by Prokofiev (1995) the role of the first intermediate hosts on the Barents Sea coast is played, respectively, by the molluscs *Nucella (Thaidus) lapillus* and *Littorina* spp., the second intermediate hosts being the mussels *Mytilus edulis*. Both hosts inhabit the lower intertidal – the upper subintertidal level, so there is no speaking about the spatial separation of their habitats. Yet, after leaving the molluscan host, the cercariae exhibit a photopositive and a geonegative response and rise to the upper water layers, i.e., leave the zone of possible contact with the mussels. It is only in a few hours, during which the newly formed local group of larvae disperses in space under the influence of hydrodynamics, that the cercariae loose their direct phototactic response and change their response to gravity to positive. Such individuals sink to the bottom and form gatherings in the "host-space". The formation of such a complex spatial-temporal organization of local cercariae hemipopulations seems to be associated with the immobility of their second intermediate host. Therefore, the dispersion over an extensive area, which takes place at the first stage of the group's existence, can be viewed as an adaptation aimed at increasing the probability of the contact with immobile mussels. It should be noted that its formation might have been associated with the special living conditions at the intertidal zone, determined by the tidal cycle and increased hydrodynamics. Anyway, in the above examples from Bartoli (see fig. 126), which describe the events in the relatively stable habitats of shallow inlets and bays, if the habitats of the first and the second intermediate hosts are the same, the local groups of larvae do not leave the bottom water layers, even if the second intermediate hosts are the sessile bivalves.

The density of local hemipopulations of cercariae is unlikely to be evenly distributed in the environment, even if the molluscan host sheds the larvae more or less evenly. It becomes more or less aggregated under the influence of environmental factors, such as, e.g., microcurrents. It may also be facilitated by the larval behavioural reactions. Thus, the combination of a photonegative and a geopositive response in the cercariae of *Maritrema subdolum* mentioned above results in the accumulation of the larvae, immediately after their leaving the mollusc, in the shadowed places of the bottom (sheltered by seaweeds, rocks). As the distribution of these shelters is patchy, the density of the forming local group of cercariae also turns out to be heterogeneous.

The formation of the spatial structure of the local hemipopulations of cercariae with an "ambuscade" type of behaviour takes place in a way somewhat different from the cases described above. Importantly, they practically lack the second dispersion stage, as the larvae cannot crawl/swim across more or less long distances. The distribution character of such cercarial groups is completely determined by the movements of the molluscan hosts. Naturally, their habitats should be at least from time to time visited by the second intermediate hosts, which is the case in the nature. For instance, all events associated with the transmission of the infective agent from the first to the second intermediate hosts in *Cercaria margaritensis* may take place on one and the same frond of the kelp *Laminaria saccarina*. In summer it is the place of the accumulation both of the bivalves *Turtonia minuta* and the gastropod *Margarites helicinus*, which are, respectively, the first and the second intermediate hosts of the parasite (see fig. 110). At present there is only a single example of a species the cercariae of which exhibit an "ambuscade" behaviour type, which overcomes the spatial separation of the habitats of the first and the second intermediate hosts. It is the above mentioned microphallid *Gynaecotyla adunca* (see p. 375), which alters the behaviour of the infected molluscs *Ilyanassa obsoleta* in such a way as to ensure the appearance of local cercariae groups both in the "host-time" and "host-space" (Curtis, 1987, 1990).

The above considerations allow us to conclude that the actual spatial-temporal structure of the local cercariae hemipopulations guarantees that the majority of their individuals find themselves in the "host-space" in "host-time". However, this is true for the majority but by far not for all individuals. As noted above, the emergence of cercariae from the infected molluscs seldom drops to zero. A certain, sometimes a significant number of the larvae is shed in the time of the day which cannot be considered as favourable for the infection of a definite host. A similar picture is observed in the manifestations of behavioural reactions by the larvae shed by one and the same molluscan host. Prokofiev (1995) showed in a series of experiments that 10-30% of the cercariae in the species *Renicola thaidus*, *Cercaria parvicaudata*, *Cryptocotyle concavum* and *Maritrema subdolum* either demonstrate response to light and gravity of another sign than the majority of individuals of the same species, or the response is absent at all. At the level of cercariae group this phenomenon, termed "the ambiguity of behavioural reaction manifestation" (Prokofiev, 1995), inevitably results in the formation of a "dominant" cloud in the places of the most probable presence of the target animal, and "satellite" clouds, dispersing in other areas. It is noteworthy that the appearance or mass recruitment of the "dominant" cloud takes place mostly in "host-time", which is not necessarily the case for "satellite" clouds. This picture recalls, to some extent, the distribution of local hemipopulations of miracidia in the environment.

Among the general reasons behind the sustenance of such spatial-temporal structure of the groups of cercariae and miracidia, the possibility of infecting the "deviationist hosts" should be noted, i.e., the individuals that belong to the host species but whose behaviour deviates from the norm. At the same time, the strategy of forming "dominant" and "satellite" cercarial clouds should be very important for the trematodes owing to the fact that many, if not the majority, of the species possess a wider specificity to the second intermediate host than to the first one. They often use the animals of several species as the second intermediate hosts, and their behaviour is sure to differ at least in some details. Usually there are the species (one or more) that play the main role in the parasites' transmission, and a certain number of secondary ones (see p. 394). The infection of the latter may be performed by the "satellite" clouds. Besides, their presence may facilitate the infection of hosts in unstable conditions. It may also be of a great importance for the speciation process by means of the involvement of new animal hosts[ii].

It remains to be said that the cercariae hemipopulations in the regions with a pronounced seasonal variability disappear from the biotopes during the unfavourable period, e.g., at the onset of the autumn cold in the northern regions. As noted above (see pp. 377 and 380), it is associated both with the ceasing of cercariae emission resulting from the drop of temperature and with the reconstructions in the reproduction of parthenogenetic generations.

3.4. Hemipopulations (Infrapopulations) of Metacercariae

The formation of local metacercariae hemipopulations in the second intermediate hosts takes place by means of a typical exogenous accumulation (see p. 329). If the host's growth is not accompanied by drastic changes in its biology, if it does not pass to another biotope, and if the superinfection does not cause its death, the numbers of the local group of larvae that it harbours would increase with age. This has often been mentioned in the literature (for reviews see: Ginetsinskaya, 1983; Galaktionov and Dobrovolskij, 1986). At the same time, a number of host-parasite systems possess mechanisms which make further infection of the host impossible after a certain limit, imposed by its parasitic capacity. This is the case for the gymnophallids. The penetration of their cercariae and the development of metacercariae is strictly confined to certain parts of the host's body (Bartoli, 1973, 1981). In particular, the larvae of *Gymnophallus fossarum* in the marine bivalves *Cardium glaucum* are located either in the central mantle part close to the hinge (the optimal microhabitat) or on its periphery (the suboptimal microhabitat) (Bartoli, 1973). If all the microhabitats are occupied, new cercariae do not penetrate, and the host has a better chance not to die of the pathogenic influence of the parasite.

The process of cercarial penetration into the second intermediate host is regulated by two mechanisms (Bartoli, 1981). In the case of limited penetration, the number of metacercariae increases up to a certain limit with the host's age and then stabilizes, as all the microhabitats are occupied. When the penetration is unlimited, the number of metacercariae grows constantly throughout the host's life. In such a case, the cercariae, which have penetrated into the host, at once leave the penetration microhabitat, freeing the space for the new larvae. In the opinion of Bartoli (1981), the choice between these two mechanisms is dictated by the conditions in the ecosystem, where the infection takes place. In sheltered lagoons, where the density of cercariae hemipopulations may be very high, the mechanism of limited penetration is switched on. It prevents the second intermediate host from dying of the superinfection. In exposed habitats with pronounced hydrodynamics cercariae hemipopulations are sparse. There the mechanism of unlimited penetration is triggered, which ensures the presence of a free microhabitat for every larva which contacts the host. As for the probability of the superinfection of the second intermediate hosts, in such ecosystems it is low.

At present it is difficult to judge the extent of the presence of the mechanisms, limiting the numbers of local metacercariae groups, as the appropriate data are absent. It is not to be ruled out that the emergence of the phenomenon of limited penetration in the gymnophallids is also associated with the limited mobility of the second intermediate host, the bivalve. It always stays in the distribution area of the "dominant" cloud of cercariae (see

p. 392). If the host leads an active life, the strategy of limited penetration seems not to pay off, as the host easily avoids the area of high density of local cercarial hemipopulations. In the numerous experiments where such hosts were infected by some plagiorchiid and strigeidid species, we did not observe any limitations of the number of penetrating larvae and the experimental animal died of the superinfection.

The heterogeneous composition of metacercariae groups in one host individual, and thus of the hemipopulation as a whole, is a consequence of an exogenous accumulation (see p. 329). Alongside with the fully formed individuals, the second intermediate hosts may contain metacercariae at different developmental stages, including young larvae that have just penetrated into the host. Such a picture is observed throughout the time period when the hemipopulation of cercariae exists in the biotope. This period may be confined to a certain season (see p. 392). For the northern regions, it is the warm season. When new infective agents are no longer supplied from the environment, all metacercariae in the hosts finally reach their invasive state, though this process seems to slow down to a certain degree at lower temperatures. As a result, in cold months the hemipopulations of metacercariae, though still heterogeneous by their age structure, develop the same level of the larval infectivity.

The individual life span of metacercariae seems to be the longest among all the life cycle stages. According to the estimates of Dönges (1969), it reaches 7-8 months even in the adolescaria of Fasciolidae and Notocotylidae, the highly specialized metacercariae of Diplostomidae retaining their infectivity for up to 5-6 years. Thus, metacercariae hemipopulation is almost always present in the second intermediate hosts regardless of seasonal or cyclic changes in the numbers of cercariae groups. At the same time, its seasonal dynamics may greatly depend on the host's life cycle. For example, at the Barents Sea coast the hemipopulation of metacercariae of *Podocotyle atomon* in gammarids is present throughout the year, but the prevalence of infection in the host population falls abruptly in spring, when heavily infected last year generations die and are substituted by the young gammarids of the new spring generation, the first in the new year (Uspenskaya, 1963; Marasaeva, 1990). As for adolescaria hemipopulations, they seem to disappear in the unfavourable season.

It should be noted that contrary to the parthenita hemipopulations, those of metacercariae are usually distributed in the populations of different species of animals, their second intermediate hosts. They may be specific, the conditions in them being the most favourable for larval development, and poorly specific, where this process is in some degree impeded. Besides, the nucleus of the hemipopulation may be located in different host species depending on the environmental conditions of different regions or even the areas of the water body or land. All this greatly complicates the actual structure of metacercariae hemipopulation in natural ecosystems, each concrete case requiring a special investigation. One thing should be kept in mind, that it is in such a way that a broad dispersion of infection is achieved (dispersion at the stage of metacercariae is the most significant in the trematode life cycle – see p. 329) and the chances of infecting the definitive host are increased.

Besides, the probability of the larvae being ingested by the animals of different species also increases. This provides an advantageous ground for the divergence processes at intraspecific and specific levels. However, there are trematode groups, mostly among the higher trematodes, with a very narrow specificity to the second intermediate hosts. Mention has already been often made of the Microphallidae, which are a good example here. They usually have 1-2 species of the second intermediate hosts, this number only rarely reaching 3-4 (Belopolskaya, 1963). It may be associated with the complexity of morphogenetic processes undergone by the metacercariae in the crustaceans (Belopolskaya, 1963, see also p. 162). A narrow host-specificity is known to be characteristic of those life cycle stages, at which an intensive morphogenesis takes place (Dubinina, 1953). Although this rule is not without exceptions (e.g., the metacercariae of *Podocotyle atomon* undergo drastic morphogenetic transformations in their development, but use at least 35 crustacean species as the second intermediate hosts – Galaktionov, 1987), it seems to apply to the microphallids as well. An inevitable decrease of the probability of infection transmission, associated with a narrow specificity, seems to be compensated by the fact that the microphallids use the most mass crustacean species as their second intermediate hosts. These crustaceans make up a large proportion in the food composition of the definitive hosts, the birds. As the number of such crustaceans is limited, one and the same species is the second intermediate host for several microphallid species. Elaborate migrations and species-specific localization in crustaceans might have originated in the microphallids as an adaptation decreasing the interspecific competition for the habitats in the body of the second intermediate host (Galaktionov, 1993).

The spatial distribution of metacercariae hemipopulations is determined by a complex of factors, both abiotic and biotic. Among the latter the primary role seems to be played by the behavioural features of the animal hosts, reflected in a specific environment. Almost as important is the spatial distribution of the cercariae hemipopulations, and, respectively, the parthenita microhemipopulations. It is the interactions of these factors that determine the actual spatial structure, which is characteristic of a metacercariae hemipopulation at a given moment of time. As in the parthenita microhemipopulations, it turns out to be heterogeneous, "spotty": areas with a high infestation of the second intermediate hosts alternate with those where the values of this parameter are minimal, or where local hemipopulations are absent.

As noted above (see p. 359), the metacercariae may change the behaviour of the host. Thus, the larvae themselves exert an influence on the formation of the spatial structure of their hemipopulations. In most cases we are concerned with non-adaptive or non-specific adaptive manipulations of the host behaviour (see p. 358). One of the most well-known variants is the alteration of the bivalves' locomotory organs by the metacercariae developing in them. Such bivalves loose, partially or completely, their ability to burrow into the sediment, becoming an easy prey for the definitive hosts, the fish or the birds (see reviews: Lauckner, 1983, 1985; Thomas and Poulin, 1998). However, specific adaptive influences have also been described in the higher trematodes. This phenomenon appears to be rather common, a relatively low number of examples known at present being associated rather with the poorly studied ecological-populational aspect of those trematodes, which are not medically or veterinary important.

We have already mentioned the *favorisation* in the *Euhaplorchis californiensis* life cycle (see p. 359). The interactions between the larvae of *Dicrocoelium dendriticum* and *Leucochloridium paradoxum* have already become a classical example. As noted in chapter 2 (see p. 272), the cercariae of the former species are shed from the molluscan host in groups enveloped in a mucus layer, the so-called "complex cysts" or "mucus balls". Each "ball" contains 100-300 cercariae. When it is ingested by an ant, the second intermediate host, the larvae encyst in the body cavity. One of them, however, attaches to the ant's suboesophageal ganglion and induces a behavioural change in the host[iii]. During the daytime the infected ant performs its usual functions of a working ant, but in the evening it crawls up a blade of grass and hangs on it, seizing it with its jaws. It stays in this state until the morning. At that time all the hemipopulation of metacercariae is concentrated in the place where the probability of getting into the grazing sheep and other herbivores, the definitive hosts of *D. dendriticum*, is the highest.

Many instances of specific adaptive influence upon the behaviour of the intermediate hosts have been described in the microphallids. According to Helluy (1982, 1983), of all the cercariae of *Microphallus papillorobustus*, penetrated into the gammarid hosts, only 1-6 encyst in the protocerebral region of the cerebral ganglion, the others locating in the host's body. When the "cerebral" metacercariae mature, normal behavioural reactions of gammarids are disrupted: the phototaxis becomes strictly positive, the geotaxis sign changes from positive to negative, the mechanisms of response to the external stimulants are disrupted. As a result, the crustacean rises to the water surface, a habitat not typical for the gammarids. There, the probability of it being eaten by the gulls, the definitive host of *M. papillorobustus*, increases greatly, which has been demonstrated experimentally (Helluy, 1984).

A similar situation has been revealed in the microphallids of the "*pygmaeus*" group from *Littorina* snails (Galaktionov and Rusanov, 1983; Sergievskii et al., 1984; Galaktionov and Marasaev, 1986; Granovitch, 1992;

Galaktionov, 1993; McCarthy et al., 2000). There, metacercariae hemipopulations as such are difficult to distinguish from the parthenita hemipopulations, as all larval development until the infective state is completed inside the daughter sporocysts and thus in the same molluscan host. It has been found that as metacercariae mature, the behaviour of periwinkles changes in such a way that they crawl to the exposed substrates (such as the surface of rocks, the tops of seaweed fronds) and stay there throughout the tidal cycle. There the probability of being eaten by the common eider and other birds, the definitive hosts of these parasites, increases. As for this group of larvae, it becomes spatially separated into two parts: the mature part localizes mainly on the exposed substrate, the immature one, mainly in sheltered habitats (e.g., between seaweed fronds, on the ground under rocks and seaweed fronds) (Fig. 127). Such distribution results from the parasitic influence on the periwinkle's locomotion. Recent laboratory and field experiments have demonstrated that the periwinkles with mature infections moved further upwards but showed less downward and horizontal movements than uninfected ones (McCarthy et al., 2000).

It is noteworthy that in both cases described the alterations in the behaviour of microphallid-infected hosts take place only after the complete development of the metacercariae. In *M. papillorobustus*, it is the "cerebral" metacercariae and those which penetrated together with them, in the microphallids of the "*pygmaeus*" group, it is the whole local group, the synchronous development of which is achieved by the nature of reproduction of the mother and daughter parthenita generations (see p. 70). It increases the probability that it will be the infective larvae that would penetrate into the host. The manipulation of the host behaviour appears to be a universal trait of the microphallids with dixenous life cycle, with the mollusc being both the first and the second intermediate host. Apart from the above microphallids of the "*pygmaeus*" group, *favorisation* of transmission from the first to the second intermediate host has also been described for *Maritrema oocysta*, *Levinseniella* sp. N 17 Deblock, 1980 from the mudsnail *Hydrobia ulvae*, and *Microphallus* sp. from the freshwater snail *Potamopyrgus antipodarum* (Huxham et al., 1995; Jokela and Lively, 1995). In both cases the molluscs with fully developed metacercariae concentrate in the microhabitat where they are most susceptible to the definitive host.

To conclude, the spatial structure of the metacercariae hemipopulations may also depend on the seasonal migrations of the second intermediate hosts. The larval groups may disappear from the biotopes where they may contact the definitive host, and, thus, the completion of the life cycle in these biotopes is temporarily rendered impossible (for details see p. 402).

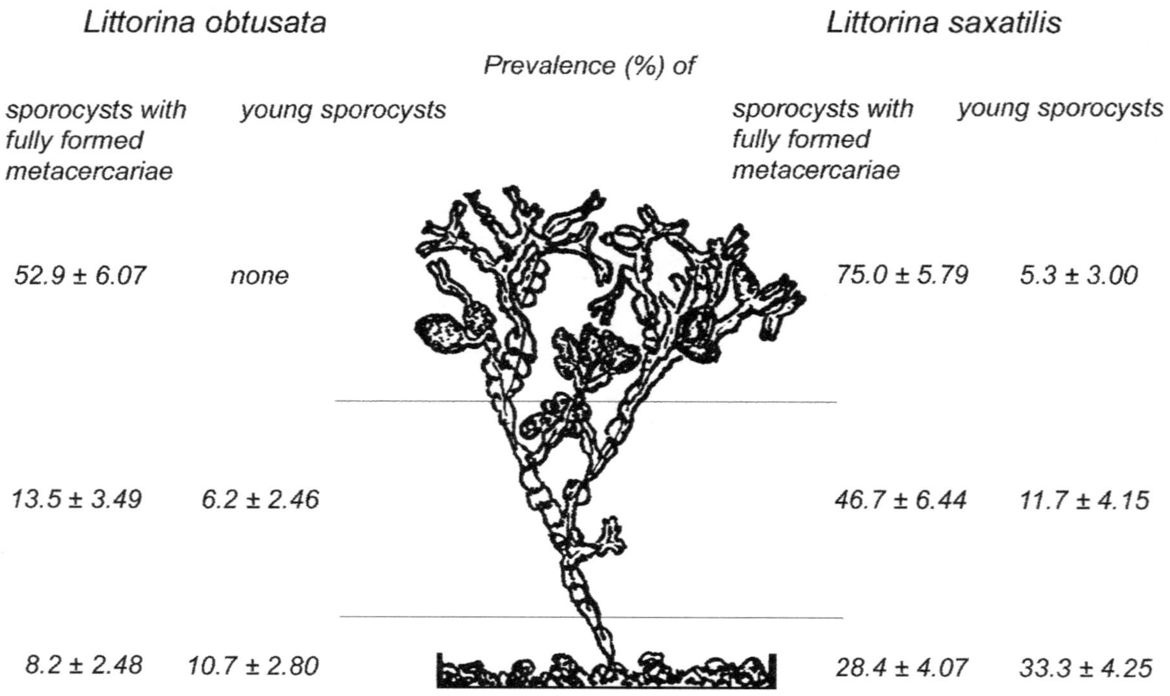

Figure 127. Distribution of periwinkles infected with "*pygmaeus*" microphallid daughter sporocysts of different age along the seaweed frond during high tide. Based on Galaktionov (1993).

3.5. Hemipopulations (Infrapopulations) of Maritae

Local hemipopulations of metacercariae give rise to local hemipopulations of maritae. The transmission of infection to the definitive hosts takes place along trophic chains. It is, in fact, a group, or block exogenous accumulation (see p. 330). Usually a group of larvae gets into one host individual. In the case of adolescariae it results from a relatively low dispersion of cercariae after they are shed by the molluscan host (Fasciolidae, Paramphistomidae), or from the accumulation of the encysted larvae on the surface of the carapace of crustaceans or the shell of molluscs (some Notocotylidae, Pronocephalidae). As shown in the previous part of the chapter, metacercariae tend to accumulate in the organism of the second intermediate host. The local maritae groups formed turn out to be quite diverse in structure in the trematodes from different taxa and with different directions of morphological evolution. The latter circumstance does not play any significant role in the determination of the structure of local metacercariae groups. Both the species with "smooth" and "metamorphic" morphogenesis are characteristically heterogeneous by their age structure (see sections 2.2.2.1. and 2.2.2.2. in chapter 2). The situation with maritae, working out various strategies of increasing reproductive potential, is quite different, which is clearly seen in the representatives of the higher orders, Strigeidida and Plagiorchiida.

A long individual life span of the maritae in the strigeidids result in the formation in the hosts of local hemipopulations, which are long-lived and heterogeneous in structure. New juveniles are more or less regularly recruited into them, and if the host is active enough, they may originate from different metacercariae hemipopulations. As a result, the groups turn out to be genetically heterogeneous, which provides abundant material for recombinant variability. Local hemipopulations of this type exist for a long time and may disperse the eggs over relatively large distances. Noteworthy, a strategy like this pays off if the host is present for a long time in the biotope, where the completion of the parasite's life cycle is possible (i.e., the first and the second intermediate hosts are present), or if it often returns there. Otherwise, the infection dissemination would be mostly "wasted".

A mass production of small eggs in combination with a short life span result in a different structure of local maritae groups in the most specialized of the plariorchiids. In less specialized forms (e.g., Plagiorchiidae, Ochetosomatidae) it seems to have many traits in common with that described for strigeidids. Life span of maritae of these plagiorchiids is rather long (exceeding 2 months in the representatives of the genus *Plagiorchis*), their individual fecundity increasing greatly with age (Krasnolobova, 1987). The greatest degree of maritae juvenilization in Lecithodendrioidea and Microphallidae, which have, in fact, an embryonic morphological-functional organization (see p. 198), limits their life up to several days. This is compensated to a great degree by a high reproductive potential of the maritae group, determined by the short period of maritogony (several hours) and a very high individual fecundity of the worms (more than 1000 eggs in the uterus at the same time – our data). These two circumstances allowed the microphallids to use migratory birds as their definitive hosts. A short presence at a given coastal area is enough for them to become infected and to disseminate the infection, eggs with miracidia. It should be emphasized that a longer life span of the maritae to a great extent looses its advantages for the population reproduction when the maritae parasitize in migratory birds.

If a constant repetitive infection of the definitive hosts takes place for a more or less long time, the age structure of the forming local maritae hemipopulations of lecithodendrioideans and microphallids becomes dynamic and diverse. Alongside with the juveniles and the worms only starting egg production, there always are mature parasites. Their proportion in the local groups is always lower than that of young worms, which can be explained by the short time of the period of active reproduction and a fast releasing of the ageing individuals from the host. The structure of local hemipopulations of maritae of the microphallids of the "*pygmaeus*" group is quite unique. The features of the parthenogenetic part of their life cycle (see p. 70) result in a multiple copying of a single genotype, which is inherited by all metacercariae in a mollusc (their number reaching 7600 – Belopolskaya, 1963). As a matter of fact, each infected mollusc contains a metcercarial clone. A considerable prevalence of the "*pygmaeus*" microphallids in the first intermediate hosts ensures an almost simultaneous infection of birds by several, and sometimes many larval clones, forming the local maritae hemipopulation. Its numbers reach hundreds of thousands individuals in the common eider (Kulachkova, 1979), its composition changing quickly owing to a short life span of mature worms and the supply of new metacercariae clones. All this allows the appearance of diverse genetic combinations resulting from the amphimixis, which leads to an increased genetic polymorphism of the populations of the microphallids of the "*pygmaeus*" group. This appears to have been the cause of an outbreak of speciation and a high level of intraspecific variability observed in these trematodes (Galaktionov, 1993; Galaktionov et al., *submit*).

Similarly to the metacercariae, the maritae hemipopulation is maintained, as a rule, by the groups in different host species. Their number is determined by the specificity, the degree of its variation correlating, according to Dubinina's concept (1953), with the nature of morphogenetic transformations in the course of maritogony. At first sight, this concept seems to be correct. For example, the microphallids, in which maritogony is minimal, also have a wide specificity (Belopolskaya, 1963). However, our data show that even within the "*pygmaeus*" group the maritae of the species *Microphallus pygmaeus* develop successfully in eiders and somewhat less successfully in gulls, the maritae of *M. piriformes*, *vice versa*, and those of

M. pseudopygmaeus develop in eiders only. The scope of the definitive hosts is also very wide in many strigeidid species (Sudarikov, 1984), despite the fact that they undergo some of the most complex and prolonged transformations in the course of maritogony among the trematodes (see p. 190). It seems evident that specificity manifestations depend not only on the nature of morphogenesis but on a number of other causes as well. A relatively narrow range of animal hosts is characteristic, e.g., of the representatives of Fasciolidae, maritogony in them being associated with migrations in the host's body. A narrow specificity of schistosomatids may be determined by a high level of immunological activity of their environment, the blood of the vertebrates. To survive there, these trematodes have formed a complex and species-specific mechanism of antigenic mimicry (for reviews see: Pearce and Sher, 1987; Wilson and Coulson, 1998).

The definitive hosts of a given parasitic species may be conditionally subdivided into three categories: specific hosts, where the worms find the most favourable developmental conditions; poorly specific, where they do reach maturity, but it takes a longer time, their life span being shorter than in specific hosts; and eliminative ones, where immature maritae die. The stable existence of trematode populations is mostly maintained by specific hosts. In concrete parasitic systems 1-2 basic species are distinguished among them, owing to their high numbers and ecological features in a given ecosystem (Kontrimavichus and Atrashkevich, 1982). It is through them that the main flow of parasites passes. As for maritae groups in poorly specific hosts, which still supply a certain number of eggs into the environment, their role may be different. If specific hosts are present in the biotope in large numbers and more or less constantly, the role of such local groups is not very important: it is not through them that the main parasitic flow passes. The situation changes if the concentration of specific hosts drops abruptly or they completely disappear (e.g., migrate to other areas, die of diseases, are extinguished by predators). In such cases, an insignificant but constant background infection is supported by these poorly specific hosts. Such a background may be the source of new potent foci after the return or restoration of the numbers of specific hosts. And, finally, the local groups in eliminative hosts completely fall out of the reproductive processes of the population.

The spatial distribution of maritae hemipopulations, as well as that of other groups of the parasitic stages of the life cycle, is determined by the biological features of the animal hosts and their reaction to the impact of environmental factors (for review see: Shaw and Dobson, 1995). This situation may be very complex and dynamic, reflecting the complex behaviour of the vertebrate animals, the definitive hosts of the trematodes. A solitary or gregarious mode of life, a confinement to a certain comparatively small area or, on the contrary, an ability to cover large distances, seasonal changes in the behaviour (flocking, migrations, seasonal variations of activity) and many other factors, taken together, determine the dispersion or concentration of the maritae hemipopulations and their temporal fluctuations.

Seasonal changes in the maritae hemipopulations, associated with migrating animals, are a very good example. In extreme instances, this group is present in a certain ecosystem only for several days a year, e.g., at the time of a short stopover of migratory birds during spring and autumn migration. This is the case with some microphallid species (for review see: Galaktionov, 1993). The maritae hemipopulations at the nesting sites of the same migratory birds in the northern regions exist for a longer time. They are present throughout almost all the summer, i.e., coincide with the time of maximum activity of the intermediate hosts. From the birds' departure in autumn to their return in spring, the hemipopulations of maritae in these areas are absent. An opposite situation is observed in the wintering places and stopover sites of the birds. For instance, according to the data of Abdul-Salam et al. (1994), the species diversity of parthenitae and cercariae in the molluscs *Clypeomorus bifasciata* in the Kuwait Bay is maximal in autumn and spring, which testifies to the presence of the maritae hemipopulations at this time. This is the case, as a mass migration of sea birds passes this region in autumn and spring.

4. GENERAL NOTES

In our opinion, the material presented in this chapter is a good illustration of the idea, outlined in the beginning, that the complex life cycles of the trematodes result in the complexly structured populations. They may be represented by different intrapopulational groups, formed by the different life cycle phases in the same locality in different seasons, contrary to free-living animals, which ecologists are accustomed to. Thus, in the northern latitudes, most trematode species are represented by the parthenita hemipopulations (mostly young or renewing groups) in the molluscs and metacercariae hemipopulations in the overwintering invertebrates, fishes and amphibians. The maritae hemipopulations disappear together with their migrating definitive hosts, migratory birds and migrating mammals. The adolescariae in the environment, and sometimes also the metacercariae in the second definitive hosts, perish. The hemipopulations of fully embryonated eggs might also be retained for some time at the bottom of the water bodies. The transmission of infection also stops: egg dissemination, cercariae emergence, recruitment of metacercariae hemipopulations. All these processes start again in spring, when the definitive hosts appear in the ecosystems. The fluctuations in the population structure within the season may be associated with the intraseasonal changes in the host's biology (changes in food composition, migration from a certain biotope, etc.) or with cyclic reconstructions in the reproduction of parthenogenetic generations. As is has often been mentioned, each specific case calls for a detailed all-embracing investigation, which would enable us to obtain a holistic view on the ways of the life cycle completion of a given trematode species in the ecosystem studied.

A remarkable plasticity of the trematode life cycle catches the eye. It is especially reflected clearly in the functioning of its phases. As a matter of fact, in every phase the process of infection transfer from host to host may follow different patterns. Free-living larvae may change their taxes and kineses under the influence of external conditions and larval age, they form "dominant" and "satellite" clouds of miracidia and cercariae, which can be redistributed in space and time. The parthenogenetic life cycle stages have a complex, and in many instances environment-dependent dynamics of demographic processes. Metacercariae usually have a wide range of hosts, whereas maritae can use both specific and non-specific hosts, their relative role in the life cycle maintenance changing under the influence of specific environmental conditions. In some cases we even come across the change in transmission ways under the influence of external factors. Thus, in warm season the cercariae of the gymnophallid *Gymnophallus choledochus* are shed from the mollusc and penetrate into the second intermediate hosts, whereas in winter they encyst inside the daughter sporocysts and develop there until they reach the infective state (Loos-Frank, 1969) (Fig. 128).

On the one hand, the above circumstances ensure a successful transmission of a complex life cycle in a highly variable environment, which is determined not only by the changing environmental factors but also by the biology, physiological state, genetic polymorphism, population parameters of the animal hosts, both intermediate and definitive. Also, a favourable ground is created for the intensification of speciation, associated with the expansion into new hosts or adaptation to different ecological groups within the same host species (see, e.g., p. 384 – the schistosomes). This plasticity may be an important component in the formation of the phenomenon of a complex parasitic life cycle, as it increases significantly, in combination with other adaptations, the probability of the infection transmission in a successive series of hosts. It must have also facilitated a remarkably broad distribution of the trematodes and their expansion into the ecosystems of various types. This will be discussed in chapter 5 (see p. 471).

In conclusion, we shall mention another evolutionary important notion. In our opinion, it is a logical extension of the populational analysis of the trematode life cycles made in this chapter. We mean the unique character of the interactions between the parthenita groups and their hosts, the molluscs. It is the parthenita microhemipopulation that, as noted above (see p. 351) possesses the most important features of a true population (the micropopulation, according to Beklemishev), such as the ability of self-reproduction and the maintenance of optimal numbers by means of a demographic process. Nothing of this kind is observed in the hemipopulations of

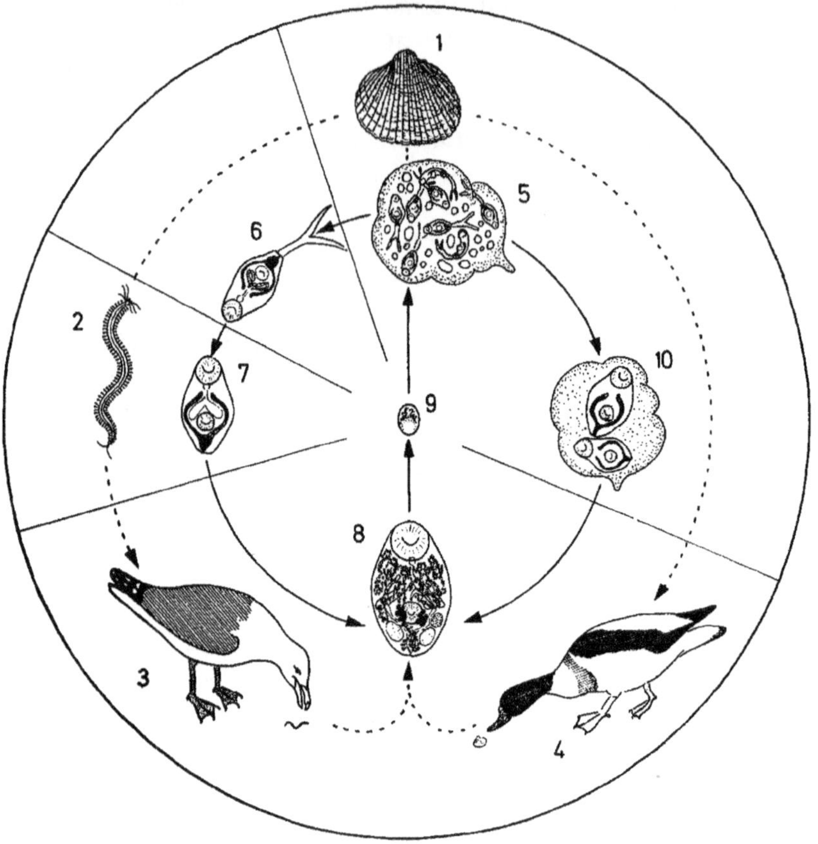

Figure 128. Alternating season-influenced transmission ways of the *Gymnophallus choledochus* life cycle. External circle – hosts: 1, the cockle *Cardium edule*; 2, the polychaete *Nereis diversicolor*; 3, the gull *Larus argentatus*; 4, the goose *Tadorna tadorna*. Internal circle – stages of the parasite life cycle: 5, the sporocyst (spring and summer); 6, the free swimming cercaria; 7, the metacercaria; 8, the marita; 9, the egg; 10, the sporocyst with metacercariae (autumn and winter). From Loos-Frank (1969), reprinted with the permission of Springer-Verlag GmbH & Co.KG.

maritae and metacercariae in the definitive and the second intermediate hosts. Such microhemipopulations appear to be characteristic of many, if not all, redial and many sporocyst-bearing trematodes (except the plagiorchiid branch), i.e., those which also comprise the representatives of the most archaic families. It also seems appropriate to recall the presence of mechanisms preventing the superinfection of the molluscs, the phenomenon of parasitic castration, the manipulation of the host's behaviour, the cercarial emergence rhythms, seasonal reconstructions in the composition of the parthenita groups and many other specific features of their functioning, which were discussed in section 3.2. Finally, the special role of the molluscs in the conservation of infection in the ecosystems should be noted. Almost all phase groups may disappear from the biotope in unfavourable seasons, except the parthenita hemipopulation. It is by means of them that the transmission of the life cycle of a trematode species resumes in the ecosystem after the onset of favourable conditions. In our opinion, all the above considerations unambiguously point to a long co-evolution of the trematode parthenitae – molluscs associations, and may provide an indirect evidence in favour of the hypothesis of the primary place of the molluscs in the life cycle of this group of parasitic flatworms.

---

[i] It should be remembered that a life cycle may comprise one ontogenesis, or, in case of metagenesis (sporozoans, cnidarians, etc.) or heterogony (myxozoans, trematodes, some crustaceans and insects), several ontogenesises.

[ii] Polymorphism (possibly, determined genetically) in the manifestation of behavioural reactions may play a much more important role in the trematode evolution that it has been discussed here. In the opinion of some researchers (Hamilton et al., 1990; Howard and Lively, 1994; Lively, 1995; Lively and Howard, 1997), the selection in the host populations should be against the common genotypes, to which the parasites are adapted. As the parasites cannot instantaneously track genetic changes in the host population, there will be a certain time lag. It is during this time that the "the ambiguity of behavioural reaction manifestation" of the larvae may play the role of utmost importance for the transmission of the parasite. In this case the phenomenon discussed may be viewed as a peculiar counter-adaptation.

[iii] This example provides yet another evidence of the existence of the group-, or kin-selection in the trematodes (see the footnote ii in chapter 2), which was also noted by Poulin (1998). Importantly, the cercaria which has penetrated into the ganglion does not develop further, and dies. As Poulin (1998) rightly remarks, this is the cost of manipulation of host behaviour by the parasite. A single larva is sacrificed for the sake of the success of a group of genetically identical individuals. Indeed, the complex cyst ingested by the ant contains a clone of larvae, all of them having been produced by a genetically identical group of daughter sporocysts, which, in their turn, have developed from a single miracidium.

Chapter 5

THE MAIN TRENDS IN TREMATODE EVOLUTION

The main feature of trematodes determining their specificity and difference from other groups of parasitic flatworms may be the phenomenon of alternation of generations relegated to different hosts (see p. 481). The differences between parthenogenetic and hermaphroditic generations are numerous and often crucial. They are manifested practically in the characteristics of the organization, biology, reproduction and even in the interactions with the host. There are several reasons for this. The main point seems to be the fact that parthenitae are the most ancient generations, which became parasitic long before the hermaphrodite one (for more details see chapter 6, p. 491). The nature of these individuals may be considered in different ways (as paedogenetic larvae or as degraded adult forms – see p. 482 for more details), but their great evolutionary age and tissue parasitism undoubtedly left a mark on all the features of their organization and biology. This can not be said about the hermaphroditic generation, which retained a great number of morphological features characteristic of flatworms. Therefore the features of the morphological evolution of these two generations are expressed in different ways. Unfortunately, we know much less about parthenitae than about cercariae and maritae, but even those fragmentary data allow us to outline the general trends of their morphological transformations in phylogeny.

1. THE MAIN TRENDS OF MORPHOLOGICAL EVOLUTION OF TREMATODES

1.1. The Parthenogenetic Generations

Parthenogenetic generations in the life cycle of trematodes are usually represented by several (more often not less than two) generations. The first one is the mother sporocyst, which is also the most ancient one. It develops with metamorphosis since its ontogenesis contains a free-living dispersive phase, the miracidium. Inhabiting different media, i.e., the external environment and the molluscan host, also predetermines different trends in the morphological evolution of the free-living and parasitic phases of the development of parthenitae according to the idea of Schmalhausen (1969).

1.1.1. The Main Trends of Morphological Evolution of Miracidia

Speaking of the regularities of the morphological evolution of miracidia, it should be noted that in the process of establishment and subsequent evolutionary transformations of the life cycles of the recent digeneans replacement of actively swimming miracidia by those protected by an egg shell (passively infecting miracidia) seems to have occurred several times and quite independently in different taxa (see pp. 2, 236). Changing the mode of existence in the environment was accompanied by the noticeable secondary simplification of larvae (see p. 236), which at the same time by convergence acquired a certain morphological similarity. This similarity does not at all reflect the real phylogenetic relations between different taxa of modern digeneans (see p. 236). The main tendencies of larval secondary simplification are manifested already in actively swimming miracidia, affecting almost all systems of organs.

Miracidia possess a characteristic epidermis combining ciliated epithelial cells (epithelial plates) and elements of definitive tegument (hypodermal ridges with cytons). In the present section we pay attention only to the main tendencies of evolutionary transformation of the epidermis.

A major tendency may be reduction of the number of epithelial plates. At the same time the number of transverse rows is also reduced: 5→4→3→2, as well as the number of cells (see p. 10). The last 2 members of this series have evidently resulted from secondary specialization. The miracidia of Gorgoderidae, though retaining an active mode of life, are characterized by a comparatively small size, which seems to have determined reduction of the number of rows of epithelial plates to three. As for the 2-row disposition of epithelial plates it occurs only in miracidia which have lost their activity in the environment. The transition to passive infection of molluscs results in a number of cases in the retention of only one row of modified plates (Halipeginae, Bunocotylinae), and sometimes in their considerable reduction (Azygiidae, Brachylaimoidea, Lecithochiriinae) (see p. 10).

The group under consideration is taxonomically heterogeneous. A simplification of locomotory ciliate apparatus in different representatives of the group must have taken place independently.

Of considerable interest are the 2 first members of the above series, with 5 and 4 transverse rows of epithelial plates. The maximum number of rows, 5, has so far been found in miracidia of only one family, Fasciolidae. At the same time, 4 rows of epithelial plates occur in miracidia of a great number of families. Most probably such miracidia are more numerous but larvae of the majority of the families still await further study. Digeneans whose miracidia possess 4 rows of epithelial plates may be divided into 3 fairly isolated phylogenetic clades. The first includes two orders, Schistosomatida and Strigeidida. The second comprises several families grouped around Echinostomatidae (Philophthalmidae, Cyclocoelidae). Paramphistomoid complex of families is also related to them (Paramphistomidae, Diplodiscidae). Finally, the third clade is represented by more specialised forms (Allocreadiidae). According to Seitner (1951), in the miracidia of *Allocreadium ictaluri* 2 plates of the 4th row become rudimentary. This seems to be a stage "transitional" to the formation of a 3-row epithelial formula. The miracidia of *Allocreadium fasciatusi* retained only 3 rows of epithelial plates (Madhavi, 1976).

The family Fasciolidae is certainly close to the second clade. This has been accepted by many investigators (Skrjabin, 1948; Yamaguti, 1958; Brooks et al., 1985a; Cribb et al., 2001b). A question arises whether it is possible to consider 5 rows of the epithelial plates of the miracidia of Fasciolidae as an archaic character or whether it is the result of secondary polymerization of locomotory structures. Unfortunately there is no unequivocal answer to this question. The main groups whose miracidia have 4 rows of epithelial plates each are treated by many investigators as ancient. Thus, Cribb et al. (2001b) unite the orders Schistosomatida and Strigeidida into one clade and treat it as a sister group with respect to all other Digenea. Gibson (1987) considers Paramphistomidae and closely allied families to be the most ancient ones. Ginetsinskaya (1968) relegates Echinostomatidae and closely allied families to the group of the archaic forms. All this suggests that the presence of 4 rows of epithelial plates in miracidia should be considered as a plesiomorphic state of this character.

Yet Ginetsinskaya (1968) considers the family Fasciolidae as the model group, recapitulating the most archaic type of the life cycle, and the presence of 5 rows of epithelial plates in miracidia as testimony of an ancient origin of this taxon. This point of view seems correct to us. Secondary polymerization of morphological structures, common in unicellular protists, does not occur in metazoan evolution (Poljansky and Raikov, 1977). On the contrary, oligomerization, i.e., reduction of the number of the homologous structures, is one of the main tendencies of morphological evolution of multicellular organisms (Dogiel, 1954). In our opinion, there is no ground to suppose that miracidia of Fasciolidae with their 5 rows of epithelial plates may be an exception from this rule. We consider the presence of 5 rows of epithelial pates as a plesiomorphic state in recent trematodes, though at dispersive stages of ancestral forms they were much more numerous. Miracidia acquiring a 4-row line-up of epithelial plates is a phenomenon secondary by nature that occurred quite independently in two large phylogenetic clades (Strigeidida – Schistosomatidae and a complex of families grouped around Fasciolidae). In the first case, ancestral forms, whose miracidia possessed 5 rows of epithelial plates, have not been retained, whereas in the second case it was Fasciolidae that have retained this plesiomorphic character.

Unfortunately, other details of structure of miracidial epithelium, especially those associated with the primordium of definitive tegument of the mother sporocyst (hypodermal ridges and their cytons) do not yield to analysis due to paucity of data available. However, one issue should be mentioned here. It was shown that in the miracidia from two families, Schistosomatidae and Paramphostomidae, hypodermal cytons deep in the body are connected by cytoplasmic bridges (Pan, 1985; Dunn et al., 1987) (Fig. 129A). Though it can not be ruled out that this feature of tegument structure is common in other trematodes (see p. 113) (not only in their miracidia), so far there are no such data. Even in the best-studied miracidia of *Fasciola hepatica* and *Philophthalmus rhionica* (Wilson, 1969a,c; Sinha, 1978; Tihomirov, 1980) nothing of the kind could be found (Fig. 129B). At the same time such an organization of the tegument is described in many species of Cestoda, not only in adult but also in larval stages (oncosphere, procercoid) (Kuperman and Davydov, 1981; Davydov, 1991; Davydov and Kuperman, 1993; Davydov et al., 1995). It is quite possible that this character reflects some structural features of the tegument inherent to ancestral forms of modern Neodermata. Cestodes and Amphilinida retained it throughout their subsequent evolution, whereas digeneans mostly lost it. The reason behind the loss of this character could be juvenilization and miniaturization that, as it has been repeatedly mentioned, accompanied digenean evolution (see pp. 27, 151, 211). If our suggestion is justified, it will be one more reason to believe that Schistosomatidae and Paramphistomidae retained many ancestral characters (Gibson, 1987; Semenov, 1991; Cribb et al., 2001b).

An important role in the realization of the main biological function of miracidia, infection of molluscs, belongs to the penetration complex (Fig. 129A,C,D,E). We shall touch upon only two of its elements: the terebratorium and the apical or penetration gland.

In the miracidia of recent digeneans, 2 types of terebratoria have been described, i.e., telescopic and spherical (see p. 15) (Fig. 129A,C). Besides, in a considerable number of species, which have passed over to the passive way of infection, terebratoria are secondarily simplified and acquire the shape of a simple conical structure.

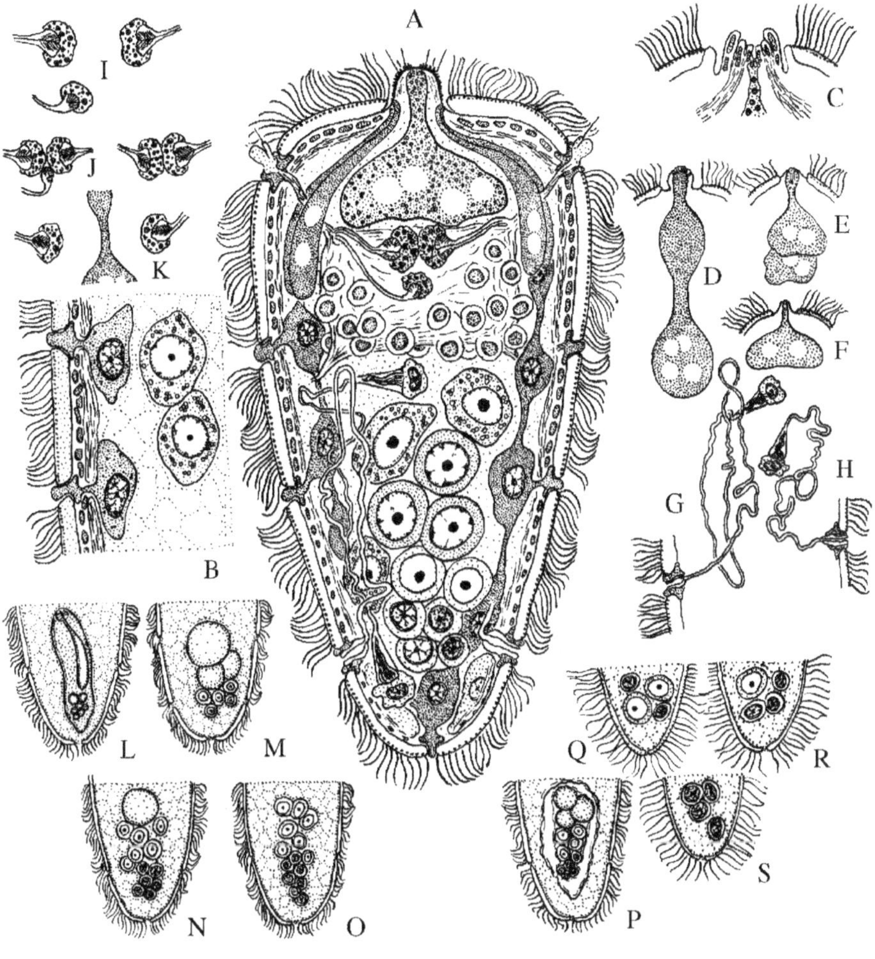

Figure 129. A generalized diagram showing the miracidium structure. A, diagram, showing a plesiomorphic state of the main morphological characters; B-S, a possible apomorphic state of the main morphological characters: B, structure of hypodermis (tegumental primordium); C, telescopic terebratorium; D-F, different variants of apical gland specialization (see text for details); G, stenostomate type of excretory system with one flame cell; H, mesostomate type of excretory system with one flame cell; I-K, different variants of location of pigmented eyespots (see text for details); L-S, different variants of germinal material organization (see text for details).

We have suggested previously (see p. 16) that telescopic terebratoria, peculiar to species belonging to a sufficiently compact group of families (Fasciolidae, Echinostomatidae, Philophthalmidae, Cyclocoelidae and possibly Cathaemasiidae and Ommatobrephidae), are the most archaic ones and reflect the plesiomorphic state of this character (Galaktionov and Dobrovolskij, 1998). At the same time the spherical terebratoria occurring in the representatives of taxa sufficiently distinct from each other (Strigeidae and

Schistosomatidae on the one hand and Paramphistomidae on the other) were considered by us as the result of secondary specialization associated with the increased role of chemical agents (secretory products) in the process of penetration. In our opinion this proposition should be reconsidered, especially since there is another point of view in the literature. Semenov (1991) considered the spherical terebratorium of Paramphistomidae as an extremely archaic character.

Telescopic terebratoria have a similar structure in all miracidia of the families Fasciolidae, Echinostomatidae, Philophthalmidae, Cyclocoelidae and possibly Cathaemasiidae and Ommatobrephidae). In Fasciolidae, though, terebratorium is somewhat simpler (Wilson, 1969a,c, 1971) than in Philophthalmidae (Tihomirov, 1980, 2000) but functionally and structurally these are sure to be homologous structures. All these families belong to the group of primitive taxa. Paramphistomidae and related groups also associated with the same complex of families. But not less ancient is the clade Strigeidida – Schistosomatida (see p. 521). These two large phylogenetic groups are sufficiently distant and possess, in effect, alternative characters (see p. 521). All the more interesting are relatively rare cases when the characters of representatives of these groups coincide. The presence of similar spherical terebratoria in Paramphistomidae, Strigeidae and Schistosomatidae is not accidental. Taking into account the fact that the terebratoria of this type are much simpler than telescopic terebratoria it can be suggested that it is they that reflect the plesiomorphic state of the above character inherited by the representatives of two large phylogenetic groups from the common ancestor. Within one of them, that is, in Fasciolidae and a group of related families, as the result of utmost specialization of muscle apparatus of the anterior end of the body telescopic terebratoria were secondarily formed. In representatives of the most specialized families (Philophthalmidae, Cyclocoelidae – Ginetsinskaya, 1968; Tihomirov, 1980, 2000) they also have a more complex structure. Thereafter, in connection with the transition of miracidia to the *per os* way of infecting molluscs, terebratoria of both types underwent secondary simplification. This process in different digenean taxa appears to have occurred quite independently. In our view, the above considerations convincingly show that the structure of the terebratorium (spherical or telescopic) is not a very valuable character for phylogenetic analysis of trematodes. This character can be used only in some families representing the lower branches of phylogenetic tree of modern digeneans.

The most important element of the penetration apparatus of miracidia is the apical gland (Fig. 129A,D,E,F). The morphological variability of these structures found in the miracidia of different Digenea is not very great. It can be reduced to several variants, which differ in the number of nuclei in the gland (see p. 17). In the majority of miracidia described to date representing both archaic and more specialized taxa (Schistosomatidae-Strigeidida, Azygiidae, a complex of families related to the Fasciolidae and

Paramphistomidae, Allocreadiidae, Opecoelidae) the apical gland contains 4 nuclei. Then miracidia with two 2-nuclei glands follow (Plagiorchiidae, Ochetosomatidae, Brachycoelidae, Heterophyidae, Opisthorchiidae). In representatives of quite different and distant families (Bucephalidae, Bunocotylinae, some Microphallidae) larvae with one 2-nuclei gland have been described. This series is brought up with miracidia whose apical gland is devoid of nuclei (Halipeginae, some Microphallidae).

There is an impression that initial stage of the above series reflects the sequential stages of juvenilization of the miracidia of the most specialized groups of trematodes. A typical 4-nuclei gland is formed in the ontogenesis of miracidia as a result of fusion of 4 separate cells, i.e., is a syncytium. The arrest of this process at an earlier stage of morphogenesis, which can be treated as one of numerous manifestations of heterochrony, gives rise to the emergence of 2-nuclei secretory cells. On the contrary, disappearance of one of these cells appears to be the result of a common reduction. The emergence of nuclear-free apical glands is, however, the result of a different process. Resorption of nuclei of specialized cells at the final stage of their differentiation is a general biological and very widely spread phenomenon. It should be noted that both arrest of syncytium formation in larval morphogenesis and reduction of part of the cells and resorption of their nuclei are carried out in different digenean families in parallel and independently of each other. All this limits the possibility of using this complex of characters for analysis of phylogenetic relations between taxa to a great extent. Actually it can be used only for a qualitative evaluation of the degree of some group specialization.

In this respect, another characteristic is of a somewhat greater interest – the shape of the apical gland and its position with respect to the ganglion (Fig. 129A). The apical gland may be compact (Schistosomatidae, a complex of families approximated to Fasciolidae, Allocreadiidae, Opecoelidae), occupying a strictly pre-ganglion position. The alternative variant is when the apical gland is elongated, its long duct sometimes forming a well expressed enlargement passes under the ganglion, whereas the swollen nuclei-containing part is always located behind the ganglion (Fig. 129D). A very elongated sac-like gland is retained in the extremely specialized larvae of Azygiidae, which have lost the period of active swimming in the environment. The same type of the apical gland seems to be inherent in miracidia of Paramphistomidae and related families. In any case judging by the data of Lengy (1960), in the miracidium of *Paramphistomum microbothrium* the apical gland consists of a slightly swollen duct and a sac-like nuclei-containing section. A considerable part of the latter is located behind the ganglion.

In each of the two clades under consideration both variants of the apical gland organization may be found. The miracidia of Schistosomatidae possess a compact gland. We treat this state of the character as plesiomorphic. The origin of the apical gland remains unknown, yet it can be supposed to be a modification of the head glands of some ancestral forms. Initially it seems to have been a compact group of small unicellular glands each of which opened outwards independently by means of a short duct. The integration of these glands into a single multinuclear complex occurred already in proto-trematodes and was retained in recent trematodes. The data of embryonic development of miracidia speak strongly in favour of this conjecture. The primordium of the apical gland is represented by 4 independent cells, which fuse and give rise to a 4-nuclear syncytium only at relatively late stages of morphogenesis (see p. 17).

The differentiation of the gland into the cyton containing nuclei and a long duct reinforced with microtubules, and moreover the emergence of a reserve swelling, the immersion of the nuclei-containing section deep into the body of the miracidium, all those are certainly secondary transformations testifying to a greater specialization of miracidia. There could be several reasons which caused those transformations. One of them may have been associated with the increasing role of the chemical agents in the process of penetration of miracidia possessing spherical terebratoria (see p. 223). In fully formed larvae the processes of secretion synthesis seem to be over, the greater part of the apical gland is filled with secretory granules. The emergence of a long duct with a large reserve swelling ensures the expansion of the gland volume and consequently the quantity of secretion stored in it. There is a distinct correlation: the modified apical glands have been so far described only in miracidia possessing spherical terebratoria. At the same time a similar association between a compact apical gland and telescopic terebratoria can not be traced as distinctly: the compact apical gland is possessed not only by the miracidia of Fasciolidae and allied families but also by the larvae of Spirorchiidae and Schistosomatidae originally possessing spherical terebratoria. This type of the apical gland is widely seen in more specialized taxa: Opecoelidae, Allocreadiidae, Encyclometridae, Bunoderidae, which have acquired a secondarily simplified conical terebratorium. It can only be supposed that in all cases like these the function of the apical gland is enhanced not at the expense of increasing its volume and the amount of secretion stored but as a result of qualitative changes in the properties of the latter, first of all, elevation of the activity level of lytic enzymes.

Of special interest for clarifying the possible phylogenetic relations between different taxa of recent trematodes is analysis of the features of the organization of the protonephridial system of miracidia (Fig. 129A,G,H). Two important characters are connected with the latter: the quantity of flame cells and the localization and passage of the excretory ducts.

Miracidia of Schistosomatidae, Spirorchiidae, Diplostomidae and Strigeidae have 2 pairs of flame cells each – the excretory formula is $2[(1) + (1)] = 4$ (Fig. 129A). In the other digeneans, if their miracidia possess protonephridia, there is only one pair of flame cells (Fig. 129G,H). Finally, in the miracidia of a number of the most specialized trematodes (Bucephalidae, Brachylaimoidea, Bunocotylinae, Lecithodendrioidea, Microphallidae) protonephridia were not found (see p. 23). In our opinion this morphological line reflects the same tendency of the excretory system evolution, which is manifested so distinctly in the hermaphroditic generation: reduction of the number of flame cells and the simplification of the excretory formula (Galaktionov and Dobrovolskij, 1987; see also p. 131). As in the case of the absence of protonephridia in the most specialized and secondary simplified larvae, a similar phenomenon is also present in the Cestoda, i.e., the oncospheres of Psedophyllidea possess a pair of flame cells with short excretory ducts, whereas the most specialized larvae of Cyclophyllidea are completely devoid of them (Rybicka, 1964; Malmberg, 1972; Kuperman, 1988).

To conclude, the presence of two pairs of flame cells appears to be a plesiomorphic character in miracidia. It is inherent in larvae of a very limited range of families – Spirorchiidae and Schistosomatidae of the order Schistosomatida and Strigeidae and Diplostomidae of the order Strigeidida. More specialized miracidia of the family Sanguinicolidae, in the order Schistosomatida, possess only two flame cells. Highly specialized larvae of the Azygiidae also possess two flame cells. This family, judging by a number of characters pertaining to body plan (see p. 522), is evidently related to the above phylogenetic clade.

Miracidia of Paramphistomidae and Fasciolidae, as well as larvae of all the related families possess only one pair of flame cells. It may be supposed that this is the result of the secondary simplification of the excretory formula of ancestral forms, which must have possessed at least 2 pairs of flame cells as do modern representatives of the Schistosomatida-Strigeidida complex which have retained the archaic state of the character under consideration.

Unfortunately, up to now nobody has ever paid attention to the location of the tubules of the miracidia excretory system. In the papers of different authors only mention of "highly twisted tubules" is made, therefore in many cases we had to resort to the analysis of the illustrative material. Dubois (1929) suggested that two types of the excretory system of cercariae, "Stenostoma" and "Mesosotoma", should be distinguished (see p. 136). The first is inherent in the archaic representatives of Digenea, the second in more specialized ones. The same two types appear to be easily distinguished in miracidia as well.

The miracidia of Diplostomidae possess a typical stenostomate excretory system (Vladimirov, 1961; Pearson, 1961; Dönges, 1964) (see fig. 1A,B, 129A). At the same time, Schistosomatidae and Spirorchiidae larvae, possessing the same excretory formula as Diplostomidae, have the excretory system of the mesostomate type.

Among miracidia possessing one pair of flame cells, two types of excretory system can be also found (Fig. 129G,H). In larvae of *Paramphis-*

tomum microbothrium the excretory channels are first directed forward and reach the level of the posterior end of the apical gland. There they form a characteristic loop, turn back and on reaching the level of excretory pores are directed forward again. The flame cells are located in the middle part of the body (Lengy, 1960). But in most of the miracidia (Fasciolidae, Echinostomatidae, Psilostomidae, Philophthalmidae, Allocreadiidae, Opecoelidae) highly twisted channels are directed forward from the excretory bladder or immediately from the excretory pore to the flame cells, usually located not far from the ganglion. In all these cases the loop characteristic of the stenostomate excretory system is absent.

The stenostomate excretory system is more archaic both in miracidia and in the hermaphroditic generation. In our opinion it also seems fairly certain that simplification of the course of the excretory system channels, i.e., formation of the mesostomate system occurred quite independently in different phylogenetic clades. Moreover, it was not connected with the simplification of the excretory formula, so as the Mesostoma type is found in miracidia both with one and two pairs of flame cells.

The location of excretory pores may turn out to be another important character associated with the excretory system (Galaktionov and Dobrovolskij, 1987, 1998). Unfortunately, data on the location of excretory pores in miracidia are scarce and therefore the significance of this character is unclear. Here we will only list the differences in the location of excretory pores known for the larvae of different species. This issue will be discussed in more detail in the section devoted to parasitic phases of parthenitae development (see p. 429). Noteworthy, the location of excretory pores is one of the key characters used in the analysis of body plan organization of the hermaphroditic generation (Galaktionov and Dobrovolskij, 1987, 1998).

The excretory pores in miracidia are located laterally and with few exceptions are located on the transverse hypodermal ridges. They may be slightly shifted either dorsally or ventrally (Spirorchiidae, Allocreadiidae, Heronimidae). Semenov (1991) noted that in miracidia possessing one pair of flame cells the excretory pores are usually localized on the very posterior transverse hypodermal ridge, between the epithelial plates of the next-to-last and the last transverse rows (in Fasciolidae between the 4th and the 5th, in the miracidia possessing 4 rows of epithelial plates between the 3rd and the 4th, in the miracidia of *Allocreadium fasciatusi*, which have lost the 4th row of epithelial plates completely, between the plates of the 2nd and the 3rd rows). Only one exception of this rule is known so far – *Philophthalmus rhionica* (Philophthalmidae). In miracidia of this species excretory channels open in the middle part of the body, between the epithelial plates of the 2nd and the 3rd rows (Tihomirov, 1980, 2000). The flame cells retain their normal position and are localized before the excretory pores, immediately behind the ganglion.

In miracidia with two pairs of flame cells a great variability of this character is observed. In representatives of two families Spirorchiidae and Clinostomidae the excretory pores are localized in the middle part of the miracidial body, between the epithelial plates of the 2nd and 3rd rows (Wall, 1941; Goodchild and Kirk, 1960; Semenov, 1991). The excretory pores of miracidia of Schistosomatidae are localized on longitudinal hypodermal ridges, between the epithelial plates of the 3rd row (Southgate and Knowles, 1977; Albaret, 1984). Finally, in the miracidia of Strigeidida the excretory pores are located on the last transverse hypodermal ridge, between the 3rd and the 4th row of epithelial plates (Pearson, 1961; Vladimirov, 1961; Dönges, 1964).

Miracidia of Sanguinicolidae belonging to the order Schistosomatida possess one pair of flame cells. Their excretory pores are located as well as in miracidia of Spirorchiidae between the epithelial plates of the 2nd and 3rd rows (McMichael-Phillips et al., 1992b). The flame cells appear to be slightly shifted backwards as to the excretory pores.

Considerable data for understanding the main tendencies in the morphological evolution of the Digenea miracidia might be afforded by a detailed analysis of organization of their nervous system and sensory structures, yet it is practically impossible because of the paucity of appropriate data. The only exception is provided by miracidia photoreceptors which may be the structures of two types: pigmented eyespots of a rhabdomeric type and ciliary non-pigmented photoreceptors (?) (see p. 20).

Pigmented eyespots have been studied best of all and in greater detail (see p. 19). However, not all miracidia possess them. They are absent in those species whose miracidia infect the mollusc passively. So far only one exception from this rule is known. Miracidia of *Tamerlania bragai* (Eucotylidae) retain pigmented eyespots, in spite of the fact that life cycle of this species is completed on land through land molluscs and the miracidia are located in egg shells (Maldonado, 1945). At the same time some miracidia leading an active life in the environment are devoid of eyespots (Schistosomatidae, Paramphistomidae, Opecoelidae, Encyclometridae).

It is not to be doubted that the presence of pigmented eyespots is a plesiomorphic state of the character. Inherent to ancestral forms, it was retained in the miracidia of most Digenea studied in this respect. The reduction of pigmented eyespots proceeded independently in different groups and was not necessarily associated with transition to passive infection of hosts. It can be illustrated with the examples of the larvae of Schistosomatidae, Paramphistomidae, Opecoelidae, Encyclometridae, on one hand, and the miracidia of Eucotylidae, on the other hand.

Loss of eyespots by miracidia seems to have accompanied digenean evolution since the moment of their establishment as an independent group. Morphological series can be presented, illustrating the gradual disappearance of pigmented eyespots (Fig. 129A,J-K). In the miracidia of *Heronimus mol-*

lis there are three eyespots – two on one side and one on the other (see p. 19) (Fig. 129I). Miracidia of Fasciolidae and Echinostomatidae also possess three eyespots, but the small eyespot most often devoid of the pigment lining (Fig. 129J). The same pattern occurs in miracidia of Allocreadiidae. In miracidia of Philophthalmidae, Diplostomidae and Strigeidae there are only two eyespots. Finally, this series is brought up by eyeless miracidia characteristic of representatives of quite different taxa.

In our opinion, occurrence of miracidia with 3 eyespots in taxa occupying the lowest levels of the phylogenetic scheme of the modern Digenea testifies to the fact that ancestral forms possessed two pairs of eyespots. A similar state may be observed by some monogeneans (both in larvae and in adult worms).

One more character worth mentioning specially is associated with the pigmented eyespots of miracidia. It is the position of eyespots as to each other in the larva. They may be located as a compact group (Fasciolidae, Echinostomatidae, Philophthalmidae, Cyclocoelidae, Allocreadiidae) or, vice versa, may be spaced out and displaced towards the ganglion side, as it happens to be in miracidia of the order Strigeidida. It is noteworthy that widely spaced of eyespots correlate exactly with the emergence of a modified apical gland consisting of a nucleus-containing part and a duct with an expanded reservoir (see p. 17) (Fig. 129K). Most probably this correlation is not random. The emergence of a duct located dorsally of the ganglion, that is, exactly where a compact group of eyespots is localized conditioned the divergence of the latter. Thus, widely spaced eyespots are the result of secondary specialization caused by the restructuring of the apical gland.

Yet an alternative treatment of this character is also possible since both in Turbellaria, including commensal ones, and in the most archaic monogeneans, the eyespots are widely spaced and do not join each other along the medial line.

The greatest difficulty is presented by the treatment of data on the organization of the germinal material of miracidia (Fig. 129L-S). It has been stated above that it is easy to distinguish two main types of the organization of the germinal material (see p. 23). The first type is possessed by miracidia which have practically completed their reproduction at the last stages of morphogenesis (Fasciolidae, Transversotrematidae and the so-called paedogenetic miracidia – see p. 23), the second, by the miracidia of all those trematodes in which realization of their reproductive function is partly or completely transferred to the parasitic phase of the development of mother parthenogenetic generation (mother sporocyst). A clear morphological series can be constructed. This series illustrates different stages of the transfer and subsequent maximizing of the reproductive function (Dobrovolskij et al., 2000).

At one end of this series there are the aforementioned "paedogenetic" miracidia (Philophthalmidae, Cyclocoelidae, some Echinostomatidae and some Allocreadiidae) in which only one individual of the next generation is formed. Miracidia of Fasciolidae and Transversotrematidae contain 1-2 embryos of the next generation and only mature germinal cells (Fig. 129M). Miracidia of Paramphistomidae, in which there are only mature germinal cells and those finishing differentiation, also seem to belong to this group. In larvae of Echinostomatidae the composition of germinal elements is supplemented with a group of undifferentiated cells (*Echinostoma paraensei*), whereas the embryos may disappear completely (*E. caproni*) (Ataev et al., 2001a,b) (Fig. 129N,O). Judging by the description, and mainly by the figure presented in the paper of Crandall (1960), organization of the germinal material in the miracidia of *Heronimus mollis* (Heronimidae) is similar to that of *Echinostoma paraensei*) (Fig. 129P). Miracidia of Strigeidida, Schistosomatida, many Plagiorchiida, Heterophyidae and Bunocotylinae containing different numbers of germinal elements possess one general character in common: as well as miracidia of *Echinostoma caproni*, they possess only undifferentiated cells and germinal cells at different stages of differentiation) (Fig. 129Q,R). This series is brought up by the larvae of some Plagiorchiida possessing only undifferentiated cells (see p. 27 and fig. 129S).

Miracidia which form the two first terms of the above series can not be called larvae in the strict sense of this word. They are not simply generatively mature organisms, these are organisms which have fully completed their reproduction (Dobrovolskij et al., 2000; see also pp. 238, 492). Miracidia can be considered true larvae if they possess only a germinal primordium represented by a group of undifferentiated cells (for instance, miracidia of *Paralepoderma cloacicola*). In such cases it is mother sporocysts that become adult organisms at the parasitic phase of development. They start reproducing after some period of morphogenetic transformations accompanied with maturation of germinal cells. The presence of intermediate terms of this row testifies, in our opinion, to the fact that "transfer" of generative function realization from the miracidium to the parasitic phase of mother sporocyst development was carried out already within the limits of digeneans itself.

It is rather difficult to define the plesiomorphic state of the character under consideration. Naturally this cannot be the first group of the series represented by the so-called "paedogenetic" miracidia. On the contrary, the two next groups could reflect this state. It is true that these two groups are not so homogenous as it may seem at first sight. Actually, between representatives of families constituting them there are considerable biological differences already seen at the parasitic stage of development of the mother sporocyst. The problem in question is the fate of germinal cells and embryos if the miracidium possesses the latter. In Fasciolidae, Transversotredmatidae, Clinostomidae, Paragonimidae well studied in this respect only one embryo is retained in the mature mother sporocyst, however many germinal cells there may be in the miracidium (see p. 34). Thus the actual fecundity of the mother sporocysts of these trematodes is lower than the potential one. The possible reasons of this phenomenon will be considered in chapter 6.

Though the fate of germinal cells forming in miracidia of Paramphistomidae was not traced in great detail, they probably do not perish at the parasitic stage of development since the number of rediae released by the mother sporocyst is close to the number of germinal cells (Dinnik, 1961).

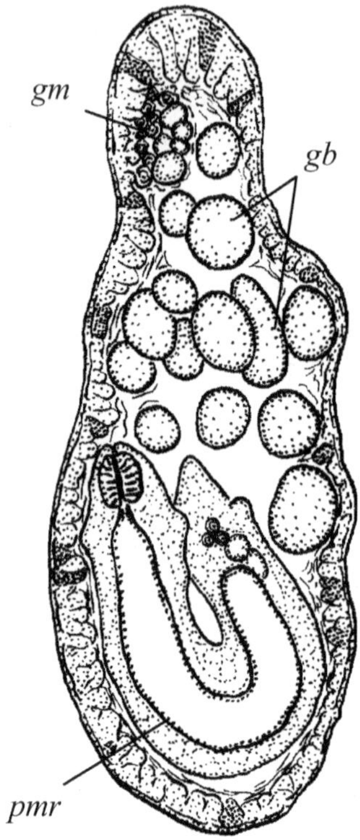

Figure 130. Mother sporocyst of *Echinostoma paraensei* (5-6 days old) with precocious mother redia. *Abbreviations*: *gb*, germinal ball; *gm*, germinal mass; *pmr*, precocious mother redia.

In this connection one should also mention *Echinosotoma paraensei*, though the germinal material of miracidia of this species, as well as other Echinostomatidae, belongs to the second type (see p. 24): undifferentiated cells which make it up, together with the embryo and mature germinal cells, proliferate and differentiate into germinal ones at the parasitic stage already (Ataev et al., 2001a,b). In the species under consideration one redia outstrips all the others in its development as early as in the mother sporocyst (Sapp et al., 1998). The authors refer to it as precocious mother redia (Fig. 130). It seems to develop from the very embryo, which is part of the germinal material of the miracidium. The mother sporocyst releases it on the 6th day after the infection of the mollusc, whereas the rest of the redia of the same generation are either represented by germinal balls or are at some intermediate stages of morphogenesis. This redia also differs from the others in its biological features: it does not leave the ventricle of the heart and is always localized close to the mother sporocyst. The similarity of the above picture with what occurs during the development of mother sporocyst of Fasciolidae (Dinnik and Dinnik, 1956; Palmieri et al., 1977; Czapski, 1978; Rakotondravao et al., 1992) was observed by the authors of the study referred to (Sapp et al., 1998). This similarity, in our opinion, reflects sufficiently ancient relations in the host-parasite system mollusc-parthenitae.

We suggest that the variant of organization of germinal material observed in miracidia of Fasciolidae, Paramphistomidae and other families constituting this "group" is most close to the plesiomorphic state. Advancing this suggestion we presume that the mother sporocyst parasitizing in the mollusc is the first and consequently the most ancient generation which started the formation of the life cycle of trematodes (a detailed foundation of this point of view and our opinion on the origin of Digenea will be discussed in chapter 6). Not "larvae", as many authors considered (Heyneman, 1960; Cable, 1965; Ginetsinskaya, 1968; Pearson, 1972) but generatively mature individuals passed over to commensal relations (James and Bowers, 1967; Dobrovolsky et al., 2000; Dobrovolskij and Galaktionov, 2003). As mentioned above, miracidia of representatives of many families which have preserved some archaic characters are just such organisms whose reproduction has been almost completed. At the early stages of formation of association "molluscan host – proto-digeneans" interrelations in this association were imperfect. This must have limited the possibilities of their reproduction: phylogenetically young parasites could not fully realize their reproductive potential. In our opinion, the limitations of actual fecundity of mother sporocysts of a whole number of archaic taxa can be considered as recapitulation of this ancient state. Subsequent evolutionary events seem to have been developing in two directions. The appearance of the so-called "paedogenetic" miracidia (Cyclocoelidae, Philophthalmidae, some Echinostomatidae, some Allocreadiidae – see p. 23) is a manifestation of one of them: the fecundity of the mother sporocyst becomes low. In such species only one individual of the next generation is developed already at the stage of the miracidium.

Reduction of actual fecundity down to the release of only one individual is observed only in "redial" trematodes whose parthenitae are characterized by a high degree of pathogenicity for their hosts (see p. 253). It cannot be ruled out that it is a kind of adaptation making it possible to slow down the increase of the number of parthenitae in the mollusc. Therefore the host has a possibility of better adaptation to the increasing parasitic load and this, in its turn, stabilizes the whole host-parasite system.

As mentioned above, another direction of evolutionary transformations of mother sporocysts was associated with the transfer of the generative function to the parasitic phase. On the one hand, it caused serious morphological transformations of mature mother sporocysts (see p. 245), on the other hand, transformation of miracidial germinal material into a simple germinal primordium represented only by non-differentiated germinal cells (*Paralepoderma cloacicola* – see p. 28).

Summing up, several important issues should be emphasized.

(1) The general tendency of miracidial morphological evolution is the simplification of their organization: reduction of the number of transverse rows of epithelial plates and their total number down to their complete disappearance, simplification of the excretory formula and replacement of the stenostomate excretory system by a simpler mesostomate one, reduction of the number and a subsequent complete reduction of pigmented eyespots, simplification of germinal material. All this is accompanied by the reduction of the size of miracidia and their cell composition.

So far, we have managed to discover only 2 exceptions from this rule: replacement of a simpler spherical terebratorium by a telescopic one, and modification of the apical gland in miracidia of Strigeidida. Both these changes concern the penetration apparatus of miracidia and are manifested in the representatives of the groups which retain quite a lot of archaic characters. Acquisition and subsequent evolution of the structures associated with penetration into the body of the mollusc could have taken place only after the ancestors of the modern digeneans turned from commensals inhabiting the host body cavities connected with the environment into parasites actively overcoming tissue barriers. Yet subsequently in miracidia of the representatives of the most specialized groups these structures have been simplified (see p. 10).

(2) The above tendency for morphological simplification of miracidia is not immediately associated with transition from active existence in the environment to passive existence protected by the eggshell. This transition only intensified the tendency under consideration that had started already in actively infecting miracidia.

(3) Unfortunately, there are considerable gaps in the knowledge of miracidia from archaic groups. The data available concern only a limited number of species. It may be due to these gaps in our knowledge that the combination of plesiomorphic and apomorphic states of almost all the main characters is "unexpected" and unique in all the miracidia described. An exception to this rule is only the shape of the apical gland and the location of pigmented eyespots (see p. 418).

The presence of different combinations of archaic characters and those testifying to the secondary specialization of their carriers appears to testify to a very early divergence of several phylogenetic branches giving rise to the recent Digenea taxa.

1.1.2. The Main Tendencies of the Morphological Evolution of the Mother Sporocyst and Daughter Generations of Parthenitae

The parasitic phase of development of the mother generation and the subsequent generations of trematode parthenitae are not well studied. The conventional opinion of parthenitae as simple organisms with a capacious brood cavity possessing a pharynx at most and a simple sac-like gut was a reason for the almost complete absence of their detailed microanatomical studies. The recent electron microscopical studies do not solve the problem since in most cases they are devoted to the description of the fine structure of some organ systems. Usually they are tegumentary structures, more rarely, the digestive system or the birth pore. The nervous and excretory systems of rediae and sporocysts are poorly studied. Since the data available in the literature are fragmentary the analysis of the possible tendencies of the morphological evolution of parthenitae is also fragmentary. We need to restrict our discussion of external morphology of rediae and sporocysts and of some details of their internal structure. First of all, it concerns the digestive and the excretory system.

We have already written that we consider parthenitae as the most ancient parasitic generations in the life cycle of digeneans which underwent the most morphological-functional restructuring under the influence of the parasitic mode of life. Though rediae and sporocysts are morphologically different (see section 1.2. of chapter 1), they are the same from an evolutionary and morphological regard. According to Ginetsinskaya (1968), they are organisms of the same kind differing only in the degree of morphological-physiological regress manifestation. This point of view goes back to Steenstrup (1842) who thought that rediae and sporocysts corresponded to each other in the life cycle of trematodes. This idea was developed and well-founded by Sewell (1922) by demonstrating the possible stages of gradual simplification of rediae to the morphological-functional state of a sporocyst. After that, this problem has been discussed in the literature (see reviews: Ginetsinskaya, 1968; Dobrovolskij et al., 1983; Pearson, 1992). However, it should be recalled that ontogenesis of the first and subsequent (if any) parthenogenetic generations is different: the mother sporocyst develops with metamorphosis, whereas direct development is characteristic of the subsequent generations. The reason for this difference and its nature will be considered later (see p. 493). It should be emphasized that it does not at all affect the correctness of the above. This circumstance allows us to consider both mother and daughter generations of parthenitae in one morphological series, reflecting only the most general tendencies of the morphological evolution of those organisms and not their phyletic relations.

The two morphological types of parthenitae deserve special mention. It is important to establish whether the redial or the sporocyst morphological type is more ancient. The more ancient type should reflect at least to some extent the patterns of organization of ancestors of modern digeneans. Up to now only "daughter" parthenitae (rediae and daughter sporocysts), i.e., parallel, "synchronic" generations were considered in the literature (Sewell, 1922; Ginetsinskaya, 1968; Dobrovolskij et al., 1983). Since the redial morphological type of this generation is inherent in all archaic taxa it was referred to as primary with respect to the sporocyst one. Mother sporocysts were usually not considered in the analysis. It seems ill-founded to us. Mother sporocysts, though they possess a certain specificity, (see p. 2) in many ways conditioned by their ancient origin, hardly differ from other parthenogenetic generations. The parthenitae are one single complex of alternating generations whose morphologic evolution is subject to the same regularities.

The generation of the mother sporocyst is always present in the life cycle of all digeneans. Yet, as mentioned above (see p. 28), the parasitic phase of the mother sporocyst development is left out in a number of species, and then this generation is represented only by the dispersive, free-living phase, the miracidium. In recent digeneans which have retained the parasitic phase of development of the mother sporocyst, the latter is always extremely specialized, which is easily accounted for by the fact that they are evolutionary very ancient organisms. The emergence of daughter generations of parthenitae is certain to have been secondary. At first, individuals released by mature mother sporocysts were shed into the environment, where the life cycle of the ancestral trematodes was completed (see p. 498). These organisms must have possessed a complete set of features inherent in Platyhelminthes in general and ancestors of the modern trematodes in particular. Therefore, it is not surprising that after the appearance in the life cycle first of the second and later of subsequent generations parasitic in the mollusc those organisms should have recapitulated some characteristics of organization and behaviour inherent in ancestral free-living forms.

From the above it follows that the redial morphological type should be considered as archaic (Fig. 131), though the rediae themselves must have emerged rather late, this phenomenon certainly being secondary. Transformation of rediae into daughter sporocysts, which occurred in representatives of a number of specialized taxa, seems to have taken place even later. From this point of view, simplification and morphological-physiological degeneration of mother and daughter sporocysts are parallel and rather independent phenomena though realized in accordance with the same laws of morphological evolution.

The morphological series (Fig. 131B,E,F,M,K,L): active rediae with well-expressed locomotory appendages → sedentary sack-like rediae →filamentous rediae → long filamentous sporocysts retaining an ability for weak undulation → immobile sac-like sporocysts → small round sporocysts, with small variations has been repeatedly suggested in the literature (Sewell, 1922; Ginetsinskaya, 1968; Dobrovolskij et al., 1983). It is quite possible that the first member of this series (the rediae of Echinostomatidae, Cathaemasiidae, Fasciolidae) are the most morphologically archaic rediae. It is also pretty certain that transition to sporocyst state occurred independently in different phylogenetic branches of the Digenea. This process was accompanied by the realization of different evolutionary tendencies.

Figure 131. Diversity of morphological types in the trematode parthenitae. A-H, rediae (A, a generalized diagram of the structure of archaic rediae – dorsal view; B-C, Echinostomatidae; D, Notocotylidae; E-F, Heterophyidae; G, Bivesiculidae; H, the mother redia of Philophthalmidae); I-N, daughter sporocysts (I, Cyathocotylidae; J, Opecoelidae; K, Plagiorchiidae; L, Lecithodendrioidea; M, Schistosomatidae (*Trichobilharzia*); N, Brachylaimoidea); O-T, mother sporocysts (O, Cyathocotylidae; P, Plagiorchiidae (*Paralepoderma*); Q, Plagiorchiidae (*Haplometra cylindracea*); R, Ochetosomatidae (*Lechriorchis*); S, Heronimidae; T, Fasciolidae).

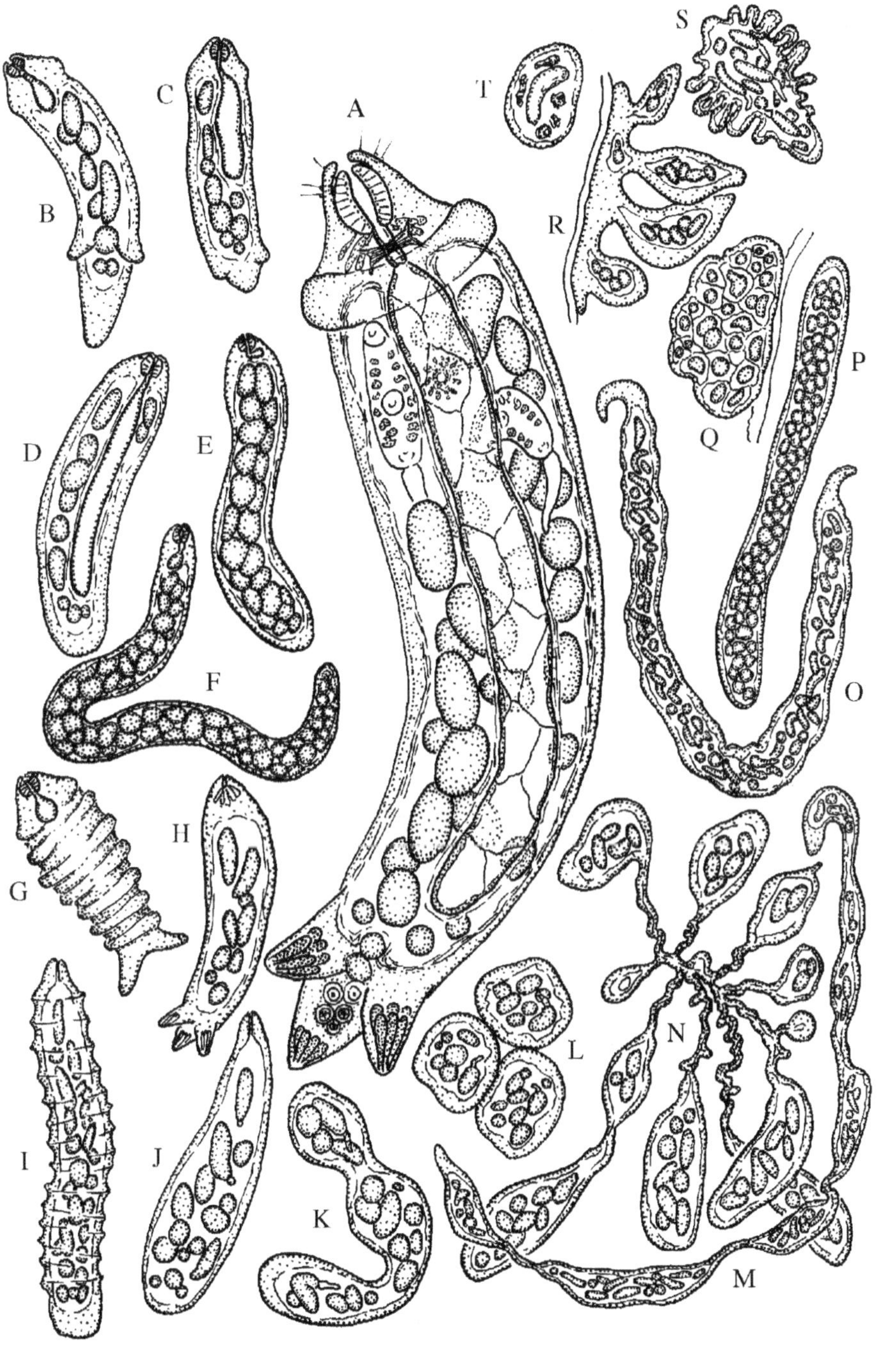
S
T
C
A
B
R
P
D
E
Q
F
O
H
G
N
L
I
J
K
M

In the line leading from the fasciolid-echinostomatid complex of families to the most specialized order Plagiorchiida, replacement of rediae by sporocysts was accompanied only by a considerable morphological simplification of the second and subsequent generations of parthenitae. Daughter sporocysts as well as their evolutionary precursors, rediae, remained unitary organisms with a single reproduction centre – one germinal mass (Fig. 131J-L). At the same time, in some representatives of two other large phylogenetic groups: redial Aporocotylidae – sporocyst-bearing Sanguinicolidae, Spirorchiidae, and Schistosomatidae (1); redial Clinostomidae – sporocyst-bearing Cyathocotylidae *sensu lato*, Diplostomidae, Strigeidae (2) long filamentous sporocysts emerge, sometimes with features of modular organization. Not one but several germinal masses are formed in them. They are evenly distributed along a very elongated body which often becomes irregularly beaded-like, especially in older individuals. Swollen parts of the body become rather isolated brood chambers (Fig. 131M).

But the true modular organisms are branching sporocysts (Fig. 131N) of species of Bucephalidae and Brachylaimoidea and the mother sporocysts of many most specialized Plagiorchiida (Fig. 131Q,R). The establishment of modular organization is accompanied by a morphological-functional disintegration of the organism which in fact loses its morphological and physiological integrity (Dobrovolskij, 1975; Dobrovolskij et al., 1983). Degeneration and death of one of its parts does not influence the other parts which continue to function normally and to give rise to individuals of the next generation.

Specific morphological transformations which the parthenitae underwent are diverse and are mostly expressed independently in different taxa. The sac-like rediae which have completely lost locomotory appendages are very common (Fig. 131D-F). They are characteristic of Aporocotylidae, Clinostomidae, Heterophyidae, Notocotylidae, Monorchiidae, Acanthocolpidae, Haploporidae, Allocreadiidae. There are more families possess rediae of this morphological type, yet the above list demonstrates very well that the families listed are very distant.

All the above also refers to the parthenitae of other morphologic types. We will limit ourselves to one example. Highly specialized small round sporocysts (Fig. 131L) are, on the one hand, characteristic of many species of Sanguinicolidae, on the other hand, of a number of representatives of Lecithodendrioidea.

During the analysis of the above morphological series illustrating the gradual simplification of rediae down to the state of sporocysts, one more feature is usually considered: the degree of development of the digestive system. The latter is indeed lost by parthenitae. Rediae possessing a long gut (Fig. 131A) are succeeded by forms in which the gut becomes noticeably shorter as related to the whole length of the body (Fig. 131B-G). It is especially evident in the rediae of Heterophyidae whose gut, very short in length, does not exceed much the diameter of the pharynx (Fig. 131E). Actually it is a complete disappearance of the organs of the digestive system that marks the transition from the redia to sporocyst morphological type. Most researchers concerned with this problem, emphasized that this transition was smooth (Sewell, 1922; Ginetsinskaya, 1968; Dobrovolskij et al., 1983).

However, the fact is that shortening and reducing the volume of the gut are not always directly connected with a change in the external morphology of rediae. These two processes are to a great extent independent and can be realized in taxa of comparatively low rank. Thus, within the family Echinostomatidae, rediae of some species possess a long intestine (*Echinoparyphium echinatoides*, *Petasiger neocommense*), whereas in others it may be very short (*Hypoderaeum conoideum*). It is by no means associated with the shape of the rediae body, the presence of a collar and posterior locomotory appendages. A similar picture is also observed in rediae with a sac-like body. In rediae of Notocotylidae the gut is very long: it almost reaches the posterior end of the body. Yet, as stated above, the gut of the rediae of all Heterophyidae is, on the contrary, very short.

It has already been noted that loss of the organs of digestive tract means transition of parthenitae to the category of sporocysts. This is a conventional and widely spread point of view. Yet in our opinion, this is a somewhat simplified approach to the problem of the definition of two morphological types of parthenitae. The ambiguity of this problem can be well illustrated by the example of Philophthalmidae. According to Tihomirov (1980) the parthenita of *Philophthalmus rhionica* inoculated by a miracidium into a mollusc is devoid of the digestive system (Fig. 131H) and due to this circumstance is referred to as a sporocyst. This is how it is most often noted. Yet, individuals of this parthenogenetic generation of *P. rhionica* possess a complete set of morphological characteristics (with the exception of the digestive system) inherent in archaic rediae (Tihomirov, 1980; Dobrovolskij et al., 1983). They possess well-developed anterior and posterior appendages, especially in young individuals, supplied with specialized glandular cells and a ventrally located birth pore. Terminally, at the anterior end there is a group of head glands which are most probably used as penetration glands when young individuals set free from the miracidium make their way into the molluscan host. There is nothing of this kind even in the most complex sporocysts. Following Tihomirov, we designate these organisms as mother rediae (Dobrovolskij et al., 1983, see p. 59). The presence or absence of the digestive system is an important but not the only characteristic determining the morphological type of parthenitae. In archaic rediae the above characters are fully present. In more specialized ones some of the characters may be reduced or lost.

The nature of these losses is treated differently by different authors (Brooks et al., 1985a; Pearson, 1992). Brooks et al. (1985a) consider that loss of the intestine by rediae and its conversion into daughter sporocysts is the result of paedomorphosis. In our opinion, this point of view is unacceptable, especially if we take into account the arguments of the above authors. It should be recollected that paedomorphosis is a process which results in the appearance of forms with "simplified" morphology corresponding to the embryonic stages of the ancestral species. Paedomorphosis is based on the phenomenon of heterochrony. However, Brooks et al. (1985a, pp. 434-5) wrote: "If the development of the digestive system were retarded in a species which normally produced rediae, the resulting equivalent larval forms would have the condition found in earlier developmental stages (i.e., the mother sporocyst). ... The resultant close similarity in form between these paedomorphic rediae and mother sporocysts might lead one to call them (*the paedomorphic rediae – K.G. & A.D.*) daughter sporocysts." One thing becomes clear from the above quotation: the organization of daughter sporocysts is compared not with the stages of rediae morphogenesis but with that of individuals of another generation, much more ancient and considerably more secondarily changed under the influence of parasitism. This is difficult to agree with. Daughter sporocysts should be compared only with rediae, whose phylogenetic offspring they are.

Daughter sporocysts, on the other hand, lost the digestive system not as the result of paedomorphosis but due to a usual reduction, the trophic function gradually passing over to the tegument (see p. 253). Reduction of the length and volume of the gut during the transition from the archaic rediae to the more specialized ones cannot be accounted for by paedomorphosis, and we agree with Pearson (1992) on this point. However, the role of heterochrony in those processes can not be ruled out. Moreover, analysis of morphogenesis of typical rediae and typical sporocysts has shown that the rudiments of the gut in rediae, irrespective of its length, always differentiate rather early. Yet in the daughter sporocyst (we have studied in detail the development of this generation in several species of Plagiorchiida), the formation of the gut is left out of morphogenesis. This is difficult to account for by reduction. Those cases when the primordium of the esophagus appears in morphogenesis but its subsequent development is inhibited and this rudiment becomes the sphincter of the birth pore, seem to be the manifestation of heterochrony (Azygiidae – Sillman, 1962). Thus, reduction and paedomorphosis could be very well superimposed, but it was reduction that gave momentum to the disappearance of the gut in daughter generations of parthenitae.

Of special interest is the excretory system of rediae and sporocysts (Fig. 132). The uniformity of the pattern of its structure is remarkable: in absolutely all the rediae and sporocysts studied in this respect it belongs to the morphological type of Mesostoma. Unfortunately we know nothing of the excretory system of mature mother sporocysts but both stenostomate and mesostomate excretory systems can be found in miracidia. In our opinion, it testifies to the fact that the excretory system of the daughter and the subsequent generations of parthenitae have undergone a noticeable morphologic

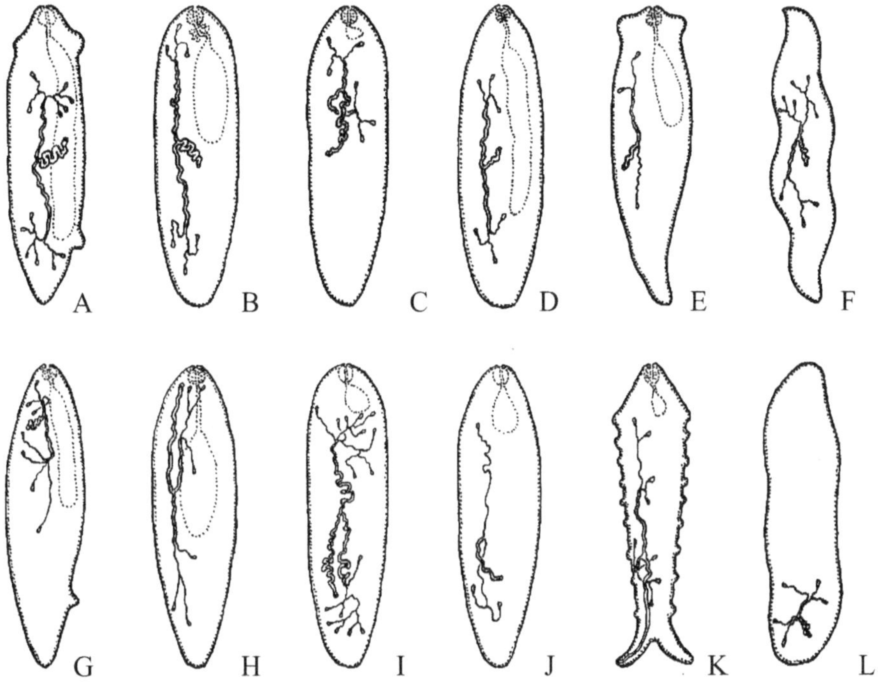

Figure 132. The excretory system of rediae and sporocysts. A, Echinostomatidae; B, Lepocreadiidae (based on Peters, 1961); C, Heterophyidae (based on Cable, 1956); D, Allocreadiidae (based on Madhavi, 1976); E, Transversotrematidae (based on Nadakal et al., 1969); F, Fellodistomidae (based on Cable, 1956); G, *Cercaria caribbea IV* (based on Cable, 1956); H, Haploporidae (based on Cable and Isseroff, 1969); I, Allocreadiidae (?) (based on Peters, 1961); J, Paragonimidae (based on Ginetsinskaya, 1968); K, Bivesiculidae (based on Pearson, 1968); L, Plagiorchiidae (based on Dobrovolskij and Reichel, 1973).

simplification as compared to the individuals of the first generation – the mother sporocyst.

The location of excretory pores deserves special analysis. The data available allow us to distinguish 3 groups of species. The most numerous one, which includes the representatives of the most different taxa (Echinostomatidae, Transversotrematidae, Heterophyidae, Lepocreadiidae, Allocreadiidae, Fellodistomidae) is characterized by the location of the excretory pores of their parthenitae (both rediae and sporocysts) almost right in the middle of the body (Fig. 132A-F). The relatively short main collecting tubes branch out, approximately at the same level, into the anterior and posterior longitudinal collecting tubules, into which capillaries of the flame cells or channels leading to groups of flame cells usually open. As a rule, the flame cells themselves are located closer to the ends of the body of the redia and the sporocyst. Yet cases are known, when the anterior and posterior groups of flame cells are displaced towards the anterior half of the body (*Cercaria caribbea XVIII* Cable, 1956 – Heterophyidae) (Cable, 1956), so that the excretory pore which retained the middle position is positioned behind them (Fig. 132C).

The second group is formed by the species in whose parthenitae the excretory pore is located closer to the posterior end of the body (Bivesiculidae, Paragonimidae, *Cercaria pomatiopsidis* Stimpson, 1865 – Allocreadiidae (?), Plagiorchiidae, Telorchiidae). A special place in this series is occupied by Bivesiculidae. The posterior end of the body of their rediae is bifurcated, the excretory pores are located at the ends of short bifurcate branches (Pearson, 1968) (Fig. 132K). This feature of the excretory system of the rediae of Bivesiculidae reminds one of the position of primary excretory pores of the cercariae of Azygiidae, Transversotrematidae and all the larvae of the hermaphroditic generation of the Schistosomatida (see p. 438).

As a rule, the main collecting tube is relatively long. The place where it branches into the anterior and posterior collecting tubules is located in the middle part of the body, as in representatives of the first group. The anterior and posterior groups of flame cells are located far from each other, closer to the ends of the body. All the elements of the excretory system are compactly situated in the posterior end of the body only in the sporocyst-bearing species (Plagiorchiidae, Telorchiidae) (Fig. 132L).

The species in whose parthenitae excretory pores are greatly shifted forwards, constitute the third group (Haploporidae, *Cercaria caribbea IV* Cable, 1956[i], Fig. 132G,H). At the same time, in *Cercaria caribbea IV* the whole excretory system is shifted to the anterior part of the body.

Such a diversity in the location of the main elements of the excretory system in the parthenita body and especially an extremely heterogeneous taxonomic composition of all the groups distinguished impede assessment of the state of the main characters. In this respect only a few suggestions can be made (Fig. 133). Taking into account the fact that both in the overwhelming majority of miracidia of archaic taxa (see p. 416), and in individuals of the hermaphroditic generation of all the species studied in this respect (see p. 438) the primary excretory pores are located in the posterior part of the body, it may be suggested that it is this position of theirs that happens to be initial for all digeneans (Fig. 133A,B,H). Accordingly, the presence of long main collecting tubes should be also considered as the plesiomorphic state. In archaic forms they often belong to a more ancient stenostomate type both in the first parthenogenetic (miracidia) and in the hermaphroditic generation. The excretory system of the stenostomate type was retained independently by both mother sporocysts (miracidia) and hermaphroditic worms. Thus, miracidia of Diplostomidae possess a typical stenostomate excretory system, whereas cercariae of the same species have a mesostomate one. An alternative situation is seen in species of the Echinostomatidae whose miracidia are characterized by the presence of a simplified mesostomate excretory system, whereas the hermaphrodite generation always possesses a typical stenostomate one.

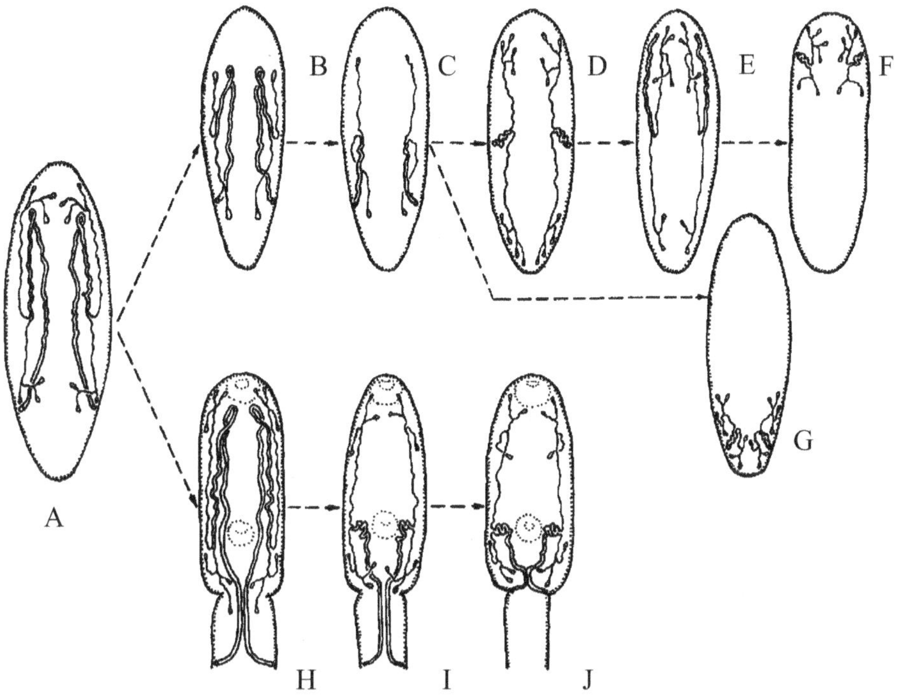

Figure 133. Possible ways of morphological evolution of the excretory system of the parthenogenetic and hermaphroditic generations of trematodes. A, the excretory system of the hypothetical ancestral form; B-C, the first parthenogenetic generation (miracidia); D-G, the daughter parthenogenetic generation (sporocysts and rediae); H-J, the hermaphroditic generation. (See text for details).

In both generations an independent realization of a tendency for shortening of the main collecting tube results in the appearance of the mesostomate excretory system (Fig. 133C-G,H-J), but the main collecting tube still remains relatively long. It is this state of the excretory system that we observe in the daughter and the subsequent generations of the parthenitae of representatives of the second group (Bivesiculidae, Paragonimidae, *Cercaria pomatiopsidis* Stimpson, 1865 – Allocreadiidae (?)).

Thereafter, in the mother and subsequent generations of parthenitae a tendency for shifting the primary excretory pores forward towards the middle part of the body appeared independently, which was accompanied by the shortening of the main collecting tubes (Fig. 133D). In the hermaphroditic generation this did not occur (see p. 438) (Fig. 133H-J). Complete shifting of protonephridia either into the anterior (see fig. 132C and fig. 133F) or the posterior part of the body (see fig. 132L and fig. 133G) must be a secondary phenomenon, associated with that of allometric growth often observed in different generations of digeneans.

A special place in parthenitae evolution was occupied by transformations associated with the reproductive function. In the overwhelming majority of cases these transformations are directed towards maximizing of the generative function.

Analysis of the evolution of parthenita reproduction is impossible within the framework of one morphologic series, as it was done by us during the analysis of rediae and sporocysts morphology. This is first of all due to the above specific features of the mother and subsequent generations of parthenitae conditioned by the difference of origin and the time of appearance of those generations (see p. 491).

We have discussed previously (see p. 418) the organization of germinal material of the free-living dispersive phase of the mother parthenogenetic generation – miracidia in which the main tendency of evolutionary transformations is a retardation of the realization of the generative function. As a result, the differentiation of germinal cells is also retarded. In miracidia of the most specialized species a simple germinal primordium appears, represented only by undifferentiated cells (see p. 421). These changes, as mentioned above, were conditioned by a transfer of the generative function to the parasitic phase of the mother sporocyst development. This transfer was accompanied by a radical restructuring of the latter.

In our opinion, the most archaic forms are the mother sporocysts of Paramphistomidae, Paragonimidae, Fasciolidae and Transversotrematidae. Morphologically, they are very simply structured sack-like organisms. The transfer of reproduction to the parasitic phase and its subsequent intensification required from the parasites the intensification of other functions as well, including the trophic one. For parasites lacking the gut it is first of all associated with the active function of tegumentary-muscular sac, all the elements of which are retained by the parasitic phase from the miracidium. In more archaic forms, whose mother sporocysts do not live long in the mollusc, the tegumentary-muscular sac is subject to a relatively fast degradation. On the contrary in long-living, actively reproducing parasites it does not degrade but develops intensively. In this case we are faced with a kind of a secondary recapitulation of the character "initially" inherent in trematodes (see p. 424).

The mother sporocysts of Strigeidida are highly integrated organisms. Relatively young, actively functioning individuals possess not only a well-developed tegumentary-muscular sac characteristic of flatworms, but very often a residual parenchyma. Even the oldest individuals, whose reproductive time is practically over, retain the main characters of a unitary organism.

The evolution of the mother sporocysts of the representatives of Plagiorchiida has taken a different course. In this case an intensified growth of soma was required since it was impossible to ensure reproduction and the first stages of development of individuals of the next generation in any other way. The subsequent events were developing quite differently since the main tendency in the morphological evolution of mother sporocysts in this phylogenetic branch happens to be morphological-physiological disintegration and passing over to the modular type of organization (see p. 247). In the most specialized forms disintegration is finally followed by a true degeneration, when the generative elements (germinal masses, single germinal cells and developing embryos of the next generation) pass over to independent parasitizing in the host (see p. 46). Actually we have returned to the initial proposition of our consideration of this series: the degradation of somatic structures sometimes resulting in the "disappearance" of the organism as such. The parasitic phase of the mother sporocyst is absent not only in cyclocoelids, phylophthalmids and a number of other archaic families, it is also absent in some Plagiorchiida (Telorchiidae, Lecithodendrioidea, Microphallidae). Something of this kind is to be also observed in some other specialized trematodes (for instance, Brachylaimoidea – see p. 70). Yet it should be recalled that whereas in primitive trematodes this is the consequence of an extremely low generative activity of mother sporocysts, which is likely to be the manifestation of their archaism and great ancientry, in some others it is a secondary adaptation ensuring maximal intensification of the generative function. These are two different phenomena. They have a different nature and are not comparable to each other in this sense. Therefore we can not agree with Pearson (1992) who having rightly noted a tendency for reduction of the parasitic phase of the mother parthenogenetic generation grouped all the above and some other examples of this kind into one continuous series.

Morphological changes in mother sporocysts are accompanied by serious transformations in the organization of their germinal material. In Paramphistomidae, Fasciolidae, Transversotrematidae, Paragonimidae and some others repeatedly mentioned above, mother sporocysts are devoid of a functioning germinal mass. In mother sporocysts of Echinostomatidae there is one small germinal mass of an attached type (Ataev et al., 2001a) (see fig. 15). According to our observations, comparatively small mother sporocysts of Bunocotylinae and Heterophyidae also possess one germinal mass. On the contrary, in large elongated mother sporocysts of Strigeidida polymerization (i.e., an increase of the number) of germinal masses occurs (Dönges, 1964; see p. 33, fig. 14).

In Plagiorchiida the same tendency dominates. In worm-like mother sporocysts of *Macrodera longicollis* we managed to find only one germinal mass with a very limited term of function. The latter feature ensures a considerable synchronization of the development of the next generation embryos (see p. 43). Transition to the modular type of organization is accompanied by the multiplication of the number of centers of reproduction of germinal cells, germinal masses. Their number may be measured by tens if not hundreds. We consider the appearance of diffuse germinal masses (e.g., in mother sporocysts of *Haplometra cylindracea* – see p. 81) as an extreme form of polymerization.

The same tendency is manifested in the most specialized Plagiorchiida in which the mother sporocyst practically disappears due to a very early degradation of somatic elements (see p. 46). In such cases we deal with the phenomenon of brood parasitism accompanied by intense proliferation of germinal elements in different parts of the body of the infected mollusc (Schell, 1965; Dobrovolskij, 1975; Galaktionov and Dobrovolskij, 1985).

The variability of the organization of the germinal material of the daughter and subsequent parthenogenetic generations is not so great. Rediae and daughter sporocysts have more frequently one, and sometimes several germinal masses. We have mentioned above that increasing the number of the latter is a secondary phenomenon. The localization of germinal masses is noticeably different. In the most archaic rediae (Echinostomatidae, Philophthalmidae) the germinal masses are sunk into the parenchyma of the posterior part of the body (see p. 68). With the expanding of the schizocoel and related reduction in the residual "parenchyma" the so-called attached germinal masses appear, which are often displaced to the lateral surface of the body (e.g., in daughter rediae of Bunocotylinae). And finally, the line is brought up by parthenitae with floating germinal masses. The appearance of centers of reproduction of germinal elements of the last type in all the known cases to date is associated with the transition from rediae to sporocysts. Besides the loss of the gut, which has been discussed above (see p. 424), this process is accompanied by other manifestations of degeneration. The maximally possible extension of the schizocoel and consequently complete reduction of the residual "parenchyma" (in mature daughter sporocysts usually only very reduced elements of tegumentary-muscular sac and endocyst are retained) results in the isolation of the germinal mass.

In all the sac-like and worm-like daughter sporocysts of Plagiorchiidae, Microphallidae and Ochetosomatidae studied, as well as in rediae there is only one germinal mass. At the same time, long filamentous daughter sporocysts of Strigeidida often acquire a great amount of germinal masses, which can be regarded as the manifestation of their secondary specialization.

1.2. The Hermaphroditic Generation

Ontogenesis of the hermaphroditic generation as well as that of the first parthenogenetic one, the mother sporocysts, includes free-living and parasitic phases. The relation between the former and the latter may be different. In the primitive digeneans (Fasciolidae, Paramphistomidae, Philophthalmidae) there are 2 free-living phases – cercaria and adolescaria and 1 parasitic phase – marita (adult). On the contrary, in the overwhelming majority of the Digenea the relation is quite reverse: the free-living phase is only cercariae, whereas metacercariae and maritae are obligatory parasitic organisms. There are rather many exceptions from this generalized scheme (for details see chapter 3). Thus, in some specialized forms free-living cercariae disappear though this phase of ontogenesis is retained (some Microphallidae and Plagiorchiidae). In Azygiidae and Bivesiculidae the phase of metacercariae is absent, whereas all blood parasites (Schistosomatida) seem to lack the marita phase (see p. 338).

All these variations and exceptions, though very important, do not change the general pattern of morphological evolution of the hermaphroditic generation. They only illustrate an unusual biological plasticity of the Digenea and the possible ways of their adaptive radiation.

1.2.1. Analysis of the Organization of the Hermaphroditic Generation

The main tendencies of cercariae morphological evolution have been considered in detail earlier (Galaktionov and Dobrovolskij, 1987). Therefore we limit ourselves to a short review of the main tendencies accompanying cercariae specialization and distinguishing the plesiomorphic states of the main characters.

The characters determining the cercariae structure can be easily divided into two groups (see p. 149). One comprises the features whose appearance was associated with the necessity of the realization of the main function of cercariae in the digenean life cycle – the transfer of the infective stage in some way from the mollusc to the next host. The majority of them are purely provisional formations acquired by larvae of the hermaphroditic generation secondarily and not retained at the subsequent stages of development. The possibilities of using the characters of this kind in evolutionary and phylogenetic research are limited and restricted to the taxa whose taxonomical level is not high. The fact is that similar provisional structures for the same function may be acquired independently and be lost as easily by representatives of different phylogenetic branches. This group first of all comprises characters associated with the main locomotory organ of cercariae: the tail, the penetration and cystogenous apparatuses.

The characters of the second group characterize such structures and organs which are initially inherent in the hermaphroditic generation and in many ways determine the characteristic features of the body plan organization (*bauplane*) of this group of parasitic flatworms. The concrete manifestation of some characters may be rather different in different species (for instance, the degree of development of the digestive system, the number of flame cells and the character of the excretory formula, the degree of development of the elements of the secondary excretory and reproductive systems). Most often this is conditioned by heterochrony which seems to play an important role in the digenean evolution. Similar to the provisional structures of the first group, the structures of this kind testify to the degree of the specialization of cercariae possessing them and in most cases cannot be used as groundwork for phylogenetic constructions. The reason for it is a great number of parallelisms in which the morphological evolution of digeneans abounds. The situation is quite different when it concerns characters of the body plan organization which are not subject to the influence of heterochrony. They are location of the primary excretory pores, stenostomate and mesostomate type of the main collecting ducts, the shape and structural features of the excretory bladder, the location of the elements of the excretory system and the boundaries between the body and the tail of the cercaria with respect to each other, the length of the intestine branches. It is these characters that can be used for phylogenetic analysis.

It was shown in chapter 1 (see section 2.1.9.) that as the result of the manifestation of heterochrony, in the course of transition from archaic taxa to more specialized one, cercariae underwent juvenilization and miniaturization (see p. 151). Externally this is manifested in the simplification of their structure. The certainty of the above tendencies does not, however, rule out the necessity of deep analysis of the origin of the main structures inherent in cercariae and evolutionary mechanisms of changes following them.

The first and the main characteristic of the cercaria is the presence of a tail. All the cases of the tail absence in larvae of the recent Digenea are certainly secondary (see some details below).

Different points of view exist concerning the origin of the tail Brooks et al. (1985a) reject of the adaptive nature of cercaria tails. To prove their views they refer to *Heronimus mollis* whose life cycle, in their opinion, is the most ancient and archaic among the Digenea. Cercariae of this species possess a well-developed motile tail though they do not leave the mollusc. Infection of final hosts (fresh water turtles) occurs when the latter eat infected molluscs (Cable and Crandall, 1956; Ulmer and Sommer, 1957; Crandall, 1960). Brooks et al. (1985a) claimed that cercariae of *H. mollis* remained in sporocysts and consequently from the point of view of adaptive evolution it is impossible to account for the appearance of a tail in those larvae. In our opinion, this assertion is not very convincing. Cercariae of *H. mollis* actually move freely in the lacunae of the haemocoel and after artificial removal from the mollusc are capable of swimming for almost 8 h (Crandall, 1960). In this case it is natural to suppose that the life cycle of *H. mollis* is secondarily reduced (a typical example of secondary two-host life cycle – see p. 341), whereas cercariae simply have not yet lost the provisional organs initially inherent in free-living larvae.

While discussing the possible variants of the primitive life cycles of digeneans Gibson (1987) wrote: "Another possibility, which I put forward as food for thought, is that the cercarial tail was not originally an organ of locomotion, since it would have been of no use for locomotion in its early developmental stage, but an organ of protection (analogous to the bladder of cyclophyllidean cysticerci) that was developed by the ancestral resting stage (proto-metacercaria) in the molluscan host prior to its ingestion by vertebrate" (Gibson, 1987, p. 450). This suggestion also has some weak spots. First, the protective function of the bladder of the cysticercus is secondary. Before this Cyclophyllidea underwent prolonged evolutionary transformations: in more primitive forms no submersion of the scolex and neck of the metacestode into the highly distended cercomer is observed. The primary function of the cercomer (in the most ancient proto-Cercomeromorpha) is that of attachment. In monogeneans it is retained (the cercomer is part of the attachment disk, so-called opisthaptor), in the rest of the representatives of this phylogenetic branch of parasitic flatworms the cercomer is either reduced (Gyrocotylida, Amphilinida) or, if it is retained it changes its function: first it is trophic and storage of nutrients, later trophic and protection, still later a possibility of proliferation is added to it, i.e., a functional substitution occurs.

Second, even supposing that the tail had initially appeared not for locomotion but for some other purposes protection would have been the last of them. All the potential ancestors of trematodes are already pre-adapted to carrying out the function of protection: they possess well-developed epidermal glands which are mostly used for this purpose (Ivanov and Mamkaev, 1973). To transform these secretory elements into a cystogenous apparatus would be easier and would require less energy expenditure than building up a great amount of superfluous tissue in the shape of a cysticercous or cystophorous tail (see p. 99). Energetically it is too costly. From the standpoint of evolutionary morphology the case is not better since the sequence of events, following Gibson's suggestions, should have been as follows. First a ring fold which would gradually enlarge forwards should have been formed, which would have been unable to perform a protective function before its edges had closed above the anterior part of the body. Thus, the problem of the functional purpose of this structure is still open. Then at the posterior end an appendage would have appeared which later would have turned into a usual locomotory tail. In the latter, practically *de novo* the locomotory muscle system should have appeared (see p. 121) since it is difficult to suppose that in such a specialized structure like a cyst-like capsule unspecialized musculature of the tegumentary-muscular sac should have been preserved.

Another thing could be supposed. Before the appearance of the protective fold trematodes already possessed some caudal appendage. Then we inevitably return to the question what kind of appendage it might have been and where from it could have appeared. No one has yet managed to think of any functionally grounded, "adaptationist", as Brooks et al. (1985a) put it, explanation. The only remaining variant seems to be that of a cercomer homologue. But the fact is that trematodes are not Cercomeromorpha that was convincingly proved by Bychowsky (1957).

The presence of a cercomer in Cercomeromorpha is a plesiomorphic character which appeared in ancestral forms at the very stages of the establishment of the group. No signs of a tail or even its primordium in the most ancient generations in the life cycle of trematodes – parthenitae – could be found. The tail as a provisional organ of adaptive nature appears much later, at the moment when necessity of a fast swimming dispersive larva appeared, maybe after primary two-host life cycles had been formed (for details see p. 491).

We have already said above (see p. 435) that the adaptive nature of this provisional organ is undoubtful. Moreover, we do not question Cable's (1965, 1974) suggestion that the tail may have formed as the result of the specialization of the posterior end of the body of free living organisms which used undulation for moving in the water column.

The groundwork of Cable's arguments are sufficiently numerous data on morphogenesis of the excretory system and cercariae tail in representatives of different groups of trematodes. The gist of his notions is well illustrated in a scheme given below (Fig. 134) borrowed from Cable's (1977) work. The tail of furcocercariae is formed as the result of the isolation of the posterior end of the growing embryo. The anterior boundary of this part which resulted from moulding, is shifted forward, which conditions the presence in the tail not only of the distal parts of the excretory system ducts but also of the last (the most posterior) primary group of flame cells. The primary excretory pores are located at the ends of furcae. Yet it should be emphasized that the group of flame cells is present in the tails of far from all furcocercariae (in Sanguinicolidae, Aporocotylidae, Bucephalidae, Gymnophallidae, some Fellodistomidae and Brachylaimoidea flame cells never enter the tail). At the same time, in the cercariae of *Heronimus mollis* possessing a simple tail with a clearly expressed tendency for megalocercous state, the latter group of flame cells is located in the tail, closer to its base (Fig. 135).

The second type of tail formation in the process of morphogenesis inherent in the representatives of such families as Fasciolidae, Paramphistomidae, Echinostomatidae, Notocotylidae, Heterophyidae, Haploporidae, Acanthcolpidae are characterized by a combination of two processes: the above moulding of the growing embryo and active cell proliferation in its posterior end. Under such circumstances the tail is no longer furcocercous, it becomes simple. The primary excretory pores are located in the middle part of the tail. The anterior boundary of the latter turns out to be slightly shifted backwards, so that flame cells do not enter the tail.

The third variant inherent in all Xiphidiocercariae, Ophthalmoxiphidiocercariae and some Cercariae Microcerca (Opecoelidae). The tail is formed only as the result of cell proliferation proceeding in the part of embryo located only between the primary excretory pores. In this case a simple tail is formed which is fully devoid of any elements of the excretory

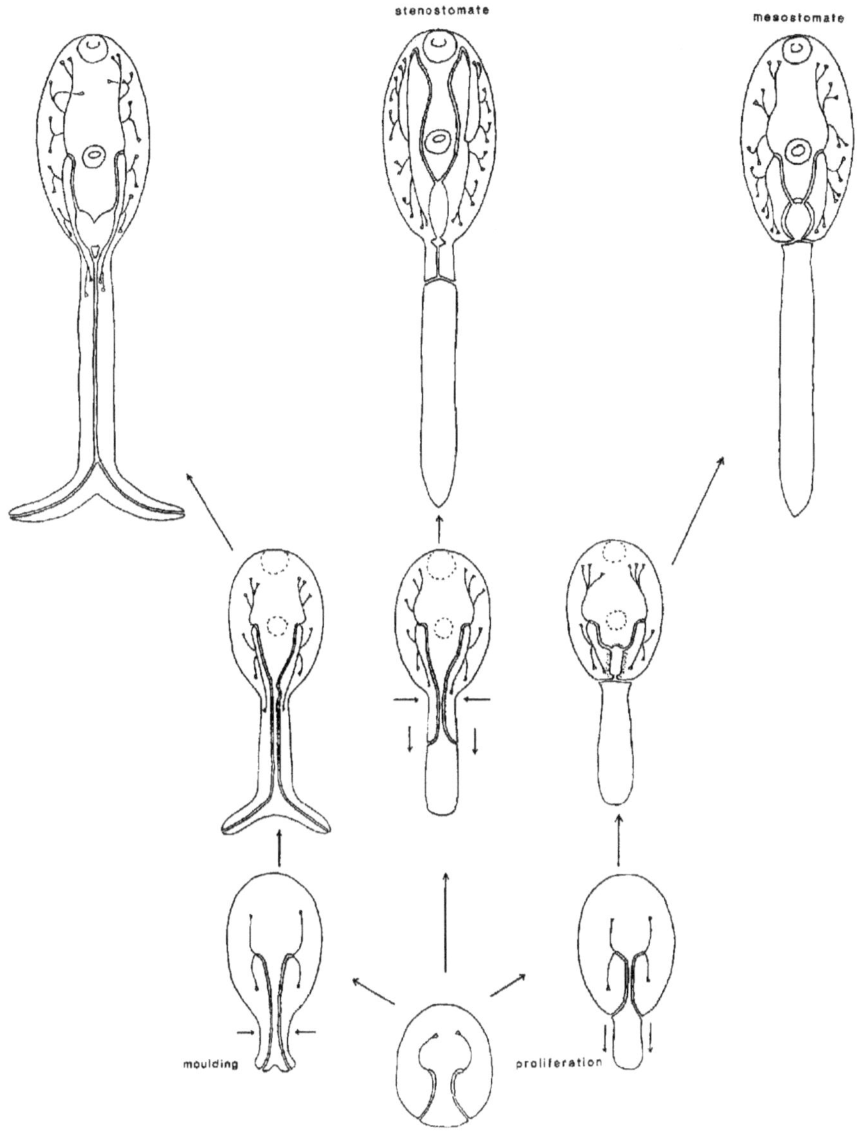

Figure 134. The development of the excretory system and the formation of the cercarial tail. After Cable (1977).

system. In Cable's opinion proliferation is a secondary phenomenon. Hence a natural conclusion suggests itself that the furcocercous type of larvae organization is the most archaic one. This well constructed logical scheme is captivating because of its simplicity, yet other treatments of the above events are possible.

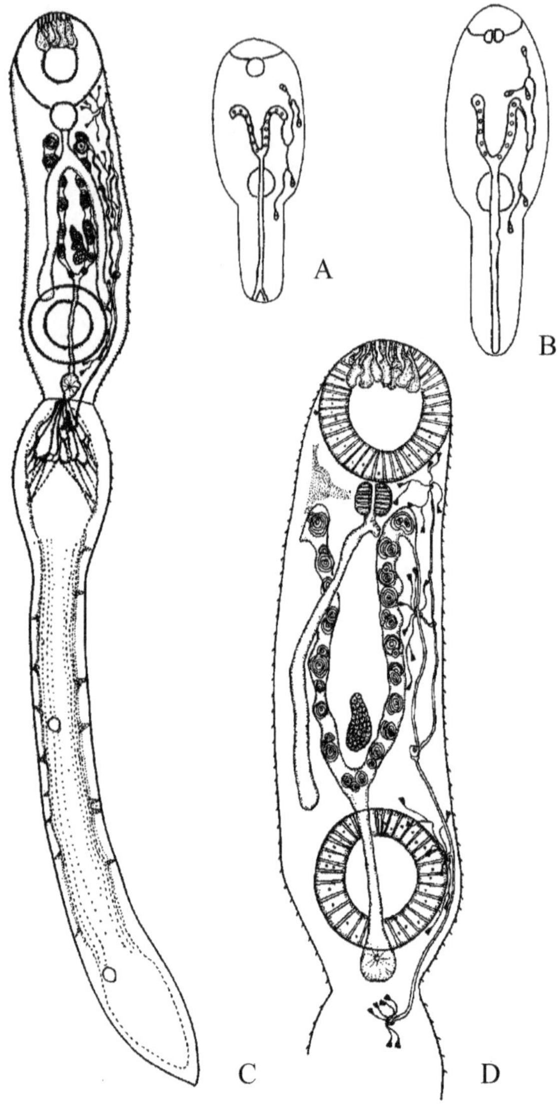

Figure 135. Development of the *Heronimus mollis* cercaria. A-B, successive developmental stages; C-D, fully formed cercaria. From Crandall (1960), reprinted with the publisher's permission.

First of all it should be noted that localization of primary excretory pores at the posterior end of the body is supposed to be primary. Such a premise seems incorrect to us. Let us consider the same data from a somewhat different point of view taking into account some features of the body plan organization of not only hermaphroditic generation but also parthenitae as well as related groups of flatworms, first of all monogeneans and gyro-

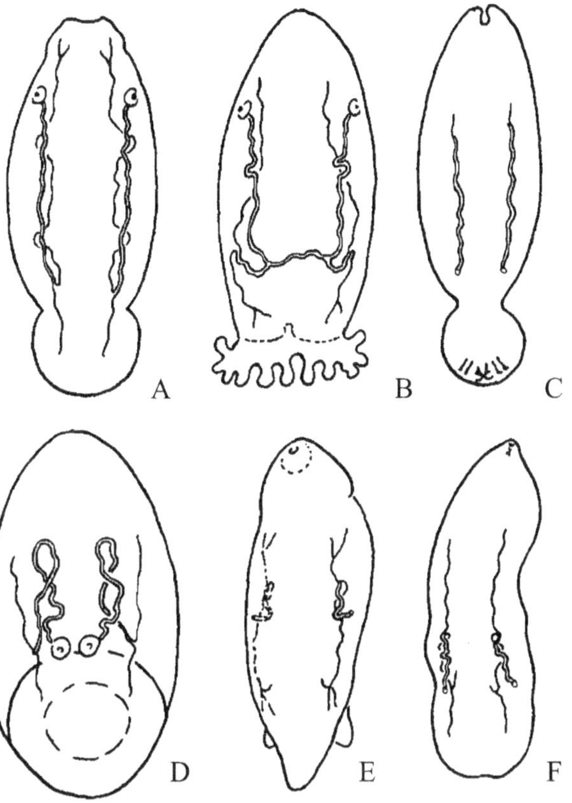

Figure 136. Location of the primary excretory pore in the body of different flatworms. A, Monogenea; B, Gyrocotylida; C, cestode procercoid; D, Aspidogastrida; E, trematode rediae; F, trematode sporocysts. From Galaktionov and Dobrovolskij (1998).

cotylids. In all the above forms the primary excretory pores open on the side of the body (Fig. 136). In monogeneans and gyrocotilids they are generally shifted forward, in procercoids of pseudophyllid cestodes they are shifted rearward but are never located on the cercomer. In fact, a very similar picture is observed in parthenitae: miracidia – mother sporocysts, redia and daughter sporocysts in which primary excretory pores, even if they are shifted to the posterior part are never localized terminally in the posterior end of the body. Between the level of their location and the posterior end there is always a well-expressed part of the body which we will further designate as the post-poral part. Correspondingly, the part of the body between the primary excretory pores and the anterior body end will be referred to as the pre-poral part. In the post-poral part of the redial body the germinal mass inbuilt into the residual parenchyma is located (see fig. 30). The excretory pores of the juvenile stages of Aspidogastrida are actually also distant from the posterior end of the body. In representatives of this group they are shifted dorsally to a great extent and are located at the level of the anterior edge of the attachment organ (adhesive disc). Though Rohde (1972) called them dorso-terminal they are situated at a distance of almost one third of the body length from the posterior edge of the body.

Importantly, that at the earliest stages of excretory system development in cercariae embryos the primary excretory pores are located closer to the equator of the yet round embryo than to its posterior pole (Fig. 137). And only later, as the result of allometric growth of the embryo they gradually shift towards the posterior end.

These considerations testify in our opinion to the fact that ancient trematodes possessed a lateral and not a terminal location of the excretory pores and a well-expressed post-poral part of the body. It is that part, together with excretory pores and distal regions of main collecting ducts of the protonephridial system that began to specialize at some moment as the locomotory part of the body and later became a cercaria tail. This state is reproduced very well in the tails of cercariae of Fasciolidae, Paramphistomidae, Echinostomatidae and other related forms. A position of the primary excretory pores at the tail (in anterior third, in anterior half, at mid-tail, etc.) discussed intensively by Pearson (1992) and Brooks et al. (1985a) could be reduced to the problem of allometric growth: in some cases pre-poral and post-poral parts of the tail grow equally intensively, and this seems to be a more archaic state, whereas in the others the intense growth falls on the post-poral part and then primary pores are close to the base of the tail to some degree, which, in our opinion, is secondary. Full suppression of the growth of the pre-poral part of the tail primordium results in the appearance of tails characteristic of xiphidiocercariae (Plagiorchiida) and other groups similar to them in this respect. In this case the boundary between the body and the tail is located at the level of excretory pores.

The initial stages of the tail formation in furcocercariae do not differ externally from similar stages of the development of the tail primordium in larval morphogenesis of Fasciolidae, Paramphistomidae and other representatives of this complex of families. Unlike what Cable (1977) showed in his scheme, the incipient tail primordium does not possess any signs of bifurcation and has a well expressed post-poral part, yet hereafter both direction of growth and character of growth processes change (see fig. 137). According to our observations the pre-poral section always grows somewhat faster and due to it the body of the larva, the tail stem, and the frontal surface of the furcae are formed. The formation of the body and tail stem proper takes place as the result of the moulding. The growth of the post-poral part proceeds not longitudinally but in the transverse direction and is mainly the result of proliferation. Actually two growth zones are formed which are lateral to the tip of the posterior part of the embryo, which ensures the formation of two diverging branches of the furca.

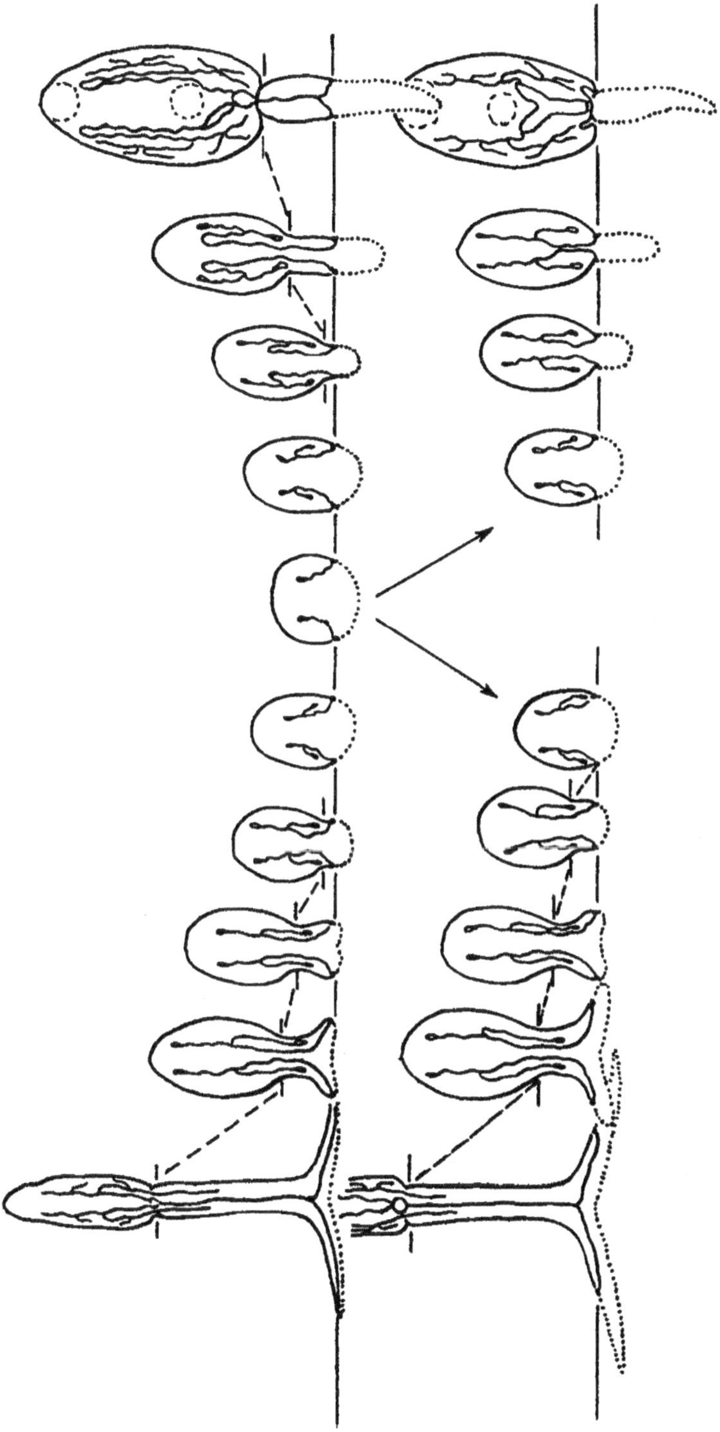

Figure 137. Comparison of the growth of the pre-poral and the post-poral body part in developing trematodes. From Galaktionov and Dobrovolskij (1998).

In representatives of a number of taxa (Bivesiculidae, Transversotrematidae, Azygiidae, all the families of order Schistosomatida, Gymnophallidae, the most archaic Brachylaimoidea) formation of the frontal and posterior surface of the furca branches proceeds approximately at the same rate. In this case the primary excretory pores are located at the ends of branches. With growth processes dominating in the post-poral part it is from this part that not only the posterior but also considerable parts of the frontal branch surface are formed. This results in a shift of the excretory pores to the middle of the frontal surface. Such relations are observed in the cercariae of all Strigeidida. Thus, in furcocercariae, as well as in cercariae with simple tails, a more intense development in ontogenesis of the post-poral part is certainly a secondary phenomenon.

If our treatment of growth processes accompanying the tail formation in cercariae ontogenesis is correct, the conclusion about the secondary nature of the furcocercous state becomes inevitable. In our opinion, it should not be surprising. The branching of the posterior end of the tail could have arisen rather late, only as the result of an specialization of the locomotory apparatus making it possible to use different forms of locomotion (swimming: only furcocercariae "learned" to swim tail backward – see p. 269), and providing a possibility for floating in the water column (see pp. 103, 267). The latter property is inherent in all furcocercariae.

Both Cable (1974) and Pearson (1972) consider that forked tails, being inherited from ancestral forms, are practically homologous in all the groups which possess them with rare exceptions (some Echinostomatidae – Prévot, 1971). In his phylogenetic scheme Cable (1974) actually contraposes all trematodes with furcocercariae to the rest of the flukes whose cercariae possess simple tails. For some reason an exception is made only for Transversotrematidae positioned not far from Notocotylidae and Pronocephalidae.

Such an unambiguous solution seems too simplified to us. All the furcocercariae can be subdivided into two groups according to the presence of flame cells in the tail stem. In all Strigeidida without exception it is the tail in which the last (sixth) primary group of flame cells is located. A similar picture is also characteristic of Bivesiculidae, Azygiidae, and Transversotrematidae. Schistosomatida demonstrate two variants of the realization of this character. In larvae of Spirorchiidae and Schistosomatidae the last primary group of flame cells is also located in the tail stem, whereas in extremely specialized Aporocotylidae and Sanguinicolidae all the flame cells are localized only in the larval bodies.

A similar picture is to be also observed in representatives of a whole number of other families (Gymnophallidae, Brachylaimidae, Fellodistomidae *pro parte*, Bucephalidae and even some Echinostomatidae – *Aporchis massiliensis*) whose cercariae also possess a furca of some shape. In our opinion, the above taxa do not form a natural group even if *Aporchis massiliensis* is crossed off this list. The excretory formula of the Gymnophallidae cercariae described to date is 2[(2 + 2) + (2 + 2)] = 16 or 2[(N × 1 + N × 1 + N × 1) + (N × 1+ N × 1)] = 2 × N × 5, which is the result of juvenilization repeatedly mentioned above (see p. 149). An incomplete number of primary groups of flame cells is associated with collecting tubules (Cable, 1953; Ching, 1965; James, 1964; Loos-Frank, 1971; Pekkarinen, 1987a,b). Excretory formulae of the same type which we will further designate as "incomplete" (see below) are also characteristic of larvae at least of two more families: Fellodistomidae (*Cercaria caribbea LX* Cable, 1956) and Leucochloridiomorphidae (Brachylaimoidea) (*Leucochloridiomorpha constantiae* – Allison, 1943; Žd'árská, 1963). On the contrary, Bucephalidae are left out of this series. Their cercariae possess six pairs of the primary groups of flame cells (*Labratrema lamirandi* – 2[(6 + 6 + 6) + (6 + 6 + 6)] = 72; *Cercaria pleuromerae* – 2[(2 + 2 + 2) + (2 + 2 + 2)] = 24) (Maillard, 1975; Wardle, 1988). All the primary groups are located in the larval body and never enter the tail. According to this feature the Bucephalidae larvae resemble cercariae of Fasciolidae, Echinostomatidae and other families of this complex.

As for the families whose cercariae possess an "incomplete" excretory formula (the terminal primary groups of flame cells are absent – 2[(N × 1 + N × 1) + (N × 1 + N × 1)] = 2 × N × 4), this is the result of heterochrony (see p. 131). Delay in the development of the excretory system occurs independently in different digenean groups, therefore any judgment of the degree of relationship of taxa according to this character is out of the question. Relegation of Aporocotylidae and Sanguinicolidae to the order Schistosomatida causes no doubt, and though cercariae of these trematodes have no flame cells in their tail stem, attribution of these families to the first group is indisputable. The situation is more complicated in Gymnophallidae and Fellodistomidae. As regards the position of Bucephalidae, the above data on the structure of excretory system in their cercariae in our opinion show that there are no close phylogenetic relationships between Bucephalidae, on the one hand, and the complex of families always possessing true furcocercariae, on the other hand.

We have discussed the tail formation in such detail because the above considerations allow us to draw two important conclusions.

(1) The events of the early stages of morphogenesis in the overwhelming majority of cercariae support the hypothesis of Cable (1965, 1974) who supposed that the posterior body part of the cercarial body became the tail. Two demonstrative arguments in favour of Cables' hypothesis are the pattern of **isolation** of the tail primordium from the rest of the embryo and the presence in the tail primordium of excretory system elements (considerable parts of the main collecting tubes and sometimes the last primary group of flame cells). At the earliest stages of morphogenesis it is moulding of the embryo and not outgrowth of some part of the body (proliferation) that re-

sults in the formation of the tail. If the tail were formed as the result of proliferation, it could be considered as a late novel provisional formation. Then the tail would play no role in determining features of the body plan organization of the hermaphroditic generation. However, this is not the case. Proliferation, though it does occur, proceeds only after the isolation of primary tail primordium. Noteworthy, the role of proliferation in tail formation is increased in the most specialized forms (Allocreadiidae, Plagiorchiida).

(2) We cannot agree with Cable's (1965, 1974) and Pearson's (1972) opinion on the primary character of bifurcated tails in the digenean larvae of the hermaphroditic generations. We consider that the appearance of a furca at the end of a tail is a late event in the evolution of the Digenea which occurred at least three times quite independently. One group is constituted by the orders Strigeidida and Schistosomatida and an adjoining group of families (see p. 521), the second is represented by Bucephalidae and, finally, the third by some Echinostomatidae (the genus *Aporchis*). A somewhat more complicated problem is presented by several other digeneans of which furcocercous larvae are typical: Gymnophallidae, Brachylaimoidea and Fellodistomidae. The first 2 families may have retained the bifurcated tail of some ancestral forms common with the first of the groups distinguished above, whereas Fellodistomidae have most likely acquired furcocercous larvae independently, since this family comprises rather a lot of species whose cercariae possess simple non-furcate tails. Incidentally, fellodistomids are often mentioned as an example of the fact that simple tails are secondary and appear as the result of furcae loss. In fact, this is only a hypothesis based on admission that bifurcated tails are primary, which, in our opinion, is erroneous.

In connection with the problems under discussion an extremely specialized monotypical family Heronimidae should be mentioned. Cercariae of *Heronimus mollis* possess a simple non-furcate tail evidently demonstrating a tendency for the megalocercous state. The last group of flame cells is located in the tail. So far, this is the only example of this kind available in the literature. The uniqueness of these larvae also lies in the fact that they possess an "incomplete" (see above) excretory formula (Crandall, 1960). Such a combination of characters is not to be found in any other digenean group. Taking into account some other characters of both hermaphroditic and the only parthenogenetic generations of this species one could suggest that Heronimidae hold a special place in the system of modern trematodes but at the same time are far from being a sister group of all the rest of the flukes as Brooks et al (1985a) claimed (see p. 507).

Thus, summing up the above, it is possible to suggest that three important characters are associated with the complex "cercariae tail-excretory system", which changed independently in the process of the digenean evolution: the shape of the tail (i), the location of primary excretory pores (ii) and the location of the last (posterior) primary groups of flame cells (iii).

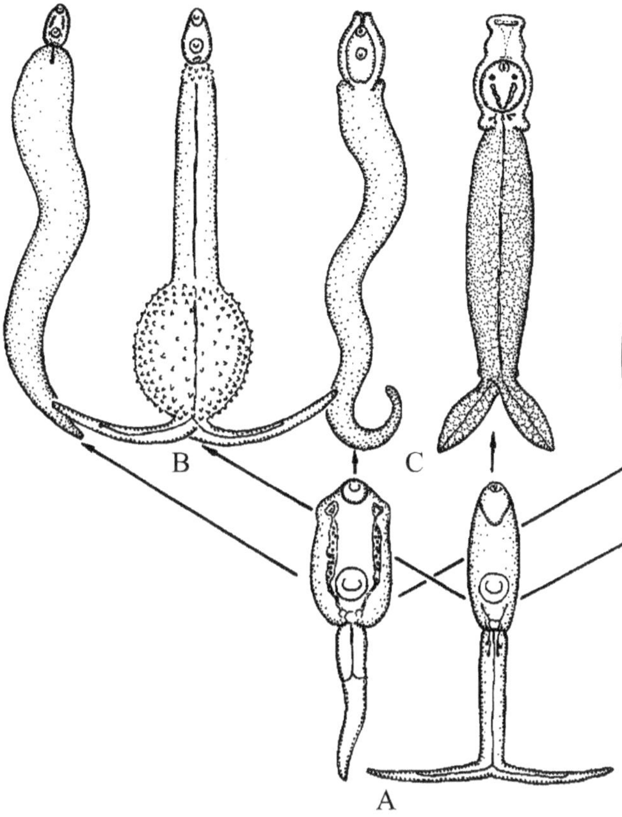

Figure 138. Parallelisms in the adaptive modifications of the two basic types of cercarial tails. A, cercariae with initial types of the tail (Echinostomatidae and Strigeidida); B, cercariae with macrocercous tails (Echinostomatidae and Strigeidida); C, cysticercous cercariae (Gorgoderidae and Bivesiculidae); D, microcercous cercariae (Opecoelidae and Brachylaimidae).

All the surprising variety of cercariae tail structure can be reduced to two main morphological types: simple and bifurcate tails. The first type seems more ancient to us and initial for larvae of the hermaphroditic generation. The appearance of a furca at the end of the tail is a character of secondary specialization. This was most probably associated with the change of behaviour of cercariae in the environment, a transition from continuous swimming to a discrete type of locomotion (intermittent swimming), which in most cases was accompanied by acquiring specific resting postures (see p. 268). Forked tails are so ideally adapted to the realization of this type of behaviour that their adaptive nature causes no doubt. It should be only noted that the appearance of such tails in the digenean evolution occurred rather early and resulted in the isolation of a large clade (see p. 522).

All the other tail modifications appeared on the basis of these two types, absolutely independently and in parallel (Fig. 138). Among the true Furcocercaria macrocercous state is to be observed, though rarely (Transversotrematidae, *Cercaria bulbocauda*, *C. clavicauda*, *C. abourda*, *C. globocaudata*, *C. notabilis*) (Miller, 1927; Szidat, 1940; Niewiadomska, 1966, 1970; Cribb, 1988). At the same time, very large tails are also inherent in the representatives of several genera in the family Echinostomatidae (*Petasiger neocommense*, *Echinochasmus milvi* and other species of this genus). They are also common enough in Heterophyidae (*Cercaria caribbea XVI*, *C. caribbea XVII*, *C. caribbea XVIII*) (Cable, 1956).

The same refers to cysticercous larvae: cercariae of Bivesiculidae and Azygiidae possess typical forked tails, whereas in cercariae of Gorgoderidae the tail belongs to a group of simple ones. A similar picture is also observed in fin folds and bristles.

The position of excretory pores in cercariae is also initially realized only in two main variants: they are localized either directly on the tail (see fig. 133H,I) or on the boundary of the body and the tail (see fig. 133J). It is the first variant, as it seems to us, that should be considered as the initial one. First, it is inherent in all the archaic taxa, possessing both furcocercariae and larvae with simple tails. Second, such location of excretory pores always correlates well with the predominance of moulding during tail formation (see p. 458).

The second variant is surely secondary. Localization of primary excretory pores on the boundary of the body and the tail is characteristic of a limited number of specialized taxa the central position among which is occupied by the order Palgiorchiida. Besides, the tail of larvae is mainly formed as an outgrowth, i.e., by means of proliferation.

The last character, the location of the most posterior (6th) primary group of flame cells, is not so easily subject to unambiguous assessment since its concrete manifestation seems to depend on one more character associated with the excretory system – the structure of the excretory formula. In those cases when cercariae possess a "complete", i.e., 6-term excretory formula, no special difficulties arise. In larvae of Azygiidae, Bivesiculidae, Transversotrematidae, Spirorchiidae and Schistosomatidae from the order Shistosomatida, Cyathocotylidae and Diplostomidae from the order Strigeidida the last 6-term primary group of flame cells is always located in the tail stem. The same is characteristic of representatives of the family Strigeidae (the order Strigeidida) in whose larvae the excretory formula is 5-term: $2[(N \times 1 + N \times 1) + (N \times 1 + N \times 1) + (N \times 1)] = 2 \times N \times 5$ (see p. 135). It seems to us that it is this state of the character under consideration that should be considered primary. To some extent it appears to recapitulate the initial stages of the tail formation in the ancestral digeneans in which the specializing locomotory part of the body included not only the post-poral but also a considerable part of the pre-poral region containing the last group of flame cells (see fig. 133H). The overwhelming majority of the recent Digenea possessing furcocercariae retained this state of the character.

On the contrary, the absence of flame cells in cercariae tails seems to be always secondary but this may be the result of absolutely different processes. In some cases (Sanguinicolidae, Aporocotylidae), as stated above, the flame cells are absent in the tail stem because of the extreme juvenilization of larvae also expressed in the simplification of their excretory formula. It cannot be ruled out that for the same reason the tail stem of cercariae is devoid of flame cells in Gymnophallidae and Brachylaimoidea. Larvae of representatives of all these families possess "incomplete" excretory formulae: 4-term $2[(N \times 1 + N \times 1) + (N \times 1 + N \times 1)] = 2 \times N \times 4$ or 5-term $2[(N \times 1 + N \times 1 + N \times 1) + (N \times 1 + N \times 1)] = 2 \times N \times 5$.

Yet in many modern digeneans flame cells in the tail of the cercariae are absent even if excretory formula is 6-term (Bucephalidae, Megaperidae, Haplosplanchnidae, Paramphistomidae, Fasciolidae, Echinostomatidae, Psilostomidae, Philophthalmidae, Notocotylidae). The same picture can be also observed in a large group Heterophyidae *sensu lato*, whose cercariae demonstrate a set of all the three most frequently occurring variations of excretory formulae: 6-, 5-, and 4-term (Cable, 1956). Though a detailed description of the structure and development of the excretory system of cercariae of Acanthocolpidae is not available, there is every reason to believe that this taxon is also characterized by the same state of the above character (Peters, 1961).

With all the dissimilarity of the above groups they have one more character in common: the initial presence in the tail of distal parts of the main collecting ducts. The primary excretory pores may be located at the ends of very long branches of the furca (Bucephalidae) (Matthews, 1973a,b, 1974; Podvyaznaya and Galaktionov, a,b, *submit*) or on the lateral surfaces of the simple tail (all the rest of the families listed above).

The absence of flame cells in the tail (see fig. 133I) in all cases like these cannot be accounted for by a retardation in the development of excretory system. This matter should probably be a shift of the "body-tail" boundary backwards, beyond the zone of flame cells location. In fact, a noticeable shortening of the posterior part of the embryo (which in the course of cercariae morphogenesis is differentiated as a tail) is observed. Yet in this case too the tail includes not only the whole post-poral but also the pre-poral part to a considerable degree.

Further realization of this tendency results in the "cercarial body-tail" boundary being shifted still further backwards (see fig. 133J). In Lepocreadiida, Allocreadiidae, Gorgoderidae, all Plagiorchiida it passes at the level of location of excretory pores. The tail is formed only at the expense of the post-poral part and does not include any elements of excretory system.

Shifting the "cercarial body-tail" boundary backwards does not rule out realization of the tendency for simplification of the excretory formula. Juvenilization results in the appearance of larvae with "incomplete" excretory formula in different families: Echinostomatidae – *Echinochasmus*, many

Heterophyidae and Opisthorchiidae, Opecoelidae, Microphallidae. The often-mentioned *Heronimus mollis* should be again discussed. We have already written (see p. 438) that larvae of this species are characterized by a combination of a simple tail, an "incomplete" excretory formula and the presence of the last primary group of flame cells in the tail (Crandall, 1960). The appearance of such a unique combination of characters may be accounted for by one fact only. In our opinion the "cercarial body-tail" boundary in the ancestors of the modern Digenea was initially still more anterior. The area of the body which was differentiated as a tail, in ancient forms initially possessing a 6-term excretory formula, also included the zone of the location of the last but one, 5th primary group of flame cells. Heronimidae retained this very ancient state of the character, which did not prevent their thorough and secondary specialization: the appearance of a schizocoel in the miracidium (see p. 527), modification of the mother sporocyst, simplification of the excretory formula, reduction of the transmission ways in the environment (see pp. 436, 527), change of a settlement site in the final host.

Not a less important feature of the Digenea than a presence of a tail in the cercariae is the presence in the hermaphrodite generation of two suckers, an oral and a ventral one. Three main groups of characters are associated with the suckers: "presence and absence", "location on the body" of cercariae and maritae and "the degree of development and specialization" which often give rise to discussion in the literature (Cable and Hunninen, 1942; Cable, 1956; Brooks et. al., 1985a; Pearson, 1992). The main subject of the discussion is determination of the plesiomorphic state of the first character. Pearson (1972, 1992) who devoted a lot of attention to this problem, states unambiguously that "… the plesiomorphic states for the Digenea are *ventral sucker absent* and *oral sucker absent*, which absences must be distinguished from secondary loss" (Pearson, 1992, p. 108). This position agrees well with the fact that both Pearson (1972, 1992) and Cable (1974) consider the family Bivesiculidae to be the most archaic group among the recent Digenea which have retained the greatest number of plesiomorphic characters. In their opinion, the absence of both oral and ventral suckers in them reflects the primary state of the character in question. We will return to the discussion of the state of Bivesiculidae among the modern Digenea (see p. 522) later. Now we only note that this assertion is not as uncontested as it may seem at first sight. Both cercariae (the body of the larva is under consideration here and not its greatly developed and extremely specialized tail), and maritae of Bivesiculidae are very juvenilized (see p. 507). In this case the absence of the suckers may be one of the numerous manifestations of heterochrony, so frequent in the Digenea. This suggestion seems plausible since there are data testifying to the fact that at least the oral sucker in some species of Bivesiculidae does exist (Gibson, 1987).

But suckers are absent not only in Bivesiculidae. There are at least two other groups of families to be mentioned in connection with the problem under consideration. First of all, it concerns the paramphistomoid complex. The presence of an enlarged ventral sucker in them is doubtless. The problem of the oral sucker is more complicated. Traditionally when describing cercariae and maritae of Paramphistomidae, Diplodiscidae and other families of this complex the researchers mention *an oral sucker* though a very modified one due to the formation of posterior-lateral diverticula (see, for instance, Skrjabin, 1949). Yet another point of view on the nature of this formation can be found in the literature. Näsmark (1937) showed rather convincingly that a muscular structure supplied with secretory cells which is situated in the beginning of the digestive system in hermaphroditic individuals of Paramphistomidae and other families of this taxonomic complex was a very modified pharynx. This point of view was shared by Pearson (1992). Moreover, he extended it to those digeneans whose cercariae belong to the group Monostomata (Notocotylidae, Pronocephalidae). Pearson suggests that in this case individuals of the hermaphroditic generation may have lost the sucker and its place and function are taken by the pharynx. Unlike the Paramphistomidae and related forms, the pharynx of representatives of Monostomata did not undergo any noticeable changes. In all the other trematodes the oral sucker both in cercariae and maritae is always present, as a usual sucker or in modified state (the anterior organ – see pp. 106, 277).

Though arguments given by Näsmark (1937) and Pearson (1992) as regards Paramphistomidae and Monostomata (Notocotylidae, Pronocephalidae) sound convincing, the final solution of the problem of homology of the anterior muscular structure to the pharynx and not to the oral sucker may be accepted only after special investigations including embryological ones.

All the above three groups (Bivesiculidae; Paramphistomidae and related taxa; Notocotylidae and Pronocephalidae) possess primary two-host life cycles and form the lower branches of the phylogenetic "tree" of the recent Digenea. From this point of view one could think that the absence of the oral sucker in them reflects the plesiomorphic state of the character. Yet most probably this is not the case. Our doubts concerning Bivesiculidae have been already stated above. As for the paramphistomoid complex of families, the absence of the oral sucker in them appears to be secondary. All the representatives of this phylogenetic group except the secondary specialized Microscaphidiidae, possess a well-developed ventral sucker located close to the posterior end of the body both in the cercariae and maritae.

The ventral sucker probably appeared in the hermaphroditic generation of trematodes later than the oral one. At present there is no direct evidence but embryological data testify in favour of this suggestion. The ventral sucker is formed later than the oral one in cercarial morphogenesis. In cercariae from many groups the ventral sucker often remains underdeveloped and sometimes is present only as a hardly distinguishable primordium (e.g.,

Cyathocotylidae, Heterophyidae). Microphallid cercariae of "Ubiquita" type lack the primordium of the ventral sucker at all, the ventral sucker being formed only at subsequent phases of development of the hermaphroditic generation. A retardation in the ventral sucker formation may reflect the succession of events in the digenean phylogeny. There is no ground to suppose that paramphistomids (*sensu lato*) have originated from some ancient forms that lacked the oral sucker and acquired the ventral one independently of all other digeneans.

Individuals of the hermaphroditic generation of Notocotylidae and Pronocephalidae lack the ventral sucker. Its functions are performed by other structures (see p. 98). Yet the retention of an archaic initial character in the representatives of these families is not to be expected: even though their life cycles are archaic, on the whole they possess a great number of characters pointing to secondary specialization.

Besides, in all rediae, even those possessing a very reduced intestine, the pharynx never opens outwards. It is surrounded with a circular fold of the tegument restricting the pre-pharyngeal cavity. Our observations of living rediae of *Cryptocotyle lingua* have shown that during their feeding on molluscan hemolymph, first the pre-pharyngeal cavity is filled, with the pharynx being closed. Then the oral opening proper (formed by the edges of the fold) closes and the pharynx opens. The food mass passes through the pharynx into the oesophagus and then into the gut. On the contrary, in rediae of several species of Echinostomatidae absorbing not only liquid food but also bits of the host tissues and sometimes younger rediae as well, capture of solid food objects proceeds differently. A widely opening fold is a little drawn back, the pharynx advancing maximally. With its help rediae first capture and hold the prey, and then push it into the gut. In this respect, feeding of rediae is reminiscent of many Turbellaria using the pharynx as an organ of food capture.

In representatives of all three groups of taxa mentioned (Bivesiculidae; Paramphistomidae and related taxa; Notocotylidae and Pronocephalidae) there is nothing of the kind. The anterior opening of the pharynx is also the oral opening since the pre-pharyngeal cavity and the circular tegumental fold are absent.

In the hermaphroditic generation of trematodes the oral sucker located at the anterior end of the body is always separated from the pharynx by a relatively short prepharynx lined with the extended external syncytial tegumental layer. In cercarial ontogenesis the primordium of this part of the digestive system appears at the same time as that of the pharynx behind the primordium of the oral sucker, i.e., rather deep in the body of the embryo. The oral sucker is symmetrical: the curve of its axis relative to the longitudinal axis of the body resulting in displacement of the oral opening subterminally to the ventral surface occurs later due to a more powerful development of myoblasts on the dorsal surface. Taken together, all these considerations make it possible to suggest that the prepharynx of hermaphroditic generation is homologous to the pre-pharyngeal cavity of rediae. As for the oral sucker, it developed at the expense of the most distal part of the circular fold provided with a rather complexly differentiated system of radial muscles (see p. 62 and fig. 27).

In our opinion all the above testifies to the loss of the oral sucker by representatives of the three groups of taxa under consideration (Bivesiculidae; Paramphistomidae and related taxa; Notocotylidae and Pronocephalidae), and not its primary absence. Pearson admits the plesiomorphic state of the organization of "... redia, which I (*i.e., Pearson – K. G. & A. D.*) have argued elsewhere is close to the adult predigenean" (Pearson, 1992; p. 115). We also proceed from this premise. But in this case, if the plesiomorphic state of the character analyzed were really characteristic of Bivesiculidae it would be also natural to expect the representatives of this family to retain the pre-pharyngeal cavity restricted by the circular fold of the tegument. Yet no structures like this have been described. Of course Bivesiculidae may be supposed to have lost not only the sucker but also the tegumental fold, which resulted in disappearance of the pre-pharyngeal cavity. But then the present state of the character under consideration should be considered not plesiomorphic, as Pearson (1972, 1992) and Cable (1974) have suggested, but an morph. Most probably it was the oral sucker that was lost, which is testified to by the above data (Gibson, 1987).

The oral sucker in the hermaphroditic generation of Bucephalidae (Gasterostomata) deserves special consideration. These digeneans possess the so-called "rhynchus" located terminally at the anterior end of the body, their oral opening being displaced to the ventral surface. Actually these are two different characters (the location of the oral sucker and that of the oral opening) which nevertheless form a united complex. Their close association is presupposed by the fact that the oral cavity was initially formed as the most anterior part of the digestive system. In fact this is a specialized part of the foregut. This has not been taken into account in most cases during the discussion of the features of the organization of the hermaphroditic generation.

There exist two points of view on the origin of the rhynchus. Brooks et al. (1985a) consider this structure to be homologous to the ventral sucker which was secondarily displaced forwards because the oral opening shifted to the ventral surface. This speculation is artificial and has not been supported. Another point of view is more wide spread (Sinitsin, 1909; Sewell, 1922; La Rue, 1926; Gibson, 1987; Pearson, 1992) – the rhynchus is homologous to the oral sucker of other digeneans, i.e., Sinitsin (1909) used information on the histological structure and function of the rhynchus and typical oral suckers to support this view. The oral opening was displaced to the ventral surface secondarily and independently. The above view is illustrated in the scheme of Pearson (1992, p. 112, Fig. 11). He noted that the three latter positions in this scheme are real, but the first one (which seems to represent the ancestor of Bucephalidae with a typical terminally positioned oral sucker and a simple sack-like gut) is hypothetical.

It appears certain that the shift of the oral opening to the ventral surface in Bucephalidae is a secondary phenomenon. On the one hand this is supported by the morphological series presented by Pearson (1992). On the other hand (which is more important in our opinion), the oral opening of the redia is always positioned strictly terminally.

On the contrary, homology of the rhynchus and the oral sucker is not as certain as may appear. The study of the morphogenesis and morphology of cercariae of *Labratrema lamirandi* and maritae of *Prosorhynchus squamatus* revealed details usually not discussed in the literature (Suloeva, 1999). The initial stages of the digestive system formation in *Labratrema lamirandi* are similar to those in other digeneans (see p. 130). The difference is that there is no oral sucker primordium in the pharyngeal region, the primordium of the prepharynx being distinct. Later a cavity is formed in the prepharynx and in the pharynx but this cavity has no connection with the outside. In the place where the prepharyx primordium is close to the tegument the latter is a little thicker, forming a small protuberance. When the oral opening is formed the prepharynx lining and the external syncytial layer of tegument are united, and the tegumental fold covering the mouth is formed. In mature maritae of *Prosorhynchus squamatus* this fold, because of specialized muscle bundles, may close the oral opening. This structure resembles the prepharyngeal cavity and the circular tegumental fold of the redia (see p. 62).

In our opinion, the aforementioned features of the anterior part of the digestive system in the hermaphroditic generation of Bucephalidae are very important. This may attest to the fact that Bucephalidae, unlike other digeneans, retained the plesiomorphic state of the anterior part of the digestive system and never possessed an oral sucker. The rhynchus is a novel structure whose origin is different from the usual suckers of other digeneans.

Pearson, when constructing the above mentioned series (Pearson, 1992, p. 112, fig. 11) considered a sucker-like rhynchus as a primitive variant. In his opinion it could have evolved in a true rhynchus only gradually. In this respect he seems to be correct. But a sucker-like structure at the anterior end of the body in Bucephalidae had nothing in common with the digestive system. Pearson (1992) convincingly showed that suckers and structures similar to them both morphologically and functionally (for instance, the "false sucker") located at the anterior end of the body, had appeared independently in different groups of parasitic flatworms. This is also true of the ventral sucker in the Digenea and attachment structures at the posterior body ending of Aspidogastrida and Cercomeromorpha (Rohde and Watson, 1995). The transformation of the anterior (but not the oral – sic!) sucker-like structure into the rhynchus of Bucephalidae must have happened because there was no connection between the structure under consideration and the functioning digestive system. This is indirectly confirmed by the fate of the oral sucker in many cercariae. The digestive system of these larvae either remains underdeveloped or performs non-characteristic functions (see pp. 130, 279). If the oral sucker is not involved in attachment of the anterior end of the body during feeding, it is frequently transformed into the so-called "anterior organ". In furcocercariae of Strigeidida and Schistosomatida it is specialized penetration-secretory complex including a glandular apparatus, an eversible proboscis and differentiated muscles. A morphologically and functionally similar proboscis was also independently formed in pleurolophocercous cercariae (Heterophyidae, Opistorchiidae). However, their "anterior organ" retains some characteristic features of a typical oral sucker and does not contain the distal parts of penetration gland ducts. What is usually referred to in the literature as the "oral sucker" in virgulate xiphidiocercariae (Lecithodendrioidea) is not an oral sucker. There is no proboscis (its functions are performed by the stylet) but also no cavity characteristic of a typical sucker. The main volume of this structure is occupied by pocket-like outgrowths of the tubular distal part of the foregut containing considerable mucous secretion, i.e., the virgula (see p. 126). The formation of this organ can be traced within the Lecithodendrioidea group (see fig. 57A).

The above examples demonstrate that during evolution oral suckers may lose their typical construction and become rhynchus-like structures in cercariae from different taxa. Importantly, at subsequent stages of ontogenesis (as a rule, in metacercariae) in all the above cases, a typical structure of the oral sucker in individuals of the hermaphroditic generation is restored when the digestive system begins to function.

In conclusion, we share Pearson's (1992) view of the independent acquisition of the oral sucker by different groups of parasitic Platyhelminthes (Neodermata). We also agree that the ancestral form of the Digenea had neither the oral nor the ventral sucker. The latter was acquired only after the individuals of hermaphroditic generation colonized vertebrates as hosts. Yet we do not agree that the absence of the suckers in Bivesiculidae reflects a plesiomorphic state. This also refers to the other two groups of taxa mentioned above – Paramphostomidae and related taxa and Notocolylidae and Pronocephalidae. The only group that may have retained the primary absence of the suckers is the Bucephalidae. Considering all the characteristics of this group, it may be considered as a group *incerta sedis* within Digenea.

The ventral sucker was acquired by the hermaphroditic generation relatively late in the evolution of the Digenea. This is supported by the sequence of formation of primordia of the suckers, the tail, and the excretory system in the course of cercarial morphogenesis. The primordium of the ventral sucker is the last to emerge. Sometimes the formation of this organ is shifted to the subsequent stages of the hermaphroditic generation development. An underdeveloped ventral sucker in cercariae of Cyathocotylidae and Heterophyidae and all the more so in Cercariae Ubiquita (some Microphallidae) are a result of heterochrony, and so is the absence of the ventral sucker in Bivesiculidae larvae (see p. 507).

Independent emergence of the oral and the ventral sucker is also supported by differences in their structure. As shown on *Philophthalmus rhionica*, the oral sucker possesses the external layer of circular muscle filaments. The latter is located immediately under the "capsule" of basal lamina separating the muscles of the sucker from the adjacent parenchyma. Such a muscle layer is absent in the ventral sucker (Tihomirov, 1980; Galaktionov and Dobrovolskij, 1987). The most superficial position is occupied by a layer of meridional muscles (transverse muscle filaments as defined by Rohde and Watson, 1995). The difference between the oral and the ventral sucker seems to arise from the specificity of their origin and function. The first, as mentioned above (see p. 452), was formed at the expense of the prepharyngeal cavity. Therefore, the oral sucker is, by its origin, the most anterior part of the digestive system. The function of the oral sucker is not only attachment of the anterior end of the body, but in absorbing liquid and semiliquid food it can close similar to the ring tegumental fold surrounding the prepharyngeal cavity in rediae (see p. 452). The external layer of circular muscles in the oral sucker (absent in the ventral one) appears to serve this purpose. The same muscles are used in the cases when villi of the host intestine are grasped by the oral sucker.

The ventral sucker formed in the course of evolution on the ventral surface of the body as an attachment organ. Its emergence should be treated as a secondary adaptation to its residence in the intestine of vertebrates, which enable maritae to resist peristalsis and the movement of rather dense food mass. It has been mentioned above that in cercariae the ventral sucker often remains underdeveloped or is absent. The ventral sucker in maritae has diverse functions. In those cases when its role as an attachment organ is limited there may be retardation in its development in the process of maritogony. For example, in strigeidid trematodes which possess an adhesive organ and an "attachment cup" on the ventral side of the body the ventral sucker is small relative to the body size of the fully formed maritae.

A similar situation also occurs in *Brandesia turgida* (Pleurogenidae) inhabiting pocket-like outgrowths in the amphibian intestine wall where they are sheltered from intestinal peristalsis. The ventral sucker remains underdeveloped in maritae of *Astiotrema trituri*. Relatively large adults of this species parasitic in the larvae of newts are so tightly enfolded in the walls of the host intestine that additional attachment is not necessary. The ventral sucker is also absent in Eucotylidae, whose maritae inhabit the kidneys of birds. Cyclocoelidae, inhabiting respiratory tract of birds and other vertebrates, also lack this organ. Sometimes the absence of the ventral sucker is compensated for by other structures acquiring the attachment function (e.g., the ventral papillae of Notocotylidae).

In all the above cases the loss of the ventral sucker is the result of usual reduction, whereas the absence of the ventral sucker in maritae of Bivesiculidae is probably the result of heterochrony. In some Microphallidae, Heterophyidae and a number of other trematodes the ventral sucker is absent or underdeveloped in cercariae but is normally differentiated at subsequent stages. On the contrary, in Bivesiculidae most maritae lack the ventral sucker as well as cercariae.

The only group in whose representatives the absence of ventral sucker can reflect the plesiomorphic state of the character is the Bucephalidae. The secondary shift of the oral opening to the ventral surface of the body may have been possible because there was no ventral sucker there.

Acquisition of the ventral sucker by the ancestors of the Digenea predetermined many structural features of the hermaphroditic generation, in particular the division of body into two regions: locomotory (pre-acetabular) and generative (post-acetabular) ones (see p. 191). This boundary is well expressed also in those cases when the ventral sucker is secondarily reduced (Eucotilidae, Cyclocoelidae).

Primary position of the ventral sucker has long been discussed. This issue is associated with a very important problem, the search for the most ancient group (sister group).

Gibson (1987) assigned this role to paramphistomoid digeneans. He viewed their amphistomate condition (subterminal position of the ventral sucker at the posterior end of the body in individuals of hermaphroditic generation) as a synapomorph of the Aspidogastrida and Digenea. Rohde and Watson (1995) hold a similar view proposing a homology of the ventral sucker of Digenea and Aspidogastrida. This approach dates back to Brooks et. al. (1985b, 1989) and Brooks (1989) who distinguished two large taxa, subphylum Cercomeria and superclass Cercomeridea. One of the main characters of Cercomeria was the presence of "... posterior adhesive organ formed by an expansion of the parenchyma into an external pad, called a cercomer" (Brooks et. al., 1985b, p. 5). Brooks et al. (1989) included in the Cercomeria both Udonellidea, Temnocephalida and the true Cercomeromorpha, as well as Aspidogastrida and Digenea. In the latter the ventral sucker was supposed to be homologous to the cercomer. These views have already been criticized in the literature (Lebedev, 1987). It has been shown that the "posterior adhesive organs" in different taxa of Cercomeria have a different structure and are not homologous (Rohde and Watson, 1995).

The only structure of individuals of the hermaphroditic generation that might correspond to the cercomere is the tail. An analogy could be seen between the place and mode of the tail formation and those of the cercomere. However, this hypothesis has already been rejected (see p. 436). Importantly, the analysis of organization of the hermaphroditic generation (see p. 438) shows that the posterior end of the **body** of cercariae and maritae is not, in fact, the posterior body end (in terms of body plan organization). The true posterior end is the end of the cercaria tail. At the phase of metacercaria it is shed. Together with it, not only the whole post-poral part disappears from the body of the marita (exceptions are few and secondary), but in the majority of the cases a considerable part of the pre-poral one. Nothing of the kind is to be found in any other group of parasitic flatworms and their close relatives. From this point of view, the body of cercariae and maritae corresponds only to the anterior part of the body of other Platyhelminthes. Accordingly, the ventral sucker, an evidently secondary acquisition of digeneans, is developed in the pre-poral part very far from the primary (in terms of body plan organization), true posterior end of the body and cannot correspond either to the (true) cercomer or terminally located sucker of udonellids.

Therefore amphistomate condition of Paramphistomidae and Aspidogastrida cannot be treated as homologies. In the former it is secondary and results from the allometric growth of the embryo. In the latter it must be primary. Accordingly, all the specific features of the organization of paramphistomids should be considered as secondary specialization (a very characteristic muscular organ located at the anterior end of the body and sometimes treated as a specialized pharynx – see p. 451; the presence of a specialized muscular structures on the oesophagus, a lymphatic system, a shift of the secondary excretory pore to the dorsal surface, the amphistomate condition itself).

Thus, localization of the ventral sucker on the ventral body surface should be considered as the plesiomorphic state of the character. However, the location of the ventral sucker may vary considerably. In most digeneans it is close to the anterior end of the body. The pre-acetabular part of the body is several times shorter than the post-acetabular one. This is typical both of many primitive digeneans (Fasciolidae, many Echinostomatidae) and of more specialized ones (Schistosomatida, Strigeidida, many plagiochiids). The shift of acetabulum to the anterior part of the body is a secondary phenomenon, conditioned by allometric growth accompanying maturation of maritae. The reason for an intensive growth of the post-acetabular part was the intensive development of the reproductive system, which in its turn was conditioned by the transition of the hermaphroditic generation to parasitism on vertebrates. It is this circumstance that allows us to consider the post-acetabular part as the generative part of the maritae body. To a considerable degree it is *de facto* a novel structure acquired by trematodes independently of other groups of parastic Platyhelminthes (Neodermata).

The above considerations could be supported by the features of the muscular and the nervous system in the forms with the most well-developed post-acetabular (generative) region. As a rule, this is characteristic of primitive Digenea (Fasciolidae, Philophthalmidae, Echinostomatidae), but sometimes is also observed in specialized species (Opisthorchiidae, Dicrocoeliidae). Circular and longitudinal muscular fibrils (in many large species also diagonal ones) in the locomotory body region of adults of these trematodes are regularly arranged and form well-developed distinct layers. Nothing of the kind is observed in the generative region of adults of the same species. In this region muscular elements of the tegumentary-muscular sac are usually weakly developed and in most cases irregularly arranged. In juvenile phases of development of hermaphroditic individuals (cercariae and metacercariae) as well as in the daughter generation of parthenitae (rediae and worm-like daughter sporocysts) the arrangement of muscular elements of the tegumentary-muscular sac is regular along the whole body.

The nervous system structure in mature maritae corresponds to the structure of the muscular apparatus. Most species studied in detail with respect to the nervous system (Niewiadomska and Moczoń, 1982, 1984, 1987; Grabda-Kazubska and Moczoń, 1988; Joffe, 1990, Kotikova and Joffe, 1990; Kotikova et al., 1990; Kotikova, 1991; Halton and Gustafsson, 1996; Reuter and Halton, 2001) possess a typical orthogon (euorthogon according to Joffe, 1990) only in the pre-acetabular (locomotory) part of the body. There are few of any circular commissures of the nervous system behind the ventral sucker. Joffe (1990) referred to this type of orthogon as to prosostomatid-like euorthogon. In *Fasciola hepatica* there is a true plexus in the place of commissures (Fig. 139). In the adults of *Podocotyle atomon* and *Dicrocoelium dendriticum* the plexus is lacking, longitudinal nervous cords often do not reach the posterior body end and are connected only by singular irregularly arranged transverse nerves (Fig. 140A).

The structure of nervous system is entirely different in species with highly juvenilized adults whose post-acetabular part is poorly developed (e.g., *Prosotocus confusus* – Pleurogenidae). Euorthogon in them reaches the posterior body end (Looss, 1894). A similar arrangement is to be found in *Prosorhynchus squamatus* (Bucephalidae) (Kotikova et al., 1984) whose body is not divided into pre- and post-acetabular regions at all (Fig. 140B).

In our opinion, it is highly significant that, as shown in *Leucochloridium* sp. (Kotikova, 1991), in metacercariae with a true euorthogon the location of longitudinal cords and transverse commissures in the post-acetabular region is also regular. Similar results have also been obtained from the studies of ontogenetic transformations of the nervous system in the hermaphroditic generation of *Diplostomum pseudospathaceum* (Niewiadomska and Moczoń, 1984, 1987). Larvae of this species possess a typical euorthogon at early stages of ontogenesis (until the stage of the 14-day-old metacercaria). After that the metacercariae of Diplostomidae demonstrate a high rate of metamorphic processes, also involving the nervous system which acquires some features of polycommissural euorthagon (following the terminology of Joffe, 1990). In contrast to prosostomatid-like euorthogon, polycommissural one retains regular arrangement of longitudinal cords and commissures in the post-acetabular region, the number of the latter significantly increasing. Importantly, reconstruction of the nervous system in developing metacercariae occurs at the same time as reconstruction of the mus-

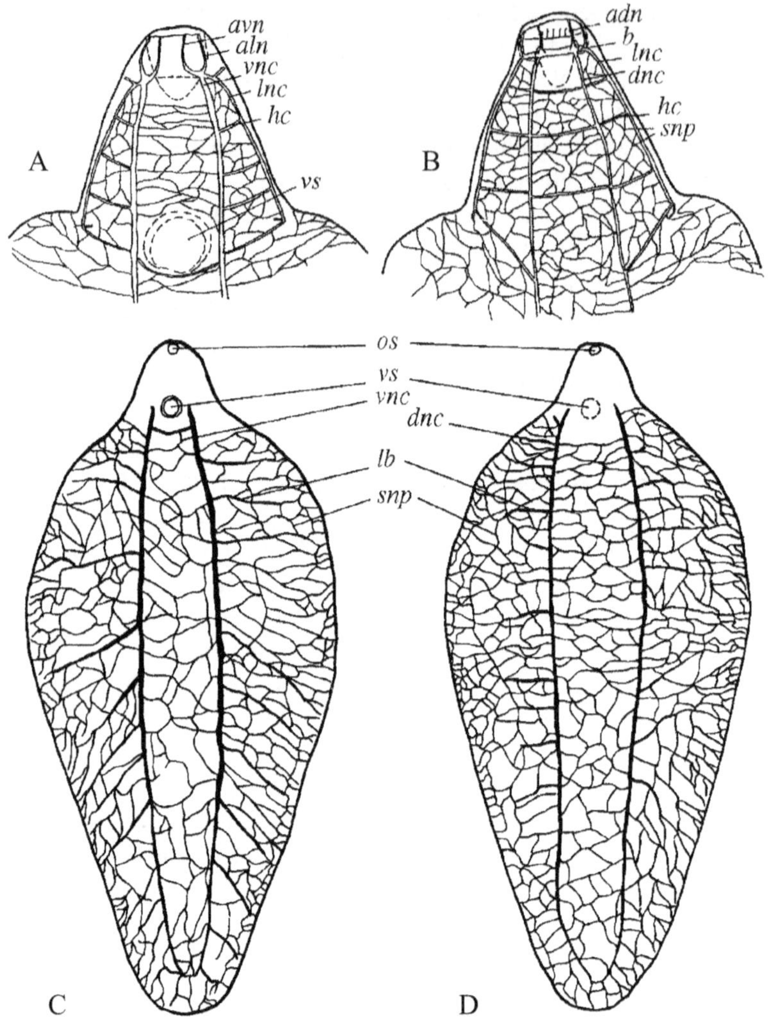

Figure 139. Nervous system of *Fasciola hepatica*. A-B, forebody; C-D, hindbody, A and C, ventral view; B and D, dorsal view. *Abbreviations*: *adn*, anterior dorsal nerve; *aln*, anterior lateral nerve; *avn*, anterior ventral nerve; *b*, brain; *dnc*, dorsal nervous cord; *hc*, hemicircular commissure; *lb*, lateral branch of the nervous cord; *lnc*, lateral nervous cord; *os*, oral sucker; *snp*, submuscular nervous plexus; *vnc*, ventral nervous cord; *vs*, ventral sucker. From Kotikova et al. (1990), with permission.

cular system (Podvyaznaya and Dobrovolskij, 2001) (see p. 176). Polycommissural euorthogon has also been described in mature *Schistosoma* (Withyachumnarkul et al., 1987) and some Brachylaimoidea (Okulewicź, 1988).

The above considerations allow the following reconstruction of the main stages of nervous system evolution in the hermaphroditic generation of

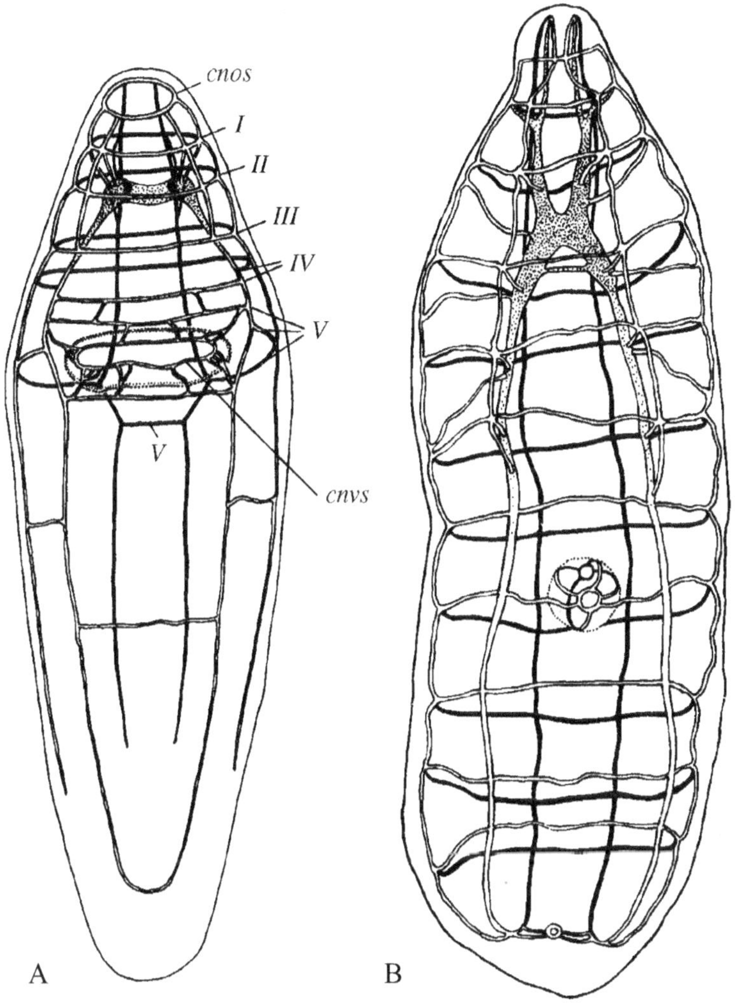

Figure 140. Different types of euorthogon. A, prosostomatid-like euorthogon of *Podocotyle atomon* (Opecoelidae), ventral view. From Kotikova et al. (1984), with permission; B, simple euorthogon of *Prosorhynchus squamatus* (Bucephalidae), ventral view. From Joffe (1990), with permission. *Abbreviations*: *I-V*, circular commissures; *cnos*, circular nerve of the oral sucker; *cnvs*, circular nerve of the ventral sucker.

Digenea. Ancestral forms must have possessed a true euorthogon. When in the process of evolution juveniles (cercariae) acquired the tail, specialized locomotory organ which is not retained in the adults and whose innervation is to a large extent autonomous (see p. 273), this acquisition involved almost no changes in the structure of the nervous system in the body region. On the contrary, the formation of the generative, post-acetabular body part (brought about by the intensive development of the reproductive system) entailed significant modifications of the nervous system. In morphologically more primitive forms these modifications result in the formation of a prosostomatid-like euorthogon (e.g., in Fasciolidae), the nervous system in the post-acetabular region recapitulating in fact proorthogon state (Joffe, 1990). Joffe (1990) considered the latter to be the initial type of the orthogonal nervous system of flatworms. Proorthogon, inherent to some Acoela, characterizes the state when the nervous cords are not completely differentiated from the plexus. In the posterior body half they do not join each other but become thinner and gradually "disappear" in the plexus. In more specialized Digenea (Schistosomatida, Strigeidida and related forms) a typical euorthogon is substituted by a polycommissural one. In secondarily juvenilized forms whose maritae are small and have a rather poorly developed generative (post-acetabular) body region, the nervous systems, like that of metacercaria, retains all the features of the evolutionary initial euorthogon.

The fact that the nervous system has evolved in parallel and independently in different clades of Trematoda in our opinion points to an early divergence in the group. This consideration is sufficient not to treat the clade Strigeidida–Schistosomatida as a sister group of all other Digenea.

1.2.2. The Main Trends of Morphological Evolution of Hermaphroditic Generation of Digenea

1.2.2.1. Heterochronies

The hermaphroditic generation of digeneans demonstrates that the evolution of larvae and sexually mature individuals of the same species may follow different trends if they inhabit environments with different conditions. This postulate of Schmalhausen (1969) has already been mentioned in the present book (see pp. 215, 408). Indeed, it seems impossible to find a common tendency in the morphological evolution of the hermaphroditic generation. At the same time, a close analysis of the data given in section 2 of chapter 1 shows that various heterochronies play an important role in this process. This was first noted by Ginetsinskaya (1968). However, manifestations of heterochronies, i.e., retardations in the development of some systems of organs accompanied by an accelerated formation of others, may be very different (sometimes opposite) at the phases of cercaria, metacercaria and marita.

Morphogenesis of plagiorchiid cercariae (with rare exceptions) is accompanied by retardation in the development of definitive systems of organs (e.g., reproductive, digestive, excretory), resulting in the juvenilization of the larvae. It is very pronounced in the representatives of highly specialized Microphallidae and Lecithodendrioidea (see pp. 131, 149). Heterochrony in metacercariae is expressed in the accelerated development of the reproductive system whose formation in fully formed larvae of the two above families is completed though it does not yet function. At the same time, their somatic organization is in fact "juvenile" and remains juvenile at the phase of marita (see p. 211).

A tendency to heterochrony is distinctly expressed in cercariae of Strigeidida. The larvae of Strigeidae and Diplostomidae are as juvenilized as those of Plagiorchiida (see p. 135). However, no juvenilization is observed in sexually mature individuals of hermaphroditic generation of Strigeidida that are the most specialized among trematodes (see p. 210). "Smooth" morphogenesis (see p. 162) appears to be insufficient to overcome the large gap between the morphological-functional organization of cercariae and maritae of strigeidids. This gap is covered by means of metamorphosis undergone by individuals of hermaphroditic generation of stigeidids at the stage of metacercaria (see p. 175). At this ontogenetic stage there are no heterochronies either. Thus, tendencies of morphological evolution of hermaphroditic generation in two main branches of higher trematodes (Plagiorchiida and Strigeidida) are similar only at the first stage of ontogenesis, the phase of the cercaria. They are different at two other stages, the phases of metacercaria and marita. This circumstance presupposes numerous differences in the strategies of life cycle transmission of these trematodes and in their association with ecosystems of different types. This will be discussed in the next section (see p. 469).

Biological significance of heterochronies lies in the fact that they facilitate the solution of tasks of functional ontogenetic stages. Juvenilization and miniaturization of individuals release energy resources which would otherwise be spent for full-scale construction of morphological organization of the phase. The resources released are channelled for the formation of provisional structures in the case of larvae and/or for an accelerated development of the reproductive system and the system of organs indispensable for the functioning of the sexually mature individual (the gut, the tegument). Juvenilization and miniaturization of cercariae bring about an increase in the fecundity of parthenogenetic generations producing a great number of small larvae. This is observed in the representatives of higher taxa Strigeidida and Plagiorchiida (see p. 70). Though the definitive organization of their cercariae is "juvenile", they possess perfect provisional structures (the tail with cross-striated muscles, the stylet, a perfect glandular apparatus – see p. 149) facilitating successful infection of the second intermediate host, i.e., the solution of the main task of the first functional ontogenetic stage.

Manifestation of heterochronies in the ontogenesis of hermaphroditic individuals becomes possible only after a certain level of relationships in the host-parasite system "the second intermediate host-the metacercaria" has been reached in the course of evolution. Cercariae encysting in the environment are never juvenilized. It is impossible since larvae encysting in the environment should possess numerous cystogenous glands and a large reserve of glycogen that would ensure the retention of adolescarial infectivity for a long time. Besides, such larvae should have a developed gut in order to start feeding immediately after entering the digestive tract of the final host. All this would entail a relatively large body, its osmoregulation requiring a complex excretory system. Even if only one link were missing from this chain of morphological-functional phenomena, the life cycle of the parasite would not be successfully completed.

When in the course of evolution the second intermediate host appeared in the life cycle, the situation became different. Formation of definitive organization (to a certain degree) and preparation for parasitism in the final host could be transferred to the phase of metacercaria, proceeding at the expense of energy resources of the second intermediate host. It is with the use of energy resources at the phase of metacercaria that the first manifestations of heterochrony in cercariae of *Echinochasmus* are associated (see p. 159). No instances of juvenilization of those echinostome cercariae that use the second intermediate host only as a "means of transport" are known. Relationships of echinostomes with the second intermediate host are at the early stages of evolution. As the relationships evolve, conditions for the juvenilization and miniaturization of cercariae become more and more favourable. Larvae of plagiorchiids and strigeidids are the most evolutionary advanced in this aspect. In the former in the course of metacercarial morphogenesis the tegument undergoes a complex differentiation aimed at intensive absorption of nutrients (see p. 160). In the latter the processes in question are facilitated by the formation (in some cases) of an organ-like capsule at the expense of host's cells (see p. 157).

As noted above, the ways of morphological evolution of Strigeidida and Plagiorchiida diverge after the ontogenetic phase of cercaria. Two variants of increasing the reproductive potential of maritae are possible. A complex morphological-functional organization of strigeidid maritae (first of all, the development of the adhesive organ) allows for an intensive inflow of nutrients from the host's organism and the production of many large yolk-rich eggs (e.g., maritae from the genus *Ichtyocotylurus* produce 10,200-23,300 eggs a day – Shigin and Shigina (1968)). On the contrary, plagiorchiid maritae, that are poorly specialized in morphological-functional respect and spend a relatively short time in the definitive host, are characterized by a mass production of small eggs with fully formed miracidia. Both strategies allow successful completion of the parasites' life cycles in ecosystems of different types (see section 2 of this chapter), which is supported by widespread distribution and species richness in the higher trematode taxa. Extreme variants of these strategies are characteristic of the most specialized Strigeidae and Diplostomidae on the one hand and Microphallidae and Lecithodendrioidea on the other hand. All other trematodes demonstrate intermediate variants with a more or less pronounced tendency to the extreme ones. Neither of the variants seems to have decisive advantages. Their diversity has been shown to reflect the diversity of ecosystems where digenean life cycles are transmitted. Therefore neither the tendency towards juvenilization nor that towards a more complex morphology of maritae has become a leading one for digeneans on the whole. On the contrary, juvenilization of cercariae is apparently advantageous for the infection of the second intermediate host since it provides an opportunity for the intensive development of provisional structures (first of all, the penetration apparatus) by means of retardation in the development of definitive systems. These advantages appear to be the reason why juvenilization has become the main tendency in the morphological evolution of the phase of cercaria. (Of course, this applies to the trematodes with the second intermediate host in the life cycle). Imaginization of cercariae is a rare phenomenon and is to be found only as an exception (see p. 151).

1.2.2.2. Temporal Polyfunctionality

When analyzing the ontogeneses of the individuals of hermaphroditic generation of trematodes, we come across one more phenomenon of morphological organization, rather widespread in parasitic flatworms. We suggest to call it "temporal polyfunctionality", i.e., the performance by one and the same organ (structure) of different functions at different ontogenetic stages. This phenomenon is often accompanied by changes (sometimes very profound) in the morphological organization of the organ (structure). It should be noted that this tendency does not apply to the organs of the reproductive system, whose differentiation is always unequivocal.

Temporal polyfunctionality is especially prominent in the dynamics of tegumentary structures. Having formed in the early cercarial embryo, the tegument later undergoes a series of morphological-physiological transformations (see p. 109). Firstly, the ultrastructure of the syncytial layer changes. Secondly, tegumental glands of different types, successively, become associated with the syncytial layer. As the result, the functional load upon the tegument differs depending on the ontogenetic stage. The tegument, naturally, retains its protective function throughout the lifetime of the hermaphroditic individual. However, the tegument of cercariae developing in the sporocysts also participates in the absorption of nutrients. Then in its syncytial layer the substance secreted by the tegumental glands is reserved, which is later spent during the free life of cercaria and its encystment. During metacercarial morphogenesis the tegument again plays the leading role in supplying the parasite with nutrients. After the completion of the main reconstructions, it acquires the organization typical for maritae, enabling young worms to pass over smoothly to the existence in the final host.

Morphological organization of the excretory bladder also undergoes reconstructions. In cercariae it mostly plays osmoregulatory function, in metacercariae it first functions as a glandular organ secreting cystogenous substance (Microphallidae) and then neutralizes and accumulates metabolites. Finally in maritae it ensures excretion. The gut is often specialized as a glandular organ before it starts to perform its characteristic function. Gut-secreted substances in metacercariae of Plagiorchiidae and Microphallidae are necessary for excystment, whereas in young maritae of *Fasciola hepatica* these substances participate in the destruction of host's cells and facilitate the penetration of the parasite via the gut wall into the abdominal cavity (see p. 312).

Some examples of temporal polyfunctionality outside Trematoda can be given, e.g., morphological-functional reconstruction of the tegument in eucestodes in the series: oncosphere → procercoid → plerocercoid → sexually mature worm (for review see: Kuperman, 1988; Davydov, 1991), and morphological reorganization of their excretory system in the course of ontogenesis. This reorganization also involves a change of functions: osmoregulation in larval stages and excretion in adult ones (Kuperman, 1988). There are more examples and their number is likely to grow with the accumulation of ultrastructural data on representatives of various groups of parasitic flatworms since most cases of temporal polyfunctionality described to-date have been revealed by ultrastructural studies.

We consider the phenomenon of temporal polyfunctionality as a specific morphological-functional adaptation to the specific conditions the larval stages develop under. In a typical case, an independent provisional structure is formed in the evolution, to atrophy after the completion of its functional activity (e.g., the tail and the stylet in cercariae). On the contrary, in the specific case considered one of the somatic systems of organs, at a certain ontogenetic stage, starts to perform a function which is not characteristic of this system in the definitive state (or else a "characteristic" function is enhanced, which is the case, e.g., with the secretory function of the metacercarial gut in Plagiorchiidae and Microphallidae). The system of organs involved in this process undergoes morphological specialization. Such a variant implies re-differentiation and thus could have originated only at the basis of a simple tissue organization, characteristic of flatworms. Metazoa, being more evolutionary advanced, no longer possess this ability since a more complex structure of somatic systems of organs calls for earlier differentiation and specialization of their cells, excluding the possibility of morphological reconstructions at a certain ontogenetic stage.

Temporal polyfunctionality in the development of somatic organs might have been one of the prerequisites of the transition to parasitism in the course of the evolution of flatworms, later facilitating the emergence of complex life cycles. These two phenomena are associated with the fact that different ontogenetic stages of one and the same individual change several environments. At first, there were two of them, the water habitat and the primary host. When flatworms became parasites, they did not possess such a perfect mechanism of protection from unfavourable environmental factors as, e.g., the cuticle of nematodes. They evolved another variant, the external syncytial layer of the tegument. In the course of evolution the tegument included different tegumental glands, differing by the nature of secreted substance (Galaktionov and Dobrovolskij, 1987). Temporal polyfunctionality was probably first manifested in the tegument. Later in evolution this tendency embraced other somatic systems of organs. The emerging morphological-functional adaptations of the kind considered have been perfected in the course of evolution. In conclusion, it should be noted that temporal polyfunctionality appears to be closely correlated with the process of juvenilization, since the ability of an organ (structure) to perform different functions at different ontogenetic stages might be a pre-requisite for retardation in the development of some somatic systems of organs. This consideration agrees well with the fact that most cases of temporal polyfunctionality are observed in some of the most juvenilized trematodes, the species from the family Microphallidae.

2. WAYS OF BIOLOGICAL RADIATION OF TREMATODES INTO DIFFERENT ECOSYSTEMS

All organisms, or groups of organisms, evolve under certain ecological conditions, adapting to them in the course of evolutionary process. It is under these conditions that their life cycles are completed. Trematodes are no exception here. Therefore, the analysis of the main evolutionary trends of trematodes necessarily entails the discussion of problems associated with their biological radiation, i.e., spreading to or invading new habitats. This process has been to a great extent determined by the changes of animal hosts in "ecological" evolution of trematodes. These changes have involved all categories of hosts, from the first intermediate to the definitive ones.

We discuss only a few examples from the enormous bulk of evidence. It is common knowledge that it is parthenitae that possess the most narrow host-specificity. It is sometimes expressed not only at the species level but even at the level of geographical races. This has been convincingly shown on the representatives of the genus *Schistosoma*, studied in most detail in this respect (see 3.2. in chapter 4). It would seem that the parthenitae of separate digenean taxa should also be rigidly associated with certain definite molluscan taxa. However, though such examples do exist, they usually apply to rather small, highly specialized groups. For instance, the parthenitae and cercariae of Bivesiculidae have been so far recorded exclusively in the *Cerithium* molluscs.

The more numerous and diverse is a digenean group, the broader is the scope of the first intermediate hosts used. The parthenitae may still be associated with certain molluscan groups, but the taxonomic rank of these groups is higher. A large superfamily Lecithodendrioidea and their close association with freshwater Prosobranchia can be considered. On the other hand, sometimes close trematode species use systematically rather distant hosts. In western Georgia we have found the cercariae of *Allassogonoporus* spp. in large numbers in *Melanopsis praemorsa* (Melanopsidae), but never in bithynids. However, the parthenitae and cercariae of this genus successfully develop in *Bithynia tentaculata* (Bithyniidae) in the central and north-western European Russia, where melanopsids are absent. Even more striking examples can be drawn from echinostomatids and schistosomatids, whose parthenitae are equally well associated with both Pulmonata and Prosobranchia. We have mentioned these two families in this connection, as their biology is incomparable in all other respects. In trematodes, groups of parasites are very often not confined to groups of hosts. Instances like these could be explained only in one way: specific as they are, trematodes readily change their molluscan hosts after incorporation into new ecosystems.

It is not so easy to overcome the specificity "barrier". However, the trematodes that do succeed may give rise to powerful outbreaks of speciation process. We are concerned with a situation like this in close species of the microphallids of the "*pygmaeus*" group. Their two-host life cycles are transmitted in the ecosystems of the intertidal zone (see below). All four European species of the group use *Littorina* snails as the first intermediate hosts. However, the intramolluscan stages of one of these species, *Microphallus pseudopygmaeus*, also develop in 8 other species of subtidal gastropods from different families and orders (Galaktionov, 1993). In the course of speciation, *M. pseudopygmaeus* appears to have evolved the mechanism permitting to overcome the barrier of specificity to the molluscan host. This allowed the expansion to almost all species of prosobranchs used as food objects by the common eider and other marine ducks (the final hosts of *M. pseudopygmaeus*). The species *M. pseudopygmaeus* is a polymorphic one. It forms a system of populations of morphometrically polymorphic metacercariae. Some of these populations may well turn out to be independent subspecies, or even species.

Even more demonstrative are maritae and metacercariae. As a rule, the former, and especially the latter do not possess narrow specificity. The fact that the change of the host often entails other changes, including the transfer of the parasites to new ecosystems, deserves special mention. This "trans-ecosystem transfer" can be seen in Lecithodendrioidea (Fig. 141). They incorporated into their life cycles aquatic larvae of insects, completing their development on land. As a result, three phylogenetically distant groups of insectivorous vertebrates have become their definitive hosts: amphibians (Anura), insectivorous birds and bats. In each of these host groups specific species and genera of the parasites have formed. In some cases the diver-

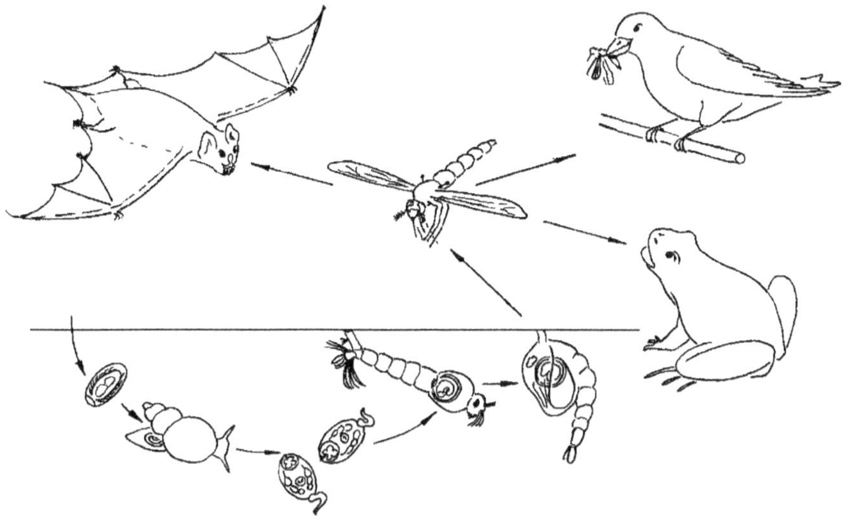

Figure 141. Diagram illustrating the "trans-ecosystem transfer" in Lecithodendrioidea (see text for explanations).

gence has gone as far as the family level (e.g., the family Allassogonoporidae is confined exclusively to bats).

These numerous "switches" between the hosts appear to have been the basis of the biological radiation of trematodes. In this way they entered new ecosystems, often with contrasting environmental conditions. The Szidat Rule, according to which more primitive parasites are associated with phylogenetically ancient groups of hosts, can not be applied to trematodes. It is true of the parasites with simple life cycles that have been evolving in more or less stable ecological conditions. Monogeneans are a good example. They have formed in aquatic environments as ectoparasites of fishes and it is with aquatic ecosystems that their further evolution was associated. Only the polystomes have managed to establish connections with other hosts (amphibians and reptiles), leaving aquatic habitats together with them. However, the main part of their life cycle (reproduction and distribution) is also timed to the presence of their hosts in water. Digeneans are different in this respect. The complication of their life cycles due to the involvement of new categories of hosts allowed them not only to occupy extremely diverse marine and freshwater ecosystems but also to radiate beyond their boundaries.

2.1. Ecological Context of Trematode Life Cycle Formation

There is no doubt that the life cycle of trematodes has formed in the aquatic environment, first the two-host and than the three-host one. This process appears to have started in the sea, marine prosobranch molluscs being the primary hosts of proto-trematodes. We also consider the opinion of Ginetsinskaya (1968) to be well-based. According to her, the primary two-host cycle, together with heterogony, was formed in the Mesozoic era, at the border between the Triassic and the Jurassic periods and was associated with the rise and, most importantly, with a broad radiation of teleost fishes. Numerous researchers have attempted more or less detailed reconstructions of the events involved in this process (Heyneman, 1960; Cable, 1965, 1974; Ginetsinskaya, 1968; James and Bowers, 1967; Pearson, 1972; Odening, 1974; Clark, 1974; Gibson, 1987; Gibson and Bray, 1994; Cribb et al., 2001a). It must have occurred in closed areas, most likely in some coastal biotopes with a high concentration of aquatic animals. Taking into account the benthic way of life of the molluscan hosts and most probably mediocre locomotory abilities of the ancient dispersive stages of proto-trematodes (for details see p. 491), the events involved in the primary life cycle formation were probably confined to the bottom water layer, i.e., in a "two-dimensional" space, actually on a plane. It was only after the proto-cercariae acquired specialized tails, more advanced locomotory organs, that the trematodes could start to colonize the water column. With the acquisition of the actively moving second intermediate host this process could be finally completed at a greater spatial scale. Thus the trematode life cycle was transferred into "three-dimensional", i.e., the bottom and the water column of the aquatic habitat. As a result, dispersion potential of proto-trematodes became much higher, enabling them to involve into their life cycles new ecological groups of hosts, both second intermediate and final ones. Mesozoic fluctuations of the ocean level resulted in the formation of extensive river basins and continental water bodies, rapidly colonized by both vertebrates and invertebrates. In particular, the intervention of prosobranchs into freshwater bodies occurred solely via the estuaries and riverbeds (Starobogatov, 1970). Both molluscs and vertebrates could have introduced trematodes into these water bodies, which gave rise to a most rich and diverse fauna.

The transition to life in freshwater has necessitated only physiological changes in trematodes, associated with the solution of osmoregulatory problems. Naturally, entrance into new ecosystems also entailed the change of hosts. A new temporal dynamics of the life cycle was formed, associated with more clearly expressed seasonal cycles of abiotic and biotic factors in relatively small continental water bodies as compared to the sea. As a result, various new adaptations emerged and speciation processes were intensified. However, none of these changes could alter the principle scheme of the trematode life cycle.

There seems to be one reason behind this stability. The spatial structure of the freshwater biotope, where the life cycle was transmitted, remained essentially the same as that of the shallow marine areas. Trematodes can adapt to existence in a large array of different ecosystems, and yet the general scheme of their life cycles remains the same (with the exception of extreme cases that will be discussed later). It was only the set of adaptations in the phases of the life cycle that was altered. This circumstance appears to be the reason why a majority of modern trematodes possesses typical three-host life cycles. Primary two-host and three-host life cycles were not destabilized even by the trans-ecosystem transfer of the trematodes becoming possible, first due to the ability to encyst on a certain substrate and then due to the emergence of the second intermediate host. Whereas initially the whole life cycle was confined to the aquatic environment, these adaptive acquisitions allowed the trematodes to colonize a new range of vertebrate hosts, i.e., reptiles, birds and land mammals, stimulating speciation (e.g., in the superfamily Lecithodendrioidea, see above – p. 468).

The stability of the three-host life cycles might seem strange, yet the explanation appears to be rather simple. The colonization of active land vertebrates is biologically very advantageous. The dispersion of infection is immeasurably enhanced. The diversity of microhabitats to be used for settling in the body of a vertebrate is also very high. The individuals of the hermaphroditic generation developing and reproducing in a warm-blooded host no longer depend on the temperature fluctuations. In most cases the parasite-exploited energy potential of the hosts is also unlimited. All these factors must have favoured the broad colonization by trematodes of land vertebrates, i.e., birds and mammals. It is with these hosts that an extensive increase of species diversity of such groups as Echinostomatidae, Notocotylidae, Heterophyidae, Strigeidida, Schistosomatida, Plagiorchiida.

However, advantageous as land vertebrates are for trematodes, the main events involved in the life cycle completion are still confined to the aquatic environment. It is there that dissemination and dispersion occur, the two stages determining, in the long run, "the success of infection" of land vertebrates. All the features of these processes have formed in the aquatic environment. They are rigidly determined by a very definite structure of the water body space. The association of parasites with aquatic or terrestrial ecosystems is determined not so much by the place inhabited by the host but by the place where dispersion occurs. The scale of dispersion determines the definite biological features of the parasites and the events in their life cycles.

2.2. Expansion into the Ecosystems of Pelagic Zone

The ability to exploit the offshore shelf and pelagic regions of oceans is impossible without specific adaptations and changes in the life cycle events. This is determined by the enormous spatial scale of this environment. A drastic decrease in the density of aquatic animals in the pelagic zone as compared to coastal and shallows ones results in a lower probability of successful infection. Biological adaptations to these conditions are directed towards increasing the longevity of invasive larvae in the environment. This has been achieved in a number of ways, including passive floating in seawater. Hemiurids, that have developed complex tails, are an excellent example here (Gibson and Bray, 1979; Køie, 1990a,b).

In our opinion, a greater spatial scale of ocean systems as compared to coastal ones required a new strategy in life cycle transmission. Dispersion abilities of cercariae and second intermediate hosts (mainly invertebrates) were no longer sufficient for ensuring the location and infection of final hosts such as pelagic carnivorous fishes. The adoption of an additional host (the third intermediate host) became necessary. Initially main functions of the additional host were transport and protection of the parasite. On the one hand, it guaranteed the prolongation of juvenile stages of the parasite and, on the other, it ensured the transport of the parasite from the dispersion site to the place of the location of the definitive hosts. In this way four-host life cycles appeared. They were originally facultative and then became obligatory.

Four-host life cycles are probably characteristic for at least some didymozoids. Young individuals of these trematodes (at the initial stages of adult morphogenesis) have been recorded in the intestines of a number of small fishes (Manter, 1934; Cable, 1956). This has led Nikolaeva (1965) and Pearson (1972) to conclude that juvenile Didymozoidae are "transported" from crustaceans to large pelagic carnivores such as tuna and sierra in the intestines of small fishes. Apparently this mode of transport is characteristic of all species in this taxon inhabiting pelagic waters.

Although the authenticity of four-host life cycles in Didymozoidae may still be in question, they are obligatory in some Hemiuroidea. Planktonic crustaceans act as hosts for hemiuroid mesocercariae (an intermediate form between cercaria and metacercaria – see pp. 175, 182). The metacercarial stage is transferred to plankton-feeding fishes, permitting the use of the large and more long-ranged carnivorous fishes as final hosts. The introduction of a mesocercaria provides for additional morphogenetic changes in the development of these trematodes.

Four-host life cycles have also arisen independently in some freshwater trematodes, such as *Strigea* and *Alaria.* Initially this might have been associated with their invasion and establishment in large freshwater basins. Adoption of new hosts (common in trematodes) promoted the incorporation of large carnivores into the life cycles. Many of these hosts such as foxes, dogs and wolves, were only peripherally associated with freshwater basins, essential for the evolution of these parasites.

The regularity described seems to be universal. Host chains utilized by the plerocercoids of some trypanorhynchs to reach their motile carnivorous hosts, sharks and rays, may include not two or three but probably dozens of hosts (Cake, 1976; Dolfus, 1976; Gaevskaya and Nigmatulin, 1981). In the life cycle of an anisakid nematode *Hysterothylacium aduncum*, whose definitive hosts are numerous species of marine fishes, a number of ctenophores, chaetognaths, polychaetes and ophiuroids may act as obligate intermediate hosts or transport hosts (Køie, 1993).

2.3. Colonization of Terrestrial and Marine Littoral Ecosystems

Evolving trematodes have also colonized an entirely different types of ecosystems, terrestrial and intertidal ones. As for the former, it should be noted that though trematodes may exist in arid conditions, they always depend on the presence of moisture (see p. 240). Both these ecosystems are characterized by two interdependent conditions, which are important. The first is the lack of the water "volume", either constant or, in the intertidal zone, temporary. On land, trematodes often experience an acute deficit of moisture. As a consequence, dispersion in such ecosystems takes place predominantly in "two-dimensional" space. Swimming dispersive stages occur in the intertidal zone, covered with water twice a day during high tides, but can not exist in terrestrial ecosystems.

Transmission of trematodes in intertidal and terrestrial ecosystems became possible only after the emergence of some adaptations that permitted trematodes to be more or less independent of the environment (Galaktionov, 1987, 1993). "Passive" infection of molluscs is probably the most important of them. This adaptation arose in trematodes that completed their life cycles in water and has eventually become widespread. The second dispersion (i.e., the dispersion of mother sporocyst larva – miracidium, see pp. 2, 236) loses the active motile phase. This fact does not exhibit a noticeable negative influence upon the parasites' dispersion potential. The main role in dispersion belongs to the activity of the infection-dispersing definitive host. An insignificant redistribution of infection in the water basin due to the swimming of the miracidia, is amply compensated for by the possibility of relocation (e.g., due to microcurrents) of the eggs containing the larvae which retain their infectivity for months (see p. 239). Therefore it is not surprising that the loss of actively swimming miracidium has taken place in many trematode groups completing their life cycles in different ecosystems. However, it was a necessary prerequisite for the penetration of free-living trematode larvae into the ecosystems with extreme environmental parameters. They only had to enhance the protective qualities of their eggshells. This is observed in Dicrocoeliidae and Brachilaimoidea, which have successfully colonized arid land areas.

The issue of the second dispersion (at the phase of cercariae) is less clear. Encystment in the environment and the formation of adolescariae could not have been a preadaptation in this case. These processes end the dispersion in water, whereas in conditions that lack moisture and the impact of numerous other unfavourable factors the larvae must be isolated from the environment from the very beginning. Therefore colonization of littoral and terrestrial ecosystems was accompanied by reconstruction of the life cycle scheme. The main tendency accompanying this reconstruction was the increasing independence of dispersive stages of the parasites from the environment.

Dicrocoeliids retain the three-host scheme of the life cycle, which probably means that their colonization of terrestrial ecosystems is a relatively recent evolutionary event. They solve the dispersion problems by means of increased isolation of their dispersive stages from the unfavourable environmental factors. Their miracidia have already been discussed. As for the cercariae of *Dicrocoelium dendriticum* (still possessing the tails), they enter the pallial cavity of the mollusc, forming the so-called "slug" balls. The latter are shed into the environment and are readily eaten by ants, acting as second intermediate hosts. The species of the genera *Corrigia* and *Eurytrema* are more specialized in this respect. Not cercariae but daughter sporocysts are released from the molluscs into the environment. They contain larvae with already very rudimentary tails. A well-developed endocyst of these parthenitae (see p. 43) protects the dispersive stages. In all three examples given above the way of infecting the second intermediate host also changes. It becomes passive and *per os*. Thus, a clear analogy with miracidia can be drawn.

Brachylaimoidea, another "terrestrial" group of trematodes, demonstrate a somewhat different trend. They have colonized land independently of dicrocoeliids. Miracidial adaptations in Brachylaimoidea and Dicrocoeliidae are the same. As for the cercariae of Brachylaimoidea, their autonomy from the environment is achieved by means of the exclusion from their life cycle of the very act of dissemination, and thus of dispersion, either active or passive. Their life cycle secondarily becomes a two-host one (*Leucochloridium*, *Hasstilesia*), the mollusc combining the functions of both the first and the second intermediate host. In the cases when a three-host life cycle and even a short period of active dispersion is retained (in primitive "terrestrial" brachylaimoideans, such as *Brachylaimus aequans*, see p. 273), the cercariae start to locomote in an entirely different way, crawling on the substrate surface (on a plane).

It is the formation of two-host life cycles that seems to be the major evolutionary trend in the trematodes colonizing non-aquatic ecosystems with extreme conditions. This is illustrated to by the fact that they are widespread in Microphallidae, another trematode group that has formed in coastal ecosystems. In this family a successive transitional series can be observed, from a typical three-host life cycle to a two-host one, when the mollusc combines the functions of the first and the second intermediate host (Belopolskaya, 1963; Deblock, 1977). This process has been accompanied by essential morphological-functional and biological reconstructions of both parthenitae and the larvae of the hermaphroditic generation (Galaktionov, 1993) (see for details pp. 65, 171).

It appears that the first stages of transition to two-host life cycles, until the emergence of "*pygmaeus*" group, did not especially increase transmission success in microphallids. This is due to certain ecological reasons that will be discussed later. Besides, species with the life cycles of "transitional" type expend a lot of energy for the formation of some provisional structures. This is especially true of the life cycle of the *Maritrema oocysta* type and, to a lesser extent, of those similar to the life cycle of *Microphallus pirum* – see p. 171. These provisional structures (first of all, the tail, the penetration complex and numerous tegumental glands – see p. 171) are indispensable for the free-living cercaria, its penetration and encystment in the second intermediate host. However, they are excessive for the larva encysting inside the daughter sporocyst. The germinal masses of the daughter sporocysts of the microphallid group under consideration, as well as those of species with three-host life cycles, are adjusted to a long-term production of embryos. A greater part of them degenerates as the body cavity of the sporocyst is filling with encysted metacercariae that have developed from the germinal balls that were the first to separate from the germinal mass. Energy expenditure for the production of embryos that would later be destroyed can be also considered as useless. Thus, two-host "transitional" life cycles did not allow microphallids either to radiate into new biotopes or to gain any special advantages in the old ones.

This can be demonstrated by the comparison of the infection of the mudsnails *Hydrobia ulvae* and *H. ventrosa* by the trematode species with two different life cycle types. The number of species with three-host life cycles and those with two-host life cycles of "transitional" type at the coast of France is, respectively, 6 and 7 (Deblock, 1980). The prevalence of species with three-host life cycles (*Maritrema subdolum* and *Microphallus claviformis*) in *H. ventrosa* turned out to be as high as that of species with two-host life cycles (*Microphallus pirum* and *M. breviatus*) even in very severe conditions of the salt marsh ponds in Iceland (Skirnisson et al., *in press*).

A "breakthrough" resulting in a broad expansion into the ecosystems with unstable environment has been made by the microphallids of "*pygmaeus*" group. The germinal mass of the daughter sporocysts of this species has a limited proliferative activity and the production of the embryos of hermaphroditic generation is not excessive (see p. 69). The development of embryos is not accompanied by the formation of any provisional structures except a rudimentary tail (see p. 174). This makes it possible to reduce energy expenses for the production of invasive metacercariae and as a result the negative impact upon the molluscan host is decreased. This consideration is supported by the results of experiments, staged under the impact of unfavourable environmental factors, that have shown the survival rate of molluscs infected by the microphallids of "*pygmaeus*" group to be as high as that of non-infected ones (see p. 355). Finally, the mechanism of transmission *favorisation* in the parasites under consideration should be noted, associated with a specific manipulation of the molluscan hosts' behaviour (see p. 358). Taken together, all these advantages have ensured the flourishing of the microphallids of "*pygmaeus*" group in ecosystems with extreme environmental conditions. The species *Microphallus helicicola* has colonized the high mountainous ecosystems of the Tien Shan (Middle Asia) (Belopolskaya and Soboleva, 1977). *M. pseudopygmaeus* is the only species of seabird trematode that completes its life cycle in the high Arctic shore ecosystems of Franz Josef Land (Galaktionov, 1996b). The microphallids of "*pygameus*" group are clearly predominant in littoral molluscs on the coasts of the northern seas (the Norwegian, Barents and White Seas) (Galaktionov and Bustness, 1999). These trematodes are also rather common in the coastal ecosystems of boreal regions. However, not this group but species with a three-host life cycle occupy a dominant position there (James, 1969; Werding, 1969; Combescot-Lang, 1976; Lauckner, 1980, 1984; Hughes and Answer, 1982; Irwin, 1983; Newell, 1986; Skirnisson and Galaktionov, 2002).

Such limited distribution of two-host life cycles appears to be associated with the negative ecological consequences of the loss of the phase of free-living cercaria. With it, the phase of the first dispersion of infection disappears as such (see p. 340). It is more significant than the loss of a stage (by far not the most important one) of the second dispersion, as in the case of the loss of the free-swimming miracidium (see above). In the case of cercaria loss, the first intermediate host also performs the functions of the second one. As a result, the range of potential definitive hosts is more limited. Besides, the activity of the second intermediate host ensures an additional dispersion of larvae, thus increasing the probability of successful life cycle completion. This seems to be the reason why two-host life cycles lacking the larvae active in the environment are not especially widespread in trematodes. Their selective advantage over the three-host life cycles is expressed only in habitats with extreme environmental parameters, such as the Arctic intertidal zone, exposed marine shores and terrestrial ecosystems.

Everything that has been said about the negative consequences of the loss of the active cercaria phase in trematodes is pertinent to the species with one-host life cycle that also lack egg dispersion by the definitive host (see p. 344). It is no wonder that life cycles like these are very rare among trematodes, even among those whose transmission takes place in the ecosystem with extreme environmental parameters (see section 4 in chapter 3). *Bunocotyle progenetica* (Hemiuroideae: Bunocotylinae) is very common in marine littoral ecosystems. Its life cycle, from the stage of mother sporocyst to sexually mature marita, is completed in mudsnails *Hydrobia ulvae* and *H. ventrosa*. Since molluscan hosts dispersing the infective agents (eggs with miracidia) have a low mobility, the life cycle of *B. progenetica* and other trematodes with one-host life cycles is completed on a limited area. A rather high density of molluscs appears to guarantee infection success in cases like this, with particular molluscs being the only hosts of the trematodes.

2.4. Patterns of Strigeidida and Plagiorchiida Transmission

The three-host life cycle scheme is predominant in trematodes. However, it should be emphasized that the evolution of both phases in the life cycle and the life cycle as such has followed very different patterns in different trematode groups. The species from various digenean taxa, whose scheme of life cycles is the same, may be adapted to transmission in entirely different ecosystems. This idea can be illustrated by comparing the representatives of the orders Strigeidida and Plagiorchiida, embracing the most evolutionary advanced trematodes. The life cycles of a majority of the representatives of these two orders follow a three-host scheme. At the same time, it has already been noted that the complex of morphological-functional and biological features of different life cycle stages in Strigeidida and Plagiorchiida turns out to be essentially different.

The eggs of strigeidids require a period of embryonation in the environment. In the course of embryonation a complex free-swimming miracidium, actively penetrating the molluscan host, is formed. The evolution of the hermaphroditic generation followed the way of maximal specialization of maritae (body division into two segments, the emergence of adhesive organ and lappets – for details see p. 210). At the same time, the cercariae exhibit a clear tendency towards juvenilization. The structure of the larvae in Diplostomidae and Strigeidae is embryonal (see p. 149). This is expressed in poorly developed definitive systems of organs, i.e., reproductive, excretory and digestive ones. As a result, there is a great difference between the levels of morphological-functional organization of cercariae and maritae. To overcome this gap, a long and complex metamorphosis at the stage of metacercariae and important morphogenetic transformations accompanying maturation of maritae in the final host are necessary. The longevity of adult worms is rather high, sometimes reaching 7-18 months (Odening, 1967; Sudarikov, 1984). The latter significantly increases the total fecundity of strigeidid maritae.

As a result of the above characteristic features of their life cycles, strigeidids are poorly adapted to life cycle completion in ecosystems with frequent and contrasting alteration of environmental conditions, such as terrestrial and intertidal ones, and occur in them very rarely. The stages of the egg and the miraciduim, functionally active in the environment, are especially vulnerable. It has already been noted that in conditions of terrestrial ecosystems a free-living miracidium cannot exist at all. At the same time some of the strigeidid adaptations (a short maritogony lasting 4-7 days and a high total fecundity of maritae) are universal, i.e., favourable for life cycle completion in ecosystems of any kind. This circumstance probably made possible the expansion of certain strigeidid species into intertidal ecosystems. However, it was in freshwater ecosystems that strigeidids reached their evolutionary success. Their life cycles, "unsuitable" for terrestrial ecosystems and rare in intertidal ecosystems, turned out to be highly favourable for freshwater ones.

The life cycles of Plagiorchiida have been evolving in an opposite direction. Their miracidium can no longer exist independently, hatching only in the intestine of the molluscan host. There are only two parthenitae generations, but the patterns of their morphological-functional organization (see pp. 40, 46, 75) allows a great increase in the reproductive capacity. The latter is also facilitated by a simplified structure of cercariae (see pp. 130-1), as energy expenses for the production of a larva are lower. The formation of definitive systems of organs is more and more shifted to the stage of the metacercaria. In the most specialized Lecithodendrioidea and Microphallidae (to a lesser extent also in Plagiorchiidae and Ochetosomatidae) the metacercaria differs from the marita only by the absence of eggs in the uterus. A simplified somatic organization of the maritae determines their short life span. This is compensated for by a high fecundity, made possible due to very small eggs and simple structure of the miracidia in them.

This direction of evolutionary transformation has yielded the life cycles of plagiorchiids. They are more "universal" than the life cycles of Strigeidida, i.e., they can be completed in ecosystems of different types. "Universal" life cycles of Plagiorchiida account for their broader distribution as compared to Strigeidida. Microphallidae and Lecithodendrioidea, which are the two most specialized and successful groups of plagiorchiids, are a good example of the above consideration. The former succeeded in colonizing the ecosystems with contrasting changes of environmental parameters (the intertidal zone, terrestrial ecosystems), whereas the life cycles of the latter are associated with freshwater basins to the same extent as strigeidids.

2.5. Concluding Remarks

Analyzing the ways of biological radiation of trematodes, one can not help pondering over the ways of formation of the phenomenon of the complex life cycle in the course of the parasites' evolution. Indeed, from a formal point of view, a complex life cycle does not appear to be justified. The completion of a simple life cycle (the chain: host – free-living larva – host) seems to be a much more plausible process than the transmission of a life cycle involving several (even if only two) host animals.

Apart from the above-mentioned favourable consequences of the involvement of the intermediate hosts (e.g., see pp. 476, 491), the disadvantages of the complex life cycle are compensated for by the numerous and diverse adaptations evolved by the life cycle phases. In fact, it was the analysis of these adaptations in trematodes, both at the organismic and populational level, that was undertaken in the preceding chapters. The main advantage of the life cycle is that it renders possible the colonization of a broad range of definitive hosts by means of a chain of intermediate ones. These definitive hosts are inaccessible for the parasites' larvae directly, especially if, as it is the case with trematodes, the role of the primary host is played by molluscs, benthic invertebrates with limited motility.

The colonization of new animal hosts (first of all, warm-blooded animals – see above) and the concurrent transition to transmission in other ecological conditions may entail further changes in the life cycle events (e.g., transition from trixenous to dixenous and even homoxenous life cycles – see chapter 3), including its secondary simplification. The variants of the latter in connection with radiation into terrestrial and marine coastal ecosystems have been discussed above. Schistosomatids are another illustrative example. It has been discussed in detail (see section 3.2. in chapter 3) that their two-host life cycles have formed on the basis of three-host ones. Metacercariae of ancestors of schistosomatids began to use the blood of vertebrates as their environment. This resulted in their progenetic development and the emergence of neotenic adults. Finally, the primary definitive host disappeared from of the life cycle, and with it, the true maritae. They were substituted by neotenic forms, maturating in the blood of the host that had initially been the second intermediate and later became the definitive one (Combes, 1991b; for details see p. 338).

Summing up, it can be concluded that in terms of biological progress, complex life cycles have been favoured and they have provided for the universal success of parasites. This is why complex life cycles are so frequently encountered among parasites and have resulted in their widespread distribution in the biosphere.

---

[i] The taxonomic position of *Cercaria caribbea IV* Cable, 1956 needs an explanation. Cable (1956) relegates this form to the group "Echinostomoid cercariae", which is incorrect. The larvae described by him are really archaic, but unlike cercariae of Echinostomatidae and Psilostomidae, possess large pigmented eyespots.

Chapter 6

EVOLUTION OF LIFE CYCLES AND PHYLOGENY OF TREMATODES

Evolution of Trematoda as a separate taxon of high level is still obscure. This is mainly due to the fact that although there are abundant publications devoted to this group of parasitic flatworms our knowledge about them is still insufficient. Trematode life cycles are studied in detail only in several model species. Unfortunately, many taxa of family level and higher are known only from total mounts of mature maritae. Our understanding of trematode morphology is therefore limited: most descriptions of the structure of sexually mature individuals of the hermaphroditic generation contain only an analysis of the topography of the main organs – the suckers, the digestive and reproductive system. Data on the structure of the excretory, nervous and muscular systems and the secretory apparatus of maritae are scarce and have been obtained from few species. Reproductively mature individuals of the trematode hermaphroditic generation are more poorly studied than turbellarians. Comparative morphology of the larvae of the hermaphroditic generation, cercariae and metacercariae, and the parthenogenetic generations, sporocysts and rediae, has not been investigated using the methodological approaches of modern evolutionary morphology.

The paucity of data on this topic makes the discussion of trematode evolution very difficult. First of all, it is difficult to speculate about the sequence of the main evolutionary stages of a complex life cycle formation in the Trematoda. Evolutionary history of this taxon started with the origin of the phenomenon of change of animal hosts and the alteration of reproduction modes accompanying transition from one phase of the life cycle to the next. Yet to establish the sequence of the main stages in trematode evolution is very important. This sequence would allow for the estimation of plesiomorphic and apomorphic states of morphological and biological characters used in phylogenetic reconstructions. A clear view on phylogeny is also necessary for the elaboration of a natural classification of the Trematoda. We had these considerations in view when writing the concluding chapter.

1. ORIGIN AND EVOLUTION OF TREMATODE LIFE CYCLES

1.1. Approaches to the Analysis of the Nature and Evolution of the Trematode Life Cycles

Two different approaches to the origin of complex life cycles of parasitic organisms were suggested in the late XIXth – early XXth century. Following Leuckart (1879, 1879-1886) some researchers thought that invertebrates were the primary hosts of helminths. Other investigators relegated the role of primary hosts to vertebrates (Claus, 1889; Mordwilko, 1908, 1909). Both points of view are still discussed in connection with the earliest stages of trematode life cycle formation.

The hypothesis that molluscs were primary hosts of Trematoda predominated in the XXth century. Some time after Leuckart (1879, 1879-1886) Sinitsin (1905, 1910, 1911, 1931a,b) suggested an original hypothesis about the formation of trematode life cycle. Sinitsin thought that heterogony was the life cycle type characteristic of Trematoda. Sinitsin also considered that the ancestors of modern trematodes colonized molluscs that became their first (primary) hosts. Similar views (though different in certain aspects) were developed by other researchers (Stunkard, 1959; Jamieson, 1966; James and Bowers, 1967; Ginetsinskaya, 1968; Rohde, 1971, 1994; Pearson, 1972; Gibson, 1987; Gibson and Bray, 1994). Many of these authors accepted that molluscs were primary hosts of trematodes and considered their life cycle as heterogony. This viewpoint was elaborated in detail by Ginetsinskaya (1968). In our opinion, her main ideas are still important and we rely on them in this chapter.

An alternative point of view (Claus, 1889; Mordwilko, 1908, 1909) according to which vertebrates were primary hosts of parasites did not gain any support as regards trematodes. Ginetsinskaya (1968) has emphasized that it is useless to search for universal answers to the questions about the origin of parasites from different taxa. Within the parasitic Platyhelminthes, now united into Neodermata, the ancestors of Cercomeromorpha (*sensu* Bychowsky, 1957) must have started parasitism with colonization of vertebrates. On the contrary, the origin of the class Trematoda was associated with the transition of ancestral forms to parasitism in molluscs.

However, ancient ideas of Claus (1889) and Mordwilko (1908, 1909) have been revived recently (Rohde, 2001b; Cribb et al., 2001a; Littlewood et al., 1999b, 2001). One of the main reasons was the acceptance of phylogenetic unity of Neodermata. Since Ehlers (1984, 1985a,b, 1986) gave good grounds for delimitation of this group, other researchers have supported common origin of all representatives of Neodermata with new data (Blair et al., 1996; Tyler and Tyler, 1997; Littlewood et al., 1999a,b, 2001). To note, monophyletic origin of a compact group of parasitic Platyhelminthes (Aspidogastrida, Trematoda and Cercomeromorpha *sensu* Bychowsky, 1957) has not been doubted (Bychowsky, 1957; Ginetsinskaya, 1968). Recent results confirm this point of view.

Modern protagonists of the views of Claus (1889) and Mordwilko (1908, 1909) consider two evolutionary events – formation of an independent phylogenetic branch Neodermata within the Platyhelminthes and the transition of ancestral Neodermata to parasitism – as closely associated. However, no detailed arguments are given in support of the origin of Neodermata from parasitic ancestors. In our view, these two evolutionary events may not be associated. The presence of a common ancestor of two groups of parasites (Trematoda and Aspidogastrida, on the one hand, and Cercomeromorpha, on the other hand) does not necessarily entail that the ancestor was a parasite, especially a parasite of vertebrates. The ancestor of modern Neodermata might have been a free-living animal or a commensal, most probably, of invertebrates. It appears to us that colonization of molluscs by ancestors of Trematoda and colonization of vertebrates by ancestors of Cercomeromorpha took place independently and probably at different geological time.

It is our opinion that another reason behind the revival of the hypothesis of the presence of a primary vertebrate host for all parasitic flatworms is a formal use of methods of cladistic analysis based on the principle of maximal parsimony. Constructing phylogenetic schemes on the basis of this principle, one should be sure that the evolution of the animal group studied indeed follows the most economic way. It means that in fact we admit the existence of a universal mechanism of evolutionary process optimization. Yet even in theory such a mechanism can not exist, the evolution of every animal group being unique.

In our view, morphological analysis of individuals from different phases and generations of the life cycle of trematodes given in chapter 5 clearly demonstrated that combinations of plesiomorphic and apomorphic states of characters in representatives of close taxa of a relatively low level (families) may be very diverse and "unexpected". To note, data obtained with the use of molecular biological methods are also contradictory. Different sequences used for analysis allow the elaboration of phylogenetic schemes differing greatly from each other (for review see: Littlewood et al. 1999a, 2001).

Finally, one more reason presupposed the reversal to the views of Claus (1889) and Mordwilko (1908, 1909). It is a simplified approach to the interpretation of the life cycle of Trematoda, considered simply as an ontogenesis. This simplification is reflected in a rhetoric question "Why are there so few species of Platyhelminthes that are parasites of invertebrates in their adult stage?" in the article devoted to the evolution of parasitism in flatworms (Littlewood et al., 1999b; p. 271). We will discuss an answer to this question later. Here we only note that the authors of this article do not consider rediae and sporocysts, actively reproducing in molluscs, as "adult stages". Then the phases parasitic in molluscs should be considered as larval and we do not have to answer the question whether the primary host was a vertebrate or an invertebrate. Such a standardized interpretation to the life cycles of Trematoda and Cercomeromorpha also results in standardized schemes of their formation in the course of evolution.

Before presenting our ideas on the origin and early evolution of trematodes, it should be explained why the above views appear questionable to us.

We set aside the question about the "sister group" of Neodermata. Various possible candidates suggested (Littlewood et al., 1999b) show that for the time being there is not a satisfactory answer. To postulate that "neoophoran turbellarians" are a "sister group" of Neodermata is a statement of the fact that parasitic flatworms originated from turbellarians.

It is more important to establish the sequence of evolutionary events associated with the origin of Neodermata and their transition to parasitism. The followers of the hypothesis of the origin of Neodermata from parasitic ancestors are correct: modern representatives of this clade are indeed highly adapted to a parasitic way of life. A high level of adaptation is first of all expressed in a specific tegument, or "neodermis". In our opinion, the most important characteristic of this structure is its dynamic cell composition, and not the syncytial organization or the presence of submerged nuclei-containing parts of cells (cytons). The cell composition changes several times during the ontogenesis of individuals of one generation, providing a background for adaptive specialization of the tegument depending on current conditions of the organism's existence (see pp. 83, 109, 152).

However, a change of epidermis cell "generations" is not an exclusive feature of Neodermata. This phenomenon is found in other Turbellaria, including those that do not form symbiotic relations with other organisms. Tyler and Tyler (1997) consider that this phenomenon brought about the formation of "neodermis" in the evolution of Neodermata ancestors. This structure later was the background for a very broad adaptive radiation. The presence of the tegument must have facilitated the transition to obligate parasitism (see p. 466). Incidentally, Littlewood et. al. (1999b) also do not rule out this possibility. As an alternative to their main hypothesis these authors mention that an already formed tegument might have been an important preadaptation for the transition of ancestors of Neodermata to obligate parasitism. We would like to emphasize that this interpretation of ancient evolutionary events does not contradict a monophyletic origin of the Neodermata.

If we accept as a working hypothesis that ancestors of Neodermata changed to parasitism once in the evolution, then indeed the only candidates for the role of primary hosts are some ancient vertebrates ("protovertebrates" according to Littlewood et. al., 1999b). However, if vertebrates were primary hosts of the Trematoda, then molluscs, involved into the life cycle much later in the evolution, are typical intermediate hosts. This supposition triggers a cascade of questions that can not be answered by this hypothesis. True intermediate hosts, with rare exceptions (some Nematoda) "transfer" larval stages of parasites' development along the food chains. This is characteristic of Amphilinida and Eucestoda within Neodermata. The second intermediate host of trematodes also plays this role.

Host-parasite system "parthenitae-mollusc" functions differently. Its main role is not the transfer of infection along the food chains but the increase in the number of infective stages exported to the environment. The emergence of this adaptation in trematodes is easy to explain if we admit that molluscs were primary hosts (see p. 491) and almost impossible to explain if we suggest that molluscs were involved in the life cycle secondarily as true intermediate hosts. All the known instances when the definitive host is infected by eating the infected mollusc are the result of secondary reduction of transmission ways of trematodes in the ecosystem, the mollusc performing the function of both the first and the second intermediate host. Such cases are typical examples of secondary two-host life cycles (see p. 340). Even though Brooks et al. (1985a) attempted to interpret the life cycle of *Heronimus mollis* (from the same typological group) as the most ancient one, neither morphology nor biology of parthenitae and larvae of this species give any grounds for this interpretation. A detailed critical analysis of this view of Brooks has already been made (Gibson, 1987; Pearson, 1992; Galaktionov and Dobrovolskij, 1998).

The working hypothesis also complicates the answer to the question about the nature of rediae and sporocysts. Supposing that trematode ancestors colonized molluscs as primary hosts, we have to consider modern rediae and sporocysts as parthenogenetic "females" that have become oviviviparous and reproduce by apomictic (diploid) parthenogenesis (for details see p. 486). But if sporocysts and rediae are larvae by their origin and appeared secondarily in the life cycle of trematodes, it is much more difficult to explain their specific reproductive mode.

The phenomenon of reproductive capacity of larvae is rather common. It is widely seen in Euscestoda within the Platyhelminthes. Metacestodes of many Cyclophyllidea demonstrate a true asexual reproduction, usually in the form of budding. It increases the number of infective stages (in this case, of metacestodes) in the intermediate host. It results eventually in intensive infection of the definitive host. However, reproduction in rediae and sporocysts can not be interpreted even as a very specialized form of budding. Bud formation always involves somatic tissue elements of the mother organism. The bud is multicellular from the beginning of its formation and usually includes specialized cells of different types. On the contrary, every individual produced by a redia or a sporocyst develops from one totipotent cell undergoing a true embryo- and morphogenesis. Thus, we either have to suppose that rediae and sporocysts of Trematoda have developed a unique type of reproduction, which is not observed in any other Metazoa, or have to revert to the idea of apomictic parthenogenesis. The latter idea seems preferable to us. All available data on the reproduction of rediae and sporocysts appear to testify that germinal cells are reproductive ones (Dobrovolskij and Ataev, 2003). The most important characteristics of reproduction in rediae and sporocysts are as follows.

1. Mature germinal cells are the only source of embryos of subsequent generations in rediae and sporocysts (see p. 75). Germinal cells themselves can not proliferate. Recruitment of their number takes place by means of specialization of non-differentiated cells capable of proliferation. Rediae and sporocysts contain such cells until the end of their reproductive period.

2. In parthenitae that have completed their morphogenesis and started reproduction the germinal mass is the only source of generative elements (non-differentiated cells, differentiating and mature germinal cells) (see section 1.3.1. in chapter 1). The presence of a single germinal mass sunk into the parenchyma at the posterior end of the parthenitae should be considered a plesiomorphic state. This is first of all characteristic of rediae (see pp. 68, 82). The germinal mass of this type is very similar to primitive germaria of some turbellarians. Germinal masses of mother and daughter sporocysts may be secondarily specialized (germinal masses freely floating in the schizocoel, diffuse germinal masses – see pp. 68, 81).

3. Typical germinal masses morphologically and functionally correspond to the ovaries of mature individuals of hermaphroditic generation. Both in germinal masses and in ovaries proliferation of non-differentiated cells occurs (oogonia in ovaries – see p. 201), followed by their physiological maturation which is not accompanied by meiosis. In fact, both in ovaries and in germinal masses primary oocytes are formed. Formation and differentiation of these cells in parthenitae and hermaphroditic individuals are similar, but their subsequent fate is different. In the proximal portions of the ducts of the female reproductive system of marita an oocyte is fertilized. Then it undergoes the first and the second meiotic divisions, to form two polar bodies and a female pronucleus. The female pronucleus conjugates with a male pronucleus derived from the spermatozoan and only after that the cleavage starts. Germinal cells of rediae and sporocysts start to cleave while they are still in the germinal mass, before they have entered the schizocoel.

4. Early embryogenetic stages of the fertilized ovum of maritae and germinal cell of parthenitae are almost identical (see pp. 83, 90, 109). Differences concern only the fate of the primary epidermis cells: in the development of miracidia they separate from the embryo's surface and surround the yolk mass, whereas the embryos of the subsequent generation of parthenitae and hermaphroditic generation retain primary epidermis (embryonic membrane) on their surface. More significant differences in the development of embryos of different generations are manifested much later, at the morphogenetic stage.

In our view, there is only one way to interpret these facts. A remarkable morphological and functional similarity of germinal masses and female gonads testifies to homology of these structures, entailing homology of cells produced by them. The same conclusion follows from the comparison of embryonic development of individuals developing from fertilized eggs and germinal cells. Since patterns of embryonic development of different generations are almost identical at early stages, these processes must be homologous, testifying to the homology of initial cells. Thus, germinal cells of rediae and sporocysts are female reproductive cells by their origin (primary oocytes that have evolved a capacity for cleavage). Reproduction by means of these cells should be considered as a true apomictic parthenogenesis according to the traditional view on this process (Maynard Smith, 1978).

However, it is emphasized in the literature that rediae and sporocysts are asexual (agamous) developmental stages. In our opinion, this issue has two aspects. The first aspect concerns the origin of this point of view. It dates back to Steenstrup (1842) who investigated the life cycle of *Fasciola hepatica* and suggested that reproduction of rediae could be considered as budding. Thus, the hypothesis of metagenesis was the first attempt to explain the life cycle of trematodes.

Brooks (1930) suggested an entirely different approach to this question. He considered the life cycle of digeneans not as the alteration of independent generations but as a prolonged ontogenesis of individuals of one generation accompanied by a complex metamorphosis. According to Brooks (1930), the multiplication of a number of germinal cells is a result of polyembryony and of "germinal lineage". Polyembryony is indeed asexual reproduction by means of division of either the zygote or already an embryo. However, neither zygotes nor dividing embryos are present in rediae and sporocysts. This is the source of contradictions inherent in the speculations of Brooks (1930).

Admitting the presence of polyembryony, we inevitably admit that germinal cells of sporocysts and rediae are genuine generative cells. The latter, if they are not zygotes, can be either oocytes (mature or maturing) or gonia (in our case, oogonia). Since no traces of meiosis have been found in rediae and sporocysts in the process of germinal cells' maturation (Dobrovolskij and Ataev, 2003), we have to accept that they are oogonia or, at best, primary oocytes. Development of generative cells (ova) without fertilization falls under the definition of parthenogenesis. Since there are no meiotic divisions, we deal with diploid (apomictic) parthenogenesis in the case considered.

To note, polyembryony certainly can not be looked upon as a reproductive mode of rediae and sporocysts. Biologically, polyembryony is asexual reproduction of individuals of subsequent generation at early stages of ontogenesis. Therefore, admitting polyembryony in digeneans we do not answer the questions about the mode of reproduction in rediae and sporocysts and the nature of these organisms.

Polyembryony in rediae and sporocysts has not yet been reliably demonstrated. Brooks (1930) and later Cort (1944) and Cort et al. (1954) considered germinal cells to be direct descendants of the primary "propagatory" cell, interpreting its subsequent reproduction as polyembryony. We have often mentioned above that germinal cells in germinal masses of rediae and sporocysts are recruited only from non-differentiated cells retaining the ability to proliferate (see p. 75, see also Dobrovolskij et al., 1983; Dobrovolskij and Ataev, 2003). There are no more reasons to consider this process as polyembryony than to call "polyembryony" the reproduction of oogonia in the ovary.

To sum up, the reproductive process of sporocysts and rediae appears to be diploid apomictic parthenogenesis. The cleavage is started by the primary oocyte whose functional maturation is not accompanied by meiosis. Since there is no meiosis even in the ovary of hermaphroditic individual reproducing by means of a true amphimixis, it is unlikely that sporocysts and rediae could have retained any traces of meiotic divisions. We find it advisable to use the term "parthenitae" for these phases of digenean life cycle. It expresses precisely the true nature of sporocysts and rediae.

It should be emphasized that the genetic consequences of apomictic parthenogenesis do not differ from those of typical asexual multiplication. In fact, it is a mechanism of producing a large number of genetic copies (clones). With this stage of reproduction in their life cycle, trematodes can produce in a short time a large number of genetically uniform cercariae, passing the infection to the next host. Genetic structure of the trematode life cycle should be considered from exactly this viewpoint in the life history and population studies of trematodes and analysis of their transmission in ecosystems

Yet when discussing zoological rather than parasitological issues (formation of Trematoda as an independent taxon, the origin of their complex life cycle, analysis of trends of morphological and biological evolution of this group), it is important to understand clearly the true nature of sporocysts and rediae. When and how did these generations of the trematode life cycle originate? What is their unique reproductive mode? Did it originate independently or as a secondary modification of reproductive mechanisms of their ancestors? In our opinion, the least contradictory answer to these and numerous other questions can be given if we admit the sexual nature of sporocysts and rediae (=parthenitae) whose ancestors lost amphimixis, acquired apomixis and vivipariry.

Host specificity is often a central point in discussions concerning the origin of the Trematoda. Supporters of the hypothesis of molluscs as primary hosts of Trematoda usually refer to two indisputable facts. The first fact is association of trematodes almost exclusively with molluscs, mostly with Gastropoda. There is only one exception, Aporocotylidae, using polychaetes as the first intermediate hosts. Trematodes whose rediae and sporocysts develop in Bivalvia and not in Gastropoda are also very rare. The second fact is a very narrow specificity of most trematodes to the first intermediate host at the species level. Specificity of maritae to vertebrates is much broader (see for details Ginetsinskaya, 1968; Gibson and Bray, 1994).

Many researchers (Ginetsinskaya, 1968; Pearson, 1972; Gibson, 1987) consider that these two facts point to a greater evolutionary age of the system molluscs-trematodes than of the system vertebrates-trematodes and thus to molluscs as primary hosts of ancient trematodes.

Opponents of this point of view recognize that much data has been accumulated in its support (Cribb et al., 2001a). However, they emphasize that trematodes changed their hosts several times in the course of evolution, changing from one group of molluscs to another (Littlewood et al., 1999b, 2001; Cribb et al., 2001a). This circumstance appears to decrease the value of the specificity character in evolutionary reconstructions. However, the same is true about the relations of trematodes and vertebrates. Furman's rule, according to which evolution of parasites and their hosts is parallel can not be applied to all groups of parasites. It appears to be true for parasites with simple life cycles without a change of hosts. On the contrary, trematodes "break" this rule and this is the basis of their broad adaptive radiation.

All the researchers appear to leave unnoticed an obligatory, without any exception, participation of molluscs in the completion of trematode life cycles. Transmission ways of parasites can be reduced due to deletion of various hosts and the life cycles can be modified due to deletion or insertion of different developmental phases (see chapter 3), yet the first intermediate host is always present. No researcher who discussed trematode evolution patterns has every doubted that of the three possible states of the character "first intermediate host" – Gastropoda, Bivalvia, Polychaeta – Gastropoda is the plesiomorphic one. This total dependence of trematodes on molluscs can not be rationally explained by the "vertebrate first" hypothesis (Cribb et al., 2001a, p. 1008). On the contrary, "scenarios invoking mollusc-first evolution" (Ibid., p. 1008) provide a logical explanation of the close association between the parasites and their hosts.

Moreover, we have not found any examples of such obligatory dependence of all the representatives of a species-rich taxon of a high level on the secondarily acquired intermediate host for transfer of infection stages to the definitive host along the food chain. Brooks and McLennan (1993) consider that host specificity, as one of the main factors of parasites' evolution is a "myth". However, in our opinion they failed to disprove that long-term association of hosts and parasites may result in a significant narrowing of host specificity exhibited by parasites ("Manter's Second Rule" – after: Brooks and McLennan, 1993). Narrow host specificity of many specialized ancient parasites is an inevitable consequence of the tighter association with the host, indispensable for formation of stable host-parasite systems. The host-parasite systems formed by molluscs and rediae or sporocysts are undoubtedly stable.

An important argument in support of the "vertebrate first" hypothesis is associated with parsimony principle. In the opinion of Cribb et al. (2001a) their hypothesis is more parsimonious than "mollusc first" scenarios. Two different problems should be discussed in this connection: monophyletic origin of Neodermata and transition of Neodermata to parasitic way of life. As noted above, adherents of the "vertebrate first" hypothesis presume that transition to parasitism, formation of a new type of epidermis ("neodermis") associated with it and separation of a monophyletic clade Neodermata are closely associated events. From these precepts, one of the two alternative hypotheses indeed has to be chosen. Moreover, if molluscs are accepted as primary hosts of Neodermata, no satisfactory explanation of the formation of Cercomeromorpha can be given. The difficulty lies not only in the number of "steps", noted by Cribb et al. (2001a), but also in that it is impossible to explain the origin of the initial type of morphological organization of Cercomeromorpha.

These and some other contradictions can be solved in two ways. The first possibility is to accept that some group of ancient Neodermata that were parasites of "provertebrata" lost the association with their vertebrate hosts and colonized a new group of hosts, molluscs. Subsequent events might have followed one of the possible well-elaborated scenarios (Ginetsinskaya, 1968; Pearson, 1972; Gibson, 1987). However, all of them imply one more "step" – a "tertiary" inclusion of the vertebrate host into the life cycle. This scenario can hardly be considered as parsimonious. The authors of the "vertebrate first" hypothesis find it hardly probable (Littlewood et al., 1999b).

In our opinion, the second possibility solves most, if not all of the problematic issues listed above. Independent transition to a parasitic way of life of two evolutionary branches of the clade Neodermata should be accepted. In a relatively small group of related forms with a tendency to a symbiotic way of life one of the groups acquired a new type of epidermis, the tegument ("neodermis"), changing dynamically throughout the lifetime. It was the most important preadaptation (see pp. 83, 109) that facilitated the transition of ancient Neodermata to parasitism and made possible their effective adaptive radiation. Those forms that did not follow this evolutionary trend remained typical "turbellarians", although specialized ones. Neodermata, on the contrary, fully used the evolutionary potential accumulated, having started the colonization of different hosts. Some of Neodermata – future Cercomeromorpha – became obligatory parasites of protovertebrates. Discussion of the main formation stages of this group is outside the scope of this book. We only note that the hypothesis of primary endoparasitism currently accepted by many researchers (Littlewood et al., 1999b, 2001; Cribb et al., 2001a; Rohde, 2001b) seems ill-founded to us. Our view on the evolution of trematodes is presented in the following section.

1.2. Main Stages in the Evolution of Trematode Life Cycles

Since there are no paleontological data, a detailed reconstruction of ancestral forms of modern trematodes, that were obligatory commensals of paleozoic gastropods, is impossible. The newest and most advanced methods do not allow us to answer the question of what group of turbellarians gave rise to the clade Neodermata. As noted above, there are many candidates for this role. Besides, both in all Neodermata and in all representatives of the class Trematoda plesiomorphic characters are "unpredictably" scattered both in different taxa and in different generations and phases of the life cycle of one species.

Nevertheless, it appears expedient to attempt at least a general reconstruction of ancient proto-trematoda in order to sum up the results of the analysis presented in chapter 5.

The first question to be addressed is what organisms colonized molluscs: juvenile developmental stages (probably, larvae) of some ancient forms or reproductively mature organisms. Many researchers think that it was larvae that colonized molluscs (Heyneman, 1960; Cable, 1965; Ginetsinskaya, 1968; Pearson, 1972). However, they do not take into account the fact that development accompanied with metamorphosis and the presence of the larva in the life cycle are rare among turbellarians. However, a dispersive larval stage is present in the life cycle of Fecampiidae. This fact was discussed by Ginetsinskaya (1968) in connection with the origin of trematodes and early stages of their life cycle evolution.

Some investigators consider the Fecampiidae as the most probable candidate for the role of the sister group of Neodermata (Kornakova and Joffe, 1999; Littlewood et al., 1999b; Cribb et al., 2001a). However, in our view the presence of larvae in parasitic Turbellaria does not prove anything. A dispersive larval phase in the organisms whose different phases of the life cycle are associated with different habitats seems quite natural. Most parasites regardless of their taxonomic position are such organisms. It is the dispersive larva that ensures the completion of their life cycles. Juvenile developmental stages are likely to have become true larvae independently in different taxa of parasitic Platyhelminthes.

James and Bowers (1967) held a different point of view. Though they did not discuss this point in detail, they emphasized that molluscs were likely to be colonized not by juveniles but by mature organisms. In our opinion, this supposition is supported well by factual material on the reproductive dynamics of the mother generation of parthenitae accumulated at present (see pp. 23, 419). As noted above, in primitive trematodes it is the miracidia that are reproductively mature organisms and not the parasitic phase of the mother sporocysts development. The mother sporocyst is capable of intensive reproduction only in specialized and evolutionary advanced digeneans.

The above considerations allow a new reconstruction of the early stages of formation of host-parasite relations between proto-trematodes and paleozoic molluscs (Dobrovolskij and Galaktionov, 2003). According to Ponder (1998), the division of molluscs into Gastropoda and Cephalopoda took place at the lower boundary of the Paleozoic, about 570 million years ago. It appears that molluscs as potential hosts were primarily colonized by reproductively mature individuals rather than by juvenile stages.

Summing up plesiomorphic characters of parthenogenetic generations of trematodes, it can be supposed that these organisms morphologically resemble the rediae of the most primitive Trematoda. Pearson (1992) admitted this possibility. He considered that it was rediae that recapitulated most fully the ancestral organization features of ancient forms (see p. 453). In our opinion, three features of body plan organization of proto-digenea are the most important: lack of body division into two parts (body – tail or body – cercomer) (i), anterior position of the mouth opening and the presence of a simple sac-like gut (ii), the presence of two primary excretory pores situated laterally and shifted to the posterior body part, and of the main collecting tubes of the excretory system of the stenostomate type (iii). The latter feature seems to contradict the data on the redial excretory system available in the literature (see p. 429). However, it should be recalled that rediae appeared secondarily in the trematode life cycle (see p. 423) and their excretory system might have been simplified. In general, the tendency to simplification is the main trend of redial morphological evolution. Rediae evolved into sporocysts several times independently in different phylogenetic branches (see p. 424).

At the early stages of formation of the association between molluscs and proto-trematodes the interrelations between the partners were probably commensalistic and not parasitic. Conditions in the mantle cavity of molluscs were favourable for completion of the worms' reproduction. The worms might have fed on slime, cell detritus and molluscan extrapallial fluid, like parthenogenetic metacercariae of modern gymnophallids *Cercaria margaritensis* (Ching, 1982; Galaktionov, 1996a; Irwin et al., 2003). Transition from commensalism to parasitism involved a complex of morphological and biological reconstructions. The most important reconstruction was probably transition to tissue parasitism. It presupposed many aspects of subsequent transformations. Releasing eggs to the environment was difficult and oviviviparity developed. At the same time transition from amphimixis to apomictic (diploid) parthenogenesis must have occurred see p. 73)

All these processes must have been accompanied by successive transformations of the reproductive system. On the one hand, it was expressed in the loss of a considerable part of the organs of reproductive system, and on the other hand, in the changes of the location of the remaining ones. The male reproductive system was completely lost. The only organ left from the female reproductive system was the germarium. A transition to true viviparity resulted in a loss of vitellaria and ducts of the reproductive system. Instead, a spacious body cavity (schizocoel) formed, located along the gut and occupying a considerable portion of the body. The germarium, transformed into the germinal mass, shifted backwards, and the birth pore shifted much forwards. Tihomirov and Seleznev (1998) suggested that the birth pore and the external uterus opening of the individuals of hermaphroditic generations were homologous; this is unlikely. The origin of the birth pore is probably secondary, associated with the emergence of the schizocoel.

Such a location of organs to each other is rather functional. It ensures an easy transportation of food to the developing embryos and a certain "age" gradient of their position in the body cavity; young embryos are situated close to the germinal mass in the posterior part of the body cavity, whereas mature ones are located in the anterior part closer to the birth pore. This must be adaptive, since unprotected embryos are not damaged by the more developed mobile individuals.

However, the parasites started to use the host's resources in the most efficient way not from the very beginning, as we can see from the mother sporocysts of archaic redial trematodes (Fasciolidae, Transversotrematidae) (see p. 245). The development of host-parasite interrelations (first, transition from active histiophagy to hematophagy, and then from gut feeding to the nutrient transport through the tegument) brought about subsequent changes in two different directions. More intensive use of the energy resources of the host allowed a gradual transfer of the generative function to the time spent by the parasite in the molluscan host and an enhancement of this function. The latter would be impossible without increasing the mass and the size of the body: mature mother sporocysts can be very large (see p. 247).

The free-living developmental phase of such proto-trematodes became the true dispersive larva, and the parasitic phase, the "mother sporocyst" in the narrow sense of the word. In fact, the emergence of the latter in the course of trematode evolution can be considered as the result of the prolongation of ontogenesis in the ancestral forms, caused by the switch to parasitism. Such ontogenetic "superstructures" (anaboly according to Sewertzoff, 1931, 1939) are quite common in the evolution of Animalia. It follows from the above argument that some archaic features of organization, recapitulating the plesiomorphic state of characters of ancestors, should be identified in miracidia, and, most of all, in the large actively swimming "larvae" that are typical of the most primitive digeneans (Fasciolidae, Paramphistomidae, Echinostomatidae).

Rediae and daughter sporocysts, characteristic of the majority of the recent species, were acquired later in the evolution of trematodes (see p. 424). Their appearance can be considered to be a secondary adaptation to efficient exploitation of the host resources.

Importantly, the transformations listed above took place not **before** but **after** the separation of proto-trematodes as an independent phylogenetic branch of parasitic flatworms. These transformations determine the evolutionary trends of the taxon Trematoda as such. In our opinion, these changes (first of all, transition to parthenogenesis) were possible due to the fact that generations reproducing by amphimixis were retained in the life cycle. Individuals reproducing in the mollusc produced juvenile organisms that during a certain evolutionary period probably reached the reproductive state in the environment. The presence of parthenogenetic generations, ensuring a considerable increase in the numbers of genetically uniform progeny, should be compensated by the presence of hermaphroditic generation in the life cycle. In the opinion of Ginetsinskaya (1968), individuals of this generation were free-living and probably had many archaic features characteristic of ancestral forms. It was at that evolutionary time that the life cycle of hypothetical proto-trematoda must have become heterogony, i.e., an obligatory alteration of generations different as to their reproductive mode (amphimixis and parthenogenesis).

No doubt the next key event in the evolution of the digenean life cycle was the emergence of the so-called definitive or final host – a vertebrate animal, colonized by the individuals of hermaphroditic generation.

Possible details of the process have been discussed in the literature (Cable, 1965, 1974; Ginetsinskaya, 1968; Pearson, 1972; Gibson, 1987). However, one aspect, directly associated with the emergence of two-host life cycle, has not been covered sufficiently. This is the appearance of a specialized larval stage, the cercaria, in the digenean life cycle. By their very nature, cercariae are the true larvae of the hermaphroditic generation. Their main function is the transfer of the infective stage to the definitive host through the external environment. We presume that originally such a transfer could have been realized directly, i.e., without an encystment stage and certainly without the involvement of the second intermediate host. The latter is likely to have been inserted into the life cycle at a much later stage (Galaktionov and Dobrovolskij, 1998). In this case the infection of the final host took place exclusively *per os*.

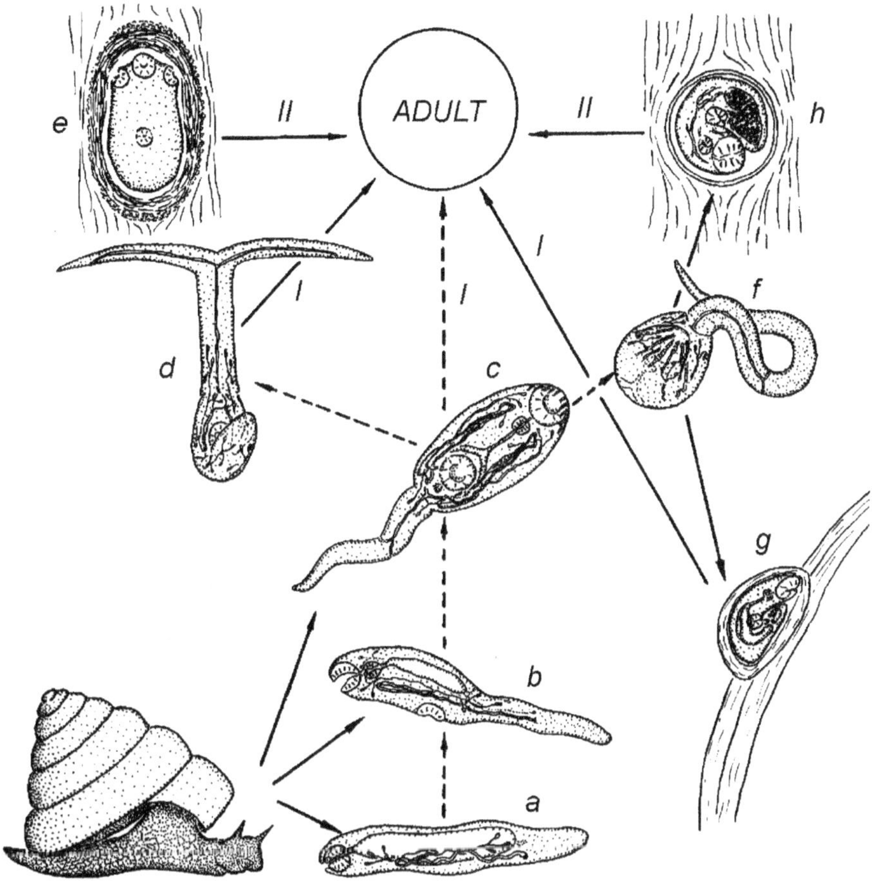

Figure 142. Possible ways of evolution of the three-host life cycle in trematodes. *a-c*, possible stages in cercarial evolution; *d*, furcocercous cercaria with intermittent swimming behaviour; *e*, metacercaria in the second intermediate host enveloped in the capsule of the host origin; *f*, monocercous cercaria with continuos type of movement; *g*, adolescaria encysted on the substrate in the environment; *h*, metacercariae encysted in a second intermediate host; *I*, primary two-host life cycle, *II*, three-host life cycle. Based on Dobrovolskij and Galaktionov (2003).

The successful infection of the primary final hosts, generally accepted as teleost fishes, largely depended on the motility of juvenile stages of parasite development that left the host. Moving in water with the use of weakly differentiated muscle elements of the tegumentary-muscular sac is possible only by means of undulation. This is precisely what Cable (1965) suggested. In our opinion, his suggestion is correct.

According to Cable (1965, 1974), it was later that the posterior part of the body became specialized as a locomotory region. This gradually transformed into the tail of recent cercaria (Fig. 142). This suggestion also seems quite sound (see p. 438).

The acquisition of the tail and the transformation of a usual juvenile developmental stage into a true dispersive larva not only increased the probability of infection transfer from the mollusc to the vertebrate animal but also entailed morphological reconstructions of hermaphroditic individuals. The posterior body part specialized into the tail. This brought about changes in the topography of the organs of reproductive system, especially of distal parts of ducts of male and female reproductive systems. External openings of these ducts inevitably shifted to the anterior body part. This process was also influenced by the presence of attachment organs, the oral and the ventral suckers, acquired by the individuals of hermaphroditic generation when they became parasites of vertebrates. The presence of the ventral sucker appears to have been associated with the division of the body into the pre-acetabular (locomotory) region and post-acetabular (generative) one. External openings of the reproductive system and correspondingly the copulative apparatus and the distal part of uterus (metraterm) functioning as a vagina shifted to the more mobile locomotory region, which facilitated cross-fertilization. As for the Laurer's canal, the duct that lost its function of a vagina, it retained its position in the generative body part.

After the two-host life cycle and the dispersive larva (cercaria) had become established, the subsequent evolution of digeneans apparently followed two main directions (Galaktionov and Dobrovolskij, 1998; Dobrovolskij and Galaktionov, 2003) (Fig. 142d,f). The efficiency of infection of the definitive host and, therefore, the completion of the life cycle, apart from other factors, significantly depended on the time that larvae spent in the environment. The life span of any endotrophic dispersive larvae, especially actively swimming ones, is always limited. In these organisms the life span can be prolonged in different ways. First of all, they can refrain from the active movement, replacing it by passive "waiting". It is precisely this strategy that has been adopted by a large group of trematodes (Fasciolidae, Philophthalmidae, Paramphistomidae, Notocotylidae) (see p. 152), whose cercariae acquired the capacity to encyst in the environment (Fig. 142g). In this case, the *per os* method of infecting the definitive host remains the same as in the most archaic trematodes. However, the role of the infective stage is played by the juveniles, the adolescariae, safely protected within a multilayered cyst, and not by actively swimming larvae. Although the life span of adolescariae in the environment is also limited, it significantly exceeds that of active cercariae.

The second way of increasing the life span of the juvenile stages involves a change in the manner of the locomotion. Continuous movement, requiring enormous energy, is replaced by intermittent swimming behaviour (see p. 267). In this case the life span of the dispersive larva is prolonged. This makes encounter with a definitive host more probable. However, an important condition should be fulfilled – cercariae should remain in the spatial zone in which they had previously moved, i.e., in the water masses inhabited by their first vertebrate hosts – fishes. Apparently, it was during the period when intermittent swimming behaviour was establishing in the Digenea, that the evolutionary path towards a furcocercous tail emerged (Fig 142d) (see p. 267).

The third major event in the evolution of digeneans and establishment of their life cycle was the replacement of primary two-host life cycles by three-host ones. This was associated with the emergence of the second intermediary host. The problem was discussed at length in the literature (Ginetsinskaya, 1968; Pearson, 1972; Gibson, 1987; Galaktionov and Dobrovolskij, 1998), but we think that one important aspect has escaped the attention of investigators.

The proposed scheme is as follows. As noted above, originally trematode larvae encysted in the environment after a short period of active swimming (Fig. 142g). They did not manifest any essential selectivity towards the substrate on which they settled. Selectivity appeared later (Pearson, 1972) (see p. 281), and ensured the transfer of the parasites along the food chains.

The next stage consists of the larval exploration of various cavities in the animal body that are directly accessible from the external environment. This requires a complex of complicated behavioural reactions, which, in turn, becomes a basis for the first manifestation of specificity (Galaktionov and Dobrovolskij, 1998). It is precisely at this evolutionary stage that we find many Echinostomatidae whose cercariae still retain, in some cases, the capacity for encystment in external environments (see p. 152).

At the same time, in some representatives (e.g., *Echinochasmus*, *Himasthla*) of the same family, the first attempts of via the skin penetration into the host can be observed (see p. 284). In fact, for the first time in digenean evolution we see a progressively specializing penetration apparatus.

Transition to tissue and internal body cavity parasitism in the second intermediate host (Fig. 142h) resulted in the subsequent simplification of the complex of cystogenous glands and, hence, reduction of the cyst. Not only did the cyst wall gradually become thinner and more permeable, but the act of encystment itself often shifted to the last stages of the morphogenesis of the developing metacercaria (see p. 152).

The argument outlined above accounts for the establishment of the three-host life cycle in the series that we conventionally denote as Fasciolidae, Echinostomatidae →Heterophyidae →Plagiorchiidae (Fig. 142f-h). Naturally, a large number of other taxa that are more or less related to these groups, belong to the same complex.

A completely different situation is observed in the representatives of another large complex of taxa – Schistosomatida, Strigeidida and related families (Fig. 142d,e). It could be concluded that, in this case, transition to the three-host life cycle and the appearance of the second intermediate host developed independently. The most archaic representatives of this group

(Azygiidae) even retain the primary two-host life cycle with a direct *per os* infection of the definitive host.

All Strigeidida are characterized by possession of a true three-host life cycle. Nevertheless, the presence of a thin poorly developed cyst around the developing metacercaria is only observed in representatives of the family Cyathocotylidae (see p. 158). It should be emphasized that the formation of this cyst proceeds is a manner different from those of other digeneans that possess encysting metacercariae. The material for cyst formation is supplied not by cystogenous cells that are specialized tegumental cytons, but by true skin glands, the ducts of which open as independent pores at the surface of the tegument. As for Diplostomidae and Strigeidae, no reliable evidence of metacercarial encystment in the second intermediate host is available in literature. Such structures, when observed, are commonly referred to as multi-layered thick cysts. These are, in fact, capsules of connective tissue formed by the host (see p. 157) (Fig. 142e).

Some researchers think that the above complex of taxa is represented by ancient groups that separated from the common stem of Trematoda very early (Cribb et al., 2001b). If it is true, an independent acquisition of the second intermediate host in the life cycle seems quite probable. Thus, three-host life cycles appeared in the evolution of trematodes at least twice.

2. THE MAIN TRENDS IN EVOLUTION OF TREMATODE LIFE CYCLES

In conclusion, we would like to discuss the main evolutionary trends of trematode life cycles (Fig. 143). It should be emphasized that we do not construct particular phylogenetic schemes but consider the main trends of transformations of life cycles. These transformations may be brought about by different reasons, some of which (e.g., trematode expansion into different ecosystems) have been discussed above (see section 2 of chapter 5).

We consider an obligatory two-host life cycle with an established heterogony as the basic life cycle type (Fig. 143,I). In our opinion, this type constitutes the boundary between the ancestral forms and modern trematodes. The definitive host was infected *per os* by swimming larvae that did not have any specific adaptations to a prolonged life span in the environment. We have attempted a reconstruction of such ancient dispersive larvae, taking into account the analysis of morphological tendencies of trematode evolution presented in chapter 5.

These larvae were probably rather large as compared to specialized cercariae of modern trematodes, their tail was simple and not very specialized (Fig. 144). Most species-specific characters of the sexually mature individuals were also characteristic of these larvae. They possessed the oral and the ventral sucker and a well-developed primordium of the digestive system. The oesophagus and the intestinal branches had a clearly expressed lumen throughout their length. Though the excretory system of stenostomate type contained many flame cells, the number of primary groups of flame cells was probably already reduced to 6. These hypothetical larvae probably lacked the specialized secretory apparatus characteristic of most modern cercariae. It is likely that the tegument ("neodermis") already performed the secretory function, which must have facilitated the transition of ancestral trematodes to parasitism in the digestive tract of vertebrates. However, larvae of the hermaphroditic generation at this stage of trematode evolution possessed neither a true cystogenous apparatus (originating as the result of specialization of a part of tegument cells) nor penetration glands (specialized head glands of the ancestral forms). As noted above (see p. 494), initially individuals of this generation reached reproductive maturity in the external environment. However, after the acquisition of the vertebrate host, reproductive activity of the parasite was shifted for the period spent in the host (a similar phenomenon had already occurred in the course of evolution when proto-trematodes colonized molluscs as hosts). Therefore, at the considered stage of life cycle evolution the larvae probably had a poorly differentiated generative primordium, if any. This early manifestation of heterochrony, which brings about retardation in the development of the reproductive system makes biological sense. Yet as noted above the larvae remain adult-like. This larval feature is partly retained in representatives of some archaic families (Echinostomatidae, Paramphistomidae, Philophthalmidae), disappearing completely in the specialized larvae of Schistosomatidae, Strigeidae, Lecithodendrioidea, Microphallidae.

The tail enabled the larvae to swim in the water column. As a result, not only the probability of infection of the definitive hosts, the fishes, increased, but also the range of hosts became much broader due to inclusion of different plankton-feeders into the transmission of the parasites. In modern trematodes life cycles this kind are not known.

Subsequent evolutionary transformations did not alter the essential life cycle scheme. The life cycle remained a two-host one. It was only the biology of dispersive stages that changed, increasing further the probability of infection of the definitive host. As noted above (see p. 496), it could be achieved in two ways.

In the ancestors of one large phylogenetic branch the larvae, retaining archaic features of organization (a large body, a simple tail, a certain similarity with adult stages) and behaviour (continuous swimming), acquired the ability to encyst in the environment. Thus the phase of adolescaria was included into the ontogenesis of the hermaphroditic generation (Fig. 143,II). Biological specialization of cercariae of the forms discussed is expressed in the ability to select the substrate for settlement and subsequent encystment. As noted above (see p. 281), they prefer substrates the final host feeds on.

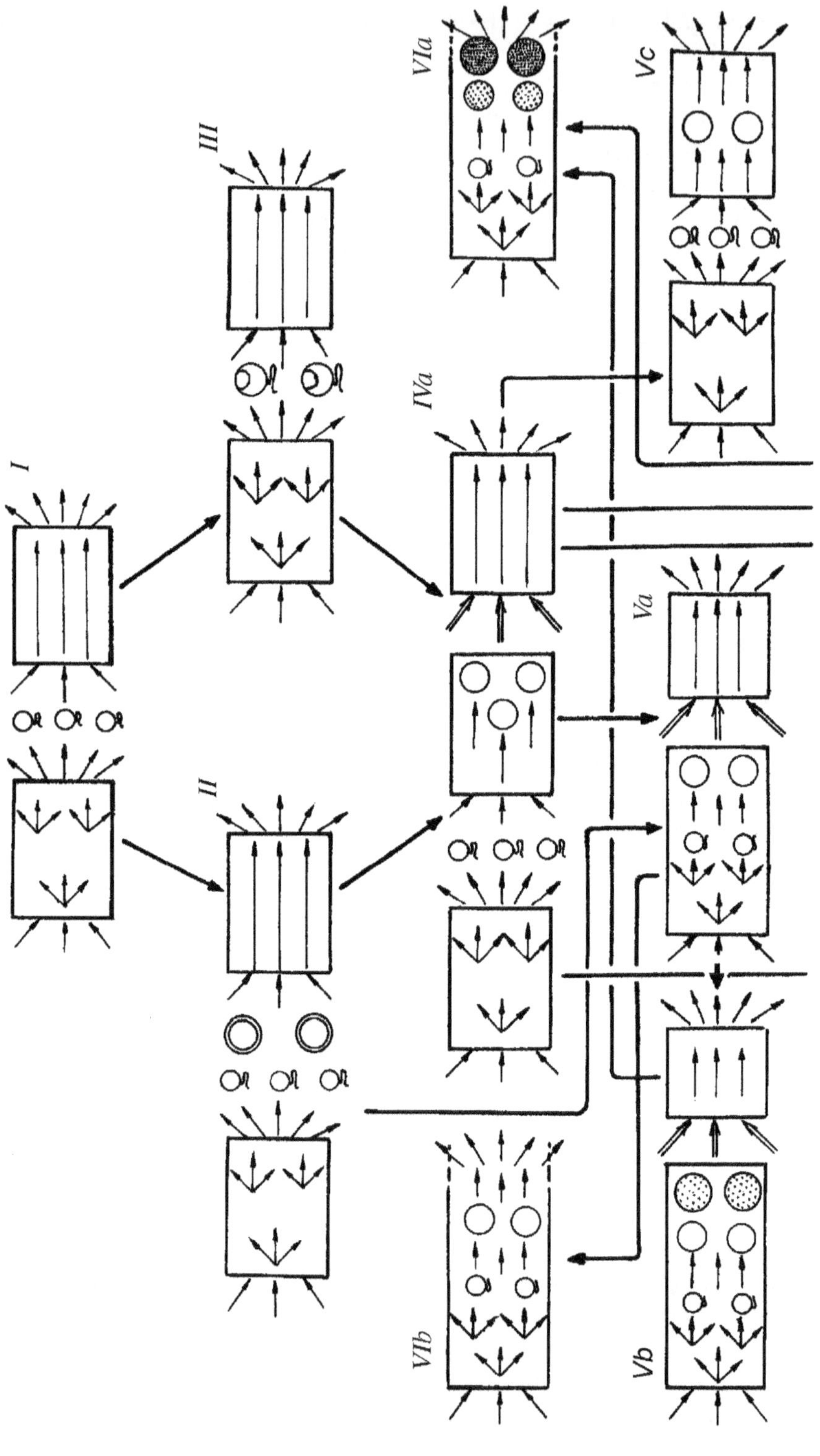
I
II
III
IVa
Va
Vb
Vc
VIa
VIb

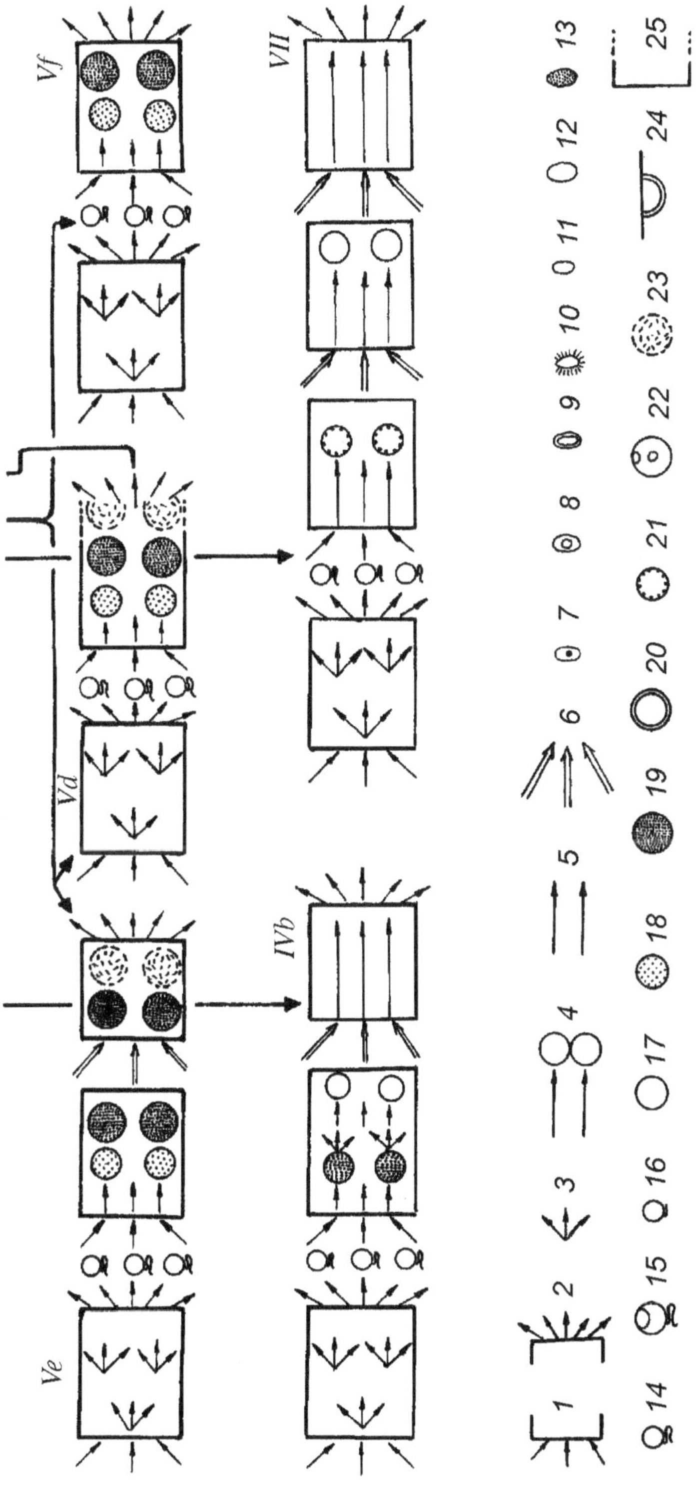
Vf
VII
Vd
IVb
Ve
1
2
3
4
5
6
7
8
9
10
11
12
13
14
15
16
17
18
19
20
21
22
23
24
25

This selectivity, accompanied by prolongation of the time spent by the infective stage (adolescariae) in the environment (as the result of encystment) further increases the probability of infection of the vertebrate host. Among modern trematodes this type of life cycle is present in the families Fasciolidae, Philophthalmidae, Paramphistomidae (and other related groups), Notocotylidae and Pronocephalidae.

In another trematode group specialization of cercarial behaviour was a dominant tendency. Prolongation of their lifespan in the environment was achieved by means of specialization of locomotion and the acquisition of intermittent type of movement. Definitive hosts were infected when they ate swimming or floating larvae, not the resting adolescariae. Archaic life cycles of this type have not been retained. However, Azygiidae and Bivesiculidae, whose cercariae infect the definitive hosts *per os*, might be direct descendant of such forms (Fig. 143,III). Only a general scheme was retained in their life cycles from the archaic state, the families having acquired many characters pointing to an extreme specialization. Both groups independently acquired cysticercous larvae. Bivesiculidae probably followed the tendency to juvenilization of both cercariae and maritae (see p. 450). On the contrary, Azygiidae, retaining large cercariae with some characters in common with

Figure 143. Possible ways of evolution of the trematode life cycle. *I*, hypothetical life cycle of ancient trematodes; *II*, primary two-host life cycle with the phase of adolescaria; *III*, primary two-host life cycle with cysticercous cercaria (Azygiidae); *IVa*, typical three-host life cycle characteristic of the majority of modern trematodes; *IVb*, three-host life cycle with additional endogenous agglomeration on the phase of metacercaria (some Gymnophallidae); *Va*, secondary two-host life cycle with the molluscan host which combines the functions of the first and the second intermediate host; *Vb*, the same, but with progenetic development of metacercariae (microphallids of the "*pygmaeus*" group); *Vc*, secondary two-host life cycle with the vertebrate host which combines the functions of the second intermediate and the definitive host (*Opisthioglyphe ranae*); *Vd*, secondary two-host life cycle with metacercaria reaching sexual maturity in the second intermediate host, dispersion of the infective eggs may be brought about by the host's death (*Paralepoderma brumpti*); *Ve*, the same, dispersion of the infective eggs by means of a non-specific predator (*Paralepoderma brumpti*); *Vf*, the same, dispersion of the infective eggs without the death of the host(blood parasites of the order Schistosomatida); *VIa*, homoxenous life cycle formed as a result of neoteny (*Paralepoderma progenetica*); *VIb*, homoxenous life cycle in which all developmental phases have been retained (*Bunocotyle progenetica*); *VII*, four-host life cycle.

Designations: (to this figure and to figures 102-109 and 111-115): *1*, exogenous accumulation; *2*, dispersion; *3*, endogenous agglomeration; *4*, metacercarial development in the second intermediate host; *5*, development of marita (maritogony); *6*, exogenous block accumulation; *7*, compound egg; *8*, embryonation of the egg; *9*, fully formed miracidium in the egg; *10*, actively swimming miracidium; *11*, young mother sporocyst; *12*, mature mother sporocyst; *13*, daughter generation of parthenitae (rediae and sporocysts); *14*, cercaria; *15*, cysticercous cercaria; *16*, secondary simplified cercariae which do not leave the molluscan host; *17*, metacercaria; *18*, metacercaria with formed reproductive system; *19*, progenetic metacercaria with functioning reproductive system; *20*, adolescaria; *21*, mesocercaria; *22*, adult (marita); *23*, death of the adult; *24*, adolescarial encystment on the surface of the host's integument; *25*, death of the host. Based on Galaktionov and Dobrovolskij (1998).

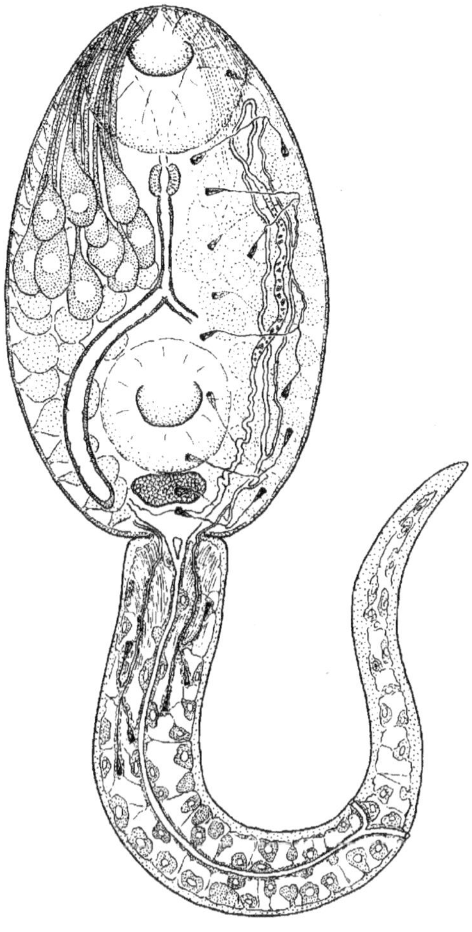

Figure 144. Hypothetical scheme of the ancient cercaria structure.

adult stages, demonstrated a high biological plasticity and gave rise to other life cycle types (see p. 504).

Life cycles of specialized Azygiidae and Bivesiculidae could also have formed in another way. Two-host life cycles of the type considered might result from a secondary loss of cystogenous apparatus and ability to form cysts. This way appears less probable but it can not be ruled out. The life cycle of *Philophthalmus rhionica* follows almost the same scheme in the cases when the cercaria that gets into the final host is not encysted yet (see p. 302). Larvae of some echinostomes also may infect the definitive host in a similar way (see p. 283).

The next stage was the formation of three-host life cycles. In most cases transition to three-host life cycles was brought about by the substitution of a free-living adolescaria by a parasitic metacercaria (Fig. 143,IVa).

Facultative three-host life cycles emerge independently in Philophthalmidae, Notocotylidae and Echinostomatidae (see pp. 283, 308, 336). However, they become obligatory only in the latter. Moreover, it is only in echinostomatids that metacercariae are true parasites, since they use the host not only as a substrate for transfer of infection (encysted larvae – metacercariae) along the food chains but also as a source of energy. In the long run, not only the life span of metacercariae in the host is prolonged but also a certain stage of morphogenesis is transferred to this phase of development of the hermaphroditic generation (see p. 159). Thus, the second intermediate host was the last to appear in the life cycle. Different researchers agree upon this point (Ginetsinskaya, 1968; Pearson, 1972; Gibson, 1987; Galaktionov and Dobrovolskij, 1998).

This type of life cycle (Fig. 143,IVa) is inherent to most modern trematodes, both those that retained a number of archaic characters (e.g., Heterophyidae) and more specialized ones (Plagiorchiida).

Three-host life cycles appeared in the course of evolution several times in different ways. As noted above, in forms whose ancestors did not acquire the ability to encyst, three-host life cycles emerged independently, though also by means of the acquisition of a second intermediate host (Cyathocotylidae, Strigeidae, Diplostomidae) (see p. 497).

At the same time, some trematodes evolved three-host life cycles in an entirely different way. For instance, in the most specialized Azygiidae a new host was not "inserted" (as it is usually the case) but "superstructured" to the life cycle. The reproductive phase of the hermaphroditic generation shifted into carnivorous pelagic elasmobranchians, whereas the former definitive host, small teleost fishes, started to function as the second intermediate one (Gibson, 1987).

Life cycles of some gymnophallids may be considered as a rare modification of the three-host life cycle (Fig. 143,IVb). These aberrant forms demonstrate a second endogenous accumulation in the metacercaria phase (see p. 73).

Further transformations of life cycles were associated with two opposite processes: transmission ways of parasites were reduced or, on the contrary, become more complicated. Reduction of the number of hosts – transition to two-host life cycles (Fig. 143,V) – was a common phenomenon. It took place several times independently in representatives of different taxa.

The main variants of two-host life cycles are as follows. First of all, transition to a two-host life cycle does not always entail the reduction of the cycle of trematode development. All the phases may be retained in the developmental cycle. In this case, two variants of a two-host life cycle are possible: either the mollusc combines the functions of the first and the second intermediate host (*Heronimus mollis* – Heronimidae, some Echinostomatidae) (Fig. 143,Va) or the host cercariae have invaded first plays the role of the second intermediate host and then becomes the final one (the genus

Opisthioglyphe – Plagiorchiidae) (Fig. 143,Vc). Two-host life cycles with the mollusc combining the functions of two hosts may originate directly on the basis of primary two-host life cycles with the adolescaria (e.g., *Catatropis* from notocotylids). Another modification of this life cycle type is observed in microphallids, whose metacercariae in fact complete morphogenesis of the reproductive system as the result of progenetic development, sometimes losing the ability to encyst (microphallids of "*pygmaeus*" group) (Fig. 143,Vb).

Another type of two-host life cycle is also based upon progenetic development of parasites. In this case the definitive host is eliminated since the metacercaria achieves reproductive maturity. Reduction of transmission ways (the loss of the definitive host) is superimposed by precocious reproductive maturation of larval stages, the phase of marita being eliminated. Egg dispersion is accomplished in different ways. In more primitive forms (*Paralepoderma brumpti* – Plagiorchiidae) egg dispersion in the environment may be brought about by the host's death (Fig. 143,Vd) or by means of a non-specific predator (Fig. 143,Ve). The latter variant appears to ensure a large-scale dispersion of infective agents. More specialized forms (Fig. 143,Vf) shed eggs into the environment constantly and dispersion does not depend on the death of the host (Schistosomatidae, Spirorchiidae, Sanguinicolidae, Transversotrematidae) (see pp. 278, 338).

The tendency to reduced transmission ways of parasites in ecosystems finds its utmost expression in facultative or obligatory one-host life cycles completed in the first intermediate host, the mollusc. Similarly to different variants of two-host life cycles, transition to a one-host life cycle may be both accompanied and not accompanied by precocious maturation and corresponding reduction of development of the hermaphroditic generation. The former variant (Fig. 143,VIa) is found in representatives of very different taxa (e.g., Plagiorchidae – *Paralepoderma progenetica*, *Plagiorchis arcuatus* (?); Azygiidae – *Proterometra dickermani*; Monorchiidae) (see p. 149). When reproductive maturity is reached, morphogenesis may stop at the phase of both metacercaria and cercaria. In Hemiuridae (Fig. 143,VIb) development of the hermaphroditic generation is completed fully (*Bunocotyle progenetica*, *Parahemiurus bennettae*) (see p. 169). In most cases dissemination occurs when the host dies or is eaten by non-specific predators. Necrophagy, observed in some molluscs, may facilitate direct transfer of the parasite from host to host. Reproductively mature cercariae of *Proterometra dickermani* leave the host and in this way egg dispersion is achieved.

The origin of one-host life cycles is diverse. In plagiorchiids and hemiuroids they are secondary, brought about by further reduction of the number of hosts in typical three-host or secondary two-host life cycles in which the molluscs started to combine the functions of two hosts. One-host life cycles in Azygiidae (with initially archaic two-host life cycles) are of a different origin (see above).

Importantly, two-host and especially one-host life cycles are mainly associated with ecosystems with extreme environmental conditions (the intertidal zone, arid terrestrial ecosystems). They are completed in fact in a "two-dimensional" space (see p. 473).

However, reduction of the number of hosts in the life cycle is not the only tendency in trematode evolution. In a number of cases usually associated with the colonization of vast spaces (see p. 471), the number of hosts can be as many as four (four-host life cycles) (Fig. 143,VII). One more phase of development appears in the life cycle, mesocercaria (see p. 175). Representatives of different taxa acquired four-host life cycles independently and in different ways. In Hemiuroidea and Didymozoidae this phenomenon probably emerged when one more link was "superstructured" into the chain of predators along which the parasites circulate (similarly to emergence of three-host life cycles in Azygiidae), whereas in *Strigea* and *Alaria* the new link was "inserted" into the chain.

The above considerations demonstrate that the evolution of trematode life cycles is accompanied by constantly emerging parallelisms. A special analysis is necessary in each particular case for correct evaluation of the importance and the nature of the phenomena observed. Attempting the construction of a "natural" trematode classification reflecting actual phyletic relations between the groups, it is necessary to take into account the features of organization and biology of all life cycle phases. Only an approach like this would allow an exact estimation of plesiomorphic states, apomorphies and homoplasies.

3. POSSIBLE APROACHES TO ESTABLISHING A NATURAL CLASSIFICATION OF TREMATODES

A natural classification of parasitic Platyhelminthes (Neodermata) is still to be elaborated. In our opinion, attempts in this direction should not be confined to the search for a "sister" group as is usually done (see p. 494). A natural classification of parasitic Platyhelminthes should be based upon extensive and detailed analysis of phylogeny of all flatworms. This issue is outside the scope of the present book. Here we discuss only possible approaches to the elaboration of a natural classification of Trematoda.

There have been numerous attempts to point out the most ancient, "primary" group of modern trematode families. Different taxa have been named as candidates for this role. In the opinion of Ginetsinskaya (1968), among modern trematodes the Fasciolidae has the greatest number of plesiomorphic characters. Cable (1974) and later Pearson (1992) considered Bivesiculidae as the most archaic group closest to the ancestral forms. Gibson (1987) suggested Paramphistomidae as a possible candidate. Brooks et al. (1985a) thought Heronimidae to be the "sister" group to all other trematodes. Cribb et al. (2001b) accepted as a basal group a large clade including a number of widely distributed and very specialized families: Strigeidae, Diplostomidae, Schistosomatidae, Sanguinicolidae.

None of the taxa of modern trematodes mentioned above could be considered as an archaic group retaining many plesiomorphic characters. We have often noted (see section 1 in chapter 5) that most "unexpected" combinations of plesiomorphic and apomorphic states of different characters can be found in representatives of almost every taxon. The search for a "sister" group inherent to cladistic approach is unproductive as applied to modern trematodes.

The monotypical family Heronimidae can not be considered as the most ancient group. *Heronimus mollis* retained very few archaic features: large cercariae, the presence in cercariae of a simple tail and the stenostomate excretory system. This species has a typical secondary two-host life cycle that must have emerged recently in the evolutionary process (see p. 436). The only other character which might be interpreted as plesiomorphic is the lack of daughter generations of parthenitae. However, the reduction of the number of parthenogenetic generations at the expense of the generations following the daughter parthenitae is a common phenomenon. For instance, it is observed in all plagiorchiids, which retain only one daughter generation of sporocysts. Since energy resources of the host are limited, it is possible that in *H. mollis*, endogenous agglomeration was suppressed when its large-tailed cercariae ceased to leave the mollusc, retaining the ability to swim (!). A very early formation of the secondary excretory pore at the dorsal surface of the posterior body end also points to a secondary specialization of these trematodes. Contrary to the view of Cable (1974), we do not consider larvae of this species as amphistomous. Their postacetabular body region is short but clearly discernible (see fig. 135).

In our opinion, members of the Bivesiculudae are also very distant from ancestral forms, even though they do possess an indisputable plesiomorphic character, the absence of the excretory bladder. It is a very primitive feature. However, certain doubts as to lack of the excretory bladder in Bivesiculidae have been noted in the literature (for review see: Gibson, 1987). Data on the presence or absence of the oral sucker in Bivesiculidae are even more contradictory. Pearson (1992) thought that the lack of the oral and the ventral sucker in individuals of the hermaphroditic generation of Bivesiculidae was a manifestation of the "turbellarian" state. However, there are data pointing to the presence of at least the oral sucker in some species (for review see: Gibson, 1987). As for the ventral sucker, its secondary reduction (enhanced by heterochrony) is a common phenomenon. Moreover, the organization of bivesiculid maritae points to an extreme juvenilization and miniaturization of these worms rather than to an initial primitive state (see

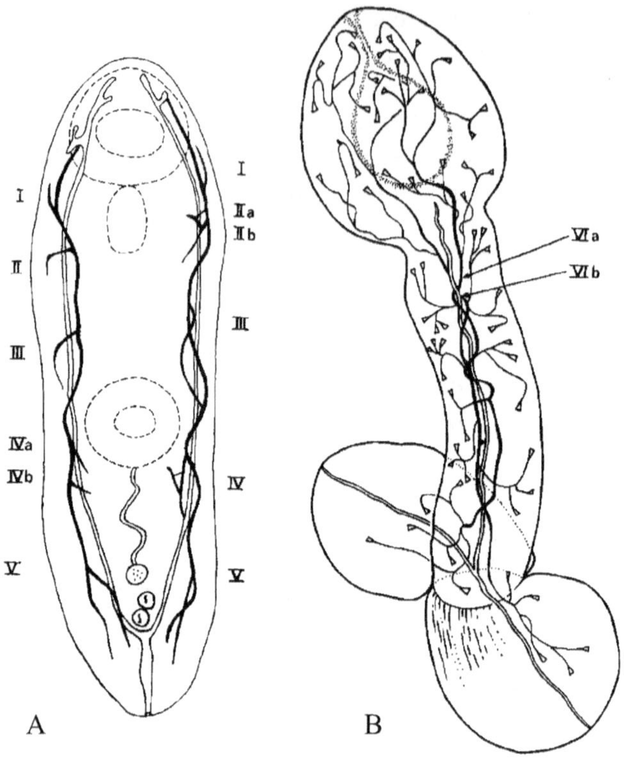

Figure 145. Excretory system of *Azygia lucei* cercaria. A, a scheme of branching of the excretory ducts; B, general view. *Abbreviations*: *I-VI*, primary groups of flame cells. From Odening (1976), reprinted with the author's permission.

p. 450). Finally, their cercariae are cysticercous, which points to a great secondary specialization (see p. 99).

Azygiidae are also very specialized. Their two-host life cycle reflects the plesiomorphic state of this character. Another archaic feature is the stenostomate excretory system, though it is present in a modified state (Fig. 145). Cercarial morphology combines primitive characters and characters pointing to secondary specialization. The former are the large size of cercariae and a complex structure of the excretory system due to multiplication of the number of flame cells in primary groups (see fig. 145). However, the number of primary groups is constant (6). The last (the 6th) group is extended into the tail as in all other true furcocercariae.

Important characters pointing to specialization of Azygiidae cercariae are an early differentiation of primordia of the reproductive system organs and a large "cysticercous" tail, whose shape and functioning may be considered as a kind of "mimicry" (see p. 283), aimed at attracting the definitive host. As noted above, azygiid cercariae are adult-like. This might be the basis of precocious development of individuals of the hermaphroditic generation. This is the case, e.g., in *Proterometra dickermani* (see p. 149). A clear tendency to three-host life cycles is another important manifestation of specialization in Azygiidae (see p. 504).

Transversotrematidae also can not be candidates for the role of the primary group. Though the representatives of this family possess a number of very ancient characters (features of reproductive dynamics of the mother sporocyst – see p. 34, the presence of rediae, large adult-like cercariae that can not encyst and lack cystogenous apparatus, the stenostomate excretory system), on the whole Transversotrematidae are one of the most specialized groups of modern trematodes. Characters pointing to deep specialization are the presence of very characteristic furcocercariae, a secondary two-host life cycle (see p. 315), progenetic development of metacercariae localized under the scales of fishes. Transversotrematidae, Bevisiculidae and Azygiidae are undoubtedly very ancient groups that diverged very early in trematode evolution. However, the divergence took place after the formation of the branch that gave rise to the clade of furcocercous trematodes.

Other trematode taxa whose representatives possess true furcocercariae in our opinion are even less suited for the role of an "archaic" group. The phenomenon of "furcocercous" cercariae and behavioural modifications associated with it are secondary (see p. 496). Specialization of "furcocercous" trematodes (Cyathocotylidae, Strigeidae, Diplostomidae, Schistosomatidae, Spirorchiidae, Sanguinicolidae) is expressed in many different characters, first of all, those of the hermaphroditic generation. These characters are: extreme miniaturization and juvenilization of cercariae, substitution of the oral sucker by the anterior organ in larvae, acquisition of the mesostomate excretory system, a complex metamorphosis at the phase of metacercariae, morphological specialization of maritae (development of the attachment cup and the adhesive organ), modifications of life cycles (transition to four-host life cycles in some Strigeidae or, on the contrary, to secondary two-host life cycles in Schistosomatidae, Spirorchiidae, Sanguinicolidae). These are the most evident characters of secondary specialization in trematodes from this phylogenetic clade. They are actually more numerous and are associated with almost all systems of organs.

Of the trematodes that retained morphologically archaic cercariae and acquired the ability to encyst in the environment, Echinostomatidae and Notocotylidae *sesu lato* are undoubtedly secondary groups. Both parthenogenetic and hermaphroditic generations of echinostomatids possess many plesiomorphic characters. The most important of them are: very archaic mother and daughter (rediae) generations of parthenitae; large, usually adult-like cercariae with a simple tail and the stenostomate excretory system , ability to encyst in the environment (retained in some species). At the same time, echinostomatids possess certain characters pointing to secondary specialization: ability to reproduce acquired by the parasitic phase of the mother sporocyst development (see pp. 24, 34) and an almost complete transition to three-host life cycle (see p. 336).

The only archaic features retained by notocotylids are their rediae and the scheme of primary two-host life cycle. Morphological and biological characters of the life cycle phases point to an extreme secondary specialization, both in the mother generation of parthenitae (see p. 24) and the hermaphroditic generation which is characterized by the mesostomate excretory system, monostomate condition and secondary attachment organs whose origin is associated with lack of the ventral sucker.

Two families are left as possible candidates for the role of the most archaic group: Fasciolidae and Paramphistomidae. Gibson (1987) favours Paramphistomidae. His main argument is the amphistomate condition of their hermaphroditic generation which he considers as a synapomorphy of aspidogastrids and digeneans. We have already discussed this issue in detail (see p. 457). In our opinion, the amphistomate condition of Paramphistomidae is secondary and not homologous to that in Aspidogastrida. Thus, those characters of organization of Paramphistomidae that could be treated as plesiomorphic ones, following our approach should be considered as characters pointing to secondary specialization.

In our opinion, Fasciolidae are the most controversial candidates for the role of "primary" group. They possess a very "unexpected" combination of characters that we consider as plesiomorphic and those pointing to secondary specialization. Fasciolid miracidia are probably the most archaic miracidia in existence. In this respect we agree with Ginetsinskaya (1968). Parthenogenetic generations of Fasciolidae are also very archaic. They produce a small number of adult-like cercariae and retain ability to self-reproduce. On the contrary, the organization of maritae, their behaviour in the organism of the definitive host (see p. 312), and the range of hosts point to a great specialization of the group.

All things considered, it seems useless to search for some single archaic "sister" group to **all** digeneans. Trematodes split into two (or more – we do not know yet) phylogenetic branches very early. No "intermediate" forms were left. Archaic characters are "unpredictably" scattered in the representatives of several taxa belonging to the lower branches of the trematode phylogenetic tree.

There have been several attempts to elaborate a natural classification of Trematoda which would reflect phylogenetic relations between the "branches". A pioneer study in this field was made by La Rue (1957) who was the first to apply for the delimitation of two large phylogenetic branches (Epitheliocystidea and Anepitheliocystidea) not morphological characters of maritae but those of cercariae, in particular, details of structure of their excretory bladder. This approach was sharply criticized by other researchers especially since reliability of facts underpinning the ideas of La Rue was questioned (see p. 138). Nevertheless, some of La Rue's groups seem natural to us (Plagiorchiata and Opisthorchiata; Azygiata, including Azygiidae and Transversotrematidae).

Cable's (1974) research is unique among the studies devoted to the natural classification of trematodes. Cable (1974) described some traits of body plan organization of the hermaphroditic generation that had been left unnoticed by other investigators and applied them in his constructions (see p. 438). The most important conclusion to be drawn from his phylogenetic scheme is that none of the groups of modern trematodes can be considered as "ancestral". Modern forms are divided into several phylogenetic branches that split a long time ago. The common ancestor at the basis of these branches is unknown. We share this point of view which is supported by our own analysis of several archaic groups (see p. 506). In our opinion, Cable's (1974) phylogenetic scheme is the best one available to-date, even though not all of his groups are convincing enough.

The phylogenetic scheme of trematodes proposed by Brooks et al. (1985a, 1989) seems the least successful to us. A detailed critical analysis of this study has already been given elsewhere (Pearson, 1992; Galaktionov and Dobrovolskij, 1998).

A very original phylogenetic scheme of trematodes was elaborated by Cribb et al. (2001b). These authors applied relatively few morphological characters, which on the one hand limited the possibilities of the analysis. On the other hand, this circumstance made it possible to ignore dubious characters that did not yield to a non-contradictory interpretation in estimations of plesiomorphic and apomorphic states. This approach allowed the authors to avoid the errors of Brooks et al. (1985a, 1989). Besides, Cribb et al. (2001b) extensively drew upon molecular data when elaborating their phylogenetic schemes. Nevertheless, the final result appears contradictory.

Ignoring concrete phyletic relations, we can see 3 distinct groups of families at the final phylogenetic scheme given by Cribb et al. (2001b) (Fig. 146). The first group comprises Strigeidae, Diplostomidae, Schistosomatidae, Sanguinicolidae and evidently related families whose representatives possess true furcocercariae (see p. 442) – Azygiidae, Bivesiculidae и Transversotrematidae. The families of the large hemiuroid-didymozooid complex are also close to this group and so are Heronimidae, Fellodistomidae and a number of families whose cercariae are unknown.

The second, central group is more heterogeneous. At its basis there is a complex of rather closely related families: Fasciolidae, Echinostomatidae, Philophthalmidae, Paramphistomidae, Haplosplanchnidae as well as a related group Notocotylidae *sensu lato*. Heterophyidae and a number of other related families are positioned very close to them. Several other families should also be placed here, though they do not form a uniform group themselves: Haploporidae, Acanthocolpidae. The second group is in fact much more numerous. We have highlighted only several key families.

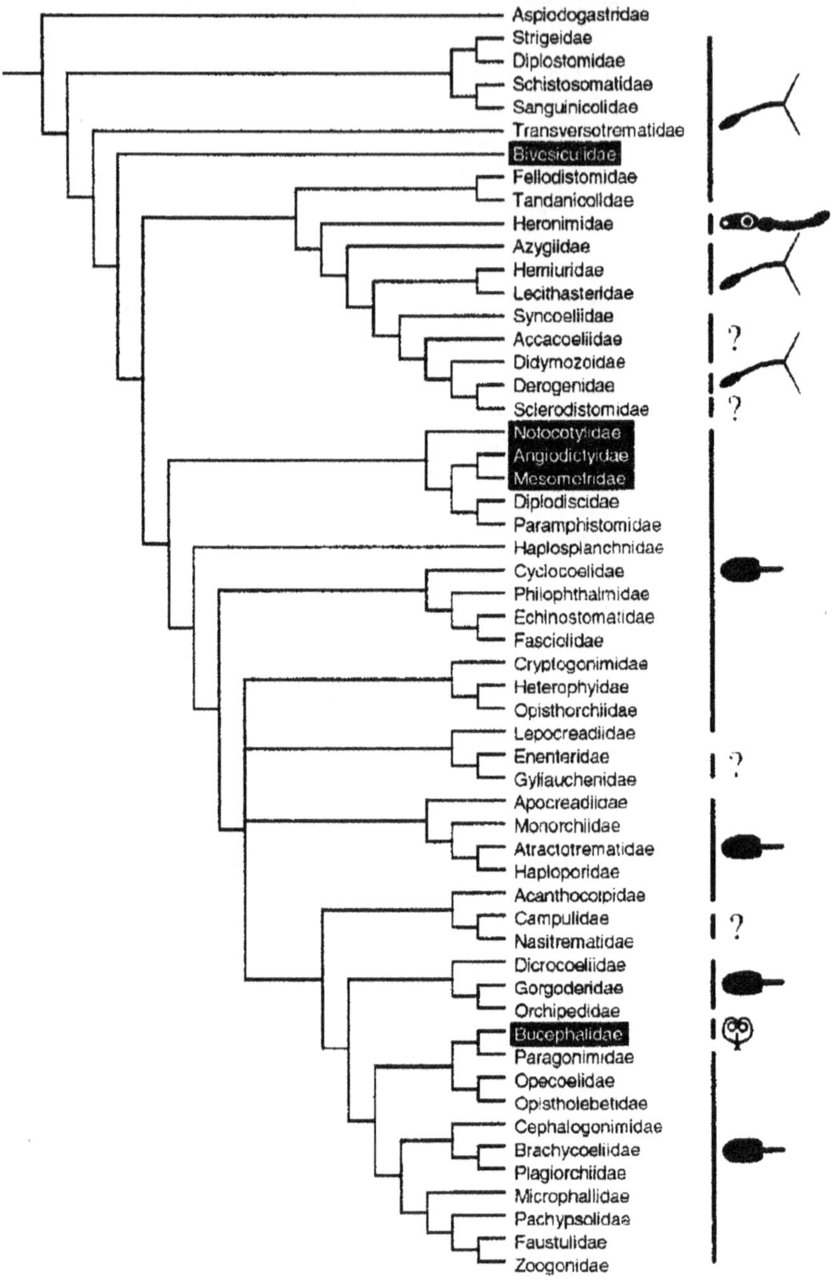

Figure 146. Digenean phylogeny based on combined morphological and molecular data sets. Distribution of complete absence of the ventral sucker at all life cycle stages (boxed taxa) and presence of fork- or simple-tailed cercariae are shown. From Cribb et al. (2001b), with permission of Taylor & Francis Publ.

The last group includes many taxa traditionally associated with the plagiorchiid complex of families: Dicrocoeliidae, Gorgoderidae, Paragonimidae, Opecoelidae, Cephalogonimidae, as well as Plagiorchiidae and Microphallidae as such. These are only some of the families that should be placed in this group.

We have presented the above (incomplete) list of families from Cribb et al. (2001b) in order to demonstrate that our grouping of these families agrees well with the major groups of taxa outlined by Cable (1974) in his phylogenetic scheme. It is unlikely that this agreement is coincidence. In our opinion, this grouping reflects actual historical relations between different groups of modern trematode taxa.

Lower part of the scheme proposed by Cribb et al. (2001b) (see fig. 146) is much less clear, inviting a long discussion. First of all, the basal position of the complex Strigeidae *sensu lato* – Schistosomatidae is problematic. The representatives of this complex (besides the above two families, it includes at least 5 more taxa – Cyathocotylidae *sensu lato*, Diplostomidae *sensu lato*, Spirorchiidae, Sanguinicolidae *sensu stricto* and Aporocotylidae) possess both very ancient plesiomorphic characters and characters pointing to a deep specialization (see chapter 5). This is true of the hermaphroditic generation to the full extent and also refers to some extent to the parthenogenetic generations.

In short, plesiomorphic and apomorphic characters of these generations are as follows. Schistosomatidae, Spirorchiidae, Sanguinicolidae *sensu stricto* and Aporocotylidae are deeply specialized blood parasites with secondarily reduced two-host life cycle (see p. 338). Strigeidae, Cyathocotylidae *sensu lato* and Diplostomidae *sensu lato* have usual three-host and even four-host life cycles. The representatives of this complex of families are sporocyst-bearing forms. There is only one exception: in Aporocotylidae the second parthenogenetic generation is represented by secondary specialized rediae. Ontogenesis of the individuals of the first parthenogenetic generation, the mother sporocyst, in all families from this complex involves a large actively swimming miracidium. Mature mother and daughter sporocysts retain a typical worm-like shape in all representatives except a number of specialized Sanguinicolidae. Moreover, in most, if not all representatives of this complex of families parthenitae of the daughter generation and subsequent ones retained the ability to produce their like. This is an ancient biological trait, lost in the more specialized representatives of digeneans. Thus, parthenogenetic generations of this complex of families are, on the whole, archaic. The only character pointing to their secondary specialization is the predominance of daughter sporocysts (not rediae) in this group of taxa.

Things are different with the hermaphroditic generation. All representatives of the complex of families considered possess furcocercariae. This fact alone can be considered as a manifestation of secondary specialization (see p. 496). With the exception of a number of the most archaic Cyathocotylidae, they all have very juvenilized cercariae. Representatives of this complex always have the mesostomate excretory system, often with an "incomplete" excretory formula (see p. 131). Besides, they possess many larval adaptive organs (the anterior organ instead of the oral sucker, a well-developed and complexly differentiated glandular penetration apparatus, the gut that functions as a secretory organ).

All forms under consideration demonstrate a complex metamorphosis completed after the penetration into the host. In Schistosomatidae, Spirorchiidae, Sanguinicolidae *sensu stricto* and Aporocotylidae metamorphosis is associated with reaching the reproductive maturity and the formation of a complex of adaptations to existence in the circulatory system. Strigeidae, Cyathocotylidae *sensu lato* and Diplostomidae *sensu lato* undergo a profound reconstruction at the metacercaria stage (see p. 175). Maritae of these three families are also deeply specialized. Their body is divided into two "segments", an adhesive organ and lappets are present, the position of reproductive system openings is changed, the cirrus sac is lost, an intensive development of the secondary excretory system takes place as early as at the stage of metacercaria. This list could be continued yet in our opinion it is sufficient to postulate a deep specialization of these trematodes. We have not found any plesiomorphic characters in the hermaphroditic generation of the families considered.

An entirely different picture is observed when we leave this "basal" group and move up the phylogenetic scheme. Transversotrematidae, the "basal" group for other representatives of the class, has already been shown to be secondary (see p. 509). Moreover, all known transversotrematids with secondary two-host, progenetic life cycles are characterized by redial state of the second parthenogenetic generation.

The latter character is retained in the subsequent series of families: Bivesiculidae, Azygiidae, Hemiuridae (more precisely, a complex of families of hemiuroid trematodes). However, sporocyst-bearing Fellodistomidae and Heronimidae fall into the middle of this series. They are also the "basal" group for Azygiidae and Hemiuridae *sensu lato* that are characterized by redial state of the second parthenogenetic generation.

This complex of families precedes forms (also characterized by redial state) that make up the majority of the second group outlined above. It is divided into several well-defined clusters: Notocotylidae–Paramphistomidae; Fasciolidae and related families; heterophyoid complex of families. Two more groups of families are associated with the latter, among them the families Lepocreadiidae, Monorchiidae and Haploporidae. Another large cluster opens with the family Acanthocolpidae. In our opinion, this cluster is rather heterogeneous: according to Cribb et al. (2001b) it includes Dicrocoeliidae, Gorgoderidae, Bucephalidae, Paragonimidae and Opecoelidae. The cluster is completed with a group of taxa related to Plagiorchiidae.

Closely following the above scheme, it is difficult to reveal any general patterns in the distribution of the main characters pointing to the level of specialization in the main taxa of modern trematodes. For instance, we have already noted that the "sporocyst-bearing" basal group Strigeidae—Schistosomatidae precedes a long series of "redial" trematodes, which ends only with the family Acanthocolpidae. However, several sporocyst-bearing families are "inserted" into this series. Sporocyst-bearing Fellodistomidae and Heronimidae precede redial Azygiidae and the families of hemiuroid complex. On the contrary, redial Paragonimidae fall into the same cluster with sporocyst-bearing Bucephalidae and Opecoelidae. Whereas it can be supposed that sporocyst-bearing forms emerged in "redial" groups independently several times, an opposite process (sporocysts becoming rediae) is unlikely. It contradicts Dollo's Law of irreversible evolution.

The distribution of the character "the excretory system type" ("stenostomate" or "mesostomate") hardly yields to analysis. All representatives of the complex Strigeidae – Schistosomatidae possess a secondarily simplified mesostomate excretory system. In the middle of the scheme, digeneans with the more archaic stenostomate type of the excretory system dominate. However, there are exceptions. For instance, Notocotylidae (with mesostomate excretory system) precede several clusters of typically "stenostomate" forms (Paramphistomidae, Haplosplanchnidae, Echinostomatidae, Fasciolidae). The same pattern is repeated in the next cluster: heterophyoid complex of families and Lepocreadiidae (with mesostomate excretory system) are in the same series with Haploporidae (with typically stenostomate excretory system). Moreover, these families occupy a basal position as to Acanthocolpidae which also possess the "stenostomate" excretory system.

A similar pattern is revealed in the distribution of other characters. For instance, the position of primary excretory pores (an important character of body plan organization) and the shape of the tail agrees well with the upper, basal part of the scheme where true furcocercariae predominate. Though the position of two groups, Fellodistomidae and Heronimidae, seems controversial to us, this is mostly due to insufficient data on development and structure of the excretory system in their cercariae. On the contrary, in the lower part of the scheme, where the most evolutionary advanced taxa are situated, the authors did not avoid contradictions. For instance, Lepocreadiidae precede Haploporidae in the phylogenetic scheme, though in the former primary excretory pores are located at the cercarial tail base (Peters, 1961) and in the latter, laterally on the tail (Cable and Isseroff, 1969). For the same reasons it is questionable to place Bucephalidae, with their highly specialized cercariae (Matthews, 1973a,b, 1974), into the group of families in whose cercariae primary excretory pores are located at the tail base (see p. 449).

Finally, the central (i.e., rather high) position in the scheme is occupied by a group of families whose representatives possess obligatory two-host life cycles with the phase of adolescaria (Notocotylidae, Paramphystomidae, Philophthalmidae, Haplosplanchnidae) or those that only start to colonize the second intermediate host (Echinostomatidae). Assuming this scheme, we have to choose between two working hypotheses. The first hypothesis postulates that the most ancient trematodes initially possessed three-host life cycles, a biological feature retained by the representatives of the basal group Strigeidae – Schistosomatidae. In this case we must admit that the ancestors of modern trematodes lost the ability to use the second intermediate host several times in the course of evolution. The larvae of Notocotylidae and subsequent families lost the penetration apparatus but acquired a secondary complex of cystogenous structures. Correspondingly, larval behaviour in the environment also altered (see p. 279).

In our opinion, this sequence of events is impossible. Therefore, we have to suppose that a series of forms substituting each other existed for a long evolutionary time. These initial forms were characterized by obligatory two-host life cycles, redial state and a number of other archaic morphological and biological features. Forms that were to give rise to clusters of modern families branched successively from the initial forms. Three-host life cycles and corresponding morphological and biological adaptations were acquired independently in every branch, and rediae were substituted by the daughter sporocysts independently.

It catches the eye that the basal position (on the whole and for separate clusters) in the scheme discussed is occupied by the most specialized groups: Strigeidae – Schistosomatidae for all other trematodes, Notocotylidae for Paramphystomidae, Cyclocoelidae for Fasciolidae. Unfortunately, there are more examples such inconsistencies.

It is interesting to compare the phylogenetic scheme of Cribb et al. (2001b) and that suggested by Tkach et al. (2001) who applied molecular data (data on the sequencing of the fragment localized at the 5′ end of the 28S rDNA). On the whole, the two schemes are similar. The basal position in both is occupied by sporocyst-bearing furcocercous groups (Schistosomatidae and Strigeidae) which are followed by two related families, Echinostomidae and Psilostomidae and by Notocotylidae. Further upwards the schemes there is a group of specialized sporocyst-bearing forms: Monorchiidae, Opecoelidae and a clade that, surprisingly, unites Encyclometridae, on the one hand, and Dicrocoeliidae and Gorgoderidae as sister groups, on the other hand. Finally, there is a group comprising *Astiotrema monticelli* and two undoubtedly related families, Nanophyetidae and Troglotrematidae. This cluster is considered as basal for two sister clusters that in the opinion of Tkach et al. (2001) make up the suborder Plagiorchiata.

Differences between the scheme compared are not essential. In fact, they share the same drawback: basal position of the most specialized taxa

(Schistosomatidae, Strigeidae, Diplostomidae). The main groups of redial and sporocyst-bearing trematodes have the same position in the scheme of Cribb et al. (2001b) and that of Tkach et al. (2001). However, the order of some groups in the latter scheme (Tkach et al., 2001) seems preferable. For instance, Echinostomatidae and Psilostomidae occupy a basal position to the Notocotylidae.

Noteworthy, Tkach et al. (2001) stated that the composition of groups in their scheme could be argued. In our opinion, some of their clusters are indeed heterogeneous. For example, the group Encyclometridae–Gorgoderidae–Dicrcoeliidae seems very artificial. Unfortunately, the life cycle has not been elucidated in any of the *Encyclometra* species. We do not even know whether these trematodes are redial or sporocyst-bearing. However, the little knowledge we have does not allow us to consider Encyclometridae as related to the Gorgoderidae and Dicrcoeliidae.

Firstly, differences between Encyclometridae and the other two families concern miracidial morphology and biology. According to our observations, miracidia of *E. colubrimurorum* have an entirely different organization than those of Gorgoderidae and especially of Dicrocoeliidae. Large miracidia of *E. colubrimurorum* actively infect the host. They possess a 4-nuclei penetration gland (see fig. 8 in Ginetsinskaya, 1968), four rows of epithelial plates and other traits characteristic of archaic larvae (see p. 408). Their development is completed in the marita uterus (see p. 219). Miracidia of Gorgoderidae possess features pointing to secondary simplification and specialization, even though they are hatched from the eggs in water (Goodchild, 1948). Miracidia of Dicrocoeliidae are even more specialized (Tang, 1950). To conclude, miracidia of each of the three families mentioned demonstrate a characteristic organization. Judging from miracidial organization, Encyclometridae seem to be related to archaic taxa, whereas Dicrocoeliidae are much closer to Plagiorchiata *sensu* Tkach et al. (2001).

There are even greater differences between the hermaphroditic generation of the three families under consideration. They can be illustrated by the analysis of the excretory system structure. The excretory system of *Encyclometra* (Odening, 1969) is unique. No such excretory system has been found in any other trematodes studied. Data on the excretory system in Gorgoderida are scarce but most probably it is stenostomate (Sinitsin, 1905; Krull, 1935a; Byrd et. al., 1940). The excretory system of Dicrocoeliidae is mesostomate. The excretory formula of *Dicrocoelium dendriticum* and *Corrigia corrigia* is $2[(2 + 2 + 2) + (2 + 2 + 2)] = 24$, i.e., it is "complete", but with a tendency towards juvenilization. In *Eurytrema pancreaticum* whose cercariae possess an "incomplete" excretory formula, this tendency is even more manifest (see p. 131).

The excretory system of Gorgoderidae contains characteristic flame cells. They are very large and mushroom shaped (Krull, 1935a; Sinitsin, 1905). Such flame cells are uncommon in trematodes. In particular, they are known from the daughter sporocysts of *Sphaerosotoma bramae* (Opecoelidae) (see fig. 35 in Ginetsinskaya, 1968), but nothing of the kind has been found in Dicrocoeliidae. Thus, morphological data available to-date demonstrate that the group of families considered is heterogeneous. A taxon of a relatively low level (superfamily – suborder) is highly unlikely to include forms that differ in respect to several important characters (the type of miracidium and its organization, the excretory system type). We do not know any examples of this kind.

As a reason for exclusion of Dicrocoeliidae out of "Plagiorchiata", Tkach et al. (2001) listed several morphological characters inherent, in their opinion, to Dicrocoeliidae and not characteristic of "all plagiorchiates" (see p. 191 in Tkach et al., 2001). Firstly, in maritae of Dicrocoeliidae the ovary is situated behind the testes, whereas in the representatives of the suborder "Plagiorchiata" it is situated before them. However, the anterior position of the ovary in maritae of Gorgoderidae did not prevent the authors (Tkach et al., 2001) from placing this family in the same group as Dicrocoeliidae. Secondly, Tkach et al. (2001) noted to the absence of the gut in cercariae of Dicrocoeliidae. This must be a misunderstanding. Branches of the gut are indeed underdeveloped (i.e., they have no lumen) in cercariae of the genus *Dicrocoelium* (see p. 130). Other cercariae from this family do have a lumen at least in the proximal part of the gut branches. Contrary to Tkach et al. (2001), the gut branches are not an "obligatory" feature in the larvae of Plagiorchiata *sensu* Tkach et al. (2001). The degree of development of the cercarial digestive system may be different not only in species belonging to different families but within one family as well. For instance, cercariae from various morphological groups of xiphidiocercariae (armatae, ornatae, virgulae) can be easily arranged into series. At one end of the series there would be forms with fully differentiated digestive system, at the other, larvae with only the pharynx (at best) left from it. The latter variant is common in families whose representatives are juvenilized: Prosthogonimidae, Allasogonoporidae, Lecithodendriidae. It is even more pronounced in Microphallidae. In their cercariae not only the formation of the reproductive system organs but also that of the ventral sucker is retarded. These highly juvenilized xiphidiocercariae are sometimes united into a separate morphological group "ubiquita" (for review see: Belopolskaya, 1963).

At present only one reliable difference between Dicrocoeliidae and Plagiorchiata *sensu* Tkach et al. (2001) is known. It concerns the morphology of the daughter sporocysts. Noteworthy, the daughter sporocysts from the genera *Corrigia* and *Eurytrema* (see fig. 24G) are highly specialized and secondarily altered. It is parthenitae from the genus *Dicrocoelium* that retain the most archaic type of organization. Their anterior body end bearing the birth pore is elongated and a long and narrow canal passes there, connecting the pore and the body cavity. A similar structure of anterior body end is observed in young daughter sporocysts of Opecoelidae (*Sphaerostomum bramae*). Regularly cylindrical daughter sporocysts of *Gorgodera amplicava* also taper towards the anterior end, as noted by Krull (1935a). At the same time, in those Plagiorchiata *sensu* Tkach et al. (2001) that have the most archaic parthenitae, daughter sporocysts are irregular in shape, sac-like and immobile. They have thin walls and a hardly discernible birth pore. The parthenitae from the genus *Dicrocoelium* are evidently less specialized and closer to the initial prototype (see p. 422).

Thus, there is no reliable morphological evidence to justify the separation of Dicrocoeliidae from a complex of taxa united under the common name of "Plagiorchiata". It is also questionable to unite forms with an essentially different excretory system (Encyclometridae, Gorgoderidae and Dicrocoeliidae) into a taxon of a relatively low level (superfamily?).

The next group distinguished by Tkach et al. (2001) is also controversial. It includes Astiotrematidae (?), Nanophyetidae and Troglotrematidae. The latter two families may indeed be related. However, Astiotrematidae is unlikely to be related to them and certainly cannot be their basal group. Tkach et al. (2001) offered two variants: either Astiotrematidae are unrelated to Plagiorchiata or they are a heterogeneous group with polyphyletic origin. In our opinion, the latter variant is correct. Unfortunately, the life cycle of only one species, *Astiotrema trituri*, has been studied in detail (Grabda, 1959; Gerasev and Dobrovolskij, 1977) but these investigations unambiguously show that *A. trituri* belongs to Plagiorchiata. Its miracidial morphology, structure and development of daughter sporocysts and cercarial morphology (*A. trituri* has large xipphidiocercariae from the group "armata") are characteristic of typical plagiorchiates. Moreover, at the scheme proposed by Tkach et al. (2001) *Astiotrema trituri* should be placed between Leptophallidae and Telorchiidae. To determine the systematic position of *A. monticelli* further investigations, especially those elucidating its life cycle, are necessary.

The above discussion demonstrates that methodical approaches currently used for analysis of trematode phylogeny fail to offer satisfactory solutions of the existing problems. Cladistic analysis, which has lately gained wide popularity, is based on the maximum parsimony principle. Its possibilities for revealing actual processes of phylogenesis of taxa of high level (groups of phyla, phyla, classes) are very limited. The "shortest trees" can hardly be considered as true phylogenetic schemes. They are in fact typical topological series demonstrating the degree of relative relatedness or unrelatedness of the taxa. The diversity of results rendered by cladistic analysis shows, in our opinion, that an objective view on phylogenesis of both parasitic Platyhelminthes and the taxon Trematoda is yet to be elaborated.

Comparison of the phylogenetic schemes given by Cribb et al. (2001b) and Tkach et al. (2001) and the one presented earlier by Brooks et al. (1985a, 1989) reveals one more tendency. The best topology results from the application of only one molecular character (Tkach et al., 2001). When data on morphology of different phases of the life cycle are used together with molecular ones, the results are much more contradictory (Cribb et al., 2001b). Finally, the scheme proposed by Brooks et al. (1985a, 1989), based upon the analysis of morphological characters only, seems unacceptable to us (for a critical review of this article see: Pearson, 1992; Galaktionov and Dobrovolskij, 1998).

In our opinion, the reason behind the contradictions is an arbitrary selection of the characters used for analysis. In fact, criteria for the selection have never been elaborated. No precise estimation of the evolutionary age of characters or the degree of their conservancy is given. Nobody has ever attempted to conduct an analysis of body plan organization of modern trematodes and to compare them with other parasitic flatworms. A strict and unambiguous estimation of plesiomorphic and apomorphic states for most morphological characters is lacking.

The above considerations made us suggest our own approach to the analysis of actual phylogeny of trematodes and the elaboration of their natural classification. General principles determining our approach to the elaboration of natural classification of trematodes have already been discussed in this book. They may be summarized as follows.

1. Ancestral forms (proto-trematodes) and the most ancient trematodes (common ancestors of most modern taxa) have not been retained (Cable, 1974).

2. Modern trematode taxa are evolutionary rather "young". The most ancient of them formed as late as in the middle or the second half of the Mesozoic and most representatives of "higher" trematodes emerged in late Mesozoic or Cainozoic.

3. The present-day parthenitae (in particular, the generation of the mother sporocyst) were the initial generation in the trematode life cycle (Ginetsinskaya, 1968; Pearson, 1972; Dobrovolskij et al., 1983).

4. The miracidium is the most ancient phase in the trematode life cycle. Miracidial morphology and biology is an important criterion for estimating the degree of specialization of taxa and establishing its phylogenetic relationships with other groups. The organization of miracidia is a very conservative character characterizing taxa of the family and superfamily level (Dobrovolskij, 1965). The miracidia of the most specialized groups are secondarily simplified and consist of only a few cells. They may be similar externally but retain some characteristic features of organization.

5. Redial state of the second parthenogenetic generation is primary. Substitution of rediae by the daughter sporocysts took place several times (at least, four) in trematode evolution.

6. Taxa of higher level should be elevated on the basis of the most ancient characters reflecting traits of body plan organization of ancestral forms. In trematodes the most important characters of this kind are: characters associated with the excretory system (the position of primary excretory

pores, the stenostomate or mesostomate type of the excretory system); the proportion between the pre-poral and post-poral parts of the cercarial body; the position of the anterior boundary of the tail primordium as to the excretory system and the mode of tail primordium formation associated with it (Cable, 1977). The shape and size of the tail, its cystocercous state, the presence of fin-folds and bristles are secondary adaptive characters. They should be applied for distinguishing taxa of relatively low level (e.g., families).

7. Characters associated with the reproductive system (the presence of certain organs and their topography), that are traditionally considered as important in the trematode classification, should be mostly used at the level of superfamilies, families and genera.

8. The phenomenon of heterochrony played a very important role in the morphological evolution of trematodes. This phenomenon entailed widespread juvenilization and miniaturization, especially among the most specialized trematodes (Galaktionov and Dobrovolskij, 1987, 1998).

9. Analyzing phylogenetic relations between taxa, it is important to consider the events in the life cycles: the presence of certain generations and phases and their order, as well as the number and order of the hosts. It should be remembered that two-host life cycles, widespread in trematodes, have different origins (see p. 334), and so do four-host life cycles (see p. 346).

10. Two different phenomena should be distinguished: reduction of transmission ways of parasites in the ecosystem (expressed as the reduction in the number of animal hosts and the emergence of secondary two-host and even one-host life cycles) and a true reduction of the life cycle (the reduction of certain phases and generations) (see p. 338). Both phenomena may be obligatory, genetically fixed characters or facultative characters, manifestations of a broad environmentally induced variation.

11. The Szidat Rule, postulating that primitive parasites are associated with ancient and archaic groups of animal hosts, can not be applied to trematodes. This is true of animal hosts of all categories. On the whole, trematode evolution is rather "autonomous". It is not closely associated with the evolution of some group of animals, including vertebrates.

Ability to encyst, on the one hand, and furcocercous cercariae with intermittent type of locomotion, on the other hand, were the autapomorphies that split trematodes into two large clades. The first clade comprises trematodes with true furcocercariae. Noteworthy, "morphological" furcocercariae, whose posterior tail part consists of two branches, emerged at least three times independently in different trematode groups. In "true" furcocercariae the distal parts of the main collecting ducts of the excretory system are initially located in the tail (always entering the furcal branches) and so is the posterior primary group of flame cells. The former criterion is universal. If the distal parts of the main collecting ducts do not enter the tail, cercariae cannot be considered as true furcocercariae. Applicability of the latter criterion depends on the degree of cercarial juvenilization (see p. 444). In small, secondarily specialized cercariae with an "incomplete" excretory formula all the flame cells are situated outside the tail stem (e.g., Aporocotylidae, Sanguinicolidae).

Trematodes with true furcocercariae fall into two groups differing as to the proportion between the pre-poral and post-poral body parts in the tail (see p. 444). It is externally expressed in different position of the primary excretory pores on the furcae. In representatives of the first group excretory pores are initially located terminally at the furcal tips. If they are slightly shifted, they may be shifted only to the posterior furcal surface. The second group comprises taxa whose larvae bear primary excretory pores on the frontal furcal surface. The first group is rather heterogeneous. It comprises very different families: Bivesiculidae, Transversotrematidae, Azygiidae, Clinostomidae and, finally, Aporocotylidae, Sanguinicolidae, Spirorchiidae and Schistosomatidae, that are often assigned the level of the order Schistosomatida. The second group is more homogeneous. It corresponds to the well-defined order Strigeidida (families Cyathocotylidae, Diplostomidae, Strigeidae). In representatives of Gymnophallidae and Brachylaimoidea primary excretory pores are located at the furcal tips but the excretory formula is "incomplete".

Azygiidae, Bivesiculidae, Transversotrematidae, Gymnophallidae and Brachylaimoidea possess the stenostomate excretory system, whereas all other families listed have the mesostomate excretory system.

In the first group Azygiidae and possibly Bivesiculidae retain primary two-host life cycle and infect the definitive host *per os*. Two-host life cycles of Transversotrematidae and all Schistosomatida are secondary, arising from progenetic development of metacercariae. All other families listed possess typical three-host life cycles. Some strigeidids and diplostomids possess secondary four-host life cycles.

Analysis reveals a very unexpected distribution of plesiomorphic and apomorphic characters across the phases of the life cycle. Representatives of most families in the furcocercous clade possess large actively infecting miracidia consisting of many cells. On the whole, they have a very similar body plan. The same organization is characteristic of miracidia of trematodes from the main groups of the second clade. Even miracidia of Azygiidae retain the major features characteristic of the group, though they infect the molluscs passively and are therefore secondarily simplified.

Data on mother sporocysts from the families listed are very scarce and do not yield to analysis. In the few cases when mother sporocysts are mentioned at all (Schistosomatidae, Cyathocotylidae, Diplostomidae), they are described as worm-like, sometimes rather large organisms with a well-developed tegumenary-muscular sac and the birth pore. Their germinal masses function for a long time, and the reproductive function is prolonged. So far only one exception is known: small mother sporocysts of Transversotrematidae. Only one individual of the next parthenogenetic generation develops in them.

Daughter generations of parthenitae in the group under consideration are represented both by rediae (Bivesiculidae, Transversotrematidae, Azygiidae, Aporocotylidae, Clinostomidae, Gymnophallidae) and by sporocysts (all other families mentioned above). Daughter sporocysts are usually worm-like, sometimes filiform. The most specialized small spherical daughter sporocysts are characteristic of some Sanguinicolidae.

Furcocercariae are a heterogeneous group morphologically. Variability in the position of excretory pores has already been discussed. The distribution of other characters is even more irregular. Cercariae from three families (Bivesiculidae, Transversotrematidae, Azygiidae and some Strigeidida) belong to the morphological group of megalocercous or cysticercous larvae. The larvae of Clinostomidae, Schistosomatida and most Strigeidida are typical furcocercariae with developed tails. A tendency towards reduction of the tail size and the loss of locomotory function by the tail is pronounced in Gymnophallidae and Brachylaimoidea.

According to the excretory system structure and some other characters, cercariae of Transversotrematidae, Azygiidae and many Cyathocotylidae are the least juvenilized. Cercariae are greatly juvenilized in most strigeidids and schistosomatids. The most juvenilized cercariae are characteristic of sanguinicolids and gymnophallids. Unfortunately, we cannot judge about the degree of juvenilization of bivesiculid cercariae due to paucity of data. Many researchers think that the structure of excretory system in bivesiculid cercariae points to the fact that bivesiculids are archaic. In our opinion, an opposite interpretation is just as probable. All parts of their protonephridia are paired, which may be a manifestation of an extreme juvenilization (also retained at the phase of marita) and hence to specialization of this family.

Absence of suckers in hermaphroditic generation in bivesiculids may also point to juvenilization. At present there are no grounds to consider the absence of suckers as a plesiomorphic character.

Azygiid cercariae lack penetration apparatus, which is not surprising. There is no penetration apparatus in transversotrematid cercariae, either. Progenetic "maritae" of transversotrematids are localized under the scales of fishes. In these cases absence of penetration apparatus is also most probably a plesiomorphic state of the character. The colonization of the second intermediate host does not necessarily entail the acquisition of true penetration glands and associated structures. Transversotrematids appear to have stopped at the same evolutionary stage as Echinostomatidae. On the contrary, Schistosomatidae and Strigeidida possess well-developed complex penetration apparatuses.

Almost all representatives of the large clade in question, including the forms with a typical three-host life cycle (see p. 157), lack the ability to encyst in the second intermediate host.

Sexually mature individuals of hermaphroditic generation of these trematodes are specialized in various degrees. Azygiids and clinostomids are the least specialized forms. Transversotrematidae and Schistosomatidae are progenetic metacercariae (see p. 338), and as such should be considered as secondarily modified forms. Moreover, schistosomatids are blood parasites, which is not characteristic of any other trematodes. Maritae of strigeidids, with their "two-segment" body and an organ of extra-intestinal digestion (adhesive organ) belong to the most specialized trematodes.

To summarize, all the families discussed above demonstrate characters pointing to deep specialization. It means that the main phylogenetic branches in the furcercarious clade split very early in the evolution. At the same time, they are undoubtedly related. Therefore they should be grouped together into a taxon of a high level. We suggest for these groups the subclass level with a provisional name "Furcocercariata".

This subclass comprises several distinct groups of different size: azygiids, transversotrematids, bivesiculids, gymnophallids, brachylaimoids, schistosomatids (clinostomids seem to be related to the latter according to cercarial morphology) and stigeidids.

Several other groups of families are probably related to "Furcocercariata". Firstly, it concerns hemuiroid complex together with Didymozoidae and Fellodistomidae. Tandanicollidae are related to the latter. All these trematodes are specialized, which makes it difficult to establish the degree of their relatedness with other "Furcocercariata". Data available on the morphology and reproductive dynamics of parthenitae and on morphogenesis and structure of cercariae are insufficient to verify the systematic position of these groups. Therefore we propose to consider them as groups of unclear systematic position (*incerta sedis*) within "Furcocercariata".

As noted above, trematodes from the second large clade used the mechanism of encystment for prolongation of the time of their existence in the environment. Modification of the tail was not necessary for this and the tail retained its initial simple structure. Only the most distant parts of the main collecting ducts enter the tail, where they form the caudal excretory canal. Primary excretory pores are situated laterally on the tail. Excretory formula in all the species studied in this respect is "complete", with all primary groups of flame cells situated in the larval body. We denote this state as a "protocercous" one.

Primary two-host life cycles with the phase of adolescaria are found in representatives of three groups of families from "protocercous" clade. The first group comprises Fasciolidae, Echinostomidae *sensu lato*, Psilostomidae and Philophthalmidae, and a closely related family Cyclocoelidae. Parasites of fishes Haploporidae, Megaperidae and Haplosplanchnidae are also situated rather close to this group. The second group comprises Paramphistomidae *sensu lato* and Mesometridae (apparently, rather closely related to the former). The third group consists of Notocotylidae *sensu lato*.

Trematodes from all the three groups mentioned are characterized by redial state of the second parthenogenetic generation. Their cercariae (with the exception of Notocolylidae *sensu lato*) possess the stenostomate excretory system.

Attempts to assign to Notocotylidae *sensu lato* the role of the basal group for all other families listed (Cribb et al., 2001b; see also p. 511, 515) seem fruitless to us. In our opinion, fasciolid–echinostomatid complex is more archaic. Notocotylidae *sensu lato* and Paramphistomidae *sensu lato* with related families possess more characters pointing to a very great specialization (see pp. 509, 510). We may distinguish either three (see above) or four groups of "protocercous" trematodes. In the former case only features of body plan organization of the hermaphroditic generation are used. In the latter case we should also take into consideration that maritae from three families (Haploporidae, Megaperidae and Haplosplanchnidae) parasitize only fishes, whereas Fasciolidae, Echinostomatidae *sensu lato*, Psilostomidae and Philophthalmidae are associated with warm-blooded vertebrates.

The family Acanthocolpidae and the heterophyoid complex of families (Heterophyidae, Opisthorchiidae, Cryptogonimidae) are also characterized by the redial state of the second parthenogenetic generation. However, they possess obligatory three-host life cycles. The body plan characteristics of these two groups are similar (they possess "protocercous" cercariae). At the same time, there is an important difference between them. Whereas Acanthocolpidae possess the stenostomate excretory system, Heterophyidae *sensu lato* possess the mesostomate one. Therefore they should be considered as independent taxa of the same level.

The families listed above possess similar characteristics of body plan organization and an obligatory redial state of the second parthenogenetic generation. We suggest that they should be grouped together into a taxon of a high level, provisionally called "Protocercariata".

The third large clade to be distinguished according to the result of analysis of body plan organization of the individuals of the hermaphroditic generation is characterized by an independent formation of the tail primordium and the organs of the excretory system (see p. 438). None of the elements of the excretory system enter the tail at any stage of ontogenesis: primary excretory pores are located at the tail base. Similarly to "Furcocercariata", this group is morphologically and biologically diverse.

In Lepocreadiidae, Allocreadiidae and Paragonimidae the second parthenogenetic generation is represented by a rediae. However, in most of the families (Gorgoderidae, Opecoelidae, Plagiorchiata *sensu* Tkach et al., 2001) parthenogenetic generations are sporocysts specialized in various degrees. Gorgoderidae and possibly also Lepocreadiidae possess the stenostomate excretory system. All other families from this clade have the mesosto mate excretory system. Life cycles of all trematodes from this clade are initially obligatory three-host ones. However, transition to various forms of two-host life cycles due to reduction of transmission ways is often observed (see p. 345).

On the whole, the third clade is very heterogeneous. It may be due to the fact that the tail primordium and the primordium of the excretory system were spatially separated several times in the course of evolution. If this is true, this group is polyphyletic. However, an alternative variant also cannot be ruled out. These forms might have had a common ancestor but several "sister" clades formed as the result of very early divergence and colonization of different ecological niches. To choose between these alternatives, it is necessary to conduct a detailed study of life cycles of these trematodes and molecular investigations of a greater number of objects. Until this has been done, we suggest to place all these trematodes into the group "Xiphidiocercariata".

In this brief review we have mentioned only the largest of trematode families. However, two more families deserve special mention. The family Bucephalidae is one of the most controversial groups of modern digeneans. They were often associated with gymnophallids and hemiuroid trematodes. However, data available in the literature (Matthews, 1973a,b, 1974, Podvyaznaya and Galaktionov, a,b, *submit*) and results of our own observations show that Bucephalidae are not related to "Furcocercariata". Cribb et al. (2001b) united bucephalids and paragonimids. In our opinion, this is not convincing. Morphological and biological characters (especially the position of the mouth opening on the body of hermaphroditic individuals) show that Bucephalidae are a very original group. Their most characteristic features are as follow.

Passively infecting miracidia of bucephalids externally resemble those of plagiorchiids. However, there is an essential difference. Bucephalid miracidia possess a long stylet located at an angle to the longitudinal body axis in the sagittal plane. Bucephalid sporocysts are very large and branch extensively. They are highly specialized and appear to possess a regenerative potential. Studies of serial histological sections of molluscs with a mature infection always reveal young parts of sporocysts, that have not or have just started reproduction, close to degenerating ones (see p. 50).

Organization of hermaphroditic generation also sets bucephalids apart from other trematodes. In particular, the position of the mouth opening and the structure of the digestive system are very original (see p. 453). Literature data (Matthews, 1973a,b, 1974) and out own observations show that at early stages of morphogenesis the tail primordium includes the distal parts of the main collecting tubes. This shows that Bucephalidae are not related to the group "Xiphidiocercariata" (Cribb et al., 2001b). At the same time, bucephalid cercariae are only slightly juvenilized. Their excretory formula consists of all the six primary groups of flame cells (see p. 445). The posterior primary group of flame cells is located in the larval body and not in the tail. Therefore Bucephalidae cannot be included into "Fucrocercariata" (La Rue, 1957; Cable, 1974).

The above combination of characters – the presence of the main collecting tubes in the tail and the absence of the posterior primary group of flame cells there – is characteristic only of "Protocercariata". However, all other traits of organization and biology of bucephalids point to a great specialization. Therefore Bucephalidae cannot be placed into "Protocercariata". In this connection it should be remembered that Poche (1926) distinguished the subclass Gasterostomata. Poche (1926) and later Odening (1961, 1974) considered the Gasterostomata as entirely different from all other trematodes. Though we don't share the latter viewpoint, in our opinion, Bucephalidae should be placed into an independent taxon of the same level as the three large clades mentioned above. It seems expedient to leave the old name Gasterostomata for this subclass.

The second group deserving special mention is the family Heronimidiae. It has often been noted in the literature that this taxon is archaic (see p. 485). The analysis of their life cycle phases (Crandall, 1960) has demonstrates that Heronimidae, though undoubtedly ancient, are a highly specialized group.

Heronimids possess a large actively infecting miracidium. It consists of many cells, has a well-developed body cavity (schizocoel) with a typical attached germinal mass in it. The germinal mass is retained for some time in young sporocysts. As they develop, they acquire a characteristic multi-branched shape (see pp. 37-8 and fig. 17). This is the mother sporocyst, the only parthenogenetic generation in the life cycle of *Heronimus mollis*. Its cercariae do not leave the mollusc, yet retain all the characters of free-swimming larvae. The cercariae are secondarily megalocercous. Their simple non-branching tail contains the distal parts of the excretory system ducts and, importantly, the posterior primary group of flame cells. Elements of the excretory system are undoubtedly present in the tail, even though some details of their structure described in Crandall (1960) are questionable. We agree with Cribb et al. (2001b) that heronimids should be placed into the furcocercous complex of species. Heronimidae as well as Fellodistomidae probably lost furcocercous state of the tail secondarily.

Concluding this brief review, we emphasize that there is no character or method that could be used alone as a panacea for elaboration of a natural classification of trematodes.

References

Abbas, M.K., Cain, G.D. Actin and intermediate sized filaments of the spines and cytoskeleton of *Schistosoma mansoni*. Z Parasitenk 1987; 73:66-74.

Abdul-Salam, J., Sreelatha, B.S., Ashkanani, H. Seasonal prevalence of trematode cercariae in *Clypeomorus bifasciata* (Gastropoda: Prosobranchia) in Kuwait Bay. Folia Parasitol 1994; 41:247-52.

Adams, J.E., Martin, W.E. Life cycle of *Himasthla rhigedana* Dietz, 1909 (Trematoda: Echinostomatidae). Trans Am Microsc Soc 1963; 82:1-6.

Adema, C.M., Locker, E.S. "Specificity and Immunobiology of Larval Digenean-Snail Associations." In *Advances in Trematode Biology*, B. Fried and T. Graczyk, eds. Boca Raton, New York: CRC Press, 1997.

Agrawal, N., Pandey, K.C. A new record of intramiracidial redia. J Helminthol 1979; 53: 130-1.

Albaret, J.L. Structures argyrophiles superficielles do miracidium de *Fascioloides magna* (Bassi, 1875) (Trematoda, Fasciolidae, Fasciolinae). Ann Parasitol Hum Comp 1985; 60:213-6.

Alderson, G.D. Ultrastructure and histochemistry of the metacercarial cyst of *Spelotrema nicolli* (Microphallidae: Trematoda). Int J Parasitol 1975; 5:659-65.

Allison, L.N. *Leucochloridiomorpha constantiae* (Mueller) (Brachylaimidae), its life cycle and taxonomic relationships among digenetic trematodes. Trans Am Microsc Soc 1943; 62:127-68.

Al-Salman, H.A., James, B.L. Ultrastructure of the tegument in daughter sporocyst and cercarial embryos of *Meiogymnophallus minutus* (Digenea: Gymnophallidae). Int J Parasitol 1988; 18:231-42.

Al-Salman, H.A., James, B.L. Ultrastructural studies on the daughter sporocysts of *Meiogymnophallus minutus* (Digenea: Gymnophallidae). Int J Parasitol 1989; 19:499-508.

Ameel, D.J., Cort, W.W., Van der Woude, A. Germinal development in the mother sporocyst and rediae of *Halipegus eccentricus* Thomas. J Parasitol 1949; 35:569-78.

Anderson, D.J., Cable, R.M. Studies on the life history *Linstowiella szidati* (Anders.). J Parasitol 1950; 36:395-410.

Anderson, M., Anderson, F. Life history of *Proterometra dickermani* Anderson, 1962. J Parasitol 1963; 49:275-80.

Ataev, G.L., Dobrovolskij, A.A. Development of the microhemipopulation of parthenitae of the trematode *Philophthalmus rhionica*. Parazitologiya 1990; 24:499-508 (in Russian).

Ataev, G.L., Avanessian, A.V., Loker, E.C., Dobrovolskij, A.A. The organization of germinal material and dynamics of mother sporocyst reproduction in the genus *Echinostoma* (Trematoda: Echinostomatidae). Parazitologiya 2001a; 35:307-19 (in Russian).

Ataev, G.L., Dobrovolskij, A.A., Avanessian, A.V., Loker, E.C. Germinal elements and their development in *Echinostoma caproni* and *Echinostoma paraensei* (Trematoda) miracidia. J Parasitol 2001b; 87:1160-4.

Ataev, G.L., Dobrovolskij, A.A., Fournier, A., Jourdane, J. Migration and development of mother sporocyst of *Echinostoma caproni* (Digenea: Echinostomatidae). J Parasitol 1997; 83:444-53.

Austin, F.G., Stirewalt, M.A., Danzinger, R.E. *Schistosoma mansoni* : stimulatory effect of rat skin lipid fractions on cercarial penetration behavior. Exp Parasitol 1972; 31:217-24.

Austin, F.G., Frappaolo, P., Gilbert, B., Landis, W., Da Rosa ,M.N., Stirewalt, M.A. Further studies of *Schistosoma mansoni* cercarial stimulation by crude egg lecithin and other lipids. Parasitology 1974; 69:455-63.

Awad, A.H.H., Probert, A.J. Scanning and transmission electron microscopy of the male reproductive system of *Schistosoma margrebowiei* Le Roux, 1933. J Helminthol 1989; 63:197-205.

Awad, A.H.H., Probert, A.J. Scanning and transmission electron microscopy of the female reproductive system of *Schistosoma margrebowiei* Le Roux, 1933. J Helminthol 1990; 64:181-92.

Azevedo, C., Corral, L. Ultrastructural and cytochemical observations of the spherites in a parasite trematode larva. J Submicrosc Cytol 1987; 19:445-64.

Bakke, T.A. Histology and biology of the larval stages of *Leucochloridium* Carus, 1835 (Trematoda, Digenea) as revealed by light and electron microscopy. Fauna Norv, Ser. A 1982; 3:41-56.

Bargues, M.D., Mas-Coma, S. Intracellular development in digenean sporocyst early stages. Research and Reviews in Parasitology 1991; 51:111-24.

Bargues, M.D., Mas-Coma, S., Llueca, M.E., Oviedo, J.A., Garulo, R., Jabbar, I., Camps, S. Early developing stages of sporocysts in species of the family Brachylaimidae (Digenea). Bull Soc Fr Parasitol 1990; 8:309.

Bartoli, P. La pénétration et l'installation des cercaires de *Gymnophallus fossarum* P. Bartoli, 1965 (Digenea, Gymnophallidae) chez *Cardium glaucum* Bruguiee. Bull Mus Hist Natur, Paris, 3e sér. N 117, janv.-fevr. 1973; Zoologie 91:319-34.

Bartoli, P. Démographie et phénomène de conpétition intraspécifique des Gymnophallidae Morosov, 1955 chez les deuxièmes hôtes intermédiaires (Trematoda, Digenea). Ann Parasitol Hum Comp 1981; 56:33-44.

Bartoli, P. Populations ou espéces? Recherches sur la signification de la transmission de Trématodes Lepocreadiinae (T. Odhner, 1905) dans deux écosystèmes marins. Ann Parasitol Hum Comp 1983; 58:117-39.

Bartoli, P. Rediscription de *Cercaria setifera* F.S. Monticelli, 1914 (nec J. Muller) (Trematoda) parasite de *Conus ventricosus* Hwass; comparaison avec quelques cercaires ophtalmotrichocerques de Méditerranée occidentale. Ann Parsitol Hum Comp 1984; 59:161-76.

Bartoli, P. Role favorisant de l'hydrodynamisme dans la contamination des deuxièmes hotes intermédiaires par les cercaires de *Maritrema missenensis* (A. Palombi, 1940) (Digenea, Microphallidae). Ann Parsitol Hum Comp 1986; 61:35-41.

Bartoli, P., Combes, C. Stratégies de dissémination des cercaires de trématodes dans un écosystème marin littoral. Acta Ecol 1986; 7:101-14

Bartoli, P., Holmes, J.C. A transmission study of two sympatric digeneans: spatial constraints and solutions. J Helminthol 1997; 64:169-75

Bartoli, P., Prévot, G. Stratégies d'infestation des hotes cibles chez les trématodes marins parasites de *Larus cachinnans michaellis* de provence. Ann Parsitol Hum Comp 1986; 61:533-52.

Basch, P.F. Why do schistosomes have separate sexes? Parasitology Today 1990; 6:160-3.

Basch, P.F., DiConza, J.J. The miracidium-sporocyst transition in *Schistosoma mansoni*: surface changes *in vitro* with ultrastructural correlation. J Parasitol 1974; 60:935-41.

Baudoin, M. Host castration as a parasitic strategy. Evolution 1975; 29:335-52.

Bayssade-Dufour, C. L'appareil sensoriel des cercaires et la systématique des Trématodes Digénétiques. Mém Mus Nat d'Hist Natur, Sér A Zool 1979; 113:1-81.

Behrens, A.C., Nollen, P.M. Responses of *Echinostoma caproni* miracidia to gravity, light and chemicals. Int J Parasitol 1992; 22:673-5.

Beklemishev, V.V. Populations and micropopulations of parasites and nidicolous. Zoologicheskii Zhurnal (*Zoological Journal*) 1959; 38:1128-37 (in Russian).

Beklemishev, V.V. Spatial and functional structure of populations. Byulleten' MOIP (Otdelenie biologii) 1960; 65:41-50 (in Russian).

Belopolskaya, M.M. "Family Microphallidae Travassos, 1920." In: *Trematody Zhivotnyh i Cheloveka (Trematodes of Animals and Man)*. K.I. Skrjabin (ed.), Vol. 21, Moscow-Leningrad, USSR: Izdatel'stvo Akademii Nauk SSSR, 1963 (in Russian).

Belopolskaya, M.M., Soboleva, T.N. The life cycle of *Microphallus helicicola* sp. nov. (Trematoda: Microphallidae Travassos, 1920). Izvestia Akademii Nauk Kazakhskoy SSR (*Proceedings of the Kazakh Academy of Sciences*), Ser Biol 1977; 4:19-24 (in Russian).

Benjamin, L.R., James, B.L. The development of the metacercaria of *Maritrema linguilla* Jag., 1908 (Digenea: Microphallidae). Parasitology 1987; 94:221-31.

Bennett, C.E. *Fasciola hepatica*: development of caecal epithelium during migration in the mouse. Exp Parasitol 1975; 37:426-41.

Bennett, C.E. *Fasciola hepatica*: development of excretory and parenchymal systems during migration in the mouse. Exp Parasitol 1977; 41:43-53.

Bennett, C.E., Threadgold, L.T. Electron microscope studies of *Fasciola hepatica*. XIII. Fine structure of newly excysted juvenile. Exp Parasitol 1973; 34:85-99.

Bennett, C.E., Threadgold, L.T. *Fasciola hepatica*: development of the tegument during migration in the mouse. Exp Parasitol 1975; 38:38-55.

Bennett, C.E., Wilson, R.A., Denison, J. An investigation of the molecular characteristics important in stimulation of an attachment response by the miracidium of *Fasciola hepatica*. Comp Biochem Physiol 1972; 41A:129-57.

Bennett, H.J. The life history of *Cotylophoron cotylophorum* a Trematode from Ruminants. Illinois Biol Monogr 1936; 14:7-120.

Bennett, H.S. Morphological aspects of extracellular polysaccharides. J Histochem Cytochem 1963; 11:13-23.

Berezantsev, Yu.A. *Trichinellosis*. Leningrad, U.S.S.R.: Medicina, 1974 (in Russian).

Berezantsev, Yu.A., Dobrovoskij A.A. Encapsulation processes of *Posthodiplostomum cuticola* (Nordmann, 1832) Dubois, 1936 metacercariae in fishes. Trudy Astrakhanskogo zapovednika (*Transactions of Astrakhansky Nature Reserve*) 1968; 11:7-12 (in Russian).

Berger, V.Ya. On the effect of parasites on the system of adaptation to salinity in the snail *Hydrobia ulvae*. Parazitologiya 1976; 10:333-7 (in Russian).

Berger, V.Ya. *Adaptations of Marine Molluscs to Changes in Salinity*. Leningrad, USSR: Nauka, 1986 (in Russian).

Berger, V.Ya., Kondratenkov, A.P. The effect of infection of *Hydrobia ulvae* with larvae of trematodes and its resistance to drying and water-freshening. Parazitologiya 1974; 8:563-4 (in Russian).

Berger, V.Ya., Kharazova, A.D. Mechanisms of salinity adaptations in marine molluscs. Hydrobiologia 1997; 355:115-26.

Berger, V.Ya., Galaktionov, K.V., Prokofiev, V.V. Influence of parasites on host adaptations to environmental factors: host-parasite system molluscs – trematode parthenitae. Parazitologiya 2001; 35:192-200 (in Russian)

Bezerra, J.C.B., Becker, W., Zelck, U.E. A comparative study of the organic-acid content of the hemolymph of *Schistosoma mansoni*-resistant and susceptible strains of *Biomphalaria glabrata*. Memorias do Instituto Oswaldo Cruz 1997; 92:421-5.

Bibby, M.C., Rees, G. The ultrastructure of the epidermis and associated structures in the metacercaria, cercaria and sporocyst of *Diplostomum phoxini* (Faust, 1918). Z Parasitenk 1971a; 37:169-86.

Bibby, M.C., Rees, G. The uptake of radioactive glucose in vivo by the metacercaria *Diplostomum phoxini* (Faust) and its conversion to glycogen. Z Parasitenk 1971b; 37:187-97.

Biguet, J., Deblock, S., Capron, A. Description d'une metacercaire progénétique du genre *Asymphylodora* Loos, 1899, découverte chez *Bythinia leachi* Sheppard dans le nord de la France. Ann Parasit Hum Comp 1956; 31:525-42.

Blair, D., Campos, A., Cummings, M.P., Laclette, J.P. Evolutionary biology of parasitic platyhelminths: the role of molecular phylogenetics. Parasitol Today 1996; 12:66-71.

Bock, D. Cercarien und ihre Parthenitae aus Süswasserschnecken des Naturreservates "Obedska Bara" bei Belgrad (Jugoslawien). Ph.D. Thesis (Druck Dorsch), Univ Hamburg, Hamburg. 1980.

Bock, D. Formation, histochemistry and ultrastructure of the metacercarial cyst wall of *Plagiorchis* species 1 (Trematoda, Plagiorchiidae). Int J Parasitol 1988; 18:379-88.

Bock, D. Hatching mechanism of the metacercaria of *Plagiorchis* species 1 (Trematoda: Plagiorchiidae). J Helminthol 1989; 63:163-71.

Bogitsh, B.J. Peptidase activity in the egg-shell-enclosed embryo of *Schistosoma mansoni* using fluorescent histochemistry. Trans Am Microsc Soc 1983; 102:169-72.

Bogitsh, B.J. An overview of surface specializations in the digenetic trematodes. Hydrobiologia 1986; 132:305-10.

Bogitsh, B.J., Carter, O.S. Developmental studies on the digestive tract of schistosomules (*Schistosoma mansoni*) grown in vitro. I. Ultrastructure. Trans Am Microsc Soc 1977; 96:219-27.

Bouix-Busson, D., Rondelaud, D., Combes, C. L'infestation de *Lymnaea glabra* Müller par *Fasciola hepatica* L. Les caractéristiques des émissions cercariennes. Ann Parasitol Hum Comp 1985; 60:11-21.

Britov, V.A. *Pathogens of trichinellosis*. Moscow, U.S.S.R.: Nauka, 1982 (in Russian)

Brooker, B.E. The sense organs of Trematode miracidia. J Linn Soc London 1972; 51(Suppl 1): 171-80.

Brooks, D.R. The phylogeny of the cercomeria (Platyhelminthes: Rhabdocoela) and general evolutionary principles. J Parasitol 1989; 75:606-16.

Brooks, D.R., O'Grady, R.T., Glem, D.R. Phylogenetic analysis of the Digenea (Plathyhelmintes: Cercomeria) with comments on their adaptive radiation. Can J Zool 1985a; 63:411-43.

Brooks, D.R., O'Grady, R.T., Glem, D.R. The phylogeny of the Cercomeria Brooks, 1982 (Plathyhelminthes). Proc Helminthol Soc Wash 1985b; 52:1-20.

Brooks, D.R., Bandoni, S.M., MacDonald, C.A., O'Grady, R.T. Aspects of the phylogeny of the Trematoda Rudolphi, 1808 (Platyhelminthes: Cercomeria). Can J Zool; 1989; 67:2609-24.

Brooks, D.R., McLennan, D.A. *Parascript: Parasites and the Language of Evolution*. Washington DC, U.S.A.: Smithsonian Institution Press, 1993.

Brooks, F.G. Studies on the germ-cell cycle of Trematodes. Am J Hyg 1930; 12:299-340.

Burton, P.R. Gametogenesis and fertilization in the frog lung fluke, *Haematoloechus medioplexus* (Trematoda: Plagiorchiidae). J Morphol 1960; 107:93-122.

Burton, P.R. In vitro uptake of radioglucose by a frog lung-fluke and correlation with the histochemical identification of glycogen. J Parasitol 1962; 48:874-82.

Burton, P.R. Fine structure of the reproductive system of a frog lung fluke. I. Mehlis' gland and associated ducts. J Parasitol 1967a; 53:540-55.

Burton, P.R. Fine structure of the reproductive system of a frog lung fluke. II. Penetration of the ovum by a spermatozoon. J Parasitol 1967b; 53:994-9.

Bush, A.O., Lafferty, K.D., Lotz, J.M., Shostak, A.W. Parasitology meets ecology on its own terms. Margolis et al. revisited. J Parasitol 1997; 83:575-83.

Bustnes, J.O., Galaktionov, K.V. Anthropogenic influences on the infestation of intertidal gastropods by seabird trematode larvae on the southern Barents Sea coast. Mar Biol 1999; 133: 449-53.

Bustnes, J.O., Galaktionov, K.V., Irwin, S.W.B. Potential threats to littoral biodiversity: is increased parasitism a consequence of human activity? Oikos 2000; 90: 189-90.

Buzzell, G.R. Composition, secretion, and fate of the glands in the miracidium and sporocysts of *Fasciola hepatica* L. J Helminthol 1983; 57:79-84.
Bychowsky, B.E. Ontogenesis and phylogenetic interrelations of parasitic flatworms. Izvestia Academii Nauk SSSR (*Proceedings of the USSR Academy of Sciences*), Ser Biol 1937; 4:1353-83 (in Russian).
Bychowsky, B.E. *Monogenetic Trematodes, their Systematics and Phylogeny*. Moscow-Leningrad, USSR: Izdatel'stvo Akademii Nauk SSSR 1957 (in Russian). English translation by W.J. Hargis and P.C. Oustinoff. Washington, USA: Graphic Arts Press, Inc., 1961.
Byrd, E.E., Maples, W.P. The intramolluscan phase in the life cycle of *Pneumatophilus leidyi* Byrd et Denton, 1937 (Trematoda: Plagiorchiidae). J Parasitol 1965; 51:23.
Byrd, E.E., Maples, W.P. Intramolluscan stages of *Dasymetra conferta* Nicoll, 1911 (Trematoda: Plagiorchiidae). J Parasitol 1969; 55:509-26.
Byrd, E.E., Venard, C.E., Reiber, R.J. The excretory system in Trematoda. I. Studies on the excretory system in the trematoda subfamily Gorgoderinae Looss 1901. J Parasitol 1940; 26:407-20.
Cable, R.M. The life cycle of *Parvatrema borinquenae* gen. et sp. nov. (Trematoda: Digenea) and the systematic position of the subfamily Gymnophallidae. J Parasitol 1953; 39:408-21.
Cable, R.M. Studies on marine digenetic trematodes of Puerto Rico. The life cycle in the family Haplosplanchnidae. J Parasitol 1954; 40:71-6.
Cable, R.M. Marine cercariae Puerto Rico. Scient Survey Puerto Rico and Virgin Islands 1956; 61:491-577.
Cable, R.M. "Thereby hangs a tail". J Parasitol 1965; 51:2-12.
Cable, R.M. "Phylogeny and Taxonomy of Trematodes with Reference to Marine Species." In *Symbiosis in the Sea*. W.E. Vernberg, ed. Columbia, South Carolina, U.S.A.: University of South Carolina Press 1974.
Cable, R.M. *An Illustrated Laboratory Manual of Parasitology*. Minneapolis, Minnesota, U.S.A.: Burgess 1977.
Cable, R.M., Crandall, R.B. Larval stages and phylogeny as exemplified by the lung fluke of turtles. Science 1956; 124:890.
Cable, R.M., Hunninen, A.V. Studies on *Deropristis inflata* (Molin), its life history and affinities to trematodes of the family Acanthocolpidae. Biol Bull Mar Biol Lab Woods Hole 1942; 82:292-312.
Cable, R.M., Isseroff, H. A protandrous haploporid cercaria, probably the larva of *Saccocoelioides sogandaresi* Lumsden, 1963. Proc Helminthol Soc Wash 1969; 36:131-5.
Cable, R.M., McLean, R.A. The occurrence of *Cercaria clausii* Monticelli, a marine rattenkoning cercaria on the west coast of Florida. Notulae Nat 1943; 129:1-7
Cake, E.W.J. A key to larval cestodes of shallow-water, benthic molluscs of the Northern Gulf of Mexico. Proc Helminthol Soc Wash 1976; 43:160-71.
Cannon, L.R. The life cycles of *Bunodera sacculata* and *B. luciopercae* (Trematoda: Allocreadiidae) in Algonquin Park, Ontario. Can J Zool 1971; 49:1417-29.
Capron, A. Schistosomiasis: forty years' war on the worm. Parasitol Today 1998; 14:379-84.
Cardell, R.R., Philpott, J.E., Philpott, D.E. The ultrastructure of the tail of the cercaria of *Himasthla quissetensis* (Miller and Northup, 1926). Trans Am Microsc Soc 1960; 79:442-50.
Carter, O.S., Bogitsh, B.J. Histologic and cytochemical observations of the effects of *Schistosoma mansoni* on *Biomphalaria glabrata*. Ann NY Acad Sci 1975; 266:380-93.
Caulfield, J.P., Yuan, H.-C., Cianci, C.M.L., Hein, A. *Schistosoma mansoni*: development of the cercarial glycocalyx. Exp Parasitol 1988; 65:10-9.
Caulfield, J.P., Cianci, C.M.L., McDiarmid, S.S., Suyemitsu, T., Schmid, K. Ultrastructure, carbohydrate, and amino acid analysis of two preparations of the cercarial glycocalyx of *Schistosoma mansoni*. J Parasitol 1987; 73:514-22.

Chapman, H.D. The functional organization and fine structure of the tail musculature of the cercariae of *Cryptocotyle lingua* and *Himasthla secunda*. Parasitology 1973; 66:487-97.

Chapman, H.D. The behaviour of the cercaria of *Cryptocotyle lingua*. Z Parasitenk 1974; 44:211-26.

Chapman, H.D., Wilson, R.A. The propulsion of the cercariae of *Himasthla secunda* (Nicoll) and *Cryptocotyle lingua*. Parasitology 1973; 67:1-15.

Chen, P. The germ cell cycle in the trematode *Paragonimus kellicotti* Ward. Trans Am Microsc Soc 1937; 56:208-36.

Cheng, T.C. Studies on the morpogenesis, development and germ cell cycle in the sporocysts and cercariae of *Glypthelmins pennsylvaniensis* Cheng, 1961 (Trematoda: Brachycoeliidae). Proc Pennsylv Acad Sci 1961; 35:10-22.

Cheng, T.C., Bier, J.W. Studies on molluscan schistosomiasis: an analysis of the development of the cercaria of *Schistosoma mansoni*. Parasitology 1972; 64:129-41.

Cheng, T.C., James, H.A. Studies on the germ cell cycle, morphogenesis and development of the cercarial stage of *Crepidostomum cornutum* (Osborn, 1903). Trans Am Microsc Soc 1960; 79:75-85.

Cheng, T.C., Snyder, R.W. Studies on host-parasite relationships between larval trematodes and their hosts. I. A review. II. The utilization of the host's glycogen by the intramolluscan larvae of *Glypthelmins pennsylvaniensis* Cheng and associated phenomena. Trans Am Microsc Soc 1962; 81:209-28.

Cheng, T.C., Snyder, R.W. Studies on host-parasite relationships between larval trematodes and their hosts. IV. A histochemical determination of glucose and its role in the metabolism of molluscan host and parasite. Trans Am Microsc Soc 1963; 82:343-6.

Chernin, E. Behavioural responses of miracidia of *Schistosoma mansoni* and other trematodes to substance emitted by snail. J Parasitol 1970; 56:287-96.

Chernin, E. Penetrative activity of *Schistosoma mansoni* miracidia stimulated by exposure to snail-conditioned water. J Parasitol 1972; 58:209-12.

Chernin, E., Dunovan, C.A. The influence of host-parasite dispersion upon the capacity of *Schistosoma mansoni* miracidia to infect *Australorbis glabratus*. Am J Trop Med Hyg 1962; 11:455-71.

Chernogorenko-Bidulina, M.N., Bliznyuk, I.D. On the life cycle of the trematode *Sphaerostoma bramae* Mull., 1776. Doklady Akademii Nauk SSSR (*Proceedings of the USSR Academy of Sciences*) 1960; 134:237-40 (in Russian).

Ching, H.L. Life cycle of *Lacunovermis conspicuus* n. gen., n. sp. and *Meiogymnophallus multigemmulus* n. gen., n. sp. (Gymnophallidae: Trematoda) from *Macoma inconspicua* and Diving Ducks from Vancouver, Canada. Proc Helminthol Soc Wash 1965; 32:53-63.

Ching, H.L. Description of germinal sacs of a gymnophallid trematode, *Cercaria margaritensis* sp. n., in the extrapallial fluid of subtidal snails (*Margarites* spp.) in British Columbia. Can J Zool 1982; 60:516-20.

Christensen, N.O. *Schistosoma mansoni*: interference with cercarial host-finding by various aquatic organisms. J Helminthol 1979; 53:7-14.

Christensen, N.O., Frandsen, F. The interaction of some environmental factors influencing *Schistosoma mansoni* cercarial host-finding. J Helminthol 1980; 54:203-5.

Christensen, N.O., Frandsen, F., Roushdy, M.Z. The influence of environmental conditions and parasite-intermediate host-related factors on the transmission of *Echinostoma liei*. Z Parasitenk 1980; 63:47-63.

Chubrik, G.K. "Fauna and Ecology of Trematode Larvae in the Molluscs of the Barents and White Seas". In *Zhiznennye Cicly Paraziticheskikh Chervi Severnih Morei (Life Cycles of Parasitic Worms in Northern Seas)*. Yu.I. Poljansky, ed., Moscow-Leningrad, USSR: Izdatel'stvo Akademii Nauk SSSR, 1966 (in Russian).

Ciordia, H. Cytological studies of the germ cell cycle of the trematode family Bucephalidae. Trans Am Microsc Soc 1956; 75:103-16.

Clark, R.B. Interpretation of life history pattern in the Digenea. Int J Parasitol 1974; 4:115-23.

Claus, C. Zur morphologischen und phylogenetischen Beurteilung des Bandwurmkörpers. Arch Zool Inst Wien 1889; 8:4-6.
Clegg, A. Secretion of lipoproteins by Mehlis' gland in *Fasciola hepatica*. Ann NY Acad Sci 1965; 118:969-86.
Clegg, J.A. Skin penetration by cercariae of the bird schistosome *Austrobilharzia terrigalensis*: the stimulatory effect of cholesterol. Parasitology 1969, 59:973-89.
Cohen, C., Reinhardt, B., Castellani, L., Norton, P., Stirewalt, W. Schistosome surface spines are "crystals" of actin. J Cell Biol 1982; 95:987-8.
Coil, W.H. Observations on egg shell formation in *Hydrophitrema gigantica* Sanders, 1960 (Hemiuridae: Digenea). Z Parasitenk 1965; 25:510-7.
Coil, W.H. The penetration of *Fascioloides magna* miracidia into the snail host *Fossaria bulimoides*. Z Parasitenk 1977; 52:53-9.
Coil, W.H. Miracidial penetration in *Fascioloides magna* (Trematoda). Z Parasitenk 1981; 65:299-307.
Combes, C. Ethological aspects of parasite transmission. Am Nat 1991a; 138:866-80.
Combes, C. The Schistosome scandal. Acta Œecol 1991b; 12:165-73.
Combes, C., Théron, A. Les densites cercariennes. Mém Mus Nat d'Hist Natur 1981; A119:186-96.
Combes, C., Fournier, A., Moné, H., Théron, A. Behaviours in trematode cercariae that enhance parasite transmission: patterns and processes. Parasitology 1994; 109S:S3-S13.
Combescot-Lang, C. Etudes des digeneans parasites de *Littorina saxatilis* (Olivi) et de leurs effect sur cet hôte. Ann Parasitol Hum Comp 1976; 51:27-36.
Conn, D.B. The role of cellular parenchyma and extracellular matrix in the histogenesis of the paruterine organ of *Mesocestoides lineatus* (Platyhelminthes: Cestoda). J Morphol 1988; 197:303-14.
Conn, D.B. The biology of flatworms (Platyhelminthes): Parenchyma cells and extracellular matrices. Trans Am Microsc Soc 1993; 112:241-61.
Conn, D.B., Rocco, L.J. Fine structure of the cellular parenchyma and extracellular matrix of *Ophiotaenia loennbergii* (Cestoda: Proteocephalidea). Acta Zool (Stockholm) 1989; 70:105-10.
Conradt, U., Peters, W. Investigations on the occurrence of pinocytosis in the tegument of *Schistocephalus solidus*. Parasitol Res 1989; 75:630-5.
Cort, W.W. The excretory system of a stylet cercaria. Univ Calif Publ Zool 1919; 19:87-94.
Cort, W.W. The germ cell cycle in the digenetic trematodes. Quart Rev Biol 1944; 19:275-84.
Cort, W.W., Ameel, D.J. Further studies on the development of the sporocyst stages of plagiorchiid trematodes. J Parasitol 1944; 30:37-56.
Cort, W.W., Ameel, D.J., Van der Woude, A. Germinal masses in redia embryos of an echinostome and a psilostome. J Parasitol 1949; 35:579-82.
Cort, W.W., Ameel, D.J., Van der Woude, A. Germinal material in the rediae of *Clinostomum marginatum* (Rud.). J Parasitol 1950; 36:157-63.
Cort, W.W., Ameel, D.J., Van der Woude, A. Development of the mother and daughter sporocysts of a snake plagiorchoid, *Lechriorchis primus* (Trematoda: Reniferidae). J Parasitol 1952; 38:187-202.
Cort, W.W., Ameel, D.J., Van der Woude, A. Germinal development in the sporocysts and rediae of the digenetic trematodes. Exp Parasitol 1954; 3:185-225.
Cort, W.W., Oliver, L. The development of the larval stages of *Plagiorchis muris* Tanabe, 1922, in the first intermediate host. J Parasitol 1943; 29:81-99.
Crabtree, J.E., Wilson, R.A. *Schistosoma mansoni*: a scanning electron microscope study of the developing schistosomulum. Parasitology 1980; 81:553-64.
Crandall, R.B. The life history and affinities of the turtle lung fluke, *Heronimus chelydrae* MacCallum, 1902. J Parasitol 1960; 46:289-307.
Cribb, T.H. The life cycle and biology of *Opecoelus variabilis*, sp. nov. (Digenea: Opecoelidae). Aust J Zool 1985; 33:715-28.

Cribb, T.H. Life cycle and biology of *Prototransversotrema steeri* Angel, 1969 (Digenea: Transversotrematidae). Aust J Zool 1988; 36:111-29

Cribb, T.H., Bray, R.A., Littlewood, D.T.J. The nature and evolution of the association among digeneans, molluscs and fishes. Int J Parasitol 2001a; 31:997-1011.

Cribb, T.H., Bray, R.A., Littlewood, D.T.J., Pichelin S.P., Herniou E.A. "The Digenea". In *Interrelationships of the Platyhelminthes.* D.T.J. Littlewood and R.A. Bray, eds. London, U.K.: Taylor & Francis 2001b.

Curtis, L.A. The influence of sex and trematode parasites on carrion response of the estuarine snail *Ilyanassa obsoleta.* Biol Bull 1985; 169:377-90.

Curtis, L.A. Vertical distribution of an estuarine snail altered by parasite. Science 1987; 235:1509-11.

Curtis, L.A. Parasitism and the movements of intertidal gastropod individuals. Biol Bull 1990; 179:105-12.

Curtis, L.A., Hurd, L.E. Age, sex, and parasites: spatial heterogeneity in a sandflat population of *Ilyanassa obsoleta.* Ecology 1983; 64:819-28.

Curtis, L.A., Kinley, J.L., Tanner, N.L. Longevity of oversized individuals: growth, parasitism, and history in an estuarine snail population. J Mar Biol Ass UK 2000; 80:811-20

Czapski, Z. New observation on the life cycle of *Fasciola hepatica* L. in *Galba truncatula* O.F. Muller, *Galba occulata* Jack. and *Galba turricola* Held. IV Intern Congr Parasitol, Short Commun 1978; Sect A:10.

Czubaj, A., Niewiadomska, K. Types of sensory cells in *Notocotylus attenuatus* (Rud., 1809) rediae (Digenea, Notocotylidae). Parasitol Res 1988; 74:243-9.

Czubaj, A., Niewiadomska, K. The muscular system of the cercaria of *Diplostomum pseudospathaceum* Niew., 1984 (Digenea): A phalloidin-rhodamine fluorescence and TEM study. Acta Parasitol 1997; 42:199-218.

Davies, T.N. *Schistosoma mansoni*: the structure and elemental composition of the preacetabular penetration gland cell secretion in the pre-emergent cercariae. Parasitology 1983; 87:55-60.

Davydov, V.G. "On Structure, Function, and Origin of Tegument in Representatives of the Cercomeromorpha". In: *Mofologicheskie Osnovy Filogenetiki Ploskih Chervei (Morphological Principles of Platyhelminthes Phylogenetics).* Y.V. Mamkaev, ed., Leningrad, Russia: Zoological Institute Publ., 1991 (in Russian).

Davydov, V.G., Kuperman, B.I. The ultrasructure of the tegument and peculiarities of the biology of *Amphlina foliacea* adult (Plathelminthes, Amphilinidea). Folia Parasitol 1993; 40:13-22.

Davydov, V.G., Korneva, J.V., Kuperman, B. I. The development of the tegument in ontogenesis of *Triaenophorus nodulosus* (Cestoda: Pseudophyllidea). Folia Parasitol 1995; 42:269-79.

Dawes, B. *The Trematoda.* Cambridge, U.K.: At the University Press, 1946.

Dawes, B. A study of the miracidium of *Fasciola hepatica* and account of the mode of penetration of the sporocyst into *Lymnaea truncatula.* In *Sobretiro del Libro Homenaje al Dr. Eduardo Caballero y Caballero.* Mexico, 1960; 95-111.

Dawes, B. On early stages of *Fasciola hepatica* penetrating into the liver of an experimental host, the mouse. A histological picture. J Helminthol 1961; Suppl.:41-52.

Dawes, B. On the growth and maturation of *Fasciola hepatica* L. in the mouse. J Helminthol 1962a; 36:11-38.

Dawes, B. Additional notes on the growths of *Fasciola hepatica* in the mouse, with some remarks about recent researches in Belgium. J Helminthol 1962b; 36:259-68.

Dawes, B. The migration of juvenile forms of *Fasciola hepatica* L. through the wall of the intestines in the mouse, with some observations on food and feeding. Parasitology 1963a; 53:109-22.

Dawes, B. Some observations of *Fasciola hepatica* L. during feeding operations in the hepatic parenchyma of the mouse, with notes on the nature of liver damage in this host. Parasitology 1963b; 53:135-43.

Dawes, B. Hyperplasia of the bile duct in fascioliasis and its relation to the problem of nutrition in the liver-fluke, *Fasciola hepatica* L. Parasitology 1963c; 53:123-33.

Dawkins, R. *The Extended Phenotype: the Gene as the Unit of Selection.* Oxford, U.K.: Oxford University Press, 1982.

Day, M.F. Changes in the epidermis of *Cryptocotyle lingua* (Creplin) during metamorphosis of cercaria to metacercaria in the second intermediate (fish) host. Parasitology 1976; 73:24.

De Jong-Brink, M. Interference of schistosome parasites with neuroendocrine mechanisms in their snail host causes physiological changes. Adv Neuroimmunol 1992; 2:199-233.

De Jong-Brink, M. How shistosomes profit from the stress responses they elicit in their hosts. Adv Parasitol 1995; 35:177-256.

Deblock, S. Contribution à l'étude des Microphallidae Travassos, 1920. XXIV. Tentative de phylogénie et de taxonomie. Bull Mus Hist Nat 1971; 3e sér, 7, Zoologie 7:353-469.

Deblock, S. Présence en France de *Bunocotyle progenetica* (Markowski, 1936), Trématode digène. Bull Soc Zool France 1974a; 99:593-600.

Deblock, S. Contribution à l'etude des Microphallidae Travassos, 1920 (Trematoda). XXVIII. *Microphallus abortivus* n. sp. espèce à cycle évolutif abrégé originaire d'Oléron. Ann Parasitol Hum Comp 1974b; 49:175-84.

Deblock, S. De l'abrégement du cycle évolutif chez les Trématodes Digènes Microphallidés. Excerta Parasitologica en memoria del Dr Eduardo Caballero y Caballero. Instituto de Biologia, publicationes especiales. Mexico, 1977; 4: 151-60.

Deblock, S. Inventaire des trématodes larvaires parasites des mollusques *Hydrobia* (Prosobranches) des côtes de France. Parassitologia 1980; 22:1-105.

Dickerman, E.E. *Paurorhynchus hiodontis* a new genus and species of trematoda (Bucephalidae, Paurorhynchinae n. subfam.) from the mooneye fishes *Hiodon tergisus*. J Parasitol 1954; 40:311-5.

Dinnik, J.A. *Paramphistomum phillerouxi* sp. nov. (Trematoda: Paramphistomatidae) and its development in *Bulinus forskalii*. J Helminthol 1961; 35:69-90.

Dinnik, J.A., Dinnik, N.N. Observations on the succession of rediae generations of *Fasciola gigantica* Cobb. in a snail host. Z Tropen Med Parasitol 1956; 7:397-419.

Dinnik, J.A., Dinnik, N.N. Development of *Carmyerius exoporus* Mapleston (Trematoda: Gastrothylacidae) in a snail host. Parasitology 1960; 50:469-80.

Dixon, K.E., Mercer, E.H. The formation of the cyst wall of the metacercaria of *Fasciola hepatica* L. Z Zellforsch 1967; 77:345-60.

Dobrovolskij, A.A. Über die Eincheitlichkeit des Bauplanes von Mirazidien der Überfamilie Plagiorchiidae. Angew Parasitol 1965a; 6:157-65.

Dobrovolskij, A.A. Some new data on the life cycle of trematode *Opisthioglyphe ranae* Flolich, 1791 (Plagiorchiidae). Helminthologia 1965b; 6:205-21 (in Russian).

Dobrovolskij, A.A. Elaboration of the life cycle of *Telorchis assula* (Dujardin, 1845) Dolfus, 1957 (Trematoda, Telorchiidae). Materialy k Nauchnoy Konferentsii Vsesoiuznogo Obshestva Gelmintologov (*Materials to Scientific Conference of All-Union Helmintological Society*), Moscow, USSR: Nauka 1967; 5:132-41 (in Russian).

Dobrovolskij, A.A. Life cycle of *Paralepoderma cloacicola* (Luhe, 1909) Dolfus, 1950 (Trematoda, Plagiorchiidae). Vestnik Leningradskogo Universiteta (*Bulletin of the Leningrad University*) 1969; 9: 28-38 (in Russian).

Dobrovolskij, A.A. Life cycle of *Macrodera longicollis* (Abildgaard, 1788) Luhe, 1909 (Trematoda, Ochetosomatidae). Vestnik Leningradskogo Universiteta (*Bulletin of the Leningrad University*) 1971; 15: 9-20 (in Russian).

Dobrovolskij, A.A. Some patterns in evolution of mother sporocysts of trematodes of the suborder Plagiorchiata. Ekologicheskaya i Eksperimentalnaya Parazitologiya (*Ecological and Experimental Parasitology*), Leningrad, USSR: Izdatel'stvo Leningradskogo Universiteta 1975; 1:96-108 (in Russian).

Dobrovolskij, A.A., Ataev, G.L. "The Nature of Reproduction of Trematodes Rediae and Sporocysts." In *Hommage à Louis Euzet - Taxonomy, Ecology, and Evolution of Metazoan Parasites*, C. Combes and J. Jourdane, eds. Perpignan, France: Presses Universitaires de Perpignan, 2003.

Dobrovolskij, A.A., Galaktionov, K.V. "Main Stages in the Evolution of Digenea Life Cycles." In *Hommage à Louis Euzet - Taxonomy, Ecology, and Evolution of Metazoan Parasites*, C. Combes and J. Jourdane, eds. Perpignan, France: Presses Universitaires de Perpignan, 2003.

Dobrovolskij, A.A., Muhamedov, G.K. Some features of reproduction in the mother sporocysts of *Xiphidiocercaria* sp. VII Odening (Trematoda, Plagiorchiata). Ekologicheskaya i Eksperimentalnaya Parazitologiya (*Ecological and Experimental Parasitology*), Leningrad, USSR: Izdatel'stvo Leningradskogo Universiteta 1979; 2:42-7 (in Russian).

Dobrovolskij, A.A., Reichel, A.C. Life cycle of *Haplometra cylindracea* Zeder, 1800 (Trematoda, Plagiorchiidae). Vestnik Leningradskogo Universiteta (*Bulletin of the Leningrad University*) 1973; 3:5-13 (in Russian).

Dobrovolskij, A.A., Galaktionov, K.V., Ataev, G.L. Peculiarities of the generative material organization and proliferation dynamics of trematode mother sporocysts. Parzitologiya 2000; 34:14-24 (In Russian).

Dobrovolskij, A.A., Galaktionov, K.V., Muhamedov, G.K., Sinha B.K., Tihomirov, I.A. Parthenogenetic generations of trematodes. Trudy Leningradskogo Obshestva Estestvoispytatelei (*Transactions of Leningrad Society of Naturalists*) 1983; 82:1-108 (in Russian).

Dobson, A.P. The population biology of parasitic-induced changes in host behaviour. Quart Rev Biol 1988; 63:139-65.

Dogiel, V.A. On the origin of migrations of helminths in the organism of their hosts. Vestnik Khasahskogo Filiala AN SSSR (*Bulletin of the Kazakh Branch of the USSR Academy of Sciences*) 1944; 1:16-19 (in Russian).

Dogiel, V.A. *A Course of General Parasitology*. Leningrad, USSR: Gosudarstvennoe Uchebno-Pedagogicheskoye Izdatel'stvo Ministerstva Prosvesheniya RSFSR, 1947 (in Russian).

Dogiel, V.A. *Oligomerization of Homologous Organs as One of the Main Directions of Animal Evolution*. Leningrad, USSR: Izdatel'stvo Leningradskogo Universiteta, 1954 (in Russian).

Dogiel, V.A. *General Parasitology*. Edinburgh. U.K.: Oliver & Boyd, 1964.

Dolfus, R.P. Sur *Monostoma petasatum* Deslongchamps, 1824 et son cycle évolutif deux hôtes. Ann Parasitol Hum Comp 1966; 41:289-99.

Dolfus, R.P. Ennumération des cestodes du plankton et des invertébrès marins. Ann Parasitol Hum Comp 1976; 51:207-20.

Dönges, J. Der Lebenszyklus von *Posthodiplostomum cuticola* (v. Nordmann, 1832) Dubois, 1936 (Trematoda, Diplostomatidae). Z Parasitenk 1964; 24:169-248.

Dönges, J. Entwicklungs- und Lebensdauer von Metacercarien. Z Parasitenk 1969; 31:340-66.

Dönges, J. Das Miracidium von *Isthmiophora melis* (Schrank, 1788) (Echinostomatidae). Okologie und Morphologie. Z Parasitenk 1973; 41:215-30.

Dubinin, V.B. New data on age-related localization of parasitic worms in the intestine of birds. Doklady Akademii Nauk SSSR (*Proceedings of the USSR Academy of Sciences*) 1941; 30:374-8 (in Russian).

Dubinina, M.N. Host-specificity in ligulid tapeworms at different phases of their life cycle. Parasitologicheskii Sbornik Zoologicheskogo Instituta Akademii Nauk SSSR (*Parasitological Proceedings of the Zoological Institute of the USSR Academy of Sciences*) 1953; 15:234-51 (in Russian).

Dubois, G. Les cercaires de la Région de Neuchatel. Bull Soc Neuchât Sci Nat 1929; 69:5-103.
Dubois, G. Monographie des Strigeida (Trematoda). Mém Soc Neuchât Sci Nat 1938; 6:1-533.
Dunn, T.S., Nizami, W.A., Hanna, R.E. Studies on the ultrastructure and histochemistry of the lymph system in three species of amphistome (Trematoda: Digenea) from the Indian water Buffalo *Bubalus bubalus*. J Helminthol 1985; 59:1-18.
Dunn, T.S., Hanna, R.E., Nizami, W.A. Ultrastructural and histochemical observations on the epidermis, presumptive tegument and glands of the miracidium of *Gigantocotyle explanatum* (Trematoda: Paramphistomatidae). Int J Parasitol 1987; 17:885-95.
Dunn, T.S., Dang, P.H., Mattison, G., Hanna, R.E.B., Nizami, W.A. Ultrastructural observations on the redial tegument of *Paramphistomum epiclitum* from the planorbid snail, *Indoplanorbis exustus*. J Helminthol 1992a; 66:167-76.
Dunn, T.S., Dang, P.H., Hanna, R.E.B., Nizami, W.A. Embryological development of the cercarial tegument of *Paramphistomum epiclitum* in the planorbid snail, *Indoplanorbis exustus*. J Helminthol 1992b; 66:243-54.
Ebrahimzaden, A. Beiträge zur Micromorphologie der Miracidien von *Schistosoma mansoni*. 1. Feinstruktur des Teguments und dessen "Begleitorganellen". Z Parasitenk 1977; 54:257-67.
Ebrahimzaden, A., Kraft, M. Ultrastrukturelle Untersuchungen zur Anatomie der Cercarien von *Schistosoma mansoni*. II. Das Exkretionssystem. Z Parsitenk 1971; 36:265-90
Ehlers, U. Phylogenetisches System der Plathelminthes. Verbandl Naturwissen Vereins Hamb 1984; 27:291-4.
Ehlers, U. *Das Phylogenetische System der Plathelminthes*. Stuttgart, D.B.R.: Gustav Fischer Verlag, 1985a.
Ehlers, U. "Phylogenetic Relationships within the Platyhelminthes." In: *The Origins and Relationships of Lower Invertebrates*. S. Conway Morris, J.D. George, R. Gibson and H.M. Platt, eds. Oxford, U.K.: Oxford University Press, 1985b.
Ehlers, U. Comments on a phylogenetic system of the Platyhelminthes. Hydrobiologia 1986; 132:1-12.
Ejsmont, L. Morphologische, systematische und entwicklungsgeschichtliche Untersuchungen an Arten des Genus *Sanguinicola* Plehn. Bull Int L'Acad Polan Sci Lettres Classe Sci Math Nat, Ser B, Sci Nat 1926: 9-10:877-966.
Eklu-Natey, D.T., Wüest, J., Swiderski, Z., Striebel, H., Huggel, H. Comparative scanning electron microscope (SEM) study of miracidia of four human schistosome species. Int J Parasitol 1985; 15:33-42.
Erasmus, D.A. *The Biology of Trematodes*. London, U.K.: Edward Arnold, 1972.
Erasmus, D.A. A comparative study of the reproductive system of mature, immature and "unisexual" female *Schistosoma mansoni*. Parasitology 1973; 67:165-83.
Erasmus, D.A. *Schistosoma mansoni*: development of the vitelline cell, its role in drug sequestration, and changes induced by Astiban. Exp Parasitol 1975; 38:240-56.
Erasmus, D.A. The host-parasite interface of trematodes. Adv Parasitol 1977; 15:201-42.
Erasmus, D.A., Popiel, I., Shaw, J.R. A comparative study of the vitelline cells in *Schistosoma mansoni, S. haematobium, S. japonicum*, and *S. matthei*. Parasitology 1982; 84:283-8.
Erwin, B.E., Halton, D.W. Fine structure observations on spermatogenesis in a progenetic trematode, *Bucephaloides gracilescens*. Int J Parasitol 1983; 13:413-26.
Esch, G.W., Gibbons, J.W., Bourque, J.E. An analysis of the relationships between stress and parasitism. Am Midl Nat 1975; 93:339-53.
Eveland, L.K., Morse, S.I. *Schistosoma mansoni*: in vitro conversion of cercariae to schistosomula. Parasitology 1975; 71:327-35.
Faust, E.C. The reaction of the miracidia of *Schistosoma japonicum* and *S. haematobium* in the presence of their intermediate hosts. J Parasitol 1924; 10:199-204.

Faust, E.C., Hoffmann, W.A. Studies on schistosomiasis mansoni in Puerto Rico. 3. Biological studies. I. The extramammalian phases of the life cycle. J Publ Helth Trop Med (Puerto Rico) 1934; 10:1-47.

Faust, E.C., Meleney, H.E. Studies of schistosomiasis japonica. II. The biology, morphology and life history of the causative organism *Schistosoma japonicum* Katsur. Amer J Hyg (Monogr. ser.) 1924; 3:1-339.

Feiler, W., Haas, W. Host-finding in *Trichobilharzia ocellata* cercariae: swimming and attachment to the host. Parasitology 1988a; 96:493-505.

Feiler, W., Haas, W. *Trichobilharzia ocellata*: chemical stimuli of duck skin for cercarial attachment. Parasitology 1988b; 96:507-17.

Fernandez, J., Esch, G.W. Guild structure of larval trematodes in the snail *Helisoma anceps*: patterns and processes at the individual host level. J Parasitol 1991; 77:528-39.

Filimonova, L.V. *Trematode Fauna of the USSR. Notocotylids.* Moscow, USSR: Nauka, 1985 (in Russian).

Fournier, A. "Photoreceptors and Photosensitivity in Platyhelminthes" In *Photoreception and Vision in Invertebrates.* M.A. Ali, ed. Plenum Publishing Corporation, 1984.

Fournier, A., Théron, A. Sectorisation morpho-anatomique et fonctionnelle du sporocyste-fils de *Schistosoma mansoni.* Z Parasitenk 1985; 71:325-36.

Fournier, A., Xia, M., Combes, C. Étude ultrastructurale du sporocyste-fils de *Schistosoma japonicum*: comparaison avec le sporocyste-fils des autres Schistosomes. Ann Sci Naturel, Zoologie, Paris, 13e Série 1986-1987; 8:5-10.

Fraenkel, G.S., Gunn, D.L. *The Orientation of Animals.* New York, U.S.A.: Dover Books, 1961.

Fried, B. Metacercarial excystment of Trematodes. Adv Parasitol 1994; 33:91-144.

Fried, B. Biology of echinostomes except Echinostoma. Adv Parasitol 2001; 49:163-215.

Fried, B., King, W. Attraction of *Echinostoma revolutum* cercariae to *Biomphalaria glabrata* dialysate. J Parasitol 1989; 75:55-7.

Fried, B., Reddy, A. Effects of snail-conditioned water from *Biomphalaria glabrata* on hatching of *Echinostoma caproni* miracidia. Parasitol Res 1999; 85:155-7.

Fried, B., Rosa-Brunet, L.C. Exposure of *Dugesia tigrina* (Turbellaria) to cercariae of *Echinostoma trivolvis* and *Echinostoma caproni* (Trematoda). J Parasitol 1991; 77:126-32.

Fripp, P.J. The sites of (1-^{14}C)-glucose assimilation in *Schistosoma haematobium.* Comp Biochem Physiol 1967; 23:893-8.

Fujino, T. "Structure and Function of the Alimentary Tract in Trematodes."*Advances in Trematode Biology*", B. Fried and T. Graczyk, eds. Boca Raton, New York, U.S.A.: CRC Press, 1997.

Fujino, T., Ishii, Y. Comparative ultrastructural topography of the gut epithelia of the lung fluke *Paragonimus* (Trematoda: Troglotrematidae). Int J Parasitol 1978; 8:139-48.

Fujino, T., Ishii, Y. Secretion, absorption and lipid excretion in the gastrodermis of the lung flukes, *Paragonimus ohirai* and *P. westermani*: ultrastructural observations. Jap J Parasitol 1988; 37:227

Fukuda, K., Fujino, T., Hirata, M. Ultrastructural characterization of tegumental spines of the adult lung fluke, *Paragonimus westermani.* Z Parasitenk 1987; 73:95-7.

Gaevskaya, A.V., Nigmatulin, Ch.M. Some ecological aspects of parasitic associations of squids *Stenoteuthis oualaniensis* (Steenstrup, 1855) in tropical Atlantic. Biologicheskie Nauki (*Biological Sciences*) 1981; 1:52-7 (in Russian).

Galaktionov, K.V. The infection of males and females of molluscs of the genus *Littorina* (Gastropoda, Prosobranchia) with parthenitae of trematodes in the Barents Sea coastal waters. Parazitologiya 1985; 19:213-9 (in Russian).

Galaktionov, K.V. "Trematode Life Cycles in Littoral Biocenoses." In *Zhiznennye Cycly Parazitov v Biotsenozakh Severnykh Morei (Parasite Life Cycles in Biocenoses of the Northern Seas).* Yu.I. Poljansky, ed. Apatity, USSR: Kola Scientific Centre of the USSR Academy of Sciences Publ., 1987 (in Russian).

Galaktionov, K.V. *Maritrema murmanica* sp. n. – a new microphalld trematode with aberrant life cycle. Parazitologiya 1989; 23:412-18 (in Russian).

Galaktionov, K.V. "Influence of Parasitizing of Microphallid Trematode Parthenitae on Resistance of Littoral Molluscs, *Littorina saxatilis* (Olivi) to Stress Factors." In *Morfologiya i Ekologiya Parazitov Morskikh Zhivotnykh (Morphology and Ecology of Parasites of Marine Animals).* A.A. Dobrovolskij, ed. Apatity, Russia: Kola Scientific Centre of the Russian Academy of Sciences Publ., 1990 (in Russian).

Galaktionov, K.V. The development of the metacercariae of *Microphallus pirum* (Syn. *Levinseniella somateria*) (Trematoda: Microphallidae). Parazitologiya 1991a; 25:116-24 (in Russian).

Galaktionov, K.V. Some evolutionary tendencies in morphogenesis of the larvae of the hermaphroditic generation of trematodes of the family Microphallidae. Parazitologiya 1991b; 25:520-5 (in Russian).

Galaktionov, K.V. Seasonal dynamics of the age composition of daughter sporocyst hemipopulations of microphallids of the *"pygmaeus"* group (Trematoda, Microphallidae) in littoral molluscs, *Littorina saxatilis,* in the Barents Sea. Parazitologiya 1992; 26:462-9 (in Russian).

Galaktionov, K.V. *Life Cycles of Trematodes as Components of Ecosystems*. Apatity, Russia: Kola Scientific Centre of the Russian Academy of Sciences Publ., 1993 (in Russian).

Galaktionov, K.V. An experimental study of the unusual life cycle of *Parvatrema* sp. (Trematoda: Gymnophallidae). Parazitologiya 1996a; 30:487-94 (in Russian).

Galaktionov, K.V. Life cycle and distribution of seabird helminths in Arctic and sub-Arctic regions. Bull Scand Soc Parasitol 1996b; 6:31-49.

Galaktionov, K.V., Bustnes, J.O. Species composition and prevalence of seabird trematode larvae in periwinkles at two littoral sites in North-Norway. Sarsia 1995; 80:187-91.

Galaktionov, K.V., Bustnes, J.O. Distribution patterns of marine bird digenean larvae in periwinkles along the southern Barents Sea coast. Dis Aquat Organisms 1999; 37:221-30.

Galaktionov, K.V., Dobrovolskij, A.A. "An Attempt at the Populational Analysis of Trematode Life Cycles by the Example of the *"Pygmaeus"* Group Microphallids (Trematoda: Microphallidae)." In *Ekologo-Parazitologicheskie Issledovaniya Severnykh Morei (Ecological-Parasitological Investigations of the Northern Seas).* Yu.I. Poljansky, ed., Apatity, USSR: Kola Scientific Centre of the USSR Academy of Sciences Publ., 1984 (in Russian).

Galaktionov, K.V., Dobrovolskij, A.A. Development and reproduction of the mother generation of parthenitae of trematodes of the genus *Microphallus* (Plagiorchiida, Microphallidae). Zoologicheskii Zhurnal (*Zoological Journal*) 1985; 64: 1468-75 (in Russian).

Galaktionov, K.V., Dobrovolskij, A.A. "Parasites in Marine Shelf Communities." In *Zhizn' i Usloviya eyo Sushchestvovaniya v Bentali Barentseva Morya (Life and Conditions of its Existence at the Bottom of the Barents Sea).* G.G. Matishov, ed. Apatity, USSR: Kola Scientific Centre of the USSR Academy of Sciences Publ., 1986 (in Russian).

Galaktionov, K.V. and A.A. Dobrovolskij. *Hermaphroditic Generation of Trematodes*. Leningrad, USSR: Nauka, 1987 (in Russian).

Galaktionov, K.V., Dobrovolskij, A.A. "Parasites in Ecosystems of the Northern Seas". In *Zhizn' i Sreda Polyarnykh Morei (Life and Environment of the Polar Seas).* G.G. Matishov, ed. Leningrad: Nauka, 1989 (in Russian).

Galaktionov, K.V., Dobrovolskij, A.A. *Origin and Evolution of Trematode Life Cycles*. St. Petersburg, Russia: Nauka, 1998 (in Russian).

Galaktionov, K.V., Krasnodembskij, E.G. Glandular formations of microphallid trematode cercariae. Vestnik Leningradskogo Universiteta (*Bulletin of the Leningrad University*) 1990; ser. 3, 2:3-10 (in Russian).

Galaktionov, K.V., Malkova, I.I. Tegument structure in microphallid trematode cercariae. Parazitologiya 1990a; 24:301-8 (in Russian).

Galaktionov, K.V., Malkova, I.I. Ultrastructure and formation of the stylet in cercariae of microphallid trematodes (Plagiorchiidae, Microphallidae). Zoologicheskii Zhurnal (*Zoological Journal*) 1990b; 69:130-5 (in Russian).

Galaktionov, K.V., Malkova, I.I. Development of the alimentary tract during morphogenesis of the metacercariae of *Levinseniella brachysoma*. J Helminthol 1993; 67:87-94.

Galaktionov, K.V., Malkova, I.I. The glands of trematode cercariae of the family Microphallidae Travassos, 1920. Int J Parasitol 1994; 24:595-604.

Galaktionov, K.V., Malkova, I.I. Changes in the excretory bladder ultrastructure during the morphogenesis of *Levinseniella brachysoma* metacercariae. J Helminthol 1995; 69:203-12.

Galaktionov, K.V., Marasaev, S.F. Patterns in distribution of molluscs of the genus *Littorina* infected by trematode parthenitae in the intertidal zone of the Barents and White Seas. Vestnik Leningradskogo Universiteta (*Bulletin of the Leningrad University*) 1986; ser. 3, 1:88-90 (in Russian).

Galaktionov, K.V., Mikhailova, N.A. "Distribution of Trematode Infection in Littoral Molluscs of the Genus *Littorina* in the Onega Bay of the White Sea. " In *Morfologiya i Ekologiya Parazitov Morskikh Zhivotnykh (Morphology and Ecology of Parasites of Marine Animals)*. A.A. Dobrovolskij, ed. Apatity, Russia: Kola Scientific Centre of the Russian Academy of Sciences Publ., 1990 (in Russian).

Galaktionov, K.V., Rusanov, N.I. "Some Ecological-Populational Aspects of Interrelations in the System Molluscs of the Genus *Littorina* – Trematode Parthenitae on the Site of the Barents Sea Coast." In *Issledovania Biologii, Morfologii i Fiziologii Gidrobiontov (Studies on Biology, Morphology, and Physiology of Water Organisms)*. G.G. Matishov, ed. Apatity, USSR: Kola Branch of the USSR Academy of Sciences Publ., 1983 (in Russian).

Galaktionov, K.V., Berger, V.Ya., Prokofiev, V.V. Comparison of resistance to environmental factors of the molluscs *Hydrobia ulvae* infected with trematode parthenitae and those free from infection. Parazitologiya 2002; 36:195-202 (in Russian).

Galaktionov, K.V., Olenev, A.V., Dobrovolskij, A.A. Two species of cyathocotylid cercariae from the freshwater snail *Melanopsis praemorsa*. Parazitologiya 1980; 14:299-307 (in Russian).

Galaktionov, K.V., Malkova, I.I., Irwin, S.W.B., Saville, D.H., Maguire, J.G. Developmental changes in the tegument of four microphallid metacercariae in their second (crustacean) intermediate hosts. J Helminthol 1996; 70:201-10.

Galaktionov, K.V., Malkova, I.I., Irwin, S.W.B., Saville, D.H., Maguire, J.G. The structure and formation of metacercarial cysts in the trematode family Microphallidae Travassos, 1920. J Helminthol 1997; 71:13-20.

Galaktionov, K.V., Bulat, S.A., Alekhina, I.A., Saville, D.H., Fitzpatrick, S.M., Irwin, S.W.B. An investigation of evolutionary relationships within "*pygmaeus*" group microphallids (Trematoda: Microphallidae) using genetic analysis and scanning electron microscopy. J Helminthol (*submit*)

Genov, T., Samnaliev, P. "Biology, Morphology and Taxonomy of *Plagiorchis elegans* (Rudolphi, 1802) (Plagiorchiidae) in Bulgaria". In: *Fauna, Taxonomiya, Ekologiya na Khelminti po Ptitsi*. I.Vasilev, ed. Sofia, Bulgaria: Bulgarian Academy of Sciences Publ., 1984.

Gerasev, P.I., Dobrovolskij, A.A. Development of the hermaphroditic generation of *Astiotrema trituri* (Trematoda, Plagiorchiidae). Parazitologicheskii Sbornik Zoologicheskogo Instituta Akademii Nauk SSSR *(Parasitological Proceedings of the Zoological Institute of the USSR Academy of Sciences)* 1977; 27:89-111 (in Russian).

Gérard, C., Moné, H., Théron, A. *Schistosoma mansoni – Biomphalaria glabrata*: dynamics of the sporocyst population in relation to the miracidial dose and host size. Can J Zool 1993; 71:1880-5.

German, S.M., Beer, S.A. A method allowing determination of viability of opisthorchis eggs based on miracidial hatching under experimental conditions. Meditsinskaya Parazitologiya i Parazitarnye Bolezni *(Medical Parasitology and Parasitic Diseases)* 1984; 1:57-60 (in Russian).

Gibson, D.I. Some ultrastructural studies on the excretory bladder of *Podocotyle staffordi* Miller, 1941 (Digenea). Bull Brit Mus Nat Hist (Zool) 1973; 24:461-5.

Gibson, D.I. Aspects of the ultrastructure of the daughter sporocyst and cercaria of *Podocotyle staffordi* Muller, 1974 (Digenea: Opecoelidae). Norw J Zool 1974; 22: 237-52.

Gibson, D.I. Questions in digenean systematics and evolution. Parasitology 1987; 95:429-60.

Gibson, D.I., Bray, R.A. The Hemiuroidea: terminology, systematics and evolution. Bull Brit Mus (Natur Hist) Zool. ser. 1979; 36:35-146.

Gibson, D.I., Bray, R.A. The evolutionary expansion and host-parasite relationships of the Digenea. Int J Parasitol 1994; 24:1213-26.

Ginetsinskaya, T.A. Life cycle and biology of the developmental stages of *Cyclocoelum microstomum* (Trematoda). Uchenye Zapiski Leningradskogo Universiteta *(Scientific Papers of the Leningrad University)*. Ser. Biol 1954a; 172:90-112 (in Russian).

Ginetsinskaya, T.A. Importance of taxes in the life activity of cercariae. Dokladyy Akademii Nauk SSSR (*Proceedings of the USSR Academy of Sciences*) 1954b; 97:369-72 (in Russian).

Ginetsinskaya, T.A. *Trematodes, Their Life Cycles, Biology and Evolution.* Leningrad, USSR: Nauka 1968. Translated in 1988 by Amerind Publ. Co. Pvt. Ltd., New Delhi.

Ginetsinskaya, T.A. "Ecology of Parasites of Invertebrates." In *Svobodnozhivushchie i Paraziticheskie Bespozvonochnye (Free-Living and Parasitic Invertebrates).* Yu.I. Poljansky, ed., Leningrad, USSR: Izdatel'stvo Leningradskogo Universiteta, 1983 (in Russian).

Ginetsinskaya, T.A., Besedina, V.V. Glycogen in trematodes and cestodes. *Materialy Nauchnoi Konferentsii VOG. II. (Materials of the Scientific Conference of the All-Union Helminthological Society. Part II.).* V.S. Ershov, ed., Moscow, USSR: Izdatel'stvo Akademii Nauk SSSR, 1965; 64-6 (in Russian).

Ginetsinskaya, T.A., Bhutta, M.Sh. Glandular formations of cercariae and the main trends of their evolution. Parazitologiya 1976; 10:338-45 (in Russian).

Ginetsinskaya, T.A., Dobrovolskij, A.A. To the fauna of trematode larvae from freshwater molluscs of the Volga delta. Part II. Echinistomatid cercariae (family Echinostomatidae). Trudy Astrakhanskogo zapovednika (*Transactions of Astrakhansky Nature Reserve*) 1964; 8:64-104 (in Russian).

Ginetsinskaya, T.A., Dobrovolskij, A.A. "Life Cycle of Trematodes as a System of Adaptations." In *Svobodnozhivushchie i Paraziticheskie Bespozvonochnye (Free-Living and Parasitic Invertebrates).* Yu.I. Poljansky, ed., Leningrad, USSR: Izdatel'stvo Leningradskogo Universiteta, 1983 (in Russian).

Ginetsinskaya, T.A., Khari, B.I. The localization of the enzyme succinate dehydrogenase and reserve nutritional substances in the body of trematode parthenitae and larvae. Ekologicheskaya i Eksperimentalnaya Parazitologiya (*Ecological and Experimental Parasitology*), Leningrad, USSR: Izdatel'stvo Leningradskogo Universiteta 1975; 1:108-24 (in Russian).

Ginetsinskaya, T.A., Saakova, E.O. Routs of migration of trematodes from the family Cyclocoelidae Koss. into the final host's organism. Doklady Akademii Nauk SSSR (*Proceedings of the USSR Academy of Sciences*) 1952; 35:1423-6 (in Russian).

Ginetsinskaya, T.A., Shtein, G.A. Patterns in the parasitic fauna of invertebrates and the application of the main rules of ecological parasitology to their infectional characteristics. Vestnik Leningradskogo Universiteta (*Bulletin of the Leningrad University*) 1961; 15:60-72 (in Russian).

Ginetsinskaya, T.A., Dobrovolskij, A.A., Mashanskii, V.F. On tissue ultrastructure in trematode parthenitae and larvae. *Materialy Nauchnoi Konferencii VOG. I. (Materials of the Scientific Conference of the All-Union Helmintological Society. Part I.).* V.S. Ershov, ed., Moscow, USSR: Izdatel'stvo Akademii Nauk SSSR, 1965; 58-60 (in Russian).

Ginetsinskaya, T.A., Dobrovolskij, A.A., Oksov, I.V. "The Role of Glycogen in Biology of Larval Stages of Trematode Development." In *Raboty po Gelmintologii (Works on Helminthology)*. V.S. Ershov, ed. Moscow, USSR: Nauka 1981 (in Russian).

Ginetsinskaya, T.A., Mashanskii, V.F., Dobrovolskij, A.A. Ultrastructure of the integuments and mode of feeding in rediae and sporocysts. Doklady Akademii Nauk SSSR 1966; 166:249-50 (in Russian).

Gladunko, I. Morphological and biological features of some representatives of the genus *Sanguinicola* Plehn, 1905. *Materialy Nauchnoi Konferencii VOG. II. (Materials of the Scientific Conference of the All-Union Helmintological Society. Part II.)*. V.S. Ershov, ed., Moscow, USSR: Izdatel'stvo Akademii Nauk SSSR, 1968; 81-7 (in Russian).

Goater, C.P., Holmes, J.C. "Parasite-mediated Natural Selection." In *Host-Parasite Evolution. General Principles and Avian Models*. D.H. Clayton and J. Moore, eds. Oxford, U.K.: Oxford University Press, 1997.

Göbel, E., Pan, J.P. Ultrastructure of the daughter sporocyst and developing cercaria of *Schistosoma japonicum* in experimentally infected snails, *Oncomelania hupensis hupensis*. Z Parasitenk 1985; 71:227-40.

Goodchild, G. Additional observation on the bionomics and life history of *Gorgodera amplicava*. J Parasitol 1948; 34:407-27.

Goodchild, C. G., Kirk, D. E. The life history of the *Spirorchis elegans* Stunkard, 1923 (Trematoda, Spirorchiidae) from the painted turtle. J Parasitol 1960; 46:219-29.

Gorbushin, A.M. The enigma of mud snail shell growth: asymmetrical competition or character displacement? Oikos 1996; 77:85-92.

Gorbushin, A.M. Field evidence of trematode-induced gigantism in *Hydrobia* spp. (Gastropoda: Prosobranchia). J Mar Biol Ass UK 1997; 77:785-800.

Gorbushin, A.M. Comparative morphological-functional analysis of the gastropod-trematode interactions. Parazitologiya 2000; 34: 502-14 (in Russian).

Gorbushin, A.M., Levakin, I.A. The effect of trematode parthenitae on the growth of *Onoba aculeus*, *Littorina saxatilis* and *L. obtusata* (Gastropoda: Prosobranchia). J Mar Biol Ass UK 1999; 79:273-9.

Grabda, B. *Astiotrema trituri* sp. n. (Trematoda: Plagiorchiidae) from the small intestine of *Triturus vulgaris* L. Bull Acad Polonica Sci Cl II 1959; 7(1):17-21.

Grabda-Kazubska, B., Moczoń, T. The nervous system of *Echinoparyphium recurvatum* (Linstow, 1973) cercaria and adult (Digenea, Echinostomatidae). Acta Parasitol Polonica 1988; 33:177-84.

Graefe, G., Hohorst, W., Drager, H. Forked tail of the cercaria of *Schistosoma mansoni*. Nature 1967; 215:207-8.

Granovitch, A.I. The effect of trematode infection on the population structure of *Littorina saxatilis* (Olivi) in the White Sea. *Proceedings of the Third International Symposium on Littorinid Biology*. J. Grahame, P.J. Mill and D.G. Reid, eds. London, U.K.: Malacological Society of London, 1992: 255-63.

Granovitch, A.I. Parasitic systems and the structure of parasite populations. Helgol Mar Res 1999; 53:9-18.

Granovitch, A.I., Johannesson, K. Digenetic trematodes in four species of *Littorina* from the west coast of Sweden. Ophelia 2000; 53:55-65.

Grant, W.C., Harkema, R., Muse, K.E. Ultrastructure of *Pharyngostomoides procyonis* Harkema 1942 (Diplostomatidae). I. Observations on the male reproductive system. J Parasitol 1976; 62:39-49.

Grant, W.C., Harkema, R., Muse, K.E. Ultrastructure of *Pharyngostomoides procyonis* Harkema 1942 (Diplostomatidae). II. The female reproductive system. J Parasitol 1977; 63:1019-30.

Gresson, R.A. Electron microscopy of the ovary of *Fasciola hepatica*. Quart J Microsc Sci 1964; 105:213-8.

Groschaft, J., Tenora, F. Trematodes of the *Plagiorchis* Lühe, 1899 (Plagiorchiidae), parasites of bats in Afganistan. Věstn Ĉs Společ Zool 1973; 37:241-9.

Guilford, H.G. Gametogenesis, egg-capsule formation, and early miracidial development in the digenetic trematode *Halipegus eccentricus*. Trans Am Microsc Soc 1961; 47:757-64.

Gupta, B.C., Guraya, S.S., Parshad, V.R. Morphological and histhochemical studies on the prostate gland of developing and adult *Paramphistomum cervi* (Digenea: Paramphistomatidae). Int J Invertebr Reprod 1983; 6:219-28.

Haas, W. Reizphysiologische Untersuchungen an Cercarien von *Diplostomum spathaceum*. Z Vergl Physiol 1969; 64:254-87.

Haas, W. Analyse der Invasionsmechanismen der Cercarien von *Diplostomum spathaceum* – I. Fixation und Penetration. Int J Parasitol 1974a; 4:311-9.

Haas, W. Analyse der Invasionsmechanismen der Cercarien von *Diplostomum spathaceum* – II. Chemische Invasionsstimuli. Int J Parasitol 1974b; 4:321-30.

Haas, W. Die Anheftung (Fixation) der Cercarie von *Schistosoma mansoni*. Einfluss natürlicher Substrate und der Temperatur. Z Parasitenk 1976; 49:63-72.

Haas, W. *Schistosoma mansoni*: cercaricidal effect of 2-tetradecenoic acid, a penetration stimulant. Exp Parasitol 1984; 58:215-22.

Haas, W. Physiological analysis of cercarial behaviour. J Parasitol 1992; 78:243-55.

Haas W. Physiological analysis of host-finding behaviour in trematode cercariae: adaptations for transmission success. Parasitology 1994; 109:S15-S29.

Haas, W. "The Behavioral Biology of Echinostomes". In *Echinostomes as Experimental Models for Biological Research*. B. Fried and T.K. Graczyk, eds. The Netherlands: Kluwer Academic Publishers, 2000.

Haas, W., Haberl, B. "Host Recognition by Trematode Miracidia and Cercariae." In *Advances in Trematode Biology* B. Fried and T.K. Graczyk, eds. Boca Raton, New York, U.S.A.: CRC Press, 1997.

Haas, W., Van de Roemer, A. Invasion of the vertebrate skin by cercariae of *Trichobilharzia ocellata*: penetration processes and stimulating cues. Parasitol Res 1998; 84:787-95.

Haas, W., Granzer, M., Brockelman, C. Finding and recognition of the bovine host by cercariae of *Schistosoma spindale*. Parasitol Res 1990a; 76:343-50.

Haas, W., Granzer, M., Brockelman, C. *Opisthorchis viverrini* : finding and recognition of the fish host by cercariae. Exp Parasitol 1990b; 71:422-31.

Haas, W., Granzer, M., Garcia, E.G. Host identification by *Schistosoma japonicum* cercariae. J Parasitol 1987; 73:568-77.

Haas, W., Gui, M., Haberl, B., Ströbel, M. Miracidia of *Schistosoma japonicum*: approach and attachment to the snail host. J Parasitol 1991; 77:509-13.

Haas, W., Haberl, B., Schmalfuss, G., Khayyal, M.T. *Schistosoma haematobium* cercarial host-finding and host-recognition differs from that *Schistosoma mansoni*. J Parasitol 1994; 80:345-53.

Haas, W., Haberl, B., Kalbe, M., Körner, M. Snail-host-finding by miracidia and cercariae: chemical host cues. Parasitol Today 1995a; 11:468-72.

Haas, W., Körner, M., Hutterer, E., Wegner, M., Haberl, B. Finding and recognition of the snail intermediate hosts by 3 species of echinostome cercariae. Parasitology 1995b; 110:133-42.

Haas, W., Diekhoff, D., Koch, K., Schmalfuss, G., Loy, Ch. *Schistosoma mansoni* cercariae: stimulation of acetabular gland secretion is adapted to the chemical composition of mammalian skin. J Parasitol 1997; 83:1079-85.

Haberl, B., Körner, M., Spengler, Y., Hertel, J., Kalbe, M., Haas, W. Host-finding in *Echinostoma caproni*: miracidia and cercariae use different signals to identify the same snail species. Parasitology 2000; 120:479-86.

Haight, M., Davidson, D., Pasternak, J. Relationship between nuclear morphology and phases of the cell cycle during cercarial development of the digenetic trematode *Trichobilharzia ocellata*. J Parasitol 1977a; 63:267-73.

Haight, M., Davidson, D., Pasternak, J. Cell cycle analysis in developing cercariae of *Trichobilharzia ocellata* (Trematoda, Schistosomatidae). J Parasitol 1977b; 63:274-81.

Hall, J.E. Studies on virgulate xiphidiocercariae from Indiana and Michigan. Am Midl Nat 1960; 63:226-45.

Halton, D.W. An unusual structural organization to the gut of a digenetic trematode, *Fellodistomum fellis*. Parasitology 1982; 85:53-60.

Halton, D.W., Gustafsson, M.K.S. Functional morphology of the platyhelminth nervous system. Parasitology 1996; 113S:S47-S72.

Halton, D.W., Johnston, B.R. Functional morphology of the metacercarial cyst of *Bucephaloides gracilescens* (Trematoda: Bucephalidae). Parasitology 1982; 85:45-52.

Halton, D.W., McCrae, J.M. Development of the tegument and alimentary tract in a digenetic trematode, *Fellodistomum fellis*. Parasitology 1985; 90:193-204.

Halton, D.W., Maule, A.G., Shaw, C."Trematode Neurobiology." In *Advances in Trematode Biology*, B. Fried and T.K. Graczyk, eds. Boca Raton, New York, U.S.A.: CRC Press, 1997.

Hamilton, W.D., Axelrod, R. and Tanese, R. Sexual reproduction as an adaptation to resist parasites (a review). Proc Nat Acad Sci USA 1990; 87:3566-73.

Hanna, R.E.B. *Fasciola hepatica*: a light and electron autoradiographic study of shell protein and glycogen synthesis by vitelline follicles in tissue slices. Exp Parasitol 1976; 39:18-28.

Hanna, R.E.B. *Fasciola hepatica*: glycocalyx replacement in the juvenile as a possible mechanism for protection against host immunity. Exp Parasitol 1980; 50:103-14.

Harris, K.R., Cheng, T.C., Cali, A. An electron microscope study of the tegument of the metacercaria and adult of *Leucochloridiomorpha constantiae* (Trematoda: Brachylaemidae). Parasitology 1974; 68:57-67.

Helluy, S. Relations hotes-parasite du trématode *Microphallus papillorobustus* (Rankin, 1940). I. Pénétration des cercaires et rapports des métacercaires avec le tissu nerveux des *Gammarus*, hôtes intermédiaires. Ann Parsitol Hum Comp 1982; 57:263-70.

Helluy, S. Relations hotes-parasite du trematode *Microphallus papillorobustus* (Rankin, 1940). II. Modifications du comportement des *Gammarus* hôtes intermédiaires et localisation des métacercaires. Ann Parsitol Hum Comp 1983; 58:1-17.

Helluy, S. Relations hotes-parasite du trematode *Microphallus papillorobustus* (Rankin, 1940). III. Facteurs impliqués dans les modifications du comportement des *Gammarus* hôtes intermédiaires et tests de prédation. Ann Parsitol Hum Comp 1984; 59:41-56.

Helluy, S., Holmes, J.C. Serotonin, octopamine, and the clinging behaviour induced by the parasite *Polymorphus paradoxus* (Acanthocephala) in *Gammarus lacustris* (Crustacea). Can J Zool 1990; 68:1214-20.

Hendow, H.T., James, B.L. Ultrastructure of the male reproductive ducts in *Maritrema linguilla* (Digenea: Microphallidae). Int J Parasitol 1988; 18:221-9.

Hendow, H.T., James, B.L. Ultrastructure of vitellarium, vitellogenesis and associated ducts in *Maritrema linguilla* (Digenea: Microphallidae). Int J Parasitol 1989; 19:489-97.

Herber, E.C. Life history studies on two trematode of the subfamily Notocotylinae. J Parasitol 1942; 28:179-96.

Heyneman, D. On the origin of complex life cycles in digenetic flukes. Libro Hamenaje al dr. E. Caballero Jubileo 1930-1960. Mexico, 1960; 133-52.

Higgins, J.C. The role of the tegument of the metacercariae stage of *Bucephalus haimeanus* (Lacaze-Duthiers, 1854) in the absorption of particulate material and small molecules in solution. Parasitology 1979; 78:99-106.

Higgins, J.C. Formation of the cyst wall and related changes in the structure of the tegument of *Bucephalus haimeanus* (Lacaze-Duthiers, 1854) during its metamorphosis from the cercarial to metacercarial stages. Parasitology 1980; 81:47-59.

Hockley, D.J. *Schistosoma mansoni*: the development of the cercarial tegument. Parasitology 1972; 64:245-52.

Hockley, D.J. Ultrastructure of the tegument of *Schistosoma*. Adv Parasitol 1973; 11:233-305.

Hockley, D.J. Blood flukes have a double outer membrane. Nature 1977; 269:147-9.

Holy, J.M., Oaks, J.A. Mechanical integration of muscle, tegument and subtegumental tissues by anchoring fibrils and microfibrils in the cestodes *Hymenolepis diminuta*. Tissue Cell 1987; 19:881-91.

Holy, J.M., Wittrock, D.D. Ultrastructure of the female reproductive organs (ovary, vitellaria, and Mehlis' gland) of *Halipegus eccentricus* (Trematoda: Derogenidae). Can J Zool 1986; 64:2203-13.

Hoole, D., Mitchell, J.B. Development of the alimentary tract of *Gorgoderina vitelliloba* during migration in *Rana temporaria*. Int J Parasitol 1983a; 13:455-62.

Hoole, D., Mitchell, J.B. *Gorgoderina vitelliloba*: an ultrastructural study on the development of the tegument from the metacercaria to the adult fluke. Parasitology 1983b; 86:323-33.

Horák, P., van der Knaap, W.P.W. Lectins in snail-trematode immune interactions: a review. Folia Parasitol 1997; 44:161-72.

Horsfall, M.W. Studies on the life history and morphology of the cystocercous cercariae. Trans Am Microsc Soc 1934; 53:311-47.

Hoskin, G.P., Cheng, T.C. *Himasthla quissetensis*: uptake and utilization of glucose by rediae as determined by autoradiography and respirometry. Exp Parasitol 1974; 35:61-7.

Howard, R.S., Lively, C.M. Parasitism, mutation, accumulation and the maintenance of sex. Nature 1994; 367:554-7.

Hsu, K.C., Lie, K.J., Bash, P.E. The life history of *Echinostoma rodriguesi* sp. n. (Trematoda: Echinostomatidae). J Parasitol 1968; 54:333-9.

Hudson, P.J., Dobson, A.P. "Host-Parasite Processes and Demographic Consequences." In: *Host-Parasite Evolution. General Principles and Avian Models.* D.H. Clayton and J. Moore, eds. Oxford, U.K.: Oxford University Press, 1997.

Hughes, R.N., Answer, P. Growth, spawning and trematode infestation of *Littorina littorea* (L.) from an exposed shore in North Wales. J Moll Stud 1982; 48:321-30.

Hulscher, J.B. Burying depth and trematode infection in *Macoma baltica*. Neth J Sea Res 1973; 6:141-56.

Hulscher, J.B. The oystercatcher *Haematopus ostralegus* as a predator of the bivalve *Macoma baltica* in the Dutch Wadden Sea. Ardea 1982; 70:89-152.

Hussey, K.L. Comparative embryological development of the excretory system in digenetic trematodes. Trans Am Microsc Soc 1941; 60:171-210.

Hutterer, E., Klaver, E., Wegner, M., Haas, W. Snails as intermediate hosts: mechanisms of host finding in cercariae. Zentrbl Bakt, Abstracts 1992; 325:43.

Huxham, M., Raffaelli, D., Pike, A.W. The effect of larval trematodes on the growth and burrowing behaviour of *Hydrobia ulvae* (Gastropoda: Prosobranchiata) in the Ythan estuary, north-east Scotland. J Exp Mar Biol Ecol 1995; 185:1-17.

Irwin, S.W.B. Incidence of trematode parasites in two populations of *Littorina saxatilis* (Olivi) from the north shore of Belfast Lough. Irish Naturalist J 1983; 21:26-9.

Irwin S.W.B. "Excystation and Cultivation of Trematodes." In *Advances in Trematode Biology*, B. Fried and T.K. Graczyk, eds. Boca Raton, New York, U.S.A.: CRC Press, 1997.

Irwin, S.W.B., Maguire, J.G. Ultrastructure of the vitelline follicles of *Gorgoderina vitelliloba* (Trematoda: Gorgoderidae). Int J Parasitol 1979; 9:47-53.

Irwin, S.W.B., Threadgold, L.T. Electron-microscope studies of *Fasciola hepatica*. VIII. The development of the vitelline cells. Exp Parasitol 1970; 28:399-411.

Irwin, S.W.B., Threadgold, L.T. Electron-microscope studies of *Fasciola hepatica*. X. Egg formation. Exp Parasitol 1972; 31:321-31.

Irwin, S.W.B., Threadgold, L.T., Howard, N.M. *Cryptocotyle lingua* (Creplin) (Digenea: Heterophyidae): observations on the morphology of the rediae, with special reference to the birth papilla and release of cercariae. Parasitology 1978; 76:193-9.

Irwin, S.W.B., Magiure, J.G., Saville, D.H. Identification of the cercarial stage of *Maritrema arenaria* (syn. *M. gratiosum*) (Trematoda: Microphallidae). J Nat Hist 1990; 24:949-54.

Irwin, S.W.B., Galaktionov, K.V., Malkova, I.I., Saville, D.H., Fitzpatrick, S.M. An ultrastructural study of reproduction in the parthenogenetic metacercariae of *Cercaria margaritensis* Ching, 1982 (Digenea: Gymnophallidae). Parasitology 2003; 126:261-71.

Ishii, Y. Studies on the development of *Fasciolopsis buski*. I. Development of the eggs outside of the host. J Med Assoc Formosa 1934; 33:349-412.

Isseroff, H. Fine structure of the eyespot in the miracidium of *Philophthalmus megalurus* (Cort, 1914). J Parasitol 1964; 50:549-54.

Isseroff, H., Cable, R.M. Fine structure of photoreceptors in larval trematodes, a comparative study. Z Zelforsch 1968; 8:511-34.

Ito, J. A monograph of cercariae in Japan and adjacent territories. Progr Med Parasitol Jap 1964; 1:395-550.

Ivanov, A.V. and Mamkaev Yu.V. *Turbellarians (Turbellaria), Their Origin and Evolution: a Phylogenetical Essay*. Leningrad, USSR: Nauka, 1973 (in Russian).

James, B.L. The life cycle of *Parvatrema homoeotecnum* sp. nov. (Trematoda: Digenea) and review of the family Gymnophallidae Morozov, 1955. Parasitology 1964; 54:1-41.

James, B.L. The occurrence of larval Digenea in ten species of intertidal prosobranch molluscs in Cardigan Bay. J Nat Hist 1968: 2: 329-43.

James, B.L. The Digenea of the intertidal prosobranch, *Littorina saxatilis* (Olivi). Z Zool, Systematik Evolutionsforsch 1969; 7:273-316.

James, B.L., Bowers, E.A. Reproduction in the daughter sporocyst of *Cercaria bucephalopsis haimeana* (Lacaze-Duthiers, 1854) (Bucephalidae) and *Cercaria dichotoma* Lebour, 1911 (non Muller) (Gymnophallidae). Parasitology 1967; 57:607-25.

James, B.L., Bowers, E.A., Richards, J.G. The ultrastructure of the daughter sporocyst of *Cercaria bucephalopsis haimaena* Lacaze-Duthiers, 1854 (Digenea: Bucephalidae) from the edible cockle, *Cardium edule* L. Parasitology 1966; 56:753-62.

Jamieson, B.G.M. Larval stages of the progenetic trematode *Parahemiurus bennettae* Jamieson, 1966 (Digena, Hemiuridae) and the evolutionary origin of cercariae. Proc R Soc Queensland 1966; 77: 81-93.

Jensen, K.T., Latama, G., Mouritsen, K.N. The effect of larval trematodes on survival rates of two species of mud snails (Hydrobiidae) experimentally exposed to desiccation, freezing and anoxia. Helgoländer Meersunters 1996; 50:317-35.

Joffe, B.I. Morphological regularities in the evolution of the nervous system in the Plathelminthes: anatomical variants of the orthogon and their dependence upon the body form. In *Parasiticheskie i Svobodnozhivushie Ploskie Chervi: Faunistica i Morfologia (Parasitic and Free-Living Plathelminthes: Faunistics and Morphology)*. Yu.V. Mamkaev and B.I. Joffe, eds. Leningrad, U.S.S.R.: Zoological Institute Publ., 1990 (in Russian).

Johansen, B.O., Jensen, A.J. Infestations of Atlantic salmon *Salmo salar*, by *Gyrodactylus salaris* in Norwegian rivers. J Fish Biol 1986; 29:233-41.

Jokela, J., Lively, C.M. Spatial variation in infection by digenetic trematodes in a population of freshwater snails (*Potamopyrgus antipodarum*). Oecologia 1995; 103:509-17.

Jokela, J., Lively, C.M., Taskinen, J., Peers, A.D. Effect of starvation on parasite-induced mortality in a freshwater snail *Potamopyrgus antipodarum*. Œcologia 1999; 119:320-5.

Jordan, H.E., Byrd, E.E. The life cycle of *Brachycoelium mesorchium* Byrd, 1937 (Trematoda: Brachycoeliinae). Z Parasitenk 1967; 29:61-84.

Justine, J.-L., Mattei, X. Ultrastructural observations on the spermatozoon, oocyte and fertilization process in *Gonapodasmius*, a gonochoristic trematode (Trematoda: Digenea: Didymozoidae). Acta Zool 1984; 65:171-7.

Kabat, A.R. Effects of trematode parasitism on reproductive output of the bivalve *Transennella tantilla*. Can J Zool 1986; 64:267-70.

Kagan, I.C. Further contributions to the life history of *Neoleucochloridium problematicum* (Magath, 1920) new. comb. (Trematoda, Brachylaemidae). Trans Am Microsc Soc 1952; 71:20-44.

Kalbe, M., Haberl, B., Haas, W. *Schistosoma mansoni* miracidial host-finding: species-specificity of an Egyptian strain. Parasitol Res 1996; 82:8-13.

Kalbe, M., Haberl, B., Haas, W. Miracidial host-finding in *Fasciola hepatica* and *Trichobilharzia ocellata* is stimulated by species-specific glycoconjugates released from the host snails. Parasitol Res 1997; 83:806-12.

Kanev, I., Vassilev, I., Bayssade-Dufour, Ch., Albaret, J.L., Cassone, J. Chétotaxie cercarienne d'*Echinostoma revolutum* (Froelich, 1802) et *E. echinatum* (Zeder, 1803) (Trematoda, Echinostomatidae). Ann Parasitol Hum Comp 1987; 62:222-34.

Karpovich, V.N. "On Possible Cycling in Number Dynamics of the Common Eider." In *Problemi Izuchenia i Ohrani Prirodi Pribelomoria (Problems of Study and Protection of the White Sea Region).* V.N. Karpovich, ed. Murmansk, U.S.S.R.: Murmanskoye Knizhnoe izdatelstvo, 1987 (in Russian).

Kawanaka, M., Hayashi, S., Ohtomo, H. Nutritional requirements of *Schistosoma japonicum* eggs. J Parasitol 1983; 69:857-61.

Kearn, G.C., Cleveland, G., Wilkins, S. An in vitro study of the forcible ejection of the metacercaria of the strigeid digenean *Apatemon (Australapatemon) minor* Yamaguti, 1933 by rapid inward expansion of the cyst wall. J Helminthol 1989; 63:63-71.

Kechemir, N., Théron, A. Existence of replicating sporocysts in the development cycle of *Schistosoma haematobium.* J Parasitol 1980; 66:1068-70.

Kendall, S.B., McCullough, F.S. The emergence of cercariae of *Fasciola hepatica* from the snail *Lymnaea truncatula.* J Helminthol 1951; 25:77-92.

Kersten, M. Time and energy budgets of oystercatchers *Haematopus ostralegus* occupying territories of different quality. Ardea 1996; 84A:291-310.

Keshavarz-Valian, H., Nollen, P.M. Responses of *Philophthalmus gralli* miracidia to gravity and light. J Parasitol 1980; 66:684-6.

Keshavarz-Valian, H., Nollen, P.M., Maynard, G. *Philophthalmus gralli*: chemosensitivity of miracidia. J Parasitol 1981; 67:527-30.

Keymer, A.E., Read, A.F. "Behavioural Ecology: the Impact of Parasitism." In *Parasite-Host Associations, Coexistence or Conflict?* C.A. Toft, A. Aeschlimann and L. Bolis, eds. Oxford, New York, Toronto, U.K.: Oxford University Press 1991.

Khari, B.I. Succinate dehydrogenase and reserve nutrients at the different life cycle stages of *Haplometra cylinracea* Zeder, 1800 (Plagiorchioidea). Ekologicheskaya i Eksperimentalnaya Parazitologiya (*Ecological and Experimental Parasitology*), Leningrad, USSR: Izdatel'stvo Leningradskogo Universiteta 1975; 1:124-32 (in Russian).

Kinoti, G.K. The attachment and penetration apparatus of the miracidium of *Schistosoma.* J Helminthol 1971; 45:229-35.

Kniskern, V.B. Studies on the trematode family Bucephalidae Poche, 1907. Part 2. The life history of *Rhipidocotyle septapapillata* Krull, 1934. Trans Am Microsc Soc 1952; 71:1-17.

Kofiadi, A.Kh. Structure of germinal masses and some problems of the dynamics of daughter sporocysts' development in Hemiuridae gen.sp. Parazitologiya 1995; 29:404-16 (in Russian).

Køie, M. On the histochemistry and ultrastructure of the redia of *Neophasis lageniformis* (Lebour, 1910) (Trematoda, Acanthocolpidae). Ophelia 1971a; 9:113-43.

Køie, M. On the histochemistry and ultrastructure of the daughter sporocyst of *Cercaria buccini* Lebour, 1911. Ophelia 1971b; 9:145-63.

Køie, M. On the histochemistry and ultrastructure of the tegument and associated structures of the cercaria of *Zoogonoides viviparus* in the first intermediate host. Ophelia 1971c; 9:165-206.

Køie, M. The ultrastructure of the caecal epithelium of the intraredial cercaria of *Neophasis lageniformis* (Lebour, 1910) (Trematoda: Acanthocolpidae). Z Parasitenk 1973; 139:405-16.

Køie, M. On the morphology and life-history of *Podocotyle reflexa* (Creplin, 1825) Odhner, 1905, and a comparison of its developmental stages with those of *P. atomon* (Rudolphi, 1802) Odhner, 1905 (Trematoda, Opecoelidae). Ophelia 1981; 20:17-43.

Køie, M. The redia, cercaria and early stages of *Aporocotyle simplex* Odhner, 1900 (Sanguinicolidae) – a digenetic trematode which has a polychaete annelid as the only intermediate host. Ophelia 1982; 21:115-45.

Køie, M. On the morphology and life-history of *Lepidapedon elongatum* (Lebour, 1908) Nicoll, 1910 (Trematoda, Lepocreadiidae). Ophelia 1985; 24:135-53.

Køie, M. Ultrastructural study of the host-parasite relationships, including encapsulation, of *Buccinum undatum* (Gastropoda, Prosobranchia) infected with daughter sporocysts of *Zoogonoides viviparus* (Trematoda, Zoogonidae). Dis Aquat Organisms 1987; 2:117-25.

Køie, M. A new cystophorous cercaria in *Lunatia pallida* (Broderip & Sowerby) (Naticidae): possibly the cercaria of *Hemiurus levinseni* Odhner, 1905 (Digenea, Hemiuridae). Ophelia 1990a; 31:77-84.

Køie, M. Redescription of the cercaria of *Lecithochirium rufoviride* (Rudolphi, 1819) Lühe, 1901 (Digenea, Hemiuridae) (= *Cercariae vaullegeardi* Pelseneer, 1906). Ophelia 1990b; 31:85-95.

Køie, M. On the morphology and life history of *Hemiurus luehei* Odhner, 1905 (Digenea: Hemiuridae). J Helminthol 1990c; 64:193-202.

Køie, M. Aspects of the life cycle and morphology of *Hysterothylacium aduncum* (Rudolphi, 1802) (Nematoda, Ascaridoidea, Anisakidae). Can J Zool 1993; 71:1289-96.

Køie, M., Frandsen, F. Stereoscan observations on the miracidium and early sporocyst of *Schistosoma mansoni*. Z Parasitenk 1976; 50:335-44.

Køie, M., Christensen, N.O., Nansen, P. Stereoscan studies of eggs, free-swimming and penetrating miracidia and early sporocysts of *Fasciola hepatica*. Z Parasitenk 1976; 51:79-90.

Køie, M., Nansen, P., Christensen, N.O. Stereoscan studies of rediae, cercariae, cysts, excysted metacercariae and migratory stages of *Fasciola hepatica*. Z Parasitenk 1977; 54:289-97.

Komiya, Y. Die Entwicklung des Exkretionssystemes einiger Trematodenlarven aus Alster und Elbe, nebst Bemerkungen über ihren Entwicklungszyklus. Z Parasitenk 1938: 10:340-85.

Kontrimavichus, V.L. Parasitism and evolution of ecosystems (ecological aspects of parasitism). Zhurnal Obshey Biologii (*Journal of General Biology*) 1982; 43:291-302 (in Russian).

Kontrimavichus, V.L., Atrashkevich, G.I. Parasitic systems and their investigation in the populational biology of helminths. Parazitologiya 1982; 16:177-87 (in Russian).

Kornakova, E.E. Structure of parenchyma of *Udonella murmanica* (Turbellaria, Udonellida) and the main trends in the evolution of the connective tissue of flatworms. Parazitologicheskii Sbornik Zoologicheskogo Instituta Academii Nauk SSSR (*Parasitological Proceedings of the Zoological Institute of the USSR Academy of Sciences*) 1989; 36:161-79 (in Russian).

Kornakova, E.E., Joffe, B.I. A new variant of the neodermatan-type spermiogenesis in a parasitic turbellarian, *Notentera ivanovi* (Plathelminthes) and the origin of Neodermata. Acta Zool (Stockholm) 1999; 80:135-51.

Körner, M., Haas, W. Chemo-orientation of echinostome cercariae towards their snail-hosts: amino acids signal a low host-specificity. Int J Parasitol 1998a; 28:511-6.

Körner, M., Haas, W. Chemo-orientation of echinostome cercariae towards their snail-hosts: the stimulating structure of amino acids and other attractants. Int J Parasitol 1998b; 28:517-25.

Korneva, Zh.V. Ultrastructural organization of parenchyma of some cestodes. Ph.D. Thesis, St. Petersburg State University: St. Petersburg, Russia, 1995 (in Russian).

Korneva, Zh.V. Cellular composition of parenchyma and extracellular matrix in ontogenesis of *Triaenophorus nodulosus* (Cestoda). Izvestia Academii Nauk (*Proceedings of the Academy of Sciences*), Ser Biol 2001; 1:11-22 (in Russian).

Kotikova, E.A. Comparative anatomical investigation of the nervous apparatus of flatworms (Plathelminthes). Ph.D. Thesis, Leningrad State University: Leningrad, USSR, 1971 (in Russian).

Kotikova, E.A. "The Orthogon of the Platyhelminthes and Main Trends of its Evolution." In: *Mofologicheskie Osnovy Filogenetiki Ploskih Chervey (Morphological Principles of Platyhelminthes Phylogenetics).* Y.V. Mamkaev, ed., Leningrad, Russia: Zoological Institute Publ., 1991 (in Russian).

Kotikova, E.A., Joffe, B.I. "Cholinesterases in the Nervous System of Trematodes *Diplodiscus subclavatus* and *Fellodistomum fellis*." In *Parasiticheskie i Svobodnozhivushie Ploskie Chervi: Faunistica i Morfologia (Parasitic and Free-Living Plathelminthes: Faunistics and Morphology).* Yu.V. Mamkaev and B.I. Joffe, eds. Leningrad, U.S.S.R.: Zoological Institute Publ., 1990 (in Russian).

Kotikova, E.A., Joffe, B.I., Resnik, G.K. "On the Nervous System of Maritae of *Fasciola hepatica* and *Dicrocoelium lanceatum*." In *Parasiticheskie i Svobodnozhivushie Ploskie Chervi: Faunistica i Morfologia (Parasitic and Free-Living Plathelminthes: Faunistics and Morphology).* Yu.V. Mamkaev and B.I. Joffe, eds. Leningrad, U.S.S.R.: Zoological Institute Publ., 1990 (in Russian).

Kotikova, E.A., Joffe, B.I., Runovskaya, I.J. The nervous system of *Prosorhynchus squamatus* (Trematoda: Bucephalidae). Parazitologiya 1984; 18:408-12 (in Russian).

Krasnodembskij, E.G. Glandular apparatus of cercariae of some trematode species. Vestnik Leningradskogo Universiteta (*Bulletin of the Leningrad University*), Ser.Biol. 1972; 2:21-30 (in Russian).

Krasnodembskij, E.G. Glandular formations in trematodes. Ph.D. Thesis, Leningrad State University: Leningrad, USSR, 1977 (in Russian).

Krasnolobova, T.A. Biological cycle of the trematode *Prosthogonimus pellucidus* (Linstow, 1873), a pathogen of birds. Helminthologia 1962-1963; 4:217-29 (in Russian).

Krasnolobova, T.A. *Trematode Fauna of the USSR. The Genus Plagiorchis.* Moscow, USSR: Nauka, 1987 (in Russian).

Kruidenier, F.J. The formation and function of mucoids in cercariae, including a study of the virgula organ. Am Midl Natur 1951; 46:660-83.

Kruidenier, F.J. Studies on the formation and function of mucoid glands in non-virgulate xiphidiocercariae. Am Midl Natur 1953; 50:382-96.

Krull, W.H. Studies on the life history of a frog bladder fluke, *Gorgodera amplicava* Looss, 1899. Papers of the Michigan Academy of Science. Arts and Letters 1935a; 20:697-710.

Krull, W.H. Studies on the life history of *Halipegus occidualis* Stafford, 1905. Am Midl Natur 1935b; 16:129-43.

Krupa, P.L., Lewis, L.M., DelVecchio, P. Electron microscopy of ouabain-inhibited potassium-dependent transport adenosine triphosphatase activity in schistosome sporocysts. Ann NY Acad Sci 1975; 266:465-79.

Kulachkova, V.G. "Mortality of the Common Eider Caused by Helminths in the Upper Part of the Kandalaksha Bay." In *Ekologiya i morphologiya gag v SSSR (Ecology and Morphology of Eiders in the USSR).* S.M Uspenskiy, ed., Moscow, USSR: Nauka 1979 (in Russian).

Kuntz, R.E. Embryonic development of the excretory system in forktailed cercariae of the Schistosomas and in a blunt-tailed Brachylaemid cercaria. Trans Am Microsc Soc 1950: 69:1-20.

Kuntz, R.E. Embryonic development of the excretory system in a Psilostome cercariae, a Gymnocephalous (Fasciolid) cercaria and in three Monostome cercariae. Trans Am Microsc Soc 1951: 70:95-118.

Kuntz, R.E. Embryonic development of the excretory system in a Pleurolophocercous cercariae, three stylet cercariae, and Microcaudate Eucotylid cercaria. Trans Am Microsc Soc 1952; 71:45-82.

Kuperman, B.I. *Functional Morphology of Lower Cestodes*. Leningrad, USSR: Nauka, 1988 (in Russian).

Kuperman B.I., Davydov V.G. The fine structure of glands in oncospheres, procercoids and plerocercoids of Pseudophyllidea (Cestoidea). Int J Parasitol 1981; 12:135-44.

Kuris, A.M. Trophic interaction: similarity of parasitic castrations to parasitoids. Quart Rev Biol 1974; 49:129-48.

Kuris, A. "Guild Structure of Larval Trematodes in Molluscan Hosts: Prevalence, Dominance and Significance of Competition." In *Parasite Communities: Patterns and Processes*. G.W. Esch, A.O. Bush and J.M. Aho, eds. London, U.K.: Chapman & Hall, 1990.

La Rue, G.R. Studies on the trematode family Strigeidae (Holostomidae) No. III. Relationships. Trans Am Microsc Soc 1926; 45:265-81.

La Rue, G.R. The classification of Digenetic Trematoda: a review and a new system. Exp Parasitol 1957; 6:306-49.

Lafferty, K.D. Foraging on pray that are modified by parasites. Am Nat 1992; 140:854-67.

Lafferty, K.D. The evolution of trophic transmission. Parasitol Today 1999; 15:111-5.

Lafferty, K.D., Morris, A.K. Altered behavior of parasitized killifish increases susceptibility to predation by bird final hosts. Ecology 1996; 77:1390-7.

Lafferty, K.D., Thomas, F., Poulin, R. "Evolution of Host Phenotype Manipulation by Parasites and its Consequences." In *Evolutionary Biology of Host-Parasite Relationships: Theory Meets Reality*. R. Poulin, S. Morand and A. Skorping, eds. Amsterdam, The Netherlands: Elsevier, 2000.

Lambert, T.C., Farley, J. The effect of parasitism by the trematode *Cryptocotyle lingua* (Creplin) on zonation and winter migration of the common periwinkle *Littorina littorea* (L.). Can J Zool 1968; 46:1139-47.

Larson, O.R., Uglem, G.L., Lee, K.L. Fine structure and permeability of the metacercarial cyst wall of *Clinostomum marginatum* (Digenea). Z Parasitenk 1988; 74:352-5.

Lauckner, G. "Diseases of Mollusca: Gastropoda." In *Diseases of marine animals. Vol. 1*. O. Kinne, ed. Chichester et al.: Wiley & Sons, 1980.

Lauckner, G. "Diseases of Mollusca: Bivalvia." In *Diseases of marine animals. Vol. 2*. O. Kinne, ed. Hamburg, D.B.R.: Biologische Anstalt Helgoland, 1983.

Lauckner, G. Impact of digenean parasitism on the fauna of a North Sea tidal flat. Helgoländer Meerunters 1984; 37:185-99.

Lauckner, G. "Diseases of Aves (Marine Birds)." In *Diseases of marine animals. Vol. 4, Part 2*. O. Kinne, ed. Hamburg, D.B.R.: Biologische Anstalt Helgoland, 1985.

Lauckner G. Ecological effects of larval trematode infestation on littoral marine invertebrate populations. Int J Parasitol 1987; 17: 391-8.

Laurie, J.S. *Himasthla quissentensis*: induced in vitro encystment of cercaria and ultrastructure of the cyst. Exp Parasitol 1974; 35:350-62.

Lebedev, B.I. Is the trematode acetabulum a cercomer? (About one platyhelminth's classification). J Parasitol 1987; 73:1250-1.

Lemly, A.D., Esch, G.W. Population biology of the trematode *Uvulifer ambloplitis* (Hughes, 1927) in the snail intermediate host, *Helisoma trivolvis*. J Parasitol 1984; 70:461-5.

Lengy, J. Study on the *Paramphistomum microbothrium* a rumen parasite of cattle in Israel. Bull Res Concil Israel 1960; 9:71-130.

Leuckart, R. *Allgemeine Naturgechichte der Parasiten*. Leipzig, Heidelberg: C.F. Winter'sche Verlagshandl, 1879.

Leuckart, R. *Die Parasiten des Menschen und die von ihnen herrührenden Krankheiten*. Bd 1, Abteil 1, Leipzig, Heidelberg: C.F. Winter'sche Verlagshandl, 1879-1886.

Lewert, R.M., Lee, C.L. Studies on the passage of helminth larvae through host tissues. I. Histochemical studies on extracellular changes caused by penetrating larvae. 2. Enzymatic activity of larvae *in vitro* and *in vivo*. J Infect Dis 1954; 95:13-51.
Lewert, R.M., Para, J., Ozcel, M.A. Miracidial uptake of glucose in intact eggs of *Schistosoma mansoni*. J Parasitol 1970; 52:95-102.
Lewis, J.W. Studies on the life history of *Brachylaimus oesophagei* Shaldybin, 1953 (Digenea: Brachylaimidae). J Helminthol 1969; 43:79-98.
Lewis, M.C., Welsford, I.G., Uglem, G.L. Cercarial emergence of *Proterometra macrostoma* and *P. edneyi* (Digenea: Azygiidae): contrasting responses to light: dark cycling. Parasitology 1989; 99:215-23.
Le Zotte, L.A. Studies on marine digenetic trematodes of Puerto Rico: the family Bevisiculidae, its biology and affinities. J Parasitol 1954; 40:148-62.
Lie, K.J. Studies on Echinostomatidae (Trematoda) in Malaya. VII. The life history of *Echinostoma lindoense* Sandground and Bonne, 1940. Trop Geogr Med 1964; 16:72-81.
Littlewood, D.T.J., Rohde, K., Clough, K.A. The interrelationship of all major groups of Platyhelminthes: phylogenetic evidence from morphology and molecules. Biol J Linn Soc 1999a; 66:75-114.
Littlewood, D.T.J., Rohde, K., Bray, R.A., Herniou, E.A. Phylogeny of the Platyhelminthes and the evolution of parasitism. Biol J Linn Soc 1999b; 68:257-87.
Littlewood, D.T.J., Cribb, T.H., Olson, P.D., Bray, R.A. Platyhelminth phylogenetics – a key to understanding parasitism? Belgium J Zool 2001; 131 (Suppl. 1):35-46.
Lively, C.M. Through the looking glass: host-parasite coevolution and sex. Bull Scand Soc Parasitol 1995; 5:19.
Lively, C.M., Howard, R.S. "Selection by Parasites for Clonal Diversity and Mixed Mating." In *Infection, Polymorphism and Evolution*. W.D. Hamilton and J.C. Howard, eds. London, U.K.: Chapman & Hall 1997.
Lo Verde, P.T. Scanning electron microscope observations on the miracidium of *Schistosoma*. Int J Parasitol 1975; 5:95-7.
Lo, C.-T. Compatibility and host-parasite relationships between species of the genus *Bulinus* (Basommatophora: Planorbidae) and an Egyptian strain of *Schistosoma haematobium* (Trematoda: Digenea). Malacologia 1972; 11:225-80.
Locker, E.S. On being a parasite in an invertebrate host: a short survival course. J Parasitol 1994; 80:728-47.
Loos-Frank, B. Experimentelle Untersuchungen über Bau, Entwicklungen und Systematik der Himasthline (Trematoda, Echinostomatidae) des Nordenseeraumes. Z Parasitenk 1967; 28:299-351.
Loos-Frank, B. Zur Kenntnis der gymnophalliden Trematoden des Nordseeraumes. I. Die Alternativ-Zyklen von *Gymnophallus choledochus*. Z Parasitenk 1969; 32:135-56.
Loos-Frank, B. Zur Kenntnis der gymnophalliden Trematoden des Nordseeraumes. IV. Übersicht über die gymnophalliden Larven aus Molluscen der Gezeitenzone. Z Parasitenk 1971; 36:206-32.
Looss, A. Die Distomen unserer Fische und Frösche. Neue Untersuchungen über Bau und Entwicklungdes Distomen – Körpers. Bibl Zool 1894; 6, H. 16:1-293.
Lowenberger, C.A., Rau, M.E. *Plagiorchis elegans*: emergence, longevity and infectivity of cercariae, and host behavioural modifications during cercarial emergence. Parasitology 1994; 109:65-72.
Lühe, M. "Parasitische Plattwürmer. I: Trematodes." In: *Die Süsswasserfauna Deutschland. Eine Exkursionsfauna, Bd. 17*, Prof. von Brauer, ed. Jena: Verlag von Gustav Fischer, 1909.
Lumsden, R.D., Hildreth, M.B. "The Fine Structure of Adult Tapeworms." In *Biology of the Eucestoda*, Vol. 1. C. Arme and W.P. Pappas, eds. London, U.K.: Acad Press, 1983.
Lyaiman, E.M. *Course of Fish Diseases*. Moscow, USSR: Pishepromisdat, 1949 (in Russian).

Machkevskii, V.K. Biology and ecology of the trematode *Proctoeces maculatus*, parasite of the Black Sea blue mussel. Ph.D. Thesis, Moscow State University: Moscow, USSR, 1984 (in Russian).

MacInnis, A.J. Responses of *Schistosoma mansoni* miracidia to chemical attractants. J Parasitol 1965; 51:731-45.

MacInnis, A.J. Identification of chemicals triggering cercarial penetration responses *Schistosoma mansoni*. Nature 1969; 224:1221-2.

Madhavi, R. Miracidium of *Allocreadium fasciatusi* Kakaji, 1969 (Trematoda: Allocreadiidae). J Parasitol 1976; 62:410-2.

Madhavi, R. Life history of *Genarchopsis goppo* Ozaki, 1925 (Trematoda: Hemiuridae) from the freshwater fish *Channa punctata*. J Helminthol 1978; 52:251-9.

Madhavi, R. Life history of *Allocreadium handiai* Pande, 1937 (Trematoda: Allocreadiidae) from the freshwater fish *Channa punctata* Bloch. Z Parasitenk 1980; 63:89-97.

Madhavi, R., Rao, K.H. *Orchispirium heterovitellatum*: chemical nature of the eggshell. Exp Parasitol 1971; 30:345-8.

Maillard, A. *Labratrema lamirandi* (Carrere, 1937) (Trematoda: Bucephalidae) parasite de *Duht labr* (L., 1758). Creation du genre *Labratrema*. Cycle évolutif. Bull Mus Nat Hist. Ser 3, Zool 1975; 6:306-44.

Makanga, B. The effect of varying the number of *Schistosoma mansoni* miracidia on the reproduction and survival of *Biomphalaria pfeifferi*. J Invertebr Pathol 1981; 37:7-10.

Maldonado, J.F. The life cycle of *Tamerlania bragai* Santos, 1934 (Eucotylidae), a kidney fluke of domestic pigeons. J Parasitol 1945; 31:306-14.

Malkova, I.I. Tail ultrastructure in microphallid cercariae (Trematoda, Microphallidae). Parazitologiya 1989; 23:159-65 (in Russian).

Malkova, I.I., Galaktionov, K.V. Electron-microscopic investigation of the excretory system of microphallid cercariae (Trematoda, Microphallidae). Parazitologiya 1989; 23:511-7 (in Russian).

Malmberg, G. On the early development of the protonephridial systems in some species belonging to the genera *Diphyllobothrium*, *Triaenophorous* and *Schistocephalus* (Cestoda, Pseudophyllidea). Zool Scripta 1972; 1:227-8.

Manter, H.W. Some digenetic trematodes from deep-water fish of Tortugas, Florida, XVI. Papers from Tortugas Labor. Carnegie Inst of Wash 1934; Publication N 435, 28:257-345.

Marasaeva, E.F. "An Ecological Analysis of Parasite Fauna of an Amphipode *Gammarus oceanicus* at the Littoral Zone of the Eastern Murman." In *Morfologiya i Ekologiya Parazitov Morskikh Zhivotnykh (Morphology and Ecology of Parasites of Marine Animals)*. A.A. Dobrovolskij, ed. Apatity: Kola Scientific Centre of the Russian Academy of Sciences Publ., 1990 (in Russian).

Margolis, L., Esch, G.M., Holmes, J.C., Kuris, A.M., Schad, G.A. The use of ecological terms in parasitology (report of an ad hoc committee of the American Society of Parasitologists). J Parasitol 1982; 68:131-3.

Markum, B.A., Nollen, P.M. The effect of light intensity on hatching of *Echinostoma caproni* miracidia. J Parasitol 1996; 82:662-3.

Martin, W.E., Bills, R.F. Trematode excretory concretions: formation and fine structure. J Parasitol 1964; 50:337-44.

Mas-Coma, S., Montoliu, I. The life cycle of *Brachylaima ruminae* n. sp. (Trematoda: Brachylaimidae), a parasite of rodents. Z Parasitenkd 1986; 72:739-53.

Mas-Coma, S., Montoliu, I. The life cycle of *Dollfusinus frontalis*, a brachylaimid trematode of small mammals (Insectivora and Rodentia). Int J Parasitol 1987; 17:1063-79.

Mas-Coma, S., Montoliu, I. Life cycle of *Pseudoleucochloridium pericardicum* n.sp. (Trematoda: Brachylaimidae), a parasite of shrews (Insectivora: Soricidae) in the Oriental Pyrenees. Research and Reviews in Parasitology 1995; 55: 155-71.

Mason, P.R., Fripp, P.J. Reactions of *Schistosoma mansoni* miracidia using dark-ground photography. J Parasitol 1976; 62:721-7.

Mason, P.R., Fripp P.J. Analysis of movements of *Schistosoma mansoni* miracidia to light. J Parasitol 1977; 63:240-6.

Matricon-Gondran, M. Origine et differenciation du tegument d'un trématode digénétiquee: etude ultrastructurele chez *Cercaria pectinata* (Larvae de *Bacciger bacciger*, Fellodistomatides). Z Parasitenk 1971; 120:488-524.

Matricon-Gondran, M. Diversite des junction "gap" dans le parenchymal d'un trématode digénien. Biol Cell 1977; 30:160.

Mattes, O. Zur Biologie der Larvenentwicklung von *Fasciola hepatica*, besonders über den Einfluss der Wasserstoffionenkonzentration auf das Ausschlüpfen der Miracidien. Zool Anz 1926; 69:138-56.

Mattes, O. Zur Frage der Wirtsauffindung der Parasiten auf Grund experimenteller Untersuchungen an Leberegel Miracidien. 38 Verhadl Deutsch Zool Gesellsch, Leipzig, 1936: 183-6.

Mattes, O. Wirtsfindung, Invasionsvorgang und Wirtspecifitat beim *Fasciola*-miracidium. Z Parasitenk 1949; 14:320-63.

Matthews, B.F. *Cercaria vaullegeardi* Pelseneer, 1906 (Digenea: Hemiuridae): development and ultrastructure. Parasitology 1981; 83:575-86.

Matthews, B.F., Matthews, R.A. The tegument in Hemiuridae (Digenea: Hemiuroidea): structure and function in the adult. J Helminthol 1988a; 62:305-16.

Matthews, B.F., Matthews, R.A. The ecsoma in Hemiuridae (Digenea: Hemiuroidea): tegumental structure and function in the mesocercaria and the metacercaria of *Lecithochirium furcolabiatum* (Jones, 1933) Dawes, 1947. J Helminthol 1988b; 62:317-30.

Matthews, B.F., Matthews, R.A. *Lecithochirium furcolabiatum* (Jones, 1933) Dawes, 1947: the miracidium and mother sporocyst. J Helminthol 1991; 65:259-69.

Matthews, R.A. The life-cycle of *Prosorhynchus crucibulum* (Rudolphi, 1819) Odhner, 1905, and a comparison of its cercaria with that of *Prosorhynchus squamatus* Odhner, 1905. Parasitology 1973a; 66:133-64.

Matthews, R.A. The life cycle of *Bucephalus haimeanus* Lacaze-Duthiers, 1854 from *Cardium edule* L. Parasitology 1973b; 67:341-50.

Matthews, R.A. The life-cycle of *Bucephaloides gracilescens* (Rudolphi, 1819) Hopkins, 1954 (Digenea: Gasterostomata). Parasitology 1974; 68:1-12.

Maynard Smith, J. *The Evolution of Sex*. Cambridge, U.K.: Cambridge University Press, 1978.

McCarthy, A.M. Phototactic responses of the cercariae of *Echinoparyphium recurvatum* during phases of sub-maximal and maximal infectivity. J Helminthol 1999; 73:63-5.

McCarthy, H.O., Fitzpatrick, S., Irwin, S.W.B. A transmissible trematode affects the direction and rhythm of movement in a marine gastropod. Animal Behav 2000; 59: 1161-6.

McLaren, D.J. *Schistosoma mansoni: the Parasite Surface in Relation to Host Immunity*. London, U.K.: John Wiley & Sons, 1980.

McLaren, D.J., Hockley, D.J. Blood flukes have a double outer membrane. Nature 1977; 269:147-9.

McManus, D.P., James, B.L. Tricarboxylic acid cycle enzymes in the digestive gland of *Littorina saxatilis rudis* (Maton) and in the daughter sporocysts of *Microphallus similis* (Jag.) (Digenea: Microphallidae). Comp Biochem Physiol 1975a; 50B:491-5.

McManus, D.P., James, B.L. Anaerobic glucose metabolism in the digestive gland of *Littorina saxatilis rudis* (Maton) and in the daughter sporocysts of *Microphallus similis* (Jag.). Comp Biochem Physiol 1975b; 51B:293-7.

McMichael-Phillips, D.F., Lewis, J.W., Thorndyke, M.C. Ultrastructure of the egg of *Sanguinicola inermis* Plehn, 1905 (Digenea: Sanguinicolidae). J Nat Hist 1992a; 26:895-904.

McMichael-Phillips, D.F., Lewis, J.W., Thorndyke, M.C. Ultrastructural studies on the miracidium of *Sanguinicola inermis* (Digenea: Sanguinicolidae). Parasitology 1992b; 105:435-43.

Meade, T.G., Garsa, J.M. Vesicle contribution to cyst wall formation of *Posthodiplostomum minimum* metacercariae. Tex J Sci 1985; 37:143-5.

Meece, J.K., Nollen, P.M. A comparison of the adult and miracidial stages of *Echinostoma paraensei* and *E. caproni*. Int J Parasitol 1996; 26:37-43.

Meuleman, E.A. Host-parasite interrelationships between freshwater pulmonate *Biomphalaria pfeifferi* and the trematode *Schistosoma mansoni*. Neth J Zool 1972; 22:355.

Meuleman, E.A., Holzmann, P.J. The development of the primitive epithelium and true tegument in the cercaria of *Schistosoma mansoni*. Z Parsitenk 1975; 45:307-18.

Meuleman, E.A., Holzmann, P.J., Peet, R.C. The development of daughter sporocysts inside the mother sporocyst of *Schistosoma mansoni* with special reference to the ultrastructure of the body wall. Z Parasitenk 1980; 61: 201-12.

Meuleman, E.A., Lyaruu, D.M., Khan, M.A., Holzmann, P.J., Sminia, T. Ultrastructural changes in the body wall of *Schistosoma mansoni* during the transformation of a miracidium into a mother sporocyst in the snail host *Biomphalaria glabrata* Pfeifferi. Z Parasitenk 1978; 56:227-42.

Miller, H.M. Furcocercous larval trematodes from San Juan Island, Washington. Parasitology 1927; 19:61-83.

Miller, H.M., Mahaffy, E.E. Reactions of *Cercaria hamata* to light and to mechanical stimuli. Biol Bull 1930; 59:95-103.

Minchella, D.J. Host life-history variation in response to parasitism. Parasitology 1985; 90:205-16.

Minchella, D.J., Loverde, P.T. A cost of increased early reproductive effort in the snail *Biomphalaria glabrata*. Am Nat 1981; 118:876-81.

Minchella, D.J., Sollenberger, K.M., Pereira de Souza, C. Distribution of schistosome genetic diversity within molluscan intermediate hosts. Parasitology 1995; 111:217-20.

Minchella, D.J., Leathers, B.K., Brown. K.M., McNair, J.M. Host and parasite counteradaptations: an example from a freshwater snails. Am Nat 1985; 126:843-54.

Mitchell, C.W. Ultrastructure of the metacercarial cyst of *Posthodiplostomum minimum* (MacCallum, 1921). J Parasitol 1974; 60:67-74.

Mitchell, C.W., Crang, R.E. *Posthodiplostomum minimum*: examination of cyst wall and metacercaria containing calcarious concretions with scanning electron microscopy and X-ray microanalysis. Exp Parasitol 1976; 40:309-13.

Mitterer, K.-E. Untersuchungen zum Schlüpfen der Miracidien des kleinen Leberegels *Dicrocoelium dendriticum*. Z Parasitenk 1975; 48:35-45.

Moczon, T., Swiderski, Z. *Schistosoma japonicum*: cytochemistry of the Mehlis' gland and of the ootype wall. Acta Parasitol 2000; 45:22-8.

Moczon, T., Swiderski, Z., Huggel, H. *Schistosoma mansoni*: The chemical nature of the secretions produced by the Mehlis' gland and ootype as revealed by cytochemical studies. Int J Parasitol 1992; 22:65-71.

Modha, J., Redman, C.A., Thornhill, J.A., Kusel, J.R. Schistosomes: unanswered questions on the basic biology of the host-parasite relationship. Parasitol Today 1998; 14:396-401.

Moore, J. Responses of an avian predator and its isopod prey to an Acanthocephalan parasite. Ecology 1983; 64:1000-15.

Moore, J., Gotelli, N.J. "A Phylogenetic Perspective on the Evolution of Altered Host Behaviours: a Critical Look at the Manipulation Hypothesis." In *Parasitism and Host Behaviour*. C.J. Barnard and J.M. Behnke, eds. London, U.K.: Taylor & Francis, 1990.

Moore, J. The behavior of parasitized animals. BioScience 1995; 45:89-96.

Mordwilko, A.K. Origin of the phenomenon of intermediate hosts. Ezhegodnik Zoologicheskogo Museya Akademii Nauk (*YB of the Zoological Museum of the Academy of Sciences*) 1908; 13:129-222 (in Russian).

Mordwilko, A.K. Über denUrsprung der Erscheinung von Zwischenwirten bei den tierischen Parasiten. Biol Centralbl 1909; 29:369-81, 395-413, 459-67.

Morris, G.P. The fine structure of the tegument and associated structures of the cercaria of *Schistosoma mansoni*. Z Parasitenk 1971; 36:15-31.

Morris, G.P. The fine structure of the caecal epithelium of *Megalodiscus temperatus*. Can J Zool 1973; 51:457-60.

Mouchet, F., Théron, A., Bremond, P., Sellin, E., Sellin, B. Pattern of cercarial emergence of *Schistosoma curassoni* from Niger and comparison with three sympatric species of schistosomes. J Parasitol 1992; 78:61-3.

Moukrim, A., Rondelaud, D. Vertical spatial behaviour patterns of *Lymnaea truncatula* in relation with origin of snails, infection with *Fasciola hepatica*, and experimental environment. Ann Parasitol Hum Comp.1992; 67:174-9.

Mouritsen, K.N., Jensen, K.T. The enigma of gigantism: effect of larval trematodes on growth, fecundity, egestion and locomotion in *Hydrobia ulvae* (Pennant) (Gastropoda: Prosobranchia). J Exp Mar Biol Ecol 1994; 181:53-66.

Muhamedov, G.K. Seasonal dynamics of the development and reproduction of the parthenogenetic generations of *Xiphidiocercaria* sp. VII Odening, 1962 (Trematoda: Plagiorchiidae) in the mollusc *Coretus corneus*. Parazitologiya 1979; 12:444-8 (in Russian).

Muhamedov, G.K. Patterns in development and reproduction of parthenogenetic generations of some trematodes. Ph.D. Thesis, Leningrad State University: Leningrad, USSR, 1981 (in Russian).

Murrills, R.J., Reader, T.A.J., Southgate, V.R. Studies on the invasion of *Notocotylus attenuatus* (Notocotylidae, Digenea) into its snail host *Lymnaea peregra*. The contents of the fully embryonated egg. Parasitology 1985a; 91:397-405.

Murrills, R.J., Reader, T.A.J., Southgate, V.R. Studies on the invasion of *Notocotylus attenuatus* (Notocotylidae, Digenea) into its snail host *Lymnaea peregra*. In vitro observations on the hatching mechanism of the egg. Parasitology 1985b; 91:545-54.

Murrills, R.J., Reader, T.A.J., Southgate, V.R. Studies on the invasion of *Notocotylus attenuatus* (Notocotylidae, Digenea) into its snail host, *Lymnaea peregra*: in vivo observations 1 and 7 days post infection. Int J Parasitol 1988; 18:161-6.

Nadakal, A.M., Mohandas, A., Sunderaraman, V. *Cercaria chakai* sp. n. (Tansversotrematidae) from Kerala, India. J Parasitol 1969; 55:1187-90.

Najim, A.T. Life history of *Giganthobilharzia huronensis* Najim, 1950. A dermatitis-producing bird blood-fluke (Trematoda, Schistosomatidae). Parasitology 1956; 46:443-69.

Näsmark, K.E. A revision of the trematode family Paramphistomatidae. Zool Bidrag fran Uppsala 1937; 16:301-565, pls 1-13.

Neill, P.J., Smith, J.H., Doughty, B.L., Kemp, M. The ultrastructure of the *Schistosoma mansoni* egg. Am J Trop Med Hyg 1988; 39:52-65.

Neuhaus, W. Entwicklung und Biologie von *Pleurogenes medians* Olss. Zool Jahrb (Syst), Jena 1940; 74:207-42.

Newell, C.R. The marine fauna and flora of the Isles of Scilly: some marine digeneans from invertebrate hosts. J Nat Hist 1986; 20: 71-7.

Niemann, G.M., Lewis, F.A. *Schistosoma mansoni*: influence of *Biomphalaria glabrata* size on susceptibility to infection and resultant cercarial production. Exp Parasitol 1990; 70:286-92.

Niewiadomska, K. A new species of furcocercaria, *Cercaria notabilis* sp. n., from the Mazurian Lakes. Acta Parasitol Polonica 1966; 14:21-5.

Niewiadomska, K. *Cercaria clavicauda* sp. n., a new species of furcocercaria from the Mazurian Lakes. Acta Parasitol Polonica 1970; 18:341-6.

Niewiadomska, K., Czubaj, A. Structure and TEM ultrastructure of the excretory system of the cercaria and daughter sporocyst of *Diplostomum pseudospathaceum* Niew., 1984 (Digenea, Diplostomidae). Acta Parasitol 1996; 41:167-83.

Niewiadomska, K., Czubaj, A. Ultrastructure of the excretory system of the metacercaria of *Diplostomum pseudospathaceum* Niew., 1984 (Digenea). Acta Parasitol 2000; 45:307-21.

Niewiadomska, K., Moczoń, T. The nervous system *of Diplostomum pseudospathaceum* Niewiadomska (Digenea, Diplostomatidae). I. Nervous system and chaetotaxy in the cercaria. Z Parasitenk 1982; 68:295-304.

Niewiadomska, K., Moczoń, T. The nervous system *of Diplostomum pseudospathaceum* Niewiadomska (Digenea, Diplostomatidae). II. Structure and development of the nervous system in metacercariae. Z Parasitenk 1984; 70:537-48.

Niewiadomska, K., Moczoń, T. The nervous system *of Diplostomum pseudospathaceum* Niewiadomska (Digenea, Diplostomatidae). III. Structure of the nervous system in the adult stage. Parasitol Res 1987; 73:217-23

Nikolaeva, V.M. On the life cycle of trematodes from the family Didymozoidae (Monticelli, 1888) Poche, 1907. Zoologicheskii Zhurnal (*Zoological Journal*) 1965; 44:1317-27 (in Russian)

Nollen, P.M. The hatching behavior of *Echinostoma trivolvis* miracidia and their responses to gravity, light and chemicals. Int J Parasitol 1994; 24:637-42.

Nollen, P.M. "Reproductive Physiology and Behavior of Digenetic Trematodes." In *Advances in Trematode Biology*, B. Fried and T. Graczyk, eds. Boca Raton, New York: CRC Press, 1997.

Odening, K. Historishe und moderne Gesichtspunkte beim Aufbau eines natürlichen Systemes der digenetischen Trematoden. Biol Beitr 1961; 1:73-90.

Odening, K. Furcocercarien (Trematoda: Strigeata and Schistosomata larvae) aus Branderburg und Sachsen. Monatsber Deutsch Acad Wiss Berlin 1962; 4:384-92.

Odening, K. Das Exkretorssystem monostomer Cecarien (Trematoda: Notocotylidae) aus Gewassern von Berlin und Umgebung. Limnologica (Berlin) 1963; 1:366-73.

Odening, K. Zur Taxonomie der Trematodenunterordnung Plagiorchiata. Monatsber Dtsch Acad Wiss Berlin 1964; 6:191-8.

Odening, K. Eine echinostome Cercarie mit 29 Kragenstacheln aus Gevässern bei Berlin und die Entwicklung ihres Exkretionssystem. Monatsberich Deutsch Akad Wissensch Berlin 1965a; 7:818-22.

Odening, K. Der Lebenszyklus von *Neodiplostomum spathoides* Dubois (Trematoda, Strigeida) im Raum Berlin nebst Beiträgen zur Entwicklungsweise verwandter Arten. Zool Jb Syst 1965b; 92:523-624.

Odening, K. Physidae und Planorbidae als Wirte in den Lebenszyklen einheimischer Notocotylidae (Trematoda, Paramphistomida). Z Parasitenk 1966; 27:210-39.

Odening, K. Die Lebenszyklen von *Strigea falconispalumbi* (Viborg), *S. strigis* (Schrank) und *S. sphaerula* (Rudolphi) (Trematoda, Strigeida) im Raum Berlin. Zool Jb Syst 1967; 94:1-67.

Odening, K. Das Exkretionssystem von *Encyclometra* bei Metazercarie und Adults und die systematische Stellung dieser Trematodengattung. Biol Rdsch 1969; 7:75-7.

Odening, K. Verwandtschaft, System und zyklo-ontogenetische Besonderheiten der Trematoden. Zool Jahr Syst 1974; 101:345-96.

Odening, K. Der Lebenszyklus von *Azygia lucii* (Trematoda) – Untersuchungen in Gebeit der DDR. Biol Zbl 1976; 95:57-94.

Okulewicź, A. Badania nad ukladem nervovym *Brachylaimus fuscatus* (Rud., 1819) (Trematoda, Brachylaimidae). Wiadomosci Parasitol 1988; 34:125-9.

Olenev, A.V. Larval trematode fauna of the freshwater mollusc *Melanopsis praemorsae* from western Georgia. II. Ekologicheskaya i Eksperimentalnaya Parazitologiya (*Ecological and Experimental Parasitology*), Leningrad, USSR: Izdatel'stvo Leningradskogo Universiteta 1979; 2:30-41 (in Russian).

Olenev, A.V., Dobrovolskij, A.A. Larval trematode fauna of the freshwater mollusc *Melanopsis praemorsae* from western Georgia. I." Ekologicheskaya i Eksperimentalnaya Parazitologiya (*Ecological and Experimental Parasitology*), Leningrad, USSR: Izdatel'stvo Leningradskogo Universiteta 1975; 1:73-96 (in Russian).

Orido, Y. Metamorphosis of the excretory system of *Paragonimus ohirai* (Trematoda), with special reference to its functional significance. J Morphol 1987a; 194:303-10.

Orido, Y. Development of the ovary and female reproductive cells of the lung fluke, *Paragonimus ohirai* (Trematoda: Troglotrematidae). J Parasitol 1987b; 73:161-71.

Orido, Y. Fertilization and oogenesis of the lung fluke, *Paragonimus ohirai* (Trematoda: Troglotrematidae). Int J Parasitol 1988; 18:973-80.

Orido, Y. Development of the excretory bladder of the lung fluke, *Paragonimus ohirai* (Trematoda: Troglotrematidae). J Parasitol 1990a; 76:205-11.

Orido, Y. Ultrastructure of the oviduct of the lung fluke, *Paragonimus ohirai* (Trematoda: Troglotrematidae). J Morphol 1990b; 204:247-55.

Orido, Y. Ultrastructure of Mehlis' gland in the lung fluke, *Paragonimus ohirai* (Trematoda: Troglotrematidae). J Morphol 1991; 207:9-16.

Orido, Y., Takamure, A., Akamatsu, T., Takashima, Y. Ultrastructure of the ootype of the lung fluke, *Paragonimus ohirai* (Trematoda: Troglotrematidae) in mated and single-infection worm. Parasitol Res 1994; 80:307-11.

Otubanjo, O.A. *Schistosoma mansoni*: the ultrastructure of the ducts of the male reproductive system. Parasitology 1980; 81:565-71.

Otubanjo, O.A. *Schistosoma mansoni*: the sustentacular cells of the testes. Parasitology 1981; 82:125-30.

Otubanjo, O.A. Scanning electron microscopic studies of the body surface and external genitalia of Dicrocoeliid Trematode, *Concinnum epomopsis* Sandrground, 1973. Z Parasitenk 1985; 71:435-504.

Page, M.R., Nadakavukaren M.J., Huizinga H.W. *Riberoia marini*: surface ultrastructure of redia, cercaria and adult. Int J Parasitol 1980; 10:5-12.

Pages, J.R., Théron, A. Analysis and comparision of cercarial emergence rhythms of *Schistosoma haematobium, S. intercalatum* and *S. bovis* and their hybrid progeny. Int J Parasitol 1990a; 20:193-7.

Pages, J.R., Théron, A. *Schistosoma intercalatum* from Cameroon and Zaire: chronobiological differentiation of cercarial emergence. J Parasitol 1990b; 76:743-5.

Palm, V. Glycogen and fat, the reserve nutritional substances in helminths and the relation of their accumulation to various biological and ecological factors. Ph.D. Thesis, Leningrad State University: Leningrad, USSR, 1967 (in Russian).

Palmieri, J.R., Sullivan J.T., Ow-Yong, C. K. Occurrence of the sporocyst generation of *Fasciola gigantica* in the mollusc *Lymnaea rubignosa*. J Parasitol 1977; 63:299-300.

Palombi, A. Il ciclo biologico di *Diphterostomum bursinae* Stoss. (Trematoda, Zoogonidae). Publ Stz Zool Napoli 1930; 10:111-49.

Pan, S.C.-T. Studies on the host-parasite relationship between *Schistosoma mansoni* and the snail *Australorbis glabratus*. Am J Trop Med Hyg 1965; 14:931-76.

Pan, S.C.-T. The fine structure of the miracidium of *Schistosoma mansoni*. J Invert Pathol 1980; 36:307-72.

Panin, V.Ya. *Dicrocoeliid Trematodes of the World Fauna*. Alma-Ata, USSR: Nuaka 1984 (in Russian).

Panin, V.Ya. "Life Cycles of Trematodes of the Family Dicrocoeliidae." In *Functionalnaya Morphologiya Lichinok Trematod i Cestod (Functional Morphology of Trematode and Cestode Larvae)*. E.V. Gvozdev, ed. Alma-Ata, USSR: Nauka, 1986 (in Russian).

Panin, V.Ya., Žd'árská Z., Nesterenko, L.T. "Ultrastructure of the Dicrocoeliid Parthenitae and Larvae." In *Functionalnaya Morphologiya Lichinok Trematod i Cestod (Functional Morphology of Trematode and Cestode Larvae)*. E.V. Gvozdev, ed. Alma-Ata, USSR: Nauka, 1986 (in Russian).

Pariselle, A., Matricon-Gondran, M. A new type of ciliated receptor in the cercariae of *Nicolla gallica* (Trematoda). Z Parasitenk 1985; 71:353-64.

Payares, G., Simpson, J.G. *Schistosoma mansoni* surface glycoproteins. Analysis of their expression and antigenicity. Eur J Biochem 1985; 153:195-201.

Payares, G., McLaren, D.J., Evans, W.H., Smithers, S.R. Changes in the surface antigen profile of *Schistosoma mansoni* during maturation from cercaria to adult worm. Parasitology 1985; 91:83-99.

Pearce, E.J., Sher, A. Mechanisms of immune evasion in schistosomiasis. Contrib Microbiol Immunol 1987; 8:219-32.

Pearson, E.J., Cheng, T.C. Studies on parasitic castration: occurrence of a gametogenesis-inhibiting factor in extract of *Zoogonus lasius* (Trematoda). J Invertebr Pathol 1985; 46:239-46.

Pearson, J.C. Studies on the life cycles and morphology of the larval stages of *Alaria arisaemoides* Augustin and *Alaria canis* La Rue (Diplostomatidae). Can J Zool 1956; 34:295-387.

Pearson, J.C. Observations on the morphology and life cycle of *Neodiplostomum intermedium* (Trematoda: Diplostomatidae). Parasitology 1961; 51:133-72.

Pearson, J.C. Observations on the morphology and life cycle of *Paucivitellosus fragilis* Coil, Reid & Kuntz, 1965 (Trematoda: Bivesiculidae). Parasitology 1968; 58:769-88.

Pearson, J.C. A phylogeny of life-cycle patterns of the Digenea. Adv Parasitol 1972; 10:153-89.

Pearson, J.C. The paranephridial system in the Digenea: occurrence and possible phylogenetic significance. In *Parasite Lives*. St.Lucia, London, New York, 1986.

Pearson, J.C. On the position of the digenean family Heronimidae: an inquiry into a cladistic classification of the Digenea. Syst Parasitol 1992; 21:81-166.

Pedersen, K.J. Fine structural observations on the turbellarians *Stenostomum* sp. and *Microstomum lineare* with special reference to the extracellular matrix and connective tissue systems. Acta Zool (Stockh.) 1983; 64:177-90.

Pekkarinen, M. The cercaria of *Lacunovermis macomae* (Lebour, 1908) (Trematoda: Gymnophallidae), and its penetration into the bivalve *Macoma baltica* (L.) in experimental conditions. Ann Zool Fennici 1987a; 24:101-21.

Pekkarinen, M. The cercaria and metacercaria of *Gymnophallus gibberosus* Loos-Frank, 1971 (Trematoda: Gymnophallidae) in *Macoma baltica* (L.) (Bivalvia) in the Baltic brackish water (southwestern Finland). Ann Zool Fennici 1987b; 24:123-36.

Pellegrino, J., De Maria, M., De Moure, M.F. Observations on the predatory activity of *Lebistes reticulatus* (Peters, 1859) on cercariae of *Schistosoma mansoni*. Am J Trop Med Hyg 1966; 15:337-41.

Peters, L.E. The Allocreadioid problem with reference to the excretory system in four types of Cercariae. Proc Helminthol Soc Wash 1961; 28:102-8.

Peters, L.E., La Bonte, R.P. Comparative morphology of four species of allocreadiid miracidia (Trematoda). J Parasitol 1965; 51:583-6.

Phifer, K. Permeation and membrane transport in animal parasites: on the mechanism of glucose uptake by *Hymenolepis diminuta*. J Parasitol 1960; 2:145-53.

Pieper, M.B. The life cycle and germ cell cycle of *Spirorchis artericola* (Ward, 1921). J Parasitol 1953; 39:310-25.

Plorin, G.G., Gilbertson, D.E. Descriptive statistics of swimming behaviour of *Schistosoma mansoni* miracidia in artificial pond water. J Parasitol 1981; 67:45-9.

Poche, F. Das System der Platodaria. Arch Natirgesch, Abt A, 1926; Hf 2:1-240, Hf 3:241-458.

Podlipaev, S.A. The influence of bird concentrations on the infection of snails *Littorina saxatilis* (Olivi) and *L. obtusata* (L.) by parthenitae of trematodes. Ekologicheskaya i Eksperimentalnaya Parazitologiya (*Ecological and Experimental Parasitology*), Leningrad, USSR: Izdatel'stvo Leningradskogo Universiteta 1979; 2:47-100 (in Russian).

Podvyaznaya, I.M. "The Morphology and Ultrastructure of the Female Reproductive System of *Prosthodendrium ascidia* (Trematoda, Lecithodendriidae)." In *Parasiticheskie i Svobodnozhivushie Ploskie Chervi: Faunistica i Morfologia (Parasitic and Free-Living*

Plathelminthes: Faunistics and Morphology). Yu.V. Mamkaev and B.I. Joffe, eds. Leningrad, U.S.S.R.: Zoological Institute Publ., 1990 (in Russian).

Podvyaznaya, I.M. The fine structure of the male reproductive system and genital atrium of bat parasite *Allassogonoporus amphoraeformis* (Trematoda: Allassogonoporidae). Parazitologiya 1996; 30:229-35 (in Russian).

Podvyaznaya, I.M. The fine structure of the tegument of cercariae and developing metacercariae of *Diplostomum chromatophorum* (Trematoda: Diplostomidae). Parazitologiya 1999; 33:507-19 (in Russian).

Podvyaznaya, I.M., Dobrovolskij, A.A. On the development of body-wall musculature in the metacercaria of *Diplostomum chromatophorum* (Trematoda: Diplostomidae). Parazitologiya 2001; 35:531-9 (in Russian).

Podvyaznaya, I.M., Galaktionov, K.V. The ultrastructure of the excretory system of the cercariae of *Bucephaloides gracilescens* (Rudolphi, 1819) Hopkins, 1954 and *Prosorhynchus squamatus* Odhner, 1905 (Digenea, Bucephalidae). J Helminthol, a *(submit)*

Podvyaznaya, I.M., Galaktionov, K.V. The development of the excretory system of the cercariae of *Bucephaloides gracilescens* (Rudolphi, 1819) Hopkins, 1954 and *Prosorhynchus squamatus* Odhner, 1905 (Digenea, Bucephalidae). J Helminthol, b *(submit)*

Pojmanska, T., Machaj, K. Differentiation of the ultrastructure of the body wall of the sporocyst of *Leucochloridium paradoxum*. Int J Parasitol 1991; 21:651-9.

Poljansky, Yu.I., Raikov, I.B. Polymerization and oligomerization in the protists' evolution. Zhurnal Obshey Biologii (*Journal of General Biology*) 1977; 37:325-35 (in Russian).

Ponder, W.F. "Classification of Mollusca." In *Molusca: The Southern Synthesis. Fauna of Australia. Vol. 5. Part A*. P.L. Beesley, G.J.B. Ross, and A. Wells, eds. Melbourne, Australia: CSIRO, 1998.

Popiel, I. The ultrastructure of the metacercaria of *Cercaria stunkardi* Palombi, 1934 (Digenea, Opecoelidae) in an experimental intermediate host, *Amphitoe rubricata.* Norw J Zool 1976; 24:353-64.

Popiel, I. The ultrastructure of the excretory bladder of the free-living cercariae and metacercariae of *Cercaria stunkardi* Palombi, 1934 (Digenea, Opecoelidae). Z Parasitenk 1977; 51:249-60.

Popiel, I. The ultrastructure of the daughter sporocyst of *Cercaria littorina saxatilis* V Popiel, 1976 (Digenea: Microphallidae). Z Parasitenk 1978a; 56:167-73.

Popiel, I. The ultrastructure of the daughter sporocyst tegument of *Cercaria vaullegeardi* Pelseneer (Digenea, Hemiuridae). Zool Scr 1978b; 7:155-7.

Popiel, I., James, B.L. Variations in the ultrastructure of the daughter sporocyst of *Microphallus pygmaeus* (Levinsen, 1881) (Digenea: Microphallidae) in chemically defined media. Parasitology 1978a; 76:349-58.

Popiel, I., James, B.L. The ultrastructure of the tegument of the daughter sporocyst of *Microphallus similis* (Jag., 1900) (Digenea: Microphallidae). Parasitology 1978b; 76:359-67.

Poulin, R. Meta-analysis of parasite-induced behavioural changes. Animal Behav 1994; 48:137-46.

Poulin, R. "Adaptive" changes in the behaviour of parasitized animals: a critical review. Int J Parasitol 1995; 25:1371-83.

Poulin, R. *Evolutionary Ecology of Parasites*. London, U.K.: Chapman & Hall, 1998.

Poulin, R., Thomas, F. Phenotypic variability induced by parasites: extent and evolutionary implications. Parasitol Today 1999; 15:28-32.

Powell, E.C. Optical and electron microscope studies on the excretory bladder of the supposed epitheliocystid cercaria of *Ochetosoma aniarum*. Z Parasitenk 1972; 40:19-30.

Powell, E.C. Studies on the excretory "bladder" and caudal ducts of the supposed anepitheliocystid cercaria of *Schistosoma mansoni*. Z Parasitenk 1973; 43:43-52.

Powell, E.C. The ultrastructure of the excretory bladder in the supposed anepitheliocystid cercaria of *Posthodiplostomum minimum* (MacCallum, 1921). Iowa State J Res 1975; 49:259-62.

Powell, E.C. Ultrastructural development of the excretory bladder in early metacercaria of *Ochetosoma aniarium* (Leidy, 1891). Proc Helminthol Soc Wash 1977; 44:136-41.

Pozdnyakov, S.E. *Trematodes of the Suborder Didimozoata*. Vladivostok, Russia: Tihookeanskii Nauchno-Issledovatelskii Ribohosiaystvennii Tsentr, 1996 (in Russian).

Prévot, G. Cycle évolutif d'*Aporchis massilliensis* Timon-David, 1955 (Digenea Echinostomatidae), parasite du goéland *Larus argentatus michellis* Naumann. Bull Soc Zool France 1971; 96:197-208.

Price, E.W. Life history of *Schistosomatium douthitti* (Cort). Amer J Hyg 1931; 13:685-727.

Prokofiev, V.V. "Ambuscade" type of cercarial behaviour in marine trematodes. Zoologichesky Zhurnal (*Zoological Journal*) 1994; 73(5):13-20 (in Russian).

Prokofiev, V.V. Cercarial biology in the littoral trematodes of the Barents and White Seas. Ph.D. Thesis, St. Petersburg State University: St. Petersburg, Russia 1995 (in Russian).

Prokofiev, V.V. Light response in cercariae of the marine littoral trematode *Renicola thaidus* (Trematoda: Renicolidae). Parazitologiya 2001; 35:429-35 (in Russian).

Prokofiev, V.V. Vertical migration of cercaria of the marine littoral trematode *Renicola thaidus* (Trematoda: Renicolidae) in the water column. Parazitologiya 2002; 36:314-21 (in Russian).

Ractliffe, L.H. Hatching of *Dicrocoelium lanceolatum* eggs. Exp Parasitol 1968; 23:67-78.

Rakotondravao, A., Moukrin, P., Hourdin, P., Rondelaud, D. Redial generations of *Fasciola gigantica* in the pulmonate snail *Lymnaea truncatula*. J Helminthol 1992; 66:159-66.

Rankin, J.S. The life cycle of the frog bladder fluke *Gorgoderina attenuatus* Stafford, 1902 (Trematoda, Gorgoderidae). Am Midl Nat 1939; 21:479-88.

Rankin, J.S. A review of the trematode genus *Halipegus* Loos, 1899, with an account of the life history of *H. amherstensis* n. sp. Trans Am Microsc Soc 1944; 63:149-64.

Raymond, K., Probert, A.J. The daily cercarial emission rhythms of *Schistosoma margrebowiei* with particular reference to dark period stimuli. J Helminthol 1991; 65:159-68.

Raymond, K., Probert, A.J. The effect of infection with *Schistosoma margrebowiei* on the growth of *Bulinus natalensis*. J Helminthol 1993; 67:10-6.

Rea, J.C., Irwin, S.W.B. Behavioural responses of the cercariae of *Cryptocotyle lingua* (Digenea: Heterophyidae) to computer-controlled shadow sequences. Parasitology 1991; 103:471-7.

Rea, J.C., Irwin, S.W.B. The ecology of host-finding behaviour and parasite transmission: past and future perspectives. Parasitology 1994; 109S:S31-9.

Rees, F.G. Studies on the germ cell cycle of the digenetic trematode *Parorchis acanthus* Nicoll. Part I. Anatomy of the genitalia and gametogenesis in the adult. Parasitology 1939; 31:417-33.

Rees, F.G. Studies on the germ cell cycle of the digenetic trematode *Parorchis acanthus* Nicoll. Part II. Structure of the miracidium and germinal development in the larval stages. Parasitology 1940; 32:372-91.

Rees, F.G. Light and electron microscope studies of the redia of *Parorchis acanthus* Nicoll. Parasitology 1966; 56:589-602.

Rees, F.G. Locomotion of the cercaria of *Parorchis acanthus*, Nicoll, and the ultrastructure of the tail. Parasitology 1971; 62:489-503.

Rees, F.G. The development of the tail and the excretory system in the cercaria of *Cryptocotyle lingua* (Creplin) (Digenea: Heterophyidae) from *Littorina littorea* (L.). Proc R Soc London 1977; Ser B195, 1121:425-52.

Rees, F.G. The morphology and ultrastructure of the female reproductive ducts in the metacercaria and adult of *Cryptocotyle lingua*, Digenea, Heterophyidae. Z Parasitenk 1979; 60:157-76.

Rees, F.G. Surface ultrastructure of the redia of *Parorchis acanthus* Nicoll (Digenea: Philophthalmidae). Z Parasitenk 1980; 63:33-46.

Rees, F.G. The ultrastructure of the epidermis of the redia of *Parorchis acanthus* Nicoll (Digenea: Philophthalmidae). Z Parasitenk 1981; 65:19-30.

Rees, F.G. The ultrastructure of the foregut of the redia of *Parorchis acanthus* Nicoll (Digenea: Philophthalmidae) from the digestive gland of *Nucella lapillus* L. Parasitology 1983a; 87:151-58.

Rees, F.G. The ultrastructure of the intestine of the redia of *Parorchis acanthus* Nicoll (Digenea: Philophthalmidae) from the digestive gland of *Nucella lapillus* L. Parasitology 1983b; 87:159-66.

Rees, F.G., Day, M.F. The origin development of the epidermis and associated structures in the cercaria of *Cryptocotyle lingua* (Creplin) (Digenea: Heterophyidae) from *Littorina littorea* (L.). Proc R Soc London 1976; Ser B192, 1108:299-321.

Reichel, A.S. Development of parthenogenetic generations of *Haplometra cylindracea* Zeder, 1800 (Plagiorchioidea). Parazitologiya 1971; 5:441-5 (in Russian).

Reuter, M., Halton, D.W. Comparative neurobiology of Platyhelminthes. In *Interrelationships of the Platyhelminthes.* D.T.J. Littlewood and R.A. Bray, eds. London, U.K.: Taylor & Francis, 2001.

Richard, J. La chétotaxie des cercaires, valuer systématique et phylétique. Mém Mus National d'Hist Nat, Ser A, Zool 1971; 67:1-179.

Richards, C.S. Genetic studies of pathologic conditions and susceptibility to infection in *Biomphalaria glabrata.* Ann NY Acad Sci 1975; 266:394-410.

Rifkin, E. An ultrastructural study of the interaction between the sporocysts and the developing cercariae of *Schistosoma mansoni.* J Parasitol 1970; 56, Sect 2:284.

Roberts, E.W. Studies on the life cycle of *Fasciola hepatica* (Linnaeus) and of its snail host, *Lymnaea (Galba) truncatula* (Muller), in the field and under controlled conditions in the laboratory. Ann Trop Med Parasitol 1950; 44:187-206.

Roberts, T.M., Linck, R.W., Chernin, E. Effector mechanism in the response of *Schistosoma mansoni* miracidia to snail-conditioned water. J Exp Zool 1980; 211:137-42.

Roberts, T.M., Ward, S., Chernin, E. Behavioural responses of *Schistosoma mansoni* miracidia in concentration gradients of snail-conditioned water. J Parasitol 1979; 65:41-9.

Robinson, R.D., Halton, D.W. Functional morphology of the tegument of *Corrigia vitta* (Trematoda: Dicrocoelidae). Z Parasitenk 1983; 69:319-33.

Robson, E.M., Threadgold, L.T. Electron microscope studies of *Fasciola hepatica.* XII. The fine structure of the gastrodermis. Exp Parasitol 1975; 37:20-36.

Robson, E.M., Williams, I.C. Relationships of some species of Digenea with the marine prosobranch *Littorina littorea* (L.). II. The effect of larval Digenea on the reproductive biology of *L. littorea.* J Helminthol 1971; 45:145-59.

Rogers, S.H., Bueding, E. Anatomical localization of glucose uptake by *Schistosoma mansoni* adults. Int J Parasitol 1975; 5:369-71.

Rohde, K. Phylogenetic origin of trematodes. Parasitol Schriftenrh 1971; 21:17-27.

Rohde, K. The Aspidogastrea, especially *Multicotyle purvisi* Dawes, 1941. Adv Parasitol 1972; 10:78-151.

Rohde, K. Ultrastructure of the protonephridial system of *Labatostoma manteri* (Trematoda, Aspidogastrea). J Submicrosc Cytol Pathol 1989; 21:599-610.

Rohde, K. The origin of parasitism in the platyhelminthes. Int J Parasitol 1994; 24:1099-115.

Rohde, K. "Protonephridia as Phylogenetic Characters." In *Interrelationships of the Platyhelminthes.* D.T.J. Littlewood and R.A. Bray, eds. London, U.K.: Taylor & Francis 2001a.

Rohde, K. "The Aspidogastrea: An Archaic Group of Platyhelminthes." In *Interrelationships of the Platyhelminthes.* D.T.J. Littlewood and R.A. Bray, eds. London, U.K.: Taylor & Francis 2001b.

Rohde, K., Watson, N. Ultrastructure of the developing protonephridial system of the cercaria *Philophthalmus* sp. (Trematoda, Digenea). Parasitol Res 1992; 78:368-75.

Rohde, K., Watson, N. Comparative ultrastructural study of the posterior suckers of four species of symbiotic Platyhelminthes, *Temnocephala* sp. (Temnocephalidae), *Udonella cali-*

gorum (Udonellidae), *Anoplodiscus cirrusspiralis* (Monogenea: Monopisthocotylea), and *Philophthalmus* sp. (Trematoda: Digenea). Folia Parasitol 1995; 42:11-28.
Rowan, W.B. The life cycle and epizootiology of the rabbit trematode *Hasstilesia tricolor* (Stiles and Hassal, 1894) Hall, 1916 (Trematoda, Brachylaemidae). Trans Am Microsc Soc 1955; 74:1-21.
Rowan, W.B. The mode of hatching of the egg of *Fasciola hepatica.* Exp Parasitol 1956; 5:118-37.
Rowan, W.B. The mode of hatching of the egg of *Fasciola hepatica.* 2. Colloidal nature of thje viscous cushion. Exp Parasitol 1957; 6:131-42.
Rusanov, N.I., Galaktionov, K.V. Seasonal dynamics of development and reproduction of the parthenitae of *Podocotyle atomon* (Rudolphi, 1802) (Trematoda: Opecoelidae) in the littoral molluscs of the Barents Sea. In *Ekologo-Parazitologicheskie Issledovaniya Severnykh Morei (Ecological-Parasitological Investigations of the Northern Seas).* Yu.I. Poljansky, ed., Apatity, USSR: Kola Scientific Centre of the USSR Academy of Sciences Publ., 1984 (in Russian).
Russell-Pinto, F., Bowers, E., James, B. Ultrastructure study of the intramolluscan stages of *Meiogymnophallus minutus* (Digenea: Gymnophallidae) in *Scrombicularia plana* (Bivalvia) from Portugal. Parasitol Res 1996; 82:428-34.
Rybicka, K. Attempt of general approach to the embryology of cyclophyllidean cestodes. Acta Parasitol Polonica 1964; 12:327-38.
Saladin, K.S. Behavioural parasitology and perspectives on miracidial host-finding. Z Parasitenk 1979; 60:197-210.
Saladin, K.S. *Schistosoma mansoni*: cercarial responses to irradiance changes. J Parasitol 1982; 68:120-4.
Samuelson, J.C., Caulfield, J.P. The cercarial glycocalyx of *Schistosoma mansoni.* J Cell Biol 1985; 100:1423-34.
Sannia, A., James, B.L. The morphology of *Cercaria cerastodermae* I nom. nov. (Monorchiidae) (= *Cercaria lepidapedon rachion* (Cobbold, 1858) sensu Lebour, 1908) a rare digenean parasite of the cockle in Britain. J Nat Hist 1978; 12:487-500.
Sapp, K.K., Locker, E.S. Mechanisms underling digenean-snail specificity: role of miracidial attachment and host plasma factors. J Parasitol 2000a; 86:1012-9.
Sapp, K.K., Locker, E.S. A comparative study of mechanisms underling digenean-snail specificity: in vitro interactions between haemocytes and digenean larvae. J Parasitol 2000b; 86:1020-9.
Sapp, K.K., Meyer, K.A., Loker, E.S. Intramolluscan development of the digenean *Echinostoma paraensei*: rapid production of a unique mother redia that adversely affects development of conspecific parasites. Invertebr Biol 1998; 117:20-8.
Saville, D.H., Irwin, S.W.B. A light and electron microscope study on in vitro excystation of *Microphallus abortivus* (Digenea: Microphallidae) metacercariae. Parasitol Res 1991; 77:82-5.
Savinov, V.A. Development of *Alaria alata* (Goeze, 1782) in a dog. Trudi VIGIS (*Transactions of the All-Union Institute of Helminthology*) 1953; 5:63-4 (in Russian).
Savinov, V.A. On some features in the strigeats stage development and on role of different hosts in it. Uchenye Zapiski Volgogradskogo Pedagogicheskogo Instituta *(Research Notes of the Volgograd Pedagogical Institute*) 1954; 15:245-306 (in Russian).
Schallig, H.D.F.H., Sassen, M.J.M., Hordijk, P.L., De Jong-Brink, M. *Trichobilharzia ocellata*: influence of infection on the fecundity of its intermediate snail host *Lymnaea stagnalis* and cercarial induction of the release of schistosomin, a snail neuropeptide antagonizing female gonadotropic hormones. Parasitology 1991; 102: 85-91.
Schell, S.C. Development of mother and daughter sporocysts of *Haplometra intestinalis* Lucker, a plagiorchioid trematode of frogs. J Parasitol 1961; 47:493-500.
Schell, S.C. Development of the sporocyst generations of *Glypthelmins quieta* (Stafford, 1900) (Trematoda, Plagiorchioidea), a parasite of frogs. J Parasitol 1962a; 48:387-94.

Schell, S.C. The life history of *Telorchis bonnerensis* Waitz (Trematoda: Reniferidae), a parasite of the long-tailed salamander *Ambystoma macrodactylum* Baird. Trans Am Microsc Soc 1962b; 81:137-46.

Schell, S.C. The life history of *Haematoloechus breviplexus* Stafford, 1902 (Trematoda: Haplometridae McMullen, 1937) with emphasis on the development of the sporocysts. J Parasitol 1965; 51:587-93.

Schell, S.C. The miracidium of *Lecithaster* Yamaguti, 1934 (Trematoda: Hemiuroidea). J Parasitol 1975; 63:562-3.

Schmalhausen, I.I. *Problems of Darwinism*. Moscow, USSR: Nauka, 1969 (in Russian).

Schuering, L. Der Lebenszyklus von *Sanguinicola inermis* Plehn. Zoologische Jarbucher Anat Ont Entw 1922-1923; 44:264-310.

Shigin, A.A. The role of trematode free-living stages in biocenoses. Parazitologiya 1978; 30:170-6 (in Russian).

Shigin, A.A. Components of freshwater biocenoses as eliminators of helminths. Itogi Nauki i Tekhniki VINITI, Zooparazitologiya (*Results of Science and Technics VINITI, Zooparasitology*) 1981; 7:89-133 (in Russian)

Shigin, A.A. *Trematode Fauna of the USSR. The Genus Diplostomum. Metacercariae*. Moscow: Nauka, 1986 (in Russian).

Shigin, A.A. *Trematode fauna of Russia and Adjacent Territories. The Genus Diplostomum. Maritae*. Moscow: Nauka, 1993 (in Russian).

Shigin, A.A. The place and role of trematodes in the biosphere. Ekologicheskoe i Taxonomicheskoe Raznoobrazie Parazitov (Trudy Instituta Parazitologii) *(Ecological and Taxonomical Diversity of Parasites (Transactions of the Institute of Parasitology))* 1997; 61:192-208 (in Russian).

Shigin, A.A., Shigina, N.G. "On productivity of trematodes of the order Strigeidida (La Rue, 1926)." In *Gelminty Cheloveka, Zhivotnyh i Rastenii i Mery Borby s nimi (Helminths of man, animals and plants and ways of struggle against them)*. V.S. Ershov, ed. Moscow: Nauka 1968 (in Russian).

Seitner, P.G. The life history of *Allocreadium ictaluri* Pearson, 1924. J Parasitol 1951; 37:223-44.

Self, J.T., Peters, L.E., Davis, C.E. The egg, miracidium and adult of *Nematobothrium texomensis* (Trematoda, Digenea). J Parasitol 1963; 49:731-6.

Semenov, O.Y. An experimental study of biology of the *Philophthalmus rhionica* Tihomirov, 1976 (Trematoda, Philophthalmidae) miracidium. Parazitologiya 1976; 10:439-43 (in Russian).

Semenov, O.Y. Vertical distribution of the miracidia of *Philophthalmus rhionica* (Trematoda, Philophthalmidae) in the water column. Parazitologiya 1980; 14:61-5 (in Russian).

Semenov, O.Y. Behaviour of the *Philophthalmus rhionica* miracidium under combined influence of light and chemical stimuli. Vestnik Leningradskogo Universiteta (*Bulletin of the Leningrad University*) 1981; 9:9-16 (in Russian).

Semenov, O.Y. *Miracidia: Morphology, Biology, Interaction with Molluscs*. Leningrad, USSR: Lenungrad University Press, 1991 (in Russian).

Sergievsky, S.O., Granovitch, A.I., Mikhailova, N.A. Irregular distribution in the littoral zone molluscs *Littorina obtusata* and *L. saxatilis* (Gastropoda: Prosobranchia) infected by trematode parthenites. Zoologicheskii Zhurnal (*Zoological Journal*) 1984; 63:929-31 (in Russian).

Sergievskii, S.O., Granovitch, A.I., Mikhailova, N.A. Effect of trematode infection on the survival of periwinkles *Littorina obtusata* and *L. saxatilis* under the conditions of extremely low salinity. Parazitologiya 1986; 20:202-7 (in Russian).

Sewertzoff, A.N. *Morphologische Gesetzmässigkeiten der Evolution*. Jena: Gustav Fischer, 1931.

Sewertzoff, A.N. *Morphological Appropriateness of Evolution*. Moscow, Leningrad, U.S.S.R.: Izdatelstvo Akademii Nauk SSSR, 1939.

Sewell, R.B.S. Cercariae indicae. Ind J Med Res 1922; Suppl 10:1-360.

Shahmatova, R.A. Trematode larvae from freshwater molluscs from the middle Volga region and an experimental study of their biology. Ph.D. Thesis, Gor'kii State University: Gor'kii, USSR 1961 (in Russian).

Sharma, P.N., Hanna, R.E.B. Ultrastructure and cytochemistry of the tegument of *Orthocoelium scoliocoelium* and *Paramphistomum cervi* (Trematoda: Digenea). J Helminthol 1988; 62:331-43.

Sharma, P.N., Ratnu, L.S. The lymphatic system of *Orthocoelium scoliocelium* (Trematoda). J Helminthol 1982; 56:59-67.

Sharma, P.N., Swarnakar, G., Hanna, R.E.B. Ultrastructure of the male reproductive system in a rumen amphistome *Cotylophoron cotylophoron*. J Helminthol 1994; 68:255-8.

Shaw, D.J., Dobson, A.P. Patterns of macroparasite abundance and aggregation in wildlife populations: a quantitative review. Parasitology 1995; 111S:S111-S33.

Sher, A., Benno, D. Decreasing immunogenicity of schistosome larvae. Parasite Immunol 1982; 4:101-7.

Shiff, C.J. Seasonal factors influencing the location of *Bulinus (Physopsis) globosus* by miracidia of *Schistosoma haematobium* in nature. J Parasitol 1974; 60:578-84.

Shiff, C.J. Molluscan defence mechanism: immunity or population biology? Parasitol Today 1994; 10:188-90.

Shiff, C.J., Graczyk, T.K. A chemokinetic response in *Schistosoma mansoni* cercariae. J Parasitol 1994; 80:879-83.

Shiff, C.J., Kriel, R.L. A water-soluble product of *Bulinus (Physopsis) globosus* by miracidia of *Schistosoma haematobium* miracidia. J Parasitol 1970; 56:281-6.

Shimazu, T., Oshima, T. Morphology and development of a lung fluke of the genus *Paragonimus* from Primorye, USSR, in snails, *Semisulcospira libertina*, in the laboratory. J Nagano-Ken Junior College 1983; 38:7-15.

Shoop, W.L. Trematode transmission patterns. J Parasitol 1988; 74:46-59.

Short, R.B., Gagne, H.T. Fine structure of a possible photoreceptor in cercariae of *Schistosoma mansoni*. J Parasitol 1975; 61:69-74.

Shostak, A.W., Esch, G.W. Photocycle-dependent emergence by cercariae of *Halipegus occidualis* from *Helisoma anceps*, with special reference to cercarial emergence patterns as adaptations for transmission. J Parasitol 1990; 76:790-5.

Sillman, E.I. The life history of the *Azygia longa* (Leidy, 1851) (Trematoda, Digenea) and notes on *A. acuminata* (Goldberger, 1911). Trans Am Microsc Soc 1962; 81:43-65.

Sindermann, C.J., Farrin, A.E. Ecological studies of *Cryptocotyle lingua* (Trematoda: Heterophyidae) whose larvae cause "pigment spots" of marine fish. Ecology 1962; 43:69-75.

Sinha, B.K. The structure of the tegument and digestive tract in *Philophthalmus rhionica* Tihomirov и *Metagonimus yokogavai* Katsurada miracidia and rediae. Ph.D. Thesis, Leningrad State University: Leningrad, USSR, 1978 (in Russian).

Sinitsin, D.F. Material on the natural history of Trematodes. Distomes of fish and frog in the environs of Warsaw. Izv Varshavsk Univ 1905; 7-9:1-96 (in Russian).

Sinitsin, D.T. Studien über die Phylogenie der Trematoden. I. Können die digenetischen Trematoden sich auf ungeschlechtlichen Wege fortpflancen? Biol Zblatt 1909; 29:664-82.

Sinitsin, D.F. Studien über die Phylogenie der Trematoden. III. *Cercaria plicata* Mihi und *Tetracotyle* Brand., als dimorphe Larven der Distomi danebest einer Hypothese über die Entstehung des Wirtswechsels bei den Trematoden. Biologicheskiy Zhurnal 1910; 1:106-45, 170-90.

Sinitsin, D.F. Parthenogenetic generation of trematodes and their progeny in the Black Sea molluscs. Zapiski Akademii Nauk 1911; 30:1-127 (in Russian).

Sinitsin, D.F. Studien über die Phylogenie der Trematoden. V. Revision of *Harmostominae* in the light of new facts from their morphology and life history. Z Parasitenk 1931a; 3:786-834.

Sinitsin, D.F. Gastropodes, digenetic trematodes and vertebrates, as a biological serves in evolution of organisms. Arch Zool Ital 1931b; 16:395-7.

Sire, C., Rognon, A., Théron, A. Failure of *Schistosoma mansoni* to reinfect *Biomphalaria glabrata* snails: Acquired humoral resistance or intra-specific larval antagonism. Parasitology 1998; 117:117-22.

Sire, C., Durand, P., Pointier, J.-P., Théron, A. Genetic diversity and recruitment pattern of *Schistosoma mansoni* in a *Biomphalaria glabrata* snail population: a field study using random-amplified polymorohic DNA markers. J Parasitol 1999; 85:436-41.

Skelly, P.J., Tielens, A.G.M., Shoemaker, C.B. Glucose transport and metabolism in mammalian-stage schistosomes. Parasitol Today 1998; 14:402-6.

Skirnisson, K., Galaktionov, K.V. Life-cycles and transmission patterns of seabird digeneans in SW Iceland. Sarsia 2002; 87:144-51.

Skirnisson, K., Galaktionov, K.V., Kozminsky, E.V. Factors influencing the distribution of digenean (Trematoda, Digenea) infection in the population of mudsnails *Hydrobia ventrosa* inhabiting saltmarsh brackish ponds of SW Iceland. J Parasitol (*in press*).

Skrjabin, K.I. "General Trematodology." In *Trematody Zhivotnyh i Cheloveka (Trematodes of Animals and Man)*. K.I. Skrjabin (ed.), Vol. 1, Izdatel'stvo Akademii Nauk SSSR, Moscow-Leningrad, 1947 (in Russian).

Skrjabin, K.I. "Family Brachylaemidae." In *Trematody Zhivotnyh i Cheloveka (Trematodes of Animals and Man)*. K.I. Skrjabin (ed.), Vol. 1, Izdatel'stvo Akademii Nauk SSSR, Moscow-Leningrad, 1947 (in Russian).

Skrjabin, K.I. "Superfamily Fascioloidea Stiles et Golder, 1910." In *Trematody Zhivotnyh i Cheloveka (Trematodes of Animals and Man)*. K.I. Skrjabin (ed.), Vol. 2, Izdatel'stvo Akademii Nauk SSSR, Moscow-Leningrad, 1948 (in Russian).

Skrjabin, K.I. "Suborder Paramphistomatata (Szidat, 1936) Skrjabin et Schulz, 1937." In *Trematody Zhivotnyh i Cheloveka (Trematodes of Animals and Man)*. K.I. Skrjabin (ed.), Vol. 3, Izdatel'stvo Akademii Nauk SSSR, Moscow-Leningrad, 1949 (in Russian).

Skrjabin, K.I., Gushanskaya, L.KH. "System of Trematodes of the Suborder Hemiurata and its General Morphological-Biological Description." In *Trematody Zhivotnyh i Cheloveka (Trematodes of Animals and Man)*. K.I. Skrjabin (ed.), Vol. 17, Izdatel'stvo Akademii Nauk SSSR, Moscow-Leningrad, 1960 (in Russian).

Smith, R.J. Ancylid snails: first intermediate host to certain trematodes with notes on ancylids as a new host for *Megalodiscus* and *Haematolloechus*. Trans Am Microsc Soc 1959; 78:228-31.

Smithers, S.R., Doenhoff, M.J. Schistosomiasis. In *Immunology of Parasitic Infections*. S. Cohen and K.S. Warren, eds. Oxford, UK: Blackwell Scientific 1982.

Smyth, J.D. *The Physiology of Trematodes*, 1st edition. San Francisco, U.S.A.: W.H. Freeman, 1966.

Smyth, J.D., Halton, D.W. *The Physiology of Trematodes*. 2nd edition. Cambridge, U.K.: Cambridge University Press, 1983.

Soboleva, T.N. "Morphology of Brachylaimoid Parthenitae and Larvae." In *Functionalnaya Morphologiya Lichinok Trematod i Cestod (Functional Morphology of Trematode and Cestode Larvae)*. E.V. Gvozdev, ed. Alma-Ata, USSR: Nauka, 1986 (in Russian).

Sokolova, I.M. Influence of trematodes on the demography of *Littorina saxatilis* (Gastropoda: Prosobranchia: Littorinidae) in the White Sea. Dis Aquat Organisms 1995; 21:91-101.

Sokolova, I.M., Bock, C., Portner, H.-O. Resistance to fresh water exposure in White Sea *Littorina* spp. I: Anaerobic metabolism and energetics. J Comp Physiol B 2000a; 170:91-103.

Sokolova, I.M., Granovitch, A.I., Berger, V.Ja., Johannesson, K. Intraspecific physiological variability of the gastropod *Littorina saxatilis* related to the vertical shore gradient in the White and North Seas. Mar Biol 2000b; 137:297-308.

Sousa, W.P. Host life history and the effect of parasite castration on growth: a field study on *Cerithidea californica* Haldeman (Gastropoda: Prosobranchia) and its trematode parasites. J Exp Mar Biol Ecol 1983; 73:273-96.

Sousa, W.P. "Spatial Scale and the Processes Structuring a Guild of Larval Trematode Parasites." In *Parasite Communities: Patterns and Processes*. G.W. Esch, A.O. Bush and J.M. Aho, eds. London, U.K.: Chapman & Hall, 1990.

Sousa, W.P., Gleason, M. Does parasitic infection compromise host survival under extreme environmental conditions? The case for *Cerithidea californica* (Gastropoda: Prosobranchia). Œcologia 1989; 80:456-64.

Southgate, V.R. Observations on the epidermis of the miracidium and on the formation of the sporocyst of *Fasciola hepatica*. Parasitology 1970; 61:177-90.

Southgate, V.R. Observations on the structure of the cercaria of *Notocotylus attenuatus* and formation of the cyst wall of the metacercariae. Z Zellforsch 1971; 120:420-49.

Southgate, V.R., Knowles, R.J. On *Schistosoma margrebowiei* Le Roux, 1933: The morphology of the egg, miracidium and cercaria, the compatibility with species of *Bulinus*, and development in *Mesocricetus auratus*. Z Parasitenk 1977; 54:233-50.

Starobogatov, Ya.I. Molluscan Fauna and Zoogeographical Classification of Continental Water Bodies. Leningrad, USSR: Nauka, 1970 (in Russian).

Steenstrup, J.J. Über den Generationswechsel oder die Fortpflanzung und Entwicklung durch abwechselnde Generationen, eine eigentümliche Form der Brutpflege in den niederen Tierklassen. Copenhagen: C.A. Reitzel, 1842.

Stein, P.C., Lumsden, R.D. The ultrastructure of developing metacercarial cyst of *Ascocotyle leighi* Burton, 1956 (Heterophyidae). Proc Helminthol Soc Wash 1971a; 38:1-10.

Stein, P.C., Lumsden, R.D. An ultrastructural and cytochemical study of metacercarial cyst development in *Ascocotyle pachycystis* Schroeder and Leigh, 1965. J Parasitol 1971b; 57:1231-46.

Stirewalt, M.A. Cercaria of schistosomule (*Schistosoma mansoni*): absence of the pericercarial envelope in vivo and early physiological and histological metamorphosis of the parasite. Exp Parasitol 1963; 13:395-406

Stirewalt, M.A. *Schistosoma mansoni*: cercaria to schistosomule. Adv Parasitol 1974; 12:115-82.

Stjernholm, R.L., Warren, K.S. *Schistosoma mansoni*: Utilization of exogenous metabolites by eggs in vitro. Exp Parasitol 1974; 36:222-32.

Strong, P., Bogitsh, B.J. Ultrastructure of lymph system of the trematode *Megalodiscus temperatus*. Trans Am Microsc Soc 1973; 92:570-8.

Strong, P., Cable, R.M. Fine structure and development of the metacercarial cyst in *Microphallus opacus*. J Parasitol 1972; 58:92-8.

Stunkard, H.W. The morphology and life history of the digenetic trematode *Asymphylodora amnicola* n. sp. the possible significance of progenesis for the phylogeny of the Digenea. Biol Bull 1959; 117:562-81.

Stunkard, H.W. The morphology and life cycle of the digenetic trematode *Himasthla littorinae* n. sp. (Echinost.). J Parasitol 1966; 52:367-72.

Styczyńska-Jurewicz, E. On the geotaxis, invasivity and span of life of *Opistioglyphe ranae* Duj. cercariae. Bull L'Acad Pol Sci 1961; 9:31-7

Styczyńska-Jurewicz, E. The life cycle of *Plagiorchis elegans* (Rud., 1802) and the revision of the genus *Plagiorchis* Lühe, 1889. Acta Parasitol Polon 1962; 10:419-45.

Sudarikov, V.E. "The Order Strigeidida (La Rue, 1926) Sudarikov, 1959." In *Trematody Zhivotnyh i Cheloveka (Trematodes of Animals and Man)*. K.I. Skrjabin (ed.), Vol. 24, Nauka, Moscow, 1971 (in Russian).

Sudarikov, V.E. *Trematode Fauna of the USSR. Strigeids.* Moscow, USSR: Nauka, 1984 (in Russian).

Sudds, R.H.J. Observations of schistosome miracidial behaviour in the presence of normal and abnormal snail host and subsequent tissue studies of the host. Elisha Sci Soc 1960; 76:121-33.

Sukhedo, M.V.K., Mettrick, D.F. The behaviour of juvenile *Fasciola hepatica.* J Parasitol 1986; 72:492-7.

Sullivan, J.T., Cheng, T.C., Howland, K.H. Studies on parasitic castration: castration of *Ilyanassa obsoleta* (Mollusca: Gastropoda) by several marine trematodes. Trans Am Microsc Soc 1985; 104:154-71.

Suloeva, T.A. The patterns of organization of different phases of the bucephalids (Trematoda, Bucephallidae) life cycle. M.S. Thesis, St. Petersburg State University: St. Petersburg, Russia, 1999 (in Russian).

Suloeva, T.A., Galaktionov, K.V., Dobrovolskij, A.A. Miracidium of *Prosorhynchus squamatus* (Trematoda, Bucephalidae). Parazitologiya (*submit*) (in Russian).

Swiderski, Z. Cytochemistry of glycogen distribution in miracidia of *Shistosoma haematobium.* Electron Micr Soc South Africa, Proc 1984; 14:133.

Swiderski, Z. Embryonic development of *Schistosoma mansoni and S. haematobium*: egg envelope formation. South African J Sci 1985; 81:48.

Swiderski, Z. Blastomere differentiation during embryonic development of *Schistosoma mansoni.* Proceedings of the 11th International Congress on Electron Microscopy; 1986 August 31 – September 7; Kyoto, 1986, 3321-2.

Swiderski, Z. Ultrastructure of shistosome eggs. Proceedings of the 6th Asia-Pacific Congress and Workshop on Electron Microscopy; 1988 July 26 – August 4; Bangkok, 1988, 555-6.

Swiderski, Z. Origin, differentiation and ultrastructure of egg envelopes surrounding the miracidia of *Schistosoma mansoni.* Acta Parasitol 1994a; 39:64-72.

Swiderski, Z. Origin, differentiation and ultrastructure of egg envelopes surrounding the coracidia of *Bothriocephalus clavibothrium* (Cestoda: Pseudophyllidea). Acta Parasitol 1994b; 39:73-81

Swiderski, Z. Homology and analogy in egg envelopes surrounding miracidia of *Schistosoma mansoni* and coracidia of *Bothriocephalus clavibothrium.* Acta Parasitol 1994c; 39:123-30.

Swiderski, Z., Moczon, T. Ultrastructure and cytochemistry of the nervous system and specialized nerve endings in the miracidia of *Schistosoma japonicum.* Proceedings of the 5th Asia-Pacific Electron Microscopy Congress; 1992; Beijng, Chine, 1992, 326-7.

Swiderski Z., Moczon T., Eklu-Natey D.T. Ultrastructure and cytochemistry of the nervous system and specialized nerve endings in the miracidia of *Schistosoma haematobium.* Proceedings of the 13th International Congress on Electron Microscopy; 1994 July 17-22; Paris. Ulis: Les Editions de Phisique, 1994, 1431-2.

Swiderski Z., Eklu-Natey D.T., Moczon T., Huggel H., Subilia L. Embryogenesis in *Shistosoma haematobium*: egg envelope formation Proceedings of the 10th International Congress on Electron Microscopy; Hamburg, 1982, 607-8.

Szidat, L. Neue Cercarienstudien. Zentbl Bakt Parasitenkde 1940; 145:438-48.

Szidat, L. Über den Entwicklungszyklus mit progenetischen Larvenstadien (Cercariaeen) von *Genarchella genarchella* Trav. (Trematoda, Hemiuridae). Z Tropenmed Parasitol 1956; 7:132-53.

Szidat, L. Über eine ungewöhnliche Form parthenogenetischer Vermehrung bei Metacercarien einer *Gymnophallus*-Art aus *Mytilus platensis*, *Gymnophallus australis* n. sp. Des Südatlantik. Z Parasitenk 1962; 22:196-213.

Taft, S.J. Aspects of larval development and histochemistry of *Ophthalmophagus singularis* (Trematoda: Cyclocoelidae). J Parasitol 1986; 72:136-41.

Takahashi, T., Mori, K., Sbigeta, Y. The phototactic, thermotactic and geotactic responses of miracidia of *Schistosoma japonicum.* Jap J Parasitol 1961; 10:686-91.

Tallmark, Bo, Norrgren G. The influence of parasitic trematodes on the ecology of *Nassarius reticulatus* (L.) in Gullmar Fjord (Sweden). Zoon 1976; 4:149-54.

Tang, C.C. Studies on the life history of *Eurytrema pancreaticum* Jans. 1889. J Parasitol 1950; 36:559-73.

Thakur, A.S., Cheng, T.C. The formation, structure and histochemistry of the metacercarial cyst of *Philophthalmus gralli* Mathis and Lager. Parasitology 1968; 58:605-18.

Théron, A. Dynamics of larval populations of *Schistosoma mansoni* in *Biomphalaria glabrata* I. Rhythmic production of cercariae in monomiracidial infection. Ann Trop Med Parasitol 1981a; 75:71-7.

Théron, A. Dynamics of larval populations of *Schistosoma mansoni* in *Biomphalaria glabrata* II. Chronobiology of the intramolluscal larval development during the shedding period. Ann Trop Med Parasitol 1981b; 75:547-54.

Théron, A. Early and late shedding patterns of *Schistosoma mansoni* cercariae: ecological significance in transmission to human and murine hosts. J Parasitol 1984; 70:652-5.

Théron, A. Hybrids between *Schistosoma mansoni* and *S. rodhaini*: characterization by cercarial emergence rhythms. Parasitology 1989; 99:225-8.

Théron, A., Combes, C. Analyse génétique du rythme d'émergence des cercaires de *Schistosoma mansoni* par croisement des souches à pics d'émission précoces ou tardifs. Comp Rend l'Acad Sci Paris 1983; 297, Sér 3:571-4.

Théron, A., Combes, C. Genetic analysis of cercarial emergence rhythms of *Schistosoma mansoni*. Behav Genetics 1988; 18:201-9.

Théron, A., Jourdane, J. Séquence de reconversion des sporocystes de *Schistosoma mansoni* producteurs de cercaries, en vue la production de nouvells générations de sporocystes. Z Parasitenk 1979; 61:63-71.

Théron, A., Gérard, C. Development of accessory sexual organs in *Biomphalaria glabrata* (Planorbidae) in relation to timing of infection by *Schistosoma mansoni*: consequences for energy itilization pattern by the parasite. J Moll Stud 1994; 60:25-31.

Théron, A., Gérard, C., Moné, H. Early enhanced growth of the digestive gland of *Biomphalaria glabrata* infected with *Schistosoma mansoni*: side effect or parasite manipulation? Parasitol Res 1992a; 78:445-50.

Théron, A., Moné, H., Gérard, C. Spatial and energy compromise between host and parasite: the *Biomphalaria glabrata – Schistosoma mansoni* system. Int J Parasitol 1992b; 22:91-4.

Théron, A., Pages, J.-R., Rognon, A. *Schistosoma mansoni*: distribution patterns of miracidia among *Biomphalaria glabrata* snail as related to host susceptibility and sporocyst regulatory processes. Exp Parasitol 1997; 85:1-9.

Thornhill, J.A., Jones, J.T., Kusel, J.R. Increased oviposition and growth in immature *Biomphalaria glabrata* after exposure to *Schistosoma mansoni*. Parasitology 1986; 93:443-50.

Thomas, A.P. The life history of the liver fluke. Quart J Micr Sci 1883; 23:90-133.

Thomas, F., Poulin, R. Manipulation of a mollusc by a trophically transmitted parasite: convergent evolution or phylogenetic inheritance? Parasitology 1998; 116:431-6.

Thomas, J.S., Pascoe, D. The digestion of exogenous carbohydrate by the daughter sporocysts of *Cercaria linearis* Lespes, 1857 and *C. stunkardi* Palombi, 1939 *in vitro*. Z Parasitenk 1973; 43:17-23.

Thomas, L.J. Life cycle of a fluke *Halipegus excentricus* n. sp. found in the ears of frogs. J Parasitol 1939; 25:207-21.

Thompson, S.N. "Physiological Alterations During Parasitism and Their Effects on Host Behaviour." In *Parasitism and Host Behaviour*. C.J. Barnard and J.M. Behnke, eds. London, U.K.: Taylor & Francis, 1990.

Thompson, S.N. "Physiology and Biochemistry of Snail-Larval Trematode Relationships." *Advances in Trematode Biology*", B. Fried and T. Graczyk, eds. Boca Raton, New York: CRC Press, 1997.

Thompson, S.N., Kavaliers, M. Phisiological bases for parasite-induced alterations of host behaviour. Parasitology 1994; 109S:S119-38.

Threadgold, L.T. *Fasciola hepatica*: the ultrastructure of the epithelium of the seminal vesicle, the ejaculatory duct and cirrus. Parasitology 1975; 71:437-43.

Threadgold, L.T. *Fasciola hepatica*: a transmission and scanning electron microscopical study of the apical surface of the gastrodermal cells. Parasitology 1978; 76:85-90.

Threadgold, L.T. *Fasciola hepatica*: stereological analysis of vitelline cell development. Exp Parasitol 1982; 54:352-65.

Threadgold, L.T. "Parasitic Platyhelminths." In *Biology of the Integument. I. Invertebrates.* J. Breiter-Hahn, A.G. Matoltsky and K.S. Richards. Berlin: Springer, 1984.

Threadgold, L.T., Gallagher, S.S.E. Electron microscope studies of *Fasciola hepatica*. I. The ultrastructure and interrelationship of the parenchymal cells. Parasitology 1966; 56:299-304.

Thulin, J. A preliminary report on the egg and miracidium of *Aporocotyle simplex*, Odhner, 1900 (Trematoda: Sanguinicolidae). Norw J Zool 1975; 23:200.

Tihomirov, I.A. Life cycle of *Philophthalmus rhionica* sp. nov. (Trematoda: Philophthalmidae). Ph.D. Thesis, Leningrad State University: Leningrad, USSR, 1980 (in Russian).

Tihomirov, I.A. Microanatomy of the miracidium of *Philophthalmus rhionica* (Trematoda: Philophthalmidae). Parazitologiya 2000; 34:210-19 (in Russian).

Tihomirov, I.A., Seleznev, K.V. Microanatomical description of the daughter redia of *Philophthalmus rhionica* (Trematoda: Philophthalmidae). Parazitologiya 1998; 32:201-12 (in Russian).

Tkach, V.V., Pawlovski, J., Mariaux, J., Swiderski, Z. "Molecular Phylogeny of the Suborder Plagiorchiata and its Position in the System of Digenea." In *Interrelationships of the Platyhelminthes.* D.T.J. Littlewood and R.A. Bray, eds. London, U.K.: Taylor & Francis 2001.

Toledo, R., Muñoz-Antoli, C., Esteban J.G. Production and chronobiology of emergence of the cercariae of *Euparyphium albuferensis* (Trematoda: Echinostomatidae). J Parasitol 1999; 85:263-7.

Touassem, R., Théron, A. Study on the intramolluscal development of *Schistosoma bovis*: demonstration of three patterns of sporocystogenesis by daughter sporocysts. Parasitology 1986; 92:337-41.

Tyler, S., Tyler, M.S. Origin of the epidermis in parasitic Platyhelminths. Int J Parasitol 1997; 27:715-38.

Uglem, G.L., Larson, O.R. Facilitated diffusion and active transport systems for glucose in metacercariae of *Clinostomum marginatum* (Digenea). Int J Parasitol 1987; 17:847-50.

Uglem, G.L., Lee, K.L. *Proterometra macrostoma* (Trematoda: Azygiidae): functional morphology of the tegument of the redia. Int J Parasitol 1985; 15:61-4.

Uglem, G.L., Read, C.P. Sugar transport and metabolism in *Schistosoma mansoni*. J Parasitol 1975; 61:390-7.

Ulmer, M.J. *Postharmostomum helicis* (Leidy, 1847) Robinson, 1949 (Trematoda), its life history and a revision of the subfamily Brachylaeminae. Parts I and II. Trans Am Microsc Soc 1951; 70: 189-238, 319-47.

Ulmer, M.J. "Site-Finding Behaviour in Helminths in Intermediate and Definitive Hosts." In *Ecology and Physiology of Parasites.* A.M. Fallis, ed. London: Adam Hilger, 1971.

Ulmer, M.J., Sommer, S.C. Development of sporocysts in the turtle lung fluke *Heronimus chelydre* MacCallum (Trematoda: Heronimidae) Proc Iowa Acad Sci 1957; 64:601-13.

Uspenskaya, A.V. *Parasite Fauna of Benthic Crustaceans of the Barents Sea.* Moscow-Leningrad, USSR: Izdatel'stvo Akademii Nauk SSSR, 1963 (in Russian).

Valle, C., Pellegrino, J., Alvarenga, N. Rhythmic emergence of *Schistosoma mansoni* from *Biomphalaria glabrata*: influence of temperature. Revista Inst Med Trop, Sao Paolo 1973; 15:195-201.

Van der Woude, A. Germ cell cycle of *Megalodiscus temporatus* (Stafford, 1905) Harwood, 1932. J Parasitol 1951; 36 (6), sect 2:14-5.

Van der Woude, A. Germ cell cycle of *Megalodiscus temporatus* (Stafford, 1905) Harwood, 1932 (Paramphistomatidae: Trematoda). Am Midl Nat 1954; 51:172-202.

Van der Woude, A., Cort, W.W., Ameel, D.J. The early development of the daughter sporocysts of Strigeoidea (Trematoda). J Parasitol 1953; 39:38-44.

Vanoverschelde, R., Vaes, F. Studies on the life cycle of *Himasthla militaris* (Trematoda: Echinostomatidae). Parasitology 1980; 81:609-17.

Vernberg, W.B., Vernberg, F.J. Influence of parasitism on thermal resistance of the mud-flat snail *Nassa obsoleta* Say. Exp Parasitol 1963; 14:330-2.

Villella, J.B. The life history of *Ectosiphonus rhomboideum* (Sinitsin, 1931) (Trematoda: Brachylaematidae) from the Short-tailed shrew *Blarina brevicaudata*. J Parasitol 1953; 39:21.

Vladimirov, V.L. Morphology and biology of the miracidium of *Posthodiplostomum cuticola* (Nordmann, 1832) Dubois, 1936, the agent of black spot disease of fish. Dokladyy Akademii Nauk SSSR (*Proceedings of the USSR Academy of Sciences*) 1961; 140:1473-6 (in Russian).

Vladimirov, V.L. *Posthodiplostomum cuticola* (Nordmann, 1832) Dubois, 1936 (morphology and biology of early developmental stages). Ph.D. Thesis, Leningrad State University, Leningrad, USSR, 1962 (in Russian).

Voge, M., Seidel, J.S. Transformation in vitro of miracidia of *Schistosoma mansoni* and *S. japonicum* into young sporocysts. J Parasitol 1972; 58:699-704.

Wajdi, N. Penetration by the miracidia of *Schistosoma mansoni* into the snail host. J Helminthol 1966; 40:235-44.

Wall, L.D. *Spirorchis parvus* – its life history and the development of its excretory system (Trematoda). Trans Am Microsc Soc 1941; 60:221-60.

Ward, R.D., Lewis, F.A., Yoshino, T.P., Dunn, T.S. *Schistosoma mansoni*: relationship between cercarial production levels and snail host susceptibility. Exp Parasitol 1988; 66:78-85.

Wardle, W.J. A bucephalid larva, *Cercaria pleuromerae* n.sp. (Trematoda: Digenea), parasitizing a deepwater bivalve from the Gulf of Mexico. J Parasitol 1988; 74:692-4.

Weber, P., Sonntag, J. The fine structure of the body wall of adult *Paragonimus uterobilateralis*. Trop Med Parasitol 1989; 40:422-8.

Werding, B. Morphologie, Entwicklung und Ökologie digener Trematoden-Larven der Strandschnecke *Littorina littorea*. Mar Biol 1969; 3:306-33.

Wesenberg-Lund, C. Contributions to the development of the Trematoda Digenea. II. The biology of the freshwater cercariae in Danish fresh-waters. D Kgl Dansk vidensk selsk 1934; 9:1-223.

West, A.F. Studies on the biology of *Philophthalmus gralli* Mathis and Leger, 1910 (Trematoda: Digenea). Am Midl Nat 1961; 66:363-83.

Wharton, D.A. The production and functional morphology of helminth egg-shells. Parasitology 1983; 86:85-97.

Wheater, P.R., Wilson, R.A. *Schistosoma mansoni*: a histological study of migration in the laboratory mouse. Parasitology 1979; 79:49-62.

Wikel, S.K., Bogitsh, B.J. *Schistosoma mansoni*: penetration apparatus and epidermis of the miracidium. Exp Parasitol 1974; 36:342-4.

Williams, J.A., Esch, G.W. Infra- and component community dynamics in the pulmonate snail *Helisoma anceps*, with special emphasis on the hemiurid trematode *Halipegus occidualis*. J Parasitol 1991; 77:246-53.

Wilson, R.A. The hatching mechanism of the egg of *Fasciola hepatica* L. Parasitology 1968; 58:79-89.

Wilson, R.A. Fine structure of the tegument of the miracidium of *Fasciola hepatica* L. J Parasitol 1969a; 55:124-33.

Wilson, R.A. Fine structure of the protonephridial system in the miracidium of *Fasciola hepatica*. Parasitology 1969b; 59:461-7.

Wilson, R.A. Fine structure and organisation of the musculature in the miracidium of *Fasciola hepatica*. J Parasitol 1969c; 55:1153-61.

Wilson, R.A. Fine structure of the nervous system and specialized nerve endings in the miracidium of *Fasciola hepatica*. Parasitology 1970; 60:399-421.

Wilson, R.A. Gland cells and secretions in the miracidium of *Fasciola hepatica*. Parasitology 1971; 63:225-31.

Wilson, R.A. "Cercariae to Liver Worms: Development and Migration in the Mammalian Host." In *The Biology of Schistosomes. From Genes to Latrines*. D. Rollinson and A.J.G. Simpson, eds. London, U.K.: Academic Press, 1987.

Wilson, R.A., Coulson, P.S. Why don't we have a schistosomiasis vaccine? Parasitol Today 1998; 14:97-9.

Wilson, R.A., Barnes, P.E. The tegument of *Schistosoma mansoni*: observations on the formation, structure and composition of cytoplasmic inclusions in relation to tegument function. Parasitology 1974; 68:239-58.

Wilson, R.A., Webster, L.A. Protonephridia. Biol Rev 1974; 49:127-60.

Wilson, R.A., Pullen, R., Denison, J. An investigation of the mechanism of infection by digenetic trematodes: the penetration of miracidium of the *Fasciola hepatica* into its snail host *Lymnaea luteola*. Parasitology 1971; 63:491-506.

Wisniewski, L.V. The development cycle of *Posthodiplostomum brevicaudatum* (v. Nordmann, 1832) Kozicka, 1958. Acta Parasitol Polon 1958; 6:251-71.

Withyachumnarkul, B., Pongsa-Asawapaiboon, A., Sobhon, P. Upatham, E.S. The neural commissural rings of the adult male *Schistosoma mansoni*. J Helminthol 1987; 61:169-72.

Wittrock, D.D. Structure and origin of the eggshell of *Quinqueserialis quinqueserialis* (Trematoda: Notocotylidae). Z Parasitenk 1982; 67:37-44.

Wittrock, D.D. Histochemical and ultrastructural studies of the prostate gland of *Quinqueserialis quinqueserialis* (Trematoda: Notocotylidae). Trans Am Microsc Soc 1986; 105:365-75.

Woodhead, A.E. Life history studies on the Trematodes family Bucephalidae I. Trans Am Microsc Soc 1929; 48:256-75.

Woodhead, A.E. Life history studies on the trematode family Bucephalidae II. Trans Am Microsc Soc 1930; 49:1-17.

Woodhead, A.E. The germ-cell cycle in the trematode family Bucephalidae. Trans Am Microsc Soc 1931; 50:169-87.

Woolhouse, M.E.J. The effect of schistosome infection on the mortality rates of *Bulinus globosus* and *Biomphalaria pfeifferi*. Ann Trop Med Parasitol 1989; 83:137-41.

Wootton, D. M. Life-history of *Allocreadium allonetenicum*, n. sp. (Allocreadiidae, Trematoda). Biol Bull 1957; 113:302-15.

Wright, C.A. Probable relationship between the Rhodometopa group of cercariae and the trematode genus *Renicola* Cohn. Nature 1953; 171, N 4357-4363:1072-3.

Wright, C.A. Studies on the life history and ecology of the trematode genus *Renicola* Cohn. Proc Zool Soc 1956; 126:1-48.

Wright, C.A. Host-location by trematode miracidia. Ann Trop Med Parasitol 1959; 53:288-92.

Wright, C.A. Miracidial responses to molluscan stimuli. Proceedings of the 1st International Congress of Parasitology 2, Rome, 1966a, 1058-9.

Wright, C.A. The pathogenesis of helminths in Mollusca. Helminthol Abstr 1966b; 35:207-24.

Wright, D.G.S Responses of the cercariae of *Trichobilharzia ocellata* to white light, monochromatic light, and irradiance reduction. Can J Zool 1974; 52:575-9.

Wu, L.-Y. A study of the life history of *Trichobilharzia cameroni* sp. nov. (Schistosomatidae). Can J Zool 1953; 31:351-3.

Xiao, X., Lu, D., Wang, T., Gao, J., Zhu, C., Zhang, B., An, J., Pong, H., Xu, M., Wu, W. Studies on the mode of human infection with *Echinochasmus liliputans*. Chin J Parasitol Parasit Dis 1995; 13:197-9.

Xiao, X., Wang, T., Wang, Y., Sheng, G., Wang, Q. In vivo and in vitro encystment of *Echinochasmus liliputans* cercariae and biological activity of the metacercariae. J Parasitol 2001; 87

Xu, Y.Z., Dresden, M.H. The hatching of schistosome eggs. Exp Parasitol 1990; 70:236-49.

Yamaguti, S. *Systema Helminthum. Part I-II. The Digenetic Trematodes of Vertebrates*. New York: Intersciences Publ., Inc., 1958.

Yamaguti, S. *Synopsis of Digenetic Trematodes of Vertebrates. Volume 1/2.* Tokyo: Kagaku, 1971.

Yasuraoka, K. Ecology of the miracidium. 2. On the behaviour to light of the miracidium of *Fasciola hepatica*. Jap J Med Sci Biol 1954; 7:181-92.

Zakikhani, M., Rau, M.E. Effects of *Plagiorchis elegans* (Digenea: Plagiorchiidae) infection on the reproduction of *Biomphalaria glabrata* (Pulmonata: Planorbidae). J Parasitol 1998; 84:927-30.

Zavarzin, A.A., Kharazova, A.D., Molitvin M.N. *Cell Biology*. St. Petersburg, Russia: Izdatel'stvo St. Peterburgskogo Universiteta, 1992 (in Russian).

Žd'árská, Z. Larvální stadia motolic z vodních plžů na území ČSSR. Československá Parasitol 1963; 10:207-62.

Žd'árská, Z. The histology and histochemistry of the cystogenic cells of the cercaria *Echinoparyphium aconiatum* Dietz, 1909. Folia Parasitol 1968; 15:213-32.

Žd'árská, Z. The gland cells in the tail of cercariae. Folia Parasitol 1969a; 16:111-20.

Žd'árská, Z. Gland cells of the cercaria of *Plagiorchis laricola* (Skrjabin, 1924) (Trematoda). Vestn Cs Spol Zool (*Acta Soc Zool Bohemoslov*) 1969b; 33:278-86.

Žd'árská, Z. The gland cells of the cercaria of *Notocotylus attenuatus* (Rudolphi, 1809) and the cyst wall of its adolescaria. Folia Parasitol 1970; 17:31-47.

Žd'árská, Z. The cystogenic gland cells of the cercaria of *Philophathalmus sp.* Folia Parasitol 1971a; 18:119-25.

Žd'árská, Z. The histology and histochemistry of the cystogenic gland cells and the cyst wall of the larval stages of *Moliniella anceps* (Molin, 1859). Folia Parasitol 1971b; 18:15-25.

Žd'árská, Z . *Evolution of Morphological Features in Larval Stages of Digenetic Trematodes*. Praha, Czechoslovakia: Academia, 1986

Žd'árská, Z. Ultrastructure of the "paletot" of *Eurytrema pancreaticum* daughter sporocysts. Folia Parasitol 1990; 37:343-5.

Žd'árská, Z., Soboleva, T.N. Ultrastructure of the metacercaria and sporocyst sac of *Leucochloridium perturbatum* Pojmanska, 1969. Folia Parasitol 1984; 31:133-9.

Žd'árská, Z., Soboleva, T.N. "Ultrastructure of the Parthenitae and Larvae of Brachylaimoids." In *Functionalnaia Morphologiia Lichinok Trematod i Cestod (Functional Morphology of Trematode and Cestode Larvae).* E.V. Gvozdev, ed. Alma-Ata, USSR: Nauka, 1986a (in Russian).

Žd'árská, Z., Soboleva, T.N. "Morphological Features of Brachylaimoid Trematodes." In *Functionalnaya Morphologiia Lichinok Trematod i Cestod (Functional Morphology of Trematode and Cestode Larvae).* E.V. Gvozdev, ed.Alma-Ata: Nauka, 1986b (in Russian).

Žd'árská, Z., Soboleva, T.N., Osipovskaya, L.L. Ultrastructure of the tegument and associated structures of *Leucochloridium paradoxum* sporocyst and metacercaria. Folia Parasitol 1982; 29:247-51.

Žd'árská, Z., Soboleva, T.N., Sterba, J., Valkounova, J. Ultrastructure of the male reproductive system of the trematode *Brachylaimus aequans*. Folia Parasitol 1991; 38:33-7.

Žd'árská, Z., Sterba, J., Soboleva, T.N., Valkounova, J. Ultrastructure of the digestive tract of *Brachylaimus aequans* (Trematoda: Brachylaimidae). Folia Parasitol 1988; 35:105-11.

Zelmer, D.A., Esch, G.W. Bridging the gap: the odonate naiad as a paratenic host for *Halipegus occidualis* (Digenea: Hemiuridae). J Parasitol 1998; 84:94-6.

Zuk, M. The role of parasites in sexual selection: current evidence and future directions. Advances in the Study of Behavior 1992; 21:39-68.

Index

GPSR Compliance

The European Union's (EU) General Product Safety Regulation (GPSR) is a set of rules that requires consumer products to be safe and our obligations to ensure this.

If you have any concerns about our products, you can contact us on ProductSafety@springernature.com

In case Publisher is established outside the EU, the EU authorized representative is:

Springer Nature Customer Service Center GmbH
Europaplatz 3
69115 Heidelberg, Germany

Batch number: 10809436

Printed by Printforce, the Netherlands